FIFTH EDITION

Calculus

FOR BUSINESS, ECONOMICS, LIFE SCIENCES, AND SOCIAL SCIENCES

College Mathematics Series

This book is part of a comprehensive series designe o accommodate a wide variety of courses. Difficulty level and topic coverag are determined by the choice of a particular book or a combination of bool from the series. Many topics in individual books are independent and may b elected in any order or omitted. All books include reviews of relevant algebr topics.

Essentials of College Mathematics: *One or two sem ers*
A brief introduction to finite mathematics and calcu that is suitable for a one- or two-semester course or a one- or two-quarte urse

College Mathematics: *Two semesters*
A more comprehensive introduction to finite mathem s and calculus that is suitable for a two-semester course or a two- or thre uarter course

Applied Mathematics: *Two or three semesters*
An even more comprehensive introduction to finite hematics and calculus for schools requiring this level and degree of cove ; additional topics are included in both finite mathematics and calculus, the definite integral is given a more formal treatment

Finite Mathematics: *One semester*
The finite mathematics portion of *College Mathematics,* an added chapter on games and decisions; may be used in combination either *Calculus* or *Applied Calculus* to create a two- or three-semester c e

Calculus: *One semester*
The calculus portion of *College Mathematics,* with an add apter on trigonometric functions

Applied Calculus: *One or two semesters*
A more extensive treatment of calculus for schools requi his level and degree of coverage; the book can be completed in two se rs, however, with appropriate topic selection—since many of the topics ndependent—the book also can be used for a strong one-semester cou

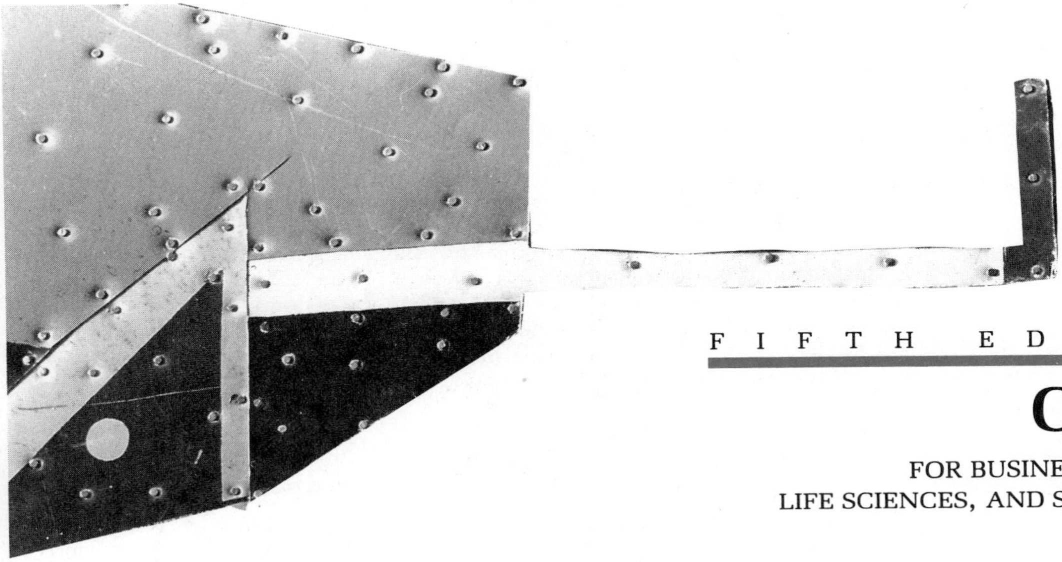

FIFTH EDITION

Calculus

FOR BUSINESS, ECONOMICS, LIFE SCIENCES, AND SOCIAL SCIENCES

RAYMOND A. BARNETT
Merritt College

MICHAEL R. ZIEGLER
Marquette University

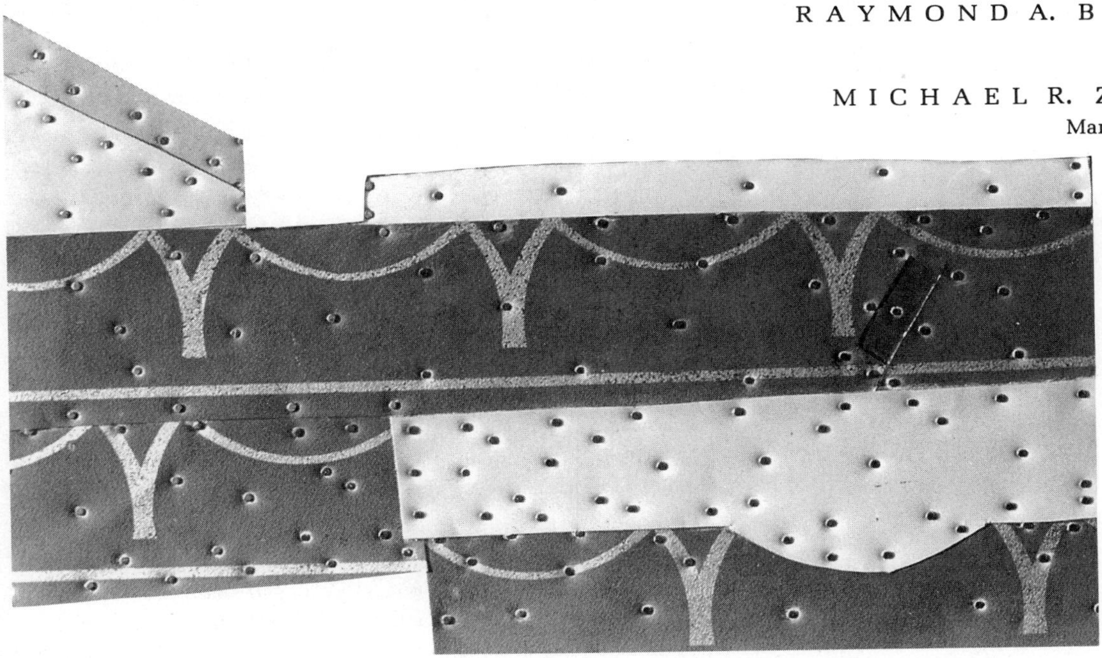

DELLEN PUBLISHING COMPANY
San Francisco, California

COLLIER MACMILLAN PUBLISHERS
London

Divisions of Macmillan, Inc.

On the cover: The detail on the cover is from a metal collage by Los Angeles artist Tony Berlant. His work can be described as part painting, part sculpture. It is resplendent with color and texture. The collage is formed by nailing fragments of printed and colored tin onto a wood backing. Berlant's work is represented by the L. A. Louver Gallery in Venice, California and the Louver Gallery in New York City. His work may also be seen in the permanent collections at the Whitney Museum of American Art in New York City, the Academy of Fine Art in Philadelphia, the Hirshorn Museum in Washington, D.C., the Art Institute of Chicago, and the Los Angeles County Museum.

Printed in the United States of America

Permissions: Dellen Publishing Company
400 Pacific Avenue
San Francisco, California 94133

Orders: Dellen Publishing Company
c/o Macmillan Publishing Company
Front and Brown Streets
Riverside, New Jersey 08075

Collier Macmillan Canada, Inc.

Library of Congress Cataloging-in-Publication Data

Barnett, Raymond A.
Calculus for business, economics, life sciences, and social sciences / Raymond A. Barnett, Michael R. Ziegler.—5th ed.
p. cm.—(College mathematics series)
Rev. ed. of: Calculus for management, life, and social sciences. c1987.
ISBN 0-02-306141-3
1. Calculus. 2. Social sciences—Mathematics. 3. Biomathematics. I. Ziegler, Michael R. II. Barnett, Raymond A. Calculus for management, life, and social sciences. III. Title. IV. Series.
QA303.B283 1990 89-48020
515—dc20 CIP

Printing: 1 2 3 4 5 6 7 8 9 Year: 9 0 1 2 3

ISBN 0-02-306141-3

CONTENTS

PREFACE

The fifth edition of *Calculus for Business, Economics, Life Sciences, and Social Sciences* is designed for a one-term course in calculus and for students who have had $1\frac{1}{2}-2$ years of high school algebra or the equivalent. It is one of six books in the authors' College Mathematics series (see page ii for a brief comparison of all six books).

This edition encompasses the experiences and recommendations of a large number of users of the last and previous editions as well as survey results from instructors, mathematics departments, course outlines, and college catalogs. The choice and independence of certain topics make the text readily adaptable to a variety of courses (see the chapter dependency chart on page xv).

◆ PRINCIPAL CHANGES FROM THE FOURTH EDITION

This edition has been divided into two parts, "Part One Preliminaries" and "Part Two Calculus." Many users have requested a more comprehensive **review of relevant topics from intermediate and college algebra** to be placed at the beginning of the book. Part One is our answer to this request. Chapters 1–3 comprise a completely rewritten and reorganized review of algebra topics. Depending on the background of a given class, this material can be totally omitted, referred to as needed, or selected topics can be covered in detail at the beginning or during the course. A few additional review topics are also included in Appendix A for convenient reference.

The following general improvements are found throughout the text: an increased emphasis on and use of **calculators; boxed definitions, results, and step-by-step processes;** and **schematic illustrations. Examples** have been improved, and many new examples have been added. **Exercise sets,** including applications, have been improved and expanded. All **exposition** has been carefully reviewed and fine-tuned or rewritten.

Specific improvements in Part Two are as follows:

1. The first four sections in the introductory **derivative** chapter of the last edition have been completely rewritten and condensed into three sections. The treatment has been both simplified and expanded. The new first section provides an **intuitive geometric introduction to limits and continuity.**

Because of numerous reviewer requests, coverage of **left-hand and right-hand limits** is again included. Because of its importance in solving optimization problems, **continuity on a closed interval** is now defined. **Increment notation** has been removed from this introductory chapter to keep the notation as simple as possible. **Increments** are now part of the section on **differentials** (Section 5-5). **Computation of difficult limits** has been reduced, and more emphasis has been placed on the **0/0 indeterminate form** because of its relevance to the definition of the derivative and its various interpretations (slope, instantaneous rate, marginal analysis, and so on). **Infinite limits** and **limits at infinity** receive more attention. **Vertical and horizontal asymptotes** are defined using these concepts.

2. The section on **curve sketching** (Section 5-3) now precedes the section on **optimization** (Section 5-4). **Solving inequalities using continuity** has become a subsection of the section on the first derivative and graphs (Section 5-1). This process of solving inequalities is fundamental to the production of **sign charts** throughout the chapter. The **curve sketching strategy** (Section 5-3) has been modified into a form that should make it easier for students to use and remember [Step 1: Use $f(x)$; Step 2: Use $f'(x)$; Step 3: Use $f''(x)$; Step 4: Graph f].
3. To satisfy many reviewer requests, **implicit differentiation** (Section 6-4) and **related rates** (Section 6-5) have been returned to the book.
4. In Chapter 7, on integration, more attention is given to **substitution techniques** and **area between the graphs of two functions.** In the applications of integration in Section 8-2, a Riemann sum approach is used to develop the formula for the **future value of a continuous income stream,** instead of the present value. The section on improper integrals in the fourth edition has been replaced by Section 8-4, on integration using tables, and a **table of integrals** is included in Appendix B.
5. In Chapter 9, on multivariable calculus, the important applications of **Cobb–Douglas production functions** are given more attention, and now include a discussion of the **marginal productivity of money.** At the request of many reviewers, our discussion of the **least-squares method** now includes two equivalent methods of determining the least-squares line: solving the system of **normal equations** or using standard formulas. To aid in this development, a section on **summation notation** has been added to Appendix A for those students who need additional experience with this notation.

◆ IMPORTANT FEATURES

Emphasis and Style

The text is **written for student comprehension.** Great care has been taken to write a book that is mathematically correct and accessible to students. Emphasis is on computational skills, ideas, and problem solving rather than mathematical theory. Most derivations and proofs are omitted except where their inclusion adds significant insight into a particular concept. General concepts and results are usually presented only after particular cases have been discussed.

Examples and Matched Problems

Over 200 completely worked examples are included. Each example is followed by a similar problem for the student to work while reading the material. This actively involves the student in the learning process. The answers to these matched problems are included at the end of each section for easy reference.

Exercise Sets

The book contains over 2,000 problems. Each exercise set is designed so that an average or below-average student will experience success and a very capable student will be challenged. Exercise sets are mostly divided into A (routine, easy mechanics), B (more difficult mechanics), and C (difficult mechanics and some theory) levels.

Applications

Enough applications are included to convince even the most skeptical student that mathematics is really useful. The majority of the applications are included at the end of exercise sets and are generally divided into business and economics, life science, and social science groupings. An instructor with students from all three disciplines can let them choose applications from their own field of interest; if most students are from one of the three areas, then special emphasis can be placed there. Most of the applications are simplified versions of actual real-world problems taken from professional journals and books. No specialized experience is required to solve any of the applications.

◆ STUDENT AIDS

1. **Think boxes** (dashed boxes) are used to enclose steps that are usually performed mentally (see Sections 1-2 and 1-3).
2. **Annotation** of examples and developments, in color type, is found throughout the text to help students through critical stages (see Sections 1-2 and 1-3).
3. **Functional use of color** improves the clarity of many illustrations, graphs, and developments, and guides students through certain critical steps (see Sections 1-2 and 1-3).
4. **Boldface type** is used to introduce new terms and highlight important comments.
5. **Screened boxes** are used to highlight important definitions, theorems, results, and step-by-step processes.
6. **Answers** to odd-numbered problems are included in the back of the book.
7. **Chapter review** sections include a review of all important terms and symbols and a comprehensive review exercise. Answers to all review exercises are included in the back of the book.
8. A **student's solutions manual** is available at a nominal cost through a book store. The manual includes detailed solutions to all odd-numbered problems and all review exercises.

9. A manual for an **Interactive Computer Applications Package (ICAP)** by Carolyn L. Meitler is available at a nominal cost through a book store. The manual contains instructions, examples, and exercises that demonstrate the use of the programs on the ICAP for *Calculus* disks. The disks containing the programs, for use on APPLE II® and IBM-PC® computers, are distributed free of charge to institutions using this book. No previous computer experience is necessary to use this package.

◆ INSTRUCTOR AIDS

1. A unique **computer-generated random test system** is available to instructors without cost. The test system utilizes an IBM-PC® computer and will produce high-quality output on a number of dot-matrix printers or on a Hewlett-Packard Laserjet 2® laser printer. The test system has been greatly expanded and now contains over 250 different problem algorithms directly related to material in the text. These carefully constructed algorithms use random number generators to produce different, yet equivalent, versions of each of these problems. An almost unlimited number of quizzes, review exercises, chapter tests, mid-terms, and final examinations, each different from the other, can be generated quickly and easily. At the same time, the system will produce answer keys and student work sheets, if desired.
2. A **printed and bound test battery** is also available to instructors without cost. The battery, organized by chapter, contains three equivalent versions (with answers) of over 250 different problems.
3. An **instructor's resource manual** provides over 100 transparency and handout masters, a detailed discussion of chapter and topic dependencies, and a detailed topic chart for comparing this book with other books in the authors' College Mathematics series.
4. An **instructor's answer manual** containing all the answers not included in the text is available to instructors without charge.
5. A **student's solutions manual** (see Student Aids) is available to instructors without charge from the publisher.
6. An **Interactive Computer Applications Package (ICAP)** by Carolyn L. Meitler (see Student Aids) is available to instructors without charge from the publisher. The programs in this package are available on disks for APPLE II® and IBM-PC® computers. Included on these disks are programs related to limit estimation, graphing functions, numerical integration, and least-squares approximations. The publisher will supply these disks without charge to institutions using this book.
7. An **interactive graphing program** for the APPLE Macintosh® computer allows a user to graph most of the mathematical functions likely to be encountered quickly, accurately, and with considerable control over axes, scales, graph size, and labeling. In addition to graphing functions, this program will perform a variety of mathematical operations related to numerical integration, root approximation, interpolating polynomials, least-square

polynomials, and approximate solutions of differential equations. Instructors will find this program useful for preparing examination material, transparency masters, and handouts. The publisher will supply this program free of charge to instructors using this book, and the program may be freely distributed to students.

◆ ERROR CHECK

Because of the careful checking and proofing by a number of mathematics instructors (acting independently), the authors and publisher believe this book to be substantially error-free. For any errors remaining, the authors would be grateful if they were sent to: Dellen Publishing Company, 400 Pacific Avenue, San Francisco, CA 94133.

◆ ACKNOWLEDGMENTS

In addition to the authors, many others are involved in the successful publication of a book. We wish to thank personally all those who reviewed the manuscript:

Chris Boldt, Eastfield College
Bob Bradshaw, Ohlone College
Charles E. Cleaver, The Citadel
Barbara Cohen, West Los Angeles College
Richard L. Conlon, University of Wisconsin—Stevens Point
Kenneth A. Dodaro, Florida State University
Martha M. Harvey, Midwestern State University
Louis F. Hoelzle, Bucks County Community College
Robert H. Johnston, Virginia Commonwealth University
Robert Krystock, Mississippi State University
Roy H. Luke, Los Angeles Pierce College
Mel Mitchell, Clarion University of Pennsylvania
Kenneth A. Peters, Jr., University of Louisville
Stephen Rodi, Austin Community College
Daniel E. Scanlon, Orange Coast College
Joan Smith, Vincennes University
Delores A. Williams, Pepperdine University
Caroline Woods, Marquette University

We also wish to thank:

John Williams for a strong and effective cover design

John Drooyan and Mark McKenna for the many sensitive and beautiful photographs throughout the book

Susan Pustejovsky and Stephen Merrill for carefully checking all examples and problems (a tedious but extremely important job)

Carolyn Meitler for assisting in the preparation of the computer-generated testing system and for developing *ICAP for Calculus.*

Brenda Rudin for assisting in the preparation of the computer-generated testing system

Janet Bollow for another outstanding book design

Phyllis Niklas for guiding the book smoothly through all publication details

Don Dellen, the publisher, who continues to provide all the support services and encouragement an author could hope for

Producing this new edition with the help of all these extremely competent people has been a most satisfying experience.

R. A. Barnett
M. R. Ziegler

Chapter Dependencies

PART ONE PRELIMINARIES

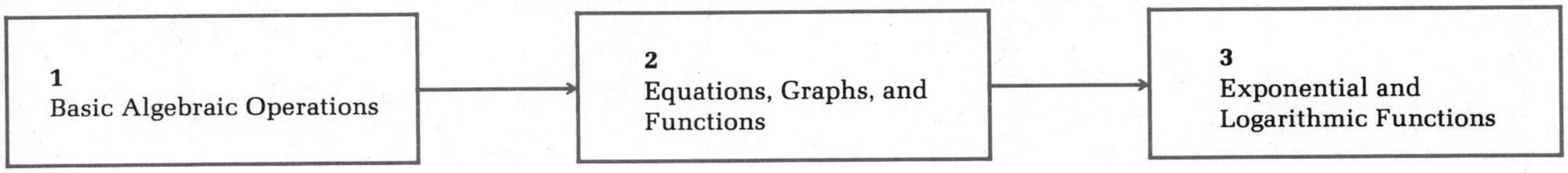

Selected topics from Part One may be referred to as needed in Part Two or reviewed systematically before starting Part Two.

PART TWO CALCULUS

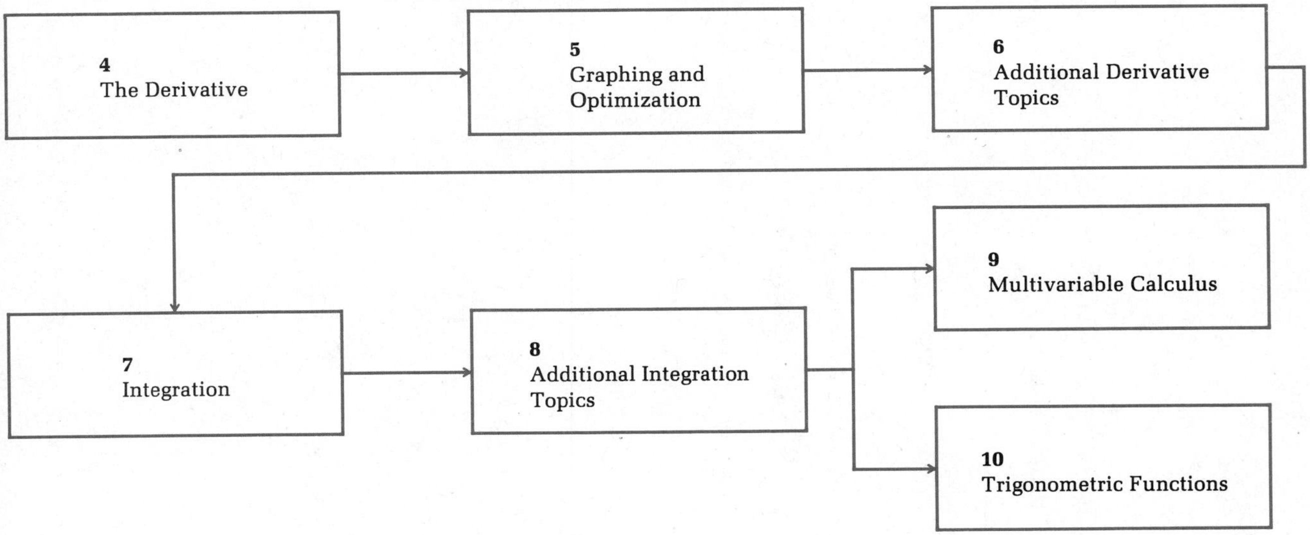

FIFTH EDITION

Calculus

FOR BUSINESS, ECONOMICS,
LIFE SCIENCES, AND SOCIAL SCIENCES

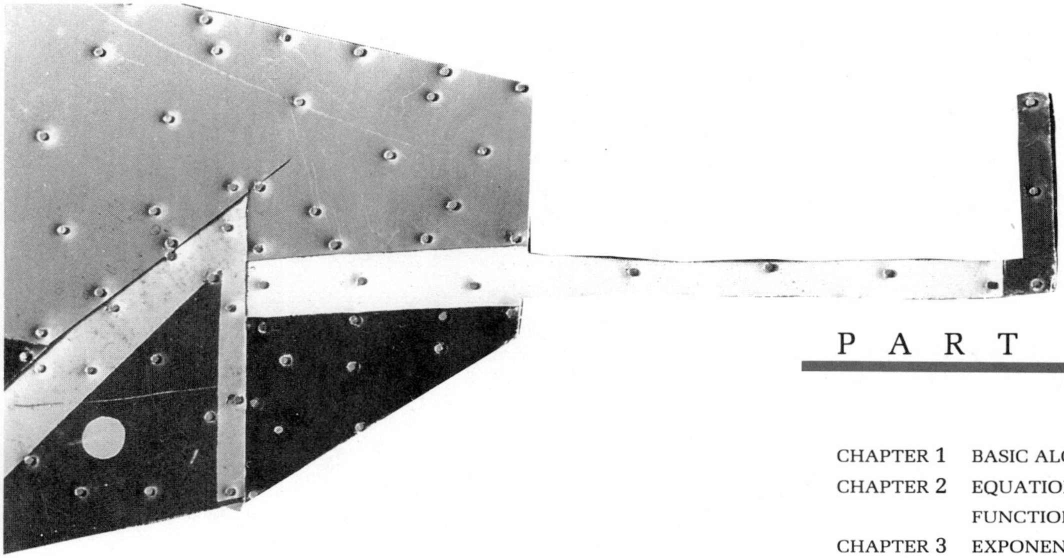

PART ONE

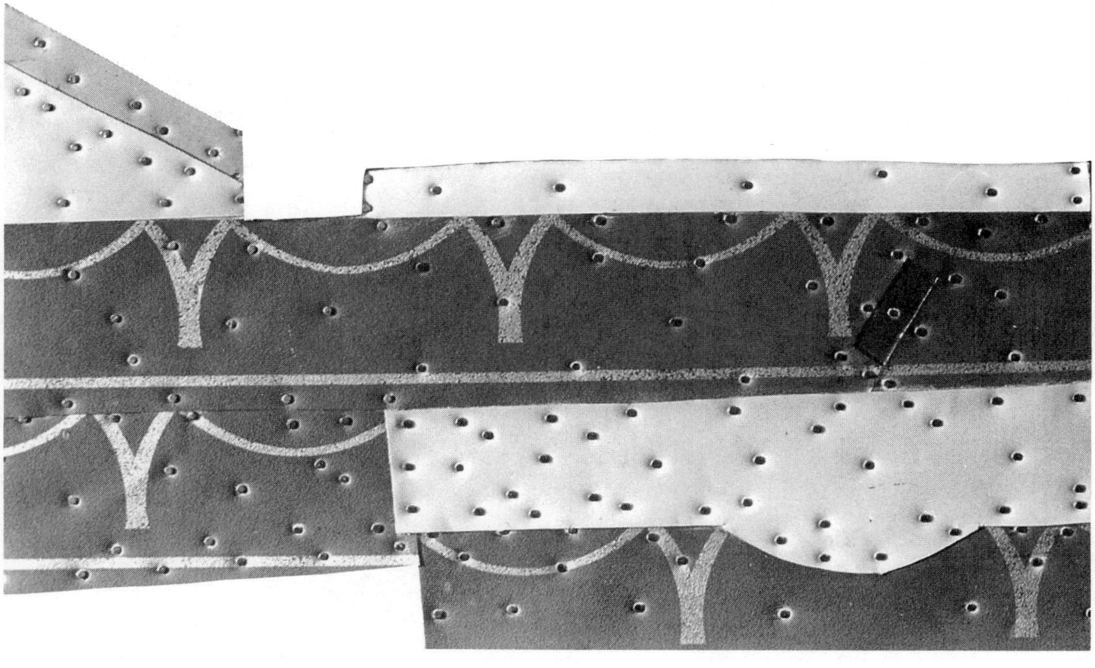

Preliminaries

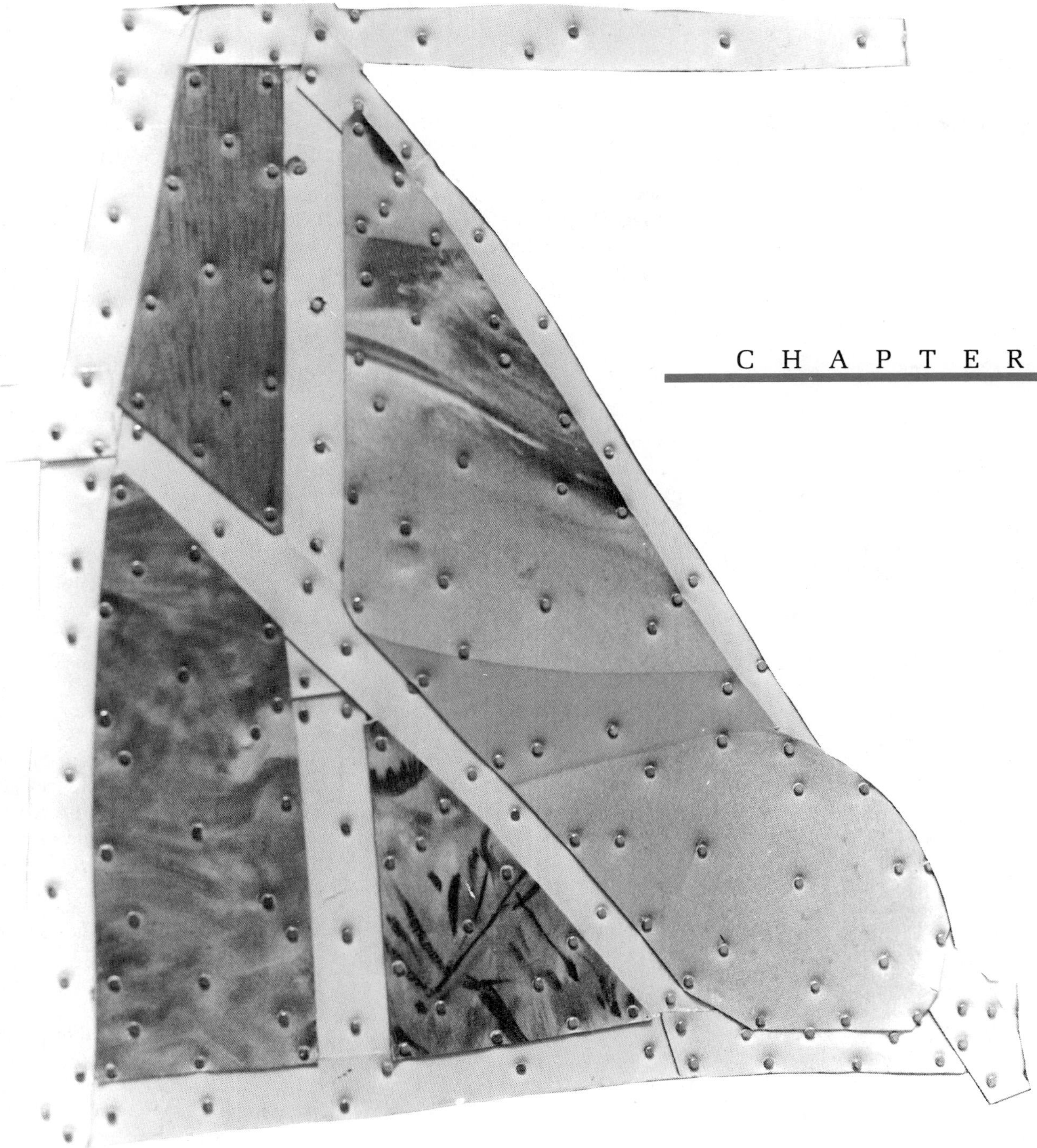

CHAPTER 1

Basic Algebraic Operations

CHAPTER 1

Contents

In this chapter we will review some important basic algebraic operations usually studied in earlier courses. The material may be studied systematically before commencing with the rest of the book or reviewed as needed.

SECTION 1-1

Algebra and Real Numbers

- THE SET OF REAL NUMBERS
- THE REAL NUMBER LINE
- BASIC REAL NUMBER PROPERTIES
- FURTHER PROPERTIES
- FRACTION PROPERTIES

The rules for manipulating and reasoning with symbols in algebra depend, in large measure, on properties of the real numbers. In this section we will look at some of the important properties of this number system. To make our discussions here and elsewhere in the text clearer and more precise, we will occasionally make use of simple *set* concepts and notation. Refer to Appendix A if you are not yet familiar with the basic ideas concerning sets.

◆ THE SET OF REAL NUMBERS

What number system have you been using most of your life? The *real number system*. Informally, a **real number** is any number that has a decimal representation. Table 1 describes the set of real numbers and some of its important subsets. Figure 1 illustrates how these sets of numbers are related.

The set of integers contains all the natural numbers and something else—their negatives and 0. The set of rational numbers contains all the integers and something else—noninteger ratios of integers. And the set of real numbers contains all the rational numbers and something else—irrational numbers.

TABLE 1

The Set of Real Numbers

SYMBOL	NAME	DESCRIPTION	EXAMPLES
N	Natural numbers	Counting numbers (also called positive integers)	$1, 2, 3, \ldots$
Z	Integers	Natural numbers, their negatives, and 0	$\ldots, -2, -1, 0, 1, 2, \ldots$
Q	Rational numbers	Numbers that can be represented as a/b, where a and b are integers and $b \neq 0$; decimal representations are repeating or terminating	$-4, 0, 1, 25, \frac{-3}{5}, \frac{2}{3}, 3.67, -0.33\overline{3}$,* $5.272\ \overline{727}$
I	Irrational numbers	Numbers that can be represented as nonrepeating and nonterminating decimal numbers	$\sqrt{2}, \pi, \sqrt[3]{7}$, 1.414 213..., 2.718 281 82...
R	Real numbers	Rational and irrational numbers	

* The overbar indicates that the number (or block of numbers) repeats indefinitely.

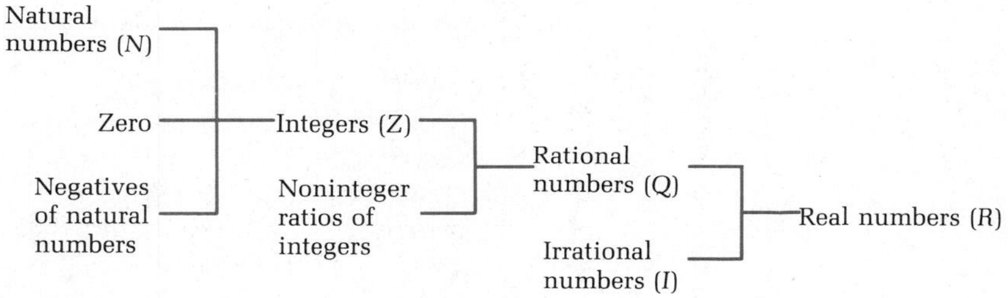

FIGURE 1
Real numbers and important subsets

◆ THE REAL NUMBER LINE

A one-to-one correspondence exists between the set of real numbers and the set of points on a line. That is, each real number corresponds to exactly one point, and each point corresponds to exactly one real number. A line with a real number associated with each point, and vice versa, as shown in Figure 2, is called a **real number line,** or simply a **real line.** Each number associated with a point is called the **coordinate** of the point.

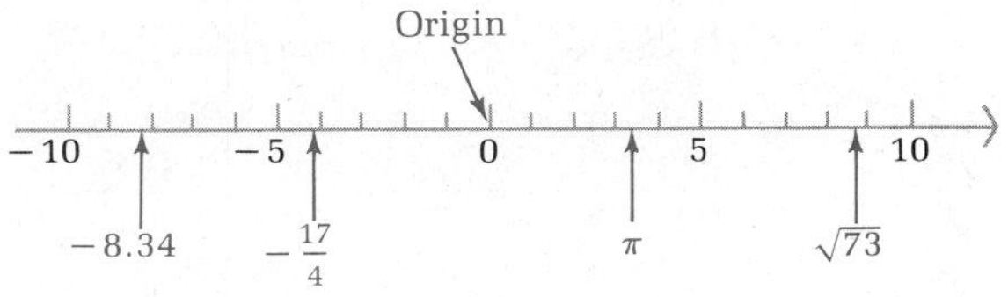

FIGURE 2
The real number line

The point with coordinate 0 is called the **origin.** The arrow on the right end of the line indicates a positive direction. The coordinates of all points to the right of the origin are called **positive real numbers,** and those to the left of the origin are called **negative real numbers.** The real number 0 is neither positive nor negative.

◆ BASIC REAL NUMBER PROPERTIES

We now take a look at some of the basic properties of the real number system that enable us to convert algebraic expressions into *equivalent forms.* These assumed basic properties, called **axioms,** become operational rules in the algebra of real numbers.

AXIOMS

Basic Properties of the Set of Real Numbers

Let a, b, and c be arbitrary elements in the set of real numbers R.

ADDITION PROPERTIES

CLOSURE: $a + b$ is a unique element in R.

ASSOCIATIVE: $(a + b) + c = a + (b + c)$

COMMUTATIVE: $a + b = b + a$

IDENTITY: 0 is the additive identity; that is, $0 + a = a + 0 = a$ for all a in R, and 0 is the only element in R with this property.

INVERSE: For each a in R, $-a$ is its unique additive inverse; that is, $a + (-a) = (-a) + a = 0$, and $-a$ is the only element in R relative to a with this property.

MULTIPLICATION PROPERTIES

CLOSURE: ab is a unique element in R.

ASSOCIATIVE: $(ab)c = a(bc)$

COMMUTATIVE: $ab = ba$

IDENTITY: 1 is the multiplicative identity; that is, $(1)a = a(1) = a$ for all a in R, and 1 is the only element in R with this property.

INVERSE: For each a in R, $a \neq 0$, $1/a$ is its unique multiplicative inverse; that is, $a(1/a) = (1/a)a = 1$, and $1/a$ is the only element in R relative to a with this property.

DISTRIBUTIVE PROPERTIES

$a(b + c) = ab + ac \qquad (a + b)c = ac + bc$

Do not be intimidated by the names of these properties. Most of the ideas presented here are quite simple. In fact, you have been using many of these properties in arithmetic for a long time.

You are already familiar with the **commutative properties** for addition and multiplication. They indicate that the order in which the addition or multiplication of two numbers is performed does not matter. For example,

$$7 + 2 = 2 + 7 \qquad \text{and} \qquad 3 \cdot 5 = 5 \cdot 3$$

Is there a commutative property relative to subtraction or division? That is, does $a - b = b - a$ or does $a \div b = b \div a$ for all real numbers a and b (division by 0 excluded)? The answer is no, since, for example,

$$8 - 6 \neq 6 - 8 \qquad \text{and} \qquad 10 \div 5 \neq 5 \div 10$$

When computing

$$3 + 2 + 6 \qquad \text{or} \qquad 3 \cdot 2 \cdot 6$$

why do we not need parentheses to indicate which two numbers are to be added or multiplied first? The answer is to be found in the **associative properties.** These properties allow us to write

$$(3 + 2) + 6 = 3 + (2 + 6) \qquad \text{and} \qquad (3 \cdot 2) \cdot 6 = 3 \cdot (2 \cdot 6)$$

so it does not matter how we group numbers relative to either operation. Is there an associative property for subtraction or division? The answer is no, since, for example,

$$(12 - 6) - 2 \neq 12 - (6 - 2) \qquad \text{and} \qquad (12 \div 6) \div 2 \neq 12 \div (6 \div 2)$$

Evaluate each side of each equation to see why.

Conclusion

Relative to addition, commutativity and associativity permit us to change the order of addition at will and insert or remove parentheses as we please. The same is true for multiplication, but not for subtraction and division.

What number added to a given number will give that number back again? What number times a given number will give that number back again? The answers are 0 and 1, respectively. Because of this, 0 and 1 are called the **identity elements** for the real numbers. Hence, for any real numbers a and b,

$$0 + 5 = 5 \qquad \text{and} \qquad (a + b) + 0 = a + b$$
$$1 \cdot 4 = 4 \qquad \text{and} \qquad (a + b) \cdot 1 = a + b$$

We now consider **inverses.** For each real number a, there is a unique real number $-a$ such that $a + (-a) = 0$. The number $-a$ is called the **additive inverse** of a, or the **negative** of a. For example, the additive inverse (or negative) of 7 is -7, since $7 + (-7) = 0$. The additive inverse (or negative) of -7 is $-(-7) = 7$, since $-7 + [-(-7)] = 0$. It is important to remember that

$-a$ is not necessarily a negative number; it is positive if a is negative and negative if a is positive.

For each number $a \neq 0$, there is a unique real number $1/a$ such that $a(1/a) = 1$. The number $1/a$ is called the **multiplicative inverse** of a, or the **reciprocal** of a. For example, the multiplicative inverse (or reciprocal) of 4 is $\frac{1}{4}$, since $4(\frac{1}{4}) = 1$. (Also note that 4 is the multiplicative inverse of $\frac{1}{4}$.)

We now turn to the **distributive properties,** which involve both multiplication and addition. Consider the following two computations:

$$5(3 + 4) = 5 \cdot 7 = 35 \qquad 5 \cdot 3 + 5 \cdot 4 = 15 + 20 = 35$$

Thus,

$$5(3 + 4) = 5 \cdot 3 + 5 \cdot 4$$

and we say that multiplication by 5 *distributes* over the sum $(3 + 4)$. In general, **multiplication distributes over addition** in the real number system. Two more illustrations are

$$9(m + n) = 9m + 9n \qquad (7 + 2)u = 7u + 2u$$

◆ EXAMPLE 1 State the real number property that justifies the indicated statement.

Statement	Property Illustrated
(A) $x(y + z) = (y + z)x$	Commutative (·)
(B) $5(2y) = (5 \cdot 2)y$	Associative (·)
(C) $2 + (y + 7) = 2 + (7 + y)$	Commutative (+)
(D) $4z + 6z = (4 + 6)z$	Distributive
(E) If $m + n = 0$, then $n = -m$.	Inverse (+)

◆

PROBLEM 1* State the real number property that justifies the indicated statement.

(A) $8 + (3 + y) = (8 + 3) + y$
(B) $(x + y) + z = z + (x + y)$
(C) $(a + b)(x + y) = a(x + y) + b(x + y)$
(D) $5xy + 0 = 5xy$
(E) If $xy = 1$, $x \neq 0$, then $y = 1/x$.

◆

◆ FURTHER PROPERTIES

Subtraction and *division* can be defined in terms of addition and multiplication, respectively:

Subtraction and Division

For all real numbers a and b:

SUBTRACTION: $a - b = a + (-b)$ $\qquad 7 - (-5) = 7 + [-(-5)] = 7 + 5 = 12$

DIVISION: $a \div b = a\left(\frac{1}{b}\right)$, $b \neq 0$ $\qquad 9 \div 4 = 9\left(\frac{1}{4}\right) = \frac{9}{4}$

* Answers to matched problems are found near the end of each section, before the exercise set.

Thus, to subtract b from a, add the negative (the additive inverse) of b to a. To divide a by b, multiply a by the reciprocal (the multiplicative inverse) of b. Note that division by 0 is not defined, since 0 does not have a reciprocal. Thus:

0 can never be used as a divisor!

The following properties of negatives (called **theorems**) can be proved using the preceding axioms and definitions.

THEOREM 1

Properties of Negatives

For all real numbers a and b:

1. $-(-a) = a$
2. $(-a)b = -(ab) = a(-b) = -ab$
3. $(-a)(-b) = ab$
4. $(-1)a = -a$
5. $\dfrac{-a}{b} = -\dfrac{a}{b} = \dfrac{a}{-b}, \quad b \neq 0$
6. $\dfrac{-a}{-b} = -\dfrac{-a}{b} = -\dfrac{a}{-b} = \dfrac{a}{b}, \quad b \neq 0$

We now state an important theorem involving 0.

THEOREM 2

Zero Properties

For all real numbers a and b:

1. $a \cdot 0 = 0$
2. $ab = 0$ if and only if $a = 0$ or $b = 0$ (or both)

◆ EXAMPLE 2 State the real number property or definition that justifies each statement.

Statement	Property Illustrated
(A) $7 - (-5) = 7 + [-(-5)]$	Definition of subtraction
(B) $-(-5) = 5$	Negatives (Theorem 1)
(C) $\dfrac{-7}{3} = -\dfrac{7}{3}$	Negatives (Theorem 1)
(D) $-\dfrac{2}{-3} = \dfrac{2}{3}$	Negatives (Theorem 1)
(E) If $(3x + 2)(x - 7) = 0$, then either $3x + 2 = 0$ or $x - 7 = 0$.	Zero (Theorem 2)

◆

PROBLEM 2 State the real number property or definition that justifies each statement.

(A) $\frac{3}{5} = 3\left(\frac{1}{5}\right)$ (B) $(-5)(2) = -(5 \cdot 2)$ (C) $(-1)3 = -3$

(D) $\frac{-7}{9} = -\frac{7}{9}$ (E) If $x + 5 = 0$, then $(x - 3)(x + 5) = 0$. ◆

◆ FRACTION PROPERTIES

Recall that the quotient $a \div b$ $(b \neq 0)$ written in the form a/b is called a **fraction.** The quantity a is called the **numerator,** and the quantity b is called the **denominator.**

Fraction Properties

For all real numbers a, b, c, d, and k (division by 0 excluded):

1. $\frac{a}{b} = \frac{c}{d}$ if and only if $ad = bc$

   $\frac{4}{6} = \frac{6}{9}$ since $4 \cdot 9 = 6 \cdot 6$

2. $\frac{ka}{kb} = \frac{a}{b}$

   $\frac{7 \cdot 3}{7 \cdot 5} = \frac{3}{5}$

3. $\frac{a}{b} \cdot \frac{c}{d} = \frac{ac}{bd}$

   $\frac{3}{5} \cdot \frac{7}{8} = \frac{3 \cdot 7}{5 \cdot 8}$

4. $\frac{a}{b} \div \frac{c}{d} = \frac{a}{b} \cdot \frac{d}{c}$

   $\frac{2}{3} \div \frac{5}{7} = \frac{2}{3} \cdot \frac{7}{5}$

5. $\frac{a}{b} + \frac{c}{b} = \frac{a + c}{b}$

   $\frac{3}{6} + \frac{5}{6} = \frac{3 + 5}{6}$

6. $\frac{a}{b} - \frac{c}{b} = \frac{a - c}{b}$

   $\frac{7}{8} - \frac{3}{8} = \frac{7 - 3}{8}$

7. $\frac{a}{b} + \frac{c}{d} = \frac{ad + bc}{bd}$

   $\frac{2}{3} + \frac{3}{5} = \frac{2 \cdot 5 + 3 \cdot 3}{3 \cdot 5}$

Answers to Matched Problems

1. (A) Associative (+) (B) Commutative (+) (C) Distributive
   (D) Identity (+) (E) Inverse (·)
2. (A) Definition of division (B) Negatives (Theorem 1)
   (C) Negatives (Theorem 1) (D) Negatives (Theorem 1)
   (E) Zero (Theorem 2)

EXERCISE 1-1

All variables represent real numbers.

A *In Problems 1–6, replace each question mark with an appropriate expression that will illustrate the use of the indicated real number property.*

1. Commutative property (·): $uv = ?$
2. Commutative property (+): $x + 7 = ?$

3. Associative property (+): $3 + (7 + y) = ?$
4. Associative property (·): $x(yz) = ?$
5. Identity property (·): $1(u + v) = ?$
6. Identity property (+): $0 + 9m = ?$

In Problems 7–18, each statement illustrates the use of one of the following properties or definitions; indicate which one:

Commutative (+, ·)	*Associative (+, ·)*	*Distributive*
Identity (+, ·)	*Inverse (+, ·)*	*Subtraction*
Division	*Negatives (Theorem 1)*	*Zero (Theorem 2)*

7. $5(8m) = (5 \cdot 8)m$
8. $a + cb = a + bc$
9. $-\frac{m}{-n} = \frac{m}{n}$
10. $5x + 7x = (5 + 7)x$
11. $7 - 11 = 7 + (-11)$
12. $(-3)(\frac{1}{-3}) = 1$
13. $9 \div (-4) = 9(\frac{1}{-4})$
14. $q + (-q) = 0$
15. $uv(w + x) = uvw + uvx$
16. $2(3x + y) + 0 = 2(3x + y)$
17. $0(m + 5) = 0$
18. $\frac{-u}{-v} = \frac{u}{v}$

B *In Problems 19–26, each statement illustrates the use of one of the following properties or definitions; indicate which one:*

Commutative (+, ·)	*Associative (+, ·)*	*Distributive*
Identity (+, ·)	*Inverse (+, ·)*	*Subtraction*
Division	*Negatives (Theorem 1)*	*Zero (Theorem 2)*

19. $(8u)(4v) = 8[u(4v)]$
20. $(7z + 4) + 2 = 2 + (7z + 4)$
21. $(x + 8)(x + 6) = (x + 8)x + (x + 8)6$
22. $(4x + 3) + (x + 2) = 4x + [3 + (x + 2)]$
23. $\frac{-4}{-(x + y)} = \frac{4}{x + y}$
24. $u(u - 2v) + v(u - 2v) = (u + v)(u - 2v)$
25. If $(x - 2)(2x + 3) = 0$, then either $x - 2 = 0$ or $2x + 3 = 0$.
26. $x(4x + 7) = 0$ if and only if $x = 0$ or $4x + 7 = 0$.
27. If $uv = 1$, does either u or v have to be 1?
28. If $uv = 0$, does either u or v have to be 0?
29. Indicate whether the following are true (T) or false (F):
    (A) All integers are natural numbers.
    (B) All rational numbers are real numbers.
    (C) All natural numbers are rational numbers.
30. Indicate whether the following are true (T) or false (F):
    (A) All natural numbers are integers.
    (B) All real numbers are irrational.
    (C) All rational numbers are real numbers.

31. Give an example of a real number that is not a rational number.
32. Give an example of a rational number that is not an integer.
33. Given the sets of numbers N (natural numbers), Z (integers), Q (rational numbers), and R (real numbers), indicate to which set(s) each of the following numbers belongs:

(A) 8 (B) $\sqrt{2}$ (C) -1.414 (D) $\frac{-5}{2}$

34. Given the sets of numbers N, Z, Q, and R (see Problem 33), indicate to which set(s) each of the following numbers belongs:

(A) -3 (B) 3.14 (C) π (D) $\frac{2}{3}$

35. Indicate true (T) or false (F), and for each false statement find real number replacements for a, b, and c that will illustrate its falseness. For all real numbers a, b, and c:

(A) $(a + b) + c = a + (b + c)$ (B) $(a - b) - c = a - (b - c)$
(C) $a(bc) = (ab)c$ (D) $(a \div b) \div c = a \div (b \div c)$

36. Indicate true (T) or false (F), and for each false statement find real number replacements for a and b that will illustrate its falseness. For all real numbers a and b:

(A) $a + b = b + a$ (B) $a - b = b - a$ (C) $ab = ba$
(D) $a \div b = b \div a$

C

37. If $c = 0.151515...$, then $100c = 15.1515...$ and

$$100c - c = 15.1515... - 0.151515...$$
$$99c = 15$$
$$c = \frac{15}{99} = \frac{5}{33}$$

Proceeding similarly, convert the repeating decimal 0.090909... into a fraction. (All repeating decimals are rational numbers, and all rational numbers have repeating decimal representations.)

38. Repeat Problem 37 for 0.181818... .
39. For a and b real numbers, justify each step below using a property given in this section:

	Statement	Reason
1.	$(a + b) + (-a) = (-a) + (a + b)$	1.
2.	$= [(-a) + a] + b$	2.
3.	$= 0 + b$	3.
4.	$= b$	4.

40. To see how the distributive property is behind the mechanics of long multiplication, compute each of the following and compare:

Long Multiplication	Use of the Distributive Property
23	$23 \cdot 12 = 23(2 + 10)$
$\times 12$	$= 23 \cdot 2 + 23 \cdot 10 =$

Use a calculator to express each number as a decimal to the capacity of your calculator. Observe the repeating decimal representation of the rational numbers and the nonrepeating decimal representation of the irrational numbers.

41. (A) $\frac{13}{6}$ (B) $\sqrt{21}$ (C) $\frac{7}{16}$ (D) $\frac{29}{111}$

42. (A) $\frac{8}{9}$ (B) $\frac{3}{11}$ (C) $\sqrt{5}$ (D) $\frac{11}{8}$

SECTION 1-2

Basic Operations on Polynomials

- NATURAL NUMBER EXPONENTS
- POLYNOMIALS
- COMBINING LIKE TERMS
- ADDITION AND SUBTRACTION
- MULTIPLICATION
- COMBINED OPERATIONS

This section covers basic operations on *polynomials*, a mathematical form encountered frequently in mathematics. Our discussion starts with a brief review of natural number exponents. (Integer and rational exponents and their properties will be discussed in detail in subsequent sections.)

◆ NATURAL NUMBER EXPONENTS

We define a **natural number exponent** as follows:

Natural Number Exponent

For n a natural number and b any real number:

$b^n = b \cdot b \cdot \cdots \cdot b$ n factors of b

$3^5 = 3 \cdot 3 \cdot 3 \cdot 3 \cdot 3$ 5 factors of 3

where n is called the **exponent** and b is called the **base.**

Along with this definition, we state the **first property of exponents:**

THEOREM 3

First Property of Exponents

For any natural numbers m and n, and any real number b:

$b^m b^n = b^{m+n}$ $(2t^4)(5t^3) = 2 \cdot 5t^{4+3}$* $= 10t^7$

* Dashed boxes are used throughout the book to represent steps that are usually performed mentally.

◆ POLYNOMIALS

Algebraic expressions are formed by using constants and variables and the algebraic operations of addition, subtraction, multiplication, division, raising to powers, and taking roots. Special types of algebraic expressions are called *polynomials*. A **polynomial in one variable** x is constructed by adding or subtracting constants and terms of the form ax^n, where a is a real number and n is a natural number. A **polynomial in two variables** x and y is constructed by adding and subtracting constants and terms of the form ax^my^n, where a is a real number and m and n are natural numbers. Polynomials in three and more variables are defined in a similar manner.

Examples of Polynomials

8	0
$3x^3 - 6x + 7$	$6x + 3$
$2x^2 - 7xy - 8y^2$	$9y^3 + 4y^2 - y + 4$
$2x - 3y + 2$	$u^5 - 3u^3v^2 + 2uv^4 - v^4$

Polynomial forms are encountered frequently in mathematics, and for their more efficient study, it is useful to classify them according to their *degree*. If a term in a polynomial has only one variable as a factor, then the **degree of the term** is the power of the variable. If two or more variables are present in a term as factors, then the **degree of the term** is the sum of the powers of the variables. The **degree of a polynomial** is the degree of the nonzero term with the highest degree in the polynomial. Any nonzero constant is defined to be a **polynomial of degree 0.** The number 0 is also a polynomial but is not assigned a degree.

◆ EXAMPLE 3

(A) The degree of the first term in $5x^3 + \sqrt{3}x - \frac{1}{2}$ is 3, the degree of the second term is 1, the degree of the third term is 0, and the degree of the whole polynomial is 3 (the same as the degree of the term with the highest degree).

(B) The degree of the first term in $8u^3v^2 - \sqrt{7}uv^2$ is 5, the degree of the second term is 3, and the degree of the whole polynomial is 5. ◆

PROBLEM 3

(A) Given the polynomial $6x^5 + 7x^3 - 2$, what is the degree of the first term? The second term? The third term? The whole polynomial?

(B) Given the polynomial $2u^4v^2 - 5uv^3$, what is the degree of the first term? The second term? The whole polynomial? ◆

In addition to classifying polynomials by degree, we also call a single-term polynomial a **monomial,** a two-term polynomial a **binomial,** and a three-term polynomial a **trinomial.**

◆ COMBINING LIKE TERMS

The concept of *coefficient* plays a central role in the process of combining *like terms.* A constant in a term of a polynomial, including the sign that precedes it, is called the **numerical coefficient,** or simply, the **coefficient,** of the term. If a constant does not appear, or only a + sign appears, the coefficient is understood to be 1. If only a − sign appears, the coefficient is understood to be −1. Thus, given the polynomial

$$5x^4 - x^3 - 3x^2 + x - 7 = 5x^4 + (-1)x^3 + (-3)x^2 + 1x + (-7)$$

the coefficient of the first term is 5, the coefficient of the second term is −1, the coefficient of the third term is −3, the coefficient of the fourth term is 1, and the coefficient of the fifth term is −7.

It is useful to state two more distributive properties of real numbers that follow from the distributive properties stated in Section 1-1.

Further Distributive Properties

1. $a(b - c) = (b - c)a = ab - ac$
2. $a(b + c + \cdots + f) = ab + ac + \cdots + af$

Two terms in a polynomial are called **like terms** if they have exactly the same variable factors to the same powers. The numerical coefficients may or may not be the same. Since constant terms involve no variables, all constant terms are like terms. If a polynomial contains two or more like terms, these terms can be combined into a single term by making use of distributive properties. The following example illustrates the reasoning behind the process:

$$\begin{aligned} 3x^2y - 5xy^2 + x^2y - 2x^2y &= 3x^2y + x^2y - 2x^2y - 5xy^2 \\ &= (3x^2y + 1x^2y - 2x^2y) - 5xy^2 \\ &= (3 + 1 - 2)x^2y - 5xy^2 \\ &= 2x^2y - 5xy^2 \end{aligned}$$

It should be clear that free use is made of the real number properties discussed earlier. The steps shown in the dashed box are usually done mentally, and the process is quickly mechanized as follows:

Like terms in a polynomial are combined by adding their numerical coefficients.

How can we simplify expressions such as $4(x - 2y) - 3(2x - 7y)$? We clear the expression of parentheses using distributive properties, and combine like terms:

$$\begin{aligned} 4(x - 2y) - 3(2x - 7y) &= 4x - 8y - 6x + 21y \\ &= -2x + 13y \end{aligned}$$

◆ EXAMPLE 4 Remove parentheses and simplify:

(A) $2(3x^2 - 2x + 5) + (x^2 + 3x - 7) = 2(3x^2 - 2x + 5) + 1(x^2 + 3x - 7)$ *Think*

$= 6x^2 - 4x + 10 + x^2 + 3x - 7$

$= 7x^2 - x + 3$

(B) $(x^3 - 2x - 6) - (2x^3 - x^2 + 2x - 3)$

$= 1(x^3 - 2x - 6) + (-1)(2x^3 - x^2 + 2x - 3)$ *Think* — Be careful with the sign here.

$= x^3 - 2x - 6 - 2x^3 + x^2 - 2x + 3$

$= -x^3 + x^2 - 4x - 3$

(C) $[3x^2 - (2x + 1)] - (x^2 - 1) = [3x^2 - 2x - 1] - (x^2 - 1)$ — Remove inner parentheses first.

$= 3x^2 - 2x - 1 - x^2 + 1$

$= 2x^2 - 2x$ ◆

PROBLEM 4 Remove parentheses and simplify:

(A) $3(u^2 - 2v^2) + (u^2 + 5v^2)$ (B) $(m^3 - 3m^2 + m - 1) - (2m^3 - m + 3)$

(C) $(x^3 - 2) - [2x^3 - (3x + 4)]$ ◆

◆ ADDITION AND SUBTRACTION

Addition and subtraction of polynomials can be thought of in terms of removing parentheses and combining like terms, as illustrated in Example 4. Horizontal and vertical arrangements are illustrated in the next two examples. You should be able to work either way, letting the situation dictate your choice.

◆ EXAMPLE 5 Add horizontally and vertically: $x^4 - 3x^3 + x^2$, $-x^3 - 2x^2 + 3x$, and $3x^2 - 4x - 5$

Solution Add horizontally:

$$(x^4 - 3x^3 + x^2) + (-x^3 - 2x^2 + 3x) + (3x^2 - 4x - 5)$$
$$= x^4 - 3x^3 + x^2 - x^3 - 2x^2 + 3x + 3x^2 - 4x - 5$$
$$= x^4 - 4x^3 + 2x^2 - x - 5$$

Or vertically, by lining up like terms and adding their coefficients:

$$\begin{array}{rrrrr} x^4 & - 3x^3 & + x^2 & & \\ & - x^3 & - 2x^2 & + 3x & \\ & & 3x^2 & - 4x & - 5 \\ \hline x^4 & - 4x^3 & + 2x^2 & - x & - 5 \end{array}$$

◆

PROBLEM 5 Add horizontally and vertically: $3x^4 - 2x^3 - 4x^2$, $x^3 - 2x^2 - 5x$, and $x^2 + 7x - 2$ ◆

◆ EXAMPLE 6 Subtract $4x^2 - 3x + 5$ from $x^2 - 8$, both horizontally and vertically.

Solution

$$(x^2 - 8) - (4x^2 - 3x + 5) = x^2 - 8 - 4x^2 + 3x - 5 = -3x^2 + 3x - 13$$

or

$$\begin{array}{r} x^2 \qquad\quad - 8 \\ -4x^2 + 3x - 5 \\ \hline -3x^2 + 3x - 13 \end{array}$$

← Change signs and add. ◆

PROBLEM 6 Subtract $2x^2 - 5x + 4$ from $5x^2 - 6$, both horizontally and vertically. ◆

◆ MULTIPLICATION

Multiplication of algebraic expressions involves the extensive use of distributive properties for real numbers, as well as other real number properties.

◆ EXAMPLE 7 Multiply: $(2x - 3)(3x^2 - 2x + 3)$

Solution

$$\begin{aligned}(2x - 3)(3x^2 - 2x + 3) &= 2x(3x^2 - 2x + 3) - 3(3x^2 - 2x + 3)\\ &= 6x^3 - 4x^2 + 6x - 9x^2 + 6x - 9\\ &= 6x^3 - 13x^2 + 12x - 9\end{aligned}$$

Or, using a vertical arrangement,

$$\begin{array}{rrrr} 3x^2 & - 2x & + 3 & \\ 2x & - 3 & & \\ \hline 6x^3 - & 4x^2 + & 6x & \\ - & 9x^2 + & 6x & - 9 \\ \hline 6x^3 - & 13x^2 + & 12x & - 9 \end{array}$$

◆

PROBLEM 7 Multiply: $(2x - 3)(2x^2 + 3x - 2)$ ◆

Thus, to multiply two polynomials, multiply each term of one by each term of the other, and combine like terms.

Products of binomial factors occur frequently, so it is useful to develop procedures that will enable us to write down their products by inspection. To find the product $(2x - 1)(3x + 2)$, we will use the popular **FOIL method.** We multiply each term of one factor by each term of the other factor as follows:

	F		O		I		L
	First product ↓		Outer product ↓		Inner product ↓		Last product ↓
$(2x - 1)(3x + 2) =$	$6x^2$	$+$	$4x$	$-$	$3x$	$-$	2

The inner and outer products are like terms and hence combine into one term. Thus,

$$(2x - 1)(3x + 2) = 6x^2 + x - 2$$

To speed up the process, we combine the inner and outer product mentally.

Products of certain binomial factors occur so frequently that it is useful to remember formulas for their products. The following formulas are easily verified by multiplying the factors on the left using the FOIL method:

Special Products

1. $(a - b)(a + b) = a^2 - b^2$ 2. $(a + b)^2 = a^2 + 2ab + b^2$
3. $(a - b)^2 = a^2 - 2ab + b^2$

◆ EXAMPLE 8 Multiply mentally:

(A) $(2x - 3y)(5x + 2y) = 10x^2 - 11xy - 6y^2$
(B) $(3a - 2b)(3a + 2b) = 9a^2 - 4b^2$
(C) $(5x - 3)^2 = 25x^2 - 30x + 9$
(D) $(m + 2n)^2 = m^2 + 4mn + 4n^2$ ◆

PROBLEM 8 Multiply mentally:

(A) $(4u - 3v)(2u + v)$ (B) $(2xy + 3)(2xy - 3)$ (C) $(m + 4n)(m - 4n)$
(D) $(2u - 3v)^2$ (E) $(6x + y)^2$ ◆

◆ COMBINED OPERATIONS

We complete this section by considering several examples that use all the operations just discussed. Note that in simplifying, we usually remove grouping symbols starting from the inside. That is, we remove parentheses ( ) first, then brackets [ ], and finally braces { }, if present. Also, multiplication and division precede addition and subtraction.

◆ EXAMPLE 9 Perform the indicated operations and simplify:

(A)
$$\begin{aligned} 3x - \{5 - 3[x - x(3 - x)]\} &= 3x - \{5 - 3[x - 3x + x^2]\} \\ &= 3x - \{5 - 3x + 9x - 3x^2\} \\ &= 3x - 5 + 3x - 9x + 3x^2 \\ &= 3x^2 - 3x - 5 \end{aligned}$$

(B)
$$\begin{aligned} (x - 2y)(2x + 3y) - (2x + y)^2 &= 2x^2 - xy - 6y^2 - (4x^2 + 4xy + y^2) \\ &= 2x^2 - xy - 6y^2 - 4x^2 - 4xy - y^2 \\ &= -2x^2 - 5xy - 7y^2 \end{aligned}$$ ◆

PROBLEM 9 Perform the indicated operations and simplify:

(A) $2t - \{7 - 2[t - t(4 + t)]\}$ (B) $(u - 3v)^2 - (2u - v)(2u + v)$ ◆

Answers to Matched Problems

3. (A) 5, 3, 0, 5 (B) 6, 4, 6
4. (A) $4u^2 - v^2$ (B) $-m^3 - 3m^2 + 2m - 4$ (C) $-x^3 + 3x + 2$

5. $3x^4 - x^3 - 5x^2 + 2x - 2$ 6. $3x^2 + 5x - 10$ 7. $4x^3 - 13x + 6$

8. (A) $8u^2 - 2uv - 3v^2$ (B) $4x^2y^2 - 9$ (C) $m^2 - 16n^2$
(D) $4u^2 - 12uv + 9v^2$ (E) $36x^2 + 12xy + y^2$

9. (A) $-2t^2 - 4t - 7$ (B) $-3u^2 - 6uv + 10v^2$

EXERCISE 1-2

A *Problems 1–8 refer to the following polynomials:*

(A) $2x - 3$ (B) $2x^2 - x + 2$ (C) $x^3 + 2x^2 - x + 3$

1. What is the degree of (C)?
2. What is the degree of (A)?
3. Add (B) and (C).
4. Add (A) and (B).
5. Subtract (B) from (C).
6. Subtract (A) from (B).
7. Multiply (B) and (C).
8. Multiply (A) and (C).

In Problems 9–30, perform the indicated operations and simplify.

9. $2(u - 1) - (3u + 2) - 2(2u - 3)$
10. $2(x - 1) + 3(2x - 3) - (4x - 5)$
11. $4a - 2a[5 - 3(a + 2)]$
12. $2y - 3y[4 - 2(y - 1)]$
13. $(a + b)(a - b)$
14. $(m - n)(m + n)$
15. $(3x - 5)(2x + 1)$
16. $(4t - 3)(t - 2)$
17. $(2x - 3y)(x + 2y)$
18. $(3x + 2y)(x - 3y)$
19. $(3y + 2)(3y - 2)$
20. $(2m - 7)(2m + 7)$
21. $(3m + 7n)(2m - 5n)$
22. $(6x - 4y)(5x + 3y)$
23. $(4m + 3n)(4m - 3n)$
24. $(3x - 2y)(3x + 2y)$
25. $(3u + 4v)^2$
26. $(4x - y)^2$
27. $(a - b)(a^2 + ab + b^2)$
28. $(a + b)(a^2 - ab + b^2)$
29. $(4x + 3y)^2$
30. $(3x + 2)^2$

B *In Problems 31–38, perform the indicated operations and simplify.*

31. $m - \{m - [m - (m - 1)]\}$
32. $2x - 3\{x + 2[x - (x + 5)] + 1\}$
33. $(x^2 - 2xy + y^2)(x^2 + 2xy + y^2)$
34. $(2x^2 + x - 2)(x^2 - 3x + 5)$
35. $(3a - b)(3a + b) - (2a - 3b)^2$
36. $(2x - 1)^2 - (3x + 2)(3x - 2)$
37. $(x - 2y)^3$
38. $(2m - n)^3$
39. Subtract the sum of the last two polynomials from the sum of the first two: $2x^2 - 4xy + y^2$, $3xy - y^2$, $x^2 - 2xy - y^2$, $-x^2 + 3xy - 2y^2$
40. Subtract the sum of the first two polynomials from the sum of the last two: $3m^2 - 2m + 5$, $4m^2 - m$, $3m^2 - 3m - 2$, $m^3 + m^2 + 2$

C *In Problems 41–44, perform the indicated operations and simplify.*

41. $(2x - 1)^3 - 2(2x - 1)^2 + 3(2x - 1) + 7$
42. $2(x - 2)^3 - (x - 2)^2 - 3(x - 2) - 4$
43. $2\{(x - 3)(x^2 - 2x + 1) - x[3 - x(x - 2)]\}$
44. $-3x\{x[x - x(2 - x)] - (x + 2)(x^2 - 3)\}$

45. If you are given two polynomials, one of degree m and the other of degree n, where $m > n$, what is the degree of their product?

46. What is the degree of the sum of the two polynomials in Problem 45?

APPLICATIONS

Business & Economics

47. *Investment.* You have $10,000 to invest, part at 9% and the rest at 12%. If x is the amount invested at 9%, write an algebraic expression that represents the total annual income from both investments. Simplify the expression.

48. *Gross receipts.* Six thousand tickets are to be sold for a concert, some for $9 each and the rest for $15 each. If x is the number of $9 tickets sold, write an algebraic expression that represents the gross receipts from ticket sales, assuming all tickets are sold. Simplify the expression.

49. *Gross receipts.* Four thousand tickets are to be sold for a musical show. If x tickets are to be sold for $10 each and three times that number for $30 each, and if the rest are sold for $50 each, write an algebraic expression that represents the gross receipts from ticket sales, assuming all tickets are sold. Simplify the expression.

50. *Investment.* A person has $100,000 to invest. If $$x$ are invested in a money market account yielding 7% and twice that amount in certificates of deposit yielding 9%, and if the rest is invested in high-grade bonds yielding 11%, write an algebraic expression that represents the total annual income from all three investments. Simplify the expression.

Life Sciences

51. *Nutrition.* Food mix A contains 2% fat, and food mix B contains 6% fat. A 10 kilogram diet mix of foods A and B is formed. If x kilograms of food A are used, write an algebraic expression that represents the total number of kilograms of fat in the final food mix. Simplify the expression.

52. *Nutrition.* Each ounce of food M contains 8 units of calcium, and each ounce of food N contains 5 units of calcium. A 160 ounce diet mix is formed using foods M and N. If x is the number of ounces of food M used, write an algebraic expression that represents the total number of units of calcium in the diet mix. Simplify the expression.

SECTION 1-3 Factoring Polynomials

- WHAT DO WE MEAN BY FACTORING?
- COMMON FACTORS
- FACTORING SECOND-DEGREE POLYNOMIALS
- MORE FACTORING

Our objective in this section is to review some of the standard factoring techniques for polynomials with integer coefficients.

◆ WHAT DO WE MEAN BY FACTORING?

If a number is written as the product of other numbers, then each number in the product is called a **factor of the original number.** Similarly, if an algebraic expression is written as the product of other algebraic expressions, then each algebraic expression in the product is called a **factor of the original algebraic expression.** For example,

$42 = 2 \cdot 3 \cdot 7$	2, 3, and 7 are factors of 42
$x^2 - 9 = (x - 3)(x + 3)$	$(x - 3)$ and $(x + 3)$ are factors of $x^2 - 9$

The process of writing a number or algebraic expression as the product of other numbers or algebraic expressions is called **factoring.** We begin our discussion of factoring with the positive integers.

An integer such as 42 can be represented in a factored form in many ways. The products

$7 \cdot 6 \qquad 21 \cdot 2 \qquad 3 \cdot 14 \qquad (\frac{1}{2})(12)(7) \qquad 2 \cdot 3 \cdot 7$

all yield 42. A particularly useful way of factoring integers greater than 1 is in terms of *prime numbers.*

Prime and Composite Numbers

An integer greater than 1 is **prime** if its only positive integer factors are itself and 1. An integer greater than 1 that is not prime is called a **composite number.** The integer 1 is neither prime nor composite.

PRIME NUMBERS:	2, 3, 5, 7, 11, 13, 17
COMPOSITE NUMBERS:	4, 6, 8, 9, 10, 12, 14

[*Note:* There is no largest prime number.]

A composite number is said to be **factored completely** if it is represented as a product of prime factors. The only factoring of 42 given above that meets this condition is $42 = 2 \cdot 3 \cdot 7$.

◆ EXAMPLE 10 Write 90 in completely factored form.

Solution

$$90 = 9 \cdot 10 = 3 \cdot 3 \cdot 2 \cdot 5 = 2 \cdot 3^2 \cdot 5$$

or

$$90 = 6 \cdot 15 = 2 \cdot 3 \cdot 3 \cdot 5 = 2 \cdot 3^2 \cdot 5$$

or

$$90 = 18 \cdot 5 = 3 \cdot 6 \cdot 5 = 3 \cdot 2 \cdot 3 \cdot 5 = 2 \cdot 3^2 \cdot 5$$

◆

Notice in Example 10 that we end up with the same set of prime factors of 90 no matter how we progress through the factoring process. This illustrates an important property of integers.

THEOREM 4

The Fundamental Theorem of Arithmetic

Each integer greater than 1 is either prime or can be expressed uniquely, except for the order of factors, as a product of prime factors.

PROBLEM 10

Write 180 in completely factored form. ◆

We also can write polynomials in completely factored form. A polynomial such as $3x^2 + 7x - 6$ can be written in factored form in many ways. The products

$$(3x - 2)(x + 3) \qquad 3x\left(x + \frac{7}{3} - \frac{2}{x}\right) \qquad 3\left(x - \frac{2}{3}\right)(x + 3)$$

all yield $3x^2 + 7x - 6$. A particularly useful way of factoring polynomials is in terms of *prime polynomials.*

Prime Polynomials

A polynomial of degree greater than 0 is said to be **prime relative to a given set of numbers:**

1. If all of its coefficients are from that set of numbers
2. If it cannot be written as a product of two polynomials, excluding 1 and itself, having coefficients from that set of numbers

Relative to the set of integers:

$x^2 - 3$ is prime

$x^2 - 4$ is not prime, since $x^2 - 4 = (x - 2)(x + 2)$

[*Note:* The set of integers is the set of numbers used most frequently in factoring polynomials.]

A nonprime polynomial is said to be **factored completely relative to a given set of numbers** if it is written as a product of prime polynomials relative to that set of numbers. The only factoring given above of $3x^2 + 7x - 6$ relative to the integers is $3x^2 + 7x - 6 = (3x - 2)(x + 3)$.

Writing polynomials in completely factored form is often a difficult task. But accomplishing it can lead to the simplification of certain algebraic expressions and to the solution of certain types of equations. The distributive properties for real numbers are central to the factoring process.

◆ COMMON FACTORS

Generally, a first step in any factoring procedure is to factor out all factors common to all terms.

◆ EXAMPLE 11 Factor out, relative to the integers, all factors common to all terms:

(A) $3x^3y - 6x^2y^2 - 3xy^3$ (B) $3y(2y + 5) + 2(2y + 5)$

Solutions (A) $3x^3y - 6x^2y^2 - 3xy^3 = \mathbf{(3xy)}x^2 - \mathbf{(3xy)}2xy - \mathbf{(3xy)}y^2$

$= 3xy(x^2 - 2xy - y^2)$

(B) $3y(2y + 5) + 2(2y + 5) = 3y\mathbf{(2y + 5)} + 2\mathbf{(2y + 5)}$

$= (3y + 2)(2y + 5)$ ◆

PROBLEM 11 Factor out, relative to the integers, all factors common to all terms:

(A) $2x^3y - 8x^2y^2 - 6xy^3$ (B) $2x(3x - 2) - 7(3x - 2)$ ◆

◆ FACTORING SECOND-DEGREE POLYNOMIALS

We now turn our attention to factoring second-degree polynomials of the form

$2x^2 - 5x - 3$ and $2x^2 + 3xy - 2y^2$

into the product of two first-degree polynomials with integer coefficients. The following example will illustrate an approach to the problem.

◆ EXAMPLE 12 Factor each polynomial, if possible, using integer coefficients:

(A) $2x^2 + 3xy - 2y^2$ (B) $x^2 - 3x + 4$ (C) $6x^2 + 5xy - 4y^2$

Solutions (A) $2x^2 + 3xy - 2y^2 = (2x + \ \ y)(x - \ \ y)$ (with ↑ ? under each blank)

Put in what we know. Signs must be opposite. (We can reverse this choice if we get $-3xy$ instead of $+3xy$ for the middle term.)

Now, what are the factors of 2 (the coefficient of y^2)?

2
1 · 2
2 · 1

The first choice gives us $-3xy$ for the middle term—close, but not there—so we reverse our choice of signs to obtain

$$2x^2 + 3xy - 2y^2 = (2x - y)(x + 2y)$$

(B) $x^2 - 3x + 4 = (x - \quad)(x - \quad)$

$$\begin{array}{cc} & \uparrow \quad \uparrow \\ & ? \quad\; ? \end{array}$$

4
2 · 2
1 · 4
4 · 1

No choice produces the middle term; hence, $x^2 - 3x + 4$ is not factorable using integer coefficients.

(C) $6x^2 + 5xy - 4y^2 = (\quad x + \quad y)(\quad x - \quad y)$

$$\uparrow \quad \uparrow \quad \uparrow \quad \uparrow$$
$$? \quad ? \quad ? \quad ?$$

The signs must be opposite in the factors, since the third term is negative. (We can reverse our choice of signs later if necessary.) We write all factors of 6 and of 4:

6	4
2 · 3	2 · 2
3 · 2	1 · 4
1 · 6	4 · 1
6 · 1	

Now try each choice on the left with each choice on the right—a total of 12 combinations that give us the first and last terms in the polynomial $6x^2 + 5xy - 4y^2$. The question is: Does any combination also give us the middle term, $5xy$? After trial and error and, perhaps, some educated guessing among the choices, we find that $3 \cdot 2$ matched with $4 \cdot 1$ gives us the correct middle term. Thus,

$$6x^2 + 5xy - 4y^2 = (3x + 4y)(2x - y)$$

If none of the 24 combinations (including reversing our sign choice) had produced the middle term, then we would conclude that the polynomial is not factorable using integer coefficients. ◆

PROBLEM 12 Factor each polynomial, if possible, using integer coefficients:

(A) $x^2 - 8x + 12$ (B) $x^2 + 2x + 5$ (C) $2x^2 + 7xy - 4y^2$
(D) $4x^2 - 15xy - 4y^2$ ◆

◆ MORE FACTORING

The factoring formulas listed below will enable us to factor certain polynomial forms that occur frequently. These formulas can be established by multiplying the factors on the right.

Special Factoring Formulas

PERFECT SQUARE:	1.	$u^2 + 2uv + v^2 = (u + v)^2$
PERFECT SQUARE:	2.	$u^2 - 2uv + v^2 = (u - v)^2$
DIFFERENCE OF SQUARES:	3.	$u^2 - v^2 = (u - v)(u + v)$
DIFFERENCE OF CUBES:	4.	$u^3 - v^3 = (u - v)(u^2 + uv + v^2)$
SUM OF CUBES:	5.	$u^3 + v^3 = (u + v)(u^2 - uv + v^2)$

◆ EXAMPLE 13

Factor completely relative to the integers:

(A) $4m^2 - 12mn + 9n^2$ (B) $x^2 - 16y^2$ (C) $z^3 - 1$ (D) $m^3 + n^3$

Solutions

(A) $4m^2 - 12mn + 9n^2 = (2m - 3n)^2$

(B) $x^2 - 16y^2 \boxed{= x^2 - (4y)^2} = (x - 4y)(x + 4y)$

(C) $z^3 - 1 = (z - 1)(z^2 + z + 1)$

(D) $m^3 + n^3 = (m + n)(m^2 - mn + n^2)$ ◆

PROBLEM 13

Factor completely relative to the integers:

(A) $x^2 + 6xy + 9y^2$ (B) $9x^2 - 4y^2$ (C) $8m^3 - 1$ (D) $x^3 + y^3z^3$ ◆

We complete this section by considering factoring that involves combinations of the preceding techniques.

Factoring Polynomials

Generally speaking, when asked to factor a polynomial, we first take out all factors common to all terms, if they are present, and then proceed as above until all factors are prime.

◆ EXAMPLE 14

Factor completely relative to the integers:

(A) $3x^3 - 48x$ (B) $3u^4 - 3u^3v - 9u^2v^2$ (C) $3m^4 - 24mn^3$
(D) $3x^4 - 5x^2 + 2$

Solutions

(A) $3x^3 - 48x = 3x(x^2 - 16) = 3x(x - 4)(x + 4)$

(B) $3u^4 - 3u^3v - 9u^2v^2 = 3u^2(u^2 - uv - 3v^2)$

(C) $3m^4 - 24mn^3 = 3m(m^3 - 8n^3) = 3m(m - 2n)(m^2 + 2mn + 4n^2)$

(D) $3x^4 - 5x^2 + 2 = (3x^2 - 2)(x^2 - 1) = (3x^2 - 2)(x - 1)(x + 1)$ ◆

PROBLEM 14

Factor completely relative to the integers:

(A) $18x^3 - 8x$ (B) $4m^3n - 2m^2n^2 + 2mn^3$ (C) $2t^4 - 16t$
(D) $2y^4 - 5y^2 - 12$ ◆

Answers to Matched Problems

10. $2^2 \cdot 3^2 \cdot 5$ 11. (A) $2xy(x^2 - 4xy - 3y^2)$ (B) $(2x - 7)(3x - 2)$
12. (A) $(x - 2)(x - 6)$ (B) Not factorable using integers
(C) $(2x - y)(x + 4y)$ (D) $(4x + y)(x - 4y)$
13. (A) $(x + 3y)^2$ (B) $(3x - 2y)(3x + 2y)$ (C) $(2m - 1)(4m^2 + 2m + 1)$
(D) $(x + yz)(x^2 - xyz + y^2z^2)$
14. (A) $2x(3x - 2)(3x + 2)$ (B) $2mn(2m^2 - mn + n^2)$
(C) $2t(t - 2)(t^2 + 2t + 4)$ (D) $(2y^2 + 3)(y - 2)(y + 2)$

EXERCISE 1-3

A

Factor out, relative to the integers, all factors common to all terms.

1. $6m^4 - 9m^3 - 3m^2$
2. $6x^4 - 8x^3 - 2x^2$
3. $8u^3v - 6u^2v^2 + 4uv^3$
4. $10x^3y + 20x^2y^2 - 15xy^3$
5. $7m(2m - 3) + 5(2m - 3)$
6. $5x(x + 1) - 3(x + 1)$
7. $a(3c + d) - 4b(3c + d)$
8. $2w(y - 2z) - x(y - 2z)$

Factor completely relative to the integers. If a polynomial is prime relative to the integers, say so.

9. $3y^2 - y - 2$
10. $2x^2 + 5x - 3$
11. $u^2 - 2uv - 15v^2$
12. $x^2 - 4xy - 12y^2$
13. $m^2 - 6m - 3$
14. $x^2 + x - 4$
15. $w^2x^2 - y^2$
16. $25m^2 - 16n^2$
17. $9m^2 - 6mn + n^2$
18. $x^2 + 10xy + 25y^2$
19. $y^2 + 16$
20. $u^2 + 81$

B

21. $4z^2 - 28z + 48$
22. $6x^2 + 48x + 72$
23. $2x^4 - 24x^3 + 40x^2$
24. $2y^3 - 22y^2 + 48y$
25. $4xy^2 - 12xy + 9x$
26. $16x^2y - 8xy + y$
27. $6m^2 - mn - 12n^2$
28. $6s^2 + 7st - 3t^2$
29. $4u^3v - uv^3$
30. $x^3y - 9xy^3$
31. $2x^3 - 2x^2 + 8x$
32. $3m^3 - 6m^2 + 15m$
33. $r^3 - t^3$
34. $m^3 + n^3$
35. $a^3 + 1$
36. $c^3 - 1$

C

37. $(x + 2)^2 - 9y^2$
38. $(a - b)^2 - 4(c - d)^2$
39. $5u^2 + 4uv - 2v^2$
40. $3x^2 - 2xy - 4y^2$
41. $6(x - y)^2 + 23(x - y) - 4$
42. $4(A + B)^2 - 5(A + B) - 6$
43. $y^4 - 3y^2 - 4$
44. $m^4 - n^4$
45. $27a^2 + a^5b^3$
46. $s^4t^4 - 8st$

SECTION 1-4 Basic Operations on Rational Expressions

- REDUCING TO LOWEST TERMS
- MULTIPLICATION AND DIVISION
- ADDITION AND SUBTRACTION
- COMPOUND FRACTIONS

We now turn our attention to fractional forms. A quotient of two algebraic expressions (division by 0 excluded) is called a **fractional expression.** If both the numerator and the denominator are polynomials, the fractional expression is called a **rational expression.** Some examples of rational expressions are

$$\frac{1}{x^3 + 2x} \qquad \frac{5}{x} \qquad \frac{x + 7}{3x^2 - 5x + 1} \qquad \frac{x^2 - 2x + 4}{1}$$

In this section we will discuss basic operations on rational expressions, including multiplication, division, addition, and subtraction.

Since variables represent real numbers in the rational expressions we will consider, the properties of real number fractions summarized in Section 1-1 will play a central role in much of the work that we will do.

Even though not always explicitly stated, we always assume that variables are restricted so that division by 0 is excluded.

◆ REDUCING TO LOWEST TERMS

Central to the process of reducing rational expressions to *lowest terms* is the *fundamental property of fractions,* which we restate here for convenient reference:

Fundamental Property of Fractions

If a, b, and k are real numbers with $b, k \neq 0$, then

$$\frac{ka}{kb} = \frac{a}{b} \qquad \frac{5 \cdot 2}{5 \cdot 7} = \frac{2}{7} \qquad \frac{x(x + 4)}{2(x + 4)} = \frac{x}{2}, \quad x \neq -4$$

Using this property from left to right to eliminate all common factors from the numerator and the denominator of a given fraction is referred to as **reducing a fraction to lowest terms.** We are actually dividing the numerator and denominator by the same nonzero common factor. Unless stated to the contrary, factors will be relative to the integers.

Using the property from right to left—that is, multiplying the numerator and denominator by the same nonzero factor—is referred to as **raising a fraction to higher terms.** We will use the property in both directions in the material that follows.

◆ EXAMPLE 15 Reduce each rational expression to lowest terms.

(A) $$\frac{6x^2 + x - 1}{2x^2 - x - 1} = \frac{(2x + 1)(3x - 1)}{(2x + 1)(x - 1)} = \frac{3x - 1}{x - 1}$$

Factor numerator and denominator completely; divide numerator and denominator by $(2x + 1)$.

(B) $\dfrac{x^4 - 8x}{3x^3 - 2x^2 - 8x} = \dfrac{x(x-2)(x^2+2x+4)}{x(x-2)(3x+4)}$

$= \dfrac{x^2+2x+4}{3x+4}$ ◆

PROBLEM 15 Reduce each rational expression to lowest terms.

(A) $\dfrac{x^2 - 6x + 9}{x^2 - 9}$ (B) $\dfrac{x^3 - 1}{x^2 - 1}$ ◆

◆ MULTIPLICATION AND DIVISION

Since we are restricting variable replacements to real numbers, multiplication and division of rational expressions follow the rules for multiplying and dividing real number fractions summarized in Section 1-1.

Multiplication and Division

If a, b, c, and d are real numbers, then:

1. $\dfrac{a}{b} \cdot \dfrac{c}{d} = \dfrac{ac}{bd}$, $b, d \neq 0$ $\qquad \dfrac{3}{5} \cdot \dfrac{x}{x+5} = \dfrac{3x}{5(x+5)}$
2. $\dfrac{a}{b} \div \dfrac{c}{d} = \dfrac{a}{b} \cdot \dfrac{d}{c}$, $b, c, d \neq 0$ $\qquad \dfrac{3}{5} \div \dfrac{x}{x+5} = \dfrac{3}{5} \cdot \dfrac{x+5}{x}$

◆ EXAMPLE 16 Perform the indicated operations and reduce to lowest terms.

(A) $\dfrac{10x^3y}{3xy + 9y} \cdot \dfrac{x^2 - 9}{4x^2 - 12x}$ Factor numerators and denominators; then divide any numerator and any denominator with a like common factor.

$= \dfrac{\overset{5x^2}{\cancel{10x^3y}}}{\underset{3 \cdot 1}{\cancel{3y(x+3)}}} \cdot \dfrac{\overset{1 \cdot 1}{\cancel{(x-3)(x+3)}}}{\underset{2 \cdot 1}{\cancel{4x(x-3)}}}$

$= \dfrac{5x^2}{6}$

(B) $\dfrac{4 - 2x}{4} \div (x - 2) = \dfrac{\overset{1}{\cancel{2}}(2-x)}{\underset{2}{\cancel{4}}} \cdot \dfrac{1}{x-2}$ $\qquad x - 2$ is the same as $\dfrac{x-2}{1}$

$= \dfrac{2 - x}{2(x-2)} = \dfrac{\overset{-1}{\cancel{-(x-2)}}}{2\underset{1}{\cancel{(x-2)}}}$ $\qquad b - a = -(a - b)$, a useful change in some problems

$= -\dfrac{1}{2}$

(C) $\dfrac{2x^3 - 2x^2y + 2xy^2}{x^3y - xy^3} \div \dfrac{x^3 + y^3}{x^2 + 2xy + y^2}$

$$= \frac{\overset{2}{\cancel{2x}}\overset{1}{\cancel{(x^2 - xy + y^2)}}}{\underset{y}{\cancel{xy}}\underset{1}{\cancel{(x + y)}}(x - y)} \cdot \frac{\overset{1}{\cancel{(x + y)^2}}}{\underset{1}{\cancel{(x + y)}}\underset{1}{\cancel{(x^2 - xy + y^2)}}}$$

$$= \frac{2}{y(x - y)}$$

◆

PROBLEM 16 Perform the indicated operations and reduce to lowest terms.

(A) $\dfrac{12x^2y^3}{2xy^2 + 6xy} \cdot \dfrac{y^2 + 6y + 9}{3y^3 + 9y^2}$ (B) $(4 - x) \div \dfrac{x^2 - 16}{5}$

(C) $\dfrac{m^3 + n^3}{2m^2 + mn - n^2} \div \dfrac{m^3n - m^2n^2 + mn^3}{2m^3n^2 - m^2n^3}$ ◆

◆ ADDITION AND SUBTRACTION

Again, because we are restricting variable replacements to real numbers, addition and subtraction of rational expressions follow the rules for adding and subtracting real number fractions.

Addition and Subtraction

For a, b, and c real numbers:

1. $\dfrac{a}{b} + \dfrac{c}{b} = \dfrac{a + c}{b}$, $b \neq 0$ $\qquad \dfrac{x}{x + 5} + \dfrac{8}{x + 5} = \dfrac{x + 8}{x + 5}$
2. $\dfrac{a}{b} - \dfrac{c}{b} = \dfrac{a - c}{b}$, $b \neq 0$ $\qquad \dfrac{x}{3x^2y^2} - \dfrac{x + 7}{3x^2y^2} = \dfrac{x - (x + 7)}{3x^2y^2}$

Thus, we add rational expressions with the same denominators by adding or subtracting their numerators and placing the result over the common denominator. If the denominators are not the same, we raise the fractions to higher terms, using the fundamental property of fractions to obtain common denominators, and then proceed as described.

Even though any common denominator will do, our work will be simplified if the *least common denominator (LCD)* is used. Often, the LCD is obvious, but if it is not, the steps in the box at the top of the next page describe how to find it.

The Least Common Denominator (LCD)

The LCD of two or more rational expressions is found as follows:

1. Factor each denominator completely.
2. Identify each different prime factor from all the denominators.
3. Form a product using each different factor to the highest power that occurs in any one denominator. This product is the LCD.

◆ EXAMPLE 17 Combine into a single fraction and reduce to lowest terms.

(A) $\frac{3}{10}+\frac{5}{6}-\frac{11}{45}$ (B) $\frac{4}{9x}-\frac{5x}{6y^2}+1$ (C) $\frac{x+3}{x^2-6x+9}-\frac{x+2}{x^2-9}-\frac{5}{3-x}$

Solutions (A) To find the LCD, factor each denominator completely:

$$\left.\begin{aligned}10&=2\cdot 5\\ 6&=2\cdot 3\\ 45&=3^2\cdot 5\end{aligned}\right\}\ \text{LCD}=2\cdot 3^2\cdot 5=90$$

Now use the fundamental property of fractions to make each denominator 90:

$$\begin{aligned}\frac{3}{10}+\frac{5}{6}-\frac{11}{45}&=\frac{9\cdot 3}{9\cdot 10}+\frac{15\cdot 5}{15\cdot 6}-\frac{2\cdot 11}{2\cdot 45}\\ &=\frac{27}{90}+\frac{75}{90}-\frac{22}{90}\\ &=\frac{27+75-22}{90}=\frac{80}{90}=\frac{8}{9}\end{aligned}$$

(B)
$$\left.\begin{aligned}9x&=3^2x\\ 6y^2&=2\cdot 3y^2\end{aligned}\right\}\ \text{LCD}=2\cdot 3^2xy^2=18xy^2$$

$$\begin{aligned}\frac{4}{9x}-\frac{5x}{6y^2}+1&=\frac{2y^2\cdot 4}{2y^2\cdot 9x}-\frac{3x\cdot 5x}{3x\cdot 6y^2}+\frac{18xy^2}{18xy^2}\\ &=\frac{8y^2-15x^2+18xy^2}{18xy^2}\end{aligned}$$

(C)
$$\frac{x+3}{x^2-6x+9}-\frac{x+2}{x^2-9}-\frac{5}{3-x}=\frac{x+3}{(x-3)^2}-\frac{x+2}{(x-3)(x+3)}+\frac{5}{x-3}$$

Note: $-\frac{5}{3-x}=\frac{5}{-(3-x)}=\frac{5}{x-3}$

We have again used the fact that $a-b=-(b-a)$.

The LCD $=(x-3)^2(x+3)$. Thus,

$$\frac{(x+3)^2}{(x-3)^2(x+3)} - \frac{(x-3)(x+2)}{(x-3)^2(x+3)} + \frac{5(x-3)(x+3)}{(x-3)^2(x+3)}$$

$$= \frac{(x^2+6x+9) - (x^2-x-6) + 5(x^2-9)}{(x-3)^2(x+3)}$$

$$= \frac{x^2+6x+9-x^2+x+6+5x^2-45}{(x-3)^2(x+3)}$$

$$= \frac{5x^2+7x-30}{(x-3)^2(x+3)}$$ ◆

PROBLEM 17 Combine into a single fraction and reduce to lowest terms.

(A) $\frac{5}{28} - \frac{1}{10} + \frac{6}{35}$ (B) $\frac{1}{4x^2} - \frac{2x+1}{3x^3} + \frac{3}{12x}$

(C) $\frac{y-3}{y^2-4} - \frac{y+2}{y^2-4y+4} - \frac{2}{2-y}$ ◆

◆ COMPOUND FRACTIONS

A fractional expression with fractions in its numerator, denominator, or both is called a **compound fraction.** It is often necessary to represent a compound fraction as a **simple fraction**—that is (in all cases we will consider), as the quotient of two polynomials. The process does not involve any new concepts. It is a matter of applying old concepts and processes in the right sequence. We will illustrate two approaches to the problem, each with its own merits, depending on the particular problem under consideration. One of the methods makes very effective use of the fundamental property of fractions in the form

$$\frac{a}{b} = \frac{ka}{kb} \qquad b, k \neq 0$$

◆ EXAMPLE 18 Express as a simple fraction: $\dfrac{\dfrac{y}{x^2} - \dfrac{x}{y^2}}{\dfrac{y}{x} - \dfrac{x}{y}}$

Solution *Method 1.* Multiply the numerator and denominator by the LCD of all fractions in the numerator and denominator—in this case, x^2y^2. (We are multiplying by a form of 1.)

$$\frac{x^2y^2\left(\dfrac{y}{x^2} - \dfrac{x}{y^2}\right)}{x^2y^2\left(\dfrac{y}{x} - \dfrac{x}{y}\right)} = \frac{x^2y^2\dfrac{y}{x^2} - x^2y^2\dfrac{x}{y^2}}{x^2y^2\dfrac{y}{x} - x^2y^2\dfrac{x}{y}} = \frac{y^3-x^3}{xy^3-x^3y}$$

$$= \frac{\overset{1}{\cancel{(y-x)}}(y^2+xy+x^2)}{xy\underset{1}{\cancel{(y-x)}}(y+x)} = \frac{y^2+xy+x^2}{xy(y+x)} \quad \text{or} \quad \frac{x^2+xy+y^2}{xy(x+y)}$$

Method 2. Write the numerator and denominator as single fractions. Then treat as a quotient.

$$\frac{\frac{y}{x^2}-\frac{x}{y^2}}{\frac{y}{x}-\frac{x}{y}} = \frac{\frac{y^3-x^3}{x^2y^2}}{\frac{y^2-x^2}{xy}} = \frac{y^3-x^3}{x^2y^2} \div \frac{y^2-x^2}{xy}$$

$$= \frac{\overset{1}{\cancel{(y-x)}}(y^2+xy+x^2)}{\underset{xy}{\cancel{x^2y^2}}} \cdot \frac{\overset{1}{\cancel{xy}}}{\underset{1}{\cancel{(y-x)}}(y+x)}$$

$$= \frac{x^2+xy+y^2}{xy(x+y)}$$

◆

PROBLEM 18 Express as a simple fraction reduced to lowest terms. Use the two methods described in Example 18.

$$\frac{\frac{a}{b}-\frac{b}{a}}{\frac{a}{b}+2+\frac{b}{a}}$$

◆

Answers to Matched Problems

15. (A) $\frac{x-3}{x+3}$ (B) $\frac{x^2+x+1}{x+1}$ 16. (A) $2x$ (B) $\frac{-5}{x+4}$ (C) mn

17. (A) $\frac{1}{4}$ (B) $\frac{3x^2-5x-4}{12x^3}$ (C) $\frac{2y^2-9y-6}{(y-2)^2(y+2)}$ 18. $\frac{a-b}{a+b}$

EXERCISE 1-4

A *Perform the indicated operations and reduce answers to lowest terms.*

1. $\frac{d^5}{3a} \div \left(\frac{d^2}{6a^2} \cdot \frac{a}{4d^3}\right)$

2. $\left(\frac{d^5}{3a} \div \frac{d^2}{6a^2}\right) \cdot \frac{a}{4d^3}$

3. $\frac{x^2}{12} + \frac{x}{18} - \frac{1}{30}$

4. $\frac{2y}{18} - \frac{-1}{28} - \frac{y}{42}$

5. $\frac{4m-3}{18m^3} + \frac{3}{4m} - \frac{2m-1}{6m^2}$

6. $\frac{3x+8}{4x^2} - \frac{2x-1}{x^3} - \frac{5}{8x}$

7. $\frac{x^2-9}{x^2-3x} \div (x^2-x-12)$

8. $\frac{2x^2+7x+3}{4x^2-1} \div (x+3)$

9. $\frac{x^2-6x+9}{x^2-x-6} \div \frac{x^2+2x-15}{x^2+2x}$

10. $\frac{m+n}{m^2-n^2} \div \frac{m^2-mn}{m^2-2mn+n^2}$

11. $\frac{3}{x^2-1} - \frac{2}{x^2-2x+1}$

12. $\frac{1}{a^2-b^2} + \frac{1}{a^2+2ab+b^2}$

13. $\frac{x+1}{x-1} - 1$

14. $m - 3 - \frac{m-1}{m-2}$

15. $\frac{3}{a-1} - \frac{2}{1-a}$

16. $\frac{5}{x-3} - \frac{2}{3-x}$

17. $\frac{2x}{x^2-y^2} + \frac{1}{x+y} - \frac{1}{x-y}$

18. $\frac{2}{y+3} - \frac{1}{y-3} + \frac{2y}{y^2-9}$

B *Perform the indicated operations and reduce answers to lowest terms. Represent any compound fractions as simple fractions reduced to lowest terms.*

19. $\frac{x^2}{x^2+2x+1} + \frac{x-1}{3x+3} - \frac{1}{6}$

20. $\frac{y}{y^2-y-2} - \frac{1}{y^2+5y-14} - \frac{2}{y^2+8y+7}$

21. $\frac{2-x}{2x+x^2} \cdot \frac{x^2+4x+4}{x^2-4}$

22. $\frac{9-m^2}{m^2+5m+6} \cdot \frac{m+2}{m-3}$

23. $\frac{c+2}{5c-5} - \frac{c-2}{3c-3} + \frac{c}{1-c}$

24. $\frac{x+7}{ax-bx} + \frac{y+9}{by-ay}$

25. $\left(\frac{x^3-y^3}{y^3} \cdot \frac{y}{x-y}\right) \div \frac{x^2+xy+y^2}{y^2}$

26. $\frac{x^2-16}{2x^2+10x+8} \div \frac{x^2-13x+36}{x^3+1}$

27. $\left(\frac{3}{x-2} - \frac{1}{x+1}\right) \div \frac{x+4}{x-2}$

28. $\left(\frac{x}{x^2-16} - \frac{1}{x+4}\right) \div \frac{4}{x+4}$

29. $\frac{1+\frac{3}{x}}{x-\frac{9}{x}}$

30. $\frac{1-\frac{y^2}{x^2}}{1-\frac{y}{x}}$

31. $\frac{\frac{1}{m^2}-1}{\frac{1}{m}+1}$

32. $\frac{\frac{1}{m}+1}{m+1}$

33. $\frac{c-d}{\frac{1}{c}-\frac{1}{d}}$

34. $\frac{\frac{1}{x}+\frac{1}{y}}{x+y}$

35. $\frac{\frac{x}{y}-2+\frac{y}{x}}{\frac{x}{y}-\frac{y}{x}}$

36. $\frac{1+\frac{2}{x}-\frac{15}{x^2}}{1+\frac{4}{x}-\frac{5}{x^2}}$

C *Represent the compound fractions as simple fractions reduced to lowest terms.*

37. $\frac{\frac{s^2}{s-t}-s}{\frac{t^2}{s-t}+t}$

38. $\frac{y-\frac{y^2}{y-x}}{1+\frac{x^2}{y^2-x^2}}$

39. $1 - \frac{1}{1-\frac{1}{1-\frac{1}{x}}}$

40. $2 - \frac{1}{1-\frac{2}{a+2}}$

SECTION 1-5 Integer Exponents and Square Root Radicals

- INTEGER EXPONENTS
- SCIENTIFIC NOTATION
- SQUARE ROOT RADICALS

We now consider basic operations on integer exponents, working with scientific notation, and basic operations on square root radicals.

◆ INTEGER EXPONENTS

Definitions for **integer exponents** are listed below.

Definition of a^n

For n an integer and a a real number:

1. For n a positive integer,

   $a^n = a \cdot a \cdot \cdots \cdot a$ $\quad$ n factors of a $\quad$ $5^4 = 5 \cdot 5 \cdot 5 \cdot 5$

2. For $n = 0$,

   $a^0 = 1 \quad a \neq 0$ $\quad$ $12^0 = 1$

   0^0 is not defined.

3. For n a negative integer,

   $$a^n = \frac{1}{a^{-n}} \quad a \neq 0 \qquad a^{-3} = \frac{1}{a^{-(-3)}} = \frac{1}{a^3}$$

   [If n is negative, then $(-n)$ is positive.]
   Note: It can be shown that for *all* integers n,

   $$a^{-n} = \frac{1}{a^n} \quad \text{and} \quad a^n = \frac{1}{a^{-n}} \quad a \neq 0 \qquad a^5 = \frac{1}{a^{-5}}, \quad a^{-5} = \frac{1}{a^5}$$

The following integer exponent properties are very useful in manipulating integer exponent forms.

Exponent Properties

For n and m integers and a and b real numbers:

1. $a^m a^n = a^{m+n}$ $\qquad$ $a^8 a^{-3} = a^{8+(-3)} = a^5$
2. $(a^n)^m = a^{mn}$ $\qquad$ $(a^{-2})^3 = a^{3(-2)} = a^{-6}$
3. $(ab)^m = a^m b^m$ $\qquad$ $(ab)^{-2} = a^{-2}b^{-2}$
4. $\left(\frac{a}{b}\right)^m = \frac{a^m}{b^m} \quad b \neq 0$ $\qquad$ $\left(\frac{a}{b}\right)^5 = \frac{a^5}{b^5}$
5. $\frac{a^m}{a^n} = a^{m-n} = \frac{1}{a^{n-m}} \quad a \neq 0$ $\qquad$ $\frac{a^{-3}}{a^7} = \frac{1}{a^{7-(-3)}} = \frac{1}{a^{10}}$

Exponent forms are frequently encountered in algebraic applications. You should sharpen your skills in using these forms by reviewing the above basic definitions and properties and the examples that follow.

◆ EXAMPLE 19 Simplify and express the answers using positive exponents only.

(A) $(2x^3)(3x^5) \boxed{= 2 \cdot 3x^{3+5}} = 6x^8$ (B) $x^5x^{-9} = x^{-4} = \dfrac{1}{x^4}$

(C) $\dfrac{x^5}{x^7} \boxed{= x^{5-7}} = x^{-2} = \dfrac{1}{x^2}$ (D) $\dfrac{x^{-3}}{y^{-4}} = \dfrac{y^4}{x^3}$

or $\dfrac{x^5}{x^7} \boxed{= \dfrac{1}{x^{7-5}}} = \dfrac{1}{x^2}$

(E) $(u^{-3}v^2)^{-2} \boxed{= (u^{-3})^{-2}(v^2)^{-2}} = u^6v^{-4} = \dfrac{u^6}{v^4}$

(F) $\left(\dfrac{y^{-5}}{y^{-2}}\right)^{-2} \boxed{= \dfrac{(y^{-5})^{-2}}{(y^{-2})^{-2}}} = \dfrac{y^{10}}{y^4} = y^6$

(G) $\dfrac{4m^{-3}n^{-5}}{6m^{-4}n^3} \boxed{= \dfrac{2m^{-3-(-4)}}{3n^{3-(-5)}}} = \dfrac{2m}{3n^8}$

(H) $\left(\dfrac{2x^{-3}x^3}{n^{-2}}\right)^{-3} = \left(\dfrac{2x^0}{n^{-2}}\right)^{-3} = \left(\dfrac{2}{n^{-2}}\right)^{-3} = \dfrac{2^{-3}}{n^6} = \dfrac{1}{2^3n^6} = \dfrac{1}{8n^6}$ ◆

PROBLEM 19 Simplify and express the answers using positive exponents only.

(A) $(3y^4)(2y^3)$ (B) m^2m^{-6} (C) $(u^3v^{-2})^{-2}$

(D) $\left(\dfrac{y^{-6}}{y^{-2}}\right)^{-1}$ (E) $\dfrac{8x^{-2}y^{-4}}{6x^{-5}y^2}$ (F) $\left(\dfrac{3m^{-3}}{2x^2x^{-2}}\right)^{-2}$ ◆

◆ SCIENTIFIC NOTATION

Writing and working with very large or very small numbers in standard decimal notation is often awkward, even with calculators. It is often convenient to represent numbers of this type in **scientific notation;** that is, as the product of a number between 1 and 10 and a power of 10.

◆ EXAMPLE 20 Decimal Fractions and Scientific Notation

$7 = 7 \times 10^0$	$0.5 = 5 \times 10^{-1}$
$67 = 6.7 \times 10$	$0.45 = 4.5 \times 10^{-1}$
$580 = 5.8 \times 10^2$	$0.003\ 2 = 3.2 \times 10^{-3}$
$43{,}000 = 4.3 \times 10^4$	$0.000\ 045 = 4.5 \times 10^{-5}$
$73{,}400{,}000 = 7.34 \times 10^7$	$0.000\ 000\ 391 = 3.91 \times 10^{-7}$

◆

Note that the power of 10 used corresponds to the number of places we move the decimal to form a number between 1 and 10. The power is positive if the decimal is moved to the left and negative if it is moved to the right. Positive exponents are associated with numbers greater than or equal to 10; negative exponents are associated with positive numbers less than 1.

PROBLEM 20 Write each number in scientific notation.

(A) 370 (B) 47,300,000,000 (C) 0.047 (D) 0.000 000 089 ◆

◆ SQUARE ROOT RADICALS

To start, we define a **square root** of a number:

Definition of Square Root

x is a **square root** of y if $x^2 = y$. 2 is a square root of 4 since $2^2 = 4$.
-2 is a square root of 4 since $(-2)^2 = 4$.

How many square roots of a real number are there? The following theorem, which we state without proof, answers the question.

THEOREM 5 **Square Roots of Real Numbers**

(A) Every positive real number has exactly two real square roots, each the negative of the other.
(B) Negative real numbers have no real number square roots (since no real number squared can be negative—think about this).
(C) The square root of 0 is 0.

Square Root Notation

For a a positive number:

$\sqrt{a}$ is the positive square root of a.
$-\sqrt{a}$ is the negative square root of a.

[*Note:* $\sqrt{-a}$ is not a real number.]

◆ EXAMPLE 21 (A) $\sqrt{4} = 2$ (B) $-\sqrt{4} = -2$ (C) $\sqrt{-4}$ is not a real number. (D) $\sqrt{0} = 0$ ◆

PROBLEM 21 Evaluate, if possible.

(A) $\sqrt{9}$ (B) $-\sqrt{9}$ (C) $\sqrt{-9}$ (D) $\sqrt{-0}$ ◆

It can be shown that if a is a positive integer that is not the square of an integer, then

$$-\sqrt{a} \quad \text{and} \quad \sqrt{a}$$

are irrational numbers. Thus,

$$-\sqrt{7} \quad \text{and} \quad \sqrt{7}$$

name irrational numbers that are, respectively, the negative and positive square roots of 7.

Properties of Radicals

For a and b nonnegative real numbers:

1. $\sqrt{a^2} = a$
2. $\sqrt{a}\,\sqrt{b} = \sqrt{ab}$
3. $\dfrac{\sqrt{a}}{\sqrt{b}} = \sqrt{\dfrac{a}{b}}$

To see that property 2 holds, let $N = \sqrt{a}$ and $M = \sqrt{b}$. Then $N^2 = a$ and $M^2 = b$. Hence,

$$\sqrt{a}\,\sqrt{b} = NM = \sqrt{(NM)^2} = \sqrt{N^2M^2} = \sqrt{ab}$$

Note how properties of exponents are used. The proof of property 3 is left as an exercise.

◆ EXAMPLE 22 (A) $\sqrt{5}\,\sqrt{10} = \sqrt{5 \cdot 10} = \sqrt{50} = \sqrt{25 \cdot 2} = \sqrt{25}\,\sqrt{2} = 5\sqrt{2}$

(B) $\dfrac{\sqrt{32}}{\sqrt{8}} = \sqrt{\dfrac{32}{8}} = \sqrt{4} = 2$ (C) $\sqrt{\dfrac{7}{4}} = \dfrac{\sqrt{7}}{\sqrt{4}} = \dfrac{\sqrt{7}}{2}$ or $\frac{1}{2}\sqrt{7}$ ◆

PROBLEM 22 Simplify as in Example 22.

(A) $\sqrt{3}\,\sqrt{6}$ (B) $\dfrac{\sqrt{18}}{\sqrt{2}}$ (C) $\sqrt{\dfrac{11}{9}}$ ◆

The foregoing definitions and theorems allow us to change algebraic expressions containing radicals to a variety of equivalent forms. One form that is often useful is called the *simplest radical form*, defined in the box on the next page.

Definition of the Simplest Radical Form

An algebraic expression that contains square root radicals is in **simplest radical form** if all three of the following conditions are satisfied:

1. No radicand (the expression within the radical sign) when expressed in completely factored form contains a factor raised to a power greater than 1.
   $\sqrt{x^3}$ violates this condition.
2. No radical appears in a denominator.
   $3/\sqrt{5}$ violates this condition.
3. No fraction appears within a radical.
   $\sqrt{\frac{2}{3}}$ violates this condition.

It should be understood that forms other than the simplest radical form may be more useful on occasion. The situation dictates what form to choose.

◆ EXAMPLE 23 Change to simplest radical form. All variables represent positive real numbers.

(A) $\sqrt{8x^3}$ (B) $\dfrac{3x}{\sqrt{3}}$ (C) $\dfrac{2\sqrt{x}-1}{\sqrt{x}+2}$ (D) $\sqrt{\dfrac{3x}{8}}$

Solutions

(A) $\sqrt{8x^3}$ violates condition 1. Separate $8x^3$ into a perfect square part (2^2x^2) and what is left over $(2x)$; then use multiplication property 2:

$$\begin{aligned}\sqrt{8x^3} &= \sqrt{(2^2x^2)(2x)}\\ &= \sqrt{2^2x^2}\,\sqrt{2x}\\ &= 2x\sqrt{2x}\end{aligned}$$

(B) $3x/\sqrt{3}$ has a radical in the denominator; hence, it violates condition 2. To remove the radical from the denominator, we multiply the numerator and denominator by $\sqrt{3}$ to obtain $\sqrt{3^2}$ in the denominator (this is called **rationalizing a denominator**):

$$\begin{aligned}\frac{3x}{\sqrt{3}} &= \frac{3x}{\sqrt{3}}\cdot\frac{\sqrt{3}}{\sqrt{3}}\\ &= \frac{3x\sqrt{3}}{\sqrt{3^2}}\\ &= \frac{3x\sqrt{3}}{3} = x\sqrt{3}\end{aligned}$$

(C) $(2\sqrt{x}-1)/(\sqrt{x}+2)$ has a radical in the denominator. Multiplying the numerator and denominator by $\sqrt{x}$ does not remove all radicals from the denominator. However, remembering that $(a+b)(a-b) = a^2 - b^2$ suggests that if we multiply the numerator and denominator by $\sqrt{x} - 2$, all radicals will disappear from the denominator (and the denominator will be rationalized).

$$\frac{2\sqrt{x} - 1}{\sqrt{x} + 2} = \frac{2\sqrt{x} - 1}{\sqrt{x} + 2} \cdot \frac{\sqrt{x} - 2}{\sqrt{x} - 2}$$

$$= \frac{2x - 5\sqrt{x} + 2}{x - 4}$$

(D) $\sqrt{3x/8}$ has a fraction within the radical; hence, it violates condition 3. To remove the fraction from the radical, we multiply the numerator and denominator of $3x/8$ by 2 to make the denominator a perfect square:

$$\sqrt{\frac{3x}{8}} = \sqrt{\frac{3x \cdot 2}{8 \cdot 2}}$$

$$= \sqrt{\frac{6x}{16}}$$

$$= \frac{\sqrt{6x}}{\sqrt{16}} = \frac{\sqrt{6x}}{4}$$

◆

PROBLEM 23 Change to simplest radical form. All variables represent positive real numbers.

(A) $\sqrt{18y^3}$ (B) $\dfrac{4xy}{\sqrt{2x}}$ (C) $\dfrac{2 + 3\sqrt{y}}{7 - \sqrt{y}}$ (D) $\sqrt{\dfrac{5y}{18x}}$ ◆

Answers to Matched Problems

19. (A) $6y^7$ (B) $\dfrac{1}{m^4}$ (C) $\dfrac{v^4}{u^6}$ (D) y^4 (E) $\dfrac{4x^3}{3y^6}$ (F) $\dfrac{4m^6}{9}$
20. (A) 3.7×10^2 (B) 4.73×10^{10} (C) 4.7×10^{-2} (D) 8.9×10^{-8}
21. (A) 3 (B) -3 (C) Not a real number (D) 0
22. (A) $3\sqrt{2}$ (B) 3 (C) $\sqrt{11}/3$ or $\frac{1}{3}\sqrt{11}$
23. (A) $3y\sqrt{2y}$ (B) $2y\sqrt{2x}$ (C) $\dfrac{14 + 23\sqrt{y} + 3y}{49 - y}$ (D) $\dfrac{\sqrt{10xy}}{6x}$ or $\dfrac{1}{6x}\sqrt{10xy}$

EXERCISE 1-5

A *Simplify and express answers using positive exponents only. Variables are restricted to avoid division by 0.*

1. $2x^{-9}$ **2.** $3y^{-5}$ **3.** $\dfrac{3}{2w^{-7}}$ **4.** $\dfrac{5}{4x^{-9}}$

5. $2x^{-8}x^5$ **6.** $3c^{-9}c^4$ **7.** $\dfrac{w^{-8}}{w^{-3}}$ **8.** $\dfrac{m^{-11}}{m^{-5}}$

9. $5v^8v^{-8}$ **10.** $7d^{-4}d^4$ **11.** $(a^{-3})^2$ **12.** $(b^4)^{-3}$

13. $(x^6y^{-3})^{-2}$ **14.** $(a^{-3}b^4)^{-3}$

Express in simplest radical form. All variables represent positive real numbers.

15. $\sqrt{x^2}$
16. $\sqrt{m^2}$
17. $\sqrt{a^5}$
18. $\sqrt{m^7}$
19. $\sqrt{18x^4}$
20. $\sqrt{8x^3}$
21. $\dfrac{1}{\sqrt{m}}$
22. $\dfrac{1}{\sqrt{A}}$
23. $\sqrt{\dfrac{2}{3}}$
24. $\sqrt{\dfrac{3}{5}}$
25. $\sqrt{\dfrac{2}{x}}$
26. $\sqrt{\dfrac{3}{y}}$

Write in scientific notation.

27. 82,300,000,000
28. 5,380,000
29. 0.783
30. 0.019
31. 0.000 034
32. 0.000 000 007 832

B *Simplify and express answers using positive exponents only. Write compound fractions as simple fractions.*

33. $(22 + 31)^0$
34. $(2x^3y^4)^0$
35. $\dfrac{10^{-3} \cdot 10^4}{10^{-11} \cdot 10^{-2}}$
36. $\dfrac{10^{-17} \cdot 10^{-5}}{10^{-3} \cdot 10^{-14}}$
37. $(5x^2y^{-3})^{-2}$
38. $(2m^{-3}n^2)^{-3}$
39. $\dfrac{8 \times 10^{-3}}{2 \times 10^{-5}}$
40. $\dfrac{18 \times 10^{12}}{6 \times 10^{-4}}$
41. $\dfrac{8x^{-3}y^{-1}}{6x^2y^{-4}}$
42. $\dfrac{9m^{-4}n^3}{12m^{-1}n^{-1}}$
43. $\left(\dfrac{6xy^{-2}}{3x^{-1}y^2}\right)^{-3}$
44. $\left(\dfrac{2x^{-3}y^2}{4xy^{-1}}\right)^{-2}$
45. $\dfrac{1 - x}{x^{-1} - 1}$
46. $\dfrac{1 + x^{-1}}{1 - x^{-2}}$
47. $\dfrac{u + v}{u^{-1} + v^{-1}}$
48. $\dfrac{x^{-1} - y^{-1}}{x - y}$

Write each problem in the form $ax^p + bx^q$ or $ax^p + bx^q + cx^r$, where a, b, and c are real numbers and p, q, and r are integers. For example,

$$\frac{2x^4 - 3x^2 + 1}{2x^3} = \frac{2x^4}{2x^3} - \frac{3x^2}{2x^3} + \frac{1}{2x^3} = x - \frac{3}{2}x^{-1} + \frac{1}{2}x^{-3}$$

49. $\dfrac{7x^5 - x^2}{4x^5}$
50. $\dfrac{5x^3 - 2}{3x^2}$
51. $\dfrac{3x^4 - 4x^2 - 1}{4x^3}$
52. $\dfrac{2x^3 - 3x^2 + x}{2x^2}$

Simplify and express answers in simplest radical form. All variables represent positive real numbers.

53. $\sqrt{18x^8y^5z^2}$
54. $\sqrt{8p^3q^2r^5}$
55. $\dfrac{12}{\sqrt{3x}}$
56. $\dfrac{10}{\sqrt{2y}}$
57. $\sqrt{\dfrac{6x}{7y}}$
58. $\sqrt{\dfrac{3m}{2n}}$
59. $\sqrt{\dfrac{4a^3}{3b}}$
60. $\sqrt{\dfrac{9m^5}{2n}}$
61. $\sqrt{18m^3n^4}\,\sqrt{2m^3n^2}$
62. $\sqrt{10x^3y}\,\sqrt{5xy}$
63. $\dfrac{\sqrt{4a^3}}{\sqrt{3b}}$
64. $\dfrac{\sqrt{9m^5}}{\sqrt{2n}}$
65. $\dfrac{5\sqrt{x}}{3 - 2\sqrt{x}}$
66. $\dfrac{3\sqrt{y}}{2\sqrt{y} - 3}$
67. $\dfrac{3\sqrt{2} - 2\sqrt{3}}{3\sqrt{3} - 2\sqrt{2}}$
68. $\dfrac{2\sqrt{5} + 3\sqrt{2}}{5\sqrt{5} + 2\sqrt{2}}$

Convert each numeral to scientific notation and simplify. Express the answer in scientific notation and in standard decimal form.

69. $\dfrac{9{,}600{,}000{,}000}{(1{,}600{,}000)(0.000\ 000\ 25)}$

70. $\dfrac{(60{,}000)(0.000\ 003)}{(0.000\ 4)(1{,}500{,}000)}$

71. $\dfrac{(1{,}250{,}000)(0.000\ 38)}{0.015\ 2}$

72. $\dfrac{(0.000\ 000\ 82)(230{,}000)}{(625{,}000)(0.008\ 2)}$

C *Simplify and write answers using positive exponents only. Write compound fractions as simple fractions.*

73. $\left[\left(\dfrac{x^{-2}y^{3}t}{x^{-3}y^{-2}t^{2}}\right)^{2}\right]^{-1}$

74. $\left[\left(\dfrac{u^{3}v^{-1}w^{-2}}{u^{-2}v^{-2}w}\right)^{-2}\right]^{2}$

75. $\left(\dfrac{2^{2}x^{2}y^{0}}{8x^{-1}}\right)^{-2}\left(\dfrac{x^{-3}}{x^{-5}}\right)^{3}$

76. $\left(\dfrac{3^{3}x^{0}y^{-2}}{2^{3}x^{3}y^{-5}}\right)^{-1}\left(\dfrac{3^{3}x^{-1}y}{2^{2}x^{2}y^{-2}}\right)^{2}$

77. $\dfrac{4(x-3)^{-4}}{8(x-3)^{-2}}$

78. $\dfrac{12(a+2b)^{-3}}{6(a+2b)^{-8}}$

79. $\dfrac{b^{-2}-c^{-2}}{b^{-3}-c^{-3}}$

80. $\dfrac{xy^{-2}-yx^{-2}}{y^{-1}-x^{-1}}$

Express in simplest radical form. All variables are restricted to avoid division by 0 and square roots of negative numbers.

81. $\dfrac{\sqrt{2x}\,\sqrt{5}}{\sqrt{20x}}$

82. $\dfrac{\sqrt{x}\,\sqrt{8y}}{\sqrt{12y}}$

83. $\dfrac{2}{\sqrt{x}-2}$

84. $\sqrt{\dfrac{1}{x-5}}$

Rationalize the numerators; that is, perform operations on the fractions that will eliminate radicals from the numerators.

85. $\dfrac{\sqrt{t}-\sqrt{x}}{t-x}$

86. $\dfrac{\sqrt{x}-\sqrt{y}}{\sqrt{x}+\sqrt{y}}$

87. $\dfrac{\sqrt{x+h}-\sqrt{x}}{h}$

88. $\dfrac{\sqrt{2+h}+\sqrt{2}}{h}$

SECTION 1-6 Rational Exponents and Radicals

- *n*TH ROOTS OF REAL NUMBERS
- RATIONAL EXPONENTS AND RADICALS
- PROPERTIES OF RADICALS

Square roots may now be generalized to *nth roots*, and the meaning of exponent may be generalized to include all rational numbers.

◆ *n*TH ROOTS OF REAL NUMBERS

Recall from Section 1-5 that r is a **square root** of b if $r^2 = b$. There is no reason to stop there. We may also say that r is a **cube root** of b if $r^3 = b$.

In general:

> For any natural number n:
>
> r is an ***n*th root** of b if $r^n = b$

How many real square roots of 16 exist? Of 7? Of -4? How many real 4th roots of 7 exist? Of -7? How many real cube roots of -8 exist? Of 11? Theorem 6 (which we state without proof) answers these questions.

THEOREM 6

Number of Real *n*th Roots of a Real Number *b*

	n EVEN	n ODD
b POSITIVE:	Two real nth roots -2 and 2 are both 4th roots of 16	One real nth root 2 is the only real cube root of 8
b NEGATIVE:	No real nth root -4 has no real square roots	One real nth root -2 is the only real cube root of -8
b ZERO:	One real nth root	One real nth root

The nth root of 0 is 0 for any natural number n.

On the basis of Theorem 6, we conclude that

7	has two real square roots, two real 4th roots, and so on.
10	has one real cube root, one real 5th root, and so on.
-13	has one real cube root, one real 5th root, and so on.
-8	has no real square roots, no real 4th roots, and so on.

◆ RATIONAL EXPONENTS AND RADICALS

We now turn to the question of what symbols to use to represent the various kinds of real nth roots. For a natural number n greater than 1 we use

$$b^{1/n} \quad \text{or} \quad \sqrt[n]{b}$$

to represent one of the **real *n*th roots of *b*.** Which one? The symbols represent the real nth root of b if n is odd and the positive real nth root of b if b is positive and n is even. The symbol $\sqrt[n]{b}$ is called an ***n*th root radical.** The number n is the

index of the radical, and the number b is called the **radicand.** Note that we write simply $\sqrt{b}$ to indicate $\sqrt[2]{b}$.

◆ EXAMPLE 24

(A) $4^{1/2} = \sqrt{4} = 2 \quad (\sqrt{4} \neq \pm 2)$ (B) $-4^{1/2} = -\sqrt{4} = -2$
(C) $(-4)^{1/2}$ and $\sqrt{-4}$ are not real numbers
(D) $8^{1/3} = \sqrt[3]{8} = 2$ (E) $(-8)^{1/3} = \sqrt[3]{-8} = -2$ ◆

PROBLEM 24

Evaluate each of the following:

(A) $16^{1/2}$ (B) $-\sqrt{16}$ (C) $\sqrt[3]{-27}$ (D) $(-9)^{1/2}$ (E) $(\sqrt[4]{81})^3$ ◆

We now define b^r for any rational number $r = m/n$.

Rational Exponents

If m and n are natural numbers without common prime factors, b is a real number, and b is nonnegative when n is even, then

$$b^{m/n} = \begin{cases} (b^{1/n})^m = (\sqrt[n]{b})^m \\ (b^m)^{1/n} = \sqrt[n]{b^m} \end{cases} \qquad \begin{array}{l} 8^{2/3} = (8^{1/3})^2 = (\sqrt[3]{8})^2 = 2^2 = 4 \\ 8^{2/3} = (8^2)^{1/3} = \sqrt[3]{8^2} = \sqrt[3]{64} = 4 \end{array}$$

and

$$b^{-m/n} = \frac{1}{b^{m/n}} \qquad b \neq 0 \qquad 8^{-2/3} = \frac{1}{8^{2/3}} = \frac{1}{4}$$

Note that the two definitions of $b^{m/n}$ are equivalent under the indicated restrictions on m, n, and b.

All the properties listed for integer exponents in Section 1-5 also hold for rational exponents, provided b is nonnegative when n is even. Unless stated to the contrary, all variables in the rest of the discussion represent positive real numbers.

◆ EXAMPLE 25

Change rational exponent form to radical form.

(A) $x^{1/7} = \sqrt[7]{x}$
(B) $(3u^2v^3)^{3/5} = \sqrt[5]{(3u^2v^3)^3}$ or $(\sqrt[5]{3u^2v^3})^3$ The first is usually preferred.
(C) $y^{-2/3} = \dfrac{1}{y^{2/3}} = \dfrac{1}{\sqrt[3]{y^2}}$ or $\sqrt[3]{y^{-2}}$ or $\sqrt[3]{\dfrac{1}{y^2}}$

Change radical form to rational exponent form.

(D) $\sqrt[5]{6} = 6^{1/5}$ (E) $-\sqrt[3]{x^2} = -x^{2/3}$
(F) $\sqrt{x^2 + y^2} = (x^2 + y^2)^{1/2}$ Note that $(x^2 + y^2)^{1/2} \neq x + y$. Why? ◆

PROBLEM 25 Convert to radical form.

(A) $u^{1/5}$ (B) $(6x^2y^5)^{2/9}$ (C) $(3xy)^{-3/5}$

Convert to rational exponent form.

(D) $\sqrt[4]{9u}$ (E) $-\sqrt[7]{(2x)^4}$ (F) $\sqrt[3]{x^3 + y^3}$ ◆

◆ EXAMPLE 26 Simplify each and express answers using positive exponents only. If rational exponents appear in final answers, convert to radical form.

(A) $(3x^{1/3})(2x^{1/2}) = 6x^{1/3+1/2} = 6x^{5/6} = 6\sqrt[6]{x^5}$

(B) $(-8)^{5/3} = [(-8)^{1/3}]^5 = (-2)^5 = -32$

(C) $(2x^{1/3}y^{-2/3})^3 = 8xy^{-2} = \dfrac{8x}{y^2}$

(D) $\left(\dfrac{4x^{1/3}}{x^{1/2}}\right)^{1/2} = \dfrac{4^{1/2}x^{1/6}}{x^{1/4}} = \dfrac{2}{x^{1/4-1/6}} = \dfrac{2}{x^{1/12}} = \dfrac{2}{\sqrt[12]{x}}$ ◆

PROBLEM 26 Simplify each and express answers using positive exponents only. If rational exponents appear in final answers, convert to radical form.

(A) $9^{3/2}$ (B) $(-27)^{4/3}$ (C) $(5y^{1/4})(2y^{1/3})$ (D) $(2x^{-3/4}y^{1/4})^4$

(E) $\left(\dfrac{8x^{1/2}}{x^{2/3}}\right)^{1/3}$ ◆

◆ EXAMPLE 27 Multiply, and express answers using positive exponents only.

(A) $3y^{2/3}(2y^{1/3} - y^2)$ (B) $(2u^{1/2} + v^{1/2})(u^{1/2} - 3v^{1/2})$

Solutions

(A) $3y^{2/3}(2y^{1/3} - y^2) = 6y^{2/3+1/3} - 3y^{2/3+2}$

$= 6y - 3y^{8/3}$

(B) $(2u^{1/2} + v^{1/2})(u^{1/2} - 3v^{1/2}) = 2u - 5u^{1/2}v^{1/2} - 3v$ ◆

PROBLEM 27 Multiply, and express answers using positive exponents only.

(A) $2c^{1/4}(5c^3 - c^{3/4})$ (B) $(7x^{1/2} - y^{1/2})(2x^{1/2} + 3y^{1/2})$ ◆

◆ PROPERTIES OF RADICALS

Changing or simplifying radical expressions is aided by several properties of radicals that follow directly from the properties of exponents considered earlier.

Properties of Radicals

If c, n, and m are natural numbers greater than or equal to 2, and if x and y are positive real numbers, then:

1. $\sqrt[n]{x^n} = x$ $\qquad \sqrt[3]{x^3} = x$
2. $\sqrt[n]{xy} = \sqrt[n]{x}\sqrt[n]{y}$ $\qquad \sqrt[5]{xy} = \sqrt[5]{x}\sqrt[5]{y}$
3. $\sqrt[n]{\dfrac{x}{y}} = \dfrac{\sqrt[n]{x}}{\sqrt[n]{y}}$ $\qquad \sqrt[4]{\dfrac{x}{y}} = \dfrac{\sqrt[4]{x}}{\sqrt[4]{y}}$
4. $\sqrt[cn]{x^{cm}} = \sqrt[n]{x^m}$ $\qquad \sqrt[12]{x^8} = \sqrt[4 \cdot 3]{x^{4 \cdot 2}} = \sqrt[3]{x^2}$

The properties of radicals provide us with the means of changing algebraic expressions containing radicals into a variety of equivalent forms. One particularly useful form is the *simplest radical form*. An algebraic expression that contains radicals is said to be in the **simplest radical form** if all four of the following conditions are satisfied:

Simplest Radical Form

1. A radicand contains no factor to a power greater than or equal to the index of the radical.
   $\sqrt[3]{x^5}$ violates this condition.
2. The power of the radicand and the index of the radical have no common factor other than 1.
   $\sqrt[6]{x^4}$ violates this condition.
3. No radical appears in a denominator.
   $y/\sqrt[3]{x}$ violates this condition.
4. No fraction appears within a radical.
   $\sqrt[4]{\frac{3}{5}}$ violates this condition.

◆ EXAMPLE 28 Write in simplest radical form.

(A) $\sqrt[3]{x^3y^6} = \sqrt[3]{(xy^2)^3} = xy^2$

or $\sqrt[3]{x^3y^6} = (x^3y^6)^{1/3} = x^{3/3}y^{6/3} = xy^2$

(B) $\sqrt[3]{32x^8y^3} = \sqrt[3]{(2^3x^6y^3)(4x^2)} = \sqrt[3]{2^3x^6y^3}\,\sqrt[3]{4x^2}$

$= 2x^2y\,\sqrt[3]{4x^2}$

(C) $\dfrac{6x^2}{\sqrt[3]{9x}} = \dfrac{6x^2}{\sqrt[3]{9x}} \cdot \dfrac{\sqrt[3]{3x^2}}{\sqrt[3]{3x^2}} = \dfrac{6x^2\sqrt[3]{3x^2}}{\sqrt[3]{3^3x^3}} = \dfrac{6x^2\sqrt[3]{3x^2}}{3x} = 2x\sqrt[3]{3x^2}$

(D) $6\sqrt[4]{\dfrac{3}{4x^3}} = 6\sqrt[4]{\dfrac{3}{2^2x^3} \cdot \dfrac{2^2x}{2^2x}} = 6\sqrt[4]{\dfrac{12x}{2^4x^4}}$

$= 6\dfrac{\sqrt[4]{12x}}{\sqrt[4]{2^4x^4}} = 6\dfrac{\sqrt[4]{12x}}{2x} = \dfrac{3\sqrt[4]{12x}}{x}$

(E) $\sqrt[6]{16x^4y^2} = \sqrt[6]{(4x^2y)^2}$

$= \sqrt[2\cdot 3]{(4x^2y)^{2\cdot 1}}$

$= \sqrt[3]{4x^2y}$ ◆

Note that in Examples 28C and 28D, we **rationalized the denominators;** that is, we performed operations to remove radicals from the denominators. This is a useful operation in some problems.

PROBLEM 28 Write in simplest radical form.

(A) $\sqrt{12x^5y^6}$ (B) $\sqrt[3]{-27x^7y^5}$ (C) $\dfrac{8y^3}{\sqrt[4]{2y}}$ (D) $4x^2\sqrt[5]{\dfrac{y^2}{2x^3}}$ (E) $\sqrt[9]{8x^6y^3}$ ◆

Answers to Matched Problems

24. (A) 4 (B) -4 (C) -3 (D) Not a real number (E) 27
25. (A) $\sqrt[5]{u}$ (B) $\sqrt[9]{(6x^2y^5)^2}$ or $(\sqrt[9]{6x^2y^5})^2$ (C) $1/\sqrt[5]{(3xy)^3}$ (D) $(9u)^{1/4}$ (E) $-(2x)^{4/7}$ (F) $(x^3 + y^3)^{1/3}$ (not $x + y$)
26. (A) 27 (B) 81 (C) $10y^{7/12} = 10\sqrt[12]{y^7}$ (D) $16y/x^3$ (E) $2/x^{1/18} = 2/\sqrt[18]{x}$
27. (A) $10c^{13/4} - 2c$ (B) $14x + 19x^{1/2}y^{1/2} - 3y$
28. (A) $2x^2y^3\sqrt{3x}$ (B) $-3x^2y\sqrt[3]{xy^2}$ (C) $4y^2\sqrt[4]{8y^3}$ (D) $2x\sqrt[5]{16x^2y^2}$ (E) $\sqrt[3]{2x^2y}$

EXERCISE 1-6

A *Change to radical form; do not simplify.*

1. $6x^{3/5}$
2. $7y^{2/5}$
3. $(4xy^3)^{2/5}$
4. $(7x^2y)^{5/7}$
5. $(x^2 + y^2)^{1/2}$
6. $x^{1/2} + y^{1/2}$

Change to rational exponent form; do not simplify.

7. $5\sqrt[4]{x^3}$
8. $7m\sqrt[5]{n^2}$
9. $\sqrt[5]{(2x^2y)^3}$
10. $\sqrt[9]{(3m^4n)^2}$
11. $\sqrt[3]{x} + \sqrt[3]{y}$
12. $\sqrt[3]{x^2 + y^3}$

Find rational number representations for each, if they exist.

13. $25^{1/2}$
14. $64^{1/3}$
15. $16^{3/2}$
16. $16^{3/4}$
17. $-36^{1/2}$
18. $-32^{3/5}$
19. $(-36)^{1/2}$
20. $(-32)^{3/5}$
21. $(\frac{4}{25})^{3/2}$
22. $(\frac{8}{27})^{2/3}$
23. $9^{-3/2}$
24. $8^{-2/3}$

Simplify each expression and write answers using positive exponents only. All variables represent positive real numbers.

25. $x^{4/5}x^{-2/5}$ 26. $y^{-3/7}y^{4/7}$ 27. $\dfrac{m^{2/3}}{m^{-1/3}}$ 28. $\dfrac{x^{1/4}}{x^{3/4}}$

29. $(8x^3y^{-6})^{1/3}$ 30. $(4u^{-2}v^4)^{1/2}$

B 31. $\left(\dfrac{4x^{-2}}{y^4}\right)^{-1/2}$ 32. $\left(\dfrac{w^4}{9x^{-2}}\right)^{-1/2}$ 33. $\dfrac{8x^{-1/3}}{12x^{1/4}}$ 34. $\dfrac{6a^{3/4}}{15a^{-1/3}}$

35. $\left(\dfrac{8x^{-4}y^3}{27x^2y^{-3}}\right)^{1/3}$ 36. $\left(\dfrac{25x^5y^{-1}}{16x^{-3}y^{-5}}\right)^{1/2}$

Multiply, and express answers using positive exponents only.

37. $3x^{3/4}(4x^{1/4} - 2x^8)$ 38. $2m^{1/3}(3m^{2/3} - m^6)$
39. $(3u^{1/2} - v^{1/2})(u^{1/2} - 4v^{1/2})$ 40. $(a^{1/2} + 2b^{1/2})(a^{1/2} - 3b^{1/2})$
41. $(5m^{1/2} + n^{1/2})(5m^{1/2} - n^{1/2})$ 42. $(2x^{1/2} - 3y^{1/2})(2x^{1/2} + 3y^{1/2})$
43. $(3x^{1/2} - y^{1/2})^2$ 44. $(x^{1/2} + 2y^{1/2})^2$

Write each problem in the form $ax^p + bx^q$, where a and b are real numbers and p and q are rational numbers. For example:

$$\frac{2x^{1/3}+4}{4x} = \frac{2x^{1/3}}{4x} + \frac{4}{4x} = \frac{1}{2}x^{1/3-1} + x^{-1} = \frac{1}{2}x^{-2/3} + x^{-1}$$

45. $\dfrac{x^{2/3}+2}{2x^{1/3}}$ 46. $\dfrac{12x^{1/2}-3}{4x^{1/2}}$ 47. $\dfrac{2x^{3/4}+3x^{1/3}}{3x}$ 48. $\dfrac{3x^{2/3}+x^{1/2}}{5x}$

49. $\dfrac{2x^{1/3}-x^{1/2}}{4x^{1/2}}$ 50. $\dfrac{x^2-4x^{1/2}}{2x^{1/3}}$

Write in simplest radical form.

51. $\sqrt[3]{16m^4n^6}$ 52. $\sqrt[3]{27x^7y^3}$ 53. $\sqrt[4]{32m^9n^7}$ 54. $\sqrt[5]{64u^{17}v^9}$

55. $\dfrac{x}{\sqrt[3]{x}}$ 56. $\dfrac{u^2}{\sqrt[3]{u^2}}$ 57. $\dfrac{4a^3b^2}{\sqrt[3]{2ab^2}}$ 58. $\dfrac{8x^3y^5}{\sqrt[3]{4x^2y}}$

59. $\sqrt[4]{\dfrac{3x^3}{4}}$ 60. $\sqrt[5]{\dfrac{3x^2}{2}}$ 61. $\sqrt[12]{(x-3)^9}$ 62. $\sqrt[8]{(t+1)^6}$

63. $\sqrt{x}\,\sqrt[3]{x^2}$ 64. $\sqrt[3]{x}\,\sqrt{x}$ 65. $\dfrac{\sqrt{x}}{\sqrt[3]{x^2}}$ 66. $\dfrac{\sqrt{x}}{\sqrt[3]{x}}$

C *Simplify by writing each expression as a simple fraction reduced to lowest terms and without negative exponents.*

67. $\dfrac{(x-1)^{1/2} - x(\frac{1}{2})(x-1)^{-1/2}}{x-1}$ 68. $\dfrac{(2x-1)^{1/2} - (x+2)(\frac{1}{2})(2x-1)^{-1/2}(2)}{2x-1}$

69. $\dfrac{(x+2)^{2/3} - x(\frac{2}{3})(x+2)^{-1/3}}{(x+2)^{4/3}}$ 70. $\dfrac{2(3x-1)^{1/3} - (2x+1)(\frac{1}{3})(3x-1)^{-2/3}(3)}{(3x-1)^{2/3}}$

CALCULATOR PROBLEMS

In Problems 71–76, evaluate using a calculator. (Refer to the instruction book for your calculator to see how exponential forms are evaluated.)

71. $22^{3/2}$ **72.** $15^{5/4}$ **73.** $827^{-3/8}$

74. $103^{-3/4}$ **75.** $37.09^{7/3}$ **76.** $2.876^{8/5}$

SECTION 1-7 Chapter Review

Important Terms and Symbols

1-1 *Algebra and Real Numbers.* Real number system; natural numbers; integers; rational numbers; irrational numbers; real numbers; real number line; coordinate; origin; positive and negative real numbers; associative, commutative, identity, and inverse properties for addition and multiplication; distributive properties; negative and reciprocal; subtraction and division; properties of negatives; zero properties; fractions and fraction properties; numerator; denominator

$N, Z, Q, I, R; \quad -a; \quad \frac{1}{a}; \quad \frac{a}{b}$

1-2 *Basic Operations on Polynomials.* Natural number exponent; base; first property of exponents; algebraic expression; polynomial; degree of a term; degree of a polynomial; monomial; binomial; trinomial; coefficient; like terms; addition and subtraction of polynomials; multiplication of polynomials; FOIL method; special products

b^n

1-3 *Factoring Polynomials.* Factor; prime number; composite number; factored completely; fundamental theorem of arithmetic; prime polynomial; common factors; factoring second-degree polynomials; perfect square; difference of squares; difference of cubes; sum of cubes

1-4 *Basic Operations on Rational Expressions.* Fractional expression; rational expression; fundamental property of fractions; reducing to lowest terms; raising to higher terms; multiplication and division of rational forms; addition and subtraction of rational forms; least common denominator; compound fraction; simple fraction

LCD

1-5 *Integer Exponents and Square Root Radicals.* Integer exponent; properties of integer exponents; scientific notation; square root; square root radical; properties of radicals; simplest radical form; rationalizing a denominator

$a^n; \quad a^{-n}; \quad a^0; \quad \sqrt{a}$

1-6 *Rational Exponents and Radicals.* nth root; nth root radical; index; radicand; rational exponent; properties of radicals; simplest radical form; rationalizing denominators

$a^{1/n}$; $\sqrt[n]{a}$; $a^{m/n}$; $\sqrt[n]{a^m}$; $a^{-m/n}$

EXERCISE 1-7

Chapter Review

Work through all the problems in this chapter review and check your answers in the back of the book. (Answers to all review problems are there.) Where weaknesses show up, review appropriate sections in the text.

A

1. Replace each question mark with an appropriate expression that will illustrate the use of the indicated real number property:
   (A) Commutative (·): $x(y + z) = ?$
   (B) Associative (+): $2 + (x + y) = ?$
   (C) Distributive: $(2 + 3)x = ?$

Problems 2–6 refer to the following polynomials:

(A) $3x - 4$ (B) $x + 2$ (C) $3x^2 + x - 8$ (D) $x^3 + 8$

2. Add all four.
3. Subtract the sum of (A) and (C) from the sum of (B) and (D).
4. Multiply (C) and (D).
5. What is the degree of (D)?
6. What is the coefficient of the second term in (C)?

Perform the indicated operations and simplify.

7. $5x^2 - 3x[4 - 3(x - 2)]$
8. $(3m - 5n)(3m + 5n)$
9. $(2x + y)(3x - 4y)$
10. $(2a - 3b)^2$

Write each polynomial in a completely factored form relative to the integers. If the polynomial is prime relative to the integers, say so.

11. $9x^2 - 12x + 4$
12. $t^2 - 4t - 6$
13. $6n^3 - 9n^2 - 15n$

Perform the indicated operations and reduce to lowest terms. Represent all compound fractions as simple fractions reduced to lowest terms.

14. $\dfrac{2}{5b} - \dfrac{4}{3a^3} - \dfrac{1}{6a^2b^2}$
15. $\dfrac{3x}{3x^2 - 12x} + \dfrac{1}{6x}$
16. $\dfrac{y - 2}{y^2 - 4y + 4} \div \dfrac{y^2 + 2y}{y^2 + 4y + 4}$
17. $\dfrac{u - \dfrac{1}{u}}{1 - \dfrac{1}{u^2}}$

Simplify Problems 18–23 and write answers using positive exponents only. All variables represent positive real numbers.

18. $6(xy^3)^5$ 19. $\dfrac{9u^8v^6}{3u^4v^8}$ 20. $(2 \times 10^5)(3 \times 10^{-3})$

21. $(x^{-3}y^2)^{-2}$ 22. $u^{5/3}u^{2/3}$ 23. $(9a^4b^{-2})^{1/2}$

24. Change to radical form: $3x^{2/5}$
25. Change to rational exponent form: $-3\sqrt[3]{(xy)^2}$

Simplify Problems 26–30 and express in simplest radical form. All variables represent positive real numbers.

26. $3x\sqrt[3]{x^5y^4}$ 27. $\sqrt{2x^2y^5}\sqrt{18x^3y^2}$ 28. $\dfrac{6ab}{\sqrt{3a}}$

29. $\dfrac{\sqrt{5}}{3-\sqrt{5}}$ 30. $\sqrt[8]{y^6}$

B *In Problems 31–36, each statement illustrates the use of one of the following real number properties or definitions. Indicate which one.*

Commutative $(+, \cdot)$	*Associative* $(+, \cdot)$	*Distributive*
Identity $(+, \cdot)$	*Inverse* $(+, \cdot)$	*Subtraction*
Division	*Negatives*	*Zero*

31. $(-7) - (-5) = (-7) + [-(-5)]$
32. $5u + (3v + 2) = (3v + 2) + 5u$
33. $(5m - 2)(2m + 3) = (5m - 2)2m + (5m - 2)3$
34. $9 \cdot (4y) = (9 \cdot 4)y$
35. $\dfrac{u}{-(v-w)} = -\dfrac{u}{v-w}$ 36. $(x - y) + 0 = (x - y)$

37. Indicate true (T) or false (F):
    (A) A natural number is a rational number.
    (B) A number with a repeating decimal representation is an irrational number.
38. Give an example of an integer that is not a natural number.
39. Given the following algebraic expressions:

    (a) $2x^2 - 3x + 5$ (b) $x^2 - \sqrt{x-3}$ (c) $x^{-3} + x^{-2} - 3x^{-1}$
    (d) $x^2 - 3xy - y^2$

    (A) Identify all second-degree polynomials.
    (B) Identify all third-degree polynomials.

Perform the indicated operations and simplify.

40. $(2x - y)(2x + y) - (2x - y)^2$
41. $(m^2 + 2mn - n^2)(m^2 - 2mn - n^2)$
42. $-2x\{(x^2 + 2)(x - 3) - x[x - x(3 - x)]\}$
43. $(x - 2y)^3$

Write in a completely factored form relative to the integers. If the polynomial is prime relative to the integers, say so.

44. $(4x - y)^2 - 9x^2$

45. $2x^2 + 4xy - 5y^2$

46. $6x^3y + 12x^2y^2 - 15xy^3$

47. $3x^3 + 24y^3$

Perform the indicated operations and reduce to lowest terms. Represent all compound fractions as simple fractions reduced to lowest terms.

48. $\dfrac{m-1}{m^2-4m+4} + \dfrac{m+3}{m^2-4} + \dfrac{2}{2-m}$

49. $\dfrac{y}{x^2} \div \left(\dfrac{x^2+3x}{2x^2+5x-3} \div \dfrac{x^3y-x^2y}{2x^2-3x+1}\right)$

50. $\dfrac{a^{-1}-b^{-1}}{ab^{-2}-ba^{-2}}$

Perform the indicated operations in Problems 51–56. Simplify, and write answers using positive exponents only. All variables represent positive real numbers.

51. $\left(\dfrac{8u^{-1}}{2^2u^2v^0}\right)^{-2}\left(\dfrac{u^{-5}}{u^{-3}}\right)^3$

52. $\dfrac{5^0}{3^2} + \dfrac{3^{-2}}{2^{-2}}$

53. $\left(\dfrac{27x^2y^{-3}}{8x^{-4}y^3}\right)^{1/3}$

54. $(a^{-1/3}b^{1/4})(9a^{1/3}b^{-1/2})^{3/2}$

55. $(x^{1/2} + y^{1/2})^2$

56. $(3x^{1/2} - y^{1/2})(2x^{1/2} + 3y^{1/2})$

57. Convert to scientific notation and simplify: $\dfrac{0.000\ 000\ 000\ 52}{(1{,}300)(0.000\ 002)}$

Perform the indicated operations in Problems 58–64 and express answers in simplest radical form. All radicands represent positive real numbers.

58. $-2x\sqrt[5]{3^6x^7y^{11}}$

59. $\dfrac{2x^2}{\sqrt[3]{4x}}$

60. $\sqrt[5]{\dfrac{3y^2}{8x^2}}$

61. $\sqrt[9]{8x^6y^{12}}$

62. $(2\sqrt{x} - 5\sqrt{y})(\sqrt{x} + \sqrt{y})$

63. $\dfrac{3\sqrt{x}}{2\sqrt{x} - \sqrt{y}}$

64. $\dfrac{2\sqrt{u} - 3\sqrt{v}}{2\sqrt{u} + 3\sqrt{v}}$

C

65. Rationalize the numerator: $\dfrac{\sqrt{t} - \sqrt{5}}{t-5}$

66. Write in the form $ax^p + bx^q$, where a and b are real numbers and p and q are rational numbers:

$$\frac{4\sqrt{x} - 3}{2\sqrt{x}}$$

67. Evaluate $x^2 - 4x + 1$ for $x = 2 - \sqrt{3}$.

68. Simplify: $x(2x - 1)(x + 3) - (x - 1)^3$

CALCULATOR PROBLEMS

Evaluate each expression using a calculator.

69. $\dfrac{(20{,}410)(0.000\ 003\ 477)}{0.000\ 000\ 022\ 09}$

70. 0.1347^5

71. $(-60.39)^{-3}$

72. $82.45^{8/3}$

73. $\sqrt[5]{0.006\ 604}$

74. $\sqrt[3]{3 + \sqrt{2}}$

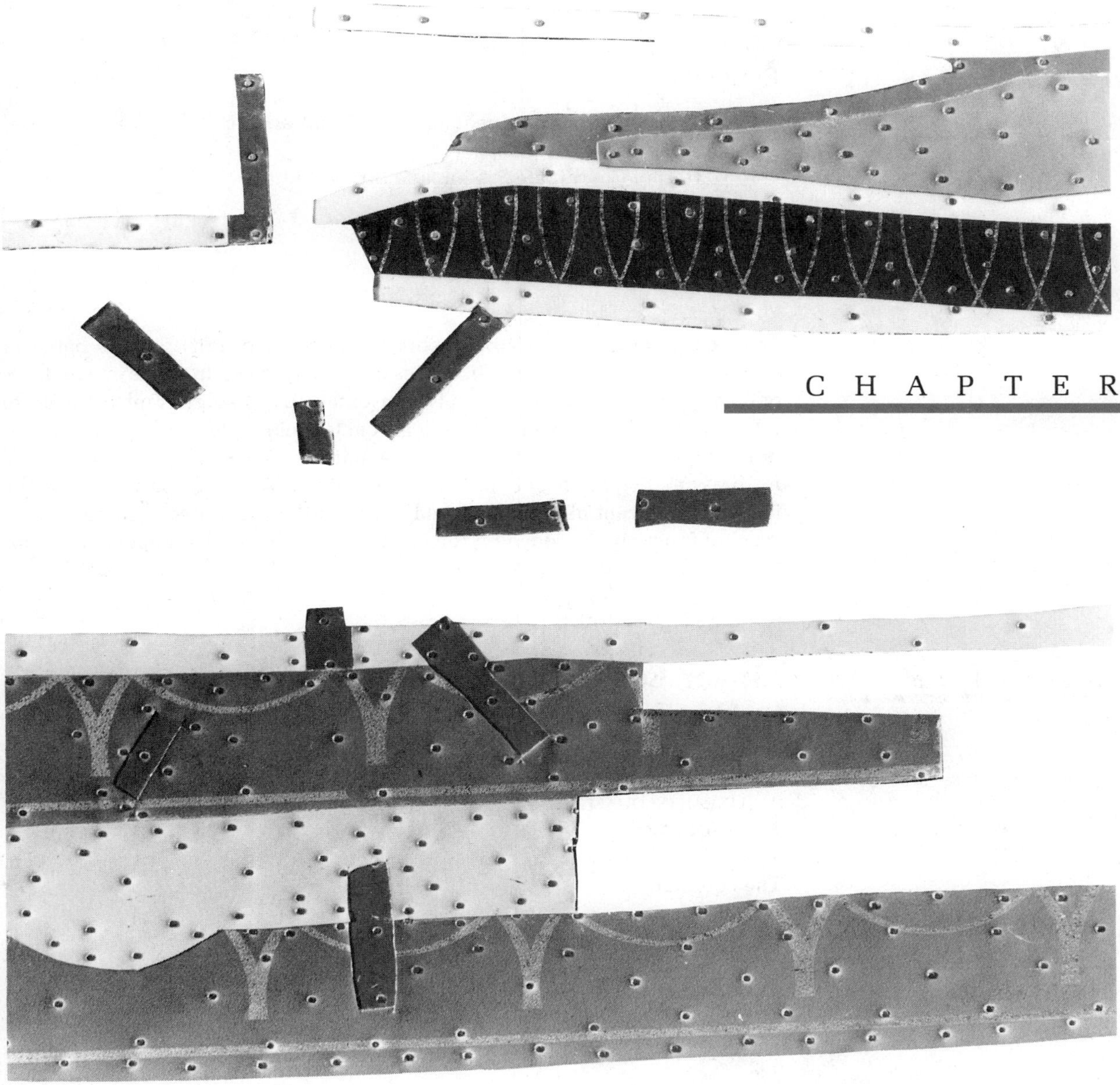

CHAPTER 2

Equations, Graphs, and Functions

CHAPTER 2

Contents

One of the important uses of algebra is the study of relationships between quantities. Many of these relationships can be expressed in terms of equations or inequalities. Additional insight is often obtained by using a graph to illustrate relationships. In this chapter, we will look at techniques for solving linear equations, linear inequalities, and quadratic equations, and we will discuss methods for drawing graphs in one and two dimensions. We will also introduce the important concept of a function and examine the properties of some specific types of functions. Throughout the chapter, we will consider applied problems that can be solved using these techniques.

SECTION 2-1

Linear Equations and Inequalities in One Variable

- LINEAR EQUATIONS
- LINEAR INEQUALITIES
- APPLICATIONS

The equation

$$3 - 2(x + 3) = \frac{x}{3} - 5$$

and the inequality

$$\frac{x}{2} + 2(3x - 1) \geq 5$$

are both first-degree in one variable. In general, a **first-degree, or linear, equation** in one variable is any equation that can be written in the form

STANDARD FORM $\qquad ax + b = 0 \qquad a \neq 0 \qquad (1)$

If the equality symbol, =, in (1) is replaced by <, >, ≤, or ≥, then the resulting expression is called a **first-degree, or linear, inequality.**

A **solution** of an equation (or inequality) involving a single variable is a number that when substituted for the variable makes the equation (or inequality)

true. The set of all solutions is called the **solution set.** When we say that we **solve an equation** (or inequality), we mean that we find its solution set.

Knowing what is meant by the solution set is one thing; finding it is another. We start by recalling the idea of equivalent equations and equivalent inequalities. If we perform an operation on an equation (or inequality) that produces another equation (or inequality) with the same solution set, then the two equations (or inequalities) are said to be **equivalent.** The basic idea in solving equations and inequalities is to perform operations on these forms that produce simpler equivalent forms, and to continue the process until we obtain an equation or inequality with an obvious solution.

◆ LINEAR EQUATIONS

The following properties of equality produce equivalent equations when applied:

Equality Properties

For a, b, and c real numbers:

ADDITION PROPERTY:	1.	If $a = b$, then $a + c = b + c$.
SUBTRACTION PROPERTY:	2.	If $a = b$, then $a - c = b - c$.
MULTIPLICATION PROPERTY:	3.	If $a = b$, then $ca = cb$, $c \neq 0$.
DIVISION PROPERTY:	4.	If $a = b$, then $\frac{a}{c} = \frac{b}{c}$, $c \neq 0$.

Several examples should remind you of the process of solving equations.

◆ EXAMPLE 1 Solve: $8x - 3(x - 4) = 3(x - 4) + 6$

Solution

$$\begin{aligned} 8x - 3(x - 4) &= 3(x - 4) + 6 \\ 8x - 3x + 12 &= 3x - 12 + 6 \\ 5x + 12 &= 3x - 6 \\ 2x &= -18 \\ x &= -9 \end{aligned}$$ ◆

PROBLEM 1 Solve: $3x - 2(2x - 5) = 2(x + 3) - 8$ ◆

◆ EXAMPLE 2 What operations can we perform on

$$\frac{x + 2}{2} - \frac{x}{3} = 5$$

to eliminate the denominators? If we can find a number that is exactly divisible by each denominator, then we can use the multiplication property of equality to clear the denominators. The LCD (least common denominator) of the fractions,

6, is exactly what we are looking for! Actually, any common denominator will do, but the LCD results in a simpler equivalent equation. Thus, we multiply both sides of the equation by 6:

$$6\left(\frac{x+2}{2} - \frac{x}{3}\right) = 6 \cdot 5$$

$$\overset{3}{\cancel{6}} \cdot \frac{(x+2)}{\underset{1}{\cancel{2}}} - \overset{2}{\cancel{6}} \cdot \frac{x}{\underset{1}{\cancel{3}}} = 30$$

$$3(x+2) - 2x = 30$$

$$3x + 6 - 2x = 30$$

$$x = 24$$

◆

PROBLEM 2 Solve: $\frac{x+1}{3} - \frac{x}{4} = \frac{1}{2}$ ◆

In many applications of algebra, formulas or equations must be changed to alternate equivalent forms. The following examples are typical.

◆ EXAMPLE 3 Solve the amount formula for simple interest, $A = P + Prt$, for:

(A) r in terms of the other variables
(B) P in terms of the other variables

Solutions (A)

$A = P + Prt$	Reverse equation.
$P + Prt = A$	Now isolate r on the left side.
$Prt = A - P$	Divide both members by Pt.
$r = \frac{A - P}{Pt}$	

(B)

$A = P + Prt$	Reverse equation.
$P + Prt = A$	Factor out P (note the use of the distributive property).
$P(1 + rt) = A$	Divide by $(1 + rt)$.
$P = \frac{A}{1 + rt}$	

◆

PROBLEM 3 Solve $M = Nt + Nr$ for:

(A) t (B) N ◆

◆ LINEAR INEQUALITIES

Before we start solving linear inequalities, let us recall what we mean by < (less than) and > (greater than). If a and b are real numbers, then we write

$a < b$ a is less than b

if there exists a positive number p such that $a + p = b$. Certainly, we would expect that if a positive number was added to any real number, the sum would be larger than the original. That is essentially what the definition states. If $a < b$, we may also write

$\boldsymbol{b > a}$ $\qquad$ b is greater than a

◆ EXAMPLE 4

(A) $3 < 5$ $\qquad$ Since $3 + 2 = 5$

(B) $-6 < -2$ $\qquad$ Since $-6 + 4 = -2$

(C) $0 > -10$ $\qquad$ Since $-10 < 0$ ◆

PROBLEM 4

Replace each question mark with either $<$ or $>$.

(A) 2 ? 8 $\qquad$ (B) -20 ? 0 $\qquad$ (C) -3 ? -30 ◆

The inequality symbols have a very clear geometric interpretation on the real number line. If $a < b$, then a is to the left of b on the number line; if $c > d$, then c is to the right of d (Fig. 1).

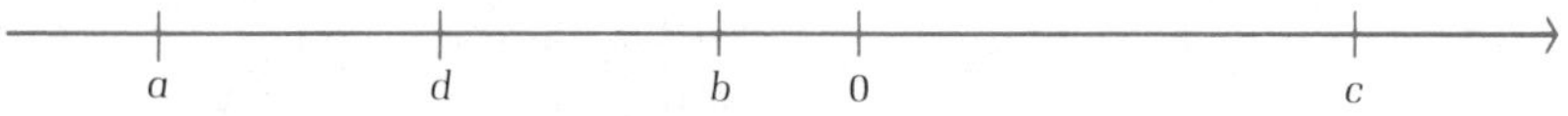

FIGURE 1
$a < b$, $c > d$

Now let us turn to the problem of solving linear inequalities in one variable. Recall that a **solution** of an inequality involving one variable is a number that, when substituted for the variable, makes the inequality true. The set of all solutions is called the **solution set.** When we say that we **solve an inequality,** we mean that we find its solution set. The procedures used to solve linear inequalities in one variable are almost the same as those used to solve linear equations in one variable but with two important exceptions (as noted below). The following properties of inequalities produce equivalent inequalities when applied:

Inequality Properties

For a, b, and c real numbers:

1. If $a > b$, then $a + c > b + c$.
2. If $a > b$, then $a - c > b - c$.
3. If $a > b$ and c is positive, then $ca > cb$. } Note difference
4. If $a > b$ and c is negative, then $ca < cb$. }
5. If $a > b$ and c is positive, then $\dfrac{a}{c} > \dfrac{b}{c}$. } Note difference
6. If $a > b$ and c is negative, then $\dfrac{a}{c} < \dfrac{b}{c}$. }

Similar properties hold if each inequality sign is reversed or if $>$ is replaced with $\geq$ **(greater than or equal to)** and $<$ is replaced with $\leq$ **(less than or equal to).** Thus, we can perform essentially the same operations on inequalities that we perform on equations, with the exception that **the sense of the inequality reverses if we multiply or divide both sides by a negative number.** Otherwise, the sense of the inequality does not change. For example, if we start with the true statement

$$-3 > -7$$

and multiply both sides by 2, we obtain

$$-6 > -14$$

and the sense of the inequality stays the same. But if we multiply both sides of $-3 > -7$ by -2, then the left side becomes 6 and the right side becomes 14, so we must write

$$6 < 14$$

to have a true statement. Thus, the sense of the inequality reverses.

If $a < b$, the double inequality $a < x < b$ means that $\boldsymbol{x > a}$ **and** $\boldsymbol{x < b}$; that is, x is between a and b. Other variations, as well as a useful interval notation, are

TABLE 1

INTERVAL NOTATION	INEQUALITY NOTATION	LINE GRAPH
$[a, b]$	$a \leq x \leq b$	
$[a, b)$	$a \leq x < b$	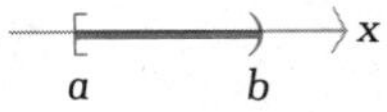
$(a, b]$	$a < x \leq b$	
(a, b)	$a < x < b$	
$(-\infty, a]$	$x \leq a$	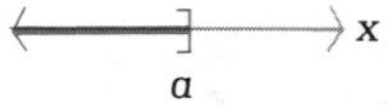
$(-\infty, a)$	$x < a$	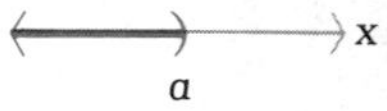
$[b, \infty)$*	$x \geq b$	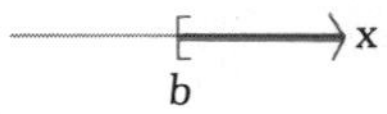
(b, ∞)	$x > b$	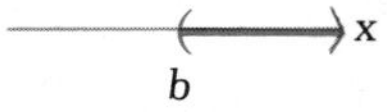

* The symbol ∞ (read "infinity") is not a number. When we write $[b, \infty)$, we are simply referring to the interval starting at b and continuing indefinitely to the right. We would never write $[b, \infty]$.

given in Table 1. Note that an end point on a line graph has a square bracket through it if it is included in the inequality and a parenthesis through it if it is not.

◆ EXAMPLE 5 Solve and graph: $2(2x + 3) < 6(x - 2) + 10$

Solution

$$\begin{aligned} 2(2x + 3) &< 6(x - 2) + 10 \\ 4x + 6 &< 6x - 12 + 10 \\ 4x + 6 &< 6x - 2 \\ -2x + 6 &< -2 \\ -2x &< -8 \\ x &> 4 \quad \text{or} \quad (4, \infty) \end{aligned}$$

Notice that the sense of the inequality reverses when we divide both sides by -2.

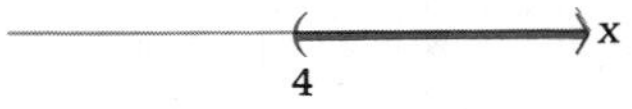

Notice that in the graph of $x > 4$, we use a parenthesis through 4, since the point 4 is not included in the graph. ◆

PROBLEM 5 Solve and graph: $3(x - 1) \leq 5(x + 2) - 5$ ◆

◆ EXAMPLE 6 Solve and graph: $-3 < 2x + 3 \leq 9$

Solution We are looking for all numbers x such that $2x + 3$ is between -3 and 9, including 9 but not -3. We proceed as above except that we try to isolate x in the middle:

$$-3 < 2x + 3 \leq 9$$

$$\boxed{-3 - 3 < 2x + 3 - 3 \leq 9 - 3}$$

$$-6 < 2x \leq 6$$

$$\boxed{\frac{-6}{2} < \frac{2x}{2} \leq \frac{6}{2}}$$

$$-3 < x \leq 3 \quad \text{or} \quad (-3, 3]$$

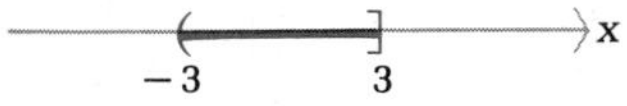

◆

PROBLEM 6 Solve and graph: $-8 \leq 3x - 5 < 7$ ◆

Note that a linear equation usually has exactly one solution, while a linear inequality usually has infinitely many solutions.

◆ APPLICATIONS

To realize the full potential of algebra, we must be able to translate real-world problems into mathematical forms. In short, we must be able to do *word problems.*

◆ EXAMPLE 7

Break-Even Analysis

It costs a record company \$9,000 to prepare a record album—recording costs, album design costs, etc. These costs represent a one-time **fixed cost.** Manufacturing, marketing, and royalty costs (all **variable costs**) are \$3.50 per album. If the album is sold to record shops for \$5 each, how many albums must be sold for the company to **break even?**

Solution Let

x = Number of records sold

C = Cost for producing x records

R = Revenue (return) on sales of x records

The company breaks even if $R = C$, with

$$\begin{aligned} C &= \text{Fixed costs} + \text{Variable costs} \\ &= \$9{,}000 + \$3.50x \\ R &= \$5x \end{aligned}$$

Find x such that $R = C$; that is, find x such that

$$\begin{aligned} 5x &= 9{,}000 + 3.5x \\ 1.5x &= 9{,}000 \\ x &= 6{,}000 \end{aligned}$$

Check For $x = 6{,}000$,

$$\begin{aligned} C &= 9{,}000 + 3.5x & \text{and} \qquad R &= 5x \\ &= 9{,}000 + 3.5(6{,}000) & &= 5(6{,}000) \\ &= \$30{,}000 & &= \$30{,}000 \end{aligned}$$

Thus, the company must sell 6,000 records to break even; any sales over 6,000 will produce a profit; any sales under 6,000 will result in a loss. ◆

PROBLEM 7

What is the break-even point in Example 7 if fixed costs are \$9,900, variable costs are \$3.70 per record, and the records are sold for \$5.50 each? ◆

Algebra has many different types of applications—so many, in fact, that no single approach applies to all. However, the following suggestions may help you get started:

Suggestions for Solving Word Problems

1. Read the problem very carefully.
2. Write down important facts and relationships.
3. Identify unknown quantities in terms of a single letter, if possible.
4. Write an equation (or inequality) relating the unknown quantities and the facts in the problem.
5. Solve the equation (or inequality).
6. Write all solutions requested in the original problem.
7. Check the solution(s) in the original problem.

◆ EXAMPLE 8

Consumer Price Index (CPI)

Table 2 lists the consumer price index (CPI) for several years. What net monthly salary in 1980 would have the same purchasing power as a net monthly salary of \$900 in 1950? Compute the answer to the nearest dollar.

Solution

To have the same purchasing power, the ratio of a salary in 1980 to a salary in 1950 would have to be the same as the ratio of the CPI in 1980 to the CPI in 1950. Thus, if x is the net monthly salary in 1980, we solve the equation

TABLE 2

CPI (1967 = 100)

YEAR	INDEX	YEAR	INDEX
1950	72	1970	116
1955	80	1975	161
1960	89	1980	247
1965	95	1985	322

$$\frac{x}{900} = \frac{247}{72}$$

$$x = 900 \cdot \frac{247}{72}$$

$$= \$3{,}088 \text{ per month}$$ ◆

PROBLEM 8

Using Table 2, what net monthly salary in 1960 would have the same purchasing power as a net monthly salary of \$2,000 in 1975? Compute the answer to the nearest dollar. ◆

Answers to Matched Problems

1. $x = 4$ 2. $x = 2$ 3. (A) $t = \dfrac{M - Nr}{N}$ (B) $N = \dfrac{M}{t + r}$

4. (A) $<$ (B) $<$ (C) $>$ 5. $x \geq -4$ or $[-4, \infty)$ 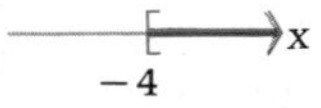

6. $-1 \leq x < 4$ or $[-1, 4)$ 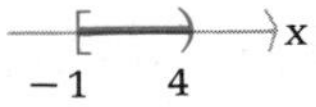 7. $x = 5{,}500$ 8. \$1,106

EXERCISE 2-1

A *Solve.*

1. $2m + 9 = 5m - 6$
2. $3y - 4 = 6y - 19$
3. $x + 5 < -4$
4. $x - 3 > -2$
5. $-3x \geq -12$
6. $-4x \leq 8$

Solve and graph.

7. $-4x - 7 > 5$
8. $-2x + 8 < 4$
9. $2 \leq x + 3 \leq 5$
10. $-3 < y - 5 < 8$

Solve.

11. $\frac{y}{7} - 1 = \frac{1}{7}$
12. $\frac{m}{5} - 2 = \frac{3}{5}$
13. $\frac{x}{3} > -2$
14. $\frac{y}{-2} \leq -1$
15. $\frac{y}{3} = 4 - \frac{y}{6}$
16. $\frac{x}{4} = 9 - \frac{x}{2}$

B

17. $10x + 25(x - 3) = 275$
18. $-3(4 - x) = 5 - (x + 1)$
19. $3 - y \leq 4(y - 3)$
20. $x - 2 \geq 2(x - 5)$
21. $\frac{x}{5} - \frac{x}{6} = \frac{6}{5}$
22. $\frac{y}{4} - \frac{y}{3} = \frac{1}{2}$
23. $\frac{m}{5} - 3 < \frac{3}{5} - m$
24. $u - \frac{2}{3} > \frac{u}{3} + 2$
25. $0.1(x - 7) + 0.05x = 0.8$
26. $0.4(u + 5) - 0.3u = 17$

Solve and graph.

27. $2 \leq 3x - 7 < 14$
28. $-4 \leq 5x + 6 < 21$
29. $-4 \leq \frac{9}{5}C + 32 \leq 68$
30. $-1 \leq \frac{2}{3}t + 5 \leq 11$

C

Solve for the indicated variable.

31. $3x - 4y = 12$, for y
32. $y = -\frac{2}{3}x + 8$, for x
33. $Ax + By = C$, for y $(B \neq 0)$
34. $y = mx + b$, for m
35. $F = \frac{9}{5}C + 32$, for C
36. $C = \frac{5}{9}(F - 32)$, for F
37. $A = Bm - Bn$, for B
38. $U = 3C - 2CD$, for C

Solve and graph.

39. $-3 \leq 4 - 7x < 18$
40. $-1 < 9 - 2u \leq 5$

APPLICATIONS

Business & Economics

41. A jazz concert brought in \$60,000 on the sale of 8,000 tickets. If the tickets sold for \$6 and \$10 each, how many of each type of ticket were sold?
42. An all-day parking meter takes only dimes and quarters. If it contains 100 coins with a total value of \$14.50, how many of each type of coin are in the meter?
43. You have \$12,000 to invest. If part is invested at 10% and the rest at 15%, how much should be invested at each rate to yield 12% on the total amount?

44. An investor has $20,000 to invest. If part is invested at 8% and the rest at 12%, how much should be invested at each rate to yield 11% on the total amount?

45. *Inflation.* If the price change of cars parallels the change in the CPI (see Table 2 in Example 8), what would a car sell for in 1980 if a comparable model sold for $3,000 in 1965?

46. *Break-even analysis.* For a business to realize a profit, it is clear that revenue R must be greater than costs C; that is, a profit will result only if $R > C$ (the company breaks even when $R = C$). A record manufacturer has a weekly cost equation $C = 300 + 1.5x$ and a revenue equation $R = 2x$, where x is the number of records produced and sold in a week. How many records must be sold for the company to make a profit?

Life Sciences

47. *Wildlife management.* A naturalist for a fish and game department estimated the total number of rainbow trout in a certain lake using the popular capture–mark–recapture technique. He netted, marked, and released 200 rainbow trout. A week later, allowing for thorough mixing, he again netted 200 trout and found 8 marked ones among them. Assuming that the proportion of marked fish in the second sample was the same as the proportion of all marked fish in the total population, estimate the number of rainbow trout in the lake.

48. *Ecology.* If the temperature for a 24 hour period at an Antarctic station ranged between $-49°F$ and $14°F$ (that is, $-49 \leq F \leq 14$), what was the range in degrees Celsius? [*Note:* $F = \frac{9}{5}C + 32$.]

Social Sciences

49. *Psychology.* The IQ (intelligence quotient) is found by dividing the mental age (MA), as indicated on standard tests, by the chronological age (CA) and multiplying by 100. For example, if a child has a mental age of 12 and a chronological age of 8, the calculated IQ is 150. If a 9-year-old girl has an IQ of 140, compute her mental age.

50. *Anthropology.* In their study of genetic groupings, anthropologists use a ratio called the *cephalic index.* This is the ratio of the width of the head to its length (looking down from above) expressed as a percentage. Symbolically,

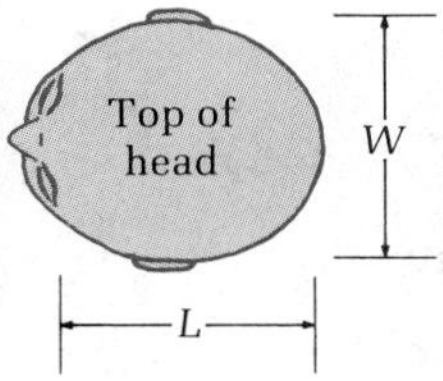

$$C = \frac{100W}{L}$$

where C is the cephalic index, W is the width, and L is the length. If an Indian tribe in Baja California (Mexico) had an average cephalic index of 66 and the average width of their heads was 6.6 inches, what was the average length of their heads?

SECTION 2-2 Quadratic Equations

- ◆ SOLUTION BY SQUARE ROOT
- ◆ SOLUTION BY FACTORING
- ◆ QUADRATIC FORMULA

A **quadratic equation** in one variable is any equation that can be written in the form

$$ax^2 + bx + c = 0 \qquad a \neq 0$$

where x is a variable and *a*, *b*, and *c* are constants. We will refer to this form as the **standard form.** The equations

$$5x^2 - 3x + 7 = 0 \qquad \text{and} \qquad 18 = 32t^2 - 12t$$

are both quadratic equations, since they are either in the standard form or can be transformed into this form.

We will restrict our review to finding real solutions to quadratic equations.

◆ SOLUTION BY SQUARE ROOT

The easiest type of quadratic equation to solve is the special form where the first-degree term is missing:

$$ax^2 + c = 0 \qquad a \neq 0$$

The method makes use of the definition of square root given in Section 1-5.

◆ EXAMPLE 9 Solve by the square root method.

(A) $x^2 - 7 = 0$ (B) $2x^2 - 10 = 0$ (C) $3x^2 + 27 = 0$

Solutions

(A) $x^2 - 7 = 0$

$x^2 = 7$ What real number squared is 7?

$x = \pm\sqrt{7}$ Short for $\sqrt{7}$ and $-\sqrt{7}$

(B) $2x^2 - 10 = 0$

$2x^2 = 10$

$x^2 = 5$ What real number squared is 5?

$x = \pm\sqrt{5}$

(C) $3x^2 + 27 = 0$

$3x^2 = -27$

$x^2 = -9$ What real number squared is -9?

No real solution, since no real number squared is negative. ◆

PROBLEM 9 Solve by the square root method.

(A) $x^2 - 6 = 0$ (B) $3x^2 - 12 = 0$ (C) $x^2 + 4 = 0$ ◆

◆ SOLUTION BY FACTORING

If the left side of a quadratic equation when written in standard form can be factored, then the equation can be solved very quickly. The method of solution by factoring rests on the following important property of real numbers (Theorem 2 in Section 1-1):

If a and b are real numbers, then $ab = 0$ if and only if $a = 0$ or $b = 0$ (or both).

◆ EXAMPLE 10 Solve by factoring using integer coefficients, if possible.

(A) $3x^2 - 6x - 24 = 0$ (B) $3y^2 = 2y$ (C) $x^2 - 2x - 1 = 0$

Solutions (A) $3x^2 - 6x - 24 = 0$ Divide both sides by 3, since 3 is a factor of each coefficient.

$x^2 - 2x - 8 = 0$ Factor the left side, if possible.

$(x - 4)(x + 2) = 0$

$x - 4 = 0$ or $x + 2 = 0$

$x = 4$ or $x = -2$

(B) $3y^2 = 2y$

$3y^2 - 2y = 0$ We lose the solution $y = 0$ if both sides are divided by y

$y(3y - 2) = 0$ ($3y^2 = 2y$ and $3y = 2$ are not equivalent).

$y = 0$ or $3y - 2 = 0$

$3y = 2$

$y = \frac{2}{3}$

(C) $x^2 - 2x - 1 = 0$

This equation cannot be factored using integer coefficients. We will solve this type of equation by another method, considered below. ◆

PROBLEM 10 Solve by factoring using integer coefficients, if possible.

(A) $2x^2 + 4x - 30 = 0$ (B) $2x^2 = 3x$ (C) $2x^2 - 8x + 3 = 0$ ◆

The factoring and square root methods are fast and easy to use when they apply. However, there are quadratic equations that look simple but cannot be solved by either method. For example, as was noted in Example 10C, the polynomial in

$$x^2 - 2x - 1 = 0$$

cannot be factored using integer coefficients. This brings us to the well-known and widely used *quadratic formula.*

◆ QUADRATIC FORMULA

There is a method called *completing the square* that will work for all quadratic equations. After briefly reviewing this method, we will then use it to develop the famous quadratic formula—a formula that will enable us to solve any quadratic equation quite mechanically.

The method of **completing the square** is based on the process of transforming a quadratic equation in standard form,

$$ax^2 + bx + c = 0$$

into the form

$$(x + A)^2 = B$$

where A and B are constants. Then, this last equation can be solved easily (if it has a real solution) by the square root method discussed above.

Consider the equation

$$x^2 - 2x - 1 = 0 \qquad (1)$$

Since the left side does not factor using integer coefficients, we add 1 to each side to remove the constant term from the left side:

$$x^2 - 2x = 1 \qquad (2)$$

Now we try to find a number that we can add to each side to make the left side a square of a first-degree polynomial. Note the following two squares:

$$(x + m)^2 = x^2 + 2mx + m^2 \qquad (x - m)^2 = x^2 - 2mx + m^2$$

In each equation, we see that the third term on the right is the square of one-half the coefficient of x in the second term on the right. To complete the square in equation (2), we add the square of one-half the coefficient of x, $(-\frac{2}{2})^2 = 1$, to each side. (This rule works only when the coefficient of x^2 is 1, that is, $a = 1$.) Thus,

$$x^2 - 2x + 1 = 1 + 1$$

The left side is the square of $x - 1$, and we write

$$(x - 1)^2 = 2$$

What number squared is 2?

$$x - 1 = \pm\sqrt{2}$$
$$x = 1 \pm \sqrt{2}$$

And equation (1) is solved!

Let us try the method on the general quadratic equation

$$ax^2 + bx + c = 0 \qquad a \neq 0 \qquad (3)$$

and solve it once and for all for x in terms of the coefficients a, b, and c. We start by multiplying both sides of (3) by $1/a$ to obtain

$$x^2 + \frac{b}{a}x + \frac{c}{a} = 0$$

Add $-c/a$ to both sides:

$$x^2 + \frac{b}{a}x = -\frac{c}{a}$$

Now we complete the square on the left side by adding the square of one-half the coefficient of x, that is, $(b/2a)^2 = b^2/4a^2$, to each side:

$$x^2 + \frac{b}{a}x + \frac{b^2}{4a^2} = \frac{b^2}{4a^2} - \frac{c}{a}$$

Writing the left side as a square and combining the right side into a single fraction, we obtain

$$\left(x + \frac{b}{2a}\right)^2 = \frac{b^2 - 4ac}{4a^2}$$

Now we solve by the square root method:

$$x + \frac{b}{2a} = \pm\sqrt{\frac{b^2 - 4ac}{4a^2}}$$

$$x = -\frac{b}{2a} \pm \frac{\sqrt{b^2 - 4ac}}{2a} \qquad \text{Since } \pm\sqrt{4a^2} = \pm 2a \text{ for any real number } a$$

When this is written as a single fraction, it becomes the **quadratic formula:**

Quadratic Formula

If $ax^2 + bx + c = 0$, $a \neq 0$, then

$$x = \frac{-b \pm \sqrt{b^2 - 4ac}}{2a}$$

TABLE 3

$b^2 - 4ac$	$ax^2 + bx + c = 0$
Positive	Two real solutions
Zero	One real solution
Negative	No real solutions

This formula is generally used to solve quadratic equations when the square root or factoring methods do not work. The quantity $b^2 - 4ac$ under the radical is called the **discriminant,** and it gives us the useful information about solutions listed in Table 3.

◆ EXAMPLE 11 Solve $x^2 - 2x - 1 = 0$ using the quadratic formula.

Solution

$$x^2 - 2x - 1 = 0$$

$$x = \frac{-b \pm \sqrt{b^2 - 4ac}}{2a} \qquad a = 1,\ b = -2,\ c = -1$$

$$= \frac{-(-2) \pm \sqrt{(-2)^2 - 4(1)(-1)}}{2(1)}$$

$$= \frac{2 \pm \sqrt{8}}{2} = \frac{2 \pm 2\sqrt{2}}{2} = 1 \pm \sqrt{2}$$

Check $x^2 - 2x - 1 = 0$

When $x = 1 + \sqrt{2}$,

$$(1 + \sqrt{2})^2 - 2(1 + \sqrt{2}) - 1 = 1 + 2\sqrt{2} + 2 - 2 - 2\sqrt{2} - 1 = 0$$

When $x = 1 - \sqrt{2}$,

$$(1 - \sqrt{2})^2 - 2(1 - \sqrt{2}) - 1 = 1 - 2\sqrt{2} + 2 - 2 + 2\sqrt{2} - 1 = 0$$

◆

PROBLEM 11 Solve $2x^2 - 4x - 3 = 0$ using the quadratic formula. ◆

If we try to solve $x^2 - 6x + 11 = 0$ using the quadratic formula, we obtain

$$x = \frac{6 \pm \sqrt{-8}}{2}$$

which is not a real number. (Why?)

Answers to Matched Problems

9. (A) $\pm\sqrt{6}$ (B) ± 2 (C) No real solution
10. (A) $-5, 3$ (B) $0, \frac{3}{2}$ (C) Cannot be factored using integer coefficients
11. $(2 \pm \sqrt{10})/2$

EXERCISE 2-2

Find only real solutions in the problems below. If there are no real solutions, say so.

A *Solve by the square root method.*

1. $x^2 - 4 = 0$
2. $x^2 - 9 = 0$
3. $2x^2 - 22 = 0$
4. $3m^2 - 21 = 0$

Solve by factoring.

5. $2u^2 - 8u - 24 = 0$
6. $3x^2 - 18x + 15 = 0$
7. $x^2 = 2x$
8. $n^2 = 3n$

Solve by using the quadratic formula.

9. $x^2 - 6x - 3 = 0$
10. $m^2 + 8m + 3 = 0$
11. $3u^2 + 12u + 6 = 0$
12. $2x^2 - 20x - 6 = 0$

B *Solve, using any method.*

13. $2x^2 = 4x$
14. $2x^2 = -3x$
15. $4u^2 - 9 = 0$
16. $9y^2 - 25 = 0$
17. $8x^2 + 20x = 12$
18. $9x^2 - 6 = 15x$
19. $x^2 = 1 - x$
20. $m^2 = 1 - 3m$
21. $2x^2 = 6x - 3$
22. $2x^2 = 4x - 1$
23. $y^2 - 4y = -8$
24. $x^2 - 2x = -3$
25. $(x + 4)^2 = 11$
26. $(y - 5)^2 = 7$

C

27. Solve $A = P(1 + r)^2$ for r in terms of A and P; that is, isolate r on the left side of the equation (with coefficient 1) and end up with an algebraic expression on the right side involving A and P but not r. Write the answer using positive square roots only.
28. Solve $x^2 + mx + n = 0$ for x in terms of m and n.

APPLICATIONS

Business & Economics

29. *Supply and demand.* The demand equation for a certain brand of popular records is $d = 3{,}000/p$. Notice that as the price (p) goes up, the number of records people are willing to buy (d) goes down, and vice versa. The supply equation is given by $s = 1{,}000p - 500$. Notice again, as the price (p) goes up, the number of records a supplier is willing to sell (s) goes up. At what price will supply equal demand; that is, at what price will $d = s$? In economic theory the price at which supply equals demand is called the **equilibrium point**—the point where the price ceases to change.
30. If P dollars is invested at $100r$ percent compounded annually, at the end of 2 years it will grow to $A = P(1 + r)^2$. At what interest rate will \$100 grow to \$144 in 2 years? [*Note:* If $A = 144$ and $P = 100$, find r.]

Life Sciences

31. *Ecology.* An important element in the erosive force of moving water is its velocity. To measure the velocity v (in feet per second) of a stream, we position a hollow L-shaped tube with one end under the water pointing upstream and the other end pointing straight up a couple of feet out of the water. The water will then be pushed up the tube a certain distance h (in feet) above the surface of the stream. Physicists have shown that $v^2 = 64h$. Approximately how fast is a stream flowing if $h = 1$ foot? If $h = 0.5$ foot?

Social Sciences

32. *Safety research.* It is of considerable importance to know the least number of feet d in which a car can be stopped, including reaction time of the driver, at various speeds v (in miles per hour). Safety research has produced the formula $d = 0.044v^2 + 1.1v$. If it took a car 550 feet to stop, estimate the car's speed at the moment the stopping process was started. You might find a calculator of help in this problem.

SECTION 2-3 Cartesian Coordinate System and Straight Lines

- ◆ CARTESIAN COORDINATE SYSTEM
- ◆ GRAPHING LINEAR EQUATIONS IN TWO VARIABLES
- ◆ SLOPE
- ◆ EQUATIONS OF LINES—SPECIAL FORMS
- ◆ APPLICATION

◆ CARTESIAN COORDINATE SYSTEM

Recall that a **Cartesian (rectangular) coordinate system** in a plane is formed by taking two mutually perpendicular real number lines **(coordinate axes)**—one horizontal and one vertical—intersecting at their origins, and then assigning unique **ordered pairs** of numbers **(coordinates)** to each point P in the plane (Fig. 2). The first coordinate **(abscissa)** is the distance of P from the vertical axis, and the second coordinate **(ordinate)** is the distance of P from the horizontal axis. In Figure 2, the coordinates of point P are (a, b). By reversing the process, each ordered pair of real numbers can be associated with a unique point in the plane. The coordinate axes divide the plane into four parts **(quadrants)**, numbered I to IV in a counterclockwise direction.

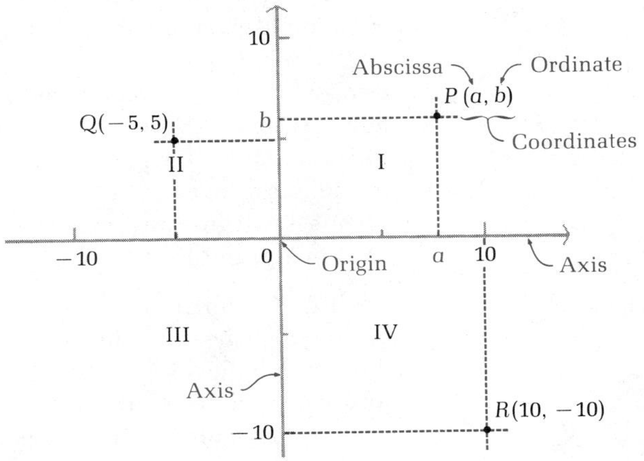

FIGURE 2
The Cartesian coordinate system

◆ GRAPHING LINEAR EQUATIONS IN TWO VARIABLES

A linear equation in two variables is an equation that can be written in the form

STANDARD FORM $\quad Ax + By = C$

with A and B not both 0. For example,

$$2x - 3y = 5 \qquad x = 7 \qquad y = \tfrac{1}{2}x - 3 \qquad y = -3$$

all can be considered linear equations in two variables. The first is in standard form, while the other three can be written in standard form as follows:

	Standard Form	
$x = 7$	$x + 0y = 7$	
$y = \frac{1}{2}x - 3$	$-\frac{1}{2}x + y = -3$	or $x - 2y = 6$
$y = -3$	$0x + y = -3$	

A **solution** of an equation in two variables is an ordered pair of real numbers that satisfy the equation. For example, $(0, -3)$ is a solution of $3x - 4y = 12$. The **solution set** of an equation in two variables is the set of all solutions of the equation. When we say that we **graph an equation** in two variables, we mean that we graph its solution set on a rectangular coordinate system.

We state the following important theorem without proof:

THEOREM 1

Graph of a Linear Equation in Two Variables

The graph of any equation of the form

STANDARD FORM $\quad Ax + By = C \quad$ (1)

where A, B, and C are constants (A and B not both 0), is a straight line. Every straight line in a Cartesian coordinate system is the graph of an equation of this type.

Also, the graph of any equation of the form

$$y = mx + b \qquad (2)$$

where m and b are constants, is a straight line. Form (2) is simply a special case of (1) for $B \neq 0$. To graph either (1) or (2), we plot any two points of their solution set and use a straightedge to draw the line through these two points. The points where the line crosses the axes—called the **intercepts**—are often the easiest to find when dealing with form (1). To find the ***y* intercept,** we let $x = 0$ and solve for y; to find the ***x* intercept,** we let $y = 0$ and solve for x. It is sometimes wise to find a third point as a check.

◆ EXAMPLE 12 (A) The graph of $3x - 4y = 12$ is

x	y
0	−3
4	0
8	3

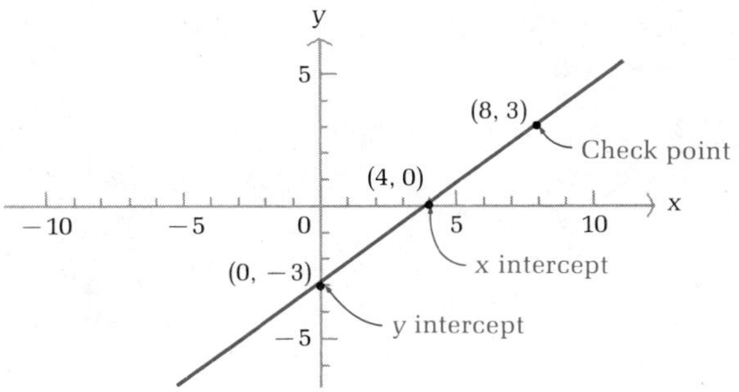

(B) The graph of $y = 2x - 1$ is

x	y
0	−1
4	7
−2	−5

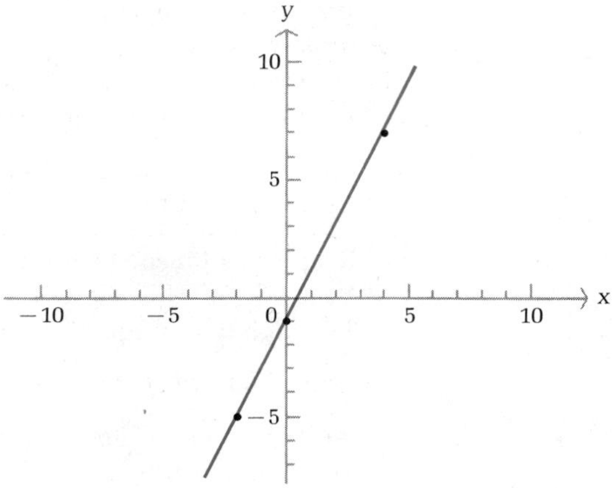

PROBLEM 12 Graph: (A) $4x - 3y = 12$ (B) $y = \frac{x}{2} + 2$

◆ SLOPE

It is very useful to have a numerical measure of the "steepness" of a line. The concept of *slope* is widely used for this purpose. The **slope** of a line through the two points (x_1, y_1) and (x_2, y_2) is given by the following formula:

Slope

$$m = \frac{y_2 - y_1}{x_2 - x_1} \qquad x_1 \neq x_2$$

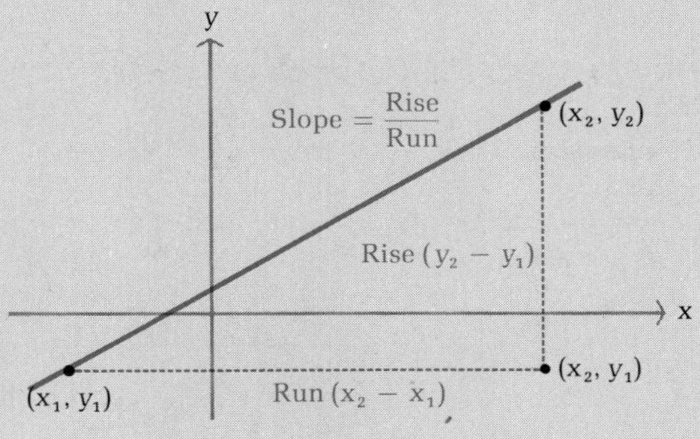

The slope of a vertical line is not defined. (Why? See Example 13B.)

◆ EXAMPLE 13 Find the slope of the line through each pair of points:

(A) $(-2, 5), (4, -7)$ (B) $(-3, -1), (-3, 5)$

Solutions (A) Let $(x_1, y_1) = (-2, 5)$ and $(x_2, y_2) = (4, -7)$. Then

$$m = \frac{y_2 - y_1}{x_2 - x_1} = \frac{-7 - 5}{4 - (-2)} = \frac{-12}{6} = -2$$

Note that we also could have let $(x_1, y_1) = (4, -7)$ and $(x_2, y_2) = (-2, 5)$, since this simply reverses the sign in both the numerator and denominator and the slope does not change:

$$m = \frac{5 - (-7)}{-2 - 4} = \frac{12}{-6} = -2$$

(B) Let $(x_1, y_1) = (-3, -1)$ and $(x_2, y_2) = (-3, 5)$. Then

$$m = \frac{y_2 - y_1}{x_2 - x_1} = \frac{5 - (-1)}{-3 - (-3)} = \frac{6}{0} \qquad \text{Not defined!}$$

Notice that $x_1 = x_2$. This is always true for a vertical line, since the abscissa (first coordinate) of every point on a vertical line is the same. Thus, the slope of a vertical line is not defined (that is, the slope does not exist). ◆

PROBLEM 13 Find the slope of the line through each pair of points:

(A) $(3, -6), (-2, 4)$ (B) $(-7, 5), (3, 5)$ ◆

In general, the slope of a line may be positive, negative, zero, or not defined. Each of these cases is interpreted geometrically in Table 4.

TABLE 4

Going from Left to Right

LINE	SLOPE	EXAMPLE
Rising	Positive	
Falling	Negative	
Horizontal	Zero	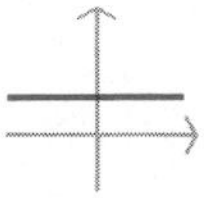
Vertical	Not defined	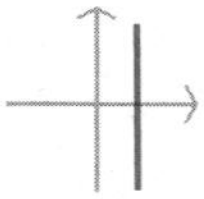

◆ EQUATIONS OF LINES—SPECIAL FORMS

The constants m and b in the equation

$$y = mx + b \tag{3}$$

have special geometric significance.

If we let $x = 0$, then $y = b$, and we observe that the graph of (3) crosses the y axis at $(0, b)$. The constant b is the *y intercept*. For example, the y intercept of the graph of $y = -4x - 1$ is -1.

To determine the geometric significance of m, we proceed as follows: If $y = mx + b$, then by setting $x = 0$ and $x = 1$, we conclude that $(0, b)$ and $(1, m + b)$ lie on its graph (a line). Hence, the slope of this graph (line) is given by:

$$\text{Slope} = \frac{y_2 - y_1}{x_2 - x_1} = \frac{(m + b) - b}{1 - 0} = m$$

Thus, m is the slope of the line given by $y = mx + b$.

Slope–Intercept Form

The equation

$$y = mx + b \qquad m = \text{Slope},\ b = y \text{ intercept} \tag{4}$$

is called the **slope–intercept form** of an equation of a line.

◆ EXAMPLE 14

(A) Find the slope and y intercept, and graph $y = -\frac{2}{3}x - 3$.
(B) Write the equation of the line with slope $\frac{2}{3}$ and y intercept -2.

Solutions

(A) Slope $= m = -\frac{2}{3}$
y intercept $= b = -3$

(B) $m = \frac{2}{3}$ and $b = -2$; thus, $y = \frac{2}{3}x - 2$

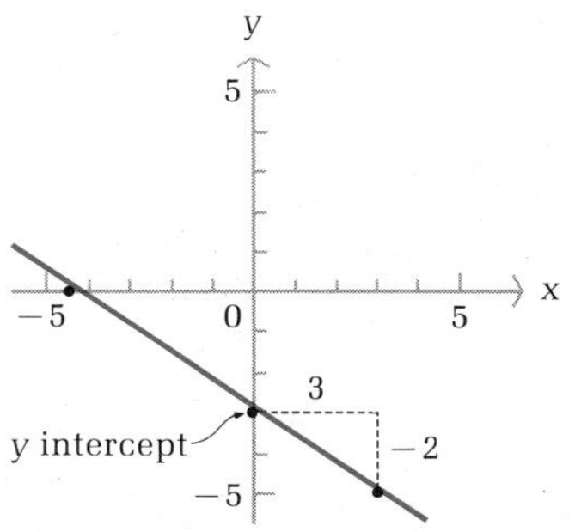

◆

PROBLEM 14 Write the equation of the line with slope $\frac{1}{2}$ and y intercept -1. Graph. ◆

Suppose a line has slope m and passes through a fixed point (x_1, y_1). If the point (x, y) is any other point on the line (Fig. 3), then

$$\frac{y - y_1}{x - x_1} = m$$

that is,

$$y - y_1 = m(x - x_1) \tag{5}$$

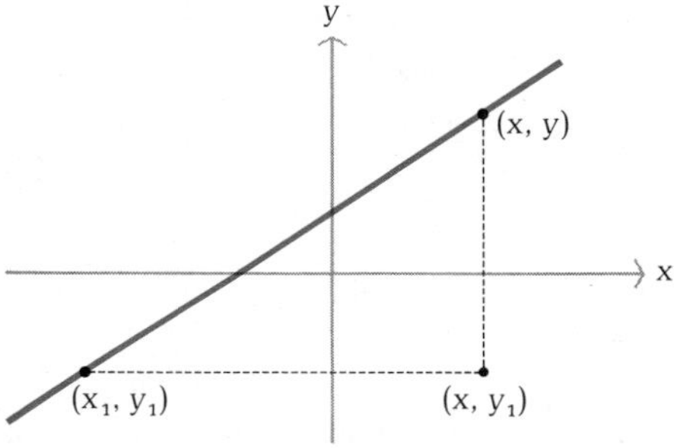

FIGURE 3

We now observe that (x_1, y_1) also satisfies equation (5) and conclude that equation (5) is an equation of a line with slope m that passes through (x_1, y_1).

Point–Slope Form

An equation of a line with slope m that passes through (x_1, y_1) is

$$y - y_1 = m(x - x_1) \tag{5}$$

which is called the **point–slope form** of an equation of a line.

The point–slope form is extremely useful, since it enables us to find an equation for a line if we know its slope and the coordinates of a point on the line or if we know the coordinates of two points on the line.

◆ EXAMPLE 15

(A) Find an equation for the line that has slope $\frac{1}{2}$ and passes through $(-4, 3)$. Write the final answer in the form $Ax + By = C$.

(B) Find an equation for the line that passes through the two points $(-3, 2)$ and $(-4, 5)$. Write the resulting equation in the form $y = mx + b$.

Solutions (A) Use $y - y_1 = m(x - x_1)$. Let $m = \frac{1}{2}$ and $(x_1, y_1) = (-4, 3)$. Then

$$y - 3 = \tfrac{1}{2}[x - (-4)]$$
$$y - 3 = \tfrac{1}{2}(x + 4) \qquad \text{Multiply by 2.}$$
$$2y - 6 = x + 4$$
$$-x + 2y = 10 \quad \text{or} \quad x - 2y = -10$$

(B) First, find the slope of the line by using the slope formula:

$$m = \frac{y_2 - y_1}{x_2 - x_1} = \frac{5 - 2}{-4 - (-3)} = \frac{3}{-1} = -3$$

Now use $y - y_1 = m(x - x_1)$ with $m = -3$ and $(x_1, y_1) = (-3, 2)$:

$$y - 2 = -3[x - (-3)]$$
$$y - 2 = -3(x + 3)$$
$$y - 2 = -3x - 9$$
$$y = -3x - 7$$

◆

PROBLEM 15 (A) Find an equation for the line that has slope $\frac{2}{3}$ and passes through $(6, -2)$. Write the resulting equation in the form $Ax + By = C$, $A > 0$.
(B) Find an equation for the line that passes through $(2, -3)$ and $(4, 3)$. Write the resulting equation in the form $y = mx + b$. ◆

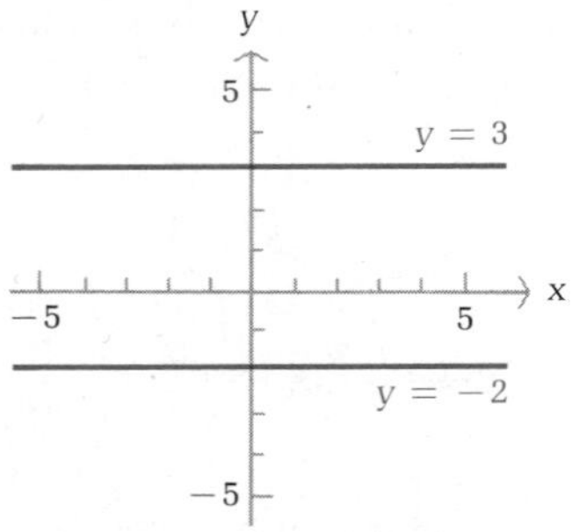

FIGURE 4

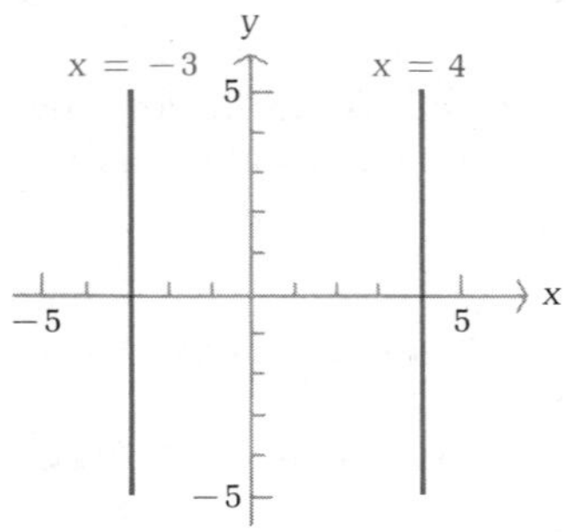

FIGURE 5

The simplest equations of a line are those for horizontal and vertical lines. A **horizontal line** has slope 0; thus, its equation is of the form

$$y = 0x + c \qquad \text{Slope} = 0,\ y \text{ intercept} = c$$

or, simply,

$$y = c$$

Figure 4 illustrates the graphs of $y = 3$ and $y = -2$.

If a line is vertical, then its slope is not defined. All x values (abscissas) of points on a vertical line are equal, while y can take on any value (Fig. 5). Thus, a **vertical line** has an equation of the form

$$x + 0y = c \qquad x \text{ intercept} = c$$

or, simply,

$$x = c$$

Figure 5 illustrates the graphs of $x = -3$ and $x = 4$.

Equations of Horizontal and Vertical Lines

HORIZONTAL LINE WITH y INTERCEPT c: $y = c$
VERTICAL LINE WITH x INTERCEPT c: $x = c$

◆ EXAMPLE 16 The equation of a horizontal line through $(-2, 3)$ is $y = 3$, and the equation of a vertical line through the same point is $x = -2$. ◆

PROBLEM 16 Find the equations of the horizontal and vertical lines through $(4, -5)$. ◆

It can be shown that if two nonvertical lines are parallel, then they have the same slope. And if two lines have the same slope, they are parallel. It also can be shown that if two nonvertical lines are perpendicular, then their slopes are the negative reciprocals of each other (that is, $m_2 = -1/m_1$, or, equivalently, $m_1 m_2 = -1$). And if the slopes of two lines are the negative reciprocals of each other, the lines are perpendicular. Symbolically:

Parallel and Perpendicular Lines

Given nonvertical lines L_1 and L_2 with slopes m_1 and m_2, respectively, then:

$L_1 \| L_2$ if and only if $m_1 = m_2$

$L_1 \perp L_2$ if and only if $m_1 m_2 = -1$ or $m_2 = -\dfrac{1}{m_1}$

[*Note:* $\|$ means "is parallel to" and $\perp$ means "is perpendicular to."]

◆ EXAMPLE 17 Given the line $x - 2y = 4$, find the equation of a line that passes through $(2, -3)$ and is:

(A) Parallel to the given line (B) Perpendicular to the given line

Write final equations in the form $y = mx + b$.

Solutions First find the slope of the given line by writing $x - 2y = 4$ in the form $y = mx + b$:

$$x - 2y = 4$$
$$y = \tfrac{1}{2}x - 2$$

The slope of the given line is $\frac{1}{2}$.

(A) The slope of a line parallel to the given line is also $\frac{1}{2}$. We have only to find the equation of a line through $(2, -3)$ with slope $\frac{1}{2}$ to solve part A:

$$\begin{aligned} y - y_1 &= m(x - x_1) \qquad m = \tfrac{1}{2} \text{ and } (x_1, y_1) = (2, -3) \\ y - (-3) &= \tfrac{1}{2}(x - 2) \\ y + 3 &= \tfrac{1}{2}x - 1 \\ y &= \tfrac{1}{2}x - 4 \end{aligned}$$

(B) The slope of the line perpendicular to the given line is the negative reciprocal of $\frac{1}{2}$; that is, -2. We have only to find the equation of a line through $(2, -3)$ with slope -2 to solve part B:

$$\begin{aligned} y - y_1 &= m(x - x_1) \qquad m = -2 \text{ and } (x_1, y_1) = (2, -3) \\ y - (-3) &= -2(x - 2) \\ y + 3 &= -2x + 4 \\ y &= -2x + 1 \end{aligned}$$

◆

PROBLEM 17

Given the line $2x = 6 - 3y$, find the equation of a line that passes through $(-3, 9)$ and is:

(A) Parallel to the given line (B) Perpendicular to the given line

Write final equations in the form $y = mx + b$. ◆

◆ APPLICATION

We will now see how equations of lines occur in certain applications.

◆ EXAMPLE 18

Cost Equation

The management of a company that manufactures roller skates has fixed costs (costs at zero output) of \$300 per day and total costs of \$4,300 per day at an output of 100 pairs of skates per day. Assume that cost C is linearly related to output x.

(A) Find the slope of the line joining the points associated with outputs of 0 and 100; that is, the line passing through $(0, 300)$ and $(100, 4{,}300)$.
(B) Find an equation of the line relating output to cost. Write the final answer in the form $C = mx + b$.
(C) Graph the cost equation from part B for $0 \leq x \leq 200$.

Solutions

(A) $m = \dfrac{y_2 - y_1}{x_2 - x_1} = \dfrac{4{,}300 - 300}{100 - 0} = \dfrac{4{,}000}{100} = 40$

(B) We must find an equation of the line that passes through $(0, 300)$ with slope 40. We use the slope–intercept form:

$$\begin{aligned} C &= mx + b \\ C &= 40x + 300 \end{aligned}$$

x	C
0	300
100	4,300
200	8,300

(C)

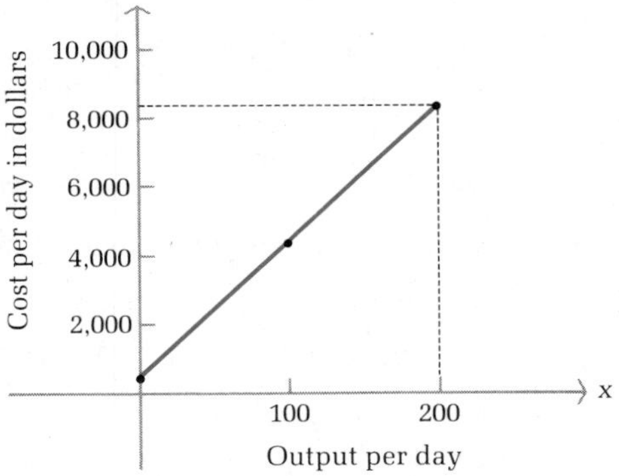

◆

PROBLEM 18 Answer parts A and B in Example 18 for fixed costs of \$250 per day and total costs of \$3,450 per day at an output of 80 pairs of skates per day. ◆

Answers to Matched Problems

12. (A)

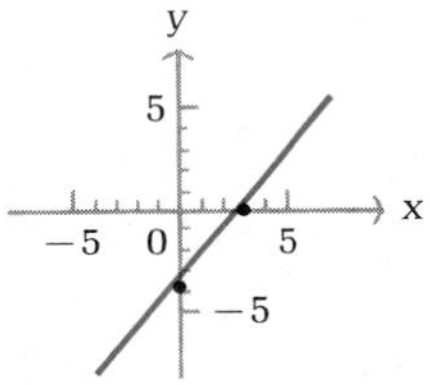

(B)

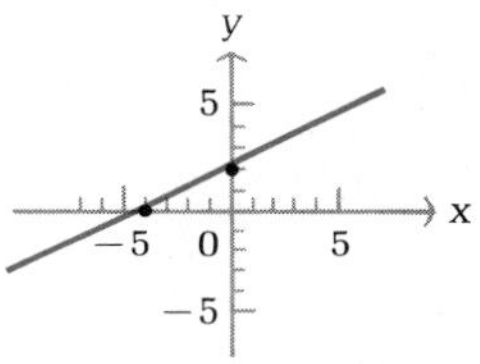

13. (A) −2
    (B) 0 (Zero is a number—it exists! It is the slope of a horizontal line.)

14. $y = \frac{1}{2}x - 1$

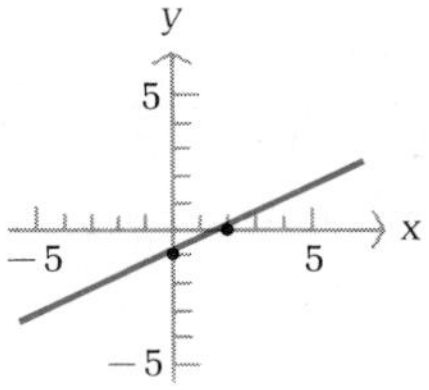

15. (A) $2x - 3y = 18$
    (B) $y = 3x - 9$
16. $y = -5$; $x = 4$
17. (A) $y = -\frac{2}{3}x + 7$
    (B) $y = \frac{3}{2}x + \frac{27}{2}$
18. (A) $m = 40$
    (B) $C = 40x + 250$

EXERCISE 2-3

A *Graph in a rectangular coordinate system.*

1. $y = 2x - 3$
2. $y = \dfrac{x}{2} + 1$
3. $2x + 3y = 12$
4. $8x - 3y = 24$

Find the slope and y intercept of the graph of each equation.

5. $y = 2x - 3$
6. $y = \dfrac{x}{2} + 1$
7. $y = -\frac{2}{3}x + 2$
8. $y = \frac{3}{4}x - 2$

Write an equation of the line with the indicated slope and y intercept.

9. Slope $= -2$
   y intercept $= 4$
10. Slope $= -\frac{2}{3}$
   y intercept $= -2$
11. Slope $= -\frac{3}{5}$
   y intercept $= 3$
12. Slope $= 1$
   y intercept $= -2$

B *Graph in a rectangular coordinate system.*

13. $y = -\frac{2}{3}x - 2$
14. $y = -\frac{3}{2}x + 1$
15. $3x - 2y = 10$
16. $5x - 6y = 15$
17. $x = 3$ and $y = -2$
18. $x = -3$ and $y = 2$

Find the slope of the graph of each equation. (First write the equation in the form $y = mx + b$.)

19. $3x + y = 5$
20. $2x - y = -3$
21. $2x + 3y = 12$
22. $3x - 2y = 10$

Write an equation of the line through each indicated point with the indicated slope. Transform the equation into the form $y = mx + b$.

23. $m = -3$; $(4, -1)$
24. $m = -2$; $(-3, 2)$
25. $m = \frac{2}{3}$; $(-6, -5)$
26. $m = \frac{1}{2}$; $(-4, 3)$

Find the slope of the line that passes through the given points.

27. $(1, 3)$ and $(7, 5)$
28. $(2, 1)$ and $(10, 5)$
29. $(-5, -2)$ and $(5, -4)$
30. $(3, 7)$ and $(-6, 4)$

Write an equation of the line through each indicated pair of points. Write the final answer in the form $Ax + By = C$, $A > 0$.

31. $(1, 3)$ and $(7, 5)$
32. $(2, 1)$ and $(10, 5)$
33. $(-5, -2)$ and $(5, -4)$
34. $(3, 7)$ and $(-6, 4)$

Write equations of the vertical and horizontal lines through each point.

35. $(3, -5)$
36. $(-2, 7)$
37. $(-1, -3)$
38. $(6, -4)$

Find an equation of the line, given the information in each problem. Write the final answer in the form $y = mx + b$.

39. Line passes through $(-2, 5)$ with slope $-\frac{1}{2}$.
40. Line passes through $(3, -1)$ with slope $-\frac{2}{3}$.
41. Line passes through $(-2, 2)$ and is:
    (A) Parallel (B) Perpendicular to $y = -\frac{1}{2}x + 5$
42. Line passes through $(-4, -3)$ and is:
    (A) Parallel (B) Perpendicular to $y = 2x - 3$
43. Line passes through $(-2, -1)$ and is:
    (A) Parallel (B) Perpendicular to $x - 2y = 4$

44. Line passes through $(-3, 2)$ and is:

(A) Parallel (B) Perpendicular to $2x + 3y = -6$

C

45. Graph $y = mx - 2$ for $m = 2$, $m = \frac{1}{2}$, $m = 0$, $m = -\frac{1}{2}$, and $m = -2$, all on the same coordinate system.

46. Graph $y = -\frac{1}{2}x + b$ for $b = -4$, $b = 0$, and $b = 4$, all on the same coordinate system.

Write an equation of the line through the indicated points. Be careful!

47. $(2, 7)$ and $(2, -3)$
48. $(-2, 3)$ and $(-2, -1)$
49. $(2, 3)$ and $(-5, 3)$
50. $(-3, -3)$ and $(0, -3)$

APPLICATIONS

Business & Economics

51. *Simple interest.* If $P (the principal) is invested at an interest rate of r, then the amount A that is due after t years is given by

$$A = Prt + P$$

If $100 is invested at 6% ($r = 0.06$), then $A = 6t + 100$, $t \geq 0$.

(A) What will $100 amount to after 5 years? After 20 years?
(B) Graph the equation for $0 \leq t \leq 20$.
(C) What is the slope of the graph? (The slope indicates the increase in the amount A for each additional year of investment.)

52. *Cost equation.* The management of a company manufacturing surfboards has fixed costs (zero output) of $200 per day and total costs of $1,400 per day at a daily output of 20 boards.

(A) Assuming the total cost per day (C) is linearly related to the total output per day (x), write an equation relating these two quantities. [*Hint:* Find an equation of the line that passes through (0, 200) and (20, 1,400).]
(B) What are the total costs for an output of 12 boards per day?
(C) Graph the equation for $0 \leq x \leq 20$.

[*Note:* The slope of the line found in part A is the increase in total cost for each additional unit produced and is called the *marginal cost.* More will be said about this concept later.]

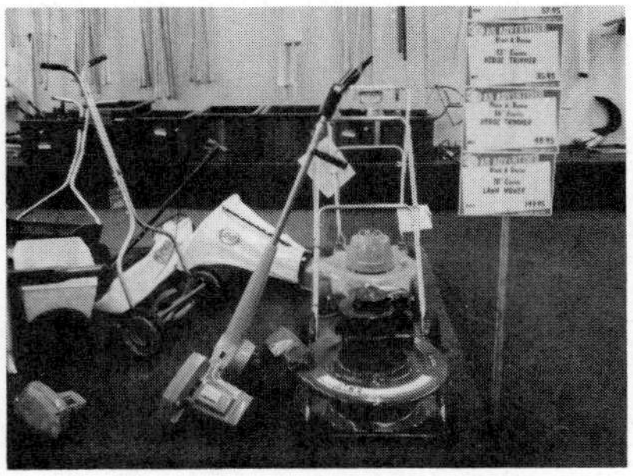

53. *Demand equation.* A manufacturing company is interested in introducing a new power mower. Its market research department gave the management the demand–price forecast listed in the table.

PRICE	ESTIMATED DEMAND
$ 70	7,800
$120	4,800
$160	2,400
$200	0

(A) Plot these points, letting d represent the number of mowers people are willing to buy (demand) at a price of $\$p$ each.

(B) Note that the points in part A lie along a straight line. Find an equation of that line.

[*Note:* The slope of the line found in part B indicates the decrease in demand for each \$1 increase in price.]

54. *Depreciation.* Office equipment was purchased for \$20,000 and is assumed to have a scrap value of \$2,000 after 10 years. If its value is depreciated linearly (for tax purposes) from \$20,000 to \$2,000:

(A) Find the linear equation that relates value (V) in dollars to time (t) in years.

(B) What would be the value of the equipment after 6 years?

(C) Graph the equation for $0 \leqslant t \leqslant 10$.

[*Note:* The slope found in part A indicates the decrease in value per year.]

Life Sciences

55. *Nutrition.* In a nutrition experiment, a biologist wants to prepare a special diet for the experimental animals. Two food mixes, A and B, are available. If mix A contains 20% protein and mix B contains 10% protein, what combination of each mix will provide exactly 20 grams of protein? Let x be the amount of A used and let y be the amount of B used. Then write a linear equation relating x, y, and 20. Graph this equation for $x \geqslant 0$ and $y \geqslant 0$.

56. *Ecology.* As one descends into the ocean, pressure increases linearly. The pressure is 15 pounds per square inch on the surface and 30 pounds per square inch 33 feet below the surface.

(A) If p is the pressure in pounds and d is the depth below the surface in feet, write an equation that expresses p in terms of d. [*Hint:* Find an equation of the line that passes through (0, 15) and (33, 30).]

(B) What is the pressure at 12,540 feet (the average depth of the ocean)?

(C) Graph the equation for $0 \leqslant d \leqslant 12{,}540$.

[*Note:* The slope found in part A indicates the change in pressure for each additional foot of depth.]

Social Sciences

57. *Psychology.* In an experiment on motivation, J. S. Brown trained a group of rats to run down a narrow passage in a cage to obtain food in a goal box. Using a harness, he then connected the rats to an overhead wire that was attached to a spring scale. A rat was placed at different distances d (in centimeters) from the goal box, and the pull p (in grams) of the rat toward the food was measured. Brown found that the relationship between these two variables was very close to being linear and could be approximated by the equation

$$p = -\tfrac{1}{5}d + 70 \qquad 30 \leqslant d \leqslant 175$$

(See J. S. Brown, *Journal of Comparative and Physiological Psychology,* 1948, 41:450–465.)

(A) What was the pull when $d = 30$? When $d = 175$?
(B) Graph the equation.
(C) What is the slope of the line?

SECTION 2-4 Functions

- DEFINITION OF A FUNCTION
- FUNCTIONS SPECIFIED BY EQUATIONS
- FUNCTION NOTATION
- APPLICATION

The function concept is one of the most important concepts in mathematics. The idea of correspondence plays a central role in its formulation. You have already had experiences with correspondences in everyday life. For example:

To each person there corresponds an annual income.

To each item in a supermarket there corresponds a price.

To each day there corresponds a maximum temperature.

For the manufacture of x items there corresponds a cost.

For the sale of x items there corresponds a revenue.

To each square there corresponds an area.

To each number there corresponds its cube.

One of the most important aspects of any science (managerial, life, social, physical, etc.) is the establishment of correspondences among various types of phenomena. Once a correspondence is known, predictions can be made. A cost analyst would like to predict costs for various levels of output in a manufacturing process; a medical researcher would like to know the correspondence between heart disease and obesity; a psychologist would like to predict the level of performance after a subject has repeated a task a given number of times; and so on.

◆ DEFINITION OF A FUNCTION

What do all the above examples have in common? Each describes the matching of elements from one set with the elements in a second set. Consider the tables of the cube, square, and square root given in Tables 5–7 at the top of the next page.

TABLE 5

DOMAIN *Number*	RANGE *Cube*
−2	−8
−1	−1
0	0
1	1
2	8

TABLE 6

DOMAIN *Number*	RANGE *Square*
−2	4
−1	1
0	0
1	1
2	4

TABLE 7

DOMAIN *Number*	RANGE *Square Root*
0	0
1	1, −1
4	2, −2
9	3, −3

Tables 5 and 6 specify functions, but Table 7 does not. Why not? The definition of the very important term *function* will explain.

Definition of a Function (Rule Form)

A **function** is a rule (process or method) that produces a correspondence between one set of elements, called the **domain,** and a second set of elements, called the **range,** such that to each element in the domain there corresponds *one and only one* element in the range.

Tables 5 and 6 specify functions, since to each domain value there corresponds exactly one range value (for example, the cube of -2 is -8 and no other number). On the other hand, Table 7 does not specify a function, since to at least one domain value there corresponds more than one range value (for example, to the domain value 9 there corresponds -3 and 3, both square roots of 9).

Since in a function elements in the range are paired with elements in the domain by some rule or process, this correspondence (pairing) can be illustrated by using ordered pairs of elements, where the first component represents a domain element and the second component represents a corresponding range element. Thus, we can write functions 1 and 2 (Tables 5 and 6) as follows:

$$\text{Function } 1 = \{(-2, -8), (-1, -1), (0, 0), (1, 1), (2, 8)\}$$

$$\text{Function } 2 = \{(-2, 4), (-1, 1), (0, 0), (1, 1), (2, 4)\}$$

In either case, notice that no two ordered pairs have the same first component and different second components. On the other hand, if we list the ordered pairs determined by Table 7, we have

$$A = \{(0, 0), (1, 1), (1, -1), (4, 2), (4, -2), (9, 3), (9, -3)\}$$

In this case, there are ordered pairs with the same first component and different second components—for example, $(1, 1)$ and $(1, -1)$ both belong to the set A—indicating once again that Table 7 does not specify a function.

This suggests an alternative but equivalent way of defining functions that produces additional insight into this concept.

Definition of a Function (Set Form)

A **function** is a set of ordered pairs with the property that no two ordered pairs have the same first element and different second elements. The set of all first components in a function is called the **domain** of the function, and the set of all second components is called the **range.**

◆ EXAMPLE 19

(A) The set $S = \{(-2, 4), (-1, 3), (0, 1), (1, 3), (2, 4)\}$ defines a function since no two ordered pairs have the same first component and different second components. The domain and range are

$$\text{Domain} = \{-2, -1, 0, 1, 2\} \qquad \text{Set of first components}$$
$$\text{Range} = \{1, 3, 4\} \qquad \text{Set of second components}$$

(B) The set $T = \{(2, 1), (1, 3), (0, 2), (2, 3), (1, 4)\}$ does not define a function since there are ordered pairs with the same first component and different second components—for example, (1, 3) and (1, 4). ◆

PROBLEM 19

Determine whether each set defines a function. If it does, then state the domain and range.

$$S = \{(1, 2), (2, 4), (3, 6), (2, 8), (1, 10)\}$$
$$T = \{(-3, 1), (-1, 2), (1, 0), (3, 2), (5, 1)\}$$ ◆

◆ FUNCTIONS SPECIFIED BY EQUATIONS

Most of the domains and ranges included in this text will be (infinite) sets of real numbers, and the rules associating range values with domain values will be equations in two variables. Consider the equation

$$y = x^2 - x \qquad x \in R \qquad \text{Recall that } R \text{ is the set of real numbers.}$$

For each **input** x we obtain one **output** y. For example:

If $x = 3$, then $y = 3^2 - 3 = 6$.

If $x = -\frac{1}{2}$, then $y = (-\frac{1}{2})^2 - (-\frac{1}{2}) = \frac{1}{4} + \frac{1}{2} = \frac{3}{4}$.

The input values are domain values, and the output values are range values. The equation (a rule) assigns each domain value x a range value y. The variable x is called an *independent variable* (since values can be "independently" assigned to x from the domain), and y is called a *dependent variable* (since the value of y "depends" on the value assigned to x). In general, any variable used as a placeholder for domain values is called an **independent variable;** any variable that is used as a placeholder for range values is called a **dependent variable.**

When does an equation specify a function?

Equations and Functions

In an equation in two variables, if there corresponds exactly one value of the dependent variable (output) to each value of the independent variable (input), then the equation specifies a function. If there is any value of the independent variable to which there corresponds more than one value of the dependent variable, then the equation does not specify a function.

◆ EXAMPLE 20 Determine which of the following equations specify functions with independent variable x.

(A) $4y - 3x = 8, \quad x \in R$ (B) $y^2 - x^2 = 9, \quad x \in R$

Solutions (A) Solving for the dependent variable y, we have

$$\begin{aligned} 4y - 3x &= 8 \qquad (1) \\ 4y &= 8 + 3x \\ y &= 2 + \tfrac{3}{4}x \end{aligned}$$

Since each input value x corresponds to exactly one output value ($y = 2 + \frac{3}{4}x$), we see that equation (1) specifies a function.

(B) Solving for the dependent variable y, we have

$$\begin{aligned} y^2 - x^2 &= 9 \qquad (2) \\ y^2 &= 9 + x^2 \\ y &= \pm\sqrt{9 + x^2} \end{aligned}$$

Since $9 + x^2$ is always a positive real number for any real number x and since each positive real number has two square roots, to each input value x there corresponds two output values ($y = -\sqrt{9 + x^2}$ and $y = \sqrt{9 + x^2}$). For example, if $x = 4$, then equation (2) is satisfied for $y = 5$ and for $y = -5$. Thus, equation (2) does not specify a function. ◆

PROBLEM 20 Determine which of the following equations specify functions with independent variable x.

(A) $y^2 - x^4 = 9, \quad x \in R$ (B) $3y - 2x = 3, \quad x \in R$ ◆

Since the graph of an equation is the graph of all the ordered pairs that satisfy the equation, it is very easy to determine whether an equation specifies a function by examining its graph. The graphs of the two equations we considered in Example 20 are shown in Figure 6. (The graph in Figure 6B was obtained using point-by-point plotting, a technique we will discuss later in this chapter.)

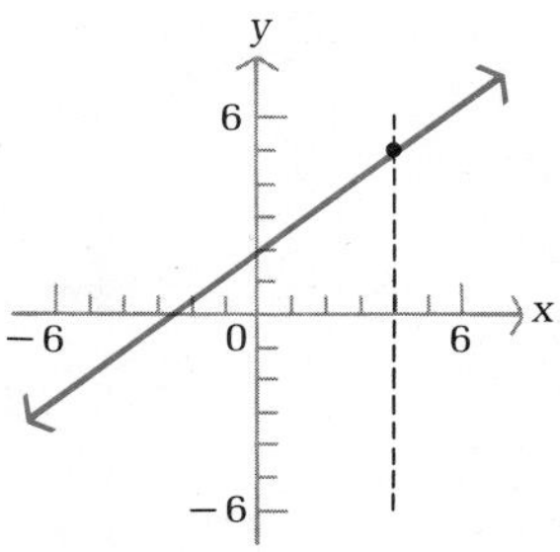

(A) $4y - 3x = 8$

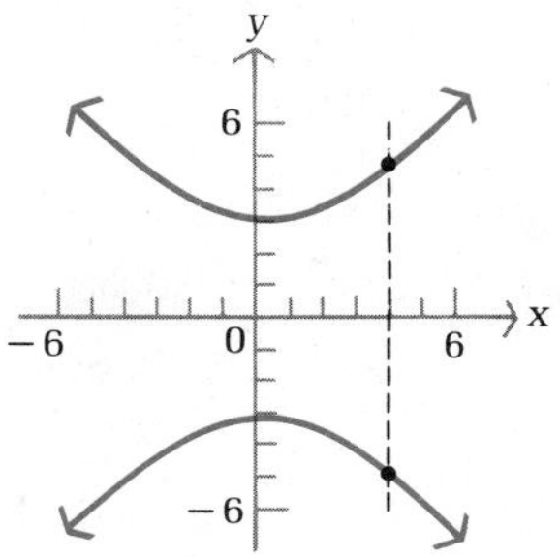

(B) $y^2 - x^2 = 9$

FIGURE 6

In Figure 6A notice that any vertical line will intersect the graph of the equation $4y - 3x = 8$ in exactly one point. This shows that to each x value there corresponds exactly one y value and confirms our conclusion that this equation specifies a function. On the other hand, Figure 6B shows that there exist vertical lines that intersect the graph of $y^2 - x^2 = 9$ in two points. This indicates that there exist x values to which there correspond two different y values and verifies our conclusion that this equation does not specify a function. These observations are generalized in Theorem 2.

THEOREM 2

Vertical Line Test for a Function

An equation specifies a function if each vertical line in the coordinate system passes through at most one point on the graph of the equation. If any vertical line passes through two or more points on the graph of an equation, then the equation does not specify a function.

In Example 20, the domains were explicitly stated along with the given equations. In many cases, this will not be done. Unless stated to the contrary, we shall adhere to the following convention regarding domains and ranges for functions specified by equations.

Agreement on Domains and Ranges

If a function is specified by an equation and the domain is not indicated, then we assume that the domain is the set of all real number replacements of the independent variable (inputs) that produce *real values* for the dependent variable (outputs). The range is the set of all outputs corresponding to input values.

◆ EXAMPLE 21 Find the domain of the function specified by the equation $y = \sqrt{x + 4}$, assuming x is the independent variable.

Solution For y to be real, $x + 4$ must be greater than or equal to 0; that is,

$$x + 4 \geqslant 0$$
$$x \geqslant -4$$

Thus,

Domain: $x \geqslant -4$ or $[-4, \infty)$ ◆

PROBLEM 21 Find the domain of the function specified by the equation $y = \sqrt{x - 2}$, assuming x is the independent variable. ◆

◆ FUNCTION NOTATION

We have just seen that a function involves two sets, a domain and a range, and a rule of correspondence that enables us to assign to each element in the domain exactly one element in the range. We use different letters to denote names for numbers; in essentially the same way, we will now use different letters to denote names for functions. For example, f and g may be used to name the functions specified by the equations $y = 2x + 1$ and $y = x^2 + 2x - 3$:

f: $y = 2x + 1$

g: $y = x^2 + 2x - 3$

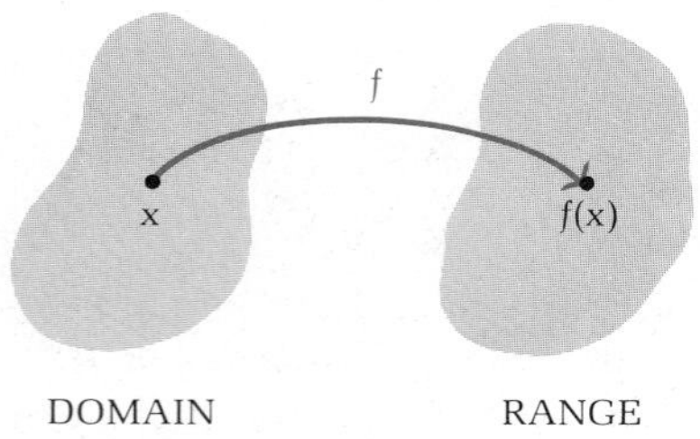

FIGURE 7

If x represents an element in the domain of a function f, then we frequently use the symbol

$$f(x)$$

in place of y to designate the number in the range of the function f to which x is paired (Fig. 7).

It is important not to think of $f(x)$ as the product of f and x. The symbol $f(x)$ is read "f of x," "f at x," or "the value of f at x." Whenever we write $y = f(x)$, we assume that the variable x is an independent variable and that both y and $f(x)$ are dependent variables. The symbol $f\colon x \rightarrow f(x)$, read "$f$ maps x into $f(x)$," is also used to indicate the relationship between the independent variable x and the dependent variable $f(x)$.

This function notation is extremely important, and its correct use should be mastered as early as possible. For example, in place of the more formal representation of the functions f and g,

f: $y = 2x + 1$ and g: $y = x^2 + 2x - 3$

we can now write

$f(x) = 2x + 1$ and $g(x) = x^2 + 2x - 3$

The function symbols $f(x)$ and $g(x)$ have certain advantages over the variable y in many situations. For example, if we write $f(3)$ and $g(-5)$, then each symbol indicates in a concise way that these are range values of particular functions associated with particular domain values. Let us find $f(3)$ and $g(-5)$.

To find $f(3)$, we replace x by 3 wherever x occurs in $f(x) = 2x + 1$ and evaluate the right side:

$$f(x) = 2x + 1$$
$$f(3) = 2 \cdot 3 + 1$$
$$= 6 + 1 = 7$$

Thus,

$f(3) = 7$ — The function f assigns the range value 7 to the domain value 3; the ordered pair (3, 7) belongs to f.

To find $g(-5)$, we replace x by -5 wherever x occurs in $g(x) = x^2 + 2x - 3$ and evaluate the right side:

$$g(x) = x^2 + 2x - 3$$
$$g(-5) = (-5)^2 + 2(-5) - 3$$
$$= 25 - 10 - 3 = 12$$

Thus,

$g(-5) = 12$ — The function g assigns the range value 12 to the domain value -5; the ordered pair $(-5, 12)$ belongs to g.

It is very important to understand and remember the definition of $f(x)$:

The Symbol $f(x)$

For any element x in the domain of the function f, the symbol $\boldsymbol{f(x)}$ represents the element in the range of f corresponding to x in the domain of f. If x is an input value, then $f(x)$ is the corresponding output value; or, symbolically, $\boldsymbol{f: x \rightarrow f(x)}$. The ordered pair $\boldsymbol{(x, f(x))}$ belongs to the function f. If x is an element that is not in the domain of f, then f is *not defined at* x and $f(x)$ *does not exist.*

Figure 8 (on the next page), which illustrates a "function machine," may give you additional insight into the nature of functions and the symbol $f(x)$. We can think of a function machine as a device that produces exactly one output (range) value for each input (domain) value on the basis of a set of instructions. (If more than one output value was produced for an input value, then the machine would not be a function machine.)

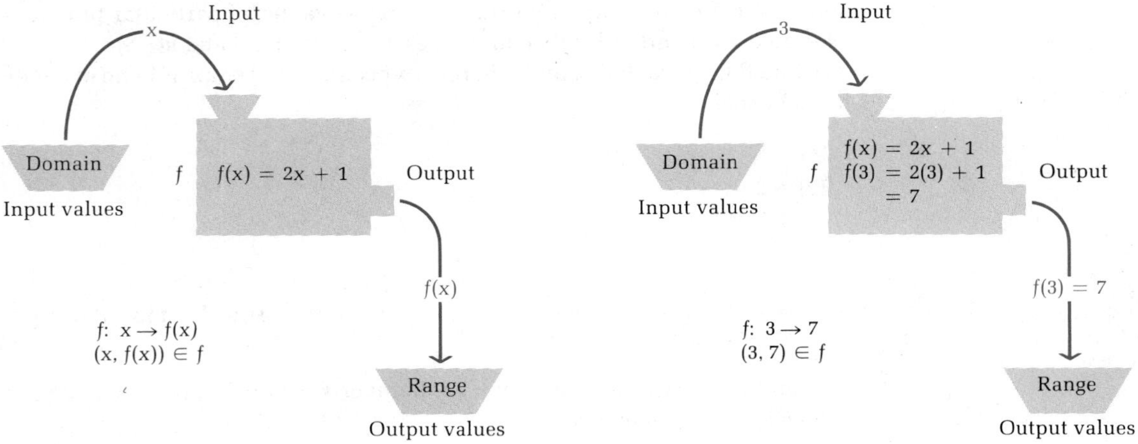

FIGURE 8
Function machine—exactly one output for each input

For the function $f(x) = 2x + 1$, the machine takes each domain value (input), multiplies it by 2, and then adds 1 to the result, to produce the range value (output). Different rules inside the function machine result in different functions.

◆ EXAMPLE 22 If

$$f(x) = \frac{12}{x - 2} \qquad g(x) = 1 - x^2 \qquad h(x) = \sqrt{x - 1}$$

then:

(A) $f(6) = \frac{12}{6 - 2} = \frac{12}{4} = 3$

(B) $g(-2) = 1 - (-2)^2 = 1 - 4 = -3$

(C) $h(-2) = \sqrt{-2 - 1} = \sqrt{-3}$

Since $\sqrt{-3}$ is not a real number, -2 is not in the domain of h and $h(-2)$ is not defined.

(D) $f(0) + g(1) - h(10) = \frac{12}{0 - 2} + (1 - 1^2) - \sqrt{10 - 1}$

$$= \frac{12}{-2} + 0 - \sqrt{9}$$

$$= -6 - 3 = -9$$ ◆

PROBLEM 22 Use the functions in Example 22 to find:

(A) $f(-2)$ (B) $g(-1)$ (C) $h(-8)$ (D) $\dfrac{f(3)}{h(5)}$ ◆

◆ **EXAMPLE 23** Find the domains of functions f, g, and h:

$$f(x) = \frac{12}{x-2} \qquad g(x) = 1 - x^2 \qquad h(x) = \sqrt{x-1}$$

Domain of f $12/(x-2)$ represents a real number for all replacements of x by real numbers except for $x = 2$ (division by 0 is not defined). Thus, $f(2)$ does not exist, and the domain of f is the set of all real numbers except 2. We often indicate this by writing

$$f(x) = \frac{12}{x-2} \qquad x \neq 2$$

Domain of g The domain is R, the set of all real numbers, since $1 - x^2$ represents a real number for all replacements of x by real numbers.

Domain of h The domain is the set of all real numbers x such that $\sqrt{x-1}$ is a real number—that is, such that

$$x - 1 \geqslant 0$$
$$x \geqslant 1 \quad \text{or} \quad [1, \infty)$$

◆

PROBLEM 23 Find the domains of functions F, G, and H:

$$F(x) = x^2 - 3x + 1 \qquad G(x) = \frac{5}{x+3} \qquad H(x) = \sqrt{2-x}$$ ◆

◆ **EXAMPLE 24** For $f(x) = x^2 - 2x + 7$, find:

(A) $f(a)$ (B) $f(a+h)$ (C) $\dfrac{f(a+h) - f(a)}{h}$

Solutions

(A) $f(a) = a^2 - 2a + 7$

(B) $f(a+h) = (a+h)^2 - 2(a+h) + 7$
$= a^2 + 2ah + h^2 - 2a - 2h + 7$

should be "plus"

(C) $$\frac{f(a+h) - f(a)}{h} = \frac{(a^2 + 2ah + h^2 - 2a - 2h + 7) - (a^2 - 2a + 7)}{h}$$
$$= \frac{2ah + h^2 - 2h}{h} = \frac{h(2a + h - 2)}{h} = 2a + h - 2$$ ◆

PROBLEM 24 Repeat Example 24 for $f(x) = x^2 - 4x + 9$. ◆

◆ APPLICATION

◆ EXAMPLE 25 — *Revenue*

The market research department of a company that manufactures memory chips for microcomputers has determined that the demand equation for 256k chips is

$$x = 10{,}000 - 50p$$

where x is the number of chips that can be sold at a price of \$p per chip.

(A) Express the revenue $R(x)$ in terms of the demand x.
(B) What is the domain of the function R?

Solutions

(A) The revenue from the sale of x units at \$p per unit is $R = xp$. To express R in terms of x, we first solve the demand equation for p in terms of x:

$$\begin{aligned} x &= 10{,}000 - 50p \\ 50p &= 10{,}000 - x \\ p &= 200 - \tfrac{1}{50}x \end{aligned}$$

Thus,

$$R(x) = xp = x(200 - \tfrac{1}{50}x) = 200x - \tfrac{1}{50}x^2$$

(B) Since neither price nor demand can be negative, x must satisfy

$$x \geq 0 \qquad \text{and} \qquad 200 - \tfrac{1}{50}x \geq 0$$

Solving the second inequality and combining it with the first, the domain of R is

$$0 \leq x \leq 10{,}000 \quad \text{or} \quad [0, 10{,}000]$$ ◆

PROBLEM 25

Repeat Example 25 for the demand equation $x = 15{,}000 - 30p$. ◆

Answers to Matched Problems

19. S does not define a function; T defines a function with domain $\{-3, -1, 1, 3, 5\}$ and range $\{0, 1, 2\}$
20. (A) Does not specify a function (B) Specifies a function
21. $x \geq 2$ (inequality notation) or $[2, \infty)$ (interval notation)
22. (A) -3 (B) 0 (C) Does not exist (D) 6
23. Domain of F: R; Domain of G: All real numbers except -3; Domain of H: $x \leq 2$ (inequality notation) or $(-\infty, 2]$ (interval notation)
24. (A) $a^2 - 4a + 9$ (B) $a^2 + 2ah + h^2 - 4a - 4h + 9$ (C) $2a + h - 4$
25. (A) $R(x) = x(500 - \frac{1}{30}x) = 500x - \frac{1}{30}x^2$
    (B) Domain: $0 \leq x \leq 15{,}000$ (inequality notation) or $[0, 15{,}000]$ (interval notation)

EXERCISE 2-4

A *Indicate whether each table specifies a function.*

1.
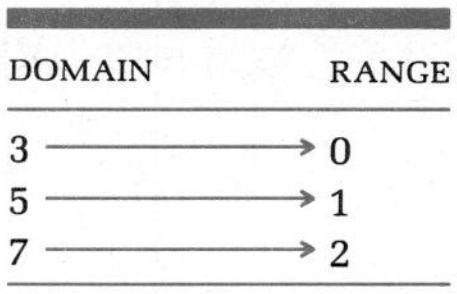

2.
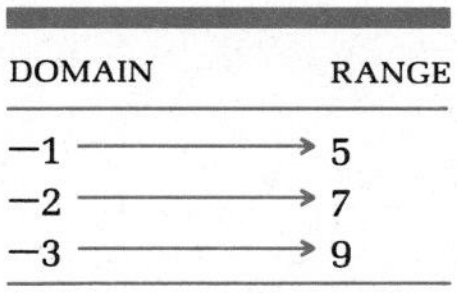

3.
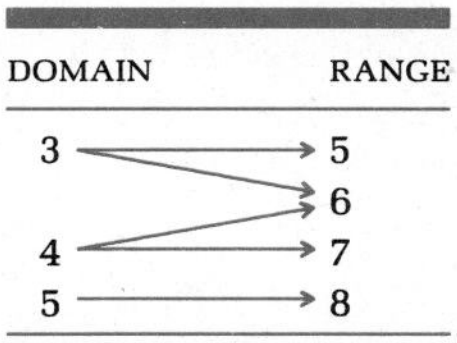

4.
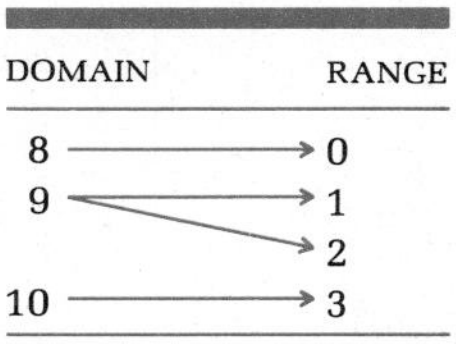

5.
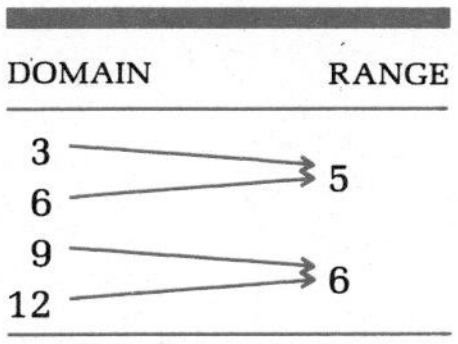

6.
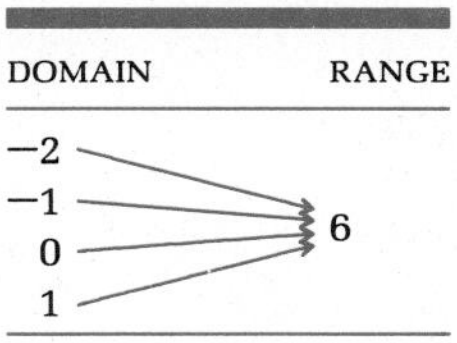

Indicate whether each set in Problems 7–12 specifies a function. Find the domain and range of each function.

7. {(1, 3), (2, 5), (4, 7), (5, 9)}
8. {(−3, 5), (−2, 4), (−1, 3), (0, 2)}
9. {(6, −6), (3, −3), (0, 0), (3, 3), (6, 6)}
10. {(−6, 6), (−3, 3), (0, 0), (3, 3), (6, 6)}
11. {(1, 3), (2, 3), (3, 3), (4, 4), (5, 4), (6, 4)}
12. {(0, 2), (1, 2), (2, 2), (0, 1), (1, 1), (2, 1)}

Indicate whether each graph specifies a function.

13.
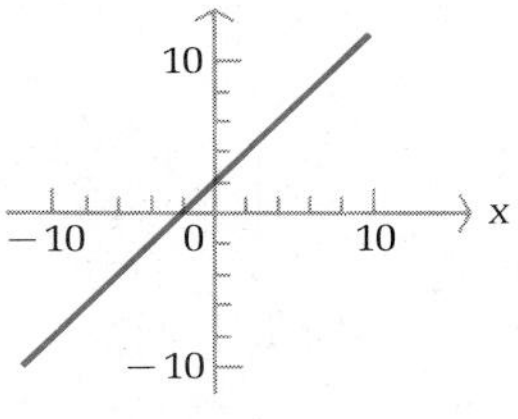

14.
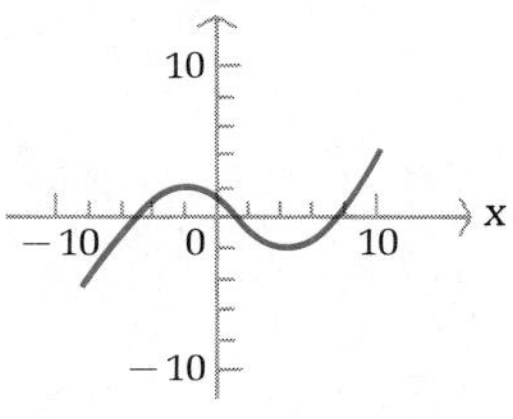

15.
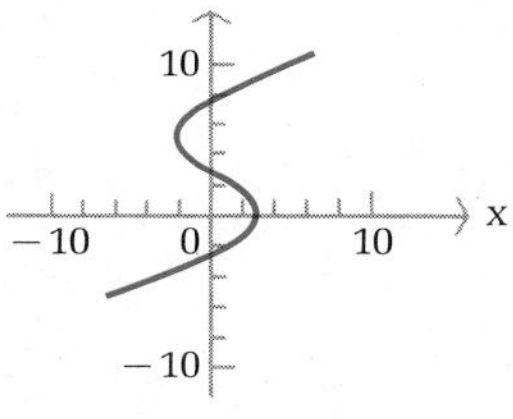

16.
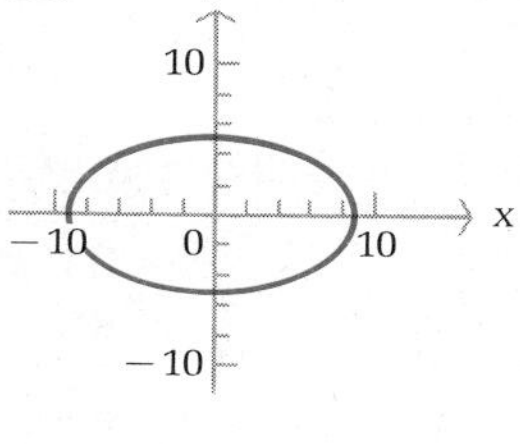

17.
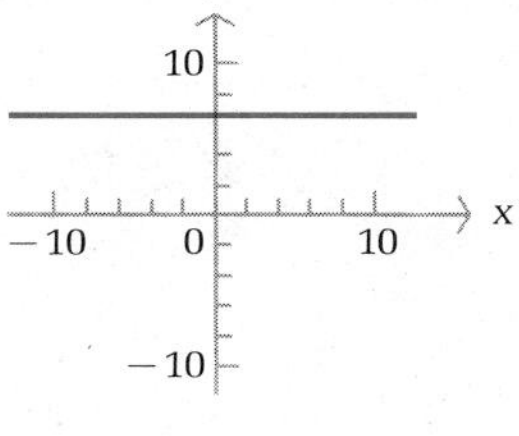

18.
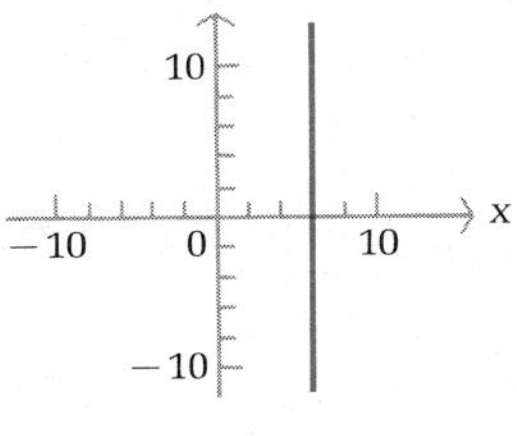

If $f(x) = 3x - 2$ and $g(x) = x - x^2$, find each of the following:

19. $f(2)$
20. $f(1)$
21. $f(-1)$
22. $f(-2)$
23. $g(3)$
24. $g(1)$
25. $f(0)$
26. $f(\frac{1}{3})$
27. $g(-3)$
28. $g(-2)$
29. $f(1) + g(2)$
30. $g(1) + f(2)$
31. $g(2) - f(2)$
32. $f(3) - g(3)$
33. $g(3) \cdot f(0)$
34. $g(0) \cdot f(-2)$
35. $\dfrac{g(-2)}{f(-2)}$
36. $\dfrac{g(-3)}{f(2)}$

B *Find the domain of each function in Problems 37–48.*

37. $f(x) = \sqrt{x}$
38. $f(x) = \dfrac{1}{\sqrt{x}}$
39. $f(x) = \dfrac{x - 3}{(x - 5)(x + 3)}$
40. $f(x) = \dfrac{x + 1}{x - 2}$
41. $f(x) = \sqrt{x + 5}$
42. $f(x) = \sqrt{7 - x}$
43. $f(x) = \dfrac{x^2 + 1}{x^2 - 1}$
44. $f(x) = \dfrac{x^2 + 5}{x^2 - 9}$
45. $f(x) = \dfrac{x}{x^2 + 3x - 4}$
46. $f(x) = \dfrac{x}{x^2 + x - 6}$
47. $f(x) = \dfrac{x + 4}{x^2 - 4x + 5}$
48. $f(x) = \dfrac{x - 7}{x^2 + 6x + 10}$

Determine which of the equations in Problems 49–58 specify functions with independent variable x. For those that do, find the domain. For those that do not, find a value of x to which there corresponds more than one value of y.

49. $4x - 5y = 20$
50. $3y - 7x = 15$
51. $x^2 - y = 1$
52. $x - y^2 = 1$
53. $x + y^2 = 10$
54. $x^2 + y = 10$
55. $xy - 4y = 1$
56. $xy + y - x = 5$
57. $x^2 + y^2 = 25$
58. $x^2 - y^2 = 16$

59. If $F(t) = 4t + 7$, find: $\dfrac{F(3 + h) - F(3)}{h}$

60. If $G(r) = 3 - 5r$, find: $\dfrac{G(2 + h) - G(2)}{h}$

61. If $g(w) = w^2 - 4$, find: $\dfrac{g(1 + h) - g(1)}{h}$

62. If $f(m) = 2m^2 + 5$, find: $\dfrac{f(4 + h) - f(4)}{h}$

63. If $Q(x) = x^2 - 5x + 1$, find: $\dfrac{Q(2 + h) - Q(2)}{h}$

64. If $P(x) = 2x^2 - 3x - 7$, find: $\dfrac{P(3 + h) - P(3)}{h}$

C *In Problems 65–72, find and simplify:* $\dfrac{f(a + h) - f(a)}{h}$

65. $f(x) = 4x - 3$
66. $f(x) = -3x + 9$
67. $f(x) = 4x^2 - 7x + 6$
68. $f(x) = 3x^2 + 5x - 8$
69. $f(x) = x^3$
70. $f(x) = x^3 - x$
71. $f(x) = \sqrt{x}$
72. $f(x) = \dfrac{1}{x}$

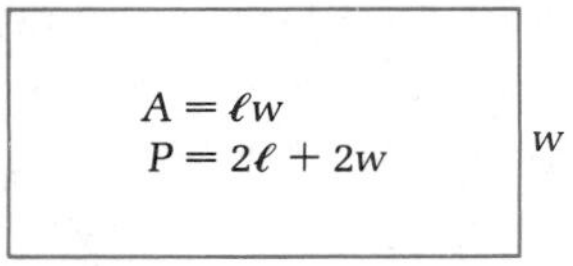

Problems 73–76 refer to the area A and perimeter P of a rectangle with length ℓ and width w (see the figure).

73. The area of a rectangle is 25 square inches. Express the perimeter $P(w)$ as a function of the width w, and state the domain of this function.

74. The area of a rectangle is 81 square inches. Express the perimeter $P(\ell)$ as a function of the length ℓ, and state the domain of this function.

75. The perimeter of a rectangle is 100 meters. Express the area $A(\ell)$ as a function of the length ℓ, and state the domain of this function.

76. The perimeter of a rectangle is 160 meters. Express the area $A(w)$ as a function of the width w, and state the domain of this function.

APPLICATIONS

Business & Economics

77. *Cost function.* The fixed costs (tooling and overhead) for manufacturing a particular stereo system are \$96,000, and the variable costs per unit (labor, material, etc.) are \$80. If x units are manufactured, express the cost $C(x)$ as a function of x. Find the cost of producing 500 stereos.

78. *Cost function.* A company that specializes in manufacturing reproductions of classic automobiles has fixed costs of \$100,000 and variable costs of \$15,000 per automobile produced. If x cars are manufactured, express the cost $C(x)$ as a function of x. Find the cost of producing 48 automobiles.

79. *Revenue function.* After extensive surveys, the research department of a stereo manufacturing company produced the demand equation

$$x = 8{,}000 - 40p$$

where x is the number of units that retailers are likely to purchase at a price of \$$p$ per unit. Express the revenue $R(x)$ in terms of the demand x. Find the domain of R.

80. *Revenue function.* Repeat Problem 79 for the demand equation

$$x = 9{,}000 - 60p$$

81. *Packaging.* A candy box is to be made out of a piece of cardboard that measures 8 by 12 inches. Equal-sized squares x inches on a side will be cut out of each corner, and then the ends and sides will be folded up to form a rectangular box.

(A) Express the volume of the box $V(x)$ in terms of x.
(B) What is the domain of the function V (determined by the physical restrictions)?
(C) Complete the table:

x	V(x)
1	
2	
3	

Notice how the volume changes with different choices of x.

82. *Packaging.* A parcel delivery service will only deliver packages with length plus girth (distance around) not exceeding 108 inches. A rectangular shipping box with square ends x inches on a side is to be used.

(A) If the full 108 inches is to be used, express the volume of the box $V(x)$ in terms of x.

(B) What is the domain of the function V (determined by the physical restrictions)?

(C) Complete the table:

x	V(x)
5	
10	
15	
20	
25	

Notice how the volume changes with different choices of x.

83. *Construction.* A veterinarian wants to construct a kennel with five individual pens, as indicated in the figure. Local ordinances require that each pen have a gate that is 3 feet wide and an area of 45 square feet. If x is the width of one pen, express the total amount of fencing $P(x)$ (excluding the gates) required for construction of the kennel as a function of x, and complete the table in the margin.

x	P(x)
3	
4	
5	
6	

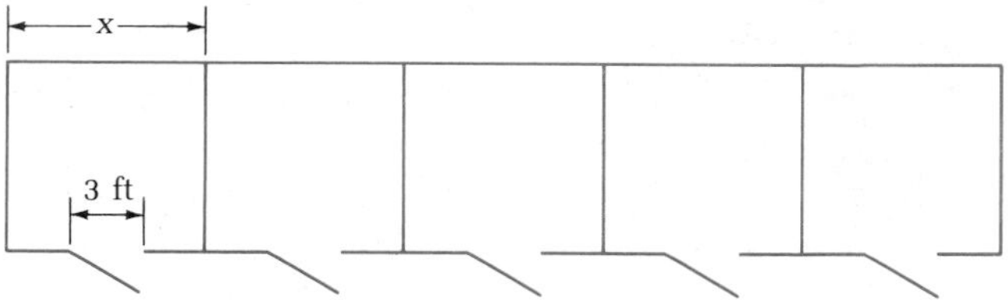

84. *Construction.* A horse breeder wants to construct three adjacent corrals, as indicated in the figure. If each corral must have an area of 600 square feet and x is the width of one corral, express the total amount of fencing $P(x)$ as a function of x, and complete the following table:

x	P(x)
10	
15	
20	
25	

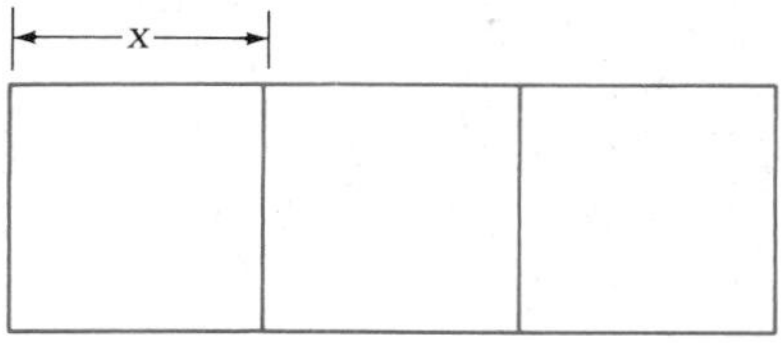

Life Sciences

85. *Muscle contraction.* In a study of the speed of muscle contraction in frogs under various loads, noted British biophysicist and Nobel prize winner A. W. Hill determined that the weight w (in grams) placed on the muscle

and the speed of contraction v (in centimeters per second) are approximately related by an equation of the form

$$(w + a)(v + b) = c$$

where a, b, and c are constants. Suppose that for a certain muscle, $a = 15$, $b = 1$, and $c = 90$. Express v as a function of w. Find the speed of contraction if a weight of 16 grams is placed on the muscle.

Social Sciences

86. *Politics.* The percentage s of seats in the House of Representatives won by Democrats and the percentage v of votes cast for Democrats (when expressed as decimal fractions) are related by the equation

$$5v - 2s = 1.4 \qquad 0 < s < 1, \quad 0.28 < v < 0.68$$

(A) Express v as a function of s, and find the percentage of votes required for the Democrats to win 51% of the seats.

(B) Express s as a function of v, and find the percentage of seats won if Democrats receive 51% of the votes.

SECTION 2-5 Linear and Quadratic Functions

- GRAPHS OF FUNCTIONS
- LINEAR FUNCTIONS AND THEIR GRAPHS
- QUADRATIC FUNCTIONS AND THEIR GRAPHS
- PIECEWISE-DEFINED FUNCTIONS
- APPLICATION: MARKET RESEARCH

GRAPHS OF FUNCTIONS

Each function that has a real number domain and range has a graph—the graph of the ordered pairs of real numbers that constitute the function. When functions are graphed, domain values are usually associated with the horizontal axis and range values with the vertical axis. Thus, the **graph of a function *f*** is the graph of the set

$$\{(x, y) \mid y = f(x), \quad x \in \text{Domain of } f\}$$

where x is the independent variable and the abscissa of a point on the graph of f, and y and $f(x)$ are dependent variables and either is the ordinate of a point on the graph of f (see Fig. 9 at the top of the next page).

The first coordinate (abscissa) of a point where the graph of a function crosses the x axis is called an ***x* intercept** of the function. The x intercepts are determined by finding the real solutions of the equation $f(x) = 0$, if any exist. The second coordinate (ordinate) of a point where the graph of a function crosses the y axis is called the ***y* intercept** of the function. The y intercept is given by $f(0)$,

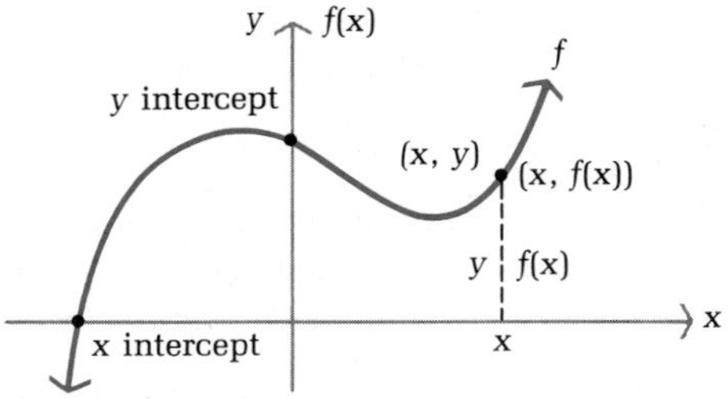

FIGURE 9

provided 0 is in the domain of f. Note that a function can have more than one x intercept, but can never have more than one y intercept (see the vertical line test in the preceding section).

◆ LINEAR FUNCTIONS AND THEIR GRAPHS

A function specified by an equation of the form

$$f(x) = mx + b$$

is called a **linear function.** In the special case $m = 0$, the function has the form

$$f(x) = b$$

and is also called a **constant function.** Graphing linear functions is equivalent to graphing the equation

$$y = mx + b \qquad \text{Slope} = m,\ y \text{ intercept} = b$$

which we discussed in detail in Section 2-3. Since the expression $mx + b$ represents a real number for all real number replacements of x, the domain of any linear function is R, the set of all real numbers. The range of a nonconstant linear function is also R, while the range of a constant function is the single real number b. See the graphs in the box.

Graph of $f(x) = mx + b$

The graph of a linear function f is a nonvertical straight line with slope m and y intercept b.

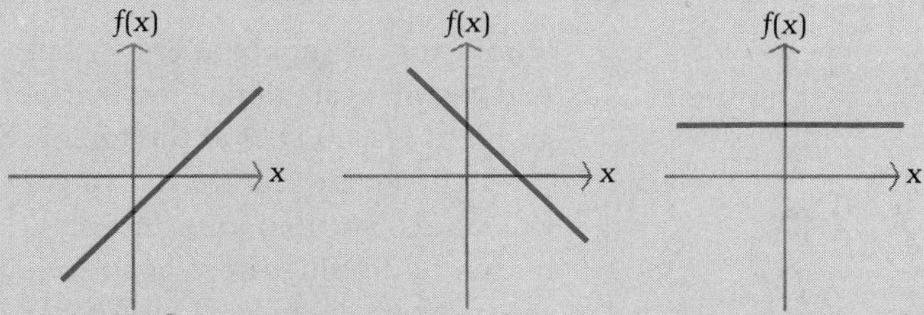

$m > 0$
Positive slope

$m < 0$
Negative slope

$m = 0$
Zero slope

◆ EXAMPLE 26 Graph the linear function defined by

$$f(x) = -\frac{x}{2} + 3$$

and indicate its slope and intercepts.

Solution

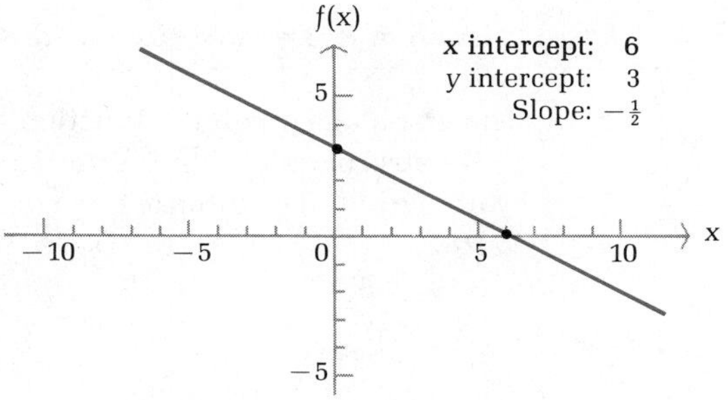

◆

PROBLEM 26 Graph the linear function defined by

$$f(x) = \frac{x}{3} + 1$$

and indicate its slope and intercepts. ◆

◆ QUADRATIC FUNCTIONS AND THEIR GRAPHS

Any function defined by an equation of the form

$$f(x) = ax^2 + bx + c \qquad a \neq 0$$

where a, b, and c are constants and x is a variable, is called a **quadratic function.**

Let us start by graphing the simple quadratic function

$$f(x) = x^2$$

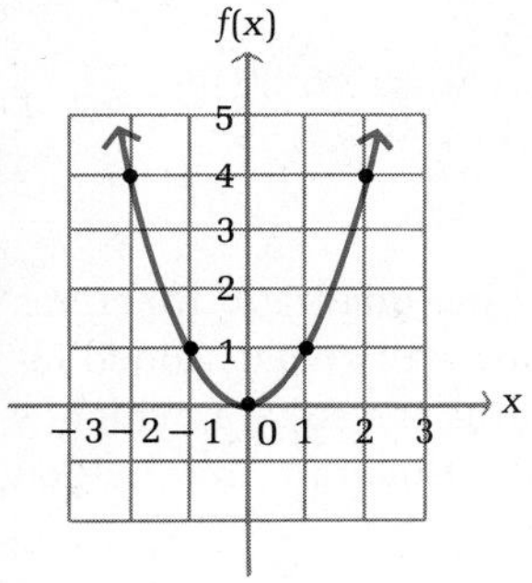

FIGURE 10

We evaluate this function for integer values from its domain, find corresponding range values, then plot the resulting ordered pairs listed in Table 8, and join these points with a smooth curve, as shown in Figure 10. The first two steps are usually done mentally or on scratch paper.

TABLE 8

Graphing $f(x) = x^2$

DOMAIN VALUES x	RANGE VALUES $y = f(x)$	ELEMENTS OF f $(x, f(x))$
-2	$y = f(-2) = (-2)^2 = 4$	$(-2, 4)$
-1	$y = f(-1) = (-1)^2 = 1$	$(-1, 1)$
0	$y = f(0) = 0^2 = 0$	$(0, 0)$
1	$y = f(1) = 1^2 = 1$	$(1, 1)$
2	$y = f(2) = 2^2 = 4$	$(2, 4)$

The curve shown in Figure 10 is called a **parabola.** It can be shown (in a course in analytic geometry) that the graph of any quadratic function is also a parabola. In general:

Graph of $f(x) = ax^2 + bx + c$, $a \neq 0$

The graph of a quadratic function f is a parabola that has its **axis** (line of symmetry) parallel to the vertical axis. It opens upward if $a > 0$ and downward if $a < 0$. The intersection point of the axis and parabola is called the **vertex.**

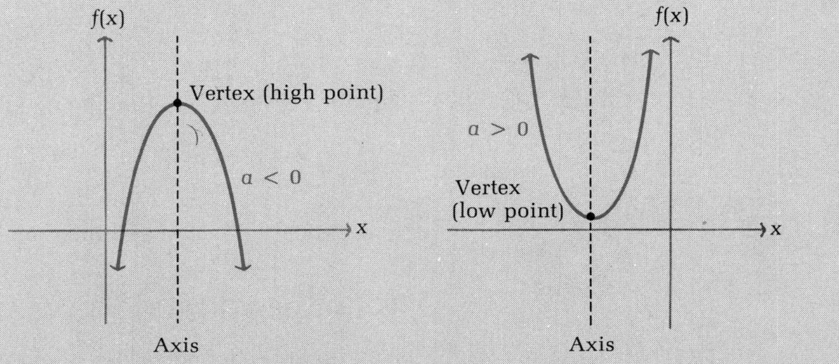

In addition to the point-by-point method of graphing quadratic functions described above, let us consider another approach that will give us added insight into these functions. (A brief review of completing the square, which is discussed in Section 2-2, may prove useful first.) We illustrate the method through an example, and then generalize the results.

Consider the quadratic function given by

$$f(x) = 2x^2 - 8x + 5$$

If we can find the vertex of the graph, then the rest of the graph can be sketched with relatively few points. In addition, we will then have found the maximum or minimum value of the function. We start by transforming the equation into the form

$$f(x) = a(x - h)^2 + k \qquad a, h, k \text{ constants}$$

by completing the square:

$$f(x) = 2x^2 - 8x + 5$$
$$= 2(x^2 - 4x) + 5$$

Factor the coefficient of x^2 out of the first two terms.

$f(x) = 2(x^2 - 4x + ?) + 5$	Complete the square within parentheses.
$= 2(x^2 - 4x + 4) + 5 - 8$	We added 4 to complete the square inside the parentheses; but because of the 2 on the outside, we have actually added 8, so we must subtract 8.
$= 2(x - 2)^2 - 3$	The transformation is complete.

Thus,

$$f(x) = \underbrace{2(x-2)^2}_{\substack{\text{Never negative} \\ \text{(Why?)}}} - 3$$

When $x = 2$, the first term on the right vanishes, and we add 0 to -3. For *any* other value of x we will add a positive number to -3, thus making $f(x)$ larger. Therefore, $f(2) = -3$ is the minimum value of $f(x)$ for all x. A very important result!

The point $(2, -3)$ is the lowest point on the parabola and is also the vertex. The vertical line $x = 2$ is the axis of the parabola. We plot the vertex and the axis and a couple of points on either side of the axis to complete the graph (Fig. 11).

x	$f(x)$
2	-3
1	-1
3	-1
0	5
4	5

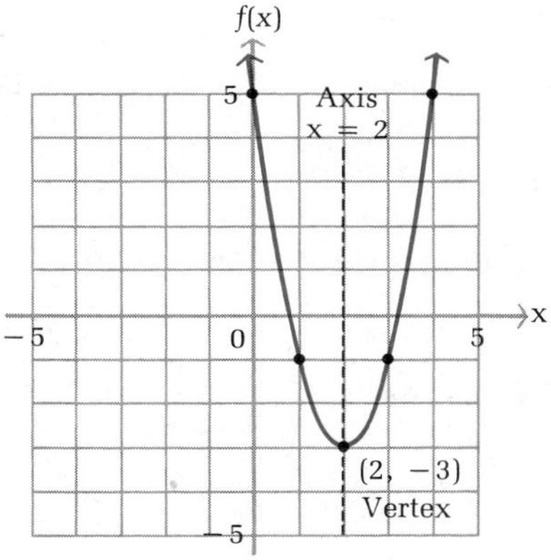

FIGURE 11

Examining the graph, we see that $f(x)$ can assume any value greater than or equal to -3, but no values less than -3. Thus,

$$\text{Range of } f = \{y \mid y \geq -3\} = [-3, \infty)$$

We also note that f has two x intercepts, which are solutions of the equation $f(x) = 0$. Since we have already completed the square, we use this method to find the x intercepts:

$$f(x) = 0$$
$$2(x - 2)^2 - 3 = 0$$

$$2(x-2)^2 = 3$$
$$(x-2)^2 = \tfrac{3}{2}$$
$$x - 2 = \pm\sqrt{\tfrac{3}{2}}$$
$$= \pm\tfrac{1}{2}\sqrt{6}$$
$$x = 2 \pm \tfrac{1}{2}\sqrt{6}$$

Thus, the x intercepts are

$$x = 2 + \tfrac{1}{2}\sqrt{6} \approx 3.22 \qquad \text{and} \qquad x = 2 - \tfrac{1}{2}\sqrt{6} \approx 0.78$$

For completeness, we observe that the y intercept is $f(0) = 5$, obtained as part of the table of values of $f(x)$.

Note all the important results we have with this approach:

1. Axis of the parabola
2. Vertex of the parabola
3. Minimum value of $f(x)$
4. Graph of $y = f(x)$
5. Range of f
6. x intercepts

If we start with the general quadratic function defined by

$$f(x) = ax^2 + bx + c \qquad a \neq 0$$

and complete the square (see Section 2-2), we obtain

$$f(x) = a\left(x + \frac{b}{2a}\right)^2 + \frac{4ac - b^2}{4a}$$

Using the same reasoning as above, we obtain the following general results:

Properties of $f(x) = ax^2 + bx + c$, $a \neq 0$

1. Axis of symmetry: $x = -\dfrac{b}{2a}$
2. Vertex: $\left(-\dfrac{b}{2a}, f\left(-\dfrac{b}{2a}\right)\right)$
3. Maximum or minimum value of $f(x)$:
$$f\left(-\frac{b}{2a}\right) = \begin{cases} \text{Minimum} & \text{if } a > 0 \\ \text{Maximum} & \text{if } a < 0 \end{cases}$$
4. Domain: All real numbers
   Range: Determine from graph
5. y intercept: $f(0) = c$
   x intercepts: Solutions of $f(x) = 0$, if any exist

To graph a quadratic function using this method, we can actually complete the square as in the preceding example or use the properties in the box. Some of you can probably more readily remember a formula, others a process. We use the properties in the box in the next example.

◆ EXAMPLE 27 Graph, finding the axis, vertex, maximum or minimum of $f(x)$, range, and intercepts:

$$f(x) = -x^2 - 4x - 5$$

Solution Note that $a = -1$, $b = -4$, and $c = -5$.

Axis of symmetry: $x = -\dfrac{b}{2a} = -\dfrac{-4}{2(-1)} = -2$

Vertex: $\left(-\dfrac{b}{2a}, f\left(-\dfrac{b}{2a}\right)\right) = (-2, f(-2)) = (-2, -1)$

Maximum value of $f(x)$ (since $a = -1 < 0$): Max $f(x) = f(-2) = -1$

To graph f, locate the axis and vertex; then plot several points on either side of the axis:

x	$f(x)$
-4	-5
-3	-2
-2	-1
-1	-2
0	-5

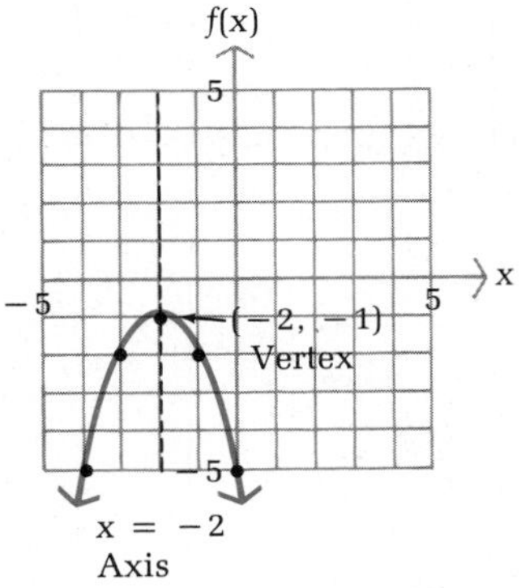

From the graph, we see that the range of f is

$$(-\infty, f(-2)] = (-\infty, -1]$$

The y intercept is $f(0) = -5$. Since the graph of f does not cross the x axis, there are no x intercepts. The equation $f(x) = 0$ has no real solutions (verify this). ◆

PROBLEM 27 Graph, finding the axis, vertex, maximum or minimum of $f(x)$, range, and intercepts:

$$f(x) = x^2 - 4x + 4$$ ◆

◆ PIECEWISE-DEFINED FUNCTIONS

The **absolute value** of a real number a, denoted $|a|$, is the (positive) distance on a number line from a to the origin. Thus, $|4| = 4$ and $|-5| = 5$. More formally, the **absolute value function** is defined by

$$f(x) = |x| = \begin{cases} -x & \text{if } x < 0 \\ x & \text{if } x \geq 0 \end{cases}$$

Notice that this function is defined by different formulas for different parts of its domain. A function whose definition involves more than one formula is called a **piecewise-defined function.** As the next example illustrates, piecewise-defined functions occur naturally in many applications.

◆ EXAMPLE 28

Service Charges

The time charges for a service call by a telephone company are \$16.20 for the first 6 minute period and \$5.40 for each additional 6 minute period (or fraction thereof). Let $C(x)$ be the total time charges for a service call that lasts x minutes. Graph C for $0 < x \leq 30$.

Solution

The total time charges are given by the following piecewise definition:

$$C(x) = \begin{cases} \$16.20 & \text{if } 0 < x \leq 6 \\ \$21.60 & \text{if } 6 < x \leq 12 \\ \$27.00 & \text{if } 12 < x \leq 18 \\ \text{And so on} \end{cases}$$

To graph C, we graph each rule in this definition for the indicated values of x:

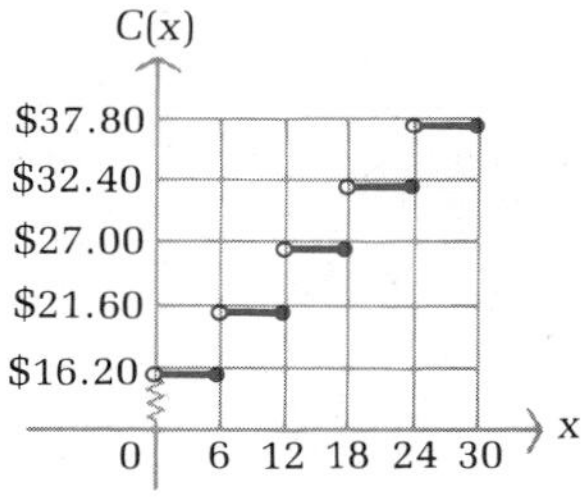

Note: A solid dot on the graph indicates that the point is part of the graph; an open dot indicates that the point is not part of the graph. ◆

PROBLEM 28

The time charges for a company that services home appliances are \$24 for the first 15 minute period and \$12 for each additional 15 minute period (or fraction thereof). Let $C(x)$ be the total time charges for a service call that lasts x minutes. Graph C for $0 < x \leq 60$. ◆

◆ APPLICATION: MARKET RESEARCH

The market research department of a company recommended to management that the company manufacture and market a promising new product. After extensive surveys, the research department backed up the recommendation with the **demand equation**

$$x = f(p) = 6{,}000 - 30p \tag{1}$$

where x is the number of units that retailers are likely to buy per month at $\$p$ per unit. Notice that as the price goes up, the number of units goes down. From the financial department, the following **cost equation** was obtained:

$$C = g(x) = 72{,}000 + 60x \tag{2}$$

where \$72,000 is the fixed cost (tooling and overhead) and \$60 is the variable cost per unit (materials, labor, marketing, transportation, storage, etc.). The **revenue equation** (the amount of money, R, received by the company for selling x units at $\$p$ per unit) is

$$R = xp \tag{3}$$

And, finally, the **profit equation** is

$$P = R - C \tag{4}$$

where P is profit, R is revenue, and C is cost.

Notice that the cost equation (2) expresses C as a function of x and the demand equation (1) expresses x as a function of p. Substituting (1) into (2), we obtain cost C as a linear function of price p:

$$\begin{aligned} C &= 72{,}000 + 60(6{,}000 - 3p) \\ &= 432{,}000 - 1{,}800p \end{aligned} \qquad \text{Linear function} \tag{5}$$

Similarly, substituting (1) into (3), we obtain revenue R as a quadratic function of price p:

$$\begin{aligned} R &= (6{,}000 - 30p)p \\ &= 6{,}000p - 30p^2 \end{aligned} \qquad \text{Quadratic function} \tag{6}$$

Now let us graph equations (5) and (6) in the same coordinate system. We obtain Figure 12 (page 106). Notice how much information is contained in this graph.

Let us compute the **break-even points;** that is, the prices at which cost equals revenue (the points of intersection of the two graphs in Figure 12). Find p so that

$$\begin{aligned} C &= R \\ 432{,}000 - 1{,}800p &= 6{,}000p - 30p^2 \\ 30p^2 - 7{,}800p + 432{,}000 &= 0 \\ p^2 - 260p + 14{,}400 &= 0 \\ p &= \frac{260 \pm \sqrt{260^2 - 4(14{,}400)}}{2} \\ &= \frac{260 \pm 100}{2} \\ &= \$80, \quad \$180 \end{aligned}$$

Solve using the quadratic formula (Section 2-2).

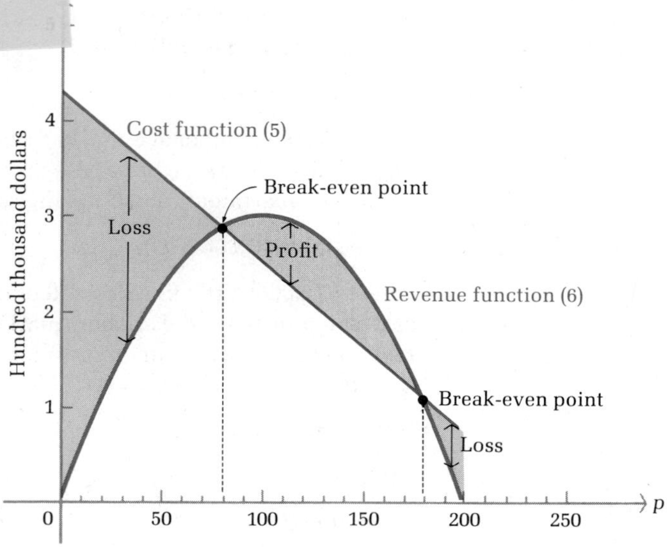

FIGURE 12

Thus, at a price of \$80 or \$180 per unit the company will break even. Between these two prices it is predicted that the company will make a profit.

At what price will a maximum profit occur? To find out, we write

$$\begin{aligned} P &= R - C \\ &= (6{,}000p - 30p^2) - (432{,}000 - 1{,}800p) \\ &= -30p^2 + 7{,}800p - 432{,}000 \end{aligned}$$

Since this is a quadratic function, the maximum profit occurs at

$$p = -\frac{b}{2a} = -\frac{7{,}800}{2(-30)} = \$130$$

Note that this is not the price at which the maximum revenue occurs. The latter occurs at $p = \$100$, as shown in Figure 12.

Answers to Matched Problems

26. Slope: $\frac{1}{3}$
y intercept: 1
x intercept: -3

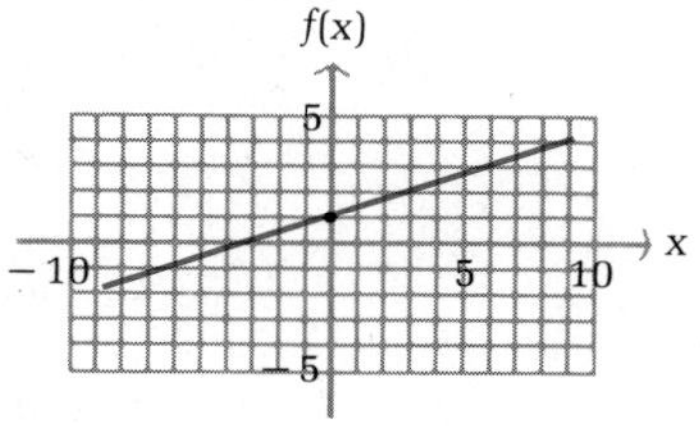

27. Minimum: $f(2) = 0$
Range: $[0, \infty)$
y intercept: 4
x intercept: 2

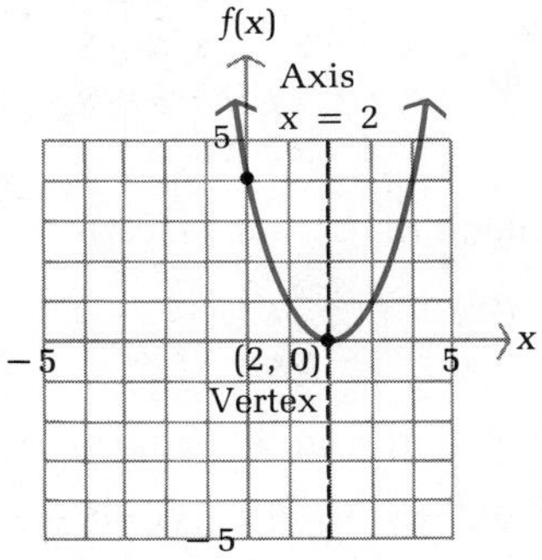

28.

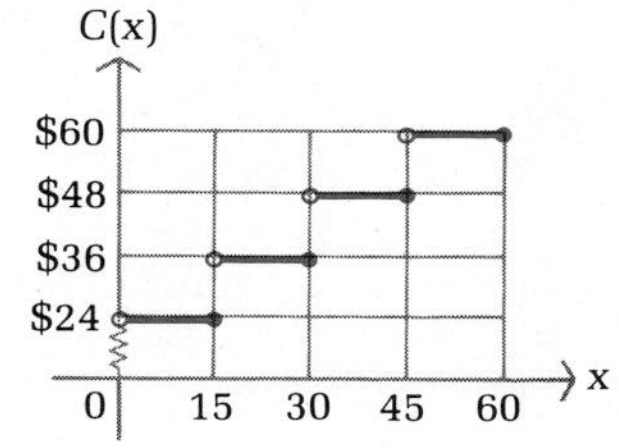

EXERCISE 2-5

A *Graph each linear function, and indicate its slope and intercepts.*

1. $f(x) = 2x - 4$
2. $g(x) = \dfrac{x}{2}$
3. $h(x) = 4 - 2x$
4. $f(x) = -\dfrac{x}{2} + 3$
5. $g(x) = -\frac{2}{3}x + 4$
6. $f(x) = 3$

Find a linear function with the indicated slope and y intercept.

7. Slope -2, y intercept 6
8. Slope 3, y intercept -5

Find a linear function whose graph passes through the indicated points.

9. $(-1, 5)$, $(5, 2)$
10. $(1, -2)$, $(7, 6)$

B *Graph, finding the axis, vertex, maximum or minimum, intercepts, and range.*

11. $f(x) = (x - 3)^2 - 1$
12. $g(x) = -(x + 2)^2 + 4$
13. $h(x) = -(x + 1)^2 + 9$
14. $k(x) = (x - 2)^2 - 16$
15. $f(x) = x^2 + 8x + 16$
16. $h(x) = x^2 - 2x - 3$
17. $f(u) = u^2 - 2u + 4$
18. $f(x) = x^2 - 10x + 25$
19. $h(x) = 2 + 4x - x^2$
20. $g(x) = -x^2 - 6x - 4$
21. $f(x) = 6x - x^2$
22. $G(x) = 16x - 2x^2$
23. $F(s) = s^2 - 4$
24. $g(t) = t^2 + 4$
25. $F(x) = 4 - x^2$
26. $G(x) = 9 - x^2$

Graph each function, and state its domain and range.

27. $f(x) = \begin{cases} 1 & \text{if } 0 \le x \le 2 \\ 3 & \text{if } 2 < x \le 3 \\ 5 & \text{if } 3 < x \le 5 \end{cases}$

28. $g(x) = \begin{cases} -2 & \text{if } -4 \le x \le -2 \\ 1 & \text{if } -2 < x < 2 \\ 2 & \text{if } 2 \le x \le 4 \end{cases}$

29. $f(x) = \begin{cases} x & \text{if } -2 \le x < 1 \\ -x + 2 & \text{if } 1 \le x \le 2 \end{cases}$

30. $f(x) = \begin{cases} x + 1 & \text{if } -1 \le x < 0 \\ -x + 1 & \text{if } 0 \le x \le 1 \end{cases}$

31. $h(x) = \begin{cases} -x^2 - 2 & \text{if } x < 0 \\ x^2 + 2 & \text{if } x > 0 \end{cases}$

32. $g(x) = \begin{cases} x^2 + 1 & \text{if } x < 0 \\ -x^2 - 1 & \text{if } x > 0 \end{cases}$

33. $G(x) = \begin{cases} -3 & \text{if } x < -3 \\ x & \text{if } -3 \le x \le 3 \\ 3 & \text{if } x > 3 \end{cases}$

34. $F(x) = \begin{cases} 1 & \text{if } x < -2 \\ 5 - x^2 & \text{if } -2 \le x \le 2 \\ 1 & \text{if } x > 2 \end{cases}$

C *Graph, finding the axis, vertex, maximum or minimum, intercepts, and range.*

35. $f(x) = x^2 - 7x + 10$
36. $g(t) = t^2 - 5t + 2$
37. $h(x) = 2 - 5x - x^2$
38. $g(t) = 4 + 3t - t^2$

Graph f and g on the same set of axes and find any points of intersection.

39. $f(x) = 2 + \frac{1}{2}x$; $g(x) = 8 - x$
40. $f(x) = -2x - 1$; $g(x) = x - 4$
41. $f(x) = x^2 - 6x + 8$; $g(x) = x - 2$
42. $f(x) = 8 + 2x - x^2$; $g(x) = x + 6$
43. $f(x) = 11 + 2x - x^2$; $g(x) = x^2 - 1$
44. $f(x) = x^2 - 4x - 10$; $g(x) = 14 - 2x - x^2$

APPLICATIONS

Business & Economics

45. *Service charges.* On weekends and holidays, an emergency plumbing repair service charges \$60 for the first 30 minute period (or fraction thereof) of a service call and \$20 for each additional 15 minute period (or fraction thereof). Let $C(x)$ be the cost of a service call that lasts x minutes. Graph $C(x)$ for $0 < x \le 90$.

46. *Delivery charges.* A nationwide package delivery service charges \$20.00 for overnight delivery of packages weighing 1 pound or less. Each additional pound (or fraction thereof) costs an additional \$2.50. Let $D(x)$ be the charge for overnight delivery of a package weighing x pounds. Graph $D(x)$ for $0 < x \le 10$.

47. *Market research.* Suppose that in the market research example in this section the demand equation (1) is changed to $x = 9{,}000 - 30p$ and the cost equation (2) is changed to $C = 90{,}000 + 30x$.

(A) Express cost C as a linear function of price p.
(B) Express revenue R as a quadratic function of price p.
(C) Graph the cost and revenue functions found in parts A and B in the same coordinate system, and identify the regions of profit and loss on your graph.
(D) Find the break-even points; that is, find the prices to the nearest dollar at which $R = C$. (A calculator might prove useful here.)
(E) Find the price that produces the maximum revenue.

48. *Market research.* Repeat Problem 47 if the demand equation (1) is changed to $x = 5{,}000 - 50p$ and the cost equation (2) is changed to $C = 40{,}000 + 12x$.

49. *Break-even analysis.* A publisher is planning to produce a new textbook. The fixed costs (reviewing, editing, typesetting, etc.) are $240,000, and the variable costs (printing, sales commissions, etc.) are $20 per book. The wholesale price (the amount received by the publisher) will be $35 per book. Let x be the number of books.

(A) Express the cost C as a linear function of x.
(B) Express the revenue R as a linear function of x.
(C) Graph the cost and revenue functions found in parts A and B on the same set of axes.
(D) Find the number of books the publisher has to sell in order to break even.

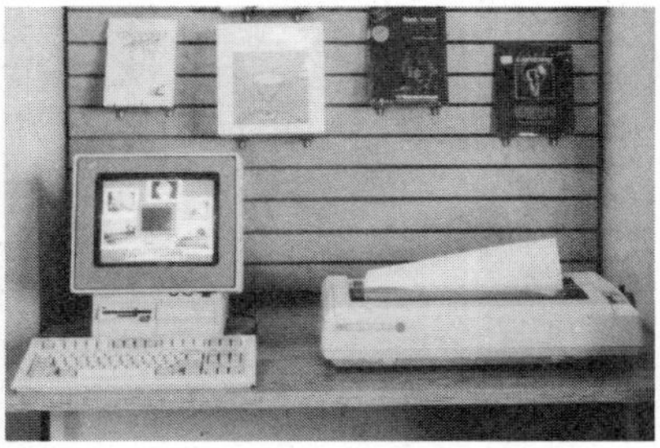

50. *Break-even analysis.* A computer software company is planning to market a new word processor for a microcomputer. The fixed costs (programming, debugging, etc.) are $300,000, and the variable costs (disk duplication, manual production, etc.) are $25 per unit. The wholesale price of the product will be $100 per unit. Let x be the number of units.

(A) Express the cost C as a linear function of x.
(B) Express the revenue R as a linear function of x.
(C) Graph the cost and revenue functions found in parts A and B on the same set of axes.
(D) Find the number of units the company has to sell in order to break even.

Life Sciences

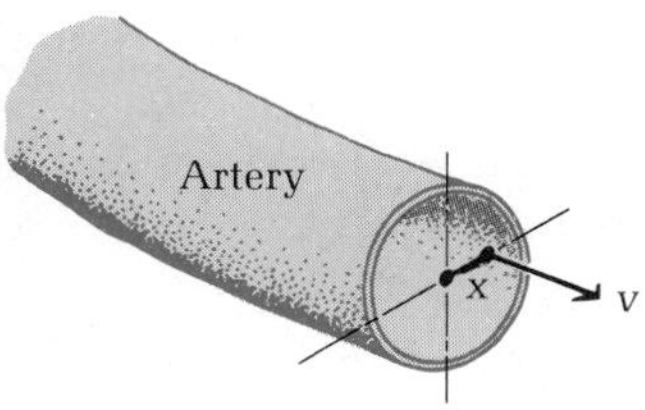

51. *Medicine.* The French physician Poiseuille was the first to discover that blood flows faster near the center of an artery than near the edge. Experimental evidence has shown that the rate of flow v (in centimeters per second) at a point x centimeters from the center of an artery (see the figure) is given by

$$v = f(x) = 1{,}000(0.04 - x^2) \qquad 0 \le x \le 0.2$$

Graph this quadratic function for the indicated values of x.

52. *Air pollution.* On an average summer day in a large city, the pollution index at 8:00 AM is 20 parts per million, and it increases linearly by 15 parts per million each hour until 3:00 PM. Let $P(x)$ be the amount of pollutants in the air x hours after 8:00 AM.

(A) Express $P(x)$ as a linear function of x.
(B) What is the air pollution index at 1:00 PM?
(C) Graph the function P for $0 \le x \le 7$.
(D) What is the slope of the graph? (The slope is the amount of increase in pollution for each additional hour of time.)

Social Sciences

53. *Psychology—sensory perception.* One of the oldest studies in psychology concerns the following question: Given a certain level of stimulation (light, sound, weight lifting, electric shock, and so on), how much should the stimulation be increased for a person to notice the difference? In the middle of the nineteenth century, E. H. Weber (a German physiologist) formulated a law that still carries his name: If Δs is the change in stimulus that will just be noticeable at a stimulus level s, then the ratio of Δs to s is a constant:

$$\frac{\Delta s}{s} = k$$

Hence, the amount of change that will be noticed is a linear function of the stimulus level, and we note that the greater the stimulus, the more it takes to notice a difference. In an experiment on weight lifting, the constant k for a given individual was found to be $\frac{1}{30}$.

(A) Find Δs (the difference that is just noticeable) at the 30 pound level; at the 90 pound level.
(B) Graph $\Delta s = s/30$ for $0 \leq s \leq 120$.
(C) What is the slope of the graph?

SECTION 2-6 Chapter Review

Important Terms and Symbols

2-1 *Linear Equations and Inequalities in One Variable.* Linear equation; linear inequality; solution; solution set; solution of an equation; equivalent equations; equality properties; inequality properties; interval notation; fixed cost; variable cost; break-even analysis

$a < b$; $a > b$; $a \leq x \leq b$; $[a, b]$; $[a, b)$; $(a, b]$; (a, b); $(-\infty; a]$; $(-\infty, a)$; $[b, \infty)$; (b, ∞)

2-2 *Quadratic Equations.* Quadratic equation; standard form; completing the square; quadratic formula; discriminant; equilibrium point

$ax^2 + bx + c = 0, a \neq 0$; $x = \dfrac{-b \pm \sqrt{b^2 - 4ac}}{2a}$; $b^2 - 4ac$

2-3 *Cartesian Coordinate System and Straight Lines.* Cartesian coordinate system; rectangular coordinate system; coordinate axes; ordered pair; coordinates; abscissa; ordinate; quadrants; solution of an equation in two variables; solution set; graph of an equation; standard form; x intercept; y intercept; slope; slope–intercept form; point–slope form; horizontal line; vertical line; parallel lines; perpendicular lines

$Ax + By = C$; $m = \dfrac{y_2 - y_1}{x_2 - x_1}$; $y = mx + b$;
$y - y_1 = m(x - x_1)$; $y = c$; $x = c$

2-4 *Functions.* Function; domain; range; input; output; independent variable; dependent variable; vertical line test; function notation; revenue function

$f(x)$; $f: x \to f(x)$; $(x, f(x))$

2-5 *Linear and Quadratic Functions.* Graph of a function; x intercept; y intercept; linear function; constant function; quadratic function; parabola; axis of a parabola; vertex of a parabola; maximum; minimum; absolute value; absolute value function; piecewise-defined function; demand equation; cost equation; revenue equation; profit equation; break-even point

$f(x) = ax + b$; $f(x) = ax^2 + bx + c, a \neq 0$; $|x|$

EXERCISE 2-6

Chapter Review

Work through all the problems in this chapter review and check your answers in the back of the book. (Answers to all review problems are there.) Where weaknesses show up, review appropriate sections in the text.

A

1. Solve: $\frac{u}{5} = \frac{u}{6} + \frac{6}{5}$
2. Solve and graph on a real number line: $2(x + 4) > 5x - 4$
3. Solve: $x^2 = 5x$
4. Graph the equation below in a rectangular coordinate system. Indicate the slope and the y intercept.

   $$y = \frac{x}{2} - 2$$

5. Write the equation of a line that passes through (4, 3) with slope $\frac{1}{2}$. Write the final answer in the form $y = mx + b$.
6. Graph $x - y = 2$ in a rectangular coordinate system. Indicate the slope.
7. For $f(x) = 2x - 1$ and $g(x) = x^2 - 2x$, find $f(-2) + g(-1)$.
8. Graph the linear function f given by the equation

   $$f(x) = \tfrac{2}{3}x - 1$$

   Indicate the slope and intercepts.
9. Find the maximum or minimum value of $f(x) = x^2 - 8x + 7$ without graphing. What are the coordinates of the vertex of the parabola?

B

10. Solve: $\frac{x}{12} - \frac{x - 3}{3} = \frac{1}{2}$

Solve and graph on a real number line.

11. $1 - \frac{x - 3}{3} \leq \frac{1}{2}$
12. $-2 \leq \frac{x}{2} - 3 < 3$

Solve for y in terms of x.

13. $2x - 3y = 6$ **14.** $xy - y = 3$

Solve Problems 15–17.

15. $3x^2 - 21 = 0$ **16.** $x^2 - x - 20 = 0$ **17.** $2x^2 = 3x + 1$

18. Graph $3x + 6y = 18$ in a rectangular coordinate system. Indicate the slope and intercepts.

19. Find an equation of the line that passes through $(-2, 3)$ and $(6, -1)$. Write the answer in the form $Ax + By = C$, $A > 0$. What is the slope of the line?

20. Write the equations of the vertical line and the horizontal line that pass through $(-5, 2)$. Graph both equations on the same coordinate system.

21. Find an equation of the line that passes through $(-2, 5)$ and $(2, -1)$. Write the answer in the form $y = mx + b$.

22. For $f(x) = 10x - 7$, $g(t) = 6 - 2t$, $F(u) = 3u^2$, and $G(v) = v - v^2$, find:

(A) $2g(-1) - 3G(-1)$ (B) $4G(-2) - g(-3)$

(C) $\dfrac{f(2) \cdot g(-4)}{G(-1)}$ (D) $\dfrac{F(-1) \cdot G(2)}{g(-1)}$

23. Find the domains of the functions f and g if

$$f(x) = 2x - x^2 \qquad g(x) = \frac{1}{x - 2}$$

24. For $f(x) = 2x - 1$, find: $\dfrac{f(3 + h) - f(3)}{h}$

25. Determine which of the following equations specify functions with independent variable x. For those that do, find the domain. For those that do not, find a value of x to which there corresponds more than one value of y.

(A) $4x - 3y = 11$ (B) $y^2 - 4x = 1$ (C) $xy + 3y + 5x = 4$

26. Graph $g(x) = 8x - 2x^2$, finding the axis, vertex, maximum or minimum, range, and intercepts.

27. Sketch the graph and find the domain and range of

$$f(x) = \begin{cases} 1 - x^2 & \text{if } -1 \leq x < 0 \\ 1 + x^2 & \text{if } 0 \leq x \leq 1 \end{cases}$$

C

28. Solve $x^2 + jx + k = 0$ for x in terms of j and k.

29. Write an equation of the line that passes through the points $(4, -3)$ and $(4, 5)$.

30. Write an equation of the line that passes through $(2, -3)$ and is:

(A) Parallel to $2x - 4y = 5$ (B) Perpendicular to $2x - 4y = 5$

Write the final answers in the form $Ax + By = C$, $A > 0$.

31. Find the domain of the function f specified by each equation.

(A) $f(x) = \frac{5}{x-3}$ (B) $f(x) = \sqrt{x-1}$

32. For $f(x) = x^2 + 7x - 9$, find and simplify: $\frac{f(a+h) - f(a)}{h}$

33. Graph $f(x) = 2x - 7$ and $g(x) = x^2 - 6x + 5$ on the same set of axes, and find any points of intersection.

APPLICATIONS

Business & Economics

34. *Investment.* An investor has \$60,000 to invest. If part is invested at 8% and the rest at 14%, how much should be invested at each rate to yield 12% on the total amount?

35. *Inflation.* If the CPI was 89 in 1960 and 247 in 1980, how much would a net salary of \$800 in 1960 have to be in 1980 in order to keep up with inflation? Set up an equation and solve.

36. *Finance.* If P dollars is invested at $100r\%$ compounded annually, at the end of 2 years it will grow to $A = P(1 + r)^2$. At what interest rate will \$1,000 grow to \$1,210 in 2 years?

37. *Linear depreciation.* A word-processing system was purchased by a company for \$12,000 and is assumed to have a salvage value of \$2,000 after 8 years (for tax purposes). If its value is depreciated linearly from \$12,000 to \$2,000:

(A) Find the linear equation that relates value V in dollars to time t in years.
(B) What would be the value of the system after 5 years?

38. *Pricing.* A sporting goods store sells a tennis racket that cost \$30 for \$48 and a pair of jogging shoes that cost \$20 for \$32.

(A) If the markup policy of the store for items that cost over \$10 is assumed to be linear and is reflected in the pricing of these two items, write an equation that relates retail price R to cost C.
(B) What should be the retail price of a pair of skis that cost \$105?

39. *Market equilibrium.* After extensive research, a retail firm has determined that the supply and demand equations for a certain product are

$$s = 10{,}000p - 25{,}000 \qquad \text{and} \qquad d = \frac{90{,}000}{p}$$

where p is the price in dollars. Find the equilibrium point (the price where supply equals demand).

40. *Break-even analysis.* A video production company is planning to produce an instructional videotape. The producer estimates that it will cost \$84,000

to shoot the video and \$15 per unit to copy and distribute the tape. The wholesale price of the tape is \$50 per unit. How many tapes must be sold for the company to break even?

41. *Market research.* The market research department of an electronics company has determined that the demand and cost equations for the production of an AM/FM clock radio are, respectively,

$$x = 500 - 10p \qquad \text{and} \quad C = 3{,}000 + 10x$$

(A) Express the revenue as a function of the price p.
(B) Express the cost as a function of the price p.
(C) Graph R and C on the same set of axes and identify the regions of profit and loss.
(D) Find the break-even points.
(E) Find the price that will produce a maximum profit.

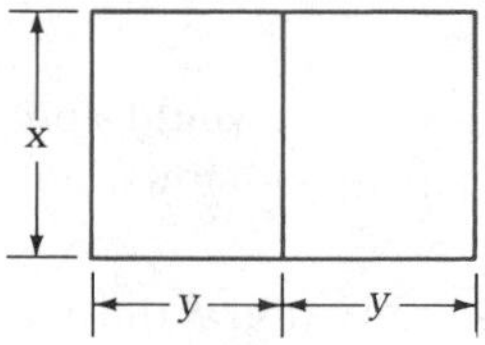

42. *Construction.* A farmer has 180 feet of fencing to be used in the construction of two identical rectangular pens sharing a common side (see the figure in the margin). Assuming all the fencing is used:

(A) Express the total area $A(x)$ enclosed by both pens as a function of the width x.
(B) From physical considerations, what is the domain of the function A?
(C) Find the dimensions of the pens that will make the total enclosed area maximum.

CHAPTER 3

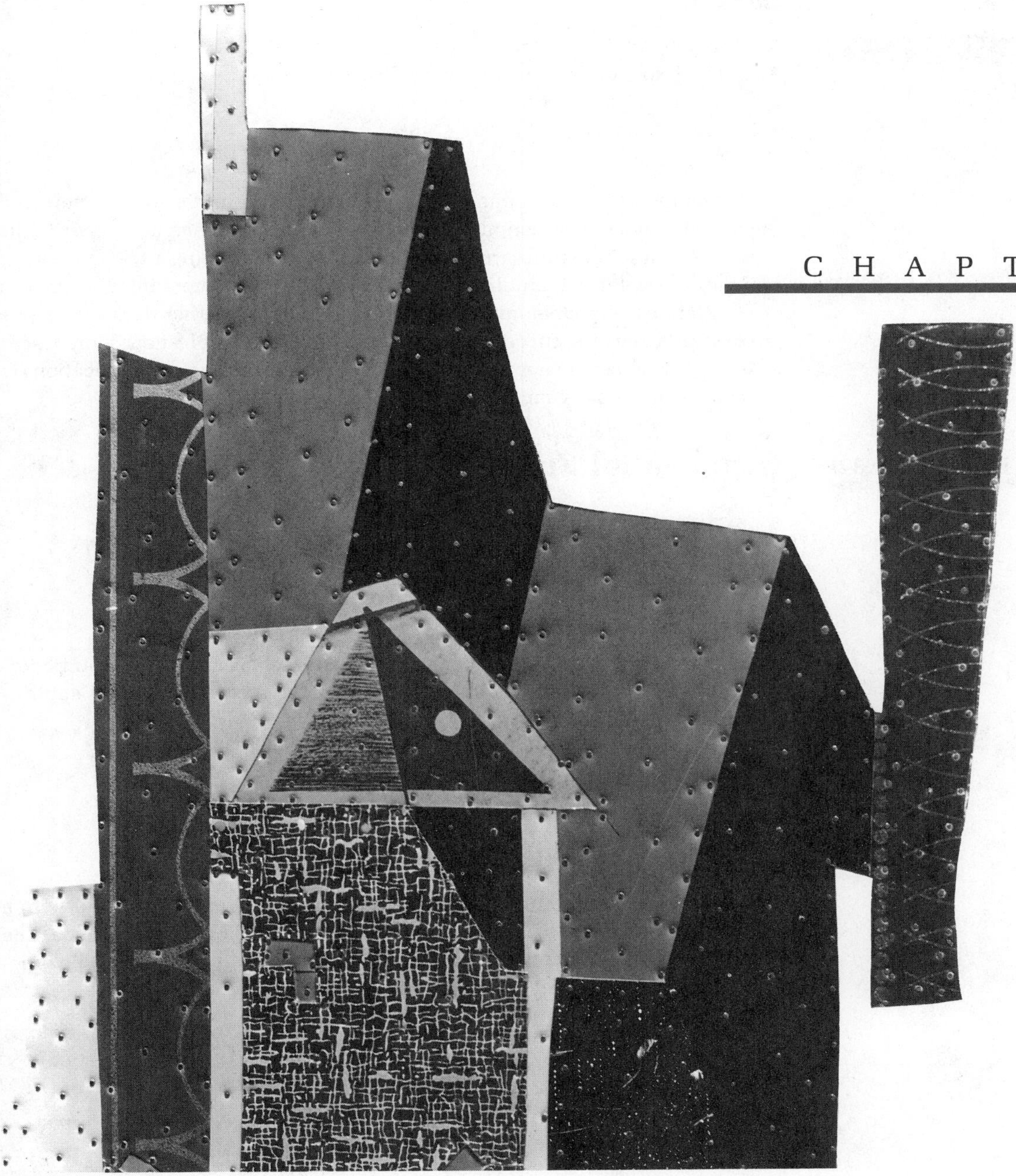

Exponential and Logarithmic Functions

CHAPTER 3

Contents

In this chapter we will define and investigate the properties of two new and important classes of functions called *exponential functions* and *logarithmic functions.* These functions are used in describing and solving a wide variety of real-world problems, including growth of money at compound interest; growth of populations of people, animals, and bacteria; radioactive decay (negative growth); and learning curves associated with the mastery of a new computer or an assembly process in a manufacturing plant. We will consider applications in these areas plus many more in sections that follow.

SECTION 3-1

Exponential Functions

- EXPONENTIAL FUNCTIONS
- BASIC EXPONENTIAL GRAPHS
- ADDITIONAL PROPERTIES
- APPLICATIONS

In this section we will define exponential functions, look at some of their important properties—including graphs—and consider several significant applications.

EXPONENTIAL FUNCTIONS

Let us start by noting that the functions f and g given by

$$f(x) = 2^x \qquad \text{and} \qquad g(x) = x^2$$

are not the same function. Whether a variable appears as an exponent with a constant base or as a base with a constant exponent, makes a big difference. The function g is a quadratic function, which we have already discussed. The function f is a new type of function called an *exponential function.*

Exponential Function

The equation

$$f(x) = b^x \qquad b > 0, b \neq 1$$

defines an **exponential function** for each different constant b, called the **base.** The independent variable x may assume any real value.

Thus, the **domain of *f*** is the set of all real numbers, and it can be shown that the **range of *f*** is the set of all positive real numbers. We require the base b to be positive to avoid nonreal numbers such as $(-2)^{1/2}$. We exclude the case $b = 1$ since $f(x) = 1^x = 1$ is a constant function, not an exponential function.

◆ BASIC EXPONENTIAL GRAPHS

Many students, if asked to graph equations such as $y = 2^x$ or $y = 2^{-x}$, would not hesitate at all. [*Note:* $2^{-x} = 1/2^x = (\frac{1}{2})^x$.] They would likely make up tables by assigning integers to x, plot the resulting points, and then join these points with a smooth curve as in Figure 1. The only catch is that 2^x has not been defined at this point for all real numbers. From Section 1-6 we know what 2^5, 2^{-3}, $2^{2/3}$, $2^{-3/5}$, $2^{1.4}$, and $2^{-3.15}$ mean (that is, 2^p, where p is a rational number), but what does

$$2^{\sqrt{2}}$$

mean? The question is not easy to answer at this time. In fact, a precise definition of $2^{\sqrt{2}}$ must wait for more advanced courses, where we can show that

$$2^x$$

names a real number for x any real number, and that the graph of $y = 2^x$ is as indicated in Figure 1A. We also can show that for x irrational, 2^x can be approximated as closely as we like by using rational number approximations for x. Since $\sqrt{2} = 1.414\ 213\ldots$, for example, the sequence

$$2^{1.4},\ 2^{1.41},\ 2^{1.414},\ \ldots$$

approximates $2^{\sqrt{2}}$, and as we move to the right, the approximation improves.

It is useful to note that the graph of

$$f(x) = b^x \qquad b > 1$$

will look very much like Figure 1A, and the graph of

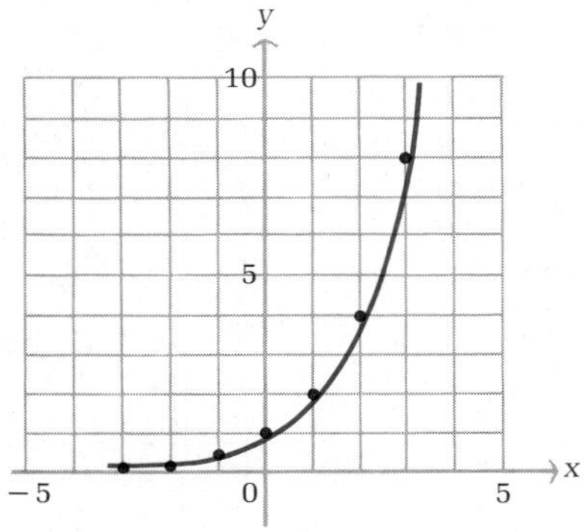

(A) $y = 2^x$

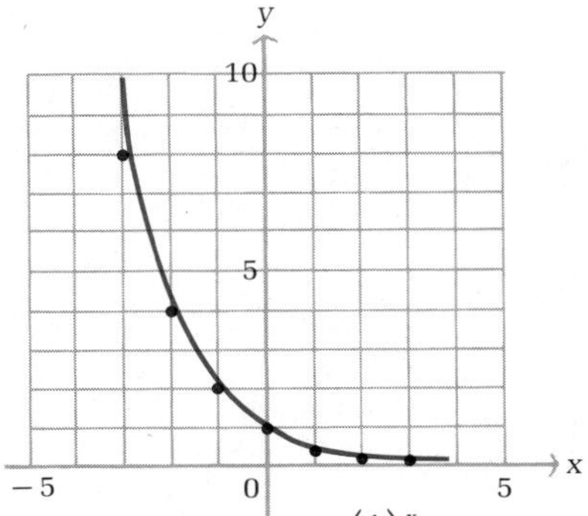

(B) $y = 2^{-x} = \left(\frac{1}{2}\right)^x$

FIGURE 1

$$f(x) = b^x \qquad 0 < b < 1$$

will look very much like Figure 1B. Observe that in both cases the graphs approach the x axis, but never touch it. The x axis is called a *horizontal asymptote* for each graph. In general, a line $y = c$ is a **horizontal asymptote** for the graph of the equation $y = f(x)$ if $f(x)$ approaches c as x increases without bound or as x decreases without bound. Asymptotes, if they exist, provide a useful aid in graphing some functions.

The graphs in Figure 1 suggest the following important general properties of exponential functions, which we state without proof:

Basic Properties of the Graph of $f(x) = b^x$, $b > 0$, $b \neq 1$

1. All graphs will pass through the point (0, 1). — $b^0 = 1$ for any permissible base b.
2. All graphs are continuous curves, with no holes or jumps.
3. The x axis is a horizontal asymptote.
4. If $b > 1$, then b^x increases as x increases.
5. If $0 < b < 1$, then b^x decreases as x increases.

The use of the $\boxed{y^x}$ key on a scientific calculator makes the plotting of accurate graphs of exponential functions almost routine. Example 1 illustrates the process.

◆ EXAMPLE 1 Graph $y = (\frac{1}{2})4^x$ for $-3 \leq x \leq 3$.

Solution

x	y
−3	0.01
−2	0.03
−1	0.13
0	0.50
1	2.00
2	8.00
3	32.00

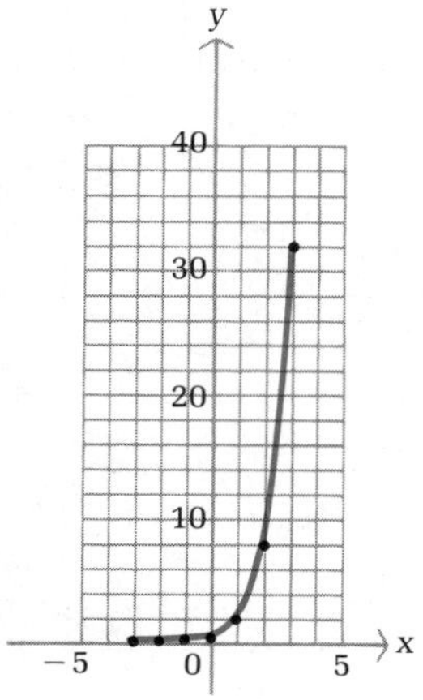

◆

PROBLEM 1 Graph $y = (\frac{1}{2})4^{-x}$ for $-3 \leq x \leq 3$. ◆

◆ ADDITIONAL PROPERTIES

Exponential functions, which include irrational exponents, obey the familiar laws of exponents we discussed earlier for rational exponents. We summarize these exponent laws and add two other important and useful properties in the following box:

Exponential Function Properties

For a and b positive, $a \neq 1$, $b \neq 1$, and x and y real:

1. Exponent laws:

$$a^x a^y = a^{x+y} \qquad \frac{a^x}{a^y} = a^{x-y} \qquad (a^x)^y = a^{xy}$$

$$(ab)^x = a^x b^x \qquad \left(\frac{a}{b}\right)^x = \frac{a^x}{b^x} \qquad \frac{4^{2y}}{4^{5y}} = 4^{2y-5y} = 4^{-3y}$$

2. $a^x = a^y$ if and only if $x = y$ — If $7^{5t+1} = 7^{3t-3}$, then $5t + 1 = 3t - 3$, and $t = -2$.

3. For $x \neq 0$,

   $a^x = b^x$ if and only if $a = b$ — If $a^5 = 2^5$, then $a = 2$.

◆ APPLICATIONS

We will now consider three applications that utilize exponential functions in their analysis: population growth (an example of exponential growth), radioactive decay (an example of negative exponential growth), and compound interest (another example of exponential growth).

Our first example involves the growth of populations, such as people, animals, insects, and bacteria. Populations tend to grow exponentially and at different rates. A convenient and easily understood measure of growth rate is the **doubling time**—that is, the time it takes for a population to double. Over short periods of time the **doubling-time growth model** is often used to model population growth:

$$P = P_0 2^{t/d}$$

where

P = Population at time t

P_0 = Population at time $t = 0$

d = Doubling time

Note that when $t = d$,

$$P = P_0 2^{d/d} = P_0 2$$

and the population is double the original, as it should be. We use this model to solve a population growth problem in Example 2.

◆ EXAMPLE 2

Population Growth

Ethiopia has a population of around 42 million people, and it is estimated that the population will double in 22 years. If population growth continues at the same rate, what will be the population (to the nearest million):

(A) 10 years from now? (B) 35 years from now?

Solutions

We use the doubling-time growth model:

$$P = P_0 2^{t/d}$$

Substituting $P_0 = 42$ and $d = 22$, we obtain

$$P = 42(2^{t/22}) \quad \text{See the graph in the margin.}$$

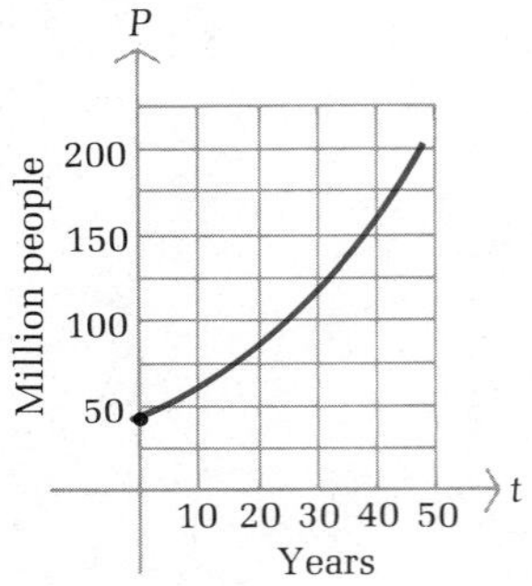

(A) Find P when $t = 10$ years:

$$P = 42(2^{10/22}) \quad \text{Use a calculator.}$$
$$P \approx 58 \text{ million people}$$

(B) Find P when $t = 35$ years:

$$P = 42(2^{36/22}) \quad \text{Use a calculator.}$$
$$P \approx 127 \text{ million people}$$ ◆

PROBLEM 2

The bacterium *Escherichia coli* (*E. coli*) is found naturally in the intestines of many mammals. In a particular laboratory experiment, the doubling time for *E. coli* is found to be 25 minutes. If the experiment starts with a population of 1,000 *E. coli* and there is no change in growth rate, how many bacteria will be present:

(A) In 10 minutes? (B) In 5 hours? ◆

Our second application involves radioactive decay (negative growth). Radioactive materials are used extensively in medical diagnosis and therapy, as power sources in satellites, and as power sources (although controversial) in many countries. If we start with an amount A_0 of a particular radioactive isotope, the amount declines exponentially in time; the rate of decay varies from isotope to isotope. A convenient and easily understood measure of the rate of decay is the **half-life** of the isotope—that is, the time it takes for half of a particular material to decay. In this section we use the following **half-life decay model:**

$$A = A_0(\tfrac{1}{2})^{t/h} = A_0 2^{-t/h}$$

where

$$A = \text{Amount at time } t$$
$$A_0 = \text{Amount at time } t = 0$$
$$h = \text{Half-life}$$

Note that when $t = h$,

$$A = A_0 2^{-h/h} = A_0 2^{-1} = \frac{A_0}{2}$$

and the amount of isotope is half the original amount, as it should be.

◆ EXAMPLE 3

Radioactive Decay

Radioactive gold-198 (^{198}Au), used in imaging the structure of the liver, has a half-life of 2.67 days. If we start with 50 milligrams of the isotope, how many milligrams will be left after:

(A) $\frac{1}{2}$ day? (B) 1 week?

Compute answers to two decimal places.

Solutions

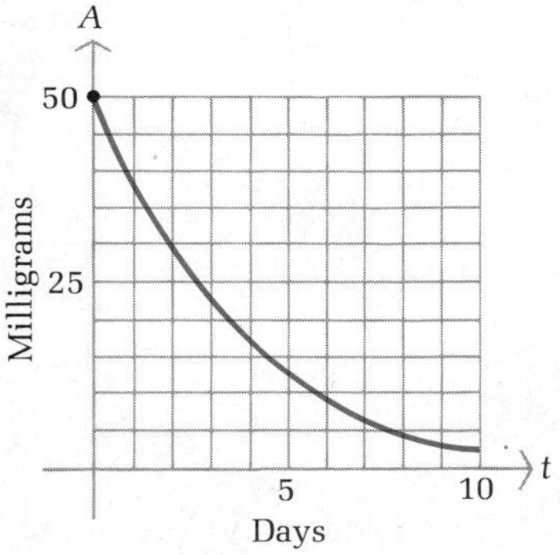

We use the half-life decay model:

$$A = A_0(\tfrac{1}{2})^{t/h} = A_0 2^{-t/h}$$

Using $A_0 = 50$ and $h = 2.67$, we obtain

$$A = 50(2^{-t/2.67}) \quad \text{See the graph in the margin.}$$

(A) Find A when $t = 0.5$ day:

$$A = 50(2^{-0.5/2.67}) \quad \text{Use a calculator.}$$
$$\approx 43.91 \text{ milligrams}$$

(B) Find A when $t = 7$ days:

$$A = 50(2^{-7/2.67}) \quad \text{Use a calculator.}$$
$$\approx 8.12 \text{ milligrams}$$

◆

PROBLEM 3

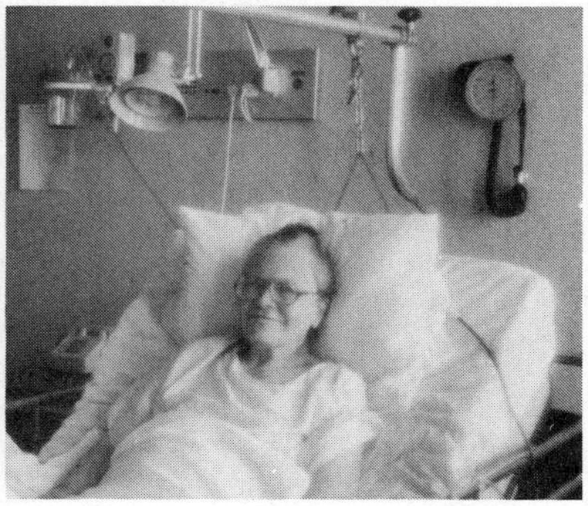

The radioactive isotope gallium-67 (^{67}Ga), used in the diagnosis of malignant tumors, has a biological half-life of 46.5 hours. If we start with 100 milligrams of the isotope, how many milligrams will be left after:

(A) 24 hours? (B) 1 week?

Compute answers to two decimal places. ◆

Our third application deals with the growth of money at compound interest. The fee paid to use another's money is called **interest.** It is usually computed as a percent (called **interest rate**) of the principal over a given period of time. If, at the end of a payment period, the interest due is reinvested at the same rate, then the interest earned as well as the principal will earn interest during the next payment period. Interest paid on interest reinvested is called **compound interest.**

Compound Interest

If a **principal** P is invested at an annual **rate** r (expressed as a decimal) compounded m times a year, then the **amount** A **(future value)** in the account at the end of t years is given by

$$A = P\left(1 + \frac{r}{m}\right)^{mt}$$

◆ EXAMPLE 4 *Compound Growth*

If $1,000 is invested in an account paying 10% compounded monthly, how much will be in the account at the end of 10 years? Compute the answer to the nearest cent.

Solution We use the compound interest formula as follows:

$$A = P\left(1 + \frac{r}{m}\right)^{mt}$$

$$= 1{,}000\left(1 + \frac{0.10}{12}\right)^{(12)(10)} \qquad \text{Use a calculator.}$$

$$= \$2{,}707.04$$

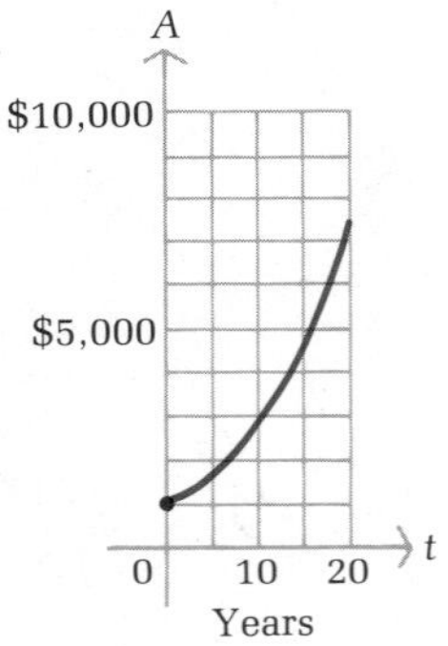

The graph of

$$A = 1{,}000\left(1 + \frac{0.10}{12}\right)^{12t}$$

for $0 \leq t \leq 20$ is shown in the margin. ◆

PROBLEM 4

If you deposit $5,000 in an account paying 9% compounded daily, how much will you have in the account in 5 years? Compute the answer to the nearest cent. ◆

Answers to Matched Problems

1. $y = (\frac{1}{2})4^{-x}$

x	y
−3	32.00
−2	8.00
−1	2.00
0	0.50
1	0.13
2	0.03
3	0.01

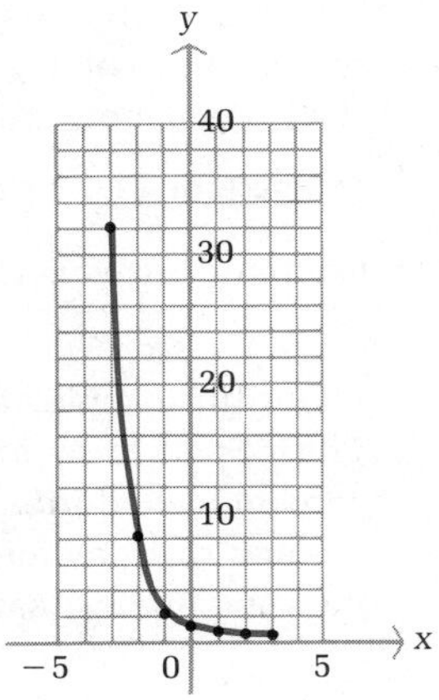

2. (A) $P \approx 1{,}320$ bacteria
   (B) $P \approx 4{,}096{,}000$ bacteria
3. (A) $A \approx 69.92$ mg
   (B) $A \approx 8.17$ mg
4. $7,841.13

EXERCISE 3-1

A *Graph each function over the indicated interval.*

1. $y = 5^x$; $[-2, 2]$

2. $y = 3^x$; $[-3, 3]$

3. $y = (\frac{1}{5})^x = 5^{-x}$; $[-2, 2]$

4. $y = (\frac{1}{3})^x = 3^{-x}$; $[-3, 3]$

5. $f(x) = -5^x$; $[-2, 2]$

6. $g(x) = -3^{-x}$; $[-3, 3]$

7. $f(x) = 4(5^x)$; $[-2, 2]$

8. $h(x) = 5(3^x)$; $[-3, 3]$

9. $y = 5^{x+2} + 4$; $[-4, 0]$

10. $y = 3^{x+3} - 5$; $[-6, 0]$

Simplify.

11. $(4^{3x})^{2y}$

12. $10^{3x-1}10^{4-x}$

13. $\dfrac{5^{x-3}}{5^{x-4}}$

14. $\dfrac{3^x}{3^{1-x}}$

15. $(2^x 3^y)^z$

16. $\left(\dfrac{4^x}{5^y}\right)^{3z}$

B *Solve for x.*

17. $10^{2-3x} = 10^{5x-6}$

18. $5^{3x} = 5^{4x-2}$

19. $4^{5x-x^2} = 4^{-6}$

20. $7^{x^2} = 7^{2x+3}$

21. $5^3 = (x + 2)^3$

22. $(1 - x)^5 = (2x - 1)^5$

23. $9^{x-1} = 3^x$

24. $2^x = 4^{x+1}$

Graph each function over the indicated interval.

25. $f(t) = 2^{t/10}$; $[-30, 30]$

26. $G(t) = 3^{t/100}$; $[-200,200]$

27. $y = 7(2^{-2x})$; $[-2, 2]$

28. $y = 11(3^{-x/2})$; $[-9, 9]$

29. $f(x) = 2^{|x|}$; $[-3, 3]$

30. $g(x) = 2^{-|x|}$; $[-3, 3]$

31. $y = 100(1.03)^x$; $[0, 20]$

32. $y = 1{,}000(1.08)^x$; $[0, 10]$

33. $y = 3^{-x^2}$; $[-2, 2]$

34. $y = 2^{-x^2}$; $[-2, 2]$

C *Simplify.*

35. $(3^x - 3^{-x})(3^x + 3^{-x})$

36. $(6^x + 6^{-x})(6^x - 6^{-x})$

37. $(3^x - 3^{-x})^2 + (3^x + 3^{-x})^2$

38. $(6^x + 6^{-x})^2 - (6^x - 6^{-x})^2$

Graph each function over the indicated interval.

39. $h(x) = x(2^x)$; $[-5, 0]$

40. $m(x) = x(3^{-x})$; $[0, 3]$

41. $g(x) = \dfrac{3^x + 3^{-x}}{2}$; $[-3, 3]$

42. $f(x) = \dfrac{2^x + 2^{-x}}{2}$; $[-3, 3]$

CALCULATOR PROBLEMS

Calculate each using a scientific calculator. Compute answers to four decimal places.

43. $3^{-\sqrt{2}}$

44. $5^{\sqrt{3}}$

45. $\pi^{-\sqrt{3}}$

46. $\pi^{\sqrt{2}}$

47. $\dfrac{3^\pi - 3^{-\pi}}{2}$

48. $\dfrac{2^\pi + 2^{-\pi}}{2}$

Business & Economics

49. *Finance.* Suppose $2,500 is invested at 7% compounded quarterly. How much money will be in the account in:

(A) $\frac{3}{4}$ year? (B) 15 years?

Compute answers to the nearest cent.

50. *Finance.* Suppose $4,000 is invested at 11% compounded weekly. How much money will be in the account in:

(A) $\frac{1}{2}$ year? (B) 10 years?

Compute answers to the nearest cent.

51. *Finance.* A person wishes to have $15,000 cash for a new car 5 years from now. How much should be placed in an account now, if the account pays 9.75% compounded weekly? Compute the answer to the nearest dollar.

52. *Finance.* A couple just had a new child. How much should they invest now at 8.25% compounded daily in order to have $40,000 for the child's education 17 years from now? Compute the answer to the nearest dollar.

Life Sciences

53. *Population growth.* If Kenya has a population of 23,000,000 people and a doubling time of 19 years, and if the growth continues at the same rate, find the population in:

(A) 10 years (B) 30 years

Compute answers to the nearest million.

54. *Bacterial growth.* If bacteria in a certain culture double every $\frac{1}{2}$ hour, write an equation that gives the number of bacteria N in the culture after t hours, assuming the culture has 100 bacteria at the start. Graph the equation for $0 \leq t \leq 5$.

55. *Radioactive tracers.* The radioactive isotope technetium-99m (^{99m}Tc) is used in imaging the brain. This isotope has a half-life of 6 hours. If 12 milligrams are used, how much will be present after:

(A) 3 hours? (B) 24 hours?

Compute answers to two decimal places.

56. *Insecticides.* The use of the insecticide DDT is no longer allowed in many countries because of its long-term adverse effects. If a farmer uses 25 pounds of active DDT, assuming its half-life is 12 years, how much will still be active after:

(A) 5 years? (B) 20 years?

Compute answers to the nearest pound.

SECTION 3-2 The Exponential Function with Base e

- BASE e EXPONENTIAL FUNCTION
- GROWTH AND DECAY APPLICATIONS REVISITED
- CONTINUOUS COMPOUND INTEREST

The number π is probably the most important irrational number you have encountered until now. In this section we will introduce another irrational number, e, that is just as important in mathematics and its applications.

BASE e EXPONENTIAL FUNCTION

Consider the following expression:

$$\left(1 + \frac{1}{m}\right)^m$$

What happens to the value of the expression as m increases without bound? (Think about this for a moment before proceeding.) Maybe you guessed that the value approaches 1 using the following reasoning: As m gets large, $1 + (1/m)$ approaches 1 (since $1/m$ approaches 0), and 1 raised to any power is 1. Let us see if this reasoning is correct by actually calculating the value of the expression for larger and larger values of m. Table 1 summarizes the results.

TABLE 1

m	$\left(1 + \frac{1}{m}\right)^m$
1	2
10	2.593 74...
100	2.704 81...
1,000	2.716 92...
10,000	2.718 14...
100,000	2.718 27...
1,000,000	2.718 28...
.	.
.	.
.	.

Interestingly, the value of $[1 + (1/m)]^m$ is never close to 1 but seems to be approaching a number close to 2.7183. As m increases without bound, the value of $[1 + (1/m)]^m$ approaches an irrational number that we call $\boldsymbol{e}$. The irrational number e to twelve decimal places is

$$e = 2.718\ 281\ 828\ 459$$

Exactly who discovered the constant e is still being debated. It is named after the great Swiss mathematician Leonhard Euler (1707–1783). This constant turns out to be an ideal base for an exponential function, because in calculus and higher mathematics many operations take on their simplest form using this base. This is why you will see e used extensively in expressions and formulas that model real-world phenomena.

Exponential Function with Base e

For x a real number, the equation

$$f(x) = e^x$$

defines the **exponential function with base e.**

The exponential function with base e is used so frequently that it is often referred to as *the* exponential function. Because of its importance, all scientific calculators have an $\boxed{e^x}$ key or its equivalent—consult your user's manual.

The important constant e, along with the two other important constants, $\sqrt{2}$ and π, are shown on the number line in Figure 2A. Using the properties of the graph of $f(x) = b^x$ discussed in Section 3-1, we obtain the graphs of $y = e^x$ and $y = (1/e)^x = e^{-x}$ shown in Figure 2B. Notice that neither graph crosses the x axis. Thus, we conclude that $e^x > 0$ and $e^{-x} > 0$ for all real numbers x, an important observation when solving certain equations involving the exponential function.

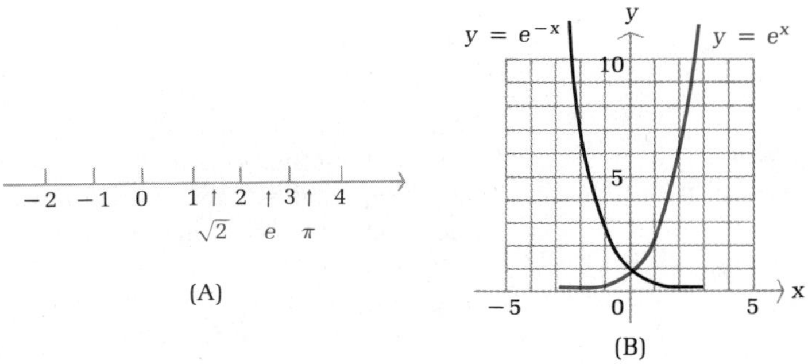

FIGURE 2

◆ GROWTH AND DECAY APPLICATIONS REVISITED

Most exponential growth and decay problems are modeled using base e exponential functions. We present two applications here and many more in Exercise 3-2.

◆ EXAMPLE 5

Exponential Growth

Cholera, an intestinal disease, is caused by a cholera bacterium that multiplies exponentially by cell division as given approximately by

$$N = N_0e^{1.386t}$$

where N is the number of bacteria present after t hours and N_0 is the number of bacteria present at $t = 0$. If we start with 10 bacteria, how many bacteria will be present:

(A) In 0.6 hour? (B) In 3.5 hours?

Solutions Substituting $N_0 = 10$ into the above equation, we obtain

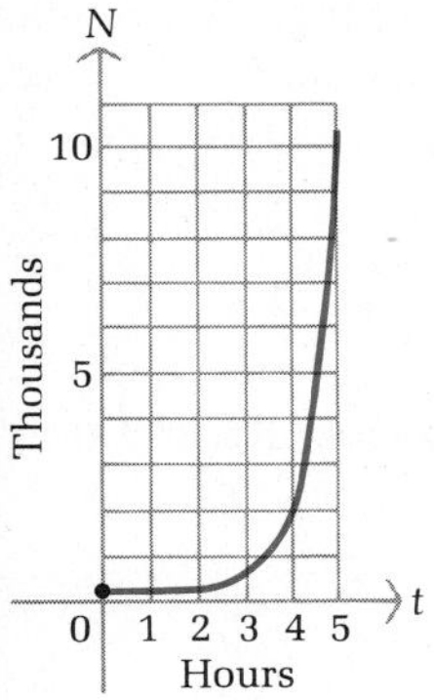

$N = 10e^{1.386t}$ The graph is shown in the margin.

(A) Solve for N when $t = 0.6$:

$$\begin{aligned} N &= 10e^{1.386(0.6)} \quad \text{Use a calculator.} \\ &= 23 \text{ bacteria} \end{aligned}$$

(B) Solve for N when $t = 3.5$:

$$\begin{aligned} N &= 10e^{1.386(3.5)} \quad \text{Use a calculator.} \\ &= 1{,}279 \text{ bacteria} \end{aligned}$$

◆

PROBLEM 5 Refer to the exponential growth model for cholera in Example 5. If we start with 50 bacteria, how many bacteria will be present:

(A) In 0.85 hour? (B) In 7.25 hours? ◆

◆ EXAMPLE 6

Exponential Decay

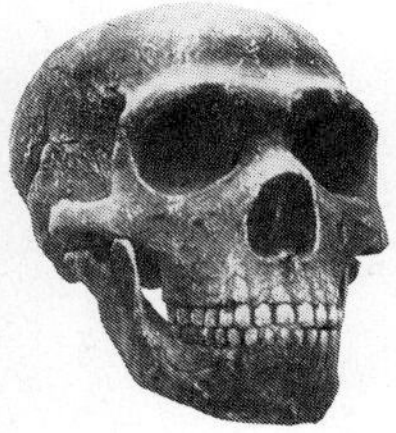

Cosmic-ray bombardment of the atmosphere produces neutrons, which in turn react with nitrogen to produce radioactive carbon-14 (^{14}C). Radioactive ^{14}C enters all living tissues through carbon dioxide, which is first absorbed by plants. As long as a plant or animal is alive, ^{14}C is maintained in the living organism at a constant level. Once the organism dies, however, ^{14}C decays according to the equation

$$A = A_0e^{-0.000124t}$$

where A is the amount present after t years and A_0 is the amount present at time $t = 0$. If 500 milligrams of ^{14}C are present in a sample from a skull at the time of death, how many milligrams will be present in the sample in:

(A) 15,000 years? (B) 45,000 years?

Compute answers to two decimal places.

Solutions Substituting $A_0 = 500$ in the decay equation, we have

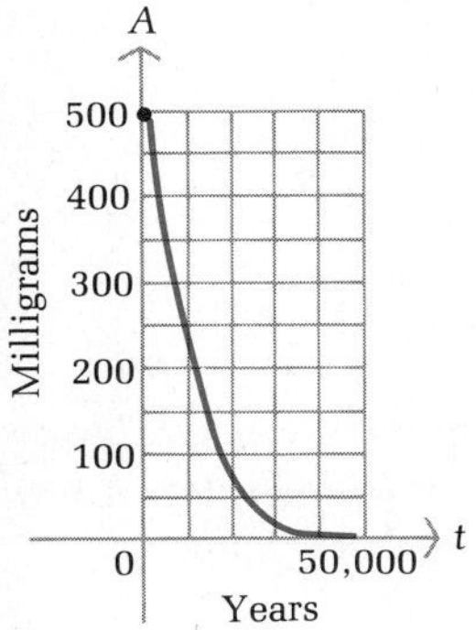

$A = 500e^{-0.000124t}$ See the graph in the margin.

(A) Solve for A when $t = 15{,}000$:

$$\begin{aligned} A &= 500e^{-0.000124(15{,}000)} \quad \text{Use a calculator.} \\ &= 77.84 \text{ milligrams} \end{aligned}$$

(B) Solve for A when $t = 45{,}000$:

$$\begin{aligned} A &= 500e^{-0.000124(45{,}000)} \quad \text{Use a calculator.} \\ &= 1.89 \text{ milligrams} \end{aligned}$$

◆

PROBLEM 6 Refer to the exponential decay model in Example 6. How many milligrams of ^{14}C would have to be present at the beginning in order to have 25 milligrams present after 18,000 years? Compute the answer to the nearest milligram. ◆

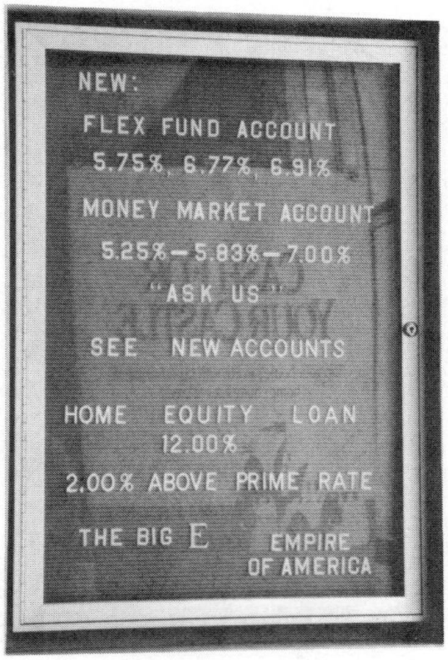

◆ CONTINUOUS COMPOUND INTEREST

The constant e occurs naturally in the study of compound interest. Returning to the compound interest formula discussed in Section 3-1,

COMPOUND INTEREST $$A = P\left(1 + \frac{r}{m}\right)^{mt}$$

recall that P is the principal invested at an annual rate r compounded m times a year and A is the amount in the account after t years. Suppose P, r, and t are held fixed and m is increased without bound. Will the amount A increase without bound or will it tend to some limiting value?

Starting with $P = \$100$, $r = 0.08$, and $t = 2$ years, we construct Table 2 for several values of m with the aid of a calculator. Notice that the largest gain appears in going from annual to semiannual compounding. Then, the gains slow down as m increases. It appears that A gets closer and closer to \$117.35 as m gets larger and larger.

TABLE 2

COMPOUNDING FREQUENCY	m	$A = 100\left(1 + \frac{0.08}{m}\right)^{2m}$
Annually	1	\$116.6400
Semiannually	2	116.9859
Quarterly	4	117.1659
Weekly	52	117.3367
Daily	365	117.3490
Hourly	8,760	117.3510

In Section 6-1, we show that

$$P\left(1 + \frac{r}{m}\right)^{mt}$$

gets closer and closer to Pe^{rt} as the number of compounding periods m gets larger and larger. The latter is referred to as the **continuous compound interest formula,** a formula that is widely used in business, banking, and economics.

Continuous Compound Interest Formula

If a principal P is invested at an annual rate r (expressed as a decimal) compounded continuously, then the amount A in the account at the end of t years is given by

$$A = Pe^{rt}$$

◆ EXAMPLE 7 What amount will an account have after 2 years if \$5,000 is invested at an annual rate of 8%:

(A) Compounded daily? (B) Compounded continuously?

Compute answers to the nearest cent.

Solutions (A) Use the compound interest formula

$$A = P\left(1 + \frac{r}{m}\right)^{mt}$$

with $P = 5{,}000$, $r = 0.08$, $m = 365$, and $t = 2$:

$$A = 5{,}000\left(1 + \frac{0.08}{365}\right)^{(365)(2)} \quad \text{Use a calculator.}$$

$$= \$5{,}867.45$$

(B) Use the continuous compound interest formula

$$A = Pe^{rt}$$

with $P = 5{,}000$, $r = 0.08$, and $t = 2$:

$$A = 5{,}000e^{(0.08)(2)} \quad \text{Use a calculator.}$$
$$= \$5{,}867.55$$

◆

PROBLEM 7 What amount will an account have after 1.5 years if \$8,000 is invested at an annual rate of 9%:

(A) Compounded weekly? (B) Compounded continuously?

Compute answers to the nearest cent. ◆

Answers to Matched Problems

5. (A) 162 bacteria (B) 1,156,054 bacteria
6. 233 mg 7. (A) \$9,155.23 (B) \$9,156.29

EXERCISE 3-2

A *Graph over the indicated interval.*

1. $y = -e^{-x}$; $[-3, 3]$
2. $y = -e^{x}$; $[-3, 3]$
3. $y = 100e^{0.1x}$; $[-5, 5]$
4. $y = 10e^{0.2x}$; $[-10, 10]$
5. $g(t) = 10e^{-0.2t}$; $[-5, 5]$
6. $f(t) = 100e^{-0.1t}$; $[-5, 5]$

B

7. $y = -3 + e^{1+x}$; $[-4, 2]$
8. $y = 2 + e^{x-2}$; $[-1, 5]$
9. $y = e^{|x|}$; $[-3, 3]$
10. $y = e^{-|x|}$; $[-3, 3]$
11. $C(x) = \dfrac{e^x + e^{-x}}{2}$; $[-5, 5]$
12. $M(x) = e^{x/2} + e^{-x/2}$; $[-5, 5]$

Simplify.

13. $e^x(e^{-x} + 1) - e^{-x}(e^x + 1)$
14. $(e^x + e^{-x})^2 + (e^x - e^{-x})^2$
15. $\dfrac{e^x(e^x + e^{-x}) - (e^x - e^{-x})e^x}{e^{2x}}$
16. $\dfrac{e^{-x}(e^x - e^{-x}) + e^{-x}(e^x + e^{-x})}{e^{-2x}}$

C *Solve each equation (remember, $e^x \neq 0$ and $e^{-x} \neq 0$).*

17. $(x - 3)e^x = 0$
18. $2xe^{-x} = 0$
19. $3xe^{-x} + x^2e^{-x} = 0$
20. $x^2e^x - 5xe^x = 0$

Graph over the indicated interval.

21. $N = \dfrac{100}{1 + e^{-t}}$; $[0, 5]$
22. $N = \dfrac{200}{1 + 3e^{-t}}$; $[0, 5]$

APPLICATIONS

Business & Economics

23. *Money growth.* If you invest $7,500 in an account paying 8.35% compounded continuously, how much money will be in the account at the end of:
(A) 5.5 years? (B) 12 years?

24. *Money growth.* If you invest $5,250 in an account paying 11.38% compounded continuously, how much money will be in the account at the end of:
(A) 6.25 years? (B) 17 years?

25. *Money growth. Barron's* (a national business and financial weekly) published the following "Top Savings Deposit Yields" for 1 year certificate of deposit accounts:
(A) Alamo Savings, 8.25% compounded quarterly
(B) Lamar Savings, 8.05% compounded continuously
Compute the value of $10,000 invested in each account at the end of 1 year.

26. *Money growth.* Refer to Problem 25. In another issue of *Barron's*, $2\frac{1}{2}$ year certificate of deposit accounts included the following:

(A) Gill Saving, 8.30% compounded continuously
(B) Richardson Savings and Loan, 8.40% compounded quarterly
(C) USA Savings, 8.25% compounded daily

Compute the value of \$1,000 invested in each account at the end of $2\frac{1}{2}$ years.

27. *Present value.* A promissory note will pay \$50,000 at maturity $5\frac{1}{2}$ years from now. How much should you be willing to pay for the note now if money is worth 10% compounded continuously?

28. *Present value.* A promissory note will pay \$30,000 at maturity 10 years from now. How much should you be willing to pay for the note now if money is worth 9% compounded continuously?

29. *Advertising.* A company is trying to introduce a new product to as many people as possible through television advertising in a large metropolitan area with 2 million possible viewers. A model for the number of people N (in millions) who are aware of the product after t days of advertising was found to be

$$N = 2(1 - e^{-0.037t})$$

Graph this function for $0 \leq t \leq 50$. What value does N tend to as t increases without bound?

30. *Learning curve.* People assigned to assemble circuit boards for a computer manufacturing company undergo on-the-job training. From past experience it was found that the learning curve for the average employee is given by

$$N = 40(1 - e^{-0.12t})$$

where N is the number of boards assembled per day after t days of training. Graph this function for $0 \leq t \leq 30$. What is the maximum number of boards an average employee can be expected to produce in 1 day?

Life Sciences

31. *Marine biology.* Marine life is dependent upon the microscopic plant life that exists in the *photic zone*, a zone that goes to a depth where about 1% of the surface light still remains. In some waters with a great deal of sediment, the photic zone may go down only 15–20 feet. In some murky harbors, the intensity of light d feet below the surface is given approximately by

$$I = I_0 e^{-0.23d}$$

What percent of the surface light will reach a depth of:

(A) 10 feet? (B) 20 feet?

32. *Marine biology.* Refer to Problem 31. Light intensity I relative to depth d (in feet) for one of the clearest bodies of water in the world, the Sargasso Sea in the West Indies, can be approximated by

$$I = I_0 e^{-0.00942d}$$

where I_0 is the intensity of light at the surface. What percent of the surface light will reach a depth of:

(A) 50 feet? (B) 100 feet?

33. *AIDS epidemic.* As of this writing, AIDS among intravenous drug users in the United States is spreading at the rate of about 46% compounded continuously. Assuming this rate does not change and there were 12,000 cases of AIDS among intravenous drug users at the end of 1987, how many cases should we expect by the end of:

(A) 1990? (B) 1995?

34. *AIDS epidemic.* At the end of 1987 there were approximately 45,000 diagnosed cases of AIDS among the general population in the United States. As of this writing, the disease is estimated to be spreading at the rate of 38% compounded continuously. Assuming this rate does not change, how many cases of AIDS should be expected by the end of:

(A) 1992? (B) 2000?

Social Sciences

35. *Population growth.* If the population in Mexico was around 100 million people in 1987 and if the population continues to grow at 2.3% compounded continuously, what will the population be in 1995? Compute the answer to the nearest million.

36. *Population growth.* If the world population was 5 billion people in 1988 and if the population continues to grow at 1.7% compounded continuously, what will the population be in 1999? Compute the answer to the nearest billion.

SECTION 3-3 Logarithmic Functions

- DEFINITION OF LOGARITHMIC FUNCTIONS
- FROM LOGARITHMIC TO EXPONENTIAL FORM AND VICE VERSA
- PROPERTIES OF LOGARITHMIC FUNCTIONS
- CALCULATOR EVALUATION OF COMMON AND NATURAL LOGARITHMS
- APPLICATION

Now we are ready to consider logarithmic functions, which are closely related to exponential functions.

◆ DEFINITION OF LOGARITHMIC FUNCTIONS

If we start with an exponential function f defined by

$$y = 2^x \tag{1}$$

and interchange the variables, we obtain an equation that defines a new relation g defined by

$$x = 2^y \tag{2}$$

Any ordered pair of numbers that belongs to f will belong to g if we interchange the order of the components. For example, (3, 8) satisfies equation (1) and (8, 3) satisfies equation (2). Thus, the domain of f becomes the range of g and the range of f becomes the domain of g. Graphing f and g on the same coordinate system (Fig. 3), we see that g is also a function. We call this new function the **logarithmic function with base 2,** and write

$$y = \log_2 x \qquad \text{if and only if} \qquad x = 2^y$$

Note that if we fold the paper along the dashed line $y = x$ in Figure 3, the two graphs match exactly.

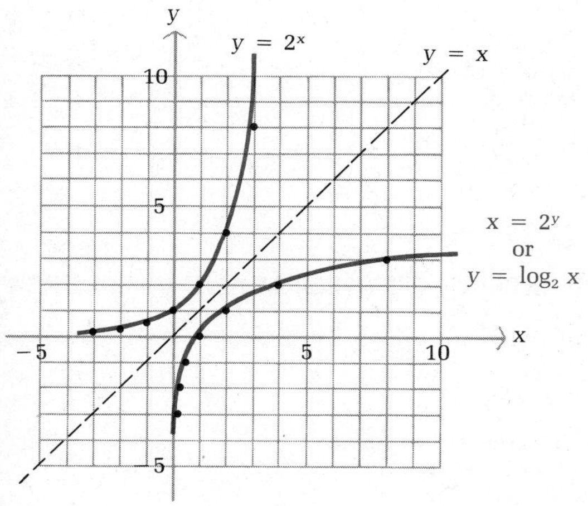

FIGURE 3

EXPONENTIAL FUNCTION		LOGARITHMIC FUNCTION	
x	$y = 2^x$	$x = 2^y$	y
−3	$\frac{1}{8}$	$\frac{1}{8}$	−3
−2	$\frac{1}{4}$	$\frac{1}{4}$	−2
−1	$\frac{1}{2}$	$\frac{1}{2}$	−1
0	1	1	0
1	2	2	1
2	4	4	2
3	8	8	3

[Ordered pairs reversed]

In general, we define the **logarithmic functions with base *b*** as follows:

Definition of Logarithmic Function

For $b > 0$ and $b \neq 1$,

$y = \log_b x$	is equivalent to	$x = b^y$
$y = \log_{10} x$	is equivalent to	$x = 10^y$
$y = \log_e x$	is equivalent to	$x = e^y$

(The log to the base b of x is the exponent to which b must be raised to obtain x.)

Remember: A logarithm is an exponent.

Since the domain of an exponential function includes all real numbers and its range is the set of positive real numbers, the **domain** of a logarithmic function is the set of all positive real numbers and its **range** is the set of all real numbers. Thus, $\log_{10} 3$ is defined, but $\log_{10} 0$ and $\log_{10}(-5)$ are not defined (3 is a logarithmic domain value, but 0 and -5 are not). Typical logarithmic curves are shown in Figure 4.

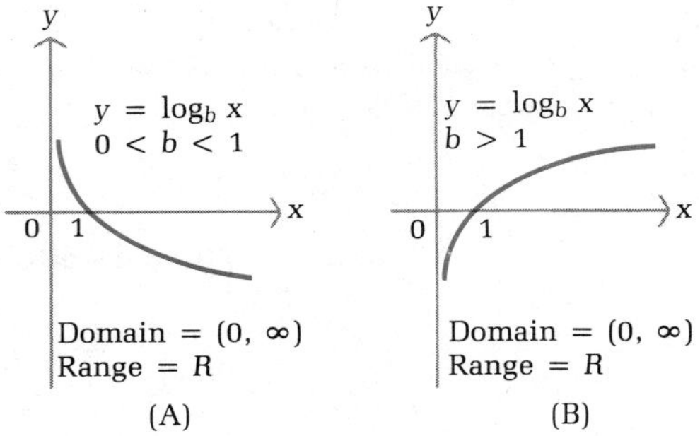

FIGURE 4
Typical logarithmic graphs

◆ FROM LOGARITHMIC TO EXPONENTIAL FORM AND VICE VERSA

We now consider the matter of converting logarithmic forms to equivalent exponential forms and vice versa.

◆ EXAMPLE 8 Change from logarithmic form to exponential form:

(A) $\log_5 25 = 2$	is equivalent to	$25 = 5^2$
(B) $\log_9 3 = \frac{1}{2}$	is equivalent to	$3 = 9^{1/2}$
(C) $\log_2(\frac{1}{4}) = -2$	is equivalent to	$\frac{1}{4} = 2^{-2}$

◆

PROBLEM 8 Change to an equivalent exponential form:

(A) $\log_3 9 = 2$ (B) $\log_4 2 = \frac{1}{2}$ (C) $\log_3(\frac{1}{9}) = -2$ ◆

◆ EXAMPLE 9 Change from exponential form to logarithmic form:

(A) $64 = 4^3$ is equivalent to $\log_4 64 = 3$
(B) $6 = \sqrt{36}$ is equivalent to $\log_{36} 6 = \frac{1}{2}$
(C) $\frac{1}{8} = 2^{-3}$ is equivalent to $\log_2(\frac{1}{8}) = -3$ ◆

PROBLEM 9 Change to an equivalent logarithmic form:

(A) $49 = 7^2$ (B) $3 = \sqrt{9}$ (C) $\frac{1}{3} = 3^{-1}$ ◆

◆ EXAMPLE 10 Find y, b, or x, as indicated.

(A) Find y: $y = \log_4 16$ (B) Find x: $\log_2 x = -3$
(C) Find y: $y = \log_8 4$ (D) Find b: $\log_b 100 = 2$

Solutions (A) $y = \log_4 16$ is equivalent to $16 = 4^y$. Thus,

$$y = 2$$

(B) $\log_2 x = -3$ is equivalent to $x = 2^{-3}$. Thus,

$$x = \frac{1}{2^3} = \frac{1}{8}$$

(C) $y = \log_8 4$ is equivalent to

$$4 = 8^y \quad \text{or} \quad 2^2 = 2^{3y}$$

Thus,

$$3y = 2$$
$$y = \frac{2}{3}$$

(D) $\log_b 100 = 2$ is equivalent to $100 = b^2$. Thus,

$b = 10$ Recall that b cannot be negative. ◆

PROBLEM 10 Find y, b, or x, as indicated.

(A) Find y: $y = \log_9 27$ (B) Find x: $\log_3 x = -1$
(C) Find b: $\log_b 1{,}000 = 3$ ◆

◆ EXAMPLE 11 Graph $y = \log_2(x + 1)$ by converting to an equivalent exponential form first. Do not use a calculator or table.

Solution Changing $y = \log_2(x + 1)$ to an equivalent exponential form, we have

$$x + 1 = 2^y \quad \text{or} \quad x = 2^y - 1$$

Even though x is the independent variable and y is the dependent variable, it is easier to assign y values and solve for x.

x	y
$-\frac{7}{8}$	-3
$-\frac{3}{4}$	-2
$-\frac{1}{2}$	-1
0	0
1	1
3	2
7	3

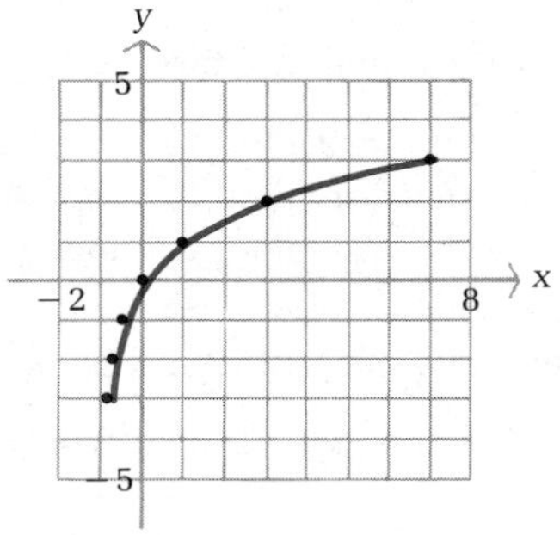

◆

PROBLEM 11 Graph $y = \log_3(x - 1)$ by converting to an equivalent exponential form first. ◆

◆ PROPERTIES OF LOGARITHMIC FUNCTIONS

Logarithmic functions have several very useful properties that follow directly from their definitions. These properties will enable us to convert multiplication problems into addition problems, division problems into subtraction problems, and power and root problems into multiplication problems. We will also be able to solve exponential equations such as $2 = 1.06^n$.

Logarithmic Properties ($b > 0$, $b \neq 1$, $M > 0$, $N > 0$)

1. $\log_b b^x = x$
2. $\log_b MN = \log_b M + \log_b N$
3. $\log_b \frac{M}{N} = \log_b M - \log_b N$
4. $\log_b M^p = p \log_b M$
5. $\log_b M = \log_b N$ if and only if $M = N$
6. $\log_b 1 = 0$

The first property follows directly from the definition of a logarithmic function. Here, we will sketch a proof for property 2. The other properties are established in a similar way. Let

$u = \log_b M$ and $v = \log_b N$

Or, in equivalent exponential form,

$M = b^u$ and $N = b^v$

Now, see if you can provide reasons for each of the following steps:

$$\log_b MN = \log_b b^u b^v = \log_b b^{u+v} = u + v = \log_b M + \log_b N$$

◆ EXAMPLE 12

(A) $\log_b \frac{wx}{yz} = \log_b wx - \log_b yz$

$= \log_b w + \log_b x - (\log_b y + \log_b z)$

$= \log_b w + \log_b x - \log_b y - \log_b z$

(B) $\log_b (wx)^{3/5} = \frac{3}{5} \log_b wx = \frac{3}{5}(\log_b w + \log_b x)$ ◆

PROBLEM 12

Write in simpler logarithmic forms, as in Example 12.

(A) $\log_b \frac{R}{ST}$ (B) $\log_b \left(\frac{R}{S}\right)^{2/3}$ ◆

The following examples and problems, though somewhat artificial, will give you additional practice in using basic logarithmic properties.

◆ EXAMPLE 13

Find x so that: $\frac{3}{2} \log_b 4 - \frac{2}{3} \log_b 8 + \log_b 2 = \log_b x$

Solution

$\frac{3}{2} \log_b 4 - \frac{2}{3} \log_b 8 + \log_b 2 = \log_b x$

$\log_b 4^{3/2} - \log_b 8^{2/3} + \log_b 2 = \log_b x$ Property 4

$\log_b 8 - \log_b 4 + \log_b 2 = \log_b x$

$\log_b \frac{8 \cdot 2}{4} = \log_b x$ Properties 2 and 3

$\log_b 4 = \log_b x$

$x = 4$ Property 5 ◆

PROBLEM 13

Find x so that: $3 \log_b 2 + \frac{1}{2} \log_b 25 - \log_b 20 = \log_b x$ ◆

◆ EXAMPLE 14

Solve: $\log_{10} x + \log_{10}(x + 1) = \log_{10} 6$

Solution

$\log_{10} x + \log_{10}(x + 1) = \log_{10} 6$

$\log_{10} x(x + 1) = \log_{10} 6$ Property 2

$x(x + 1) = 6$ Property 5

$x^2 + x - 6 = 0$ Solve by factoring.

$(x + 3)(x - 2) = 0$

$x = -3, 2$

We must exclude $x = -3$, since the domain of the function $\log_{10}(x + 1)$ is $x > -1$ or $(-1, \infty)$; hence, $x = 2$ is the only solution. ◆

PROBLEM 14

Solve: $\log_3 x + \log_3(x - 3) = \log_3 10$ ◆

◆ CALCULATOR EVALUATION OF COMMON AND NATURAL LOGARITHMS

Of all possible logarithmic bases, the base e and the base 10 are used almost exclusively. Before we can use logarithms in certain practical problems, we need to be able to approximate the logarithm of any positive number either to base 10 or to base e. And conversely, if we are given the logarithm of a number to base 10 or base e, we need to be able to approximate the number. Historically, tables were used for this purpose, but now calculators make computations faster and far more accurate.

Common logarithms (also called **Briggsian logarithms**) are logarithms with base 10. **Natural logarithms** (also called **Napierian logarithms**) are logarithms with base e. Most scientific calculators have a key labeled "log" (or "LOG") and a key labeled "ln" (or "LN"). The former represents a common (base 10) logarithm and the latter a natural (base e) logarithm. In fact, "log" and "ln" are both used extensively in mathematical literature, and whenever you see either used in this book without a base indicated they will be interpreted as follows:

Logarithmic Notation

COMMON LOGARITHM: $\log x = \log_{10} x$
NATURAL LOGARITHM: $\ln x = \log_e x$

Finding the common or natural logarithm using a scientific calculator is very easy: You simply enter a number from the domain of the function and push [log] or [ln].

◆ EXAMPLE 15 Use a scientific calculator to evaluate each to six decimal places:

(A) log 3,184 (B) ln 0.000 349 (C) log(−3.24)

Solutions

	Enter	Press	Display
(A)	3184	[log]	3.502973
(B)	0.000 349	[ln]	−7.960439
(C)	−3.24	[log]	Error

An error is indicated in part C because −3.24 is not in the domain of the log function. [*Note:* The manner in which error messages are displayed varies from one brand of calculator to the next.] ◆

PROBLEM 15 Use a scientific calculator to evaluate each to six decimal places:

(A) log 0.013 529 (B) ln 28.693 28 (C) ln(−0.438) ◆

We now turn to the second problem mentioned above: Given the logarithm of a number, find the number. We make direct use of the logarithmic–exponential relationships, which follow from the definition of logarithmic function given at the beginning of this section.

Logarithmic–Exponential Relationships

$\log x = y$	is equivalent to	$x = 10^y$
$\ln x = y$	is equivalent to	$x = e^y$

◆ EXAMPLE 16 Find x to four decimal places, given the indicated logarithm:

(A) $\log x = -2.315$ (B) $\ln x = 2.386$

Solutions (A) $\log x = -2.315$ Change to equivalent exponential form.

$x = 10^{-2.315}$ Evaluate with a calculator.

$x = 0.0048$

(B) $\ln x = 2.386$ Change to equivalent exponential form.

$x = e^{2.386}$ Evaluate with a calculator.

$x = 10.8699$ ◆

PROBLEM 16 Find x to four decimal places, given the indicated logarithm:

(A) $\ln x = -5.062$ (B) $\log x = 2.0821$ ◆

◆ APPLICATION

A convenient and easily understood way of comparing different investments is to use their **doubling times**—the length of time it takes the value of an investment to double. Logarithm properties, as you will see in Example 17, provide us with just the right tool for solving some doubling-time problems.

◆ EXAMPLE 17 How long (to the next whole year) will it take money to double if it is invested at 20% compounded annually?

Solution We use the compound interest formula discussed in Section 3-1:

$$A = P\left(1 + \frac{r}{m}\right)^{mt} \quad \text{Compound interest}$$

The problem is to find t, given $r = 0.20$, $m = 1$, and $A = 2P$; that is,

$$2P = P(1 + 0.2)^t$$
$$2 = 1.2^t$$
$$1.2^t = 2$$

$$\ln 1.2^t = \ln 2$$ Solve for t by taking the natural or common logarithm of both sides (we choose the natural logarithm).

$$t \ln 1.2 = \ln 2$$ Property 4

$$t = \frac{\ln 2}{\ln 1.2} \qquad \text{Use a calculator. [Note: } (\ln 2)/(\ln 1.2) \neq \ln 2 - \ln 1.2]$$

$$= 3.8 \text{ years}$$

$$\approx 4 \text{ years} \qquad \text{To the next whole year}$$

When interest is paid at the end of 3 years, the money will not be doubled; when paid at the end of 4 years, the money will be slightly more than doubled. ◆

PROBLEM 17 How long (to the next whole year) will it take money to double if it is invested at 13% compounded annually? ◆

It is interesting and instructive to graph the doubling times for various rates compounded annually. We proceed as follows:

$$A = P(1 + r)^t$$

$$2P = P(1 + r)^t$$

$$2 = (1 + r)^t$$

$$(1 + r)^t = 2$$

$$\ln(1 + r)^t = \ln 2$$

$$t \ln(1 + r) = \ln 2$$

$$t = \frac{\ln 2}{\ln(1 + r)}$$

Figure 5 shows the graph of this equation (doubling time in years) for interest rates compounded annually from 1% to 70% (expressed as decimals). Note the dramatic change in doubling time as rates change from 1% to 20% (from 0.01 to 0.20).

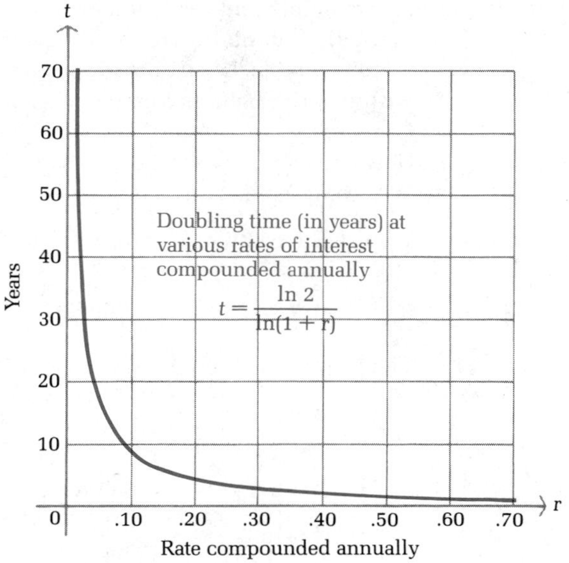

FIGURE 5

Answers to Matched Problems

8. (A) $9 = 3^2$ (B) $2 = 4^{1/2}$ (C) $\frac{1}{9} = 3^{-2}$
9. (A) $\log_7 49 = 2$ (B) $\log_9 3 = \frac{1}{2}$ (C) $\log_3(\frac{1}{3}) = -1$
10. (A) $y = \frac{3}{2}$ (B) $x = \frac{1}{3}$ (C) $b = 10$
11. $y = \log_3(x - 1)$ is equivalent to $x = 3^y + 1$

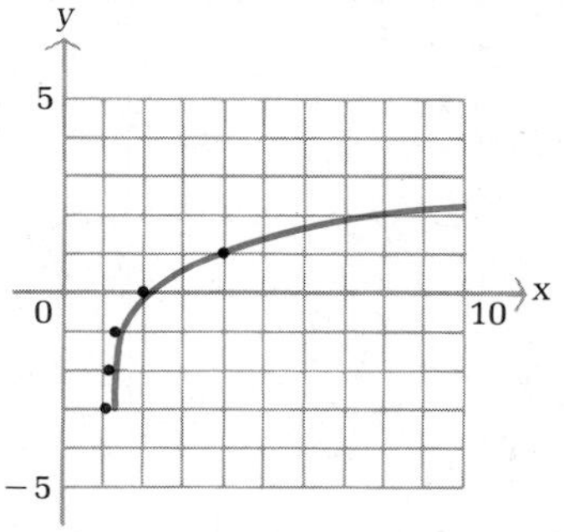

12. (A) $\log_b R - \log_b S - \log_b T$ (B) $\frac{2}{3}(\log_b R - \log_b S)$
13. $x = 2$ 14. $x = 5$
15. (A) −1.868 734 (B) 3.356 663 (C) Not defined
16. (A) 0.0063 (B) 120.8092 17. 6 yr

EXERCISE 3-3

A *Rewrite in exponential form.*

1. $\log_3 27 = 3$ 2. $\log_2 32 = 5$ 3. $\log_{10} 1 = 0$
4. $\log_e 1 = 0$ 5. $\log_4 8 = \frac{3}{2}$ 6. $\log_9 27 = \frac{3}{2}$

Rewrite in logarithmic form.

7. $49 = 7^2$ 8. $36 = 6^2$ 9. $8 = 4^{3/2}$
10. $9 = 27^{2/3}$ 11. $A = b^u$ 12. $M = b^x$

Evaluate each of the following:

13. $\log_{10} 10^3$ 14. $\log_{10} 10^{-5}$ 15. $\log_2 2^{-3}$
16. $\log_3 3^5$ 17. $\log_{10} 1{,}000$ 18. $\log_6 36$

Write in terms of simpler logarithmic forms, as in Example 12.

19. $\log_b \dfrac{P}{Q}$ 20. $\log_b FG$ 21. $\log_b L^5$
22. $\log_b w^{15}$ 23. $\log_b \dfrac{p}{qrs}$ 24. $\log_b PQR$

B *Find x, y, or b.*

25. $\log_3 x = 2$ 26. $\log_2 x = 2$ 27. $\log_7 49 = y$
28. $\log_3 27 = y$ 29. $\log_b 10^{-4} = -4$ 30. $\log_b e^{-2} = -2$

31. $\log_4 x = \frac{1}{2}$
32. $\log_{25} x = \frac{1}{2}$
33. $\log_{1/3} 9 = y$
34. $\log_{49}(\frac{1}{7}) = y$
35. $\log_b 1{,}000 = \frac{3}{2}$
36. $\log_b 4 = \frac{2}{3}$

Write in terms of simpler logarithmic forms, going as far as you can with logarithmic properties (see Example 12).

37. $\log_b \dfrac{x^5}{y^3}$
38. $\log_b x^2y^3$
39. $\log_b \sqrt[3]{N}$
40. $\log_b \sqrt[5]{Q}$
41. $\log_b(x^2 \sqrt[3]{y})$
42. $\log_b \sqrt[3]{\dfrac{x^2}{y}}$
43. $\log_b(50 \cdot 2^{-0.2t})$
44. $\log_b(100 \cdot 1.06^t)$
45. $\log_b P(1 + r)^t$
46. $\log_e Ae^{-0.3t}$
47. $\log_e 100e^{-0.01t}$
48. $\log_{10}(67 \cdot 10^{-0.12x})$

Find x.

49. $\log_b x = \frac{2}{3} \log_b 8 + \frac{1}{2} \log_b 9 - \log_b 6$
50. $\log_b x = \frac{2}{3} \log_b 27 + 2 \log_b 2 - \log_b 3$
51. $\log_b x = \frac{3}{2} \log_b 4 - \frac{2}{3} \log_b 8 + 2 \log_b 2$
52. $\log_b x = 3 \log_b 2 + \frac{1}{2} \log_b 25 - \log_b 20$
53. $\log_b x + \log_b(x - 4) = \log_b 21$
54. $\log_b(x + 2) + \log_b x = \log_b 24$
55. $\log_{10}(x - 1) - \log_{10}(x + 1) = 1$
56. $\log_{10}(x + 6) - \log_{10}(x - 3) = 1$

Graph by converting to exponential form first.

57. $y = \log_2(x - 2)$
58. $y = \log_3(x + 2)$

In Problems 59 and 60, evaluate to five decimal places using a scientific calculator.

59. (A) log 3,527.2 (B) log 0.006 913 2
(C) ln 277.63 (D) ln 0.040 883
60. (A) log 72.604 (B) log 0.033 041
(C) ln 40,257 (D) ln 0.005 926 3

In Problems 61 and 62, find x to four decimal places.

61. (A) $\log x = 1.1285$ (B) $\log x = -2.0497$
(C) $\ln x = 2.7763$ (D) $\ln x = -1.8879$
62. (A) $\log x = 2.0832$ (B) $\log x = -1.1577$
(C) $\ln x = 3.1336$ (D) $\ln x = -4.3281$

Evaluate each of the following to three decimal places using a calculator.

63. $n = \dfrac{\log 2}{\log 1.15}$
64. $n = \dfrac{\log 2}{\log 1.12}$
65. $n = \dfrac{\ln 3}{\ln 1.15}$
66. $n = \dfrac{\ln 4}{\ln 1.2}$
67. $x = \dfrac{\ln 0.5}{-0.21}$
68. $x = \dfrac{\ln 0.1}{-0.0025}$

Graph, using a calculator.

69. $y = \ln x$
70. $y = -\ln x$
71. $y = |\ln x|$
72. $y = \ln|x|$
73. $y = 2\ln(x + 2)$
74. $y = 2\ln x + 2$
75. $y = 4\ln x - 3$
76. $y = 4\ln(x - 3)$

C

77. Find the logarithm of 1 for any permissible base.
78. Why is 1 not a suitable logarithmic base? [*Hint:* Try to find $\log_1 8$.]
79. Write $\log_{10} y - \log_{10} c = 0.8x$ in an exponential form that is free of logarithms.
80. Write $\log_e x - \log_e 25 = 0.2t$ in an exponential form that is free of logarithms.

APPLICATIONS

Business & Economics

81. *Doubling time.* How long (to the next whole year) will it take money to double if it is invested at 6% interest compounded annually?
82. *Doubling time.* How long (to the next whole year) will it take money to double if it is invested at 3% interest compounded annually?
83. *Tripling time.* Write a formula similar to the doubling-time formula in Figure 5 for the tripling time of money invested at 100r% interest compounded annually.
84. *Tripling time.* How long (to the next whole year) will it take money to triple if it is invested at 15% interest compounded annually?

Life Sciences

85. *Sound intensity—decibels.* Because of the extraordinary range of sensitivity of the human ear (a range of over 1,000 million millions to 1), it is helpful to use a logarithmic scale, rather than an absolute scale, to measure sound intensity over this range. The unit of measure is called the *decibel*, after the inventor of the telephone, Alexander Graham Bell. If we let N be the number of decibels, I the power of the sound in question (in watts per square centimeter), and I_0 the power of sound just below the threshold of hearing (approximately 10^{-16} watt per square centimeter), then

$$I = I_0 10^{N/10}$$

Show that this formula can be written in the form

$$N = 10 \log \frac{I}{I_0}$$

86. *Sound intensity—decibels.* Use the formula in Problem 85 (with $I_0 = 10^{-16}$ watt/cm^2) to find the decibel ratings of the following sounds:
(A) Whisper: 10^{-13} watt/cm^2
(B) Normal conversation: 3.16×10^{-10} watt/cm^2
(C) Heavy traffic: 10^{-8} watt/cm^2
(D) Jet plane with afterburner: 10^{-1} watt/cm^2

Social Sciences

87. *World population.* If the world population is now 4 billion (4×10^9) people and if it continues to grow at 2% per year compounded annually, how long will it be before there is only 1 square yard of land per person? (The earth contains approximately 1.68×10^{14} square yards of land.)

88. *Archaeology—carbon-14 dating.* The radioactive carbon-14 (^{14}C) in an organism at the time of its death decays according to the equation

$$A = A_0 e^{-0.000124t}$$

where t is time in years and A_0 is the amount of ^{14}C present at time $t = 0$. (See Example 6 in Section 3-2.) Estimate the age of a skull uncovered in an archaeological site if 10% of the original amount of ^{14}C is still present. [*Hint:* Find t such that $A = 0.1A_0$.]

SECTION 3-4 Chapter Review

Important Terms and Symbols

3-1 *Exponential Functions.* Exponential function; base; domain; range; basic graphs; horizontal asymptote; basic properties; exponential growth; exponential decay; doubling-time growth model; half-life decay model; compound interest

$$b^x; \quad P = P_0 2^{t/d}; \quad A = A_0(\tfrac{1}{2})^{t/h}; \quad A = P\left(1 + \frac{r}{m}\right)^{mt}$$

3-2 *The Exponential Function with Base e.* Irrational number e; exponential function with base e; exponential growth; exponential decay; continuous compound interest

$$N = N_0 e^{kt}; \quad A = A_0 e^{-kt}; \quad A = Pe^{rt}$$

3-3 *Logarithmic Functions.* Logarithmic function; base; domain; range; exponential form; properties; common logarithm; natural logarithm; calculator evaluation; doubling time

$$\log_b x; \quad \log x; \quad \ln x$$

EXERCISE 3-4 Chapter Review

Work through all the problems in this chapter review and check your answers in the back of the book. (Answers to all review problems are there.) Where weaknesses show up, review appropriate sections in the text.

A

1. Write in logarithmic form using base e: $u = e^v$

2. Write in logarithmic form using base 10: $x = 10^y$

3. Write in exponential form using base e: $\ln M = N$
4. Write in exponential form using base 10: $\log u = v$

Simplify.

5. $\dfrac{5^{x+4}}{5^{4-x}}$

6. $\left(\dfrac{e^u}{e^{-u}}\right)^u$

Solve for x exactly without the use of a calculator.

7. $\log_3 x = 2$
8. $\log_x 36 = 2$
9. $\log_2 16 = x$

Solve for x to three decimal places.

10. $10^x = 143.7$
11. $e^x = 503{,}000$
12. $\log x = 3.105$
13. $\ln x = -1.147$

B *Solve for x exactly without the use of a calculator.*

14. $\log(x + 5) = \log(2x - 3)$
15. $2 \ln(x - 1) = \ln(x^2 - 5)$
16. $9^{x-1} = 3^{1+x}$
17. $e^{2x} = e^{x^2-3}$
18. $2x^2e^x = 3xe^x$
19. $\log_{1/3} 9 = x$
20. $\log_x 8 = -3$
21. $\log_9 x = \frac{3}{2}$

Solve for x to four decimal places.

22. $x = 3(e^{1.49})$
23. $x = 230(10^{-0.161})$
24. $\log x = -2.0144$
25. $\ln x = 0.3618$
26. $35 = 7(3^x)$
27. $0.01 = e^{-0.05x}$
28. $8{,}000 = 4{,}000(1.08^x)$
29. $5^{2x-3} = 7.08$

Simplify.

30. $e^x(e^{-x} + 1) - (e^x + 1)(e^{-x} - 1)$
31. $(e^x - e^{-x})^2 - (e^x + e^{-x})(e^x - e^{-x})$

Graph over the indicated interval.

32. $y = 2^{x-1}$; $[-2, 4]$
33. $f(t) = 10e^{-0.08t}$; $t \geq 0$
34. $y = \ln(x + 1)$; $(-1, 10]$

C *Solve Problems 35–38 exactly without the use of a calculator.*

35. $\log x - \log 3 = \log 4 - \log(x + 4)$
36. $\ln(2x - 2) - \ln(x - 1) = \ln x$
37. $\ln(x + 3) - \ln x = 2 \ln 2$
38. $\log 3x^2 = 2 + \log 9x$

39. Write $\ln y = -5t + \ln c$ in an exponential form free of logarithms. Then solve for y in terms of the remaining variables.
40. Explain why 1 cannot be used as a logarithmic base.

The two formulas below will be of use in some of the problems that follow:

$$A = P\left(1 + \frac{r}{m}\right)^{mt} \qquad \text{Compound interest}$$

$$A = Pe^{rt} \qquad \text{Continuous compound interest}$$

Business & Economics

41. *Money growth.* If \$5,000 is invested at 12% compounded weekly, how much (to the nearest cent) will be in the account 6 years from now?

42. *Money growth.* If \$5,000 is invested at 12% compounded continuously, how much (to the nearest cent) will be in the account 6 years from now?

43. *Finance.* Find the tripling time (to the next whole year) for money invested at 15% compounded annually.

44. *Finance.* Find the doubling time (to two decimal places) for money invested at 10% compounded continuously.

Life Sciences

45. *Medicine.* One leukemic cell injected into a healthy mouse will divide into 2 cells in about $\frac{1}{2}$ day. At the end of the day these 2 cells will divide into 4. This doubling continues until 1 billion cells are formed; then the animal dies with leukemic cells in every part of the body.

(A) Write an equation that will give the number N of leukemic cells at the end of t days.

(B) When, to the nearest day, will the mouse die?

46. *Marine biology.* The intensity of light entering water is reduced according to the exponential equation

$$I = I_0e^{-kd}$$

where I is the intensity d feet below the surface, I_0 is the intensity at the surface, and k is the coefficient of extinction. Measurements in the Sargasso Sea in the West Indies have indicated that half of the surface light reaches a depth of 73.6 feet. Find k (to five decimal places), and find the depth (to the nearest foot) at which 1% of the surface light remains.

Social Sciences

47. *Population growth.* Many countries have a population growth rate of 3% (or more) per year. At this rate, how many years (to the nearest tenth of a year) will it take a population to double? Use the annual compounding growth model $P = P_0(1 + r)^t$.

48. *Population growth.* Repeat Problem 47 using the continuous compounding growth model $P = P_0e^{rt}$.

PART TWO

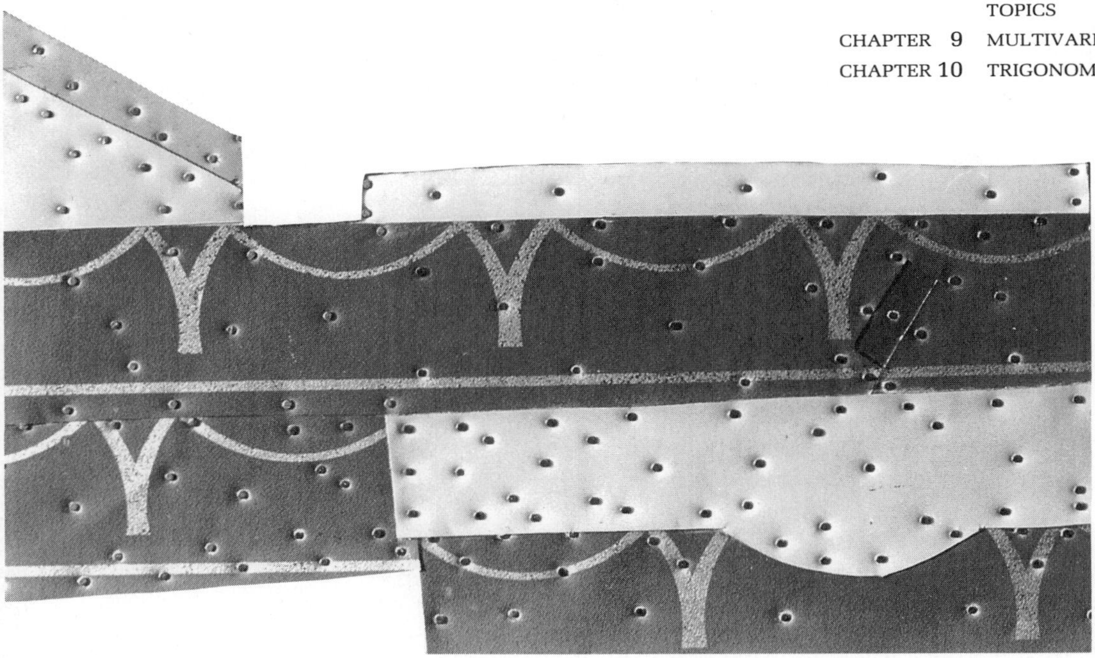

Calculus

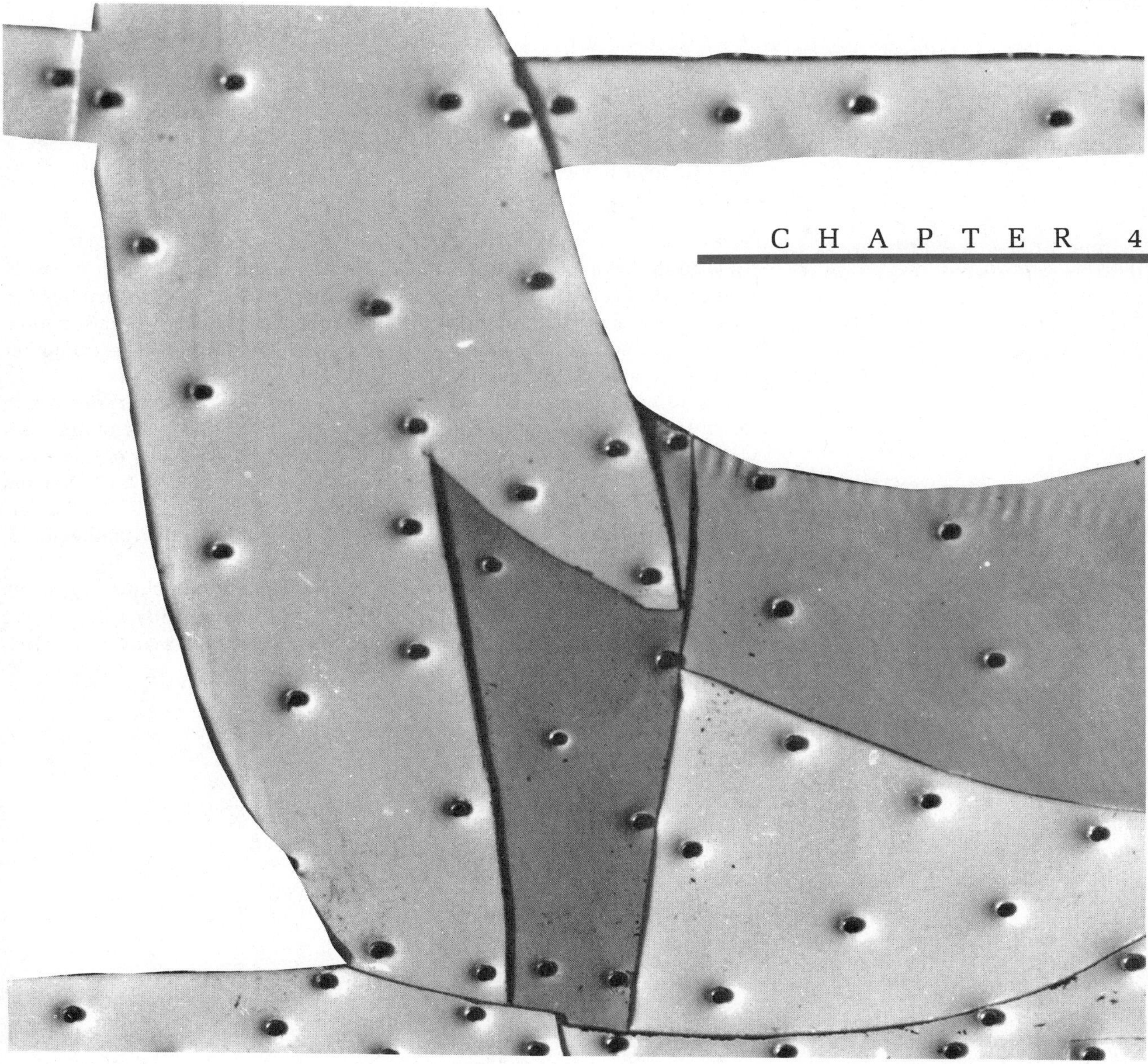

CHAPTER 4

The Derivative

CHAPTER 4

Contents

How do algebra and calculus differ? The two words *static* and *dynamic* probably come as close as any in expressing the difference between the two disciplines. In algebra, we solve equations for a particular value of a variable—a static notion. In calculus, we are interested in how a change in one variable affects another variable —a dynamic notion.

Figure 1 illustrates three basic problems in calculus. It may surprise you to learn that all three problems—as different as they appear—are mathematically related. The solutions to these problems and the discovery of their relationship required the creation of a new kind of mathematics. Isaac Newton (1642–1727) of England and Gottfried Wilhelm von Leibniz (1646–1716) of Germany simultaneously and independently developed this new mathematics, called **the calculus**—it was an idea whose time had come.

In addition to solving the problems described in Figure 1, calculus will enable us to solve many other important problems. Until fairly recently, calculus was used primarily in the physical sciences, but now people in many other disciplines are finding it a useful tool.

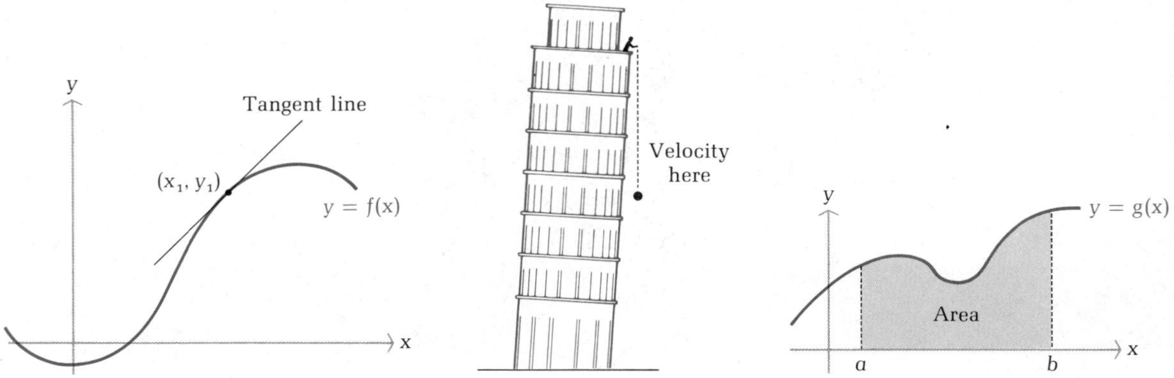

(A) Find the equation of the tangent line at (x_1, y_1) given $y = f(x)$

(B) Find the instantaneous velocity of a falling object

(C) Find the indicated area bounded by $y = g(x)$, $x = a$, $x = b$, and the x axis

FIGURE 1

SECTION 4-1

Limits and Continuity— A Geometric Introduction

- FUNCTIONS AND GRAPHS—A BRIEF REVIEW
- LIMITS
- CONTINUITY
- CONTINUITY PROPERTIES
- APPLICATION

Basic to the study of calculus is the concept of *limit*. This concept helps us to describe in a precise way the behavior of $f(x)$ when x is close to, but not equal to, a particular number c, or when x increases or decreases without bound. Our discussion here will concentrate on concept development and understanding rather than on formal mathematical detail.

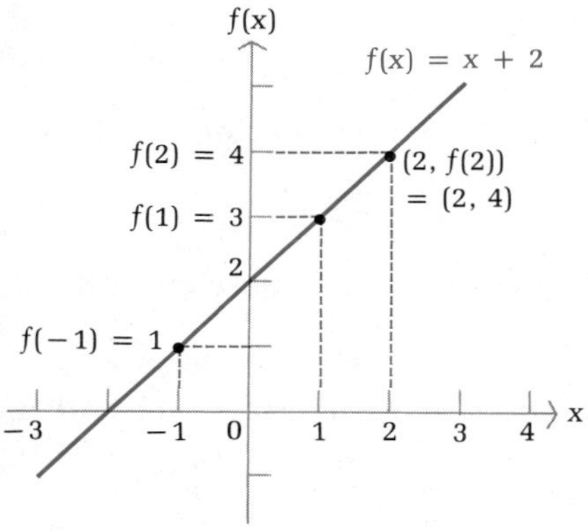

FIGURE 2

◆ FUNCTIONS AND GRAPHS—A BRIEF REVIEW

The graph of the function $y = f(x) = x + 2$ is the graph of the set of all ordered pairs $(x, f(x))$. For example, if $x = 2$, then $f(2) = 4$ and $(2, f(2)) = (2, 4)$ is a point on the graph of f. Figure 2 shows $(-1, f(-1))$, $(1, f(1))$, and $(2, f(2))$ plotted on the graph of f. Notice that the domain values -1, 1, and 2 are associated with the x axis, and the range values $f(-1) = 1$, $f(1) = 3$, and $f(2) = 4$ are associated with the y axis.

Given x, it is sometimes useful to be able to read $f(x)$ directly from the graph of f. Example 1 reviews this process.

◆ EXAMPLE 1

Complete the table below using the given graph of the function g.

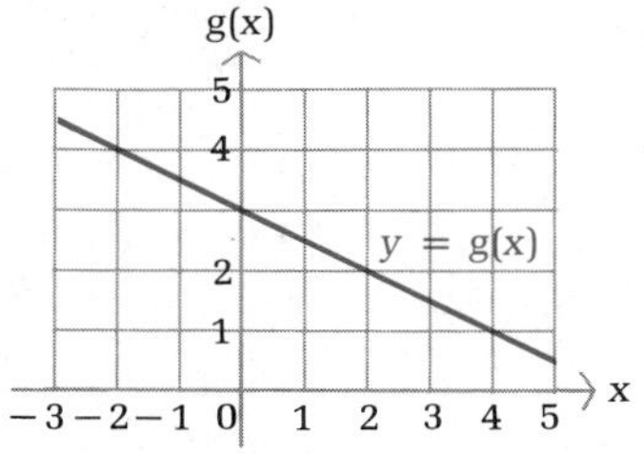

x	g(x)
−2	
1	
3	
4	

Solution

To determine $g(x)$, proceed vertically from the x value on the x axis to the graph of g, then horizontally to the corresponding y value, $g(x)$, on the y axis (as indicated by the dashed lines):

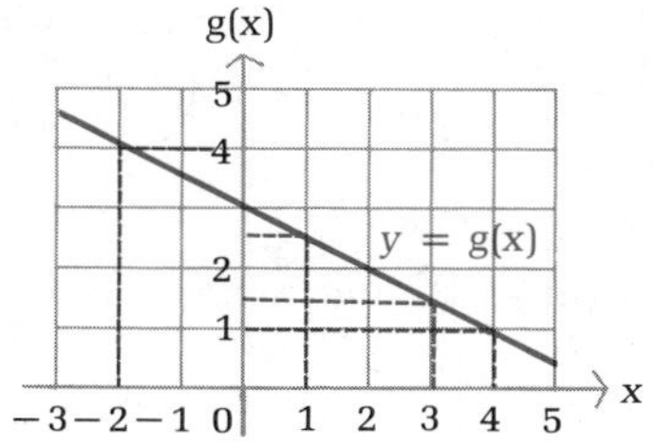

x	g(x)
−2	4.0
1	2.5
3	1.5
4	1.0

◆

PROBLEM 1 Complete the table below using the given graph of the function h.

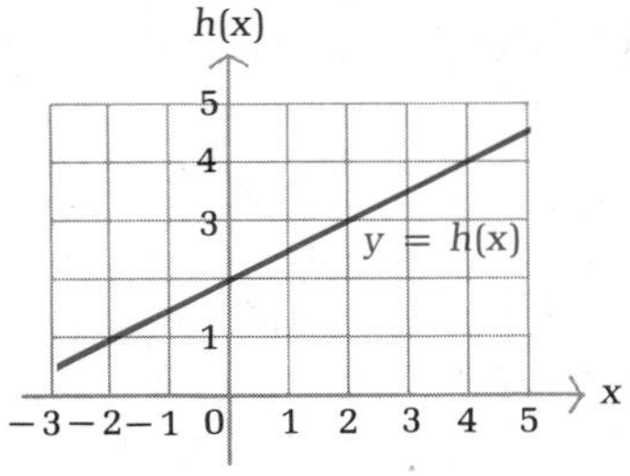

x	$h(x)$
−2	
−1	
0	
1	
2	
3	
4	

◆

◆ LIMITS

We introduce the important notion of *limit* through two examples, after which the limit concept will be defined.

◆ EXAMPLE 2 Let $f(x) = x + 2$. What happens to $f(x)$ when x is chosen closer and closer to 2, but not equal to 2?

Solution We construct a table of values of $f(x)$ for some values of x close to 2 and on either side of 2:

x approaches 2 from the left → 2 ← x approaches 2 from the right

x	1.5	1.8	1.9	1.99	1.999 → 2 ← 2.001	2.01	2.1	2.2	2.5
$f(x)$	3.5	3.8	3.9	3.99	3.999 → ? ← 4.001	4.01	4.1	4.2	4.5

f(x) approaches 4 → 4 ← f(x) approaches 4

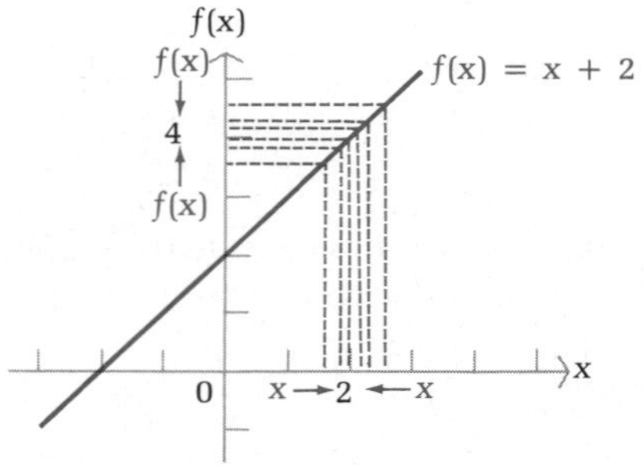

We also draw a graph of f for x near 2, as shown in the margin. Referring to the table and the graph, we see that $f(x)$ approaches 4 as x approaches 2 from either side of 2. We say that "the limit of $x + 2$ as x approaches 2 is 4" and write

$$\lim_{x \to 2} (x + 2) = 4 \qquad \text{or} \qquad x + 2 \to 4 \quad \text{as} \quad x \to 2$$

This means that we can make $f(x) = x + 2$ as close to 4 as we like by restricting x to a sufficiently small interval about 2 but excluding 2.

Also note that $f(2) = 4$. Thus, the value of the function at 2 and the limit of the function at 2 are the same. That is,

$$\lim_{x \to 2} (x + 2) = f(2)$$

Geometrically, this means there is no break, or hole, in the graph of f at $x = 2$.

◆

PROBLEM 2 Let $f(x) = x + 1$.

(A) Complete the following table:

x	0.9	0.99	0.999 → 1 ← 1.001	1.01	1.1
$f(x)$	?	?	? → ? ← ?	?	?

(B) Referring to the table in part A, find

$$\lim_{x \to 1} (x + 1)$$

(That is, what value does $x + 1$ approach as x approaches 1 from either side of 1, but is not equal to 1?)

(C) Graph $f(x) = x + 1$ for $-1 \leq x \leq 4$.

(D) Referring to the graph in part C, find

$$\lim_{x \to 0} (x + 1) \qquad \text{and} \qquad \lim_{x \to 3} (x + 1)$$ ◆

Example 2 and Problem 2 were fairly obvious. The next example is a little less obvious.

◆ EXAMPLE 3 Let

$$g(x) = \frac{x^2 - 4}{x - 2} \qquad x \neq 2$$

Even though the function is not defined when $x = 2$ (both the numerator and denominator are 0), we can still ask how $g(x)$ behaves when x is near 2, but not equal to 2. Can you guess what happens to $g(x)$ as x approaches 2? The numerator tending to 0 is a force pushing the fraction toward 0. The denominator tending to 0 is another force pushing the fraction toward larger values. How do these two forces balance out?

Solution We could proceed by constructing a table of values to the left and to the right of 2, as in Example 2. Instead, we take advantage of some algebraic simplification using factoring:

$$g(x) = \frac{x^2 - 4}{x - 2} = \frac{(x - 2)(x + 2)}{x - 2} = x + 2 \qquad x \neq 2$$

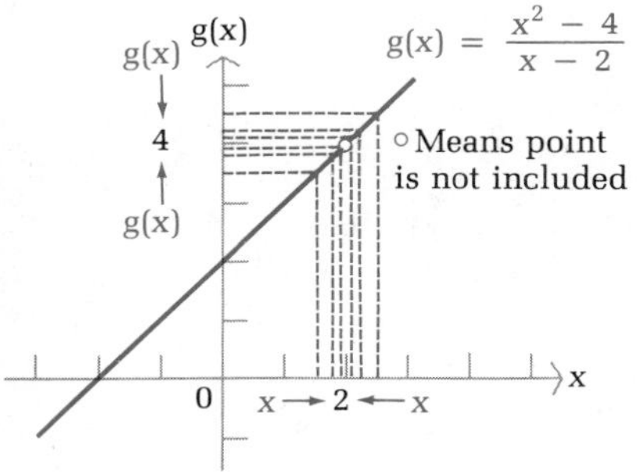

Thus, we see that the graph of function g is the same as the graph of function f in Example 2, except that the graph of g has a hole at the point with coordinates (2, 4), as shown in the margin.

Since the behavior of $(x^2 - 4)/(x - 2)$ for x near 2, but not equal to 2, is the same as the behavior of $x + 2$ for x near 2, but not equal to 2, we have

$$\lim_{x \to 2} \frac{x^2 - 4}{x - 2} = \lim_{x \to 2} (x + 2) = 4$$

And we see that the limit of the function g at 2 exists even though the function is not defined there (the graph has a hole at $x = 2$). ◆

PROBLEM 3 Let

$$g(x) = \frac{x^2 - 1}{x - 1} \qquad x \neq 1$$

[Before proceeding with parts A and B, try to guess what g(x) approaches as x approaches 1, but does not equal 1. Experiment a little with a calculator.]

(A) Graph g for $-1 \leq x \leq 4$. [*Hint:* Proceed as in Example 3.]
(B) Using the graph in part A, find

$$\lim_{x \to 1} \frac{x^2 - 1}{x - 1}$$

◆

We now present an informal definition of the important concept of limit:

Limit

We write

$$\lim_{x \to c} f(x) = L \qquad \text{or} \qquad f(x) \to L \quad \text{as} \quad x \to c$$

if the functional value $f(x)$ is close to the single real number L whenever x is close to, but not equal to, c (on either side of c).

[*Note:* The existence of a limit at c has nothing to do with the value of the function at c. In fact, c may not even be in the domain of f (see Examples 2 and 3). However, the function must be defined on both sides of c.]

The next example involves the **absolute value function:**

$$|x| = \begin{cases} x & \text{if } x > 0 \\ 0 & \text{if } x = 0 \\ -x & \text{if } x < 0 \end{cases} \qquad \begin{aligned} |3| &= 3 \\ |0| &= 0 \\ |-2| &= -(-2) = 2 \end{aligned}$$

◆ EXAMPLE 4 Let $h(x) = |x|/x$. Explore the behavior of $h(x)$ for x near 0, but not equal to 0, using a table and a graph. Find $\lim_{x \to 0} h(x)$, if it exists.*

* "$\lim_{x \to 0} h(x)$" is another way of writing "$\displaystyle\lim_{x \to 0} h(x)$."

Solution The function h is defined for all real numbers except 0. For example,

$$h(-2) = \frac{|-2|}{-2} = \frac{2}{-2} = -1$$

$$h(0) = \frac{|0|}{0} = \frac{0}{0} \qquad \text{Not defined}$$

$$h(2) = \frac{|2|}{2} = \frac{2}{2} = 1$$

In general, $h(x)$ is -1 for all negative x and 1 for all positive x. The following table and graph illustrate the behavior of $h(x)$ for x near 0:

			x approaches 0 from the left →			0	← x approaches 0 from the right				
x	-2	-1	-0.1	-0.01	$-0.001 \rightarrow$	0	$\leftarrow 0.001$	0.01	0.1	1	2
$h(x)$	-1	-1	-1	-1	$-1 \rightarrow -1$		$1 \leftarrow 1$	1	1	1	1
					$h(x) = -1 \rightarrow -1$		$1 \leftarrow h(x) = 1$				

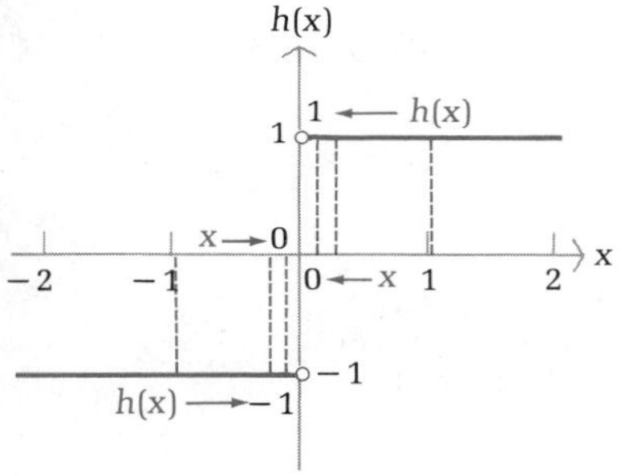

When x is near 0 (on either side of 0), is $h(x)$ near one specific number? The answer is "No," because $h(x)$ is -1 for $x < 0$ and 1 for $x > 0$. Consequently, we say that

$$\lim_{x \to 0} \frac{|x|}{x} \quad \text{does not exist}$$

Thus, neither $h(x)$ nor the limit of $h(x)$ exist at $x = 0$. However, the limit from the left and the limit from the right both exist at 0, but they are not equal. (We will discuss this further below.) ◆

PROBLEM 4 Graph

$$h(x) = \frac{x-2}{|x-2|} \qquad -1 \le x \le 5$$

and find $\lim_{x \to 2} h(x)$, if it exists. ◆

In Example 4, we mentioned the "limit from the left" and the "limit from the right." These phrases suggest the useful notion of **one-sided limits**, which we now define.

We write

$$\lim_{x \to c^-} f(x) = K \qquad c^- \text{ means } x < c \quad (c^- \neq -c)$$

and call K the **limit from the left** (or **left-hand limit**) if $f(x)$ is close to K whenever x is close to c, but to the left of c on the real number line. We write

$$\lim_{x \to c^+} f(x) = L \qquad c^+ \text{ means } x > c \quad (c^+ \neq +c)$$

and call L the **limit from the right** (or **right-hand limit**) if $f(x)$ is close to L whenever x is close to c, but to the right of c on the real number line.

We now make the following important observation:

On the Existence of a Limit

In order for a limit to exist, the limit from the left and the limit from the right must exist and be equal.

In Example 4,

$$\lim_{x\to 0^-} \frac{|x|}{x} = -1 \qquad \text{and} \qquad \lim_{x\to 0^+} \frac{|x|}{x} = 1$$

Since the left- and right-hand limits are not the same,

$$\lim_{x\to 0} \frac{|x|}{x} \quad \text{does not exist}$$

◆ EXAMPLE 5

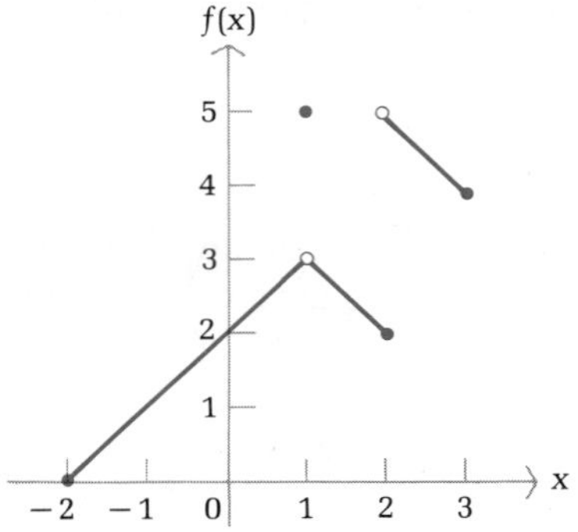

Given the graph of the function f shown in the margin, we discuss the behavior of $f(x)$ for x near -1, 1, 2, and 3:

(A) Behavior of $f(x)$ for x near -1:

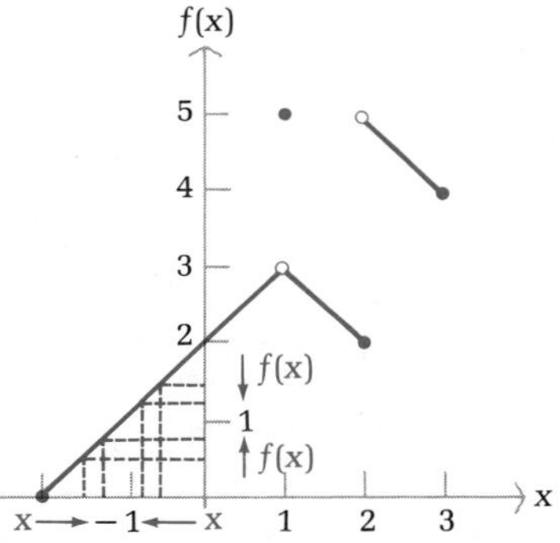

$$\lim_{x\to -1^-} f(x) = 1$$
$$\lim_{x\to -1^+} f(x) = 1$$
$$\lim_{x\to -1} f(x) = 1$$
$$f(-1) = 1$$

(B) Behavior of $f(x)$ for x near 1:

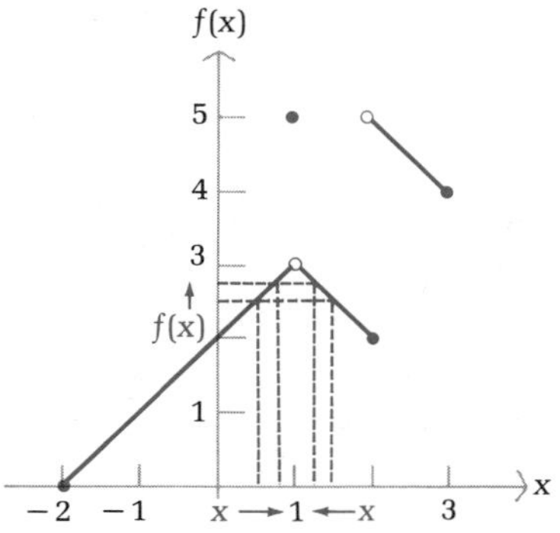

$$\lim_{x\to 1^-} f(x) = 3$$
$$\lim_{x\to 1^+} f(x) = 3$$
$$\lim_{x\to 1} f(x) = 3$$
$$f(1) = 5$$

(C) Behavior of $f(x)$ for x near 2:

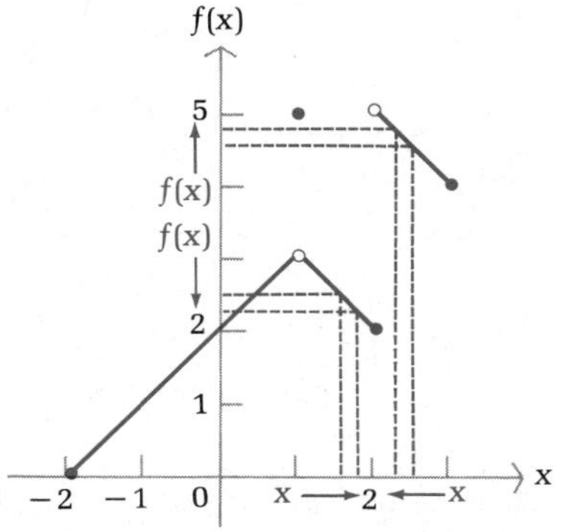

$\lim_{x \to 2^-} f(x) = 2$

$\lim_{x \to 2^+} f(x) = 5$

$\lim_{x \to 2} f(x)$ does not exist

$f(2) = 2$

(D) Behavior of $f(x)$ for x near 3:

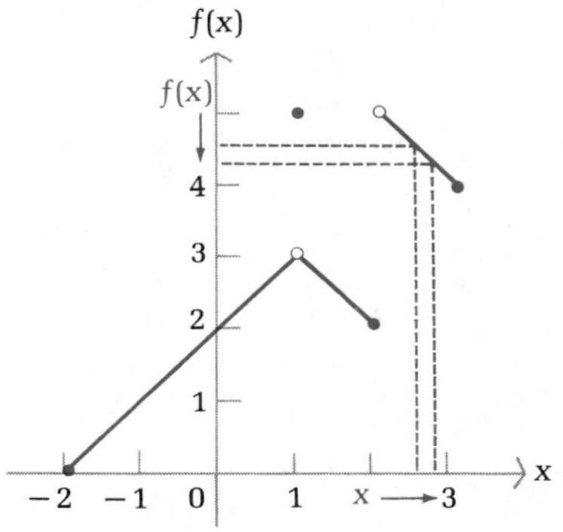

$\lim_{x \to 3^-} f(x) = 4$

$\lim_{x \to 3^+} f(x)$ does not exist

f is not defined for $x > 3$

$\lim_{x \to 3} f(x)$ does not exist

$f(3) = 4$

◆

PROBLEM 5

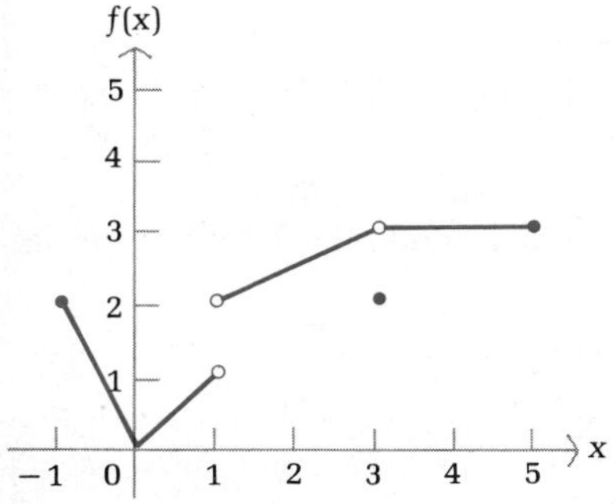

Given the graph of the function f shown in the margin, discuss the following, as we did in Example 5:

(A) Behavior of $f(x)$ for x near -1
(B) Behavior of $f(x)$ for x near 0
(C) Behavior of $f(x)$ for x near 1
(D) Behavior of $f(x)$ for x near 3

◆

◆ CONTINUITY

Compare the graphs from Examples 2–4, which are repeated in Figure 3 (at the top of the next page). Notice that two of the graphs are broken; that is, they cannot be drawn without lifting a pen off the paper. Informally, a function is *continuous over an interval* if its graph over the interval can be drawn without removing a pen from the paper. A function whose graph is broken (disconnected) at $x = c$ is said to be *discontinuous* at $x = c$. Function f (Figure 3A) is continuous for all x. Function g (Figure 3B) is discontinuous at $x = 2$, but is continuous over any interval that does not include 2. Function h (Figure 3C) is discontinuous at $x = 0$, but is continuous over any interval that does not include 0.

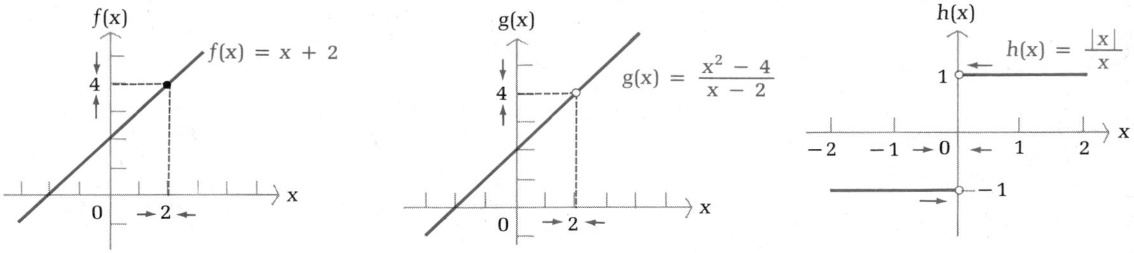

(A) $\lim_{x \to 2} f(x) = 4$
$f(2) = 4$

(B) $\lim_{x \to 2} g(x) = 4$
$g(2)$ is not defined

(C) $\lim_{x \to 0} h(x)$ does not exist
$h(0)$ is not defined

FIGURE 3

Most graphs of natural phenomena are continuous, whereas many graphs in business and economics applications have discontinuities. Figure 4A illustrates temperature variation over a 24 hour period—a continuous phenomenon. Figure 4B illustrates warehouse inventory over a 1 week period—a discontinuous phenomenon.

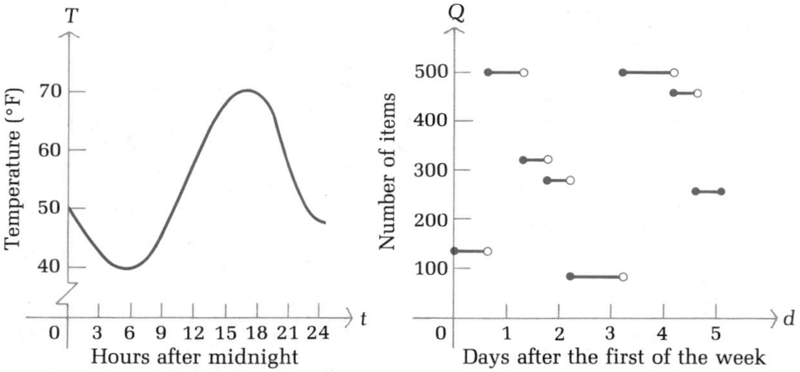

(A) Temperature for a 24 hour period

(B) Inventory in a warehouse during 1 week

FIGURE 4

The preceding discussion, examples, and figures lead to the following formal definition of continuity:

Continuity

A function f is **continuous at the point $x = c$** if

1. $\lim_{x \to c} f(x)$ exists 2. $f(c)$ exists 3. $\lim_{x \to c} f(x) = f(c)$

A function is **continuous on the open interval (a, b)** if it is continuous at each point on the interval.

If one or more of the three conditions in the definition fails, then the function is **discontinuous** at $x = c$.

◆ EXAMPLE 6 Using the definition of continuity, discuss the continuity of each function at the indicated point(s).

(A) $f(x) = x + 2$ at $x = 2$ (B) $g(x) = \dfrac{x^2 - 4}{x - 2}$ at $x = 2$

(C) $h(x) = \dfrac{|x|}{x}$ at $x = 0$ and at $x = 1$

Solutions

(A) f is continuous at $x = 2$, since

$$\lim_{x \to 2} f(x) = 4 = f(2)$$

(See Example 2 and Figure 3A.)

(B) g is not continuous at $x = 2$, since $g(2) = 0/0$ is not defined. (See Example 3 and Figure 3B.)

(C) h is not continuous at $x = 0$, since $h(0) = |0|/0$ is not defined; also, $\lim_{x \to 0} h(x)$ does not exist.

h is continuous at $x = 1$, since

$$\lim_{x \to 1} \frac{|x|}{x} = 1 = h(1)$$

(See Example 4 and Figure 3C.) ◆

PROBLEM 6 Using the definition of continuity, discuss the continuity of each function at the indicated point(s). (Compare with Problems 2–4.)

(A) $f(x) = x + 1$ at $x = 1$ (B) $g(x) = \dfrac{x^2 - 1}{x - 1}$ at $x = 1$

(C) $h(x) = \dfrac{x - 2}{|x - 2|}$ at $x = 2$ and at $x = 0$ ◆

We can also talk about one-sided continuity, just as we talked about one-sided limits. For example, a function is said to be **continuous on the right** at $x = c$ if $\lim_{x \to c^+} f(x) = f(c)$ and **continuous on the left** at $x = c$ if $\lim_{x \to c^-} f(x) = f(c)$. A function is **continuous on the closed interval [*a*, *b*]** if it is continuous on the open interval (a, b) and is continuous on the right at a and continuous on the left at b.

Figure 5A (at the top of the next page) illustrates a function that is continuous on the closed interval $[-1, 1]$. Figure 5B illustrates a function that is continuous on a half-closed interval $[0, \infty)$.

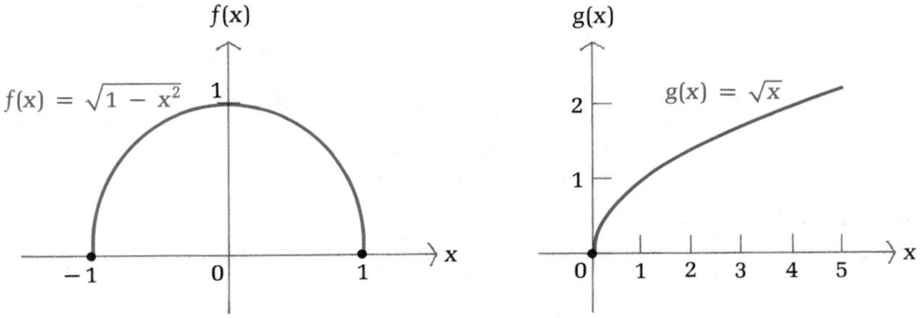

(A) f is continuous on the closed interval $[-1, 1]$

(B) g is continuous on the half-closed interval $[0, \infty)$

FIGURE 5
Continuity on closed and half-closed intervals

◆ CONTINUITY PROPERTIES

Functions have useful continuity properties:

> **If two functions are continuous on the same interval, then their sum, difference, product, and quotient are continuous on the same interval, except for values of x that make a denominator 0.**

These properties, along with Theorem 1 below, enable us to determine intervals of continuity for some important classes of functions without having to look at their graphs or use the three conditions in the definition.

THEOREM 1

Continuity Properties of Some Specific Functions

(A) A constant function $f(x) = k$, where k is a constant, is continuous for all x.
$f(x) = 7$ is continuous for all x.

(B) For n a positive integer, $f(x) = x^n$ is continuous for all x.
$f(x) = x^5$ is continuous for all x.

(C) A polynomial function is continuous for all x.
$2x^3 - 3x^2 + x - 5$ is continuous for all x.

(D) A rational function is continuous for all x except those values that make a denominator 0.
$\dfrac{x^2 + 1}{x - 1}$ is continuous for all x except $x = 1$, a value that makes the denominator 0.

(E) For n an odd positive integer greater than 1, $\sqrt[n]{f(x)}$ is continuous wherever $f(x)$ is continuous.
$\sqrt[3]{x^2}$ is continuous for all x.

(F) For n an even positive integer, $\sqrt[n]{f(x)}$ is continuous wherever $f(x)$ is continuous and nonnegative.
$\sqrt[4]{x}$ is continuous on the interval $[0, \infty)$.

Notice that Theorem 1C follows from parts A, B, and the general continuity properties stated above. Also, note that part D follows from part C and the general continuity properties, since a rational function is a function that can be expressed as the quotient of two polynomials.

◆ EXAMPLE 7 Using Theorem 1 and the general properties of continuity, determine where each function is continuous.

(A) $f(x) = x^2 - 2x + 1$

(B) $f(x) = \dfrac{x}{(x+2)(x-3)}$

(C) $f(x) = \sqrt[3]{x^2 - 4}$

(D) $f(x) = \sqrt{x - 2}$

Solutions

(A) Since f is a polynomial function, f is continuous for all x.

(B) Since f is a rational function, f is continuous for all x except -2 and 3 (values that make the denominator 0).

(C) The polynomial function $x^2 - 4$ is continuous for all x. Since $n = 3$ is odd, f is continuous for all x.

(D) The polynomial function $x - 2$ is continuous for all x and nonnegative for $x \geq 2$. Since $n = 2$ is even, f is continuous for $x \geq 2$, or on the interval $[2, \infty)$. ◆

PROBLEM 7 Using Theorem 1 and the general properties of continuity, determine where each function is continuous.

(A) $f(x) = x^4 + 2x^2 + 1$

(B) $f(x) = \dfrac{x^2}{(x+1)(x-4)}$

(C) $f(x) = \sqrt{x - 4}$

(D) $f(x) = \sqrt[3]{x^3 + 1}$ ◆

◆ APPLICATION

A bicycle messenger service in a downtown financial district uses the weight of a package to determine the charge for a delivery. The charge is \$12 for the first pound (or any fraction thereof) and \$1 for each additional pound (or fraction thereof) up to 10 pounds. If $C(x)$ is the charge for delivering a package weighing x pounds, then

$$C(x) = \begin{cases} \$12 & \text{if } 0 < x \leq 1 \\ \$13 & \text{if } 1 < x \leq 2 \\ \$14 & \text{if } 2 < x \leq 3 \\ \text{and so on} & \end{cases}$$

The function C is graphed for $0 < x \leq 3$ in Figure 6 (at the top of the next page).

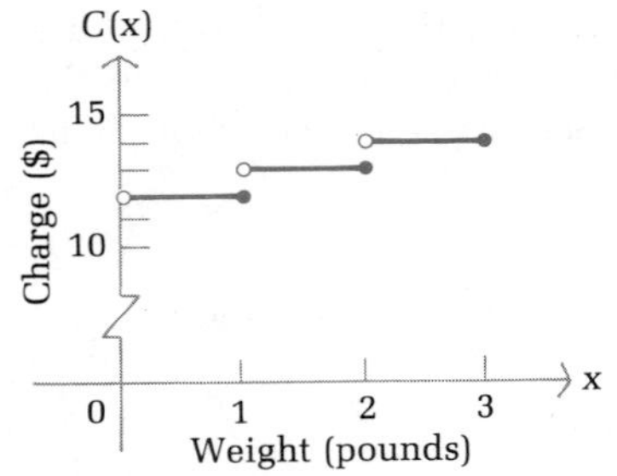

FIGURE 6
Delivery cost function

The following are a few observations about limits and continuity relative to the cost function C over the interval $0 < x \leq 3$:

1. $\lim_{x \to 1.5^-} C(x) = 13$

   $\lim_{x \to 1.5^+} C(x) = 13$

   $\lim_{x \to 1.5} C(x) = 13$

   $C(1.5) = 13$

2. $\lim_{x \to 1^-} C(x) = 12$

   $\lim_{x \to 1^+} C(x) = 13$

   $\lim_{x \to 1} C(x)$ does not exist

   $C(1) = 12$

3. $\lim_{x \to 2^-} C(x) = 13$

   $\lim_{x \to 2^+} C(x) = 14$

   $\lim_{x \to 2} C(x)$ does not exist

   $C(2) = 13$

4. C is continuous at $x = 1.5$, since

   $$\lim_{x \to 1.5} C(x) = C(1.5) = 13$$

5. C is discontinuous at 1 and 2, since

   $$\lim_{x \to 1} C(x) \quad \text{does not exist}$$

   $$\lim_{x \to 2} C(x) \quad \text{does not exist}$$

6. C is continuous from the left, but not from the right, at 1:

   $$\lim_{x \to 1^-} C(x) = 12 = C(1)$$

   $$\lim_{x \to 1^+} C(x) = 13 \neq C(1) = 12$$

7. C is continuous on the intervals $(0, 1]$, $(1, 2]$, and $(2, 3]$.

Answers to Matched Problems

1.

x	-2	-1	0	1	2	3	4
$h(x)$	1.0	1.5	2.0	2.5	3.0	3.5	4.0

2. (A)

x	0.9	0.99	$0.999 \to 1 \leftarrow 1.001$	1.01	1.1
$f(x)$	1.9	1.99	$1.999 \to 2 \leftarrow 2.001$	2.01	2.1

(B) $\lim_{x \to 1} (x + 1) = 2$

(C)

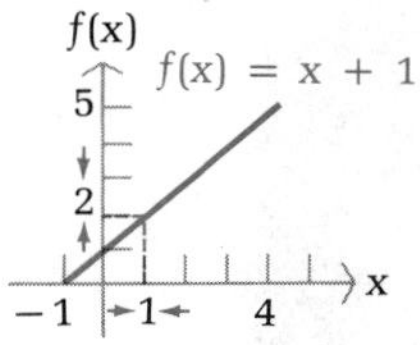

(D) $\lim_{x \to 0} (x + 1) = 1;$

$\lim_{x \to 3} (x + 1) = 4$

3. (A)

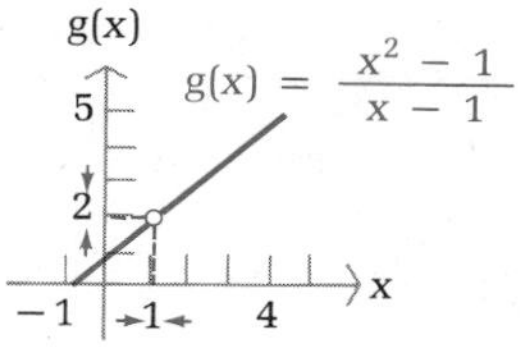

(Note that 1 is not in the domain of g.)

(B) $\lim_{x \to 1} \frac{x^2 - 1}{x - 1} = \lim_{x \to 1} (x + 1) = 2$

4.

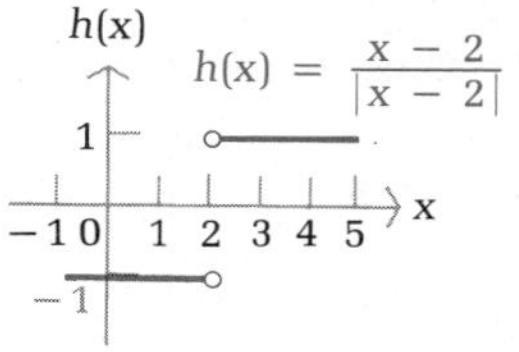

$\lim_{x \to 2} \frac{x - 2}{|x - 2|}$ does not exist

5. (A) $\lim_{x \to -1^-} f(x)$ does not exist

$\lim_{x \to -1^+} f(x) = 2$

$\lim_{x \to -1} f(x)$ does not exist

$f(-1) = 2$

(B) $\lim_{x \to 0^-} f(x) = 0$

$\lim_{x \to 0^+} f(x) = 0$

$\lim_{x \to 0} f(x) = 0$

$f(0) = 0$

(C) $\lim_{x \to 1^-} f(x) = 1$

$\lim_{x \to 1^+} f(x) = 2$

$\lim_{x \to 1} f(x)$ does not exist

$f(1)$ not defined

(D) $\lim_{x \to 3^-} f(x) = 3$

$\lim_{x \to 3^+} f(x) = 3$

$\lim_{x \to 3} f(x) = 3$

$f(3) = 2$

6. (A) f is continuous at $x = 1$, since $\lim_{x \to 1} f(x) = 2 = f(1)$. (See answer to Problem 2 above.)
   (B) g is not continuous at $x = 1$, since $g(1) = 0/0$ is not defined. (See answer to Problem 3 above.)
   (C) h is not continuous at $x = 2$, since $h(2) = 0/|0| = 0/0$ is not defined; also, $\lim_{x \to 2} h(x)$ does not exist. h is continuous at $x = 0$, since $\lim_{x \to 0} h(x) = -1 = h(0)$. (See answer to Problem 4 above.)
7. (A) Since f is a polynomial function, it is continuous for all x.
   (B) Since f is a rational function, f is continuous for all x except -1 and 4 (values that make the denominator 0).
   (C) The polynomial function $x - 4$ is continuous for all x and nonnegative for $x \geqslant 4$. Since $n = 2$ is even, f is continuous for $x \geqslant 4$, or on the interval $[4, \infty)$.
   (D) The polynomial function $x^3 + 1$ is continuous for all x. Since $n = 3$ is odd, f is continuous for all x.

EXERCISE 4-1

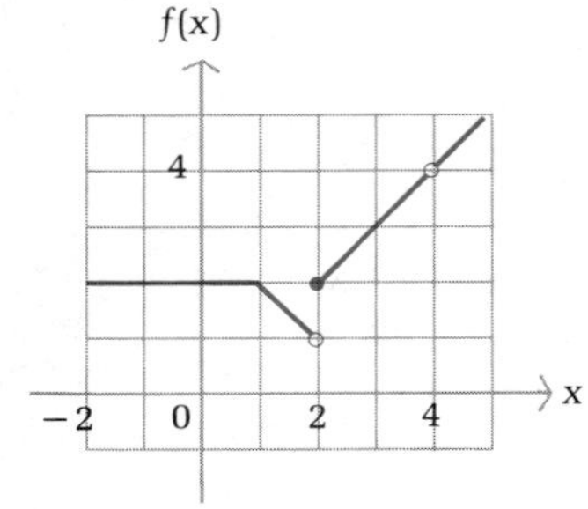

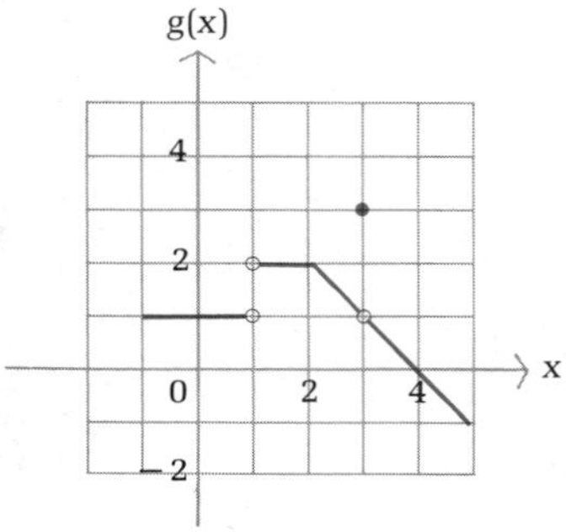

A

1. Use the graph of the function f shown in the margin to estimate each limit, if it exists.
   (A) $\lim_{x \to 0} f(x)$
   (B) $\lim_{x \to 1} f(x)$
   (C) $\lim_{x \to 2} f(x)$
   (D) $\lim_{x \to 4} f(x)$

2. Use the graph of the function g shown in the margin to estimate each limit, if it exists.
   (A) $\lim_{x \to 0} g(x)$
   (B) $\lim_{x \to 1} g(x)$
   (C) $\lim_{x \to 2} g(x)$
   (D) $\lim_{x \to 3} g(x)$

3. Use the graph of the function f in Problem 1 to estimate each value, if it is defined:
   (A) $f(0)$ (B) $f(1)$ (C) $f(2)$ (D) $f(4)$

4. Use the graph of the function g in Problem 2 to estimate each value, if it is defined:
   (A) $g(0)$ (B) $g(1)$ (C) $g(2)$ (D) $g(3)$

5. Referring to the graph of function f in Problem 1, for which of the values $c = 0$, 1, 2, or 4 does $\lim_{x \to c} f(x) = f(c)$?

6. Referring to the graph of function g in Problem 2, for which of the values $c = 0$, 1, 2, or 3 does $\lim_{x \to c} g(x) = g(c)$?

7. For which of the following values of x is the function f in Problem 1 discontinuous?

   0, 1, 2, 4

8. For which of the following values of x is the function g in Problem 2 discontinuous?

   0, 1, 2, 3

Use Theorem 1 to determine where each function in Problems 9–14 is continuous.

9. $f(x) = 2x - 3$
10. $g(x) = 3 - 5x$
11. $h(x) = \dfrac{2}{x - 5}$
12. $k(x) = \dfrac{x}{x + 3}$
13. $g(x) = \dfrac{x - 5}{(x - 3)(x + 2)}$
14. $F(x) = \dfrac{1}{x(x + 7)}$

B *Problems 15–28 refer to the function f shown in the following graph. Use the graph to estimate limits.*

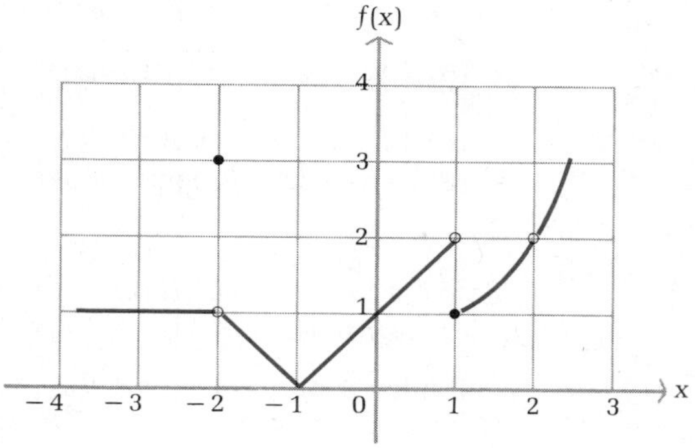

15. (A) $\lim_{x\to 0^-} f(x)$ (B) $\lim_{x\to 0^+} f(x)$ (C) $\lim_{x\to 0} f(x)$

16. (A) $\lim_{x\to -1^-} f(x)$ (B) $\lim_{x\to -1^+} f(x)$ (C) $\lim_{x\to -1} f(x)$

17. (A) $\lim_{x\to 1^-} f(x)$ (B) $\lim_{x\to 1^+} f(x)$ (C) $\lim_{x\to 1} f(x)$

18. (A) $\lim_{x\to 2^-} f(x)$ (B) $\lim_{x\to 2^+} f(x)$ (C) $\lim_{x\to 2} f(x)$

19. (A) $\lim_{x\to -2^-} f(x)$ (B) $\lim_{x\to -2^+} f(x)$ (C) $\lim_{x\to -2} f(x)$

20. (A) $\lim_{x\to 0.5^-} f(x)$ (B) $\lim_{x\to 0.5^+} f(x)$ (C) $\lim_{x\to 0.5} f(x)$

21. (A) $\lim_{x\to 0} f(x)$ (B) $f(0) = ?$ (C) Is f continuous at $x = 0$?

22. (A) $\lim_{x\to -1} f(x)$ (B) $f(-1) = ?$ (C) Is f continuous at $x = -1$?

23. (A) $\lim_{x\to 1} f(x)$ (B) $f(1) = ?$ (C) Is f continuous at $x = 1$?

24. (A) $\lim_{x\to 2} f(x)$ (B) $f(2) = ?$ (C) Is f continuous at $x = 2$?

25. (A) $\lim_{x\to -2} f(x)$ (B) $f(-2) = ?$ (C) Is f continuous at $x = -2$?

26. (A) $\lim_{x\to 0.5} f(x)$ (B) $f(0.5) = ?$ (C) Is f continuous at $x = 0.5$?

27. Function f is discontinuous at which of the following values of x?

$-3, \quad -2, \quad -1$

28. Function f is discontinuous at which of the following values of x?

$0, \quad 1, \quad 2$

29. Given the following function f:

$$f(x) = \begin{cases} 2 & \text{if } x \text{ is an integer} \\ 1 & \text{if } x \text{ is not an integer} \end{cases}$$

(A) Graph f. (B) $\lim_{x\to 2} f(x) = ?$ (C) $f(2) = ?$

(D) Is f continuous at $x = 2$? (E) Where is f discontinuous?

30. Given the following function g:

$$g(x) = \begin{cases} -1 & \text{if } x \text{ is an even integer} \\ 1 & \text{if } x \text{ is not an even integer} \end{cases}$$

(A) Graph g. (B) $\lim_{x \to 1} g(x) = ?$ (C) $g(1) = ?$

(D) Is g continuous at $x = 1$?
(E) Where is g discontinuous?

In Problems 31–38, find all limits that exist using intuition, graphing, or a little algebra, as needed.

31. $\lim_{x \to 2} (2x + 1)$
32. $\lim_{x \to 1} (3x - 2)$
33. $\lim_{x \to 2} 7$
34. $\lim_{x \to 5} 9$
35. $\lim_{x \to -3} \frac{x^2 - 9}{x + 3}$
36. $\lim_{x \to -5} \frac{x^2 - 25}{x + 5}$
37. $\lim_{x \to 1^+} \frac{|x - 1|}{x - 1}$
38. $\lim_{x \to 3^-} \frac{x - 3}{|x - 3|}$

Use Theorem 1 to determine where each function in Problems 39–46 is continuous. Express the answer in interval notation.

39. $F(x) = 2x^8 - 3x^4 + 5$
40. $h(x) = \frac{x^4 - 3x + 5}{x^2 + 2x}$
41. $g(x) = \sqrt{x - 5}$
42. $f(x) = \sqrt{3 - x}$
43. $K(x) = \sqrt[3]{x - 5}$
44. $H(x) = \sqrt[3]{3 - x}$
45. $f(x) = \frac{x^2 - 1}{x^2 - 3x + 2}$
46. $k(x) = \frac{x^2 - 4}{x^2 + x - 2}$

Complete the following table for each function in Problems 47–50:

x	0.9	0.99	0.999 → 1 ← 1.001	1.01	1.1
f(x)			→ ? ←		

From the completed table, guess the following (a calculator will be helpful for some):

(A) $\lim_{x \to 1^-} f(x)$ (B) $\lim_{x \to 1^+} f(x)$ (C) $\lim_{x \to 1} f(x)$

47. $f(x) = \frac{|x - 1|}{x - 1}$
48. $f(x) = \frac{x - 1}{|x| - 1}$
49. $f(x) = \frac{x^3 - 1}{x - 1}$
50. $f(x) = \frac{x^4 - 1}{x - 1}$

In Problems 51–56, graph f and locate all points of discontinuity.

51. $f(x) = \begin{cases} 1 + x & \text{if } x < 1 \\ 5 - x & \text{if } x \geq 1 \end{cases}$

52. $f(x) = \begin{cases} x^2 & \text{if } x \leq 1 \\ 2x & \text{if } x > 1 \end{cases}$

53. $f(x) = \begin{cases} 1 + x & \text{if } x \leq 2 \\ 5 - x & \text{if } x > 2 \end{cases}$

54. $f(x) = \begin{cases} x^2 & \text{if } x \leq 2 \\ 2x & \text{if } x > 2 \end{cases}$

55. $f(x) = \begin{cases} -x & \text{if } x < 0 \\ 1 & \text{if } x = 0 \\ x & \text{if } x > 0 \end{cases}$

56. $f(x) = \begin{cases} 1 & \text{if } x < 0 \\ 0 & \text{if } x = 0 \\ 1 + x & \text{if } x > 0 \end{cases}$

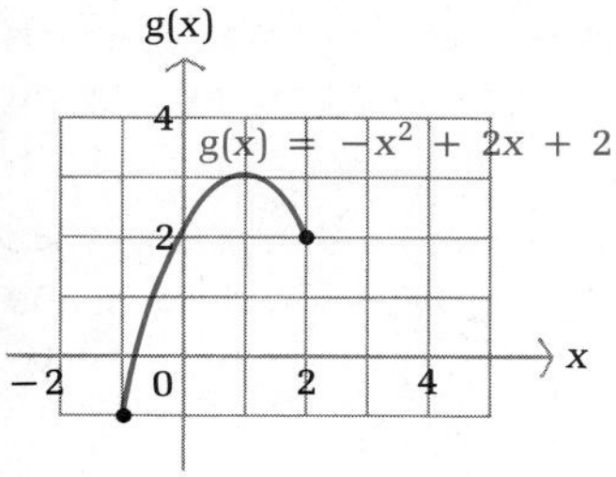

C

57. Use the graph of the function g shown in the margin to answer the following questions:
(A) Is g continuous on the open interval $(-1, 2)$?
(B) Is g continuous from the right at $x = -1$? That is, does $\lim_{x \to -1^+} g(x) = g(-1)$?
(C) Is g continuous from the left at $x = 2$? That is, does $\lim_{x \to 2^-} g(x) = g(2)$?
(D) Is g continuous on the closed interval $[-1, 2]$?

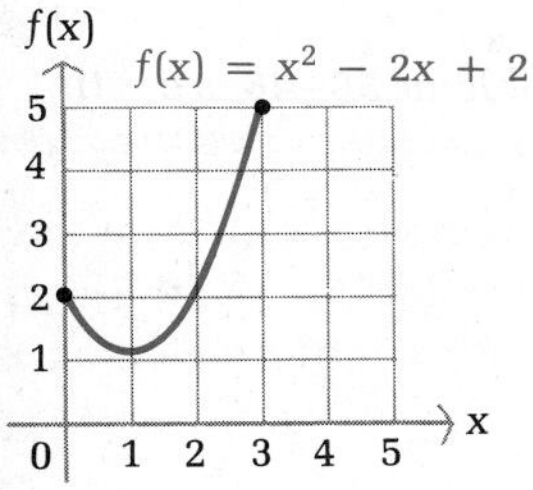

58. Use the graph of the function f shown in the margin to answer the following questions:
(A) Is f continuous on the open interval $(0, 3)$?
(B) Is f continuous from the right at $x = 0$? That is, does $\lim_{x \to 0^+} f(x) = f(0)$?
(C) Is f continuous from the left at $x = 3$? That is, does $\lim_{x \to 3^-} f(x) = f(3)$?
(D) Is f continuous on the closed interval $[0, 3]$?

Problems 59 and 60 refer to the ***greatest integer function,*** *which is denoted by* $[\![x]\!]$ *and is defined as follows:*

$[\![x]\!]$ = *Greatest integer* $\leq x$

For example,

$[\![-3.6]\!]$ = *Greatest integer* $\leq -3.6 = -4$
$[\![2]\!]$ = *Greatest integer* $\leq 2 = 2$
$[\![2.5]\!]$ = *Greatest integer* $\leq 2.5 = 2$

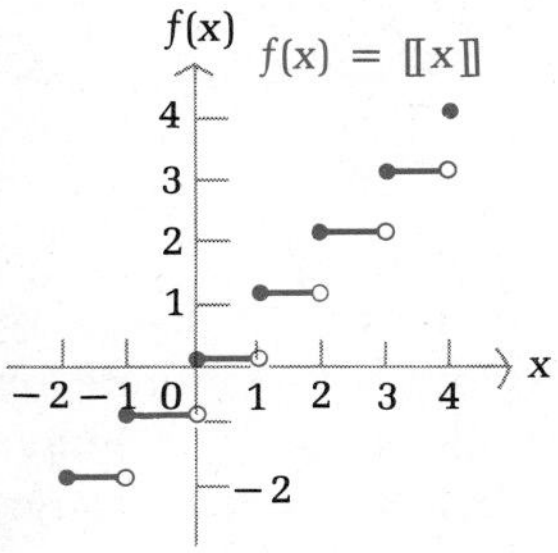

The graph of $f(x) = [\![x]\!]$ *is shown in the margin. There, we can see that*

$[\![x]\!] = -2$ *for* $-2 \leq x < -1$
$[\![x]\!] = -1$ *for* $-1 \leq x < 0$
$[\![x]\!] = 0$ *for* $0 \leq x < 1$
$[\![x]\!] = 1$ *for* $1 \leq x < 2$
$[\![x]\!] = 2$ *for* $2 \leq x < 3$

and so on.

59. (A) Is f continuous from the right at $x = 0$?
(B) Is f continuous from the left at $x = 0$?
(C) Is f continuous on the open interval $(0, 1)$?
(D) Is f continuous on the closed interval $[0, 1]$?
(E) Is f continuous on the half-closed interval $[0, 1)$?

60. (A) Is f continuous from the right at $x = 2$?
(B) Is f continuous from the left at $x = 2$?
(C) Is f continuous on the open interval $(1, 2)$?
(D) Is f continuous on the closed interval $[1, 2]$?
(E) Is f continuous on the half-closed interval $[1, 2)$?

Use your intuition and current knowledge about continuous functions to answer Problems 61 and 62. (The ideas in these two problems will be generalized in later sections.)

61. If a function f is continuous and never equal to 0 on the interval $(1, 5)$, and if $f(2) = 3$, can $f(4)$ be negative? Can $f(x)$ be negative for any x on the interval $(1, 5)$?

62. If a function g is continuous and never equal to 0 on the interval $(-1, 8)$, and if $g(0) = -3$, can $g(1)$ be positive? Is $g(x)$ negative for all x on the interval $(-1, 8)$?

APPLICATIONS

Business & Economics

63. *Postal rates.* First-class postage in 1989 was $0.25 for the first ounce (or any fraction thereof) and $0.20 for each additional ounce (or fraction thereof) up to 11 ounces. If $P(x)$ is the amount of postage for a letter weighing x ounces, then we can write

$$P(x) = \begin{cases} \$0.25 & \text{if } 0 < x \leq 1 \\ \$0.45 & \text{if } 1 < x \leq 2 \\ \$0.65 & \text{if } 2 < x \leq 3 \\ \text{and so on} \end{cases}$$

(A) Graph P for $0 < x \leq 5$.
(B) Find $\lim_{x \to 4.5} P(x)$ and $P(4.5)$.
(C) Find $\lim_{x \to 4} P(x)$ and $P(4)$.
(D) Is P continuous at $x = 4.5$? At $x = 4$?

64. *Telephone rates.* A person placing a station-to-station call on Saturday from San Francisco to New York is charged $0.30 for the first minute (or any fraction thereof) and $0.20 for each additional minute (or fraction thereof). If the length of a call is x minutes, then the long-distance charge $R(x)$ is

$$R(x) = \begin{cases} \$0.30 & \text{if } 0 < x \leq 1 \\ \$0.50 & \text{if } 1 < x \leq 2 \\ \$0.70 & \text{if } 2 < x \leq 3 \\ \text{and so on} \end{cases}$$

(A) Graph R for $0 < x \leq 6$.
(B) Find $\lim_{x \to 2.5} R(x)$ and $R(2.5)$.
(C) Find $\lim_{x \to 2} R(x)$ and $R(2)$.
(D) Is R continuous at $x = 2.5$? At $x = 2$?

65. *Income.* A personal computer salesperson receives a base salary of \$1,000 per month and a commission of 5% of all sales over \$10,000 during the month. If the monthly sales are \$20,000 or more, the salesperson is given an additional \$500 bonus. Let $E(s)$ represent the person's earnings during the month as a function of the monthly sales s.

(A) Graph $E(s)$ for $0 \leq s \leq 30{,}000$.
(B) Find $\lim_{s \to 10{,}000} E(s)$ and $E(10{,}000)$.
(C) Find $\lim_{s \to 20{,}000} E(s)$ and $E(20{,}000)$.
(D) Is E continuous at $s = 10{,}000$? At $s = 20{,}000$?

66. *Equipment rental.* An office equipment rental and leasing company rents electric typewriters for \$10 per day (and any fraction thereof) or for \$50 per 7 day week. Let $C(x)$ be the cost of renting a typewriter for x days.

(A) Graph $C(x)$ for $0 \leq x \leq 10$.
(B) Find $\lim_{x \to 4.5} C(x)$ and $C(4.5)$.
(C) Find $\lim_{x \to 8} C(x)$ and $C(8)$.
(D) Is C continuous at $x = 4.5$? At $x = 8$?

Life Sciences

67. *Animal supply.* A medical laboratory raises its own rabbits. The number of rabbits $N(t)$ available at any time t depends on the number of births and deaths. When a birth or death occurs, the function N generally has a discontinuity, as shown in the figure.

(A) Where is the function N discontinuous?
(B) $\lim_{t \to t_5} N(t) = ?$; $N(t_5) = ?$ (C) $\lim_{t \to t_3} N(t) = ?$; $N(t_3) = ?$

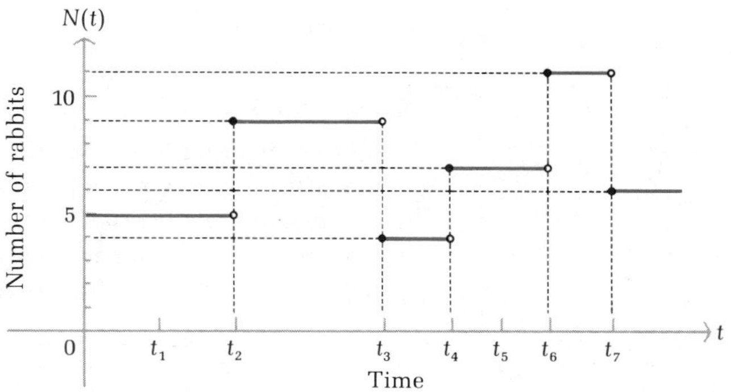

Social Sciences

68. *Learning.* The graph shown on the next page might represent the history of a particular person learning the material on limits and continuity in this book. At time t_2, the student's mind goes blank during a quiz. At time t_4, the instructor explains a concept particularly well, and suddenly, a big jump in understanding takes place.

(A) Where is the function p discontinuous?
(B) $\lim_{t\to t_1} p(t) = ?$; $p(t_1) = ?$
(C) $\lim_{t\to t_2} p(t) = ?$; $p(t_2) = ?$
(D) $\lim_{t\to t_4} p(t) = ?$; $p(t_4) = ?$

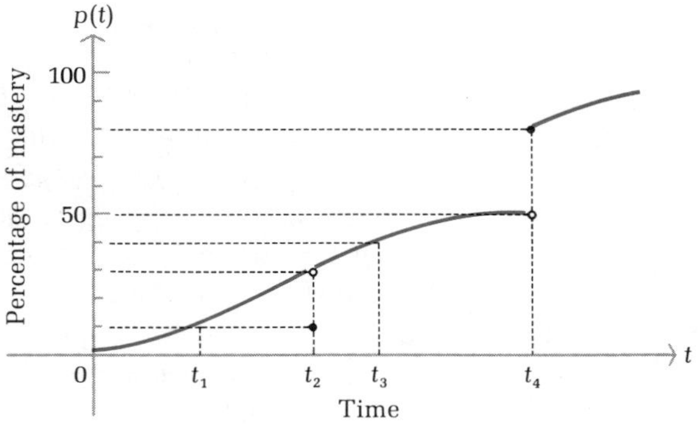

SECTION 4-2

Computation of Limits

- LIMITS AT POINTS OF CONTINUITY
- LIMITS AND INFINITY
- LIMIT PROPERTIES
- LIMITS AT INFINITY
- SUMMARY OF LIMIT FORMS

To introduce the concept of limit in the preceding section, we relied heavily on tables, graphs, and intuition. But we will find it very useful to be able to compute limits of functions without having to resort to graphs or tables. In this section, we will introduce and use various properties of limits to find limits of a wide variety of functions, in the same way we used continuity properties earlier. We will also discuss the use of the infinity symbol, ∞, relative to limits.

◆ LIMITS AT POINTS OF CONTINUITY

If we know that a function is continuous at a point, then, using the definition of continuity at a point from the preceding section, it is easy to find the limit of the function at that point—we simply evaluate the function at that point.

Limit at a Point of Continuity

If a function f is continuous at $x = c$, then

$\lim_{x \to c} f(x) = f(c)$ For example,

$\lim_{x \to 2} (x^2 - x + 1) = 2^2 - 2 + 1 = 3$

(since a polynomial function is continuous for all x).

◆ EXAMPLE 8 Find the following limits:

(A) $\lim_{x \to 3} \dfrac{x^2 - 3x}{x + 7}$ (B) $\lim_{x \to 5} \sqrt{x - 2}$ (C) $\lim_{x \to 3} 5$

Solutions (A) Since $(x^2 - 3x)/(x + 7)$ is continuous for all x except $x = -7$, it is continuous at $x = 3$. Thus,

$$\lim_{x \to 3} \frac{x^2 - 3x}{x + 7} = \frac{3^2 - 3(3)}{3 + 7} = \frac{9 - 9}{10} = \frac{0}{10} = 0$$

(B) Since $x - 2$ is continuous for all x and nonnegative for $x \geq 2$, $\sqrt{x - 2}$ is continuous for all $x \geq 2$. In particular, $\sqrt{x - 2}$ is continuous for $x = 5$. Thus,

$$\lim_{x \to 5} \sqrt{x - 2} = \sqrt{5 - 2} = \sqrt{3}$$

(C) Since $f(x) = 5$ is a constant function, it is continuous for all x. In particular, it is continuous when $x = 3$. Thus,

$$\lim_{x \to 3} f(x) = f(3) = 5$$

That is, $\lim_{x \to 3} 5 = 5$. ◆

PROBLEM 8 Find the following limits:

(A) $\lim_{x \to 4} (x + \sqrt{x})$ (B) $\lim_{x \to -7} \sqrt[3]{1 - x}$ (C) $\lim_{x \to -2} 10$ ◆

◆ LIMITS AND INFINITY

How does $f(x) = 1/x$ behave for x near 0? When a number gets small (close to 0), the reciprocal of its absolute value gets large. For example, when $x = 0.001$, $1/x = 1{,}000$; when $x = -0.001$, $1/x = -1{,}000$. Thus, as x approaches 0 from the right, $f(x)$ is positive and increases without bound. We indicate this behavior by writing

$$\lim_{x \to 0^+} \frac{1}{x} = \infty$$

Similarly, as x approaches 0 from the left, $f(x)$ is negative and decreases without bound. We indicate this behavior by writing

$$\lim_{x \to 0^-} \frac{1}{x} = -\infty$$

In neither case does the limit exist, since neither ∞ nor $-\infty$ are numbers. Nevertheless, the notation tells us something about the behavior of $f(x)$ for x near 0. The graph of f is illustrated in Figure 7A. Figure 7B illustrates the behavior of $g(x) = 1/x^2$ for x near 0. Note that for $g(x)$ we can write

$$\lim_{x \to 0} \frac{1}{x^2} = \infty$$

since $g(x)$ remains positive and increases without bound as x approaches 0 from either side.

For both functions f and g, the line $x = 0$ (the vertical axis) is called a *vertical asymptote.* In general, a line $x = a$ is a **vertical asymptote** for the graph of $y = f(x)$ if $f(x)$ either increases or decreases without bound as x approaches a

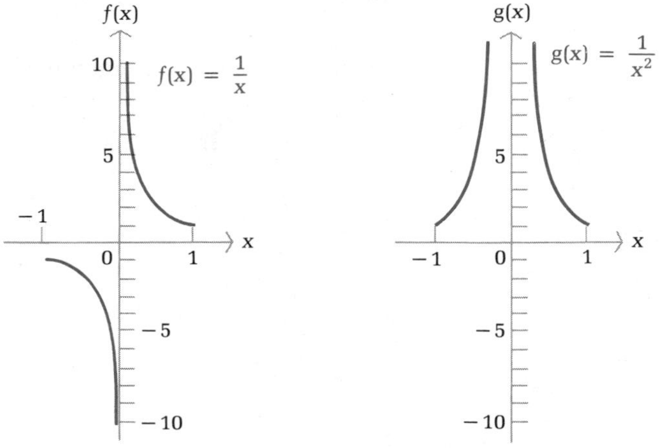

(A) $\lim_{x \to 0^-} \frac{1}{x} = -\infty$

$\lim_{x \to 0^+} \frac{1}{x} = \infty$

$\lim_{x \to 0} \frac{1}{x}$ does not exist

(B) $\lim_{x \to 0^-} \frac{1}{x^2} = \infty$

$\lim_{x \to 0^+} \frac{1}{x^2} = \infty$

$\lim_{x \to 0} \frac{1}{x^2} = \infty$

FIGURE 7

from either the left or the right. Symbolically, $x = a$ is a vertical asymptote if any of the following hold:

$$\lim_{x \to a^-} f(x) = -\infty \text{ (or } \infty)$$

$$\lim_{x \to a^+} f(x) = -\infty \text{ (or } \infty)$$

$$\lim_{x \to a} f(x) = -\infty \text{ (or } \infty)$$

◆ LIMIT PROPERTIES

We now turn to some basic properties of limits that will enable us to evaluate many limits when either the function involved is not continuous at the point or it is not certain whether the function is continuous at the point. [Remember, if a function f is continuous at $x = c$, then we can find $\lim_{x \to c} f(x)$ by evaluating the function f at $x = c$.]

THEOREM 2

Properties of Limits

Let f and g be two functions, and assume that

$$\lim_{x \to c} f(x) = L \qquad \lim_{x \to c} g(x) = M$$

where L and M are real numbers (both limits exist). Then,

1. $\lim_{x \to c} [f(x) + g(x)] = \lim_{x \to c} f(x) + \lim_{x \to c} g(x) = L + M$
2. $\lim_{x \to c} [f(x) - g(x)] = \lim_{x \to c} f(x) - \lim_{x \to c} g(x) = L - M$
3. $\lim_{x \to c} kf(x) = k \lim_{x \to c} f(x) = kL \qquad$ for any constant k
4. $\lim_{x \to c} [f(x) \cdot g(x)] = \left[\lim_{x \to c} f(x)\right]\left[\lim_{x \to c} g(x)\right] = LM$
5. $\lim_{x \to c} \dfrac{f(x)}{g(x)} = \dfrac{\lim_{x \to c} f(x)}{\lim_{x \to c} g(x)} = \dfrac{L}{M} \qquad$ if $M \neq 0$
6. $\lim_{x \to c} \sqrt[n]{f(x)} = \sqrt[n]{\lim_{x \to c} f(x)} = \sqrt[n]{L} \qquad L \geq 0$ for n even

Suppose

$$\lim_{x \to 2} f(x) = 3 \qquad \text{and} \qquad \lim_{x \to 2} g(x) = 7$$

then no other information is needed about the functions f and g to determine that

$$\lim_{x \to 2} [f(x) - g(x)] = \lim_{x \to 2} f(x) - \lim_{x \to 2} g(x) = 3 - 7 = -4$$

$$\lim_{x \to 2} 9f(x) = 9 \lim_{x \to 2} f(x) = 9 \cdot 3 = 27$$

$$\lim_{x \to 2} \frac{f(x)}{g(x)} = \frac{\lim_{x \to 2} f(x)}{\lim_{x \to 2} g(x)} = \frac{3}{7}$$

$$\lim_{x \to 2} \sqrt{f(x)} = \sqrt{\lim_{x \to 2} f(x)} = \sqrt{3}$$

Now, suppose

$$\lim_{x \to c} f(x) = 0 \qquad \text{and} \qquad \lim_{x \to c} g(x) = 0$$

then finding

$$\lim_{x \to c} \frac{f(x)}{g(x)} \tag{1}$$

may present some difficulties, since limit property 5 (the limit of a quotient) does not apply when $\lim_{x \to c} g(x) = 0$. We often have to use algebraic manipulation or other devices to determine the outcome. Recall from Examples 3 and 4 (in Section 4-1) that

$$\lim_{x \to 2} \frac{x^2 - 4}{x - 2} = \lim_{x \to 2} \frac{(x-2)(x+2)}{x - 2} = \lim_{x \to 2} (x + 2) = 4$$

and

$$\lim_{x \to 0} \frac{|x|}{x} \quad \text{does not exist}$$

From these two examples, it is clear that knowing only that $\lim_{x \to c} f(x) = 0$ and $\lim_{x \to c} g(x) = 0$ is not enough to determine limit (1). Depending on the choice of functions f and g, the limit (1) may or may not exist. Consequently, if we are given (1) and $\lim_{x \to c} f(x) = 0$ and $\lim_{x \to c} g(x) = 0$, then (1) is said to be **indeterminate**, or, more specifically, a **0/0 indeterminate form.**

The 0/0 indeterminate form may seem unusual at first—a form one might not encounter too often—but it turns out that this is not the case. This form turns up in a number of very important developments in calculus. Example 9 provides a preview of a form that will be considered in detail in the following sections.

◆ EXAMPLE 9 Find $\lim_{h \to 0} \frac{f(2 + h) - f(2)}{h}$ for:

(A) $f(x) = 3x - 4$ (B) $f(x) = \sqrt{x}$ (C) $f(x) = |x - 2|$

Solutions (A) $\lim_{h \to 0} \frac{f(2 + h) - f(2)}{h}$ $\qquad f(x) = 3x - 4$

$$= \lim_{h \to 0} \frac{[3(2 + h) - 4] - [3(2) - 4]}{h}$$

Since this is a 0/0 indeterminate form and property 5 in Theorem 2 does not apply, we proceed with algebraic simplification.

$$= \lim_{h \to 0} \frac{6 + 3h - 4 - 6 + 4}{h}$$

$$= \lim_{h \to 0} \frac{3h}{h} = \lim_{h \to 0} 3 = 3 \qquad h \neq 0$$

(B) $$\lim_{h \to 0} \frac{f(2 + h) - f(2)}{h}$$ $f(x) = \sqrt{x}$

$$= \lim_{h \to 0} \frac{\sqrt{2 + h} - \sqrt{2}}{h}$$

This is a 0/0 indeterminate form, so property 5 in Theorem 2 does not apply. Rationalizing the numerator will be of help.

$$= \lim_{h \to 0} \frac{\sqrt{2 + h} - \sqrt{2}}{h} \cdot \frac{\sqrt{2 + h} + \sqrt{2}}{\sqrt{2 + h} + \sqrt{2}}$$ $(A - B)(A + B) = A^2 - B^2$

$$= \lim_{h \to 0} \frac{2 + h - 2}{h(\sqrt{2 + h} + \sqrt{2})}$$

$$= \lim_{h \to 0} \frac{1}{\sqrt{2 + h} + \sqrt{2}}$$ Use properties 1, 5, and 6 ($h \neq 0$).

$$= \frac{1}{\sqrt{2} + \sqrt{2}} = \frac{1}{2\sqrt{2}}$$

(C) $$\lim_{h \to 0} \frac{f(2 + h) - f(2)}{h}$$ $f(x) = |x - 2|$

$$= \lim_{h \to 0} \frac{|(2 + h) - 2| - |2 - 2|}{h}$$

Since this is a 0/0 indeterminate form and property 5 in Theorem 2 does not apply, we proceed with algebraic simplification.

$$= \lim_{h \to 0} \frac{|h|}{h} \quad \text{does not exist}$$ See Example 4. ◆

PROBLEM 9 Find $\lim_{h \to 0} \frac{g(1 + h) - g(1)}{h}$ for:

(A) $g(x) = 1 - x$ (B) $g(x) = \sqrt{x}$ (C) $g(x) = |x - 1|$ ◆

◆ LIMITS AT INFINITY

Earlier in this section, we discussed the behavior of $f(x) = 1/x$ for x near 0. We now investigate the behavior of $f(x)$ as x increases or decreases without bound. When the absolute value of a number increases, its reciprocal gets smaller. For example, if $x = 10{,}000$, then $1/x = 0.0001$; if $x = -10{,}000$, then $1/x = -0.0001$. Thus, as x increases without bound, indicated by $x \to \infty$, $f(x)$ approaches 0. We indicate this behavior by writing

$$\lim_{x \to \infty} \frac{1}{x} = 0$$

Similarly, as x decreases without bound, indicated by $x \to -\infty$, $f(x)$ approaches 0. We indicate this behavior by writing

$$\lim_{x \to -\infty} \frac{1}{x} = 0$$

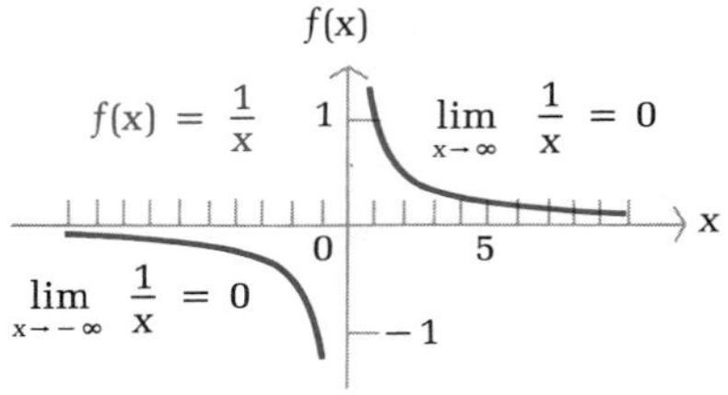

FIGURE 8

Figure 8 illustrates the behavior of $f(x) = 1/x$ as x increases and decreases without bound.

It is important to understand that the symbol ∞ does not represent an actual number that x is approaching, but is used to indicate only that the value of x is increasing with no upper limit on its size. In particular, the statement "$x = \infty$" is meaningless, since ∞ is not a symbol for a real number.

For the function f in Figure 8, the line $y = 0$ (the x axis) is called a *horizontal asymptote*. In general, a line $y = b$ is a **horizontal asymptote** for the graph of $y = f(x)$ if $f(x)$ approaches b as x either increases or decreases without bound. Symbolically, $y = b$ is a horizontal asymptote if either

$$\lim_{x \to -\infty} f(x) = b \qquad \text{or} \qquad \lim_{x \to \infty} f(x) = b$$

In the first case, the graph of f will be close to the horizontal line $y = b$ for large (in absolute value) negative x. In the second case, the graph will be close to the horizontal line $y = b$ for large positive x.

How can we evaluate limits of the form $\lim_{x \to -\infty} f(x)$ and $\lim_{x \to \infty} f(x)$ without drawing graphs or performing complicated calculations? Fortunately, **limit properties 1–6, listed earlier in this section in Theorem 2, are valid if we replace the statement $x \to c$ with $x \to -\infty$ or $x \to \infty$.** These properties, together with Theorem 3, enable us to evaluate limits at infinity for many functions.

THEOREM 3

Two Special Limits at Infinity

If p is a positive real number, and k is any real constant, then

$$\lim_{x \to -\infty} \frac{k}{x^p} = 0 \qquad \text{and} \qquad \lim_{x \to \infty} \frac{k}{x^p} = 0$$

provided that x^p names a real number for negative values of x.

Figure 9 illustrates Theorem 3 for several values of p. (Note that the x axis is a horizontal asymptote in all three cases.)

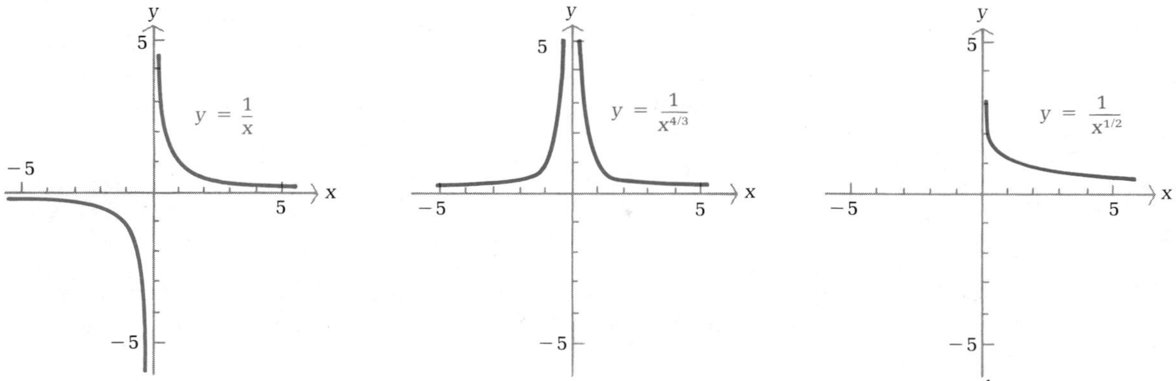

(A) $\lim_{x\to\infty} \frac{1}{x} = 0, \ \lim_{x\to-\infty} \frac{1}{x} = 0$

(B) $\lim_{x\to\infty} \frac{1}{x^{4/3}} = 0, \ \lim_{x\to-\infty} \frac{1}{x^{4/3}} = 0$

(C) $\lim_{x\to\infty} \frac{1}{x^{1/2}} = 0$

$x^{1/2}$ is not defined for negative values of x

FIGURE 9

Now, consider the following limit of a polynomial:

$$\lim_{x\to\infty} (2x^3 - 4x^2 - 9x - 100)$$

At first glance, it is not clear which part of the polynomial, $-4x^2 - 9x - 100$ or $2x^3$, dominates as x increases without bound. To help resolve the conflict, we factor out the highest power of x (x^3):

$$\lim_{x\to\infty} x^3\left(2 - \frac{4}{x} - \frac{9}{x^2} - \frac{100}{x^3}\right)$$

Looking at the factors separately, we have

$$\lim_{x\to\infty} x^3 = \infty$$

and (using Theorems 2 and 3)

$$\lim_{x\to\infty}\left(2 - \frac{4}{x} - \frac{9}{x^2} - \frac{100}{x^3}\right) \boxed{= 2 - 0 - 0 - 0} = 2$$

Now, intuitively, we would like to write

$$\lim_{x\to\infty} x^3\left(2 - \frac{4}{x} - \frac{9}{x^2} - \frac{100}{x^3}\right) = \infty$$

because if one factor is increasing without bound and at the same time the other factor is approaching a positive constant, it seems reasonable to think that their product must increase without bound. This reasoning turns out to be correct; that is:

If $\quad \lim_{x\to\infty} f(x) = \infty \quad$ **and** $\quad \lim_{x\to\infty} g(x) = L > 0$

then $\quad \lim_{x\to\infty} [f(x) \cdot g(x)] = \infty$

This statement also holds if $x \to \infty$ is replaced with $x \to -\infty$ or $x \to c$, for c any real number. (Other variations of this statement are also possible.)

We are now ready to turn our attention to the limits of rational functions as x increases without bound.

◆ EXAMPLE 10 Find the following limits, if they exist.

(A) $\lim_{x\to\infty} \frac{5x^2+3}{3x^2-2}$ (B) $\lim_{x\to\infty} \frac{3x^4-x^2+1}{8x^6-10}$ (C) $\lim_{x\to\infty} \frac{2x^5-x^3-1}{6x^3+2x^2-7}$

Solutions In all three cases, we begin (as in the polynomial example above) by factoring out the highest power of x from the numerator and the highest power of x from the denominator.

(A) $$\lim_{x\to\infty} \frac{5x^2+3}{3x^2-2} = \lim_{x\to\infty} \frac{x^2\left(5+\frac{3}{x^2}\right)}{x^2\left(3-\frac{2}{x^2}\right)} = \lim_{x\to\infty} \frac{5+\frac{3}{x^2}}{3-\frac{2}{x^2}}$$ Use Theorems 2 and 3.

$$= \frac{5+0}{3-0} = \frac{5}{3}$$

(B) $$\lim_{x\to\infty} \frac{3x^4-x^2+1}{8x^6-10} = \lim_{x\to\infty} \frac{x^4\left(3-\frac{1}{x^2}+\frac{1}{x^4}\right)}{x^6\left(8-\frac{10}{x^6}\right)}$$ Use Theorems 2 and 3.

$$= \lim_{x\to\infty} \frac{1}{x^2} \cdot \lim_{x\to\infty} \frac{3-\frac{1}{x^2}+\frac{1}{x^4}}{8-\frac{10}{x^6}} = 0 \cdot \frac{3-0+0}{8-0} = 0 \cdot \frac{3}{8} = 0$$

(C) $$\lim_{x\to\infty} \frac{2x^5-x^3-1}{6x^3+2x^2-7} = \lim_{x\to\infty} \frac{x^5\left(2-\frac{1}{x^2}-\frac{1}{x^5}\right)}{x^3\left(6+\frac{2}{x}-\frac{7}{x^3}\right)}$$

$$= \lim_{x\to\infty} x^2 \left(\frac{2-\frac{1}{x^2}-\frac{1}{x^5}}{6+\frac{2}{x}-\frac{7}{x^3}}\right)$$

$= \infty$ Since $\lim_{x\to\infty} x^2 = \infty$ and $\lim_{x\to\infty} \frac{2-\frac{1}{x^2}-\frac{1}{x^5}}{6+\frac{2}{x}-\frac{7}{x^3}} = \frac{2}{6}$ ◆

PROBLEM 10 Find the following limits, if they exist:

(A) $\lim_{x\to\infty} \frac{4x^3-5x+8}{2x^4-7}$ (B) $\lim_{x\to\infty} \frac{5x^6+3x}{2x^5-x-5}$ (C) $\lim_{x\to\infty} \frac{2x^3-x+7}{4x^3+3x^2-100}$ ◆

◆ SUMMARY OF LIMIT FORMS

Table 1 provides a summary of some of the limit forms we have considered.

TABLE 1

Summary of Some Important Limit Forms

NOTATION	DESCRIPTION	EXAMPLE
$\lim_{x\to c} f(x) = L$	$f(x)$ approaches L as x approaches c from either side of c, but is not equal to c	$\lim_{x\to 2} \frac{x^2-4}{x-2} = 4$
$\lim_{x\to c^-} f(x) = L$	$f(x)$ approaches L as x approaches c from the left	$\lim_{x\to 0^-} \frac{\lvert x\rvert}{x} = -1$
$\lim_{x\to c^+} f(x) = L$	$f(x)$ approaches L as x approaches c from the right	$\lim_{x\to 0^+} \frac{\lvert x\rvert}{x} = 1$
$\lim_{x\to c^-} f(x) = -\infty$	$f(x)$ decreases without bound as x approaches c from the left	$\lim_{x\to 0^-} \frac{1}{x} = -\infty$
$\lim_{x\to c^+} f(x) = \infty$	$f(x)$ increases without bound as x approaches c from the right	$\lim_{x\to 0^+} \frac{1}{x} = \infty$
$\lim_{x\to c} f(x) = \infty$	$f(x)$ increases without bound as x approaches c from either side of c, but is not equal to c	$\lim_{x\to 0} \frac{1}{x^2} = \infty$
$\lim_{x\to \infty} f(x) = L$	$f(x)$ approaches L as x increases without bound	$\lim_{x\to \infty} \frac{2x^2+1}{3x^2-1} = \frac{2}{3}$
$\lim_{x\to \infty} f(x) = \infty$	$f(x)$ increases without bound as x increases without bound	$\lim_{x\to \infty} \frac{2x^3}{x^2-7} = \infty$

Answers to Matched Problems

8. (A) $\lim_{x\to 4} (x + \sqrt{x}) = 4 + \sqrt{4} = 6$ (B) $\lim_{x\to -7} \sqrt[3]{1-x} = \sqrt[3]{1-(-7)} = 2$
   (C) $\lim_{x\to -2} 10 = 10$
9. (A) -1 (B) $\frac{1}{2}$ (C) Does not exist
10. (A) 0 (B) ∞ (does not exist) (C) $\frac{1}{2}$

EXERCISE 4-2

A *Given* $\lim_{x\to 3} f(x) = 5$ *and* $\lim_{x\to 3} g(x) = 9$, *find the indicated limits in Problems 1–10.*

1. $\lim_{x\to 3} [f(x) - g(x)]$
2. $\lim_{x\to 3} [f(x) + g(x)]$
3. $\lim_{x\to 3} 4g(x)$
4. $\lim_{x\to 3} (-2)f(x)$
5. $\lim_{x\to 3} \frac{f(x)}{g(x)}$
6. $\lim_{x\to 3} [f(x) \cdot g(x)]$
7. $\lim_{x\to 3} \sqrt{f(x)}$
8. $\lim_{x\to 3} \sqrt{g(x)}$
9. $\lim_{x\to 3} \frac{f(x)+g(x)}{2f(x)}$
10. $\lim_{x\to 3} \frac{g(x)-f(x)}{3g(x)}$

Find each limit in Problems 11–26, if it exists. (Algebraic manipulation is useful in some cases.)

11. $\lim_{x\to 5} (2x^2 - 3)$

12. $\lim_{x\to 2} (x^2 - 8x + 2)$

13. $\lim_{x\to 4} (x^2 - 5x)$

14. $\lim_{x\to -2} (3x^3 - 9)$

15. $\lim_{x\to 2} \dfrac{5x}{2 + x^2}$

16. $\lim_{x\to 10} \dfrac{2x + 5}{3x - 5}$

17. $\lim_{x\to 2} (x + 1)^3(2x - 1)^2$

18. $\lim_{x\to 3} (x + 2)^2(2x - 4)$

19. $\lim_{x\to 0} \dfrac{x^2 - 3x}{x}$

20. $\lim_{x\to 0} \dfrac{2x^2 + 5x}{x}$

21. $\lim_{x\to\infty} \dfrac{3}{x^2}$

22. $\lim_{x\to\infty} \dfrac{6}{x^4}$

23. $\lim_{x\to\infty} \left(5 - \dfrac{3}{x} + \dfrac{2}{x^2}\right)$

24. $\lim_{x\to\infty} \left(4 + \dfrac{1}{x^2} - \dfrac{3}{x^4}\right)$

25. $\lim_{h\to 0} \dfrac{2(3 + h) - 2(3)}{h}$

26. $\lim_{h\to 0} \dfrac{[(4 + h) - 2] - (4 - 2)}{h}$

B *Find each limit in Problems 27–54, if it exists. (Use algebraic manipulation where necessary. Also, use $-\infty$ or ∞ where appropriate.)*

27. $\lim_{x\to 1} (3x^4 - 2x^2 + x - 2)$

28. $\lim_{x\to -1} (5x^3 - 3x^2 - 5x + 3)$

29. $\lim_{x\to 1} \dfrac{x - 2}{x^2 - 2x}$

30. $\lim_{x\to 1} \dfrac{x + 3}{x^2 + 3x}$

31. $\lim_{x\to 2} \dfrac{x - 2}{x^2 - 2x}$

32. $\lim_{x\to -3} \dfrac{x + 3}{x^2 + 3x}$

33. $\lim_{x\to\infty} \dfrac{x - 2}{x^2 - 2x}$

34. $\lim_{x\to\infty} \dfrac{x + 3}{x^2 + 3x}$

35. $\lim_{x\to 2} \dfrac{x^2 - x - 6}{x + 2}$

36. $\lim_{x\to 3} \dfrac{x^2 + x - 6}{x + 3}$

37. $\lim_{x\to -2} \dfrac{x^2 - x - 6}{x + 2}$

38. $\lim_{x\to -3} \dfrac{x^2 + x - 6}{x + 3}$

39. $\lim_{x\to\infty} \dfrac{x^2 - x - 6}{x + 2}$

40. $\lim_{x\to\infty} \dfrac{x^2 + x - 6}{x + 3}$

41. $\lim_{x\to\infty} \dfrac{2x + 4}{x}$

42. $\lim_{x\to\infty} \dfrac{3x^2 + 5}{x^2}$

43. $\lim_{x\to\infty} \dfrac{3x^3 - x + 1}{5x^3 - 7}$

44. $\lim_{x\to\infty} \dfrac{3x^4 - x^3 + 5}{2x^4 - 10}$

45. $\lim_{h\to 0} \dfrac{(2 + h)^2 - 2^2}{h}$

46. $\lim_{h\to 0} \dfrac{(3 + h)^2 - 3^2}{h}$

47. $\lim_{x\to 3} \left(\dfrac{x}{x + 3} + \dfrac{x - 3}{x^2 - 9}\right)$

48. $\lim_{x\to 2} \left(\dfrac{1}{x + 2} + \dfrac{x - 2}{x^2 - 4}\right)$

49. $\lim_{x \to 0^-} \dfrac{x - 2}{x^2 - 2x}$

50. $\lim_{x \to 0^-} \dfrac{x + 3}{x^2 + 3x}$

51. $\lim_{x \to 0^+} \dfrac{x - 2}{x^2 - 2x}$

52. $\lim_{x \to 0^+} \dfrac{x + 3}{x^2 + 3x}$

53. $\lim_{x \to 0} \dfrac{x - 2}{x^2 - 2x}$

54. $\lim_{x \to 0} \dfrac{x + 3}{x^2 + 3x}$

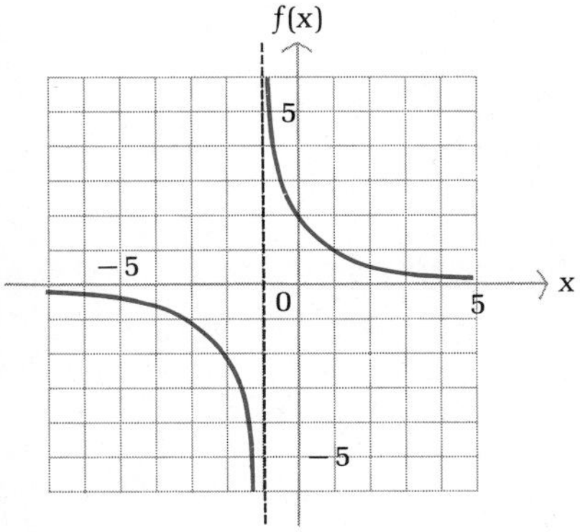

Use the graph of the function f shown in the margin to answer Problems 55 and 56.

55. (A) $\lim_{x \to \infty} f(x) = ?$ (B) $\lim_{x \to \infty} f(x) = ?$
(C) Write the equation of any horizontal asymptote.

56. (A) $\lim_{x \to -1^-} f(x) = ?$ (B) $\lim_{x \to -1^+} f(x) = ?$ (C) $\lim_{x \to -1} f(x) = ?$
(D) Write the equation of any vertical asymptote.

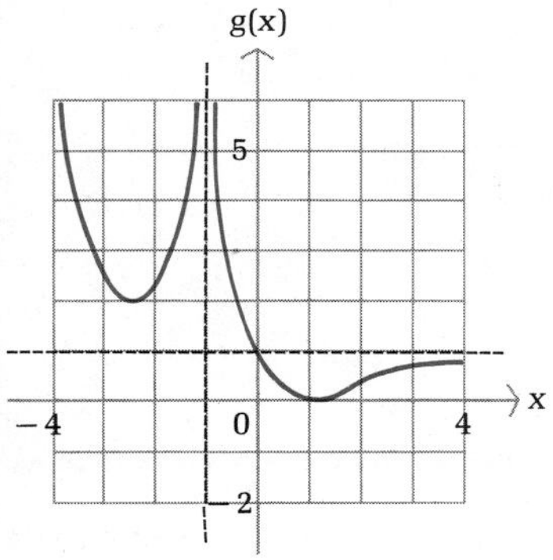

Use the graph of the function g shown in the margin to answer Problems 57 and 58.

57. (A) $\lim_{x \to -1^-} g(x) = ?$ (B) $\lim_{x \to -1^+} g(x) = ?$ (C) $\lim_{x \to -1} g(x) = ?$
(D) Write the equation of any vertical asymptote.

58. (A) $\lim_{x \to \infty} g(x) = ?$ (B) $\lim_{x \to -\infty} g(x) = ?$
(C) Write the equation of any horizontal asymptote.

Compute the following limit for each function in Problems 59–68:

$$\lim_{h \to 0} \frac{f(2 + h) - f(2)}{h}$$

59. $f(x) = 3x + 1$
60. $f(x) = 5x - 1$
61. $f(x) = x^2 + 1$
62. $f(x) = x^2 - 2$
63. $f(x) = 5$
64. $f(x) = -2$
65. $f(x) = \sqrt{x} - 2$
66. $f(x) = 1 + \sqrt{x}$
67. $f(x) = |x - 2| - 3$
68. $f(x) = 2 + |x - 2|$

C *Find each limit in Problems 69–82, if it exists. (Use algebraic manipulation where necessary and $-\infty$ or ∞ where appropriate.) [Recall: $a^3 - b^3 = (a - b)(a^2 + ab + b^2)$ and $a^3 + b^3 = (a + b)(a^2 - ab + b^2)$.]*

69. $\lim_{x \to 1} \sqrt{x^2 + 2x}$

70. $\lim_{x \to 1} \sqrt{2x^4 + 1}$

71. $\lim_{x \to 4} \sqrt[3]{x^2 - 3x}$

72. $\lim_{x \to 2} \sqrt[3]{x^3 - 2x}$

73. $\lim_{x \to 2} \dfrac{4}{(x - 2)^2}$

74. $\lim_{x \to 0} \dfrac{3}{|x|}$

75. $\lim_{x \to \infty} \left(\dfrac{1}{x^2} + \dfrac{1}{\sqrt{x}}\right)$

76. $\lim_{x \to \infty} \left(\dfrac{1}{\sqrt[3]{x}} - \dfrac{2}{x^3}\right)$

77. $\lim_{x \to 0} \left(\sqrt{x^2 + 9} - \dfrac{x^2 + 3x}{x}\right)$

78. $\lim_{x \to 1} \left(\dfrac{x^2 - 1}{x - 1} + \sqrt{x^2 + 3}\right)$

79. $\lim_{x \to 4} \frac{\sqrt{x} - 2}{x - 4}$

80. $\lim_{x \to 0} \frac{\sqrt{x + 4} - 2}{x}$

81. $\lim_{x \to 2} \frac{x^3 - 8}{x - 2}$

82. $\lim_{x \to -1} \frac{x^2 - 1}{x^3 + 1}$

83. Find each limit, if it exists. (Use $-\infty$ or ∞, as appropriate.)

(A) $\lim_{x \to -2^-} \frac{2}{x + 2}$ (B) $\lim_{x \to -2^+} \frac{2}{x + 2}$ (C) $\lim_{x \to -2} \frac{2}{x + 2}$

(D) Is $x = -2$ a vertical asymptote?

84. Find each limit, if it exists. (Use $-\infty$ or ∞, as appropriate.)

(A) $\lim_{x \to 1^-} \frac{1}{x - 1}$ (B) $\lim_{x \to 1^+} \frac{1}{x - 1}$ (C) $\lim_{x \to 1} \frac{1}{x - 1}$

(D) Is $x = 1$ a vertical asymptote?

Find each limit in Problems 85–88, where a is a real constant.

85. $\lim_{h \to 0} \frac{(a + h)^2 - a^2}{h}$

86. $\lim_{h \to 0} \frac{[3(a + h) - 2] - (3a - 2)}{h}$

87. $\lim_{h \to 0} \frac{\sqrt{a + h} - \sqrt{a}}{h}, \quad a > 0$

88. $\lim_{h \to 0} \frac{\frac{1}{a + h} - \frac{1}{a}}{h}, \quad a \neq 0$

APPLICATIONS

Business & Economics

89. *Average cost.* The cost equation for manufacturing a particular compact disk album is

$$C(x) = 20{,}000 + 3x$$

where x is the number of disks produced. The average cost per disk, denoted by $\overline{C}(x)$, is found by dividing $C(x)$ by x:

$$\overline{C}(x) = \frac{C(x)}{x} = \frac{20{,}000 + 3x}{x}$$

If only 10 disks were manufactured, for example, the average cost per disk would be $2,003. Find:

(A) $\overline{C}(1{,}000)$ (B) $\overline{C}(100{,}000)$ (C) $\lim_{x \to 10{,}000} \overline{C}(x)$ (D) $\lim_{x \to \infty} \overline{C}(x)$

90. *Employee training.* A company producing computer components has established that on the average, a new employee can assemble $N(t)$ components per day after t days of on-the-job training, as given by

$$N(t) = \frac{100t}{t + 9}$$

(See the figure below.) Find:

(A) $N(1)$ (B) $N(11)$ (C) $\lim_{t\to 11} N(t)$ (D) $\lim_{t\to\infty} N(t)$

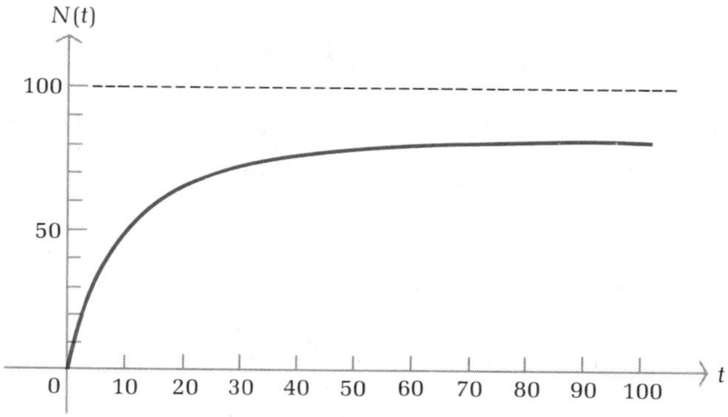

91. *Compound interest.* If \$100 is invested at 8% compounded n times per year, then the amount in the account $A(n)$ at the end of 1 year is given by

$$A(n) = 100\left(1 + \frac{0.08}{n}\right)^n$$

COMPOUNDED	n	$A(n)$
Annually	1	\$108.00
Semiannually	2	\$108.16
Quarterly	4	\$108.24
Monthly	12	
Weekly	52	
Daily	365	
Hourly	8,760	

(A) Use a calculator with a y^x key to complete the table in the margin. (Give each entry to the nearest cent.)

(B) Using the results of part A, guess the following limit:

$$\lim_{n\to\infty} A(n) = ?$$

(This problem leads to the important concept of compounding continuously, which will be discussed in detail in Section 6-1.)

92. *Pollution.* In Silicon Valley (in California), a number of computer-related manufacturing firms were found to be contaminating underground water supplies with toxic chemicals stored in leaking underground containers. A water quality control agency ordered the companies to take immediate corrective action and to contribute to a monetary pool for testing and cleanup of the underground contamination. Suppose the required monetary pool (in millions of dollars) for the testing and cleanup is estimated by

$$P(x) = \frac{2x}{1 - x}$$

where x is the percentage (expressed as a decimal fraction) of the total contaminant removed.

PERCENTAGE REMOVED	POOL REQUIRED
0.50 (50%)	\$2 million
0.60 (60%)	\$3 million
0.70 (70%)	
0.80 (80%)	
0.90 (90%)	
0.95 (95%)	
0.99 (99%)	

(A) Complete the table in the margin.

(B) Find $\lim_{x\to 0.80} P(x)$.

(C) What happens to the required monetary pool as the desired percentage of contaminant removed approaches 100% (x approaches 1 from the left)?

Life Sciences

93. *Medicine.* A drug is injected into the bloodstream of a patient through her right arm. The concentration of the drug in the bloodstream of the left arm t hours after the injection is given by

$$C(t) = \frac{0.14t}{t^2 + 1}$$

(See the figure below.) Find:

(A) $C(0.5)$ (B) $C(1)$ (C) $\lim_{t \to 1} C(t)$ (D) $\lim_{t \to \infty} C(t)$

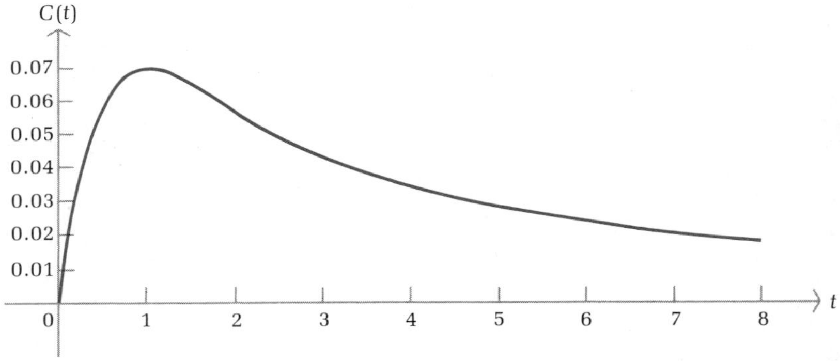

94. *Physiology.* In a study on the speed of muscle contraction in frogs under various loads, researchers W. O. Fems and J. Marsh found that the speed of contraction decreases with increasing loads. More precisely, they found that the relationship between speed of contraction S (in centimeters per second) and load w (in grams) is given approximately by

$$S(w) = \frac{26 + 0.06w}{w} \qquad w > 5$$

Find:

(A) $S(10)$ (B) $S(50)$ (C) $\lim_{w \to 50} S(w)$ (D) $\lim_{w \to \infty} S(w)$

Social Sciences

95. *Psychology—learning theory.* In 1917, L. L. Thurstone, a pioneer in quantitative learning theory, proposed the function

$$f(x) = \frac{a(x + c)}{(x + c) + b}$$

to describe the number of successful acts per unit time that a person could accomplish after x practice sessions. Suppose that for a particular person enrolling in a typing school,

$$f(x) = \frac{60(x + 1)}{x + 5}$$

where $f(x)$ is the number of words per minute the person is able to type after x weeks of lessons. (See the figure below.) Find:

(A) $f(3)$ (B) $f(10)$ (C) $\lim_{x\to 10} f(x)$ (D) $\lim_{x\to\infty} f(x)$

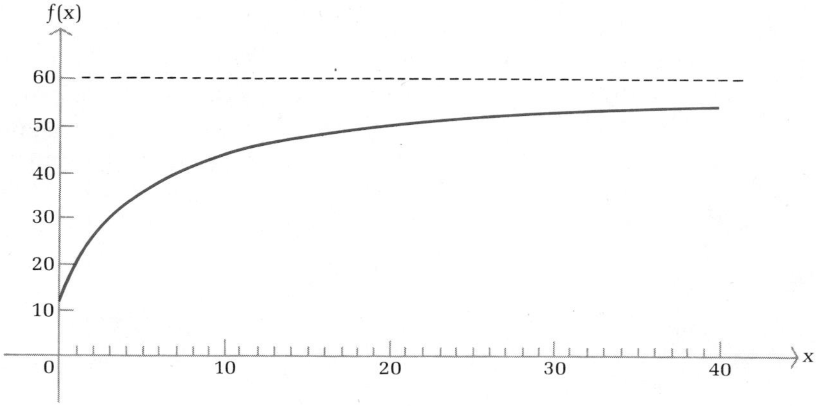

96. *Psychology—retention.* An experiment on retention is conducted in a psychology class. Each student in the class is given 1 day to memorize the same list of 30 special characters. The lists are turned in at the end of the day, and for each succeeding day for 30 days each student is asked to turn in a list of as many of the symbols as can be recalled. Averages are taken, and it is found that

$$N(t) = \frac{5t + 20}{t} \qquad t \geq 1$$

provides a good approximation of the average number of symbols, $N(t)$, retained after t days. (See the figure below.) Find:

(A) $N(2)$ (B) $N(10)$ (C) $\lim_{t\to 10} N(t)$ (D) $\lim_{t\to\infty} N(t)$

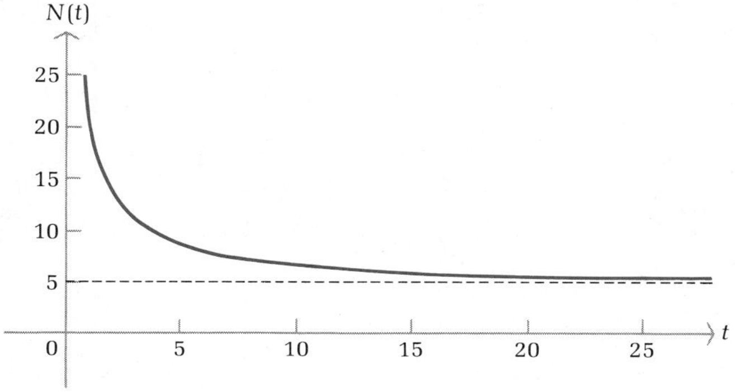

SECTION 4-3 The Derivative

- ◆ TANGENT LINES
- ◆ AVERAGE AND INSTANTANEOUS RATES OF CHANGE
- ◆ THE DERIVATIVE
- ◆ NONEXISTENCE OF THE DERIVATIVE
- ◆ SUMMARY

We now use the concept of limit to solve two of the important problems illustrated in Figure 1 at the beginning of this chapter. These two problems are repeated in Figure 10. The solution of these two apparently unrelated problems involves a common concept called *the derivative*, which we will be able to define after we have solved the problems illustrated in Figure 10.

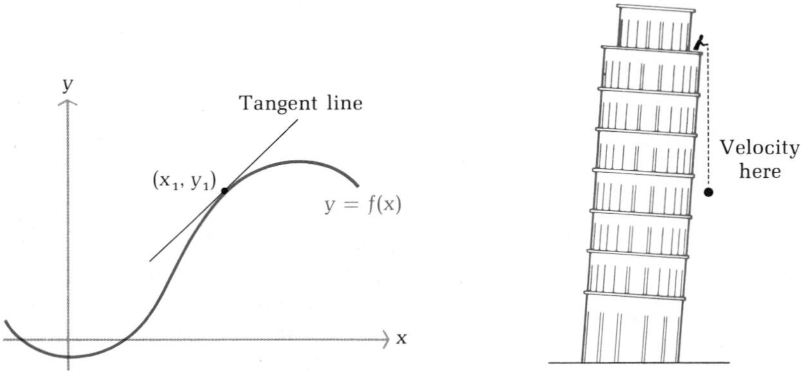

(A) Find the equation of the tangent line at (x_1, y_1) given $y = f(x)$

(B) Find the instantaneous velocity of a falling object

FIGURE 10

◆ TANGENT LINES

From plane geometry, we know that a tangent to a circle is a line that passes through one and only one point on the circle. This definition is not satisfactory for graphs of functions in general, as shown in Figure 11.

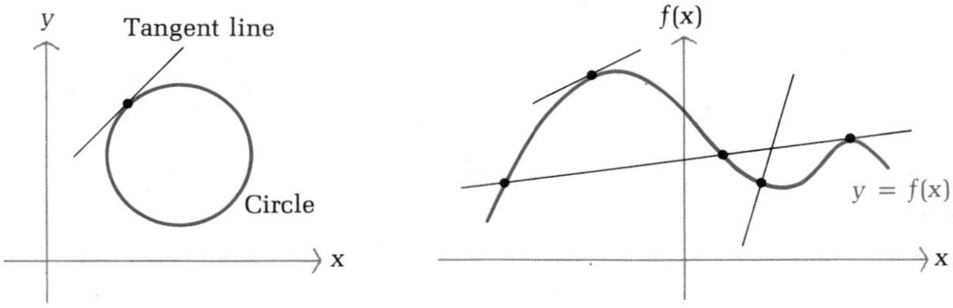

(A) (B)

FIGURE 11

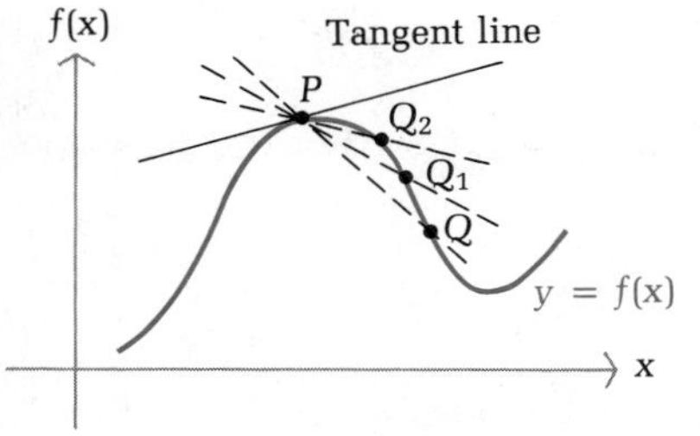

FIGURE 12

To define a tangent line for the graph of a function f at a point P on its graph, we proceed as follows. Select another point Q on the graph of f, and draw a line through P and Q. This line is called a **secant line.** Let Q approach P along the graph, as indicated by Q_1 and Q_2 in Figure 12. If the slopes of the corresponding secant lines have a limit m as Q approaches P, then m is defined to be the *slope of the tangent line* at P and the *slope of the graph of f* at P. (These definitions will be made more precise below.) If we know the slope of the tangent line and the coordinates of P, we can write the equation of the tangent line at P using the point–slope formula for a line (from Section 2-3).

To find the slope of the tangent line at a particular point $P(x_1, f(x_1))$ on the graph of f, we locate any other point Q on the graph, as indicated in Figure 13. Since h can be positive or negative (but not equal to 0), Q can be on the right or on the left of P (but cannot coincide with P). Both cases are illustrated in Figure 13.

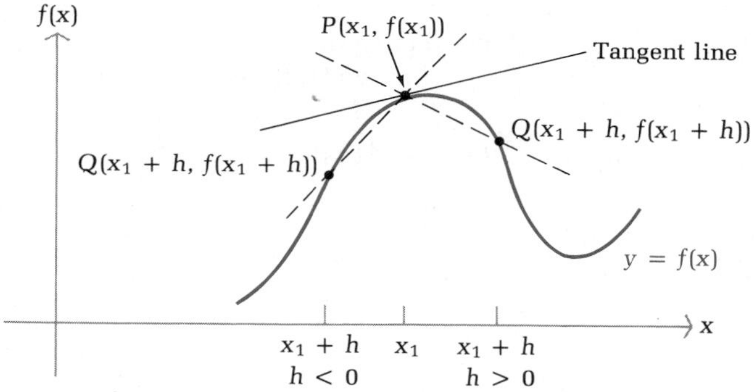

FIGURE 13

We now write the slope of the secant line through P and Q, using the slope formula, $m = (y_2 - y_1)/(x_2 - x_1)$, from Section 2-3:

$$\text{Slope of secant line} = \frac{f(x_1 + h) - f(x_1)}{(x_1 + h) - x_1}$$

$$= \frac{f(x_1 + h) - f(x_1)}{h} \qquad (1)$$

We refer to (1) as a **difference quotient.** Now, as h approaches 0 from either side of 0, but not equal to 0, Q will approach P along the graph of f (assuming f is continuous at $x = x_1$). If the limit of the difference quotient (1) exists as $h \to 0$, we define this limit to be the slope of the tangent line to the graph of f at the point P.

It is useful to note that the limit used to define the slope of a tangent line (and the slope of a graph) is usually a 0/0 indeterminate form—one of the reasons we placed special emphasis on this form earlier.

Tangent Line

Given the graph of $y = f(x)$, the **tangent line** at $x = x_1$ is the line that passes through the point $(x_1, f(x_1))$ with slope

$$\textbf{Slope of tangent line} = \lim_{h\to 0} \frac{f(x_1 + h) - f(x_1)}{h}$$

if the limit exists. The slope of the tangent line is also referred to as the **slope of the graph** at $(x_1, f(x_1))$.

◆ EXAMPLE 11 Given $f(x) = x^2$:

(A) Find the slope of the tangent line at $x = 1$.
(B) Find the equation of the tangent line at $x = 1$; that is, through $(1, f(1))$.
(C) Sketch the graph of f, the tangent line at $(1, f(1))$, and the secant line passing through $(1, f(1))$ and $(2, f(2))$.

Solutions (A) Slope of the tangent line:

Step 1. Write the difference quotient and simplify:

$$\frac{f(1+h) - f(1)}{h} = \frac{(1+h)^2 - 1^2}{h}$$

This is the slope of a secant line passing through $(1, f(1))$ and $(1 + h, f(1 + h))$.

$$= \frac{1 + 2h + h^2 - 1}{h}$$

$$= \frac{2h + h^2}{h} = \frac{h(2+h)}{h} = 2 + h \qquad h \neq 0$$

Step 2. Find the limit of the difference quotient:

$$\text{Slope of tangent line} = \lim_{h\to 0} \frac{f(1+h) - f(1)}{h}$$

$$= \lim_{h\to 0} (2 + h) = 2$$

This is also the slope of the graph of $f(x) = x^2$ at $(1, f(1))$.

(B) Equation of the tangent line: The tangent line passes through $(1, f(1)) = (1, 1)$ with slope $m = 2$ (from part A). The point–slope formula from Section 2-3 gives its equation:

$$y - y_1 = m(x - x_1) \qquad \text{Point–slope formula}$$
$$y - 1 = 2(x - 1)$$
$$y = 2x - 1 \qquad \text{Tangent line equation}$$

(C)

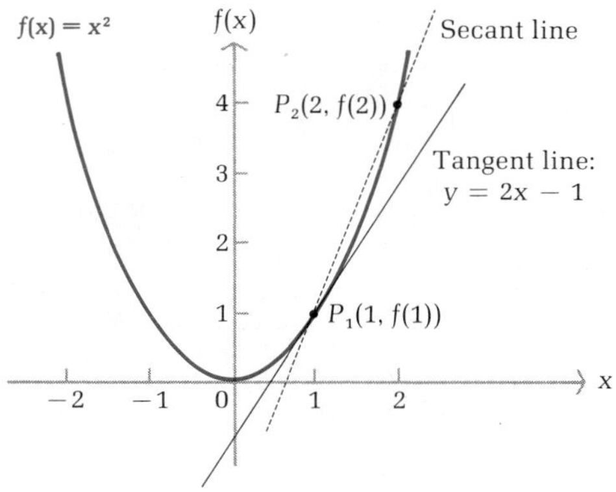

◆

PROBLEM 11 Find the slope of the tangent line for the graph of $f(x) = x^2$ at $x = 2$, and write the equation of the tangent line in the form $y = mx + b$. ◆

◆ AVERAGE AND INSTANTANEOUS RATES OF CHANGE

We now turn to the second problem stated at the beginning of this section.

◆ EXAMPLE 12 A small steel ball dropped from a tower will fall a distance of y feet in x seconds, as given approximately by the formula (from physics)

$$y = f(x) = 16x^2$$

Figure 14 shows the position of the ball on a coordinate line (positive direction down) at the end of 0, 1, 2, and 3 seconds. Note that the distances fallen between these times increases with time; thus, the velocity of the ball is increasing as it falls. Our main objective in this example is to find the velocity of the ball at a given instant, say, at the end of 2 seconds. (*Velocity* is the rate of change of distance with respect to time.)

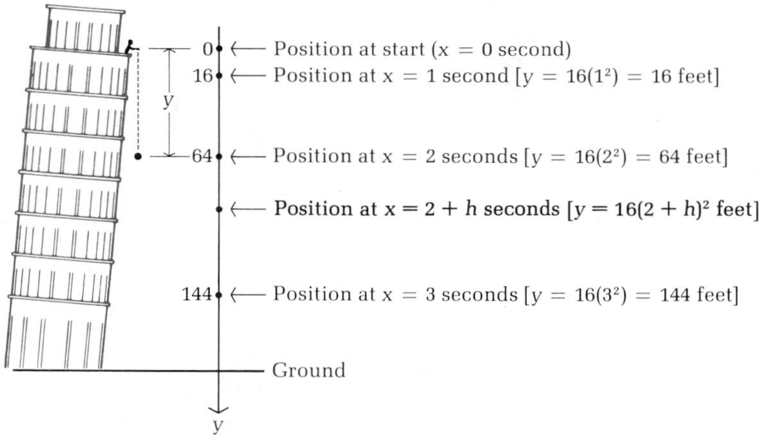

FIGURE 14
Note: Positive y direction is down.

(A) Find the average velocity from $x = 2$ seconds to $x = 3$ seconds.
(B) Find the average velocity from $x = 2$ seconds to $x = 2 + h$ seconds, $h \neq 0$.
(C) Find the limit of the expression from part B as $h \to 0$, if it exists. (What do you think the limit represents?)

Solutions

(A) Average velocity from $x = 2$ seconds to $x = 3$ seconds: Recall the formula $d = rt$, which can be written in the form

$$r = \frac{d}{t} = \frac{\text{Distance covered}}{\text{Elapsed time}} = \text{Average rate}$$

For example, if a person drives from San Francisco to Los Angeles (a distance of about 420 miles) in 7 hours, then the average velocity is

$$r = \frac{d}{t} = \frac{420}{7} = 60 \text{ miles per hour}$$

Sometimes the person will be traveling faster and sometimes slower, but the average velocity is 60 miles per hour. In our present problem, the average velocity of the steel ball is given by

$$\begin{aligned}\text{Average velocity} &= \frac{\text{Distance covered}}{\text{Elapsed time}}\\ &= \frac{f(3) - f(2)}{3 - 2}\\ &= \frac{16 \cdot 3^2 - 16 \cdot 2^2}{1} = 80 \text{ feet per second}\end{aligned}$$

(B) Average velocity from $x = 2$ seconds to $x = 2 + h$ seconds, $h \neq 0$: Proceeding as in part A,

$$\begin{aligned}\text{Average velocity} &= \frac{\text{Distance covered}}{\text{Elapsed time}}\\ &= \frac{f(2 + h) - f(2)}{h} && \text{Difference quotient}\\ &= \frac{16(2 + h)^2 - 16(2)^2}{h} && \text{0/0 indeterminate form requires simplification (for part C).}\\ &= \frac{64 + 64h + 16h^2 - 64}{h}\\ &= \frac{h(64 + 16h)}{h} = 64 + 16h && h \neq 0\end{aligned}$$

Note that if $h = 1$, the average velocity is 80 feet per second (our result in part A); if $h = 0.5$, the average velocity is 72 feet per second; if $h = -0.01$, the average velocity is 63.84 feet per second; and so on. The closer to 0 that h gets (on either side of 0), the closer the average velocity is to 64 feet per second.

(C) Limit of the expression from part B as $h \to 0$:

$$\lim_{h \to 0} \frac{f(2+h) - f(2)}{h} = \lim_{h \to 0} (64 + 16h)$$
$$= 64 \text{ feet per second}$$ ◆

We call the answer to Example 12C, 64 feet per second, the *instantaneous velocity* (or the *instantaneous rate of change*) at $x = 2$ seconds, and we have solved the second basic problem stated at the beginning of this chapter!

Average and Instantaneous Rates of Change

For $y = f(x)$, the **average rate of change from $x = x_1$ to $x = x_1 + h$** is

$$\frac{f(x_1 + h) - f(x_1)}{h} \qquad h \neq 0$$

and the **instantaneous rate of change at $x = x_1$** is

$$\lim_{h \to 0} \frac{f(x_1 + h) - f(x_1)}{h}$$

if the limit exists.

Note that the limit used to define instantaneous rate of change is usually a 0/0 indeterminate form—another reason we placed special emphasis on this form earlier.

PROBLEM 12 For the falling steel ball in Example 12, find:

(A) The average velocity from $x = 1$ second to $x = 2$ seconds.
(B) The average velocity from $x = 1$ second to $x = 1 + h$ seconds, $h \neq 0$.
(C) The instantaneous velocity (instantaneous rate of change) at $x = 1$. ◆

◆ THE DERIVATIVE

In the last two examples, we found that the special limit

$$\lim_{h \to 0} \frac{f(x_1 + h) - f(x_1)}{h} \qquad (2)$$

if it exists, gives us the slope of the tangent line to the graph of $y = f(x)$ at $(x_1, f(x_1))$ and also the instantaneous rate of change of y with respect to x at $x = x_1$. Many other applications give rise to this form. In fact, the limit (2) is of such basic importance to calculus and to the applications of calculus that we give it a special name and study it in detail. To keep form (2) simple and general, we drop the subscript on x_1 and think of the difference quotient

$$\frac{f(x+h)-f(x)}{h}$$

as a function of h, with x held fixed as we let h tend to 0. We are now ready to define one of the basic concepts in calculus, *the derivative:*

The Derivative

For $y = f(x)$, we define the **derivative of f at x,** denoted by **$f'(x)$,** to be

$$f'(x) = \lim_{h \to 0} \frac{f(x+h)-f(x)}{h} \qquad \text{if the limit exists}$$

If $f'(x)$ exists for each x in the open interval (a, b), then f is said to be **differentiable** over (a, b).

(Differentiability from the left or from the right is defined using $h \to 0^-$ or $h \to 0^+$, respectively, in place of $h \to 0$ in the above definition.)

The process of finding the derivative of a function is called **differentiation.** That is, the derivative of a function is obtained by **differentiating** the function.

The Function $f'(x)$

The derivative of a function f is a new function f' that gives us, among other things, the slope of the tangent line to the graph of $y = f(x)$ for each x and the instantaneous rate of change of $y = f(x)$ with respect to x. The domain of f' is a subset of the domain of f.

◆ EXAMPLE 13 Find $f'(x)$, the derivative of f at x, for $f(x) = 4x - x^2$.

Solution To find $f'(x)$, we use a two-step process:

Step 1. Form the difference quotient and simplify:

$$\begin{aligned}\frac{f(x+h)-f(x)}{h} &= \frac{[4(x+h)-(x+h)^2]-(4x-x^2)}{h}\\ &= \frac{4x+4h-x^2-2xh-h^2-4x+x^2}{h}\\ &= \frac{4h-2xh-h^2}{h}\\ &= \frac{h(4-2x-h)}{h}\\ &= 4-2x-h \qquad h \neq 0\end{aligned}$$

Step 2. Find the limit of the difference quotient:

$$f'(x) = \lim_{h \to 0} \frac{f(x+h) - f(x)}{h}$$

$$= \lim_{h \to 0} (4 - 2x - h) = 4 - 2x$$

Thus, if $f(x) = 4x - x^2$, then $f'(x) = 4 - 2x$. The derivative f' is a new function derived from the function f. ◆

PROBLEM 13 Find $f'(x)$, the derivative of f at x, for $f(x) = 8x - 2x^2$. ◆

◆ EXAMPLE 14 In Example 13, we started with the function specified by $f(x) = 4x - x^2$ and found the derivative of f at x to be $f'(x) = 4 - 2x$. Thus, the slope of a tangent line to the graph of f at any point $(x, f(x))$ on the graph of f is

$$m = f'(x) = 4 - 2x$$

(A) Find the slope of the graph of f at $x = 0$, 2, and 3.
(B) Graph $y = f(x) = 4x - x^2$, and use the slopes found in part A to make a rough sketch of the tangent lines to the graph at $x = 0$, 2, and 3.

Solutions (A) Using $f'(x) = 4 - 2x$, we have

$m = f'(0) = 4 - 2(0) = 4$	Slope at $x = 0$
$m = f'(2) = 4 - 2(2) = 0$	Slope at $x = 2$
$m = f'(3) = 4 - 2(3) = -2$	Slope at $x = 3$

(B)

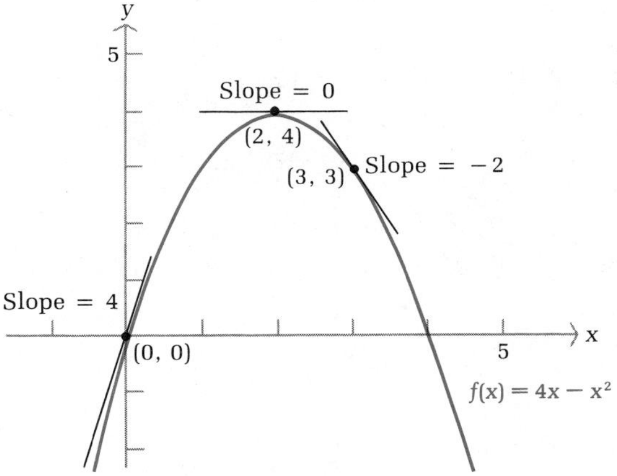

◆

PROBLEM 14 In Problem 13, we started with the function specified by $f(x) = 8x - 2x^2$. Using the derivative found there:

(A) Find the slope of the graph of f at $x = 1$, 2, and 4.
(B) Graph $y = f(x) = 8x - 2x^2$, and use the slopes from part A to make a rough sketch of the tangent lines to the graph at $x = 1$, 2, and 4. ◆

◆ EXAMPLE 15 Find $f'(x)$, the derivative of f at x, for $f(x) = \sqrt{x} + 2$.

Solution To find $f'(x)$, we find

$$\lim_{h \to 0} \frac{f(x+h) - f(x)}{h}$$

using the two-step process.

Step 1. Form the difference quotient and simplify:

$$\frac{f(x+h) - f(x)}{h} = \frac{(\sqrt{x+h} + 2) - (\sqrt{x} + 2)}{h}$$
$$= \frac{\sqrt{x+h} - \sqrt{x}}{h}$$

Since this is a 0/0 indeterminate form, we change the form by rationalizing the numerator:

$$\frac{\sqrt{x+h} - \sqrt{x}}{h} \cdot \frac{\sqrt{x+h} + \sqrt{x}}{\sqrt{x+h} + \sqrt{x}} = \frac{x + h - x}{h(\sqrt{x+h} + \sqrt{x})}$$
$$= \frac{h}{h(\sqrt{x+h} + \sqrt{x})}$$
$$= \frac{1}{\sqrt{x+h} + \sqrt{x}} \qquad h \neq 0$$

Step 2. Find the limit of the difference quotient:

$$f'(x) = \lim_{h \to 0} \frac{f(x+h) - f(x)}{h}$$
$$= \lim_{h \to 0} \frac{1}{\sqrt{x+h} + \sqrt{x}}$$
$$= \frac{1}{\sqrt{x} + \sqrt{x}} = \frac{1}{2\sqrt{x}} \qquad x > 0$$

Thus, the derivative of $f(x) = \sqrt{x} + 2$ is $f'(x) = 1/(2\sqrt{x})$, a new function. The domain of f is $[0, \infty)$. Since $f'(0)$ is not defined, the domain of f' is $(0, \infty)$, a subset of the domain of f. ◆

PROBLEM 15 Find $f'(x)$ for $f(x) = x^{-1}$. ◆

◆ NONEXISTENCE OF THE DERIVATIVE

The existence of a derivative at $x = a$ depends on the existence of a limit at $x = a$; that is, on the existence of

$$f'(a) = \lim_{h \to 0} \frac{f(a+h) - f(a)}{h} \tag{3}$$

If the limit does not exist at $x = a$, we say that the function f is **nondifferentiable at $x = a$**, or **$f'(a)$ does not exist.**

How can we recognize the points on the graph of f where $f'(a)$ does not exist? It is impossible to describe all the ways that the limit in (3) can fail to exist. However, we can illustrate some common situations where $f'(a)$ does fail to exist:

1. If f is not continuous at $x = a$, then $f'(a)$ does not exist (Figure 15A). Or, equivalently, **if f is differentiable at $x = a$, then f must be continuous at $x = a$.**
2. If the graph of f has a sharp corner at $x = a$, then $f'(a)$ does not exist and the graph has no tangent line at $x = a$ (Figure 15B). (In Figure 15B, the left- and right-hand derivatives exist but are not equal.)
3. If the graph of f has a vertical tangent line at $x = a$, then $f'(a)$ does not exist (Figure 15C and D).

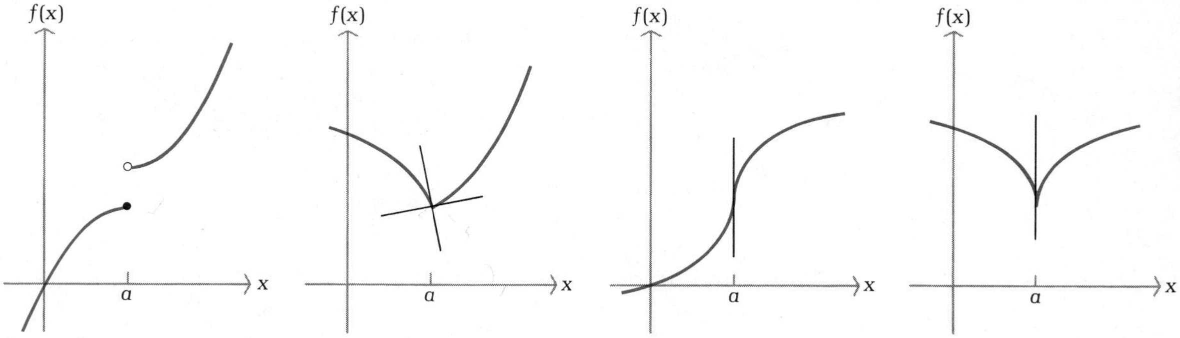

(A) Not continuous at $x = a$ (B) Graph has sharp corner at $x = a$ (C) Vertical tangent at $x = a$ (D) Vertical tangent at $x = a$

FIGURE 15
The function f is nondifferentiable at $x = a$.

◆ SUMMARY

The concept of the derivative is a very powerful mathematical idea, and its applications are many and varied. In the next three sections we will develop formulas and general properties of derivatives that will enable us to find the derivatives of many functions without having to go through the two-step limiting process each time.

Answers to Matched Problems

11. $y = 4x - 4$
12. (A) 48 ft/sec (B) $32 + 16h$ (C) 32 ft/sec
13. $f'(x) = 8 - 4x$
14. (A) $f'(1) = 4$, $f'(2) = 0$, $f'(4) = -8$ (B)

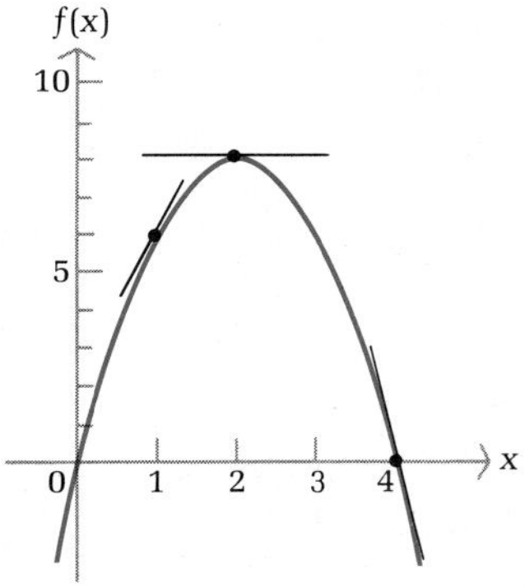

15. $f'(x) = -1/x^2$, or $-x^{-2}$

EXERCISE 4-3

A

Use $f(x) = 3x - 2$ in Problems 1 and 2 to find the given difference quotients and limits.

1. (A) $\dfrac{f(1+h) - f(1)}{h}$ (B) $\lim_{h \to 0} \dfrac{f(1+h) - f(1)}{h}$
2. (A) $\dfrac{f(2+h) - f(2)}{h}$ (B) $\lim_{h \to 0} \dfrac{f(2+h) - f(2)}{h}$

Use $f(x) = 2x^2$ in Problems 3 and 4 to find the given difference quotients and limits.

3. (A) $\dfrac{f(2+h) - f(2)}{h}$ (B) $\lim_{h \to 0} \dfrac{f(2+h) - f(2)}{h}$
4. (A) $\dfrac{f(3+h) - f(3)}{h}$ (B) $\lim_{h \to 0} \dfrac{f(3+h) - f(3)}{h}$

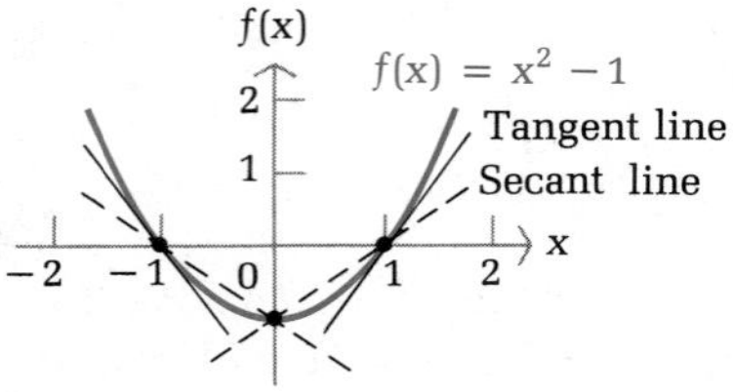

Solve Problems 5 and 6 for the graph of $y = f(x) = x^2 - 1$ (see the figure in the margin).

5. (A) Find the slope of the secant line joining $(0, f(0))$ and $(1, f(1))$.
   (B) Find the slope of the secant line joining $(1, f(1))$ and $(1 + h, f(1 + h))$.
   (C) Find the slope of the tangent line at $x = 1$.
6. (A) Find the slope of the secant line joining $(-1, f(-1))$ and $(0, f(0))$.
   (B) Find the slope of the secant line joining $(-1, f(-1))$ and $(-1 + h, f(-1 + h))$.
   (C) Find the slope of the tangent line at $x = -1$.

In Problems 7–10, find $f'(x)$ using the two-step process:

Step 1. *Simplify:* $\dfrac{f(x+h)-f(x)}{h}$

Step 2. *Evaluate:* $\lim\limits_{h\to 0}\dfrac{f(x+h)-f(x)}{h}$

Then find $f'(1)$, $f'(2)$, and $f'(3)$.

7. $f(x) = 2x - 3$

8. $f(x) = 4x + 3$

9. $f(x) = 2 - x^2$

10. $f(x) = 2x^2 + 5$

B

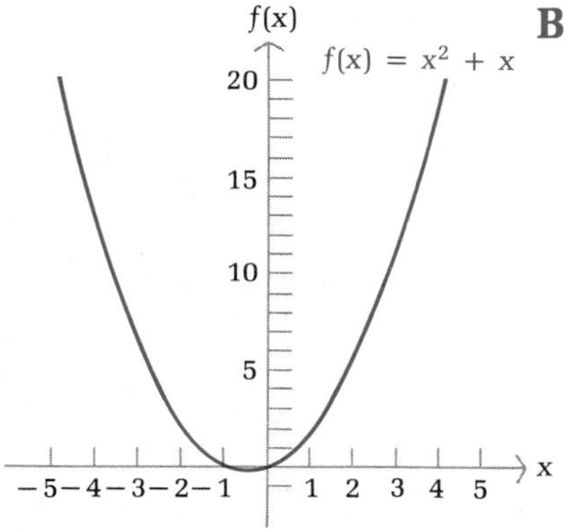

Problems 11 and 12 refer to the graph of $y = f(x) = x^2 + x$ shown in the margin.

11. (A) Find the slope of the secant line joining $(1, f(1))$ and $(3, f(3))$.
(B) Find the slope of the secant line joining $(1, f(1))$ and $(1 + h, f(1 + h))$.
(C) Find the slope of the tangent line at $(1, f(1))$.
(D) Find the equation of the tangent line at $(1, f(1))$.

12. (A) Find the slope of the secant line joining $(2, f(2))$ and $(4, f(4))$.
(B) Find the slope of the secant line joining $(2, f(2))$ and $(2 + h, f(2 + h))$.
(C) Find the slope of the tangent line at $(2, f(2))$.
(D) Find the equation of the tangent line at $(2, f(2))$.

In Problems 13 and 14, suppose an object moves along the y axis so that its location is $y = f(x) = x^2 + x$ at time x (y is in meters and x is in seconds). Find:

13. (A) The average velocity (the average rate of change of y with respect to x) for x changing from 1 to 3 seconds
(B) The average velocity for x changing from 1 to $1 + h$ seconds
(C) The instantaneous velocity at $x = 1$ second

14. (A) The average velocity (the average rate of change of y with respect to x) for x changing from 2 to 4 seconds
(B) The average velocity for x changing from 2 to $2 + h$ seconds
(C) The instantaneous velocity at $x = 2$ seconds

In Problems 15–20, find $f'(x)$ using the two-step limiting process. Then find $f'(1)$, $f'(2)$, and $f'(3)$.

15. $f(x) = 6x - x^2$

16. $f(x) = 2x - 3x^2$

17. $f(x) = \sqrt{x} - 3$

18. $f(x) = 2 - \sqrt{x}$

19. $f(x) = \dfrac{-1}{x}$

20. $f(x) = \dfrac{1}{x+1}$

Problems 21–28 refer to the function F in the graph below. Use the graph to determine whether F′(x) exists at the indicated value of x.

21. $x = a$ **22.** $x = b$ **23.** $x = c$ **24.** $x = d$
25. $x = e$ **26.** $x = f$ **27.** $x = g$ **28.** $x = h$

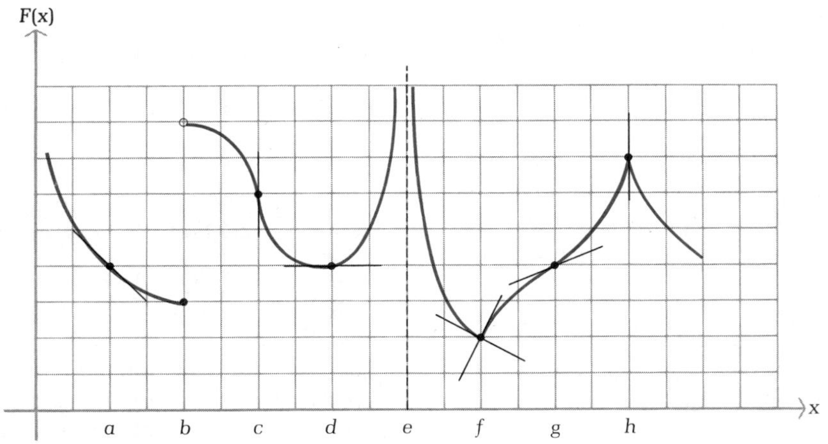

29. Given $f(x) = x^2 - 4x$:
(A) Find $f'(x)$.
(B) Find the slopes of the tangent lines to the graph of f at $x = 0, 2$, and 4.
(C) Graph f, and sketch in the tangent lines at $x = 0, 2$, and 4.

30. Given $f(x) = x^2 + 2x$:
(A) Find $f'(x)$.
(B) Find the slopes of the tangent lines to the graph of f at $x = -2, -1$, and 1.
(C) Graph f, and sketch in the tangent lines at $x = -2, -1$, and 1.

31. If an object moves along a line so that it is at $y = f(x) = 4x^2 - 2x$ at time x (in seconds), find the instantaneous velocity function $v = f'(x)$, and find the velocity at times $x = 1, 3$, and 5 seconds (y is measured in feet).

32. Repeat Problem 31 with $f(x) = 8x^2 - 4x$.

C *In Problems 33 and 34, sketch the graph of f and determine where f is nondifferentiable.*

33. $f(x) = \begin{cases} 2x & \text{if } x < 1 \\ 2 & \text{if } x \geq 1 \end{cases}$ **34.** $f(x) = \begin{cases} 2x & \text{if } x < 2 \\ 6 - x & \text{if } x \geq 2 \end{cases}$

In Problems 35–38, determine whether f is differentiable at x = 0 by considering

$$\lim_{h \to 0} \frac{f(0+h) - f(0)}{h}$$

35. $f(x) = |x|$ **36.** $f(x) = 1 - |x|$ **37.** $f(x) = x^{1/3}$ **38.** $f(x) = x^{2/3}$

39. Show that $f(x) = 2x - x^2$ is differentiable over the closed interval [0, 2] by showing that each of the following limits exists:

(A) $\lim_{h \to 0} \frac{f(x+h) - f(x)}{h}, \quad 0 < x < 2$ (B) $\lim_{h \to 0^+} \frac{f(0+h) - f(0)}{h}, \quad x = 0$

(C) $\lim_{h \to 0^-} \frac{f(2+h) - f(2)}{h}, \quad x = 2$

40. Show that $f(x) = \sqrt{x}$ is differentiable over the open interval $(0, \infty)$ but not over the half-closed interval $[0, \infty)$ by considering

$$\lim_{h \to 0} \frac{f(x+h) - f(x)}{h} \qquad 0 < x < \infty$$

and

$$\lim_{h \to 0^+} \frac{f(0+h) - f(0)}{h} \qquad x = 0$$

APPLICATIONS

Business & Economics

Problems 41 and 42 provide a first look at the concept of "marginal analysis," an important topic in business and economics that is developed further in succeeding sections and in detail in Section 4-7.

41. *Cost analysis.* The total cost per day, $C(x)$ (in hundreds of dollars), for manufacturing x windsurfing boards is given by

$$C(x) = 3 + 10x - x^2 \qquad 0 \leq x \leq 5$$

(A) Find: $\frac{C(4) - C(3)}{4 - 3}$

(This is the increase in cost for a 1 unit increase in production at the level of production of 3 boards per day.)

(B) Find: $\frac{C(3+h) - C(3)}{h}$

(C) Find: $C'(3)$
(This is called the *marginal cost* at the level of production of 3 boards per day; it is the instantaneous rate of change of cost relative to production and approximates the result found in part A.

(D) Find $C'(x)$, the *marginal cost function*.

(E) Using $C'(x)$, find $C'(1)$, $C'(2)$, $C'(3)$, and $C'(4)$, and interpret. [*Note:* As production levels increase, the rate of change of cost relative to production (marginal cost) decreases.]

42. *Cost analysis.* Repeat Problem 41 for

$$C(x) = 5 + 12x - x^2 \qquad 0 \leq x \leq 5$$

SECTION 4-4 Derivatives of Constants, Power Forms, and Sums

- DERIVATIVE OF A CONSTANT
- POWER RULE
- DERIVATIVE OF A CONSTANT TIMES A FUNCTION
- DERIVATIVES OF SUMS AND DIFFERENCES
- APPLICATIONS

In the preceding section, we defined the derivative of f at x as

$$f'(x) = \lim_{h \to 0} \frac{f(x+h) - f(x)}{h}$$

if the limit exists, and we used this definition and a two-step process to find the derivatives of several functions. In this and the next two sections, we will develop some rules based on this definition that will enable us to determine the derivatives of a rather large class of functions without having to go through the two-step process each time.

Before starting on these rules, we list some symbols that are widely used to represent derivatives:

Derivative Notation

Given $y = f(x)$, then

$$f'(x) \qquad y' \qquad \frac{dy}{dx} \qquad D_x f(x)$$

all represent the derivative of f at x.

Each of these symbols for the derivative has its particular advantage in certain situations. All of them will become familiar to you after a little experience.

◆ DERIVATIVE OF A CONSTANT

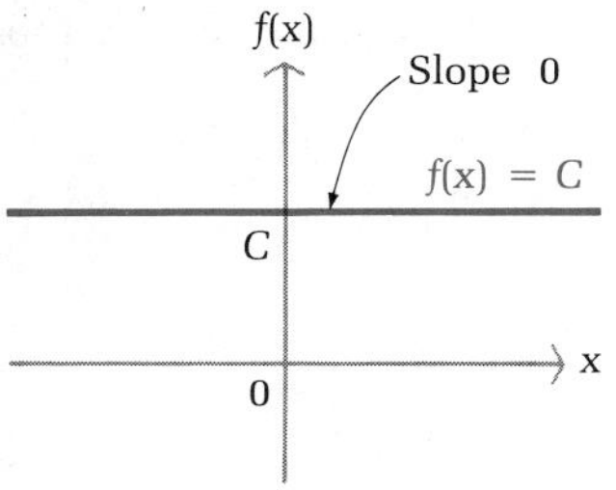

FIGURE 16

Suppose

$$f(x) = C \qquad C \text{ a constant} \qquad \text{A constant function}$$

Geometrically, the graph of $f(x) = C$ is a horizontal straight line with slope 0 (see Figure 16); hence, we would expect $f'(x) = 0$. We will show that this is actually the case using the definition of the derivative and the two-step process introduced earlier. We want to find

$$f'(x) = \lim_{h \to 0} \frac{f(x+h) - f(x)}{h} \qquad \text{Definition of } f'(x)$$

Step 1. $\dfrac{f(x+h) - f(x)}{h} = \dfrac{C - C}{h} = \dfrac{0}{h} = 0 \qquad h \neq 0$

Step 2. $\lim_{h \to 0} 0 = 0$

Thus,

$$f'(x) = 0$$

We conclude that:

The derivative of any constant is 0.

Derivative of a Constant

If $y = f(x) = C$, then

$$f'(x) = 0$$

Also, $y' = 0$, $dy/dx = 0$, and $D_x\, C = 0$.

[*Note:* When we write $D_x\, C = 0$, we mean $D_x\, f(x) = 0$, where $f(x) = C$.]

◆ EXAMPLE 16 (A) If $f(x) = 3$, then $f'(x) = 0$. (B) If $y = -1.4$, then $y' = 0$.
(C) If $y = \pi$, then $dy/dx = 0$. (D) $D_x\,(23) = 0$ ◆

PROBLEM 16 Find:

(A) $f'(x)$ for $f(x) = -24$ (B) y' for $y = 12$
(C) dy/dx for $y = -\sqrt{7}$ (D) $D_x(-\pi)$ ◆

◆ POWER RULE

Using the definition of derivative and the two-step process introduced in the preceding section, we can show that:

If $f(x) = x$, then $f'(x) = 1$.
If $f(x) = x^2$, then $f'(x) = 2x$.
If $f(x) = x^3$, then $f'(x) = 3x^2$.
If $f(x) = x^4$, then $f'(x) = 4x^3$.

In general, for any positive integer n:

If $f(x) = x^n$, then $f'(x) = nx^{n-1}$. (1)

In fact, more advanced techniques can be used to show that (1) holds for *any* real number n. We will assume this general result for the remainder of this book.

Power Rule

If $y = f(x) = x^n$, where n is a real number, then

$$f'(x) = nx^{n-1}$$

◆ EXAMPLE 17

(A) If $f(x) = x^5$, then $f'(x) = 5x^{5-1} = 5x^4$.
(B) If $y = x^{25}$, then $y' = 25x^{25-1} = 25x^{24}$.
(C) If $y = x^{-3}$, then $dy/dx = -3x^{-3-1} = -3x^{-4}$.
(D) $D_x\, x^{5/3} = \frac{5}{3}x^{(5/3)-1} = \frac{5}{3}x^{2/3}$ ◆

PROBLEM 17

Find:

(A) $f'(x)$ for $f(x) = x^6$ (B) y' for $y = x^{30}$
(C) dy/dx for $y = x^{-2}$ (D) $D_x\, x^{3/2}$ ◆

In some cases, properties of exponents must be used to rewrite an expression before the power rule is applied.

◆ EXAMPLE 18

(A) If $f(x) = 1/x^4$, then we can write $f(x) = x^{-4}$ and

$$f'(x) = -4x^{-4-1} = -4x^{-5} \quad \text{or} \quad \frac{-4}{x^5}$$

(B) If $y = \sqrt{x}$, then we can write $y = x^{1/2}$ and

$$y' = \frac{1}{2}x^{(1/2)-1} = \frac{1}{2}x^{-1/2} \quad \text{or} \quad \frac{1}{2\sqrt{x}}$$

(C) $$D_x \frac{1}{\sqrt[3]{x}} = D_x\, x^{-1/3} = -\frac{1}{3}x^{(-1/3)-1} = -\frac{1}{3}x^{-4/3} \quad \text{or} \quad \frac{-1}{3\sqrt[3]{x^4}}$$ ◆

PROBLEM 18 Find:

(A) $f'(x)$ for $f(x) = \frac{1}{x}$ (B) y' for $y = \sqrt[3]{x^2}$ (C) $D_x \frac{1}{\sqrt{x}}$ ◆

◆ DERIVATIVE OF A CONSTANT TIMES A FUNCTION

Let $f(x) = ku(x)$; where k is a constant and u is differentiable at x. Then, using the two-step process, we have the following:

Step 1. $$\frac{f(x+h) - f(x)}{h} = \frac{ku(x+h) - ku(x)}{h} = k\left[\frac{u(x+h) - u(x)}{h}\right]$$

Step 2. $$\lim_{h\to 0} \frac{f(x+h) - f(x)}{h} = \lim_{h\to 0} k\left[\frac{u(x+h) - u(x)}{h}\right] \qquad \lim_{x\to c} kg(x) = k \lim_{x\to c} g(x)$$

$$= k \lim_{h\to 0}\left[\frac{u(x+h) - u(x)}{h}\right] \qquad \text{Definition of } u'(x)$$

$$= ku'(x)$$

Thus:

The derivative of a constant times a differentiable function is the constant times the derivative of the function.

Constant Times a Function Rule

If $y = f(x) = ku(x)$, then

$$f'(x) = ku'(x)$$

Also, $y' = ku'$, $dy/dx = k\, du/dx$, and $D_x\, ku(x) = k\, D_x\, u(x)$.

◆ EXAMPLE 19 (A) If $f(x) = 3x^2$, then $f'(x) = 3 \cdot 2x^{2-1} = 6x$.

(B) If $y = \frac{x^3}{6} = \frac{1}{6}x^3$, then $\frac{dy}{dx} = \frac{1}{6} \cdot 3x^{3-1} = \frac{1}{2}x^2$.

(C) If $y = \frac{1}{2x^4} = \frac{1}{2}x^{-4}$, then $y' = \frac{1}{2}(-4x^{-4-1}) = -2x^{-5}$ or $\frac{-2}{x^5}$.

(D) $D_x \frac{4}{\sqrt{x^3}} = D_x \frac{4}{x^{3/2}} = D_x\, 4x^{-3/2} = 4\left[-\frac{3}{2}x^{(-3/2)-1}\right]$

$= -6x^{-5/2}$ or $-\frac{6}{\sqrt{x^5}}$ ◆

PROBLEM 19 Find:

(A) $f'(x)$ for $f(x) = 4x^5$ (B) $\frac{dy}{dx}$ for $y = \frac{x^4}{12}$

(C) y' for $y = \frac{1}{3x^3}$ (D) $D_x \frac{9}{\sqrt[3]{x}}$ ◆

◆ DERIVATIVES OF SUMS AND DIFFERENCES

Let $f(x) = u(x) + v(x)$, where $u'(x)$ and $v'(x)$ exist. Then, using the two-step process, we have the following:

Step 1.

$$\frac{f(x+h) - f(x)}{h} = \frac{[u(x+h) + v(x+h)] - [u(x) + v(x)]}{h}$$

$$= \frac{u(x+h) + v(x+h) - u(x) - v(x)}{h}$$

$$= \frac{u(x+h) - u(x)}{h} + \frac{v(x+h) - v(x)}{h}$$

Step 2.

$$\lim_{h \to 0} \frac{f(x+h) - f(x)}{h} = \lim_{h \to 0} \left[\frac{u(x+h) - u(x)}{h} + \frac{v(x+h) - v(x)}{h} \right]$$

$$\lim_{x \to c} [g(x) + h(x)] = \lim_{x \to c} g(x) + \lim_{x \to c} h(x)$$

$$= \lim_{h \to 0} \frac{u(x+h) - u(x)}{h} + \lim_{h \to 0} \frac{v(x+h) - v(x)}{h}$$

$$= u'(x) + v'(x)$$

Thus:

The derivative of the sum of two differentiable functions is the sum of the derivatives.

Similarly, we can show that:

The derivative of the difference of two differentiable functions is the difference of the derivatives.

Together, we then have the **sum and difference rule** for differentiation:

Sum and Difference Rule

If $y = f(x) = u(x) \pm v(x)$, then

$f'(x) = u'(x) \pm v'(x)$

[*Note:* This rule generalizes to the sum and difference of any given number of functions.]

With this and the other rules stated previously, we will be able to compute the derivatives of all polynomials and a variety of other functions.

◆ EXAMPLE 20

(A) If $f(x) = 3x^2 + 2x$, then

$$f'(x) \boxed{= (3x^2)' + (2x)' = 3(2x) + 2(1)} = 6x + 2$$

(B) If $y = 4 + 2x^3 - 3x^{-1}$, then

$$y' \boxed{= (4)' + (2x^3)' - (3x^{-1})' = 0 + 2(3x^2) - 3(-1)x^{-2}} = 6x^2 + 3x^{-2}$$

(C) If $y = \sqrt[3]{x} - 3x$, then

$$\frac{dy}{dx} = \frac{d}{dx} x^{1/3} - \frac{d}{dx} 3x = \frac{1}{3} x^{-2/3} - 3$$

(D) $$D_x\left(\frac{5}{3x^2} - \frac{2}{x^4} + \frac{x^3}{9}\right) \boxed{= D_x \frac{5}{3} x^{-2} - D_x\, 2x^{-4} + D_x \frac{1}{9} x^3}$$

$$= \frac{5}{3}(-2)x^{-3} - 2(-4)x^{-5} + \frac{1}{9} \cdot 3x^2 = -\frac{10}{3x^3} + \frac{8}{x^5} + \frac{1}{3}x^2$$ ◆

PROBLEM 20

Find:

(A) $f'(x)$ for $f(x) = 3x^4 - 2x^3 + x^2 - 5x + 7$

(B) y' for $y = 3 - 7x^{-2}$ (C) $\dfrac{dy}{dx}$ for $y = 5x^3 - \sqrt[4]{x}$

(D) $D_x\left(-\dfrac{3}{4x} + \dfrac{4}{x^3} - \dfrac{x^4}{8}\right)$ ◆

◆ APPLICATIONS

◆ EXAMPLE 21
Instantaneous Velocity

An object moves along the y axis (marked in feet) so that its position at time x (in seconds) is

$$f(x) = x^3 - 6x^2 + 9x$$

(A) Find the instantaneous velocity function v.
(B) Find the velocity at $x = 2$ and $x = 5$ seconds.
(C) Find the time(s) when the velocity is 0.

Solutions

(A) $v = f'(x) \boxed{= (x^3)' - (6x^2)' + (9x)'} = 3x^2 - 12x + 9$

(B) $f'(2) = 3(2)^2 - 12(2) + 9 = -3$ feet per second
$f'(5) = 3(5)^2 - 12(5) + 9 = 24$ feet per second

(C) $$\begin{aligned} v = f'(x) = 3x^2 - 12x + 9 &= 0 \\ 3(x^2 - 4x + 3) &= 0 \\ 3(x - 1)(x - 3) &= 0 \\ x &= 1, 3 \end{aligned}$$

Thus, $v = 0$ at $x = 1$ and $x = 3$ seconds. ◆

PROBLEM 21 Repeat Example 21 for $f(x) = x^3 - 15x^2 + 72x$. ◆

◆ EXAMPLE 22 Let $f(x) = x^4 - 8x^2 + 10$.

Tangents

(A) Find $f'(x)$.
(B) Find the equation of the tangent line at $x = 1$.
(C) Find the values of x where the tangent line is horizontal.

Solutions

(A) $f'(x) = (x^4)' - (8x^2)' + (10)'$
$= 4x^3 - 16x$

(B) $y - y_1 = m(x - x_1)$ $\quad y_1 = f(x_1) = f(1) = (1)^4 - 8(1)^2 + 10 = 3$
$y - 3 = -12(x - 1)$ $\quad m = f'(x_1) = f'(1) = 4(1)^3 - 16(1) = -12$
$y = -12x + 15$ $\quad$ Tangent line at $x = 1$

(C) Since a horizontal line has 0 slope, we must solve $f'(x) = 0$ for x:

$$f'(x) = 4x^3 - 16x = 0$$
$$4x(x^2 - 4) = 0$$
$$4x(x - 2)(x + 2) = 0$$
$$x = 0, 2, -2$$

Thus, the tangent line to the graph of f will be horizontal at $x = -2$, $x = 0$, and $x = 2$. (In the next chapter, we will see how this information is used to help sketch the graph of f.) ◆

PROBLEM 22 Repeat Example 22 for $f(x) = x^4 - 4x^3 + 7$. ◆

In business and economics, one is often interested in the rate at which something is taking place. A manufacturer, for example, is not only interested in the total cost $C(x)$ at certain production levels x, but is also interested in the rate of change of costs at various production levels.

In economics, the word **marginal** refers to a rate of change; that is, to a derivative. Thus, if

$C(x)$ = Total cost of producing x units during some unit of time

then

$C'(x)$ = Marginal cost
= Rate of change in cost per unit change in production at an output level of x units

Just as with instantaneous velocity, marginal cost $C'(x)$ is an instantaneous rate of change of total costs $C(x)$ with respect to production at a production level of x units. If this rate remains constant as production is increased by 1 unit, then it represents the change in cost for a 1 unit change in production. If the rate does not remain constant, then the instantaneous rate is an approximation of what actually happens during the next unit change in production.

The Marginal Cost Function

If the marginal cost function $C'(x)$ is a constant function, then $C'(x)$ represents the cost of producing 1 more unit at any production level x. If $C'(x)$ is not a constant function, then $C'(x)$ approximates the cost of producing 1 more unit at a production level of x units.

Example 23 should help to clarify these ideas.

◆ EXAMPLE 23 *Marginal Cost*

Suppose the total cost $C(x)$, in thousands of dollars, for manufacturing x sailboats per year is given by the function

$$C(x) = 575 + 25x - \frac{x^2}{4} \qquad 0 \leq x \leq 50$$

shown in the figure below.

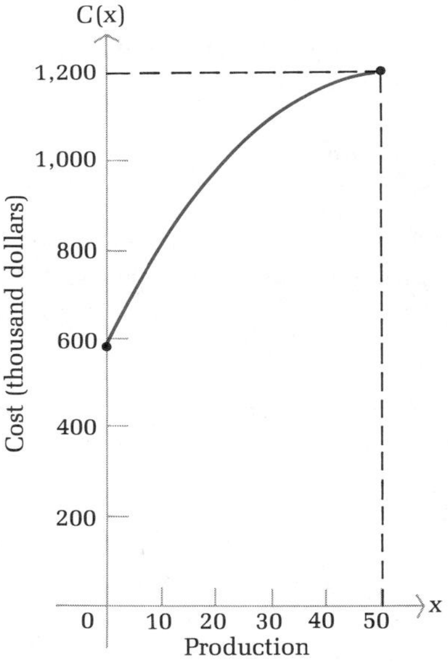

(A) Find the marginal cost at a production level of x boats.
(B) Find the marginal cost at a production level of 40 boats, and interpret the result.
(C) Find the actual cost of producing the 41st boat, and compare this cost with the result found in part B.
(D) Find $C'(30)$, and interpret the result.

Solutions (A) The marginal cost at a production level of x boats is

$$C'(x) = (575)' + (25x)' - \left(\frac{x^2}{4}\right)' = 25 - \frac{x}{2}$$

(B) The marginal cost at a production level of 40 boats is

$$C'(40) = 25 - \frac{40}{2} = 5 \quad \text{or} \quad \$5{,}000 \text{ per boat}$$

At a production level of 40 boats, the rate of change of total cost relative to production is $5,000 per boat. Thus, the cost of producing 1 more boat at a production level of 40 boats is approximately $5,000.

(C) The actual cost of producing the 41st boat is

$$\begin{pmatrix}\text{Total cost of}\\ \text{producing}\\ \text{41 boats}\end{pmatrix} - \begin{pmatrix}\text{Total cost of}\\ \text{producing}\\ \text{40 boats}\end{pmatrix}$$
$$= C(41) - C(40)$$
$$= 1{,}179.75 - 1{,}175.00 = 4.75 \quad \text{or} \quad \$4{,}750$$

The marginal cost of $5,000 per boat found in part B is a close approximation to this value.

(D) $$C'(30) = 25 - \frac{30}{2} = 10 \quad \text{or} \quad \$10{,}000 \text{ per boat}$$

At a production level of 30 boats, the rate of change of total cost relative to production is $10,000 per boat. Thus, the cost of producing 1 more boat at this level of production is approximately $10,000. ◆

In Example 23, we observe that as production goes up, the marginal cost goes down.

PROBLEM 23 Suppose the total cost C(x), in thousands of dollars, for manufacturing x sailboats per year is given by the function

$$C(x) = 500 + 24x - \frac{x^2}{5} \qquad 0 \leq x \leq 50$$

(A) Find the marginal cost at a production level of x boats.
(B) Find the marginal cost at a production level of 35 boats, and interpret the result.
(C) Find the actual cost of producing the 36th boat, and compare this cost with the result found in part B.
(D) Find $C'(40)$, and interpret the result. ◆

Answers to Matched Problems

16. All are 0. 17. (A) $6x^5$ (B) $30x^{29}$ (C) $-2x^{-3}$ (D) $\frac{3}{2}x^{1/2}$
18. (A) $-x^{-2}$ or $-1/x^2$ (B) $\frac{2}{3}x^{-1/3}$ or $2/(3\sqrt[3]{x})$
(C) $-\frac{1}{2}x^{-3/2}$ or $-1/(2\sqrt{x^3})$
19. (A) $20x^4$ (B) $x^3/3$ (C) $-x^{-4}$ or $-1/x^4$ (D) $-3x^{-4/3}$ or $-3/\sqrt[3]{x^4}$
20. (A) $12x^3 - 6x^2 + 2x - 5$ (B) $14x^{-3}$
(C) $15x^2 - \frac{1}{4}x^{-3/4}$ (D) $3/(4x^2) - (12/x^4) - (x^3/2)$
21. (A) $v = 3x^2 - 30x + 72$ (B) $f'(2) = 24$ ft/sec; $f'(5) = -3$ ft/sec
(C) $x = 4$ and $x = 6$ sec
22. (A) $f'(x) = 4x^3 - 12x^2$ (B) $y = -8x + 12$ (C) $x = 0$ and $x = 3$
23. (A) $C'(x) = 24 - (2x/5)$
(B) $C'(35) = 10$ or \$10,000 per boat; at a production level of 35 boats, the rate of change of total cost relative to production is \$10,000 per boat; thus, the cost of producing 1 more boat at this level of production is approx. \$10,000
(C) $C(36) - C(35) = 9.8$ or \$9,800; the marginal cost of \$10,000 per boat found in part B is a close approximation to this value
(D) $C'(40) = 8$ or \$8,000 per boat; at a production level of 40 boats, the rate of change of total cost relative to production is \$8,000 per boat; thus, the cost of producing 1 more boat at this level of production is approx. \$8,000

EXERCISE 4-4

Find each of the following:

A

1. $f'(x)$ for $f(x) = 12$
2. $\frac{dy}{dx}$ for $y = -\sqrt{3}$
3. $D_x\, 23$
4. y' for $y = \pi$
5. $\frac{dy}{dx}$ for $y = x^{12}$
6. $D_x\, x^5$
7. $f'(x)$ for $f(x) = x$
8. y' for $y = x^7$
9. y' for $y = x^{-7}$
10. $f'(x)$ for $f(x) = x^{-11}$
11. $\frac{dy}{dx}$ for $y = x^{5/2}$
12. $D_x\, x^{7/3}$
13. $D_x \frac{1}{x^5}$
14. $f'(x)$ for $f(x) = \frac{1}{x^9}$
15. $f'(x)$ for $f(x) = 2x^4$
16. $\frac{dy}{dx}$ for $y = -3x$
17. $D_x(\frac{1}{3}x^6)$
18. y' for $y = \frac{1}{2}x^4$
19. $\frac{dy}{dx}$ for $y = \frac{x^5}{15}$
20. $f'(x)$ for $f(x) = \frac{x^6}{24}$

B

21. $D_x(2x^{-5})$
22. y' for $y = -4x^{-1}$
23. $f'(x)$ for $f(x) = \frac{4}{x^4}$
24. $\frac{dy}{dx}$ for $y = \frac{-3}{x^6}$

25. $D_x \dfrac{-1}{2x^2}$

26. y' for $y = \dfrac{1}{6x^3}$

27. $f'(x)$ for $f(x) = -3x^{1/3}$

28. $\dfrac{dy}{dx}$ for $y = -8x^{1/4}$

29. $D_x(2x^2 - 3x + 4)$

30. y' for $y = 3x^2 + 4x - 7$

31. $\dfrac{dy}{dx}$ for $y = 3x^5 - 2x^3 + 5$

32. $f'(x)$ for $f(x) = 2x^3 - 6x + 5$

33. $D_x(3x^{-4} + 2x^{-2})$

34. y' for $y = 2x^{-3} - 4x^{-1}$

35. $\dfrac{dy}{dx}$ for $y = \dfrac{1}{2x} - \dfrac{2}{3x^3}$

36. $f'(x)$ for $f(x) = \dfrac{3}{4x^3} + \dfrac{1}{2x^5}$

37. $D_x(3x^{2/3} - 5x^{1/3})$

38. $D_x(8x^{3/4} + 4x^{-1/4})$

39. $D_x\left(\dfrac{3}{x^{3/5}} - \dfrac{6}{x^{1/2}}\right)$

40. $D_x\left(\dfrac{5}{x^{1/5}} - \dfrac{8}{x^{3/2}}\right)$

41. $D_x \dfrac{1}{\sqrt[3]{x}}$

42. y' for $y = \dfrac{10}{\sqrt[5]{x}}$

43. $\dfrac{dy}{dx}$ for $y = \dfrac{12}{\sqrt{x}} - 3x^{-2} + x$

44. $f'(x)$ for $f(x) = 2x^{-3} - \dfrac{6}{\sqrt[3]{x^2}} + 7$

For Problems 45–48, find:

(A) $f'(x)$
(B) The slope of the graph of f at $x = 2$ and $x = 4$.
(C) The equations of the tangent lines at $x = 2$ and $x = 4$.
(D) The value(s) of x where the tangent line is horizontal.

45. $f(x) = 6x - x^2$

46. $f(x) = 2x^2 + 8x$

47. $f(x) = 3x^4 - 6x^2 - 7$

48. $f(x) = x^4 - 32x^2 + 10$

If an object moves along the y axis (marked in feet) so that its position at time x (in seconds) is given by the indicated function in Problems 49–52, find:

(A) The instantaneous velocity function $v = f'(x)$.
(B) The velocity when $x = 0$ and $x = 3$ seconds.
(C) The time(s) when $v = 0$.

49. $f(x) = 176x - 16x^2$

50. $f(x) = 80x - 10x^2$

51. $f(x) = x^3 - 9x^2 + 15x$

52. $f(x) = x^3 - 9x^2 + 24x$

C *In Problems 53–56, find each derivative.*

53. $f'(x)$ for $f(x) = \dfrac{10x + 20}{x}$

54. $\dfrac{dy}{dx}$ for $y = \dfrac{x^2 + 25}{x^2}$

55. $D_x \dfrac{x^4 - 3x^3 + 5}{x^2}$

56. y' for $y = \dfrac{2x^5 - 4x^3 + 2x}{x^3}$

In Problems 57 and 58, use the definition of derivative and the two-step process to verify each statement.

57. $D_x\, x^3 = 3x^2$ **58.** $D_x\, x^4 = 4x^3$

APPLICATIONS

Business & Economics

59. *Marginal cost.* The total cost of producing x tennis rackets per day is

$$C(x) = 800 + 60x - \frac{x^2}{4} \qquad 0 \leq x \leq 120$$

(A) Find the marginal cost at a production level of x rackets.
(B) Find the marginal cost at a production level of 60 rackets, and interpret the result.
(C) Find the actual cost of producing the 61st racket, and compare this cost with the result found in part B.
(D) Find $C'(80)$, and interpret the result.

60. *Marginal cost.* The total cost of producing x portable radios per day is

$$C(x) = 1{,}000 + 100x - \frac{x^2}{2} \qquad 0 \leq x \leq 100$$

(A) Find the marginal cost at a production level of x radios.
(B) Find the marginal cost at a production level of 80 radios, and interpret the result.
(C) Find the actual cost of producing the 81st radio, and compare this cost with the result found in part B.
(D) Find $C'(50)$, and interpret the result.

61. *Advertising.* Using past records, it is estimated that a company will sell $N(x)$ units of a product after spending $x thousand on advertising, as given by

$$N(x) = 60x - x^2 \qquad 5 \leq x \leq 30$$

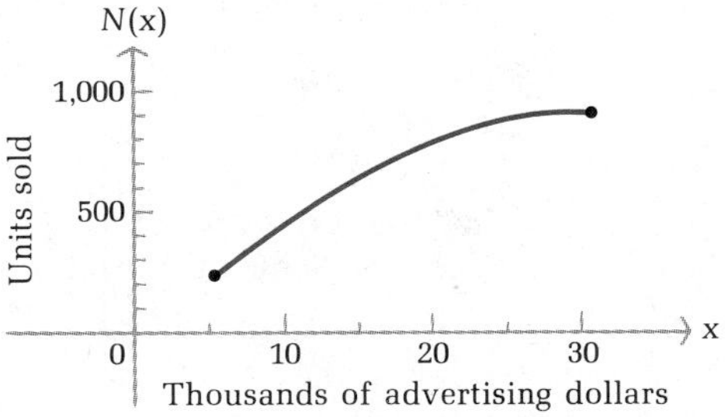

(A) Find $N'(x)$, the instantaneous rate of change of sales with respect to the amount of money spent on advertising at the $x thousand level.
(B) Find $N'(10)$ and $N'(20)$, and interpret the results.

62. *Demand function.* Suppose that in a given gourmet food store, people are willing to buy $D(x)$ pounds of chocolate candy per day at $\$x$ per quarter pound, as given by the demand function

$$D(x) = 100 - x^2 \qquad 1 \leq x \leq 10$$

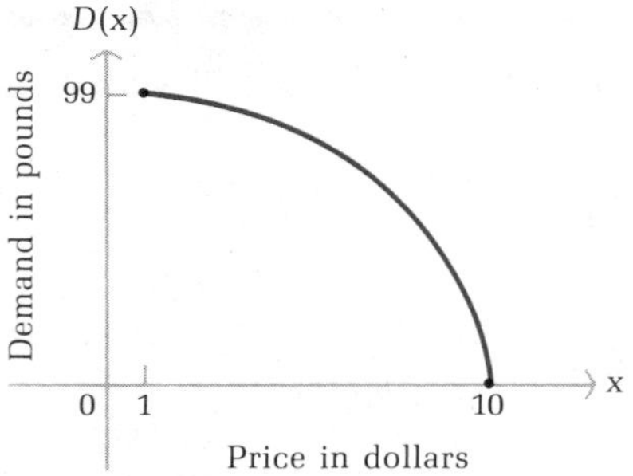

(A) Find $D'(x)$, the instantaneous rate of change of demand with respect to price at the $\$x$ price level.
(B) Find $D'(2)$ and $D'(8)$, and interpret the results.

Life Sciences

63. *Medicine.* A person x inches tall has a pulse rate of y beats per minute, as given approximately by

$$y = 590x^{-1/2} \qquad 30 \leq x \leq 75$$

What is the instantaneous rate of change of pulse rate at the:

(A) 36 inch level? (B) 64 inch level?

64. *Ecology.* A coal-burning electrical generating plant emits sulfur dioxide into the surrounding air. The concentration $C(x)$, in parts per million, is given approximately by

$$C(x) = \frac{0.1}{x^2}$$

where x is the distance from the plant in miles. Find the instantaneous rate of change of concentration at:

(A) $x = 1$ mile (B) $x = 2$ miles

Social Sciences

65. *Learning.* Suppose a person learns y items in x hours, as given by

$$y = 50\sqrt{x} \qquad 0 \leq x \leq 9$$

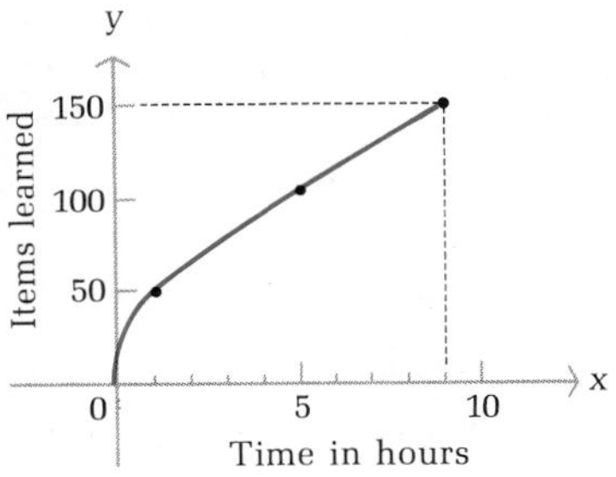

(see the figure in the margin). Find the rate of learning at the end of:

(A) 1 hour (B) 9 hours

66. *Learning.* If a person learns y items in x hours, as given by

$$y = 21\sqrt[3]{x^2} \qquad 0 \leq x \leq 8$$

find the rate of learning at the end of:

(A) 1 hour (B) 8 hours

SECTION 4-5 Derivatives of Products and Quotients

- DERIVATIVES OF PRODUCTS
- DERIVATIVES OF QUOTIENTS

The derivative rules discussed in the preceding section added substantially to our ability to compute and apply derivatives to many practical problems. In this and the next section, we will add a few more rules that will increase this ability even further.

DERIVATIVES OF PRODUCTS

In Section 4-4, we found that the derivative of a sum is the sum of the derivatives. Is the derivative of a product the product of the derivatives? Let us take a look at a simple example. Consider

$$f(x) = u(x)v(x) = (x^2 - 3x)(2x^3 - 1) \tag{1}$$

where $u(x) = x^2 - 3x$ and $v(x) = 2x^3 - 1$. The product of the derivatives is

$$u'(x)v'(x) = (2x - 3)6x^2 = 12x^3 - 18x^2 \tag{2}$$

To see if this is equal to the derivative of the product, we multiply the right side of (1) and use the derivative formulas we already know:

$$f(x) = (x^2 - 3x)(2x^3 - 1) = 2x^5 - 6x^4 - x^2 + 3x$$

Thus,

$$f'(x) = 10x^4 - 24x^3 - 2x + 3 \tag{3}$$

Since (2) and (3) are not equal, we conclude that the derivative of a product is *not* the product of the derivatives. There is a product rule for derivatives, but it is slightly more complicated than you might expect.

Using the definition of derivative and the two-step process, we can show that:

The derivative of the product of two functions is the first function times the derivative of the second function plus the second function times the derivative of the first function.

That is:

Product Rule

If

$$y = f(x) = F(x)S(x)$$

and if $F'(x)$ and $S'(x)$ exist, then

$$f'(x) = F(x)S'(x) + S(x)F'(x)$$

Also,

$$y' = FS' + SF' \qquad \frac{dy}{dx} = F\frac{dS}{dx} + S\frac{dF}{dx}$$

$$D_x[F(x)S(x)] = F(x)\, D_x\, S(x) + S(x)\, D_x\, F(x)$$

◆ EXAMPLE 24 Use two different methods to find $f'(x)$ for $f(x) = 2x^2(3x^4 - 2)$.

Solution *Method 1.* Use the product rule:

$$\begin{aligned} f'(x) &= 2x^2(3x^4 - 2)' + (3x^4 - 2)(2x^2)' \\ &= 2x^2(12x^3) + (3x^4 - 2)(4x) \\ &= 24x^5 + 12x^5 - 8x \\ &= 36x^5 - 8x \end{aligned}$$

First times derivative of second plus second times derivative of first

Method 2. Multiply first; then take derivatives:

$$f(x) = 2x^2(3x^4 - 2) = 6x^6 - 4x^2$$
$$f'(x) = 36x^5 - 8x$$

◆

PROBLEM 24 Use two different methods to find $f'(x)$ for $f(x) = 3x^3(2x^2 - 3x + 1)$. ◆

At this point, all the products we will encounter can be differentiated by either of the methods illustrated in Example 24. In the next and later sections, we will see that there are situations where the product rule must be used. Unless instructed otherwise, you should use the product rule to differentiate all products in this section to gain experience with the use of this important differentiation rule.

◆ EXAMPLE 25 Let $f(x) = (2x - 9)(x^2 + 6)$.

(A) Find the equation of the line tangent to the graph of $f(x)$ at $x = 3$.
(B) Find the value(s) of x where the tangent line is horizontal.

Solutions (A) First, find $f'(x)$:

$$f'(x) = (2x - 9)(x^2 + 6)' + (x^2 + 6)(2x - 9)'$$
$$= (2x - 9)(2x) + (x^2 + 6)(2)$$

Now, find the equation of the tangent line at $x = 3$:

$$y - y_1 = m(x - x_1) \qquad y_1 = f(x_1) = f(3) = -45$$
$$y - (-45) = 12(x - 3) \qquad m = f'(x_1) = f'(3) = 12$$
$$y = 12x - 81 \qquad \text{Tangent line at } x = 3$$

(B) The tangent line is horizontal at any value of x such that $f'(x) = 0$, so

$$f'(x) = (2x - 9)2x + (x^2 + 6)2 = 0$$
$$6x^2 - 18x + 12 = 0$$
$$x^2 - 3x + 2 = 0$$
$$(x - 1)(x - 2) = 0$$
$$x = 1, 2$$

The tangent line is horizontal at $x = 1$ and at $x = 2$. ◆

PROBLEM 25 Repeat Example 25 for $f(x) = (2x + 9)(x^2 - 12)$. ◆

As Example 25 illustrates, the way we write $f'(x)$ depends on what we want to do with it. If we are interested only in evaluating $f'(x)$ at specified values of x, the form in part A is sufficient. However, if we want to solve $f'(x) = 0$, we must multiply and collect like terms, as we did in part B.

◆ DERIVATIVES OF QUOTIENTS

As in the case with a product, the derivative of a quotient of two functions is *not* the quotient of the derivatives of the two functions.

Let

$$f(x) = \frac{T(x)}{B(x)} \qquad \text{where } T'(x) \text{ and } B'(x) \text{ exist}$$

Starting with the definition of a derivative, it can be shown that

$$f'(x) = \frac{B(x)T'(x) - T(x)B'(x)}{[B(x)]^2}$$

Thus:

The derivative of the quotient of two functions is the bottom function times the derivative of the top function minus the top function times the derivative of the bottom function, all over the bottom function squared.

Quotient Rule

If

$$y = f(x) = \frac{T(x)}{B(x)}$$

and if $T'(x)$ and $B'(x)$ exist, then

$$f'(x) = \frac{B(x)T'(x) - T(x)B'(x)}{[B(x)]^2}$$

Also,

$$y' = \frac{BT' - TB'}{B^2} \qquad \frac{dy}{dx} = \frac{B\dfrac{dT}{dx} - T\dfrac{dB}{dx}}{B^2}$$

$$D_x \frac{T(x)}{B(x)} = \frac{B(x)\, D_x\, T(x) - T(x)\, D_x\, B(x)}{[B(x)]^2}$$

◆ EXAMPLE 26

(A) If $f(x) = \dfrac{x^2}{2x - 1}$, find $f'(x)$.

(B) Find: $D_x \dfrac{x^2 - x}{x^3 + 1}$

(C) Find $D_x \dfrac{x^2 - 3}{x^2}$ by using the quotient rule and also by splitting the fraction into two fractions.

Solutions

(A) $$f'(x) = \frac{(2x - 1)(x^2)' - x^2(2x - 1)'}{(2x - 1)^2}$$

The bottom times the derivative of the top minus the top times the derivative of the bottom, all over the square of the bottom

$$= \frac{(2x - 1)(2x) - x^2(2)}{(2x - 1)^2}$$

$$= \frac{4x^2 - 2x - 2x^2}{(2x - 1)^2}$$

$$= \frac{2x^2 - 2x}{(2x - 1)^2}$$

(B) $$D_x \frac{x^2 - x}{x^3 + 1} = \frac{(x^3 + 1)\, D_x(x^2 - x) - (x^2 - x)\, D_x(x^3 + 1)}{(x^3 + 1)^2}$$

$$= \frac{(x^3 + 1)(2x - 1) - (x^2 - x)(3x^2)}{(x^3 + 1)^2}$$

$$= \frac{2x^4 - x^3 + 2x - 1 - 3x^4 + 3x^3}{(x^3 + 1)^2}$$

$$= \frac{-x^4 + 2x^3 + 2x - 1}{(x^3 + 1)^2}$$

(C) *Method 1.* Use the quotient rule:

$$D_x \frac{x^2 - 3}{x^2} = \frac{x^2 D_x(x^2 - 3) - (x^2 - 3) D_x x^2}{(x^2)^2}$$

$$= \frac{x^2(2x) - (x^2 - 3)2x}{x^4}$$

$$= \frac{2x^3 - 2x^3 + 6x}{x^4} = \frac{6x}{x^4} = \frac{6}{x^3}$$

Method 2. Split into two fractions:

$$\frac{x^2 - 3}{x^2} = \frac{x^2}{x^2} - \frac{3}{x^2} = 1 - 3x^{-2}$$

$$D_x(1 - 3x^{-2}) = 0 - 3(-2)x^{-3} = \frac{6}{x^3}$$

Comparing methods 1 and 2, we see that it often pays to change an expression algebraically before blindly using a differentiation formula. ◆

PROBLEM 26

Find:

(A) $f'(x)$ for $f(x) = \frac{2x}{x^2 + 3}$ (B) y' for $y = \frac{x^3 - 3x}{x^2 - 4}$

(C) $D_x \frac{2 + x^3}{x^3}$ two ways ◆

◆ EXAMPLE 27

Sales Analysis

When a successful home video game is first introduced, the monthly sales generally increase rapidly for a period of time, and then begin to decrease. Suppose that the monthly sales $S(t)$, in thousands of games, t months after the game is introduced are given by

$$S(t) = \frac{200t}{t^2 + 100}$$

(A) Find $S'(t)$. (B) Find $S(5)$ and $S'(5)$, and interpret the results.
(C) Find $S(30)$ and $S'(30)$, and interpret the results.

Solutions

(A) $$S'(t) = \frac{(t^2 + 100)(200t)' - 200t(t^2 + 100)'}{(t^2 + 100)^2}$$

$$= \frac{(t^2 + 100)200 - 200t(2t)}{(t^2 + 100)^2}$$

$$= \frac{200t^2 + 20{,}000 - 400t^2}{(t^2 + 100)^2}$$

$$= \frac{20{,}000 - 200t^2}{(t^2 + 100)^2}$$

(B) $S(5) = \dfrac{200(5)}{5^2 + 100} = 8$ and $S'(5) = \dfrac{20{,}000 - 200(5)^2}{(5^2 + 100)^2} = 0.96$

The sales for the 5th month are 8,000 units. At this point in time, sales are increasing at the rate of 0.96(1,000) = 960 units per month.

(C) $S(30) = \dfrac{200(30)}{30^2 + 100} = 6$ and $S'(30) = \dfrac{20{,}000 - 200(30)^2}{(30^2 + 100)^2} = -0.16$

The sales for the 30th month are 6,000 units. At this point in time, sales are decreasing at the rate of 0.16(1,000) = 160 units per month. ◆

The function $S(t)$ in Example 27 is graphed in Figure 17. Notice that the maximum monthly sales seem to occur during the 10th month. In the next chapter, we will see how the derivative $S'(t)$ is used to help sketch the graph of $S(t)$ and to find the maximum monthly sales.

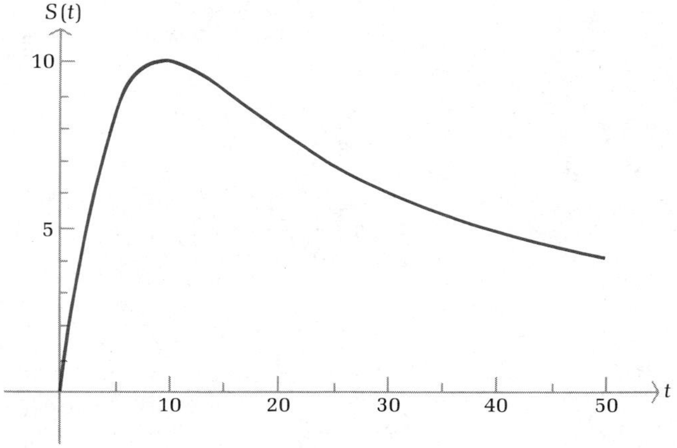

FIGURE 17

PROBLEM 27 Refer to Example 27. Suppose that the monthly sales $S(t)$, in thousands of games, t months after the game is introduced are given by

$$S(t) = \frac{200t}{t^2 + 64}$$

(A) Find $S'(t)$. (B) Find $S(4)$ and $S'(4)$, and interpret the results.
(C) Find $S(24)$ and $S'(24)$, and interpret the results. ◆

Answers to Matched Problems

24. $30x^4 - 36x^3 + 9x^2$ 25. (A) $y = 84x - 297$ (B) $x = -4, x = 1$

26. (A) $\dfrac{(x^2 + 3)2 - (2x)(2x)}{(x^2 + 3)^2} = \dfrac{6 - 2x^2}{(x^2 + 3)^2}$

(B) $\dfrac{(x^2 - 4)(3x^2 - 3) - (x^3 - 3x)(2x)}{(x^2 - 4)^2} = \dfrac{x^4 - 9x^2 + 12}{(x^2 - 4)^2}$ (C) $-\dfrac{6}{x^4}$

27. (A) $S'(t) = \dfrac{12{,}800 - 200t^2}{(t^2 + 64)^2}$

(B) $S(4) = 10$; $S'(4) = 1.5$; at $t = 4$ months, monthly sales are 10,000 and increasing at 1,500 games per month

(C) $S(24) = 7.5$; $S'(24) = -0.25$; at $t = 24$ months, monthly sales are 7,500 and decreasing at 250 games per month

EXERCISE 4-5

A *For f(x) as given, find f′(x) and simplify.*

1. $f(x) = 2x^3(x^2 - 2)$
2. $f(x) = 5x^2(x^3 + 2)$
3. $f(x) = (x - 3)(2x - 1)$
4. $f(x) = (3x + 2)(4x - 5)$
5. $f(x) = \dfrac{x}{x - 3}$
6. $f(x) = \dfrac{3x}{2x + 1}$
7. $f(x) = \dfrac{2x + 3}{x - 2}$
8. $f(x) = \dfrac{3x - 4}{2x + 3}$
9. $f(x) = (x^2 + 1)(2x - 3)$
10. $f(x) = (3x + 5)(x^2 - 3)$
11. $f(x) = \dfrac{x^2 + 1}{2x - 3}$
12. $f(x) = \dfrac{3x + 5}{x^2 - 3}$
13. $f(x) = (x^2 + 2)(x^2 - 3)$
14. $f(x) = (x^2 - 4)(x^2 + 5)$
15. $f(x) = \dfrac{x^2 + 2}{x^2 - 3}$
16. $f(x) = \dfrac{x^2 - 4}{x^2 + 5}$

B *Find each of the following and simplify:*

17. $f'(x)$ for $f(x) = (2x + 1)(x^2 - 3x)$
18. y' for $y = (x^3 + 2x^2)(3x - 1)$
19. $\dfrac{dy}{dx}$ for $y = (2x - x^2)(5x + 2)$
20. $D_x[(3 - x^3)(x^2 - x)]$
21. y' for $y = \dfrac{5x - 3}{x^2 + 2x}$
22. $f'(x)$ for $f(x) = \dfrac{3x^2}{2x - 1}$
23. $D_x \dfrac{x^2 - 3x + 1}{x^2 - 1}$
24. $\dfrac{dy}{dx}$ for $y = \dfrac{x^4 - x^3}{3x - 1}$

In Problems 25–28, find f′(x) and find the equation of the line tangent to the graph of f at x = 2.

25. $f(x) = (1 + 3x)(5 - 2x)$
26. $f(x) = (7 - 3x)(1 + 2x)$
27. $f(x) = \dfrac{x - 8}{3x - 4}$
28. $f(x) = \dfrac{2x - 5}{2x - 3}$

In Problems 29–32, find $f'(x)$ and find the value(s) of x where $f'(x) = 0$.

29. $f(x) = (2x - 15)(x^2 + 18)$

30. $f(x) = (2x - 3)(x^2 - 6)$

31. $f(x) = \frac{x}{x^2 + 1}$

32. $f(x) = \frac{x}{x^2 + 9}$

In Problems 33–36, find $f'(x)$ two ways; by using the product or quotient rule and by simplifying first.

33. $f(x) = x^3(x^4 - 1)$

34. $f(x) = x^4(x^3 - 1)$

35. $f(x) = \frac{x^3 + 9}{x^3}$

36. $f(x) = \frac{x^4 + 4}{x^4}$

C *Find each of the following. Do not simplify.*

37. $f'(x)$ for $f(x) = (2x^4 - 3x^3 + x)(x^2 - x + 5)$

38. $\frac{dy}{dx}$ for $y = (x^2 - 3x + 1)(x^3 + 2x^2 - x)$

39. $D_x \frac{3x^2 - 2x + 3}{4x^2 + 5x - 1}$

40. y' for $y = \frac{x^3 - 3x + 4}{2x^2 + 3x - 2}$

41. $\frac{dy}{dx}$ for $y = 9x^{1/3}(x^3 + 5)$

42. $D_x[(4x^{1/2} - 1)(3x^{1/3} + 2)]$

43. $f'(x)$ for $f(x) = \frac{6\sqrt[3]{x}}{x^2 - 3}$

44. y' for $y = \frac{2\sqrt{x}}{x^2 - 3x + 1}$

45. $D_x \frac{x^3 - 2x^2}{\sqrt[3]{x^2}}$

46. $\frac{dy}{dx}$ for $y = \frac{x^2 - 3x + 1}{\sqrt[4]{x}}$

47. $f'(x)$ for $f(x) = \frac{(2x^2 - 1)(x^2 + 3)}{x^2 + 1}$

48. y' for $y = \frac{2x - 1}{(x^3 + 2)(x^2 - 3)}$

APPLICATIONS

Business & Economics

49. *Sales analysis.* The monthly sales S (in thousands) for a record album are given by

$$S(t) = \frac{200t}{t^2 + 36}$$

where t is the number of months since the release of the album.

(A) Find $S'(t)$, the instantaneous rate of change of monthly sales with respect to time.
(B) Find $S(2)$ and $S'(2)$, and interpret the results.
(C) Find $S(8)$ and $S'(8)$, and interpret the results.

50. *Sales analysis.* A communications company has installed a cable television system in a city. The total number N (in thousands) of subscribers t months after the installation of the system is given by

$$N(t) = \frac{200t}{t+5}$$

(A) Find $N'(t)$, the instantaneous rate of change of the total number of subscribers with respect to time.
(B) Find $N(5)$ and $N'(5)$, and interpret the results.
(C) Find $N(15)$ and $N'(15)$, and interpret the results.

51. *Price–demand function.* According to classical economic theory, the demand $d(x)$ for a commodity in a free market decreases as the price x increases. Suppose that the number $d(x)$ of transistor radios people are willing to buy per week in a given city at a price $\$x$ is given by

$$d(x) = \frac{50{,}000}{x^2 + 10x + 25} \qquad 4 \leq x \leq 15$$

(A) Find $d'(x)$, the instantaneous rate of change of demand with respect to price change.
(B) Find $d'(5)$ and $d'(10)$, and interpret the results.

52. *Employee training.* A company producing computer components has established that on the average, a new employee can assemble $N(t)$ components per day after t days of on-the-job training, as given by

$$N(t) = \frac{100t}{t+9}$$

(A) Find $N'(t)$, the instantaneous rate of change of units assembled with respect to time.
(B) Find $N'(1)$ and $N'(11)$, and interpret the results.

Life Sciences

53. *Medicine.* A drug is injected into the bloodstream of a patient through her right arm. The concentration of the drug in the bloodstream of the left arm t hours after the injection is given by

$$C(t) = \frac{0.14t}{t^2+1}$$

(A) Find $C'(t)$, the instantaneous rate of change of drug concentration with respect to time.
(B) Find $C'(0.5)$ and $C'(3)$, and interpret the results.

54. *Drug sensitivity.* One hour after x milligrams of a particular drug are given to a person, the change in body temperature $T(x)$, in degrees Fahrenheit, is given approximately by

$$T(x) = x^2\left(1 - \frac{x}{9}\right) \qquad 0 \leq x \leq 7$$

The rate at which T changes with respect to the size of the dosage x, $T'(x)$, is called the *sensitivity* of the body to the dosage.

(A) Find $T'(x)$, using the product rule.
(B) Find $T'(1)$, $T'(3)$, and $T'(6)$.

Social Sciences

55. *Learning.* In the early days of quantitative learning theory (around 1917), L. L. Thurstone found that a given person successfully accomplished $N(x)$ acts after x practice acts, as given by

$$N(x) = \frac{100x + 200}{x + 32}$$

(A) Find the instantaneous rate of change of learning, $N'(x)$, with respect to the number of practice acts x.
(B) Find $N'(4)$ and $N'(68)$.

SECTION 4-6 Chain Rule: Power Form

- CHAIN RULE: POWER FORM
- COMBINING RULES OF DIFFERENTIATION

◆ CHAIN RULE: POWER FORM

We have already made extensive use of the power rule,

$$D_x\, x^n = nx^{n-1} \qquad n \text{ any real number} \tag{1}$$

Now we want to generalize this rule so that we can differentiate functions of the form $[u(x)]^n$. Is rule (1) still valid if we replace x with a function $u(x)$? We begin by considering a simple example. Let $u(x) = 2x$ and $n = 4$. Then

$$[u(x)]^n = (2x)^4 = 2^4x^4 = 16x^4$$

and

$$D_x[u(x)]^n = D_x\, 16x^4 = 64x^3 \tag{2}$$

But

$$n[u(x)]^{n-1} = 4(2x)^3 = 32x^3 \tag{3}$$

Comparing (2) and (3), we see that

$$D_x[u(x)]^n \neq n[u(x)]^{n-1}$$

for this particular choice of $u(x)$ and n. (In fact, it can be shown that the only time this last equation is valid is if $u(x) = x + k$, k a constant.) Thus, we cannot generalize the power rule by simply substituting $u(x)$ for x in rule (1).

How can we find a formula for $D_x[u(x)]^n$ where $u(x)$ is an arbitrary differentiable function? Let us first find $D_x[u(x)]^2$ and $D_x[u(x)]^3$ to see if a general pattern emerges. Since $[u(x)]^2 = u(x)u(x)$, we use the product rule with $F(x) = u(x)$ and $S(x) = u(x)$ to write

$$\begin{aligned} D_x[u(x)]^2 &= D_x[u(x)u(x)] = u(x)u'(x) + u(x)u'(x) \\ &= 2u(x)u'(x) \end{aligned} \tag{4}$$

Since $[u(x)]^3 = [u(x)]^2u(x)$, we now use the product rule with $F(x) = [u(x)]^2$ and (4) to write

$$\begin{aligned} D_x[u(x)]^3 &= D_x[u(x)]^2u(x) = [u(x)]^2 D_x\, u(x) + u(x)\, D_x[u(x)]^2 \\ &= [u(x)]^2u'(x) + u(x)[2u(x)u'(x)] \\ &= 3[u(x)]^2u'(x) \end{aligned}$$

Continuing in this fashion, it can be shown that

$$D_x[u(x)]^n = n[u(x)]^{n-1}u'(x) \qquad n \text{ a positive integer} \tag{5}$$

Using more advanced techniques, formula (5) can be established for all real numbers n. Thus, we have the **general power rule.**

General Power Rule

If n is any real number, then

$$D_x[u(x)]^n = n[u(x)]^{n-1}u'(x)$$

provided $u'(x)$ exists. This rule is often written more compactly as

$$D_x\, u^n = nu^{n-1}\frac{du}{dx} \qquad u = u(x)$$

The general power rule is a special case of a very important and useful differentiation rule called the **chain rule.** In essence, the chain rule will enable us to differentiate a composition form $f[g(x)]$ if we know how to differentiate $f(x)$ and $g(x)$. We defer a complete discussion of the chain rule until Chapter 6.

◆ EXAMPLE 28 Find $f'(x)$:

(A) $f(x) = (3x + 1)^4$ (B) $f(x) = (x^3 + 4)^7$

(C) $f(x) = \dfrac{1}{(x^2 + x + 4)^3}$ (D) $f(x) = \sqrt{3 - x}$

Solutions (A) $f(x) = (3x + 1)^4$ Let $u = 3x + 1$, $n = 4$.

$$f'(x) = 4(3x+1)^3 D_x(3x+1) \qquad nu^{n-1}\frac{du}{dx}$$

$$= 4(3x+1)^3 3 \qquad \frac{du}{dx} = 3$$

$$= 12(3x+1)^3$$

(B) $f(x) = (x^3 + 4)^7$ Let $u = (x^3 + 4)$, $n = 7$.

$$f'(x) = 7(x^3+4)^6 D_x(x^3+4) \qquad nu^{n-1}\frac{du}{dx}$$

$$= 7(x^3+4)^6 3x^2 \qquad \frac{du}{dx} = 3x^2$$

$$= 21x^2(x^3+4)^6$$

(C) $f(x) = \dfrac{1}{(x^2+x+4)^3} = (x^2+x+4)^{-3}$ Let $u = x^2 + x + 4$, $n = -3$.

$$f'(x) = -3(x^2+x+4)^{-4} D_x(x^2+x+4) \qquad nu^{n-1}\frac{du}{dx}$$

$$= -3(x^2+x+4)^{-4}(2x+1) \qquad \frac{du}{dx} = 2x+1$$

$$= \frac{-3(2x+1)}{(x^2+x+4)^4}$$

(D) $f(x) = \sqrt{3-x} = (3-x)^{1/2}$ Let $u = 3 - x$, $n = \frac{1}{2}$.

$$f'(x) = \frac{1}{2}(3-x)^{-1/2} D_x(3-x) \qquad nu^{n-1}\frac{du}{dx}$$

$$= \frac{1}{2}(3-x)^{-1/2}(-1) \qquad \frac{du}{dx} = -1$$

$$= -\frac{1}{2(3-x)^{1/2}} \quad \text{or} \quad -\frac{1}{2\sqrt{3-x}}$$

◆

PROBLEM 28 Find $f'(x)$:

(A) $f(x) = (5x+2)^3$ (B) $f(x) = (x^4 - 5)^5$

(C) $f(x) = \dfrac{1}{(x^2+4)^2}$ (D) $f(x) = \sqrt{4-x}$

◆

Notice that we used two steps to differentiate each function in Example 28. First, we applied the general power rule; then we found du/dx. As you gain experience with the general power rule, you may want to combine these two steps. If you do this, be certain to multiply by du/dx. For example,

$D_x(x^5+1)^4 = 4(x^5+1)^3 5x^4$ Correct

$D_x(x^5+1)^4 \neq 4(x^5+1)^3$ $du/dx = 5x^4$ is missing

If we let $u(x) = x$, then $du/dx = 1$, and the general power rule reduces to the (ordinary) power rule discussed in Section 4-4. Compare the following:

$D_x\, x^n = nx^{n-1}$ Yes—power rule

$D_x\, u^n = nu^{n-1}\dfrac{du}{dx}$ Yes—general power rule

$D_x\, u^n \neq nu^{n-1}$ Unless $u(x) = x + k$ so that $du/dx = 1$

◆ COMBINING RULES OF DIFFERENTIATION

The following examples illustrate the use of the general power rule in combination with other rules of differentiation.

◆ EXAMPLE 29 Find the equation of the line tangent to the graph of f at $x = 2$ for $f(x) = x^2\sqrt{2x+12}$.

Solution

$$f(x) = x^2\sqrt{2x+12} = x^2(2x+12)^{1/2}$$

Apply the product rule with $F(x) = x^2$ and $S(x) = (2x+12)^{1/2}$.

$$f'(x) = x^2\, D_x(2x+12)^{1/2} + (2x+12)^{1/2}\, D_x\, x^2$$
$$= x^2[\tfrac{1}{2}(2x+12)^{-1/2}](2) + (2x+12)^{1/2}(2x)$$
$$= \frac{x^2}{\sqrt{2x+12}} + 2x\sqrt{2x+12}$$

Use the general power rule to differentiate $(2x+12)^{1/2}$ and the ordinary power rule to differentiate x^2.

$$f'(2) = \frac{4}{\sqrt{16}} + 4\sqrt{16} = 1 + 16 = 17$$

$$f(2) = 4\sqrt{16} = 16$$

$(x_1, y_1) = (2, f(2)) = (2, 16)$ Point

$m = f'(2) = 17$ Slope

$y - 16 = 17(x-2)$ $y - y_1 = m(x - x_1)$

$y = 17x - 18$ Tangent line ◆

PROBLEM 29 Find the equation of the line tangent to the graph of f at $x = 3$ for $f(x) = x\sqrt{15-2x}$. ◆

◆ EXAMPLE 30 Find the value(s) of x where the tangent line is horizontal for

$$f(x) = \frac{x^3}{(2-3x)^5}$$

Solution Use the quotient rule with $T(x) = x^3$ and $B(x) = (2-3x)^5$:

$$\begin{aligned} f'(x) &= \frac{(2-3x)^5 D_x x^3 - x^3 D_x(2-3x)^5}{[(2-3x)^5]^2} \\ &= \frac{(2-3x)^5 3x^2 - x^3 5(2-3x)^4(-3)}{(2-3x)^{10}} \\ &= \frac{(2-3x)^4 3x^2[(2-3x)+5x]}{(2-3x)^{10}} \\ &= \frac{3x^2(2+2x)}{(2-3x)^6} = \frac{6x^2(x+1)}{(2-3x)^6} \end{aligned}$$

Use the ordinary power rule to differentiate x^3 and the general power rule to differentiate $(2-3x)^5$.

Since a fraction is 0 when the numerator is 0 and the denominator is not, we see that $f'(x) = 0$ at $x = -1$ and $x = 0$. Thus, the graph of f will have horizontal tangent lines at $x = -1$ and $x = 0$. ◆

PROBLEM 30 Find the value(s) of x where the tangent line is horizontal for

$$f(x) = \frac{x^3}{(3x-2)^2}$$

◆

◆ EXAMPLE 31 Starting with the function f in Example 30, write f as a product and then differentiate.

Solution

$$f(x) = \frac{x^3}{(2-3x)^5} = x^3(2-3x)^{-5}$$

$$\begin{aligned} f'(x) &= x^3 D_x(2-3x)^{-5} + (2-3x)^{-5} D_x x^3 \\ &= x^3(-5)(2-3x)^{-6}(-3) + (2-3x)^{-5} 3x^2 \\ &= 15x^3(2-3x)^{-6} + 3x^2(2-3x)^{-5} \end{aligned}$$

At this point, we have an unsimplified form for $f'(x)$. This may be satisfactory for some purposes, but not for others. For example, if we need to solve the equation $f'(x) = 0$, we must simplify algebraically:

$$\begin{aligned} f'(x) &= \frac{15x^3}{(2-3x)^6} + \frac{3x^2}{(2-3x)^5} = \frac{15x^3}{(2-3x)^6} + \frac{3x^2(2-3x)}{(2-3x)^6} \\ &= \frac{15x^3 + 3x^2(2-3x)}{(2-3x)^6} = \frac{3x^2(5x+2-3x)}{(2-3x)^6} \\ &= \frac{3x^2(2+2x)}{(2-3x)^6} = \frac{6x^2(1+x)}{(2-3x)^6} \end{aligned}$$

◆

PROBLEM 31 Refer to the function f in Problem 30, above. Write f as a product and then differentiate. Do not simplify. ◆

As Example 31 illustrates, any quotient can be converted to a product and differentiated by the product rule. However, if the derivative must be simplified, it is usually easier to use the quotient rule. (Compare the algebraic simplifications in Example 31 with those in Example 30.) There is one special case where using negative exponents is the preferred method—a fraction whose numerator is a constant.

◆ EXAMPLE 32 Find $f'(x)$ two ways for: $f(x) = \dfrac{4}{(x^2+9)^3}$

Solution *Method 1.* Use the quotient rule:

$$f'(x) = \frac{(x^2+9)^3 D_x 4 - 4 D_x(x^2+9)^3}{[(x^2+9)^3]^2}$$

$$= \frac{(x^2+9)^3(0) - 4[3(x^2+9)^2(2x)]}{(x^2+9)^6}$$

$$= \frac{-24x(x^2+9)^2}{(x^2+9)^6} = \frac{-24x}{(x^2+9)^4}$$

Method 2. Rewrite as a product, and use the general power rule:

$$f(x) = \frac{4}{(x^2+9)^3} = 4(x^2+9)^{-3}$$

$$f'(x) = 4(-3)(x^2+9)^{-4}(2x)$$

$$= \frac{-24x}{(x^2+9)^4}$$

Which method do you prefer? ◆

PROBLEM 32 Find $f'(x)$ two ways for: $f(x) = \dfrac{5}{(x^3+1)^2}$ ◆

Answers to Matched Problems

28. (A) $15(5x+2)^2$ (B) $20x^3(x^4-5)^4$ (C) $-4x/(x^2+4)^3$ (D) $-1/(2\sqrt{4-x})$
29. $y = 2x + 3$ 30. $x = 0, x = 2$
31. $-6x^3(3x-2)^{-3} + 3x^2(3x-2)^{-2}$ 32. $-30x^2/(x^3+1)^3$

EXERCISE 4-6

A *In Problems 1–12, find $f'(x)$ using the general power rule and simplify.*

1. $f(x) = (2x+5)^3$ **2.** $f(x) = (3x-7)^5$ **3.** $f(x) = (5-2x)^4$
4. $f(x) = (9-5x)^2$ **5.** $f(x) = (3x^2+5)^5$ **6.** $f(x) = (5x^2-3)^6$

7. $f(x) = (x^3 - 2x^2 + 2)^8$ 8. $f(x) = (2x^2 + x + 1)^7$ 9. $f(x) = (2x - 5)^{1/2}$
10. $f(x) = (4x + 3)^{1/2}$ 11. $f(x) = (x^4 + 1)^{-2}$ 12. $f(x) = (x^5 + 2)^{-3}$

In Problems 13–16, find $f'(x)$ and the equation of the line tangent to the graph of f at the indicated value of x. Find the value(s) of x where the tangent line is horizontal.

13. $f(x) = (2x - 1)^3$; $x = 1$
14. $f(x) = (3x - 1)^4$; $x = 1$
15. $f(x) = (4x - 3)^{1/2}$; $x = 3$
16. $f(x) = (2x + 8)^{1/2}$; $x = 4$

B *In Problems 17–34, find dy/dx using the general power rule.*

17. $y = 3(x^2 - 2)^4$
18. $y = 2(x^3 + 6)^5$
19. $y = 2(x^2 + 3x)^{-3}$
20. $y = 3(x^3 + x^2)^{-2}$
21. $y = \sqrt{x^2 + 8}$
22. $y = \sqrt[3]{3x - 7}$
23. $y = \sqrt[3]{3x + 4}$
24. $y = \sqrt{2x - 5}$
25. $y = (x^2 - 4x + 2)^{1/2}$
26. $y = (2x^2 + 2x - 3)^{1/2}$
27. $y = \dfrac{1}{2x + 4}$
28. $y = \dfrac{1}{3x - 7}$
29. $y = \dfrac{1}{(x^3 + 4)^5}$
30. $y = \dfrac{1}{(x^2 - 3)^6}$
31. $y = \dfrac{1}{4x^2 - 4x + 1}$
32. $y = \dfrac{1}{2x^2 - 3x + 1}$
33. $y = \dfrac{4}{\sqrt{x^2 - 3x}}$
34. $y = \dfrac{3}{\sqrt[3]{x - x^2}}$

In Problems 35–40, find $f'(x)$, and find the equation of the line tangent to the graph of f at the indicated value of x.

35. $f(x) = x(4 - x)^3$; $x = 2$
36. $f(x) = x^2(1 - x)^4$; $x = 2$
37. $f(x) = \dfrac{x}{(2x - 5)^3}$; $x = 3$
38. $f(x) = \dfrac{x^4}{(3x - 8)^2}$; $x = 4$
39. $f(x) = x\sqrt{2x + 2}$; $x = 1$
40. $f(x) = x\sqrt{x - 6}$; $x = 7$

In Problems 41–46, find $f'(x)$, and find the value(s) of x where the tangent line is horizontal.

41. $f(x) = x^2(x - 5)^3$
42. $f(x) = x^3(x - 7)^4$
43. $f(x) = \dfrac{x}{(2x + 5)^2}$
44. $f(x) = \dfrac{x - 1}{(x - 3)^3}$
45. $f(x) = \sqrt{x^2 - 8x + 20}$
46. $f(x) = \sqrt{x^2 + 4x + 5}$

C *In Problems 47–58, find each derivative and simplify.*

47. $D_x[3x(x^2 + 1)^3]$
48. $D_x[2x^2(x^3 - 3)^4]$
49. $D_x \dfrac{(x^3 - 7)^4}{2x^3}$
50. $D_x \dfrac{3x^2}{(x^2 + 5)^3}$
51. $D_x[(2x - 3)^2(2x^2 + 1)^3]$
52. $D_x[(x^2 - 1)^3(x^2 - 2)^2]$
53. $D_x(4x^2\sqrt{x^2 - 1})$
54. $D_x(3x\sqrt{2x^2 + 3})$

55. $D_x \dfrac{2x}{\sqrt{x-3}}$

56. $D_x \dfrac{x^2}{\sqrt{x^2+1}}$

57. $D_x \sqrt{(2x-1)^3(x^2+3)^4}$

58. $D_x \sqrt{\dfrac{4x+1}{2x^2+1}}$

APPLICATIONS

Business & Economics

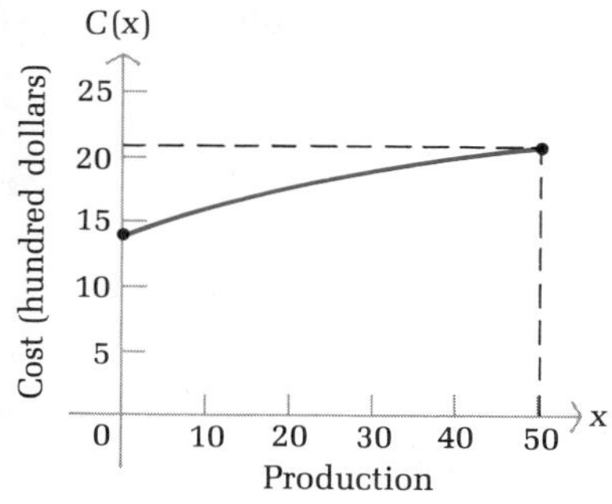

59. *Marginal cost.* The total cost (in hundreds of dollars) of producing x calculators per day is

$$C(x) = 10 + \sqrt{2x+16} \qquad 0 \le x \le 50$$

(see the figure in the margin).

(A) Find the marginal cost at a production level of x calculators.
(B) Find $C'(24)$ and $C'(42)$, and interpret the results.

60. *Marginal cost.* The total cost (in thousands of dollars) of producing x cameras per week is

$$C(x) = 6 + \sqrt{4x+4} \qquad 0 \le x \le 30$$

(A) Find the marginal cost at a production level of x cameras.
(B) Find $C'(15)$ and $C'(24)$, and interpret the results.

61. *Compound interest.* If $1,000 is invested at an interest rate of i compounded monthly, the amount in the account at the end of 4 years is given by

$$A = 1{,}000(1 + \tfrac{1}{12}i)^{48}$$

Find dA/di.

62. *Compound interest.* If $100 is invested at an interest rate of i compounded semiannually, the amount in the account at the end of 5 years is given by

$$A = 100(1 + \tfrac{1}{2}i)^{10}$$

Find dA/di.

Life Sciences

63. *Bacteria growth.* The number y of bacteria in a certain colony after x days is given approximately by

$$y = (3 \times 10^6)\left[1 - \frac{1}{\sqrt[3]{(x^2-1)^2}}\right]$$

Find dy/dx.

64. *Pollution.* A small lake in a resort area became contaminated with harmful bacteria because of excessive septic tank seepage. After treating the lake with a bactericide, the Department of Public Health estimated the bacteria concentration (number per cubic centimeter) after t days to be given by

$$C(t) = 500(8-t)^2 \qquad 0 \le t \le 7$$

(A) Find $C'(t)$ using the general power rule.
(B) Find $C'(1)$ and $C'(6)$, and interpret the results.

Social Sciences

65. *Learning.* In 1930, L. L. Thurstone developed the following formula to indicate how learning time T depends on the length of a list n:

$$T = f(n) = \frac{c}{k}\, n\sqrt{n - a}$$

where a, c, and k are empirical constants. Suppose that for a particular person, time T (in minutes) for learning a list of length n is

$$T = f(n) = 2n\sqrt{n - 2}$$

(A) Find dT/dn, the instantaneous rate of change in time with respect to n.
(B) Find $f'(11)$ and $f'(27)$, and interpret the results.

SECTION 4-7 Marginal Analysis in Business and Economics

- MARGINAL COST, REVENUE, AND PROFIT
- APPLICATION
- MARGINAL AVERAGE COST, REVENUE, AND PROFIT

◆ MARGINAL COST, REVENUE, AND PROFIT

One important use of calculus in business and economics is in *marginal analysis*. We introduced the concept of *marginal cost* earlier. There is no reason to stop there. Economists also talk about *marginal revenue* and *marginal profit*. Recall that the word "marginal" refers to an instantaneous rate of change—that is, a derivative. Thus, we define the following:

Marginal Cost, Revenue, and Profit

If x is the number of units of a product produced in some time interval, then

Total cost $= C(x)$ Total revenue $= R(x)$
Marginal cost $= C'(x)$ **Marginal revenue** $= R'(x)$

Total profit $= P(x) = R(x) - C(x)$
Marginal profit $= P'(x) = R'(x) - C'(x)$
$=$ (Marginal revenue) $-$ (Marginal cost)

Marginal cost (or revenue or profit) is the instantaneous rate of change of cost (or revenue or profit) relative to production at a given production level.

Marginal functions have several important economic interpretations. Earlier in this chapter, we discussed interpretations of marginal cost, which we summarize here. Similar interpretations can be made for marginal revenue and marginal profit.

Returning to the definition of a derivative, we observe the following (assuming the limit exists):

$$C'(x) = \lim_{h \to 0} \frac{C(x+h) - C(x)}{h} \qquad \text{Marginal cost}$$

$$C'(x) \approx \frac{C(x+h) - C(x)}{h} \qquad h \neq 0$$

$$C'(x) \approx \frac{C(x+1) - C(x)}{1}$$

$= C(x + 1) - C(x)$

= Exact change in total cost for 1 unit change in production at the x level of production

= Exact cost of producing the $(x + 1)$st item at the x level of production

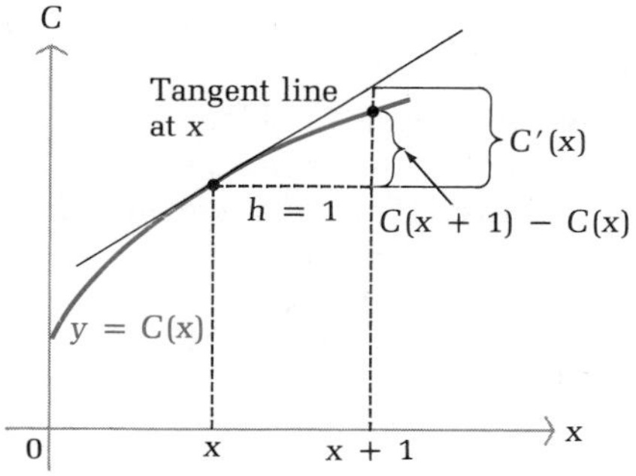

FIGURE 18
$C'(x) \approx C(x + 1) - C(x)$

Thus, at a production level of x units:

Marginal cost $C'(x)$ approximates the change in total cost that results from a 1 unit change in production.

In other words,

Marginal cost $C'(x)$ approximates the cost of producing the $(x + 1)$st item.

These observations are illustrated in Figure 18. Note that if $h = 1$ as shown in the figure, then $C(x + 1) - C(x)$ is the exact change in total cost per unit change in production at a production level of x units. The marginal cost, $C'(x)$, is the slope of the tangent line, and is approximately equal to the change in total cost C per unit change in production at a production level of x units.

◆ APPLICATION

We now present an example in market research to show how marginal cost, revenue, and profit are tied together.

◆ EXAMPLE 33

Production Strategy

The market research department of a company recommends that the company manufacture and market a new transistor radio. After suitable test marketing, the research department presents the following **demand equation:**

$$x = 10{,}000 - 1{,}000p \qquad x \text{ is demand at \$}p\text{ per radio} \tag{1}$$

or

$$p = 10 - \frac{x}{1{,}000} \tag{2}$$

where x is the number of radios retailers are likely to buy per week at $\$p$ per radio. Equation (2) is simply equation (1) solved for p in terms of x. Notice that as price goes up, demand goes down.

The financial department provides the following **cost equation:**

$$C(x) = 7{,}000 + 2x \tag{3}$$

where \$7,000 is the estimated fixed costs (tooling and overhead), and \$2 is the estimated variable costs (cost per unit for materials, labor, marketing, transportation, storage, etc.).

The **marginal cost** is

$$C'(x) = 2$$

Since this is a constant, it costs an additional \$2 to produce 1 more radio at all production levels.

The **revenue** (the amount of money R received by the company for manufacturing and selling x units at $\$p$ per unit) is

$$R = (\text{Number of units sold})(\text{Price per unit}) = xp$$

In general, the revenue R can be expressed in terms of p by using equation (1) or in terms of x by using equation (2). In marginal analysis (problems involving marginal cost, marginal revenue, or marginal profit), cost, revenue, and profit must be expressed in terms of the number of units x. Thus, the **revenue equation** in terms of x is

$$R(x) = xp = x\left(10 - \frac{x}{1{,}000}\right) \quad \text{Using equation (2)} \tag{4}$$

$$= 10x - \frac{x^2}{1{,}000}$$

The **marginal revenue** is

$$R'(x) = 10 - \frac{x}{500}$$

For production levels of $x = 2{,}000$, 5,000, and 7,000, we have

$$R'(2{,}000) = 6 \qquad R'(5{,}000) = 0 \qquad R'(7{,}000) = -4$$

This means that at production levels of 2,000, 5,000, and 7,000, the respective approximate changes in revenue per unit change in production are \$6, \$0, and −\$4. That is, at the 2,000 output level, revenue increases as production increases; at the 5,000 output level, revenue does not change with a "small" change in production; and at the 7,000 output level, revenue decreases with an increase in production.

When we graph $R(x)$ and $C(x)$ in the same coordinate system, we obtain Figure 19.

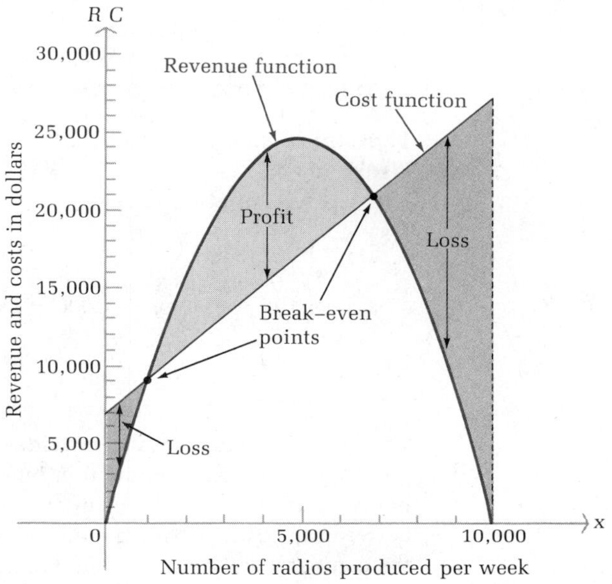

FIGURE 19

The **break-even points** (the points where revenue equals cost) are obtained as follows:

$$C(x) = R(x)$$

$$7{,}000 + 2x = 10x - \frac{x^2}{1{,}000}$$

$$\frac{x^2}{1{,}000} - 8x + 7{,}000 = 0$$

$$x^2 - 8{,}000x + 7{,}000{,}000 = 0$$ Solve using the quadratic formula (see Section 2-2).

$$x = \frac{8{,}000 \pm \sqrt{8{,}000^2 - 4(7{,}000{,}000)}}{2}$$

$$= \frac{8{,}000 \pm \sqrt{36{,}000{,}000}}{2}$$

$$= \frac{8{,}000 \pm 6{,}000}{2}$$

$$= 1{,}000, \quad 7{,}000$$

$$R(1{,}000) = 10(1{,}000) - \frac{1{,}000^2}{1{,}000} = 9{,}000$$

$$C(1{,}000) = 7{,}000 + 2(1{,}000) = 9{,}000$$

$$R(7{,}000) = 10(7{,}000) - \frac{7{,}000^2}{1{,}000} = 21{,}000$$

$$C(7{,}000) = 7{,}000 + 2(7{,}000) = 21{,}000$$

Thus, the break-even points are (1,000, 9,000) and (7,000, 21,000), as shown in Figure 19.

The **profit equation** is

$$\begin{aligned} P(x) &= R(x) - C(x) \\ &= \left(10x - \frac{x^2}{1{,}000}\right) - (7{,}000 + 2x) \\ &= -\frac{x^2}{1{,}000} + 8x - 7{,}000 \end{aligned}$$

The graph in Figure 19 also provides some useful information concerning the profit equation. At a production level of 1,000 or 7,000, revenue equals cost; hence, profit is 0 and the company will break even. For any production level between 1,000 and 7,000, revenue is greater than cost; hence, $P(x)$ is positive and the company will make a profit. For production levels less than 1,000 or greater than 7,000, revenue is less than cost; hence, $P(x)$ is negative and the company will have a loss.

The **marginal profit** is

$$P'(x) = -\frac{x}{500} + 8$$

For production levels of 1,000, 4,000, and 6,000, we have

$$P'(1{,}000) = 6 \qquad P'(4{,}000) = 0 \qquad P'(6{,}000) = -4$$

This means that at production levels of 1,000, 4,000, and 6,000, the respective approximate changes in profit per unit change in production are \$6, \$0, and −\$4. That is, at the 1,000 output level, profit will be increased if production is increased; at the 4,000 output level, profit does not change for "small" changes in production; and at the 6,000 output level, profits will decrease if production is increased. It seems the best production level to produce a maximum profit is 4,000.

Example 33 warrants careful study, since a number of important ideas in economics and calculus are involved. In the next chapter, we will develop a systematic procedure for finding the production level (and, using the demand equation, the selling price) that will maximize profit. ◆

PROBLEM 33 Refer to the revenue and profit equations in Example 33.

(A) Find $R'(3{,}000)$ and $R'(6{,}000)$, and interpret the results.
(B) Find $P'(2{,}000)$ and $P'(7{,}000)$, and interpret the results. ◆

◆ MARGINAL AVERAGE COST, REVENUE, AND PROFIT

Sometimes, it is desirable to carry out marginal analysis relative to **average cost (cost per unit), average revenue (revenue per unit), and average profit (profit per unit).** The relevant definitions are summarized in the following box:

Marginal Average Cost, Revenue, and Profit

If x is the number of units of a product produced in some time interval, then

COST PER UNIT:

$$\textbf{Average cost} = \overline{C}(x) = \frac{C(x)}{x}$$

$$\textbf{Marginal average cost} = \overline{C}'(x) = D_x\,\overline{C}(x)$$

REVENUE PER UNIT:

$$\textbf{Average revenue} = \overline{R}(x) = \frac{R(x)}{x}$$

$$\textbf{Marginal average revenue} = \overline{R}'(x) = D_x\,\overline{R}(x)$$

PROFIT PER UNIT:

$$\textbf{Average profit} = \overline{P}(x) = \frac{P(x)}{x}$$

$$\textbf{Marginal average profit} = \overline{P}'(x) = D_x\,\overline{P}(x)$$

As was the case with marginal cost:

The marginal average cost approximates the change in average cost that results from a unit increase in production.

Similar statements can be made for marginal average revenue and marginal average profit.

◆ EXAMPLE 34

Cost Analysis

A small machine shop manufactures drill bits used in the petroleum industry. The shop manager estimates that the total daily cost (in dollars) of producing x bits is

$$C(x) = 1{,}000 + 25x - \frac{x^2}{10} \qquad \text{Total cost function}$$

Thus,

$$\overline{C}(x) = \frac{C(x)}{x} = \frac{1{,}000}{x} + 25 - \frac{x}{10} \qquad \text{Average cost function}$$

$$\overline{C}'(x) = D_x\,\overline{C}(x) = -\frac{1{,}000}{x^2} - \frac{1}{10} \qquad \text{Marginal average cost function}$$

$$\overline{C}(10) = \frac{1{,}000}{10} + 25 - \frac{10}{10}$$ Average cost per unit if 10 units are produced

$$= \$124$$

$$\overline{C}'(10) = -\frac{1{,}000}{100} - \frac{1}{10}$$ A unit increase in production will decrease the average cost per unit by approximately $10.10 at a production level of 10 units. ◆

$$= -\$10.10$$

PROBLEM 34 Consider the cost function $C(x) = 7{,}000 + 2x$ from Example 33.

(A) Find $\overline{C}(x)$ and $\overline{C}'(x)$.
(B) Find $\overline{C}(1{,}000)$ and $\overline{C}'(1{,}000)$, and interpret the results. ◆

Answers to Matched Problems

33. (A) $R'(3{,}000) = 4$ (at a production level of 3,000, a unit increase in production will increase revenue by approx. $4); $R'(6{,}000) = -2$ (at a production level of 6,000, a unit increase in production will decrease revenue by approx. $2)
(B) $P'(2{,}000) = 4$ (at a production level of 2,000, a unit increase in production will increase profit by approx. $4); $P'(7{,}000) = -6$ (at a production level of 7,000, a unit increase in production will decrease profit by approx. $6)

34. (A) $\overline{C}(x) = \frac{7{,}000}{x} + 2$, $\overline{C}'(x) = -\frac{7{,}000}{x^2}$
(B) $\overline{C}(1{,}000) = 9$ (at a production level of 1,000, the average cost per unit is $9); $\overline{C}'(1{,}000) = -0.007$ (at a production level of 1,000, a unit increase in production will decrease the average cost per unit by approx. 0.7¢)

EXERCISE 4-7

APPLICATIONS

Business & Economics

1. *Cost analysis.* The total cost (in dollars) of producing x food processors is

$$C(x) = 2{,}000 + 50x - \frac{x^2}{2}$$

(A) Find the exact cost of producing the 21st food processor.
(B) Use the marginal cost to approximate the cost of producing the 21st food processor.

2. *Cost analysis.* The total cost (in dollars) of producing x electric guitars is

$$C(x) = 1{,}000 + 100x - \frac{x^2}{4}$$

(A) Find the exact cost of producing the 51st guitar.
(B) Use the marginal cost to approximate the cost of producing the 51st guitar.

3. *Cost analysis.* The total cost (in dollars) of manufacturing x auto body frames is

$$C(x) = 60{,}000 + 300x$$

(A) Find the average cost per unit if 500 frames are produced.
(B) Find the marginal average cost at a production level of 500 units, and interpret the results.

4. *Cost analysis.* The total cost (in dollars) of printing x dictionaries is

$$C(x) = 20{,}000 + 10x$$

(A) Find the average cost per unit if 1,000 dictionaries are produced.
(B) Find the marginal average cost at a production level of 1,000 units, and interpret the results.

5. *Revenue analysis.* The total revenue (in dollars) from the sale of x clock radios is

$$R(x) = 100x - \frac{x^2}{40}$$

Evaluate the marginal revenue at the given values of x, and interpret the results.

(A) $x = 1{,}600$ (B) $x = 2{,}500$

6. *Revenue analysis.* The total revenue (in dollars) from the sale of x steam irons is

$$R(x) = 50x - \frac{x^2}{20}$$

Evaluate the marginal revenue at the given values of x, and interpret the results.

(A) $x = 400$ (B) $x = 650$

7. *Profit analysis.* The total profit (in dollars) from the sale of x skateboards is

$$P(x) = 30x - \frac{x^2}{2} - 250$$

(A) Find the exact profit from the sale of the 26th skateboard.
(B) Use the marginal profit to approximate the profit from the sale of the 26th skateboard.

8. *Profit analysis.* The total profit (in dollars) from the sale of x portable stereos is

$$P(x) = 22x - \frac{x^2}{10} - 400$$

(A) Find the exact profit from the sale of the 41st stereo.
(B) Use the marginal profit to approximate the profit from the sale of the 41st stereo.

9. *Profit analysis.* The total profit (in dollars) from the sale of x video cassettes is

$$P(x) = 5x - \frac{x^2}{200} - 450$$

Evaluate the marginal profit at the given values of x, and interpret the results.

(A) $x = 450$ (B) $x = 750$

10. *Profit analysis.* The total profit (in dollars) from the sale of x cameras is

$$P(x) = 12x - \frac{x^2}{50} - 1{,}000$$

Evaluate the marginal profit at the given values of x, and interpret the results.

(A) $x = 200$ (B) $x = 350$

11. *Profit analysis.* Refer to the profit equation in Problem 9.
(A) Find the average profit per unit if 150 cassettes are produced.
(B) Find the marginal average profit at a production level of 150 units, and interpret the results.

12. *Profit analysis.* Refer to the profit equation in Problem 10.
(A) Find the average profit per unit if 200 cameras are produced.
(B) Find the marginal average profit at a production level of 200 units, and interpret the results.

13. *Revenue, cost, and profit.* In Example 33, suppose we have the demand equation

$$x = 6{,}000 - 30p \qquad \text{or} \qquad p = 200 - \frac{x}{30}$$

and the cost equation

$$C(x) = 72{,}000 + 60x$$

(A) Find the marginal cost.
(B) Find the revenue equation in terms of x.
(C) Find the marginal revenue.
(D) Find $R'(1{,}500)$ and $R'(4{,}500)$, and interpret the results.

(E) Graph the cost function and the revenue function on the same coordinate system for $0 \leq x \leq 6{,}000$. Find the break-even points, and indicate regions of loss and profit.
(F) Find the profit equation in terms of x.
(G) Find the marginal profit.
(H) Find $P'(1{,}500)$ and $P'(3{,}000)$, and interpret the results.

14. *Revenue, cost, and profit.* In Example 33, suppose we have the demand equation

$$x = 9{,}000 - 30p \qquad \text{or} \qquad p = 300 - \frac{x}{30}$$

and the cost equation

$$C(x) = 150{,}000 + 30x$$

(A) Find the marginal cost.
(B) Find the revenue equation in terms of x.
(C) Find the marginal revenue.
(D) Find $R'(3{,}000)$ and $R'(6{,}000)$, and interpret the results.
(E) Graph the cost function and the revenue function on the same coordinate system for $0 \leq x \leq 9{,}000$. Find the break-even points, and indicate regions of loss and profit.
(F) Find the profit equation in terms of x.
(G) Find the marginal profit.
(H) Find $P'(1{,}500)$ and $P'(4{,}500)$, and interpret the results.

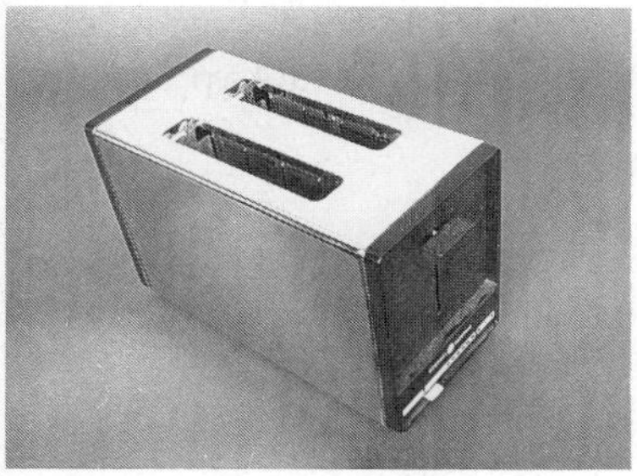

15. *Revenue, cost, and profit.* A company is planning to manufacture and market a new two-slice electric toaster. After conducting extensive market surveys, the research department provides the following estimates: a weekly demand of 200 toasters at a price of \$16 per toaster and a weekly demand of 300 toasters at a price of \$14 per toaster. The financial department estimates that weekly fixed costs will be \$1,400 and variable costs (cost per unit) will be \$4.

(A) Assume that the demand equation is linear. Use the research department's estimates to find the demand equation.
(B) Find the revenue equation in terms of x.
(C) Assume that the cost equation is linear. Use the financial department's estimates to find the cost equation.
(D) Graph the cost function and the revenue function on the same coordinate system for $0 \leq x \leq 1{,}000$. Find the break-even points, and indicate regions of loss and profit.
(E) Find the profit equation in terms of x.
(F) Evaluate the marginal profit at $x = 250$ and $x = 475$, and interpret the results.

16. *Revenue, cost, and profit.* The company in Problem 15 is also planning to manufacture and market a four-slice toaster. For this toaster, the research department's estimates are a weekly demand of 300 toasters at a price of \$25 per toaster and a weekly demand of 400 toasters at a price of \$20. The financial department's estimates are fixed weekly costs of \$5,000 and variable costs of \$5 per toaster. Assume the demand and cost equations are linear (see Problem 15, parts A and C).

(A) Use the research department's estimates to find the demand equation.
(B) Find the revenue equation in terms of x.
(C) Use the financial department's estimates to find the cost equation in terms of x.
(D) Graph the cost function and the revenue function on the same coordinate system for $0 \leq x \leq 800$. Find the break-even points, and indicate regions of loss and profit.
(E) Find the profit equation in terms of x.
(F) Evaluate the marginal profit at $x = 325$ and $x = 425$, and interpret the results.

SECTION 4-8 Chapter Review

Important Terms and Symbols

4-1 *Limits and Continuity—A Geometric Introduction.* Limit as x approaches c; one-sided limits; limit as x approaches c from the left; left-hand limit; limit as x approaches c from the right; right-hand limit; continuous curve; continuity at a point; discontinuity at a point; continuity on an open interval; continuity on a closed or half-closed interval; continuity properties

$$\lim_{x \to c} f(x); \quad \lim_{x \to c^-} f(x); \quad \lim_{x \to c^+} f(x)$$

4-2 *Computation of Limits.* Limits at points of continuity; limits and infinity; vertical asymptote; limit properties; 0/0 indeterminate form; limits at infinity; horizontal asymptote

$$\lim_{x \to c} f(x) = f(c) \quad \text{if } f \text{ is continuous at } x = c;$$

$$\lim_{x \to c} f(x) = \infty; \quad \lim_{x \to c} f(x) = -\infty; \quad \lim_{x \to \infty} f(x); \quad \lim_{x \to -\infty} f(x)$$

4-3 *The Derivative.* Secant lines; tangent lines; difference quotient; slope of a tangent line; slope of a graph; average rate of change; instantaneous rate of change; the derivative; differentiable at a point; nondifferentiable at a point; differentiable over an open interval; differentiable over a closed or half-closed interval; differentiation

$$f'(x) = \lim_{h \to 0} \frac{f(x + h) - f(x)}{h}$$

4-4 *Derivatives of Constants, Power Forms, and Sums.* Derivative notation; derivative of a constant; power rule; constant times a function rule; sum and difference rule; marginal cost

$$f'(x); \quad y'; \quad \frac{dy}{dx}; \quad D_x f(x)$$

4-5 *Derivatives of Products and Quotients.* Product rule; quotient rule

4-6 *Chain Rule: Power Form.* General power rule; chain rule; combining rules of differentiation

4-7 *Marginal Analysis in Business and Economics.* Marginal cost; marginal revenue; marginal profit; demand equation; cost equation; revenue equation; break-even points; profit equation; average cost; marginal average cost; average revenue; marginal average revenue; average profit; marginal average profit

$$C'(x); \quad \overline{C}(x); \quad \overline{C}'(x); \quad R'(x); \quad \overline{R}(x); \quad \overline{R}'(x); \quad P'(x); \quad \overline{P}(x); \quad \overline{P}'(x)$$

Summary of Rules of Differentiation

$$D_x k = 0$$

$$D_x x^n = nx^{n-1}$$

$$D_x kf(x) = kf'(x)$$

$$D_x[u(x) \pm v(x)] = u'(x) \pm v'(x)$$

$$D_x[F(x)S(x)] = F(x)S'(x) + S(x)F'(x)$$

$$D_x \frac{T(x)}{B(x)} = \frac{B(x)T'(x) - T(x)B'(x)}{[B(x)]^2}$$

$$D_x[u(x)]^n = n[u(x)]^{n-1}u'(x)$$

EXERCISE 4-8

Chapter Review

Work through all the problems in this chapter review and check your answers in the back of the book. (Answers to all review problems are there.) Where weaknesses show up, review appropriate sections in the text.

A *In Problems 1–10, find $f'(x)$ for $f(x)$ as given.*

1. $f(x) = 3x^4 - 2x^2 + 1$

2. $f(x) = 2x^{1/2} - 3x$

3. $f(x) = 5$

4. $f(x) = \dfrac{1}{2x^2} + \dfrac{x^2}{2}$

5. $f(x) = (2x - 1)(3x + 2)$

6. $f(x) = (x^2 - 1)(x^3 - 3)$

7. $f(x) = \dfrac{2x}{x^2 + 2}$

8. $f(x) = \dfrac{1}{3x + 2}$

9. $f(x) = (2x - 3)^3$

10. $f(x) = (x^2 + 2)^{-2}$

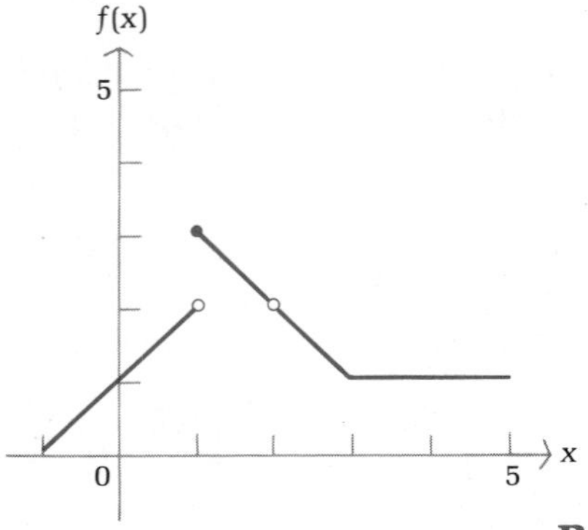

In Problems 11–13, use the graph of the function f shown in the margin to answer each question.

11. (A) $\lim_{x\to 1} f(x) = ?$ (B) $f(1) = ?$
(C) Is f continuous at $x = 1$?

12. (A) $\lim_{x\to 2} f(x) = ?$ (B) $f(2) = ?$
(C) Is f continuous at $x = 2$?

13. (A) $\lim_{x\to 3} f(x) = ?$ (B) $f(3) = ?$
(C) Is f continuous at $x = 3$?

B

In Problems 14–21, find the indicated derivative.

14. $\dfrac{dy}{dx}$ for $y = 3x^4 - 2x^{-3} + 5$

15. y' for $y = (2x^2 - 3x + 2)(x^2 + 2x - 1)$

16. $f'(x)$ for $f(x) = \dfrac{2x - 3}{(x - 1)^2}$

17. y' for $y = 2\sqrt{x} + \dfrac{4}{\sqrt{x}}$

18. $D_x[(x^2 - 1)(2x + 1)^2]$

19. $D_x \sqrt[3]{x^3 - 5}$

20. $\dfrac{dy}{dx}$ for $y = \dfrac{3x^2 + 4}{x^2}$

21. $D_x \dfrac{(x^2 + 2)^4}{2x - 3}$

22. For $y = f(x) = x^2 + 4$, find:
(A) The slope of the graph at $x = 1$
(B) The equation of the tangent line at $x = 1$ in the form $y = mx + b$

23. Repeat Problem 22 for $f(x) = x^3(x + 1)^2$.

In Problems 24–27, find the value(s) of x where the tangent line is horizontal.

24. $f(x) = 10x - x^2$

25. $f(x) = (x + 3)(x^2 - 45)$

26. $f(x) = \dfrac{x}{x^2 + 4}$

27. $f(x) = x^2(2x - 15)^3$

28. If an object moves along the y axis (scale in feet) so that it is at $y = f(x) = 16x^2 - 4x$ at time x (in seconds), find:
(A) The instantaneous velocity function
(B) The velocity at time $x = 3$ seconds

29. An object moves along the y axis (scale in feet) so that at time x (in seconds) it is at $y = f(x) = 96x - 16x^2$. Find:
(A) The instantaneous velocity function
(B) The time(s) when the velocity is 0

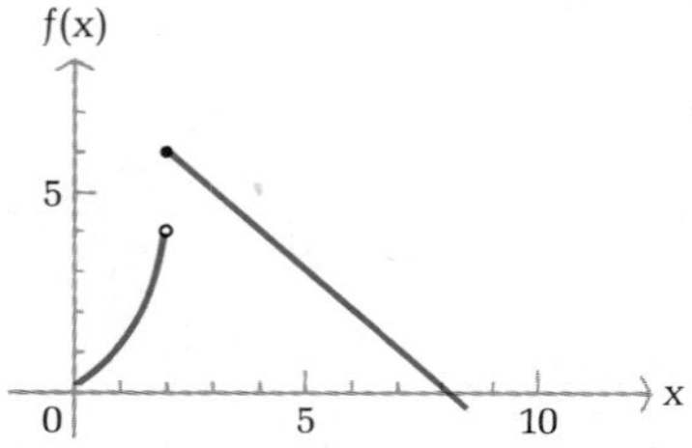

$$f(x) = \begin{cases} x^2 & 0 \le x < 2 \\ 8 - x & x \ge 2 \end{cases}$$

Problems 30 and 31 refer to the function f described in the figure in the margin.

30. (A) $\lim_{x\to 2^-} f(x) = ?$ (B) $\lim_{x\to 2^+} f(x) = ?$ (C) $\lim_{x\to 2} f(x) = ?$
(D) $f(2) = ?$ (E) Is f continuous at $x = 2$?

31. (A) $\lim_{x \to 5^-} f(x) = ?$ (B) $\lim_{x \to 5^+} f(x) = ?$ (C) $\lim_{x \to 5} f(x) = ?$
(D) $f(5) = ?$ (E) Is f continuous at $x = 5$?

32. Find each limit. (Use $-\infty$ or ∞, if appropriate.)

(A) $\lim_{x \to 0^-} \frac{1}{|x|}$ (B) $\lim_{x \to 0^+} \frac{1}{|x|}$ (C) $\lim_{x \to 0} \frac{1}{|x|}$

33. Find each limit. (Use $-\infty$ or ∞, if appropriate.)

(A) $\lim_{x \to 1^-} \frac{1}{x-1}$ (B) $\lim_{x \to 1^+} \frac{1}{x-1}$ (C) $\lim_{x \to 1} \frac{1}{x-1}$

In Problems 34–38, determine where f is continuous. Express the answer in interval notation.

34. $f(x) = 2x^2 - 3x + 1$

35. $f(x) = \frac{1}{x+5}$

36. $f(x) = \frac{x-3}{x^2 - x - 6}$

37. $f(x) = \sqrt{x-3}$

38. $f(x) = \sqrt[3]{1-x^2}$

In Problems 39–52, find each limit, if it exists. (Use $-\infty$ or ∞, if appropriate.)

39. $\lim_{x \to 3} \frac{2x-3}{x+5}$

40. $\lim_{x \to 3} (2x^2 - x + 1)$

41. $\lim_{x \to 0} \frac{2x}{3x^2 - 2x}$

42. $\lim_{h \to 0} \frac{[(2+h)^2 - 1] - [2^2 - 1]}{h}$

43. $\lim_{h \to 0} \frac{f(2+h) - f(2)}{h}$ for $f(x) = x^2 + 4$

44. $\lim_{x \to 3} \frac{x-3}{x^2 - 9}$

45. $\lim_{x \to -3} \frac{x-3}{x^2 - 9}$

46. $\lim_{x \to 7} \frac{\sqrt{x} - \sqrt{7}}{x - 7}$

47. $\lim_{x \to -2} \sqrt{\frac{x^2 + 4}{2 - x}}$

48. $\lim_{x \to \infty} \left(3 + \frac{1}{x^{1/3}} + \frac{2}{x^3}\right)$

49. $\lim_{x \to \infty} (3x^3 - 2x^2 - 10x - 100)$

50. $\lim_{x \to \infty} \frac{2x^2 + 3}{3x^2 + 2}$

51. $\lim_{x \to \infty} \frac{2x + 3}{3x^2 + 2}$

52. $\lim_{x \to \infty} \frac{2x^2 + 3}{3x + 2}$

In Problems 53 and 54, use the definition of the derivative to find $f'(x)$.

53. $f(x) = x^2 - x$

54. $f(x) = \sqrt{x} - 3$

C *Problems 55–58 refer to the function f in the figure. Determine whether f is differentiable at the indicated value of x.*

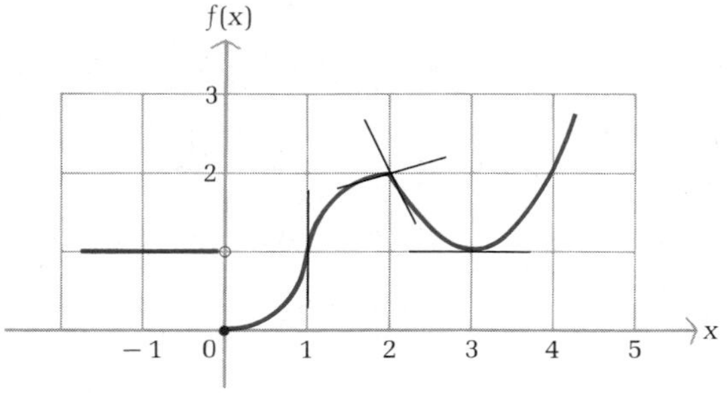

55. $x = 0$ **56.** $x = 1$ **57.** $x = 2$ **58.** $x = 3$

In Problems 59 and 60, graph f and find all discontinuities.

59. $f(x) = \begin{cases} 4 - x^2 & \text{if } x < 0 \\ 2 + x^2 & \text{if } x \geq 0 \end{cases}$

60. $f(x) = \begin{cases} 4 - x^2 & \text{if } x < 1 \\ 3x & \text{if } x \geq 1 \end{cases}$

In Problems 61–64, find f′(x) and simplify.

61. $f(x) = (x - 4)^4(x + 3)^3$

62. $f(x) = \dfrac{x^5}{(2x + 1)^4}$

63. $f(x) = \dfrac{\sqrt{x^2 - 1}}{x}$

64. $f(x) = \dfrac{x}{\sqrt{x^2 + 4}}$

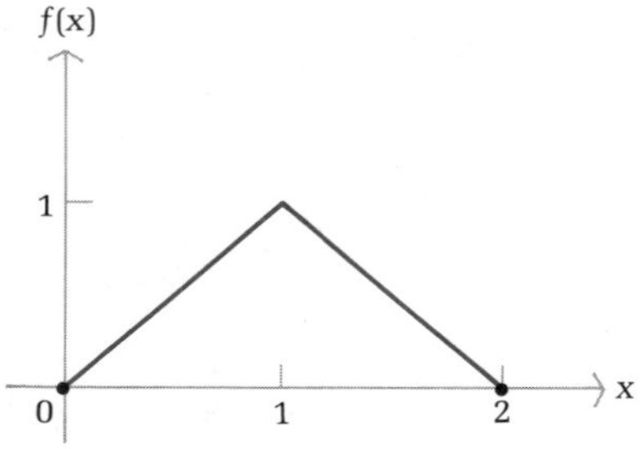

Answer the questions in Problems 65–67 for the function f shown in the figure in the margin and given below:

$$f(x) = 1 - |x - 1| \qquad 0 \leq x \leq 2$$

65. (A) $\lim_{x \to 1^-} f(x) = ?$ (B) $\lim_{x \to 1^+} f(x) = ?$

(C) $\lim_{x \to 1} f(x) = ?$ (D) Is f continuous at $x = 1$?

66. (A) Is f continuous on the open interval (0, 2)?
(B) Is f continuous from the right at $x = 0$?
(C) Is f continuous from the left at $x = 2$?
(D) Is f continuous on the closed interval [0, 2]?

67. (A) $\lim_{h \to 0^-} \dfrac{f(1 + h) - f(1)}{h} = ?$ (B) $\lim_{h \to 0^+} \dfrac{f(1 + h) - f(1)}{h} = ?$

(C) $\lim_{h \to 0} \dfrac{f(1 + h) - f(1)}{h} = ?$ (D) Does $f'(1)$ exist?

APPLICATIONS

Business & Economics

68. *Cost analysis.* The total cost (in dollars) of producing x television sets is

$$C(x) = 10{,}000 + 200x - 0.1x^2$$

(A) Find the exact cost of producing the 101st television set.

(B) Use the marginal cost to approximate the cost of producing the 101st television set.

69. *Marginal analysis.* Let

$$p = 20 - x \qquad \text{and} \qquad C(x) = 2x + 56 \qquad 0 \leq x \leq 20$$

be the demand equation and the cost function, respectively, for a certain commodity.

(A) Find the marginal cost, average cost, and marginal average cost functions.

(B) Express the revenue in terms of x, and find the marginal revenue, average revenue, and marginal average revenue functions.

(C) Find the profit, marginal profit, average profit, and marginal average profit functions.

(D) Find the break-even point(s).

(E) Evaluate the marginal profit at $x = 7$, 9, and 11, and interpret the results.

(F) Graph $R = R(x)$ and $C = C(x)$ on the same coordinate system, and locate regions of profit and loss.

70. *Employee training.* A company producing computer components has established that on the average, a new employee can assemble $N(t)$ components per day after t days of on-the-job training, as given by

$$N(t) = \frac{40t}{t + 2}$$

(A) Find the average rate of change of $N(t)$ from 3 days to 6 days.

(B) Find the instantaneous rate of change of $N(t)$ at 3 days.

(C) Find $\lim_{t\to\infty} N(t)$.

Life Sciences

71. *Pollution.* A sewage treatment plant disposes of its effluent through a pipeline that extends 1 mile toward the center of a large lake. The concentration of effluent $C(x)$, in parts per million, x meters from the end of the pipe is given approximately by

$$C(x) = 500(x + 1)^{-2}$$

What is the instantaneous rate of change of concentration at 9 meters? At 99 meters?

Social Sciences

72. *Learning.* If a person learns N items in t hours, as given by

$$N(t) = 20\sqrt{t}$$

find the rate of learning after:

(A) 1 hour (B) 4 hours

CHAPTER 5

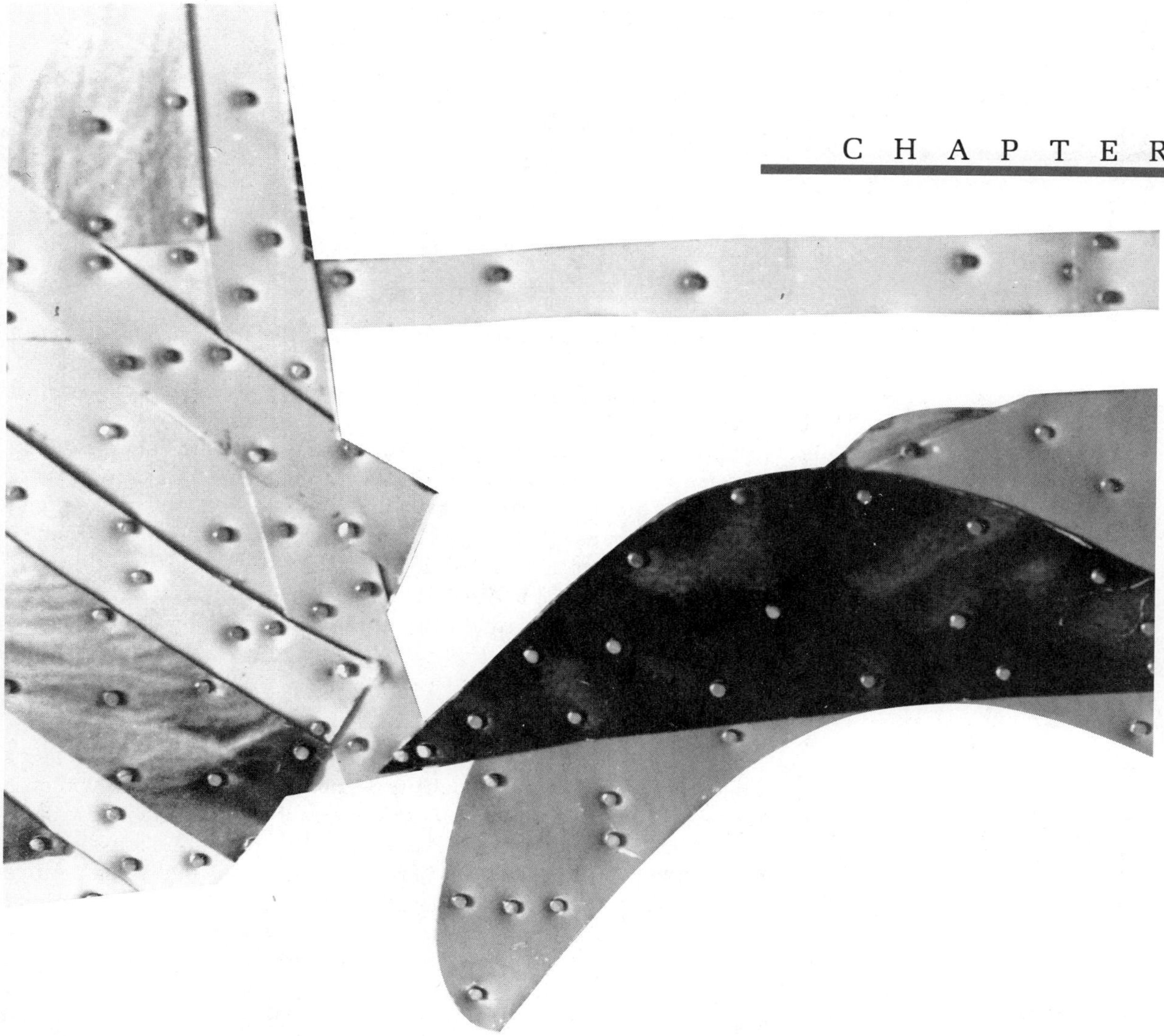

Graphing and Optimization

CHAPTER 5

Contents

SECTION 5-1 First Derivative and Graphs

- SOLVING INEQUALITIES USING CONTINUITY PROPERTIES
- INCREASING AND DECREASING FUNCTIONS
- CRITICAL VALUES AND LOCAL EXTREMA
- FIRST-DERIVATIVE TEST

Since the derivative is associated with the slope of the graph of a function at a point, we might expect that it is also associated with other properties of a graph. As we will see in this and the next section, the derivative can tell us a great deal about the shape of the graph of a function. In addition, this investigation will lead to methods for finding absolute maximum and minimum values for functions that do not require graphing. Manufacturing companies can use these methods to find production levels that will minimize cost or maximize profit. Pharmacologists can use them to find levels of drug dosages that will produce maximum sensitivity to a drug. And so on.

We digress for a moment to discuss the use of continuity and *sign charts* in solving inequalities, a process that will see frequent use in this and succeeding sections.

◆ SOLVING INEQUALITIES USING CONTINUITY PROPERTIES

In our informal discussion of continuity in Section 4-1, we said that a function is continuous over an interval if we can draw its graph over the interval without lifting a pencil from the paper. Suppose a function f is continuous over the interval $(1, 8)$ and $f(x) \neq 0$ for any x in $(1, 8)$. Also suppose $f(2) = 5$, a positive number. Is it possible for $f(x)$ to be negative for any x in $(1, 8)$? The answer is "no." If $f(7)$ were -3, for example, as shown in Figure 1, how would it be possible to join the points $(2, 5)$ and $(7, -3)$ with the graph of a continuous function without crossing the x axis between 1 and 8 at least once? [Crossing the x axis would violate our assumption that $f(x) \neq 0$ for any x in $(1, 8)$.] Thus, we conclude that $f(x)$ must be positive for all x in $(1, 8)$. If $f(2)$ were negative, then, using the same type of reasoning, $f(x)$ would have to be negative over the whole interval $(1, 8)$.

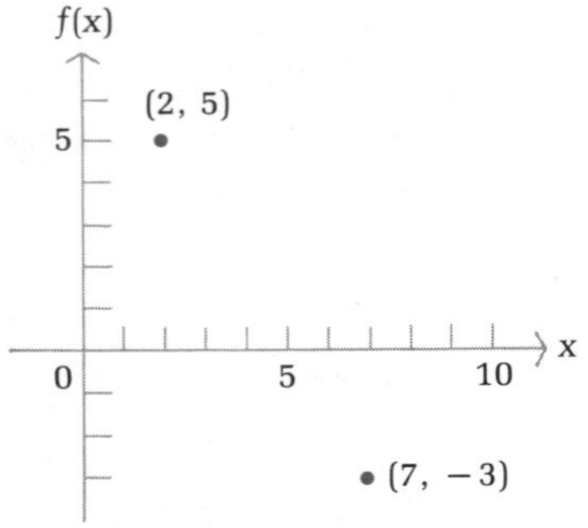

FIGURE 1

In general, **if f is continuous and $f(x) \neq 0$ on the interval (a, b), then $f(x)$ cannot change sign on (a, b).** This is the essence of Theorem 1.

THEOREM 1

Sign Properties on an Interval (a, b)

If f is continuous on (a,b) and $f(x) \neq 0$ for all x in (a, b), then either $f(x) > 0$ for all x in (a,b) or $f(x) < 0$ for all x in (a,b).

Theorem 1 provides the basis for an effective method of solving many types of inequalities. Example 1 illustrates the process.

◆ EXAMPLE 1 Solve: $\dfrac{x+1}{x-2} > 0$

Solution We start by using the left side of the inequality to form the function f:

$$f(x) = \frac{x+1}{x-2}$$

The rational function f is discontinuous at $x = 2$, and $f(x) = 0$ for $x = -1$ (a fraction is 0 when the numerator is 0 and the denominator is not 0). We plot $x = 2$ and $x = -1$, which we call *partition numbers*, on a real number line:

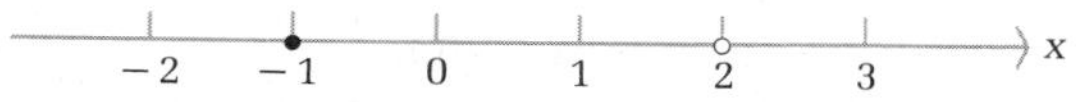

(Note that the dot at 2 is open, because the function is not defined at $x = 2$.) The partition numbers 2 and -1 determine three open intervals: $(-\infty, -1)$, $(-1, 2)$, and $(2, \infty)$. The function f is continuous and nonzero on each of these intervals. From Theorem 1 we know that $f(x)$ does not change sign on any of these intervals. Thus, we can find the sign of $f(x)$ on each of these intervals by selecting a **test number** in each interval and evaluating $f(x)$ at that number. Since any number in each subinterval will do, we choose test numbers that are easy to evaluate: -2, 0, and 3. The table in the margin shows the results.

x	-2	0	3
$f(x)$	$\frac{1}{4}$ +	$-\frac{1}{2}$ −	4 +

Test numbers

The sign of $f(x)$ at each test number is the same as the sign of $f(x)$ over the interval containing that test number. Using this information, we construct a **sign chart** for $f(x)$:

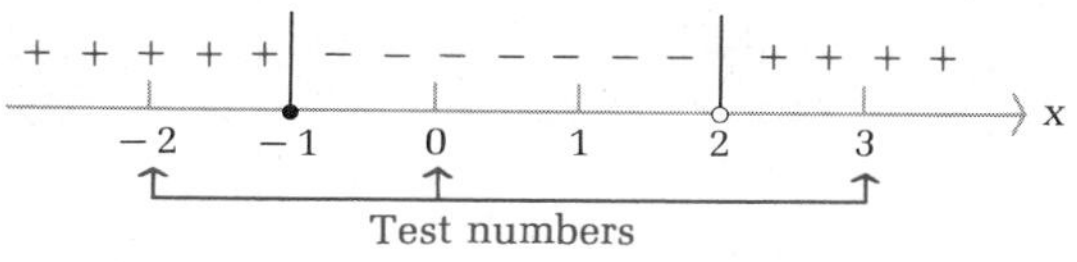

Now using the sign chart, we can easily write the solution for the given nonlinear inequality: $f(x) > 0$ for

$x < -1$ or $x > 2$	Inequality notation
$(-\infty, -1) \cup (2, \infty)$	Interval notation

◆

Most of the inequalities we will encounter will involve strict inequalities ($>$ or $<$). If it is necessary to solve inequalities of the form $\geqslant$ or $\leqslant$, we simply include the end point of any interval if it is a zero of f [that is, if it is a value of x such that $f(x) = 0$]. For example, referring to the sign chart in Example 1, the solution of the inequality

$\dfrac{x+1}{x-2} \geqslant 0$ is

$x \leqslant -1$ or $x > 2$	Inequality notation
$(-\infty, -1] \cup (2, \infty)$	Interval notation

In general, given a function f, we will call all values x such that f is discontinuous at x or $f(x) = 0$ **partition numbers. Partition numbers determine open intervals where $f(x)$ does not change sign.** By using a test number from each interval, we can construct a sign chart for $f(x)$ on the real number line. It is then an easy matter to determine where $f(x) < 0$ or $f(x) > 0$; that is, to solve the inequality $f(x) < 0$ or $f(x) > 0$.

We summarize the procedure for constructing sign charts in the following box:

Constructing Sign Charts

Given a function f:

Step 1. Find all partition numbers. That is:

(A) Find all numbers where f is discontinuous. (Rational functions are discontinuous for values of x that make a denominator 0.)
(B) Find all numbers where $f(x) = 0$. (For a rational function, this occurs where the numerator is 0 and the denominator is not 0.)

Step 2. Plot the numbers found in step 1 on a real number line, dividing the number line into intervals.

Step 3. Select a test number in each open interval determined in step 2, and evaluate $f(x)$ at each test number to determine whether $f(x)$ is positive (+) or negative (−) in each interval.

Step 4. Construct a sign chart using the real number line in step 2. This will show the sign of $f(x)$ on each open interval

[*Note:* From the sign chart, it is easy to find the solution for the inequality $f(x) < 0$ or $f(x) > 0$.]

PROBLEM 1 Solve: $\dfrac{x^2 - 1}{x - 3} < 0$

◆

◆ INCREASING AND DECREASING FUNCTIONS

Graphs of functions generally have *rising* or *falling* sections as we scan the graphs from left to right. It would be an aid to graphing if we could determine where these sections occur. Suppose the graph of a function f is as indicated in Figure 2. As we look from left to right, we see that on the interval (a, b) the graph of f is *rising*, $f(x)$ is *increasing*,* and the slope of the graph is positive $[f'(x) > 0]$. On the other hand, on the interval (b, c) the graph of f is *falling*, $f(x)$ is *decreasing*, and the slope of the graph is negative $[f'(x) < 0]$. At $x = b$, the graph of f changes direction (from rising to falling), $f(x)$ changes from increasing to decreasing, the slope of the graph is 0 $[f'(b) = 0]$, and the tangent line is horizontal.

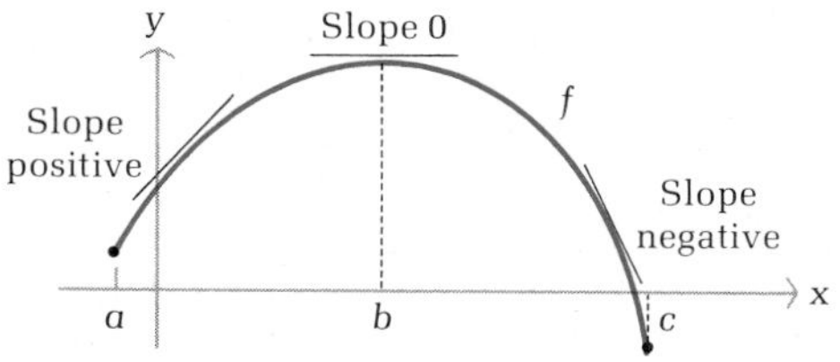

FIGURE 2

In general, if $f'(x) > 0$ (is positive) on the interval (a, b), then $f(x)$ increases (↗) and the graph of f rises as we move from left to right over the interval; if $f'(x) < 0$ (is negative) on an interval (a, b), then $f(x)$ decreases (↘) and the graph of f falls as we move from left to right over the interval. We summarize these important results in the box.

Increasing and Decreasing Functions

For the interval (a, b):

$f'(x)$	$f(x)$	GRAPH OF f	EXAMPLES
+	Increases ↗	Rises ↗	
−	Decreases ↘	Falls ↘	

◆ EXAMPLE 2 Given $f(x) = 8x - x^2$:

(A) Which values of x correspond to horizontal tangent lines?
(B) For which values of x is $f(x)$ increasing? Decreasing?
(C) Sketch a graph of f. Add horizontal tangent lines.

---

* Formally, we say that $f(x)$ is **increasing** on an interval (a, b) if $f(x_2) > f(x_1)$ whenever $a < x_1 < x_2 < b$; f is **decreasing** on (a, b) if $f(x_2) < f(x_1)$ whenever $a < x_1 < x_2 < b$.

Solutions (A) $f'(x) = 8 - 2x = 0$

$$x = 4$$

Thus, a horizontal tangent line exists at $x = 4$ only.

(B) As in Example 1, we will construct a sign chart for $f'(x)$ to determine which values of x make $f'(x) > 0$ and which values make $f'(x) < 0$. To do this, we must first determine the partition numbers (values of x for which $f'(x) = 0$ or for which $f'(x)$ is discontinuous). From part A we know that $f'(x) = 8 - 2x = 0$ at $x = 4$. Since $f'(x) = 8 - 2x$ is a polynomial, it is continuous for all x. Thus, 4 is the only partition number. We construct a sign chart for the intervals $(-\infty, 4)$ and $(4, \infty)$, using test numbers 3 and 5:

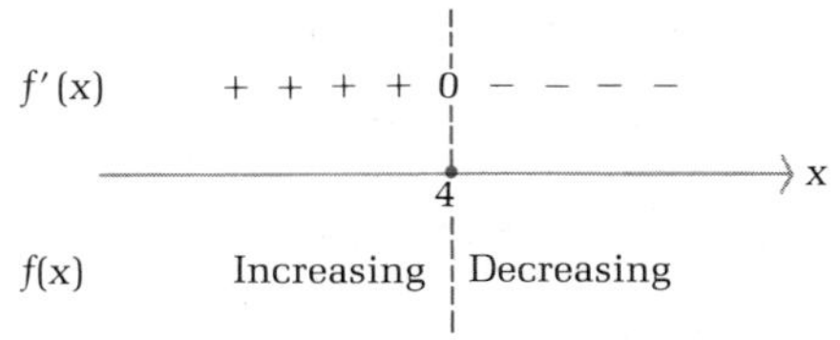

Test Numbers

x	$f'(x)$	
3	2	(+)
5	−2	(−)

Thus, $f(x)$ is increasing on $(-\infty, 4)$ and decreasing on $(4, \infty)$.

x	$f(x)$
0	0
2	12
4	16
6	12
8	0

(C)

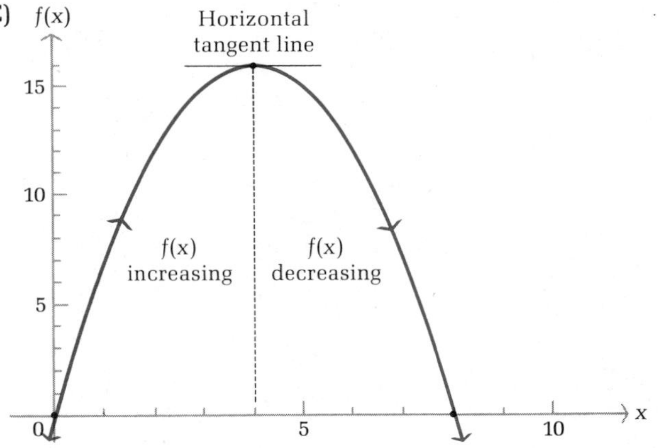

◆

PROBLEM 2 Repeat Example 2 for $f(x) = x^2 - 6x + 10$. ◆

◆ EXAMPLE 3 Determine the intervals where f is increasing and those where f is decreasing for:

(A) $f(x) = 1 + x^3$ (B) $f(x) = (1 - x)^{1/3}$ (C) $f(x) = \dfrac{1}{x - 2}$

Solutions (A) $f(x) = 1 + x^3$ $\quad f'(x) = 3x^2 = 0$

$$x = 0$$

Sign chart for $f'(x) = 3x^2$ (partition number is 0):

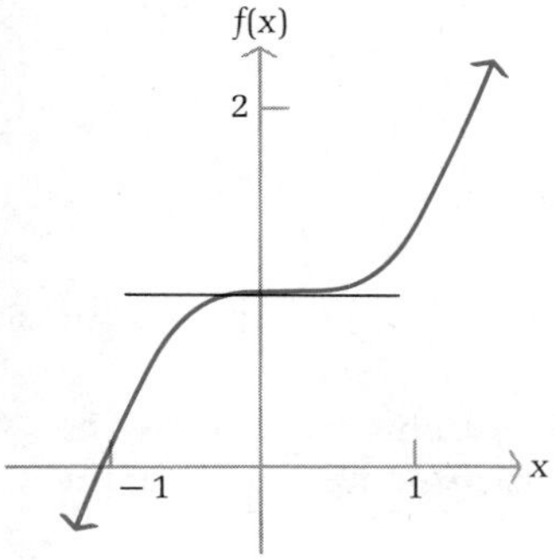

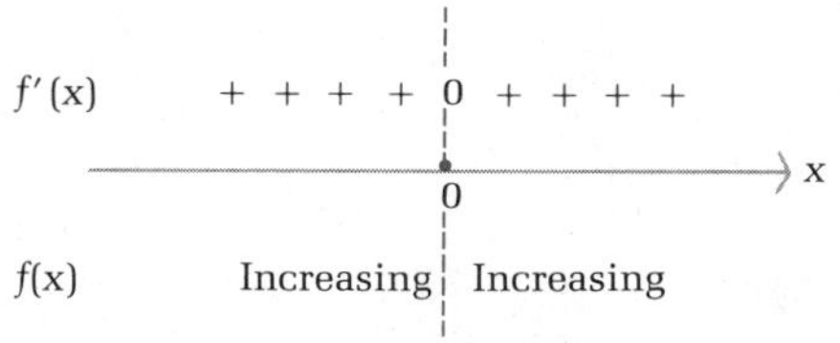

Test Numbers

x	$f'(x)$
−1	3 (+)
1	3 (+)

The sign chart indicates that $f(x)$ is increasing on $(-\infty, 0)$ and $(0, \infty)$. Since f is continuous at $x = 0$, it follows that $f(x)$ is increasing for all x. The graph of f is shown in the margin.

(B) $f(x) = (1 - x)^{1/3}$ $\quad f'(x) = -\frac{1}{3}(1 - x)^{-2/3} = \frac{-1}{3(1 - x)^{2/3}}$

To find partition numbers for $f'(x)$, we note that f' is continuous for all x except for values of x for which the denominator is 0; that is, f' is discontinuous at $x = 1$. Since the numerator is the constant -1, $f'(x) \neq 0$ for any value of x. Thus, we plot the partition number $x = 1$ on the real number line and use the abbreviation ND to note the fact that $f'(x)$ is *not defined* at $x = 1$.

Sign chart for $f'(x) = -1/[3(1 - x)^{2/3}]$ (partition number is 1):

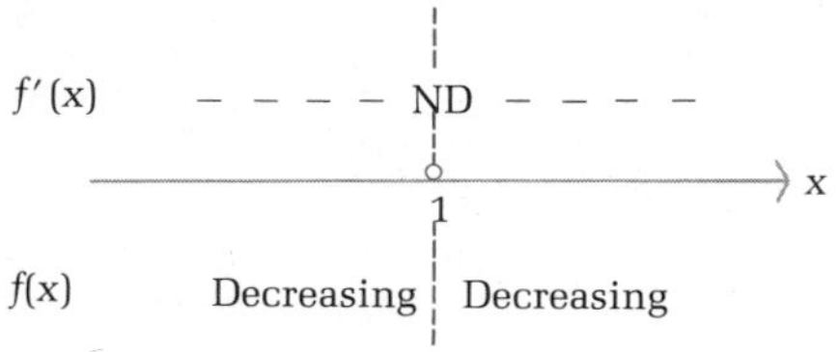

Test Numbers

x	$f'(x)$
0	$-\frac{1}{3}$ (−)
2	$-\frac{1}{3}$ (−)

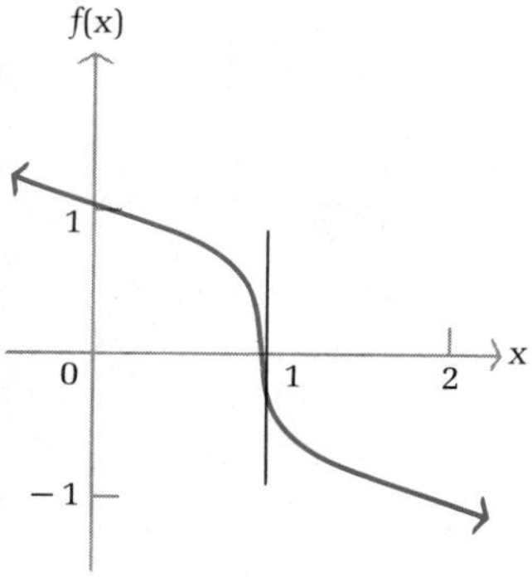

The sign chart indicates that f is decreasing on $(-\infty, 1)$ and $(1, \infty)$. Since f is continuous at $x = 1$, it follows that $f(x)$ is decreasing for all x. Thus, **a continuous function can be decreasing (or increasing) on an interval containing values of x where $f'(x)$ does not exist.** The graph of f is shown in the margin. Notice that the undefined derivative at $x = 1$ results in a vertical tangent line at $x = 1$. In general, **a vertical tangent will occur at $x = c$ if f is continuous at $x = c$ and $|f'(x)|$ becomes larger and larger as x approaches c.**

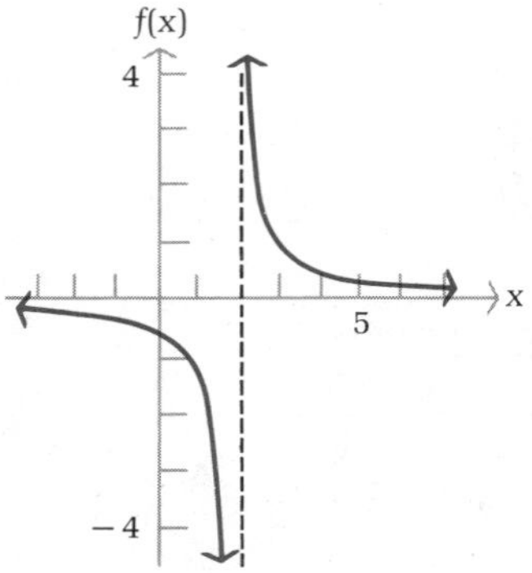

(C) $f(x) = \dfrac{1}{x-2}$ $\quad f'(x) = \dfrac{-1}{(x-2)^2}$

Sign chart for $f'(x) = -1/(x-2)^2$ (partition number is 2):

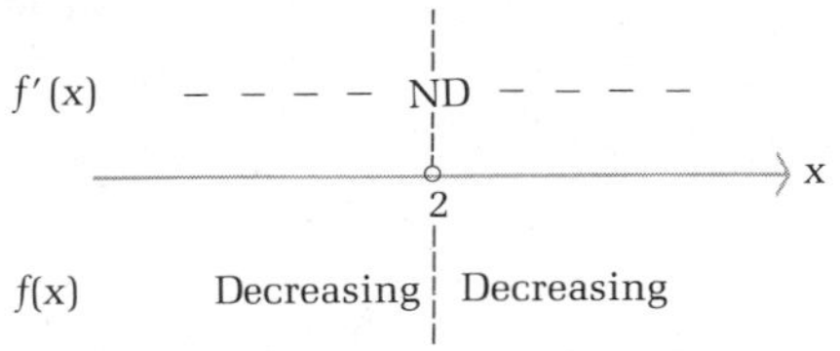

Test Numbers

x	$f'(x)$
1	−1 (−)
3	−1 (−)

Thus, f is decreasing on $(-\infty, 2)$ and $(2, \infty)$. See the graph of f in the margin. ◆

The values where a function is increasing or decreasing must always be expressed in terms of open intervals that are subsets of the domain of the function.

PROBLEM 3 Determine the intervals where f is increasing and those where f is decreasing for:

(A) $f(x) = 1 - x^3$ (B) $f(x) = (1 + x)^{1/3}$ (C) $f(x) = \dfrac{1}{x}$ ◆

◆ CRITICAL VALUES AND LOCAL EXTREMA

When the graph of a continuous function changes from rising to falling, a high point, or *local maximum*, occurs; and when the graph changes from falling to rising, a low point, or *local minimum*, occurs. In Figure 3, high points occur at c_3 and c_6, and low points occur at c_2 and c_4. In general, we call $f(c)$ a **local maximum** if there exists an interval (m, n) containing c such that

$$f(x) \le f(c)$$

for all x in (m, n).

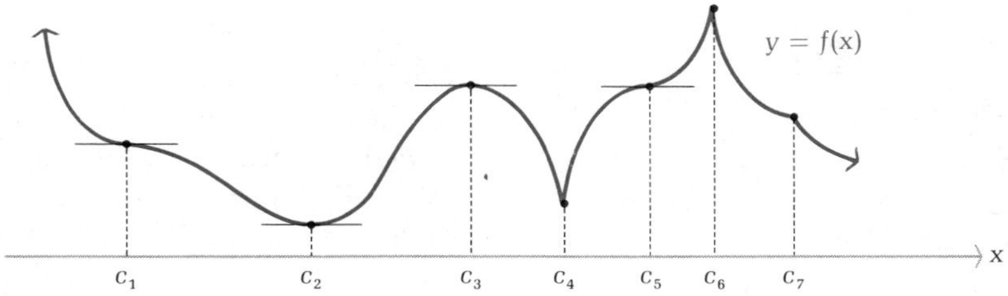

FIGURE 3

The quantity $f(c)$ is called a **local minimum** if there exists an interval (m, n) containing c such that

$$f(x) \geqslant f(c)$$

for all x in (m, n). The quantity $f(c)$ is called a **local extremum** if it is either a local maximum or a local minimum. Thus, in Figure 3, we see that local maxima occur at c_3 and c_6, local minima occur at c_2 and c_4, and all four of these points are local extrema.

How can we locate local maxima and minima if we are given the equation for a function and not its graph? Figure 3 suggests an approach. It appears that local maxima and minima occur among those values of x such that $f'(x) = 0$ or $f'(x)$ does not exist; that is, among the values $c_1, c_2, c_3, c_4, c_5, c_6$, and c_7. [Recall from Section 4-3 that $f'(x)$ is not defined at points on the graph of f where there is a sharp corner or a vertical tangent line.]

Critical Values of f

The values of x in the domain of f where $f'(x) = 0$ or $f'(x)$ does not exist are called the **critical values** of f.

It is possible to prove the following theorem:

THEOREM 2

Existence of Local Extrema

If f is continuous on the interval (a, b) and $f(c)$ is a local extremum, then either $f'(c) = 0$ or $f'(c)$ does not exist (is not defined).

Theorem 2 states that a local extremum can occur only at a critical value, but it does not imply that every critical value produces a local extremum. In Figure 3, c_1 and c_5 are critical values (the slope is 0), but the function does not have a local maximum or local minimum at either of these values.

Our strategy for finding local extrema is now clear. We find all critical values for f and test each one to see if it produces a local maximum, a local minimum, or neither.

◆ FIRST-DERIVATIVE TEST

If $f'(x)$ exists on both sides of a critical value c, then the sign of $f'(x)$ can be used to determine whether the point $(c, f(c))$ is a local maximum, a local minimum, or neither. The various possibilities are summarized in the box and illustrated in Figure 4.

First-Derivative Test for Local Extrema

Let c be a critical value of f [$f(c)$ defined and either $f'(c) = 0$ or $f'(c)$ not defined]. Construct a sign chart for $f'(x)$ close to and on either side of c.

SIGN CHART | $f(c)$

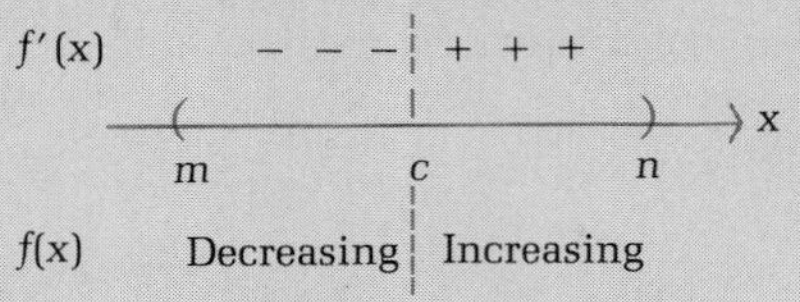

$f(c)$ is a local minimum.
If $f'(x)$ changes from negative to positive at c, then $f(c)$ is a local minimum.

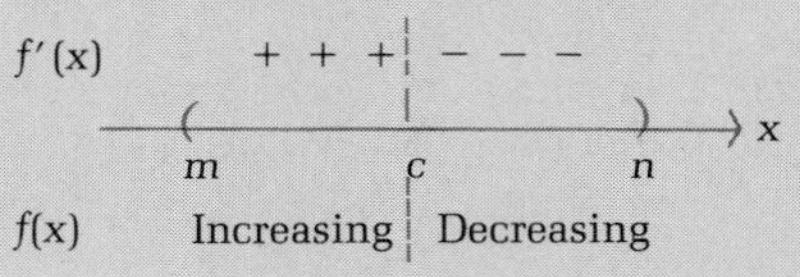

$f(c)$ is a local maximum.
If $f'(x)$ changes from positive to negative at c, then $f(c)$ is a local maximum.

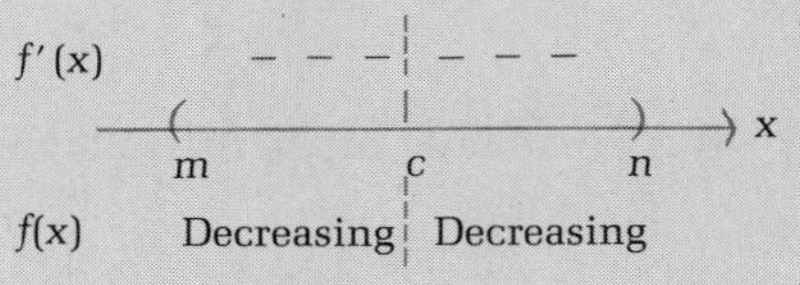

$f(c)$ is not a local extremum.
If $f'(x)$ does not change sign at c, then $f(c)$ is neither a local maximum nor a local minimum.

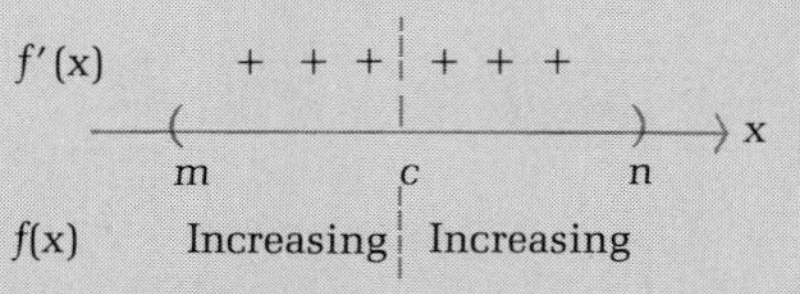

$f(c)$ is not a local extremum.
If $f'(x)$ does not change sign at c, then $f(c)$ is neither a local maximum nor a local minimum.

◆ EXAMPLE 4 Given $f(x) = x^3 - 6x^2 + 9x + 1$:

(A) Find the critical values of f. (B) Find the local maxima and minima.
(C) Sketch the graph of f.

Solutions (A) Find all numbers x in the domain of f where $f'(x) = 0$ or $f'(x)$ does not exist.

$$f'(x) = 3x^2 - 12x + 9 = 0$$
$$3(x^2 - 4x + 3) = 0$$
$$3(x - 1)(x - 3) = 0$$
$$x = 1 \quad \text{or} \quad x = 3$$

$f'(x)$ exists for all x; the critical values are $x = 1$ and $x = 3$.

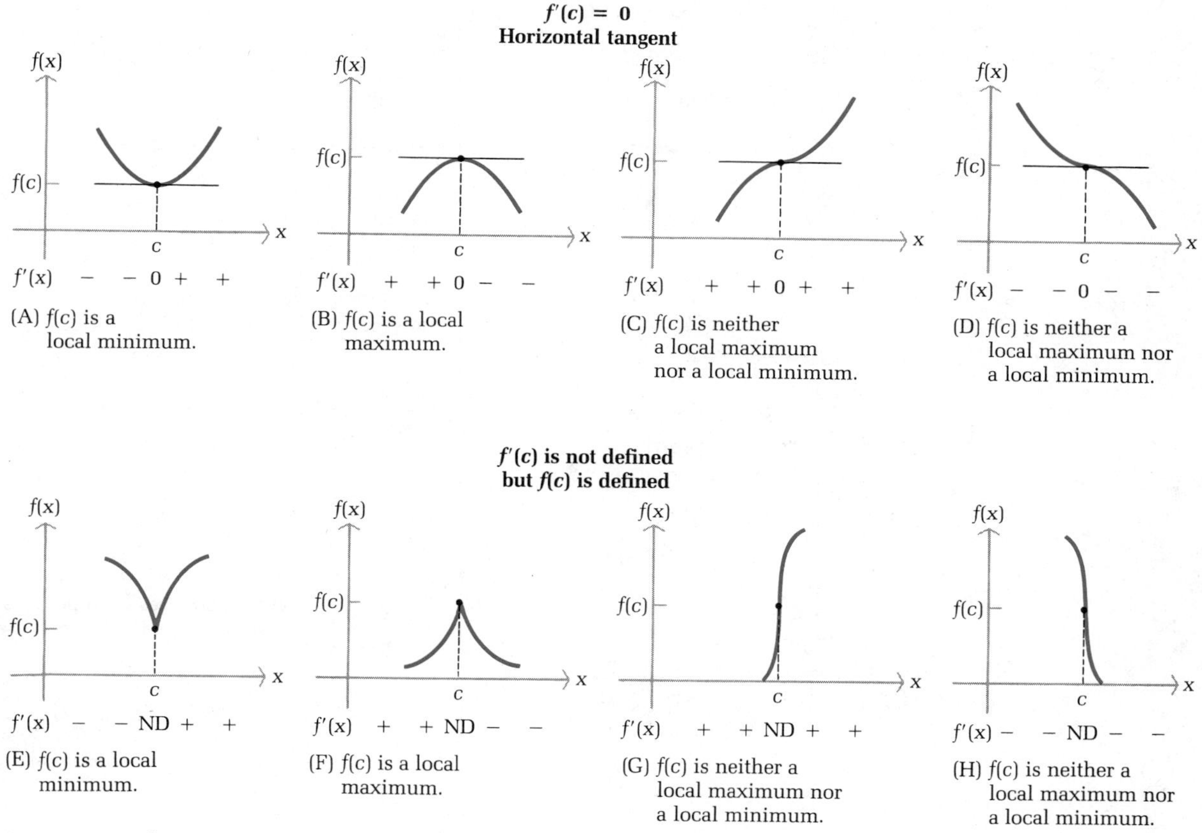

FIGURE 4
Local extrema

(B) The easiest way to apply the first-derivative test for local maxima and minima is to construct a sign chart for $f'(x)$ for all x. Partition numbers for $f'(x)$ are $x = 1$ and $x = 3$ (which also happen to be critical values for f).

Sign chart for $f'(x) = 3(x - 1)(x - 3)$:

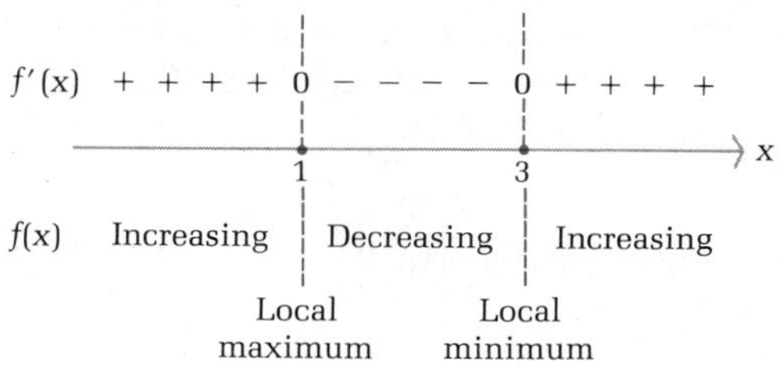

Test Numbers

x	$f'(x)$
0	9 (+)
2	−3 (−)
4	9 (+)

The sign chart indicates that f increases on $(-\infty, 1)$, has a local maximum at $x = 1$, decreases on $(1, 3)$, has a local minimum at $x = 3$, and increases on $(3, \infty)$. These facts are summarized in the following table:

x	$f'(x)$	$f(x)$	GRAPH OF f
$(-\infty, 1)$	+	Increasing	Rising
$x = 1$	0	Local maximum	Horizontal tangent
$(1, 3)$	−	Decreasing	Falling
$x = 3$	0	Local minimum	Horizontal tangent
$(3, \infty)$	+	Increasing	Rising

(C) We sketch a graph of f using the information from part B and point-by-point plotting.

x	$f(x)$
0	1
1	5
2	3
3	1
4	5

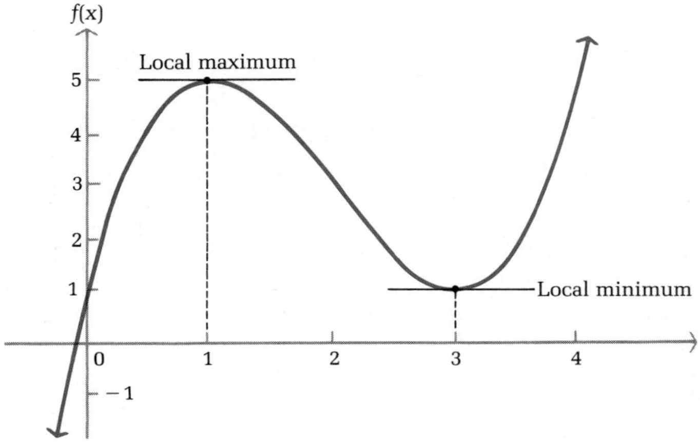

PROBLEM 4 Given $f(x) = x^3 - 9x^2 + 24x - 10$:

(A) Find the critical values of f.
(B) Find the local maxima and minima.
(C) Sketch a graph of f.

In Example 4, the function f had local extrema at both of its critical values. However, as was noted earlier, not every critical value of a function will produce a local extremum. For example, consider the function discussed in Example 3B:

$$f(x) = (1 - x)^{1/3} \quad \text{and} \quad f'(x) = \frac{-1}{3(1 - x)^{2/3}}$$

Since $f(1)$ exists and $f'(1)$ does not exist, $x = 1$ is a critical value for this function. However, the sign chart for $f'(x)$ shows that $f'(x)$ does not change sign at $x = 1$:

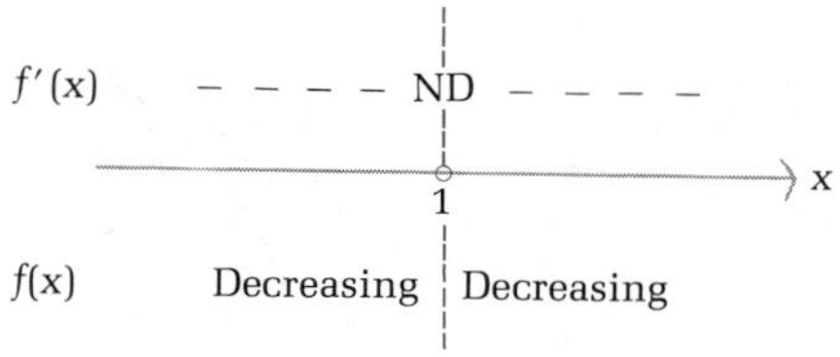

Test Numbers

x	$f'(x)$	
0	$-\frac{1}{3}$	(−)
2	$-\frac{1}{3}$	(−)

Thus, f does not have a local maximum nor a local minimum at $x = 1$.

Finally, it is important to remember that a critical value must be in the domain of the function. Refer to the function discussed in Example 3C:

$$f(x) = \frac{1}{x - 2} \quad \text{and} \quad f'(x) = \frac{-1}{(x - 2)^2}$$

The derivative is not defined at $x = 2$, but neither is the function. Thus, $x = 2$ is not a critical value for f (in fact, this function does not have any critical values). Nevertheless, $x = 2$ is a partition number for $f'(x)$ and must be included in the sign chart for $f'(x)$.

In general, every critical value for $f(x)$ is a partition number for $f'(x)$, but some partition numbers for $f'(x)$ may not be critical values for f. [Critical values have the added requirement that $f(x)$ must be defined at that value.]

Answers to Matched Problems

1. $-\infty < x < -1$ or $1 < x < 3$; $(-\infty, -1) \cup (1, 3)$
2. (A) Horizontal tangent line at $x = 3$.
   (B) Decreasing on $(-\infty, 3)$; increasing on $(3, \infty)$
   (C)

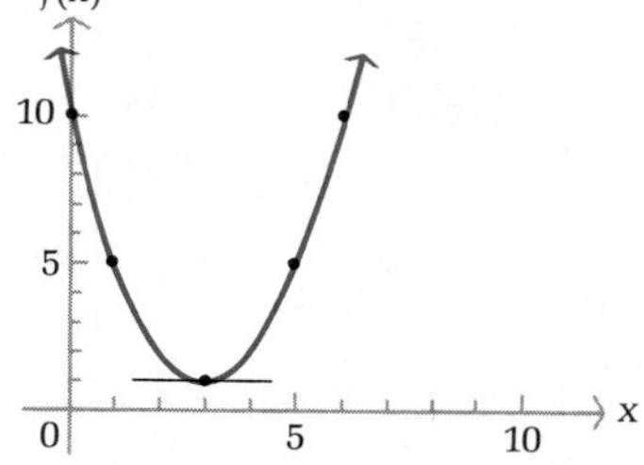

3. (A) Decreasing for all x (B) Increasing for all x
   (C) Decreasing on $(-\infty, 0)$ and $(0, \infty)$

4. (A) Critical values: $x = 2$, $x = 4$
(B) Local maximum at $x = 2$; local minimum at $x = 4$
(C)

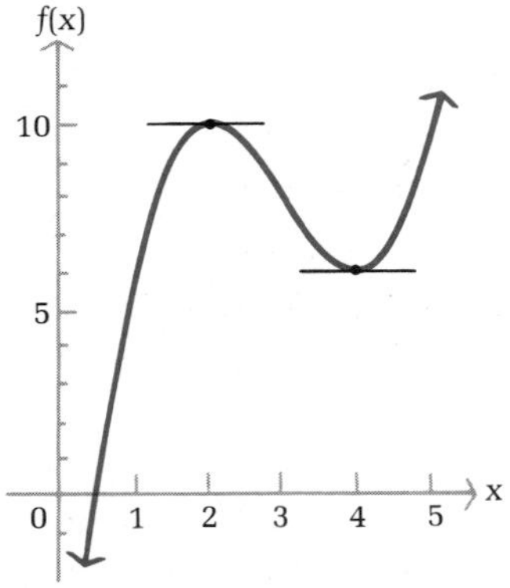

EXERCISE 5-1

A *Problems 1–6 refer to the following graph of $y = f(x)$:*

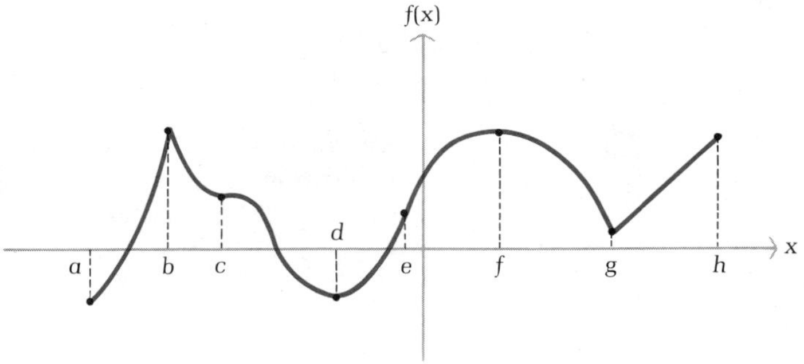

1. Identify the intervals over which $f(x)$ is increasing.
2. Identify the intervals over which $f(x)$ is decreasing.
3. Identify the points where $f'(x) = 0$.
4. Identify the points where $f'(x)$ does not exist.
5. Identify the points where f has a local maximum.
6. Identify the points where f has a local minimum.

B *Solve each inequality using a sign chart. Express answers in inequality and interval notation.*

7. $x^2 - x - 12 < 0$
8. $x^2 - 2x - 8 < 0$
9. $x^2 + 21 > 10x$
10. $x^2 + 7x > -10$
11. $\dfrac{x^2 + 5x}{x - 3} > 0$
12. $\dfrac{x - 4}{x^2 + 2x} < 0$

Find the intervals where $f(x)$ is increasing, the intervals where $f(x)$ is decreasing, and the local extrema.

13. $f(x) = x^2 - 16x + 12$
14. $f(x) = x^2 + 6x + 7$
15. $f(x) = 4 + 10x - x^2$
16. $f(x) = 5 + 8x - 2x^2$
17. $f(x) = 2x^3 + 4$
18. $f(x) = 2 - 3x^3$
19. $f(x) = 2 - 6x - 2x^3$
20. $f(x) = x^3 + 9x + 7$
21. $f(x) = x^3 - 12x + 8$
22. $f(x) = 3x - x^3$
23. $f(x) = x^3 - 3x^2 - 24x + 7$
24. $f(x) = x^3 + 3x^2 - 9x + 5$
25. $f(x) = 2x^2 - x^4$
26. $f(x) = x^4 - 8x^2 + 3$

Find the intervals where $f(x)$ is increasing, the intervals where $f(x)$ is decreasing, and sketch the graph. Add horizontal tangent lines.

27. $f(x) = 4 + 8x - x^2$
28. $f(x) = 2x^2 - 8x + 9$
29. $f(x) = x^3 - 3x + 1$
30. $f(x) = x^3 - 12x + 2$
31. $f(x) = 10 - 12x + 6x^2 - x^3$
32. $f(x) = x^3 + 3x^2 + 3x$

C Find the critical values, the intervals where $f(x)$ is increasing, the intervals where $f(x)$ is decreasing, and the local extrema. Do not graph.

33. $f(x) = \dfrac{x - 1}{x + 2}$
34. $f(x) = \dfrac{x + 2}{x - 3}$
35. $f(x) = x + \dfrac{4}{x}$
36. $f(x) = \dfrac{9}{x} + x$
37. $f(x) = 1 + \dfrac{1}{x} + \dfrac{1}{x^2}$
38. $f(x) = 3 - \dfrac{4}{x} - \dfrac{2}{x^2}$
39. $f(x) = \dfrac{x^2}{x - 2}$
40. $f(x) = \dfrac{x^2}{x + 1}$
41. $f(x) = x^4(x - 6)^2$
42. $f(x) = x^3(x - 5)^2$
43. $f(x) = 3(x - 2)^{2/3} + 4$
44. $f(x) = 6(4 - x)^{2/3} + 4$
45. $f(x) = 2\sqrt{x} - x, \quad x > 0$
46. $f(x) = x - 4\sqrt{x}, \quad x > 0$

APPLICATIONS

Business & Economics

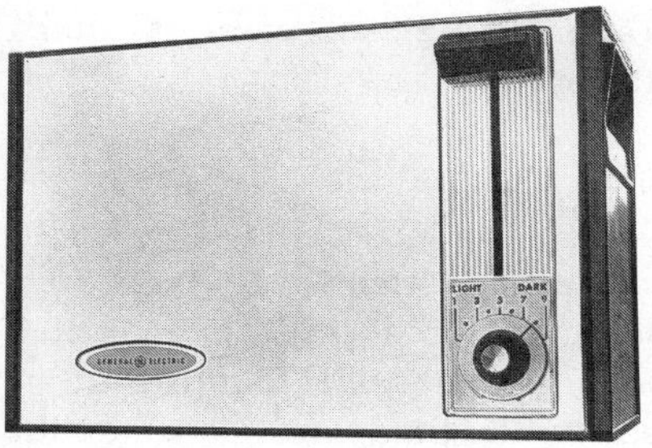

47. *Average cost.* A manufacturer has the following costs in producing x toasters in one day for $0 < x < 150$: fixed costs, \$320; unit production cost, \$20 per toaster; equipment maintenance and repairs, $x^2/20$ dollars. Thus, the cost of manufacturing x toasters in one day is given by

$$C(x) = \frac{x^2}{20} + 20x + 320 \qquad 0 < x < 150$$

and the average cost per toaster is given by

$$\overline{C}(x) = \frac{C(x)}{x} = \frac{x}{20} + 20 + \frac{320}{x} \qquad 0 < x < 150$$

Find the critical values for $\overline{C}(x)$, the intervals where the average cost per toaster is decreasing, the intervals where the average cost per toaster is increasing, and the local extrema. Do not graph.

48. *Average cost.* A manufacturer has the following costs in producing x blenders in one day for $0 < x < 200$: fixed costs, \$450; unit production cost, \$60 per blender; equipment maintenance and repairs, $x^2/18$ dollars.

(A) What is the average cost $\overline{C}(x)$ per blender if x blenders are produced in one day?

(B) Find the critical values for $\overline{C}(x)$, the intervals where the average cost per blender is decreasing, the intervals where the average cost per blender is increasing, and the local extrema. Do not graph.

49. *Marginal analysis.* Show that profit will be increasing over production intervals (a, b) for which marginal revenue is greater than marginal cost. [*Hint:* $P(x) = R(x) - C(x)$]

50. *Marginal analysis.* Show that profit will be decreasing over production intervals (a, b) for which marginal revenue is less than marginal cost.

Life Sciences

51. *Medicine.* A drug is injected into the bloodstream of a patient through the right arm. The concentration of the drug in the bloodstream of the left arm t hours after the injection is approximated by

$$C(t) = \frac{0.14t}{t^2 + 1} \qquad 0 < t < 24$$

Find the critical values for $C(t)$, the intervals where the concentration of the drug is increasing, the intervals where the concentration of the drug is decreasing, and the local extrema. Do not graph.

52. *Medicine.* The concentration $C(t)$, in milligrams per cubic centimeter, of a particular drug in a patient's bloodstream is given by

$$C(t) = \frac{0.16t}{t^2 + 4t + 4} \qquad 0 < t < 12$$

where t is the number of hours after the drug is taken orally. Find the critical values for $C(t)$, the intervals where the concentration of the drug is increasing, the intervals where the concentration of the drug is decreasing, and the local extrema. Do not graph.

Social Sciences

53. *Politics.* Public awareness of a Congressional candidate before and after a successful campaign was approximated by

$$P(t) = \frac{8.4t}{t^2 + 49} + 0.1 \qquad 0 < t < 24$$

where t is time (in months) after the campaign started and $P(t)$ is the fraction of people in the Congressional district who could recall the candidate's (and later, Congressman's) name. Find the critical values for $P(t)$, the time intervals where the fraction is increasing, the time intervals where the fraction is decreasing, and the local extrema. Do not graph.

SECTION 5-2 Second Derivative and Graphs

- CONCAVITY
- INFLECTION POINTS
- SECOND-DERIVATIVE TEST
- APPLICATION

In the preceding section, we saw that the derivative can be used to determine when a graph is rising and falling. Now we want to see what the *second derivative* (the derivative of the derivative) can tell us about the shape of a graph.

CONCAVITY

Consider the functions

$$f(x) = x^2 \qquad \text{and} \qquad g(x) = \sqrt{x}$$

for x in the interval $(0, \infty)$. Since

$$f'(x) = 2x > 0 \qquad \text{for } 0 < x < \infty$$

and

$$g'(x) = \frac{1}{2\sqrt{x}} > 0 \qquad \text{for } 0 < x < \infty$$

both functions are increasing on $(0, \infty)$.

Notice the different shapes of the graphs of f and g, shown in Figure 5. Even though the graph of each function is rising and each graph starts at (0, 0) and goes through (1, 1), the graphs are quite dissimilar. The graph of f opens upward, while the graph of g opens downward. We say that the graph of f is *concave upward*, and the graph of g is *concave downward*. It will help us draw a graph if we can determine the concavity of the graph before we draw it. How can we find a mathematical formulation of concavity?

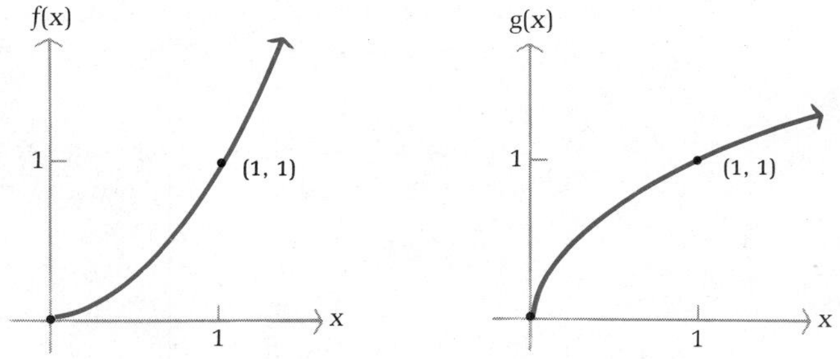

(A) $f(x) = x^2$ (B) $g(x) = \sqrt{x}$

FIGURE 5

It will be instructive to examine the slopes of f and g at various points on their graphs (see Figure 6). We can make two observations about each graph. Looking at the graph of f in Figure 6A, we see that $f'(x)$ (the slope of the tangent line) is *increasing* and that the graph lies *above* each tangent line. Looking at Figure 6B, we see that $g'(x)$ is *decreasing* and that the graph lies *below* each tangent line.

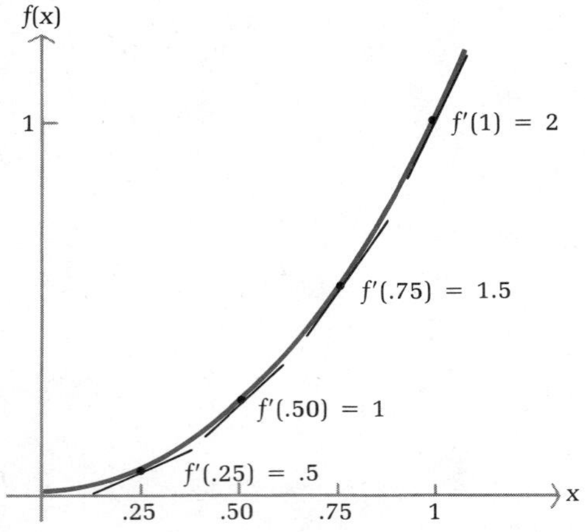

(A) $f(x) = x^2$

(B) $g(x) = \sqrt{x}$

FIGURE 6

With these ideas in mind, we state the general definition of concavity:

The graph of a function f is concave upward on the interval (a, b) if $f'(x)$ is *increasing* on (a, b) and is concave downward on the interval (a, b) if $f'(x)$ is *decreasing* on (a, b).

Geometrically, the graph is concave upward on (a, b) if it lies above its tangent lines in (a, b) and is concave downward on (a, b) if it lies below its tangent lines in (a, b).

How can we determine when $f'(x)$ is increasing or decreasing? In the preceding section, we used the derivative of a function to determine when that function is increasing or decreasing. Thus, to determine when the function $f'(x)$ is increasing or decreasing, we use the derivative of $f'(x)$. The derivative of the derivative of a function is called the *second derivative* of the function. Various notations for the second derivative are given in the following box:

Second Derivative

For $y = f(x)$, the **second derivative** of f, provided it exists, is

$$f''(x) = D_x f'(x)$$

Other notations for $f''(x)$ are

$$\frac{d^2y}{dx^2} \qquad y'' \qquad D_x^2 f(x)$$

Returning to the functions f and g discussed at the beginning of this section, we have

$$f(x) = x^2 \qquad g(x) = \sqrt{x} = x^{1/2}$$

$$f'(x) = 2x \qquad g'(x) = \frac{1}{2} x^{-1/2} = \frac{1}{2\sqrt{x}}$$

$$f''(x) = D_x\, 2x = 2 \qquad g''(x) = D_x \frac{1}{2} x^{-1/2} = -\frac{1}{4} x^{-3/2} = -\frac{1}{4\sqrt{x^3}}$$

For $x > 0$, we see that $f''(x) > 0$; thus, $f'(x)$ is increasing and the graph of f is concave upward (see Fig. 6A). For $x > 0$, we also see that $g''(x) < 0$; thus, $g'(x)$ is decreasing and the graph of g is concave downward (see Fig. 6B). These ideas are summarized in the following box:

Concavity

For the interval (a, b):

$f''(x)$	$f'(x)$	GRAPH OF $y = f(x)$	EXAMPLE
+	Increasing	Concave upward	
−	Decreasing	Concave downward	

Be careful not to confuse concavity with falling and rising. As Figure 7 (on the next page) illustrates, a graph that is concave upward on an interval may be falling, rising, or both falling and rising on that interval. A similar statement holds for a graph that is concave downward.

$f''(x) > 0$ over (a, b)
Concave upward

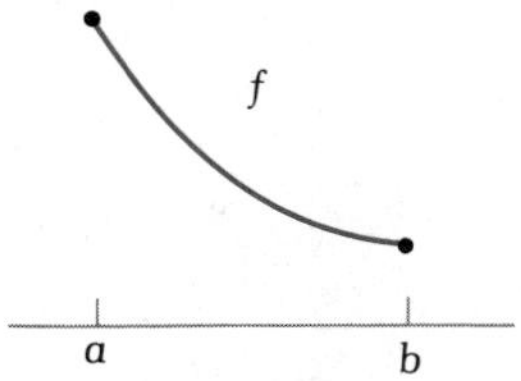

(A) $f'(x)$ is negative and increasing. Graph of f is falling.

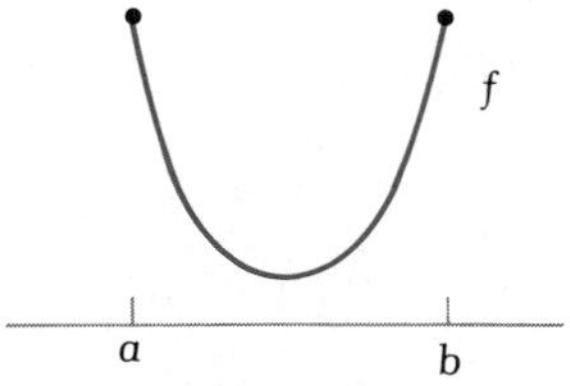

(B) $f'(x)$ increases from negative to positive. Graph of f falls, then rises.

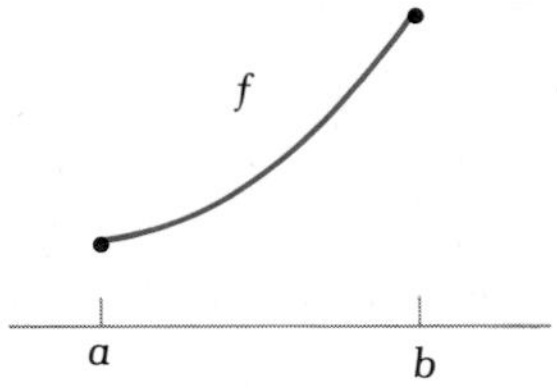

(C) $f'(x)$ is positive and increasing. Graph of f is rising.

$f''(x) < 0$ over (a, b)
Concave downward

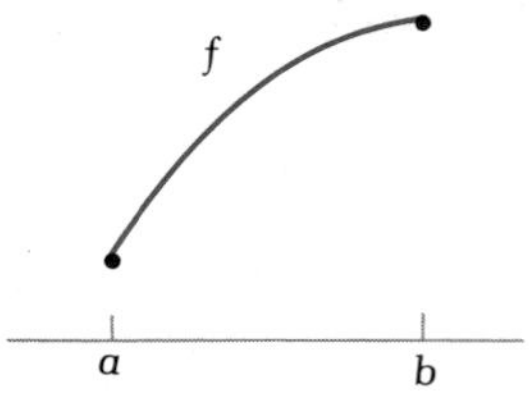

(D) $f'(x)$ is positive and decreasing. Graph of f is rising.

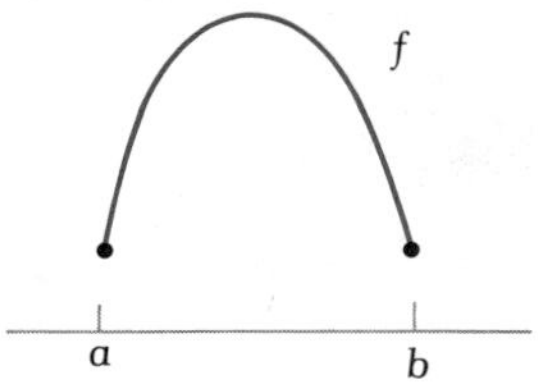

(E) $f'(x)$ decreases from positive to negative. Graph of f rises, then falls.

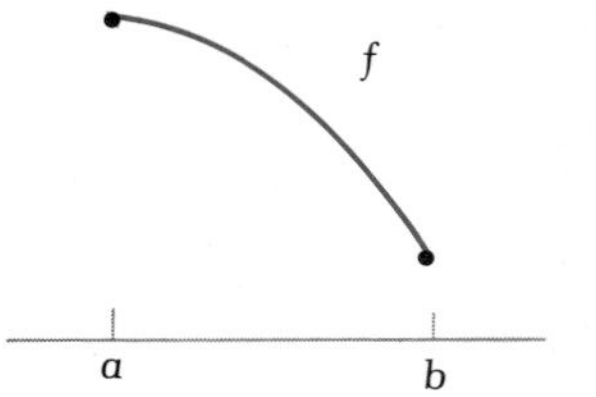

(F) $f'(x)$ is negative and decreasing. Graph of f is falling.

FIGURE 7
Concavity

◆ EXAMPLE 5 Let $f(x) = x^3$. Find the intervals where the graph of f is concave upward and the intervals where the graph of f is concave downward. Sketch a graph of f.

Solution To determine concavity, we must determine the sign of $f''(x)$.

$$f(x) = x^3 \qquad f'(x) = 3x^2 \qquad f''(x) = 6x$$

Sign chart for $f''(x) = 6x$ (partition number is 0):

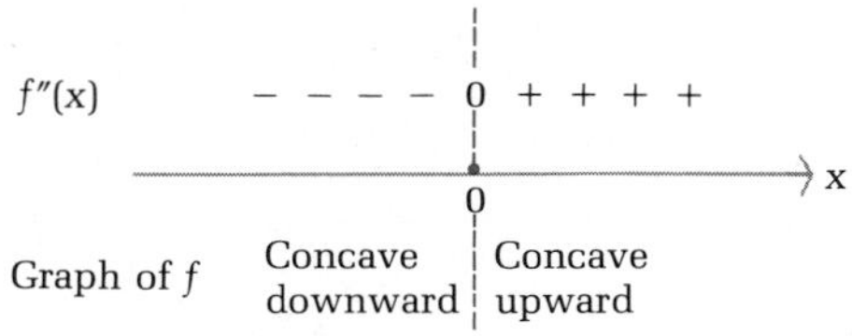

Test Numbers

x	$f''(x)$
-1	-6 (−)
1	6 (+)

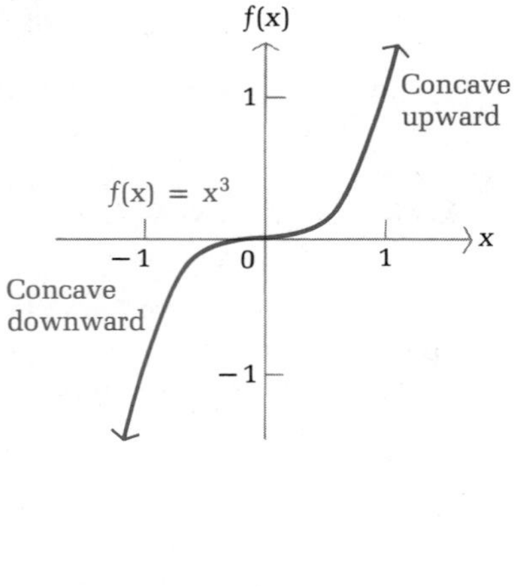

Thus, the graph of f is concave downward on $(-\infty, 0)$ and concave upward on $(0, \infty)$. The graph of f (without going through other graphing details) is shown in the figure in the margin. ◆

PROBLEM 5 Repeat Example 5 for $f(x) = 1 - x^3$. ◆

The graph in Example 5 changes from concave downward to concave upward at the point (0, 0). This point is called an *inflection point.*

◆ INFLECTION POINTS

In general, an **inflection point** is a point on the graph of a function where the concavity changes (from upward to downward or from downward to upward). In order for the concavity to change at a point, $f''(x)$ must change sign at that point. Reasoning as we did in the previous section, we conclude that the inflection points must occur at points where $f''(x) = 0$ or $f''(x)$ does not exist [but $f(x)$ must exist]. Figure 8 illustrates several typical cases.

If $f'(c)$ exists and $f''(x)$ changes sign at $x = c$, then the tangent line at an inflection point $(c, f(c))$ will always lie below the graph on the side that is concave upward and above the graph on the side that is concave downward (see Figs. 8A, B, and C).

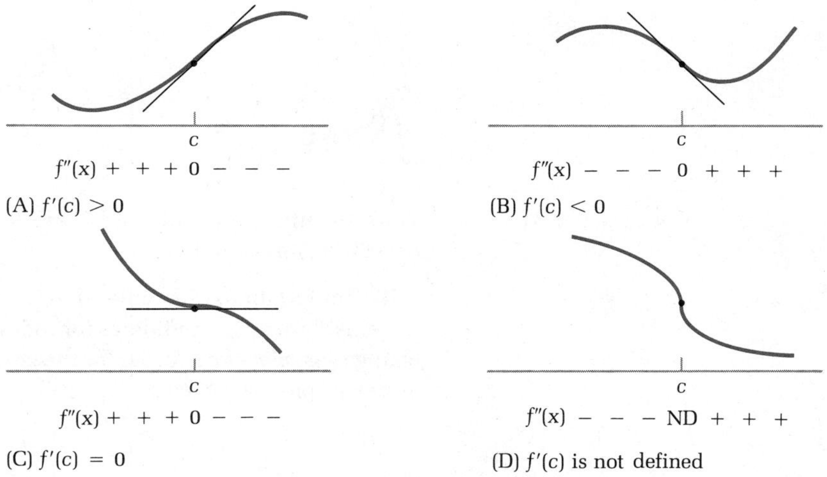

FIGURE 8
Inflection points

◆ EXAMPLE 6 Find the inflection points of $f(x) = x^3 - 6x^2 + 9x + 1$.

Solution Since inflection points occur at values of x where $f''(x)$ changes sign, we construct a sign chart for $f''(x)$.

$$f(x) = x^3 - 6x^2 + 9x + 1$$
$$f'(x) = 3x^2 - 12x + 9$$
$$f''(x) = 6x - 12 = 6(x - 2)$$

Sign chart for $f''(x) = 6(x - 2)$ (partition number is 2):

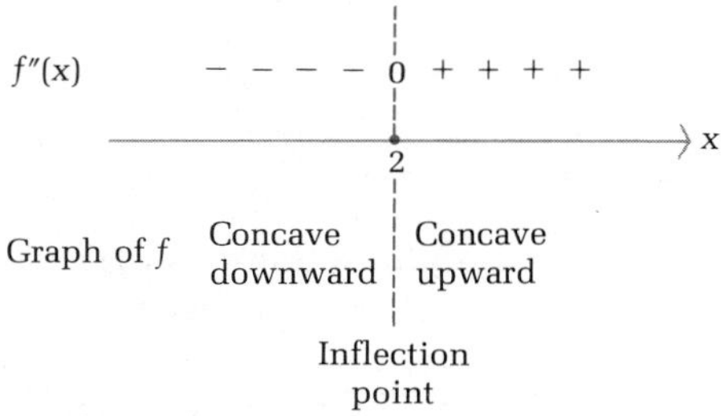

Test Numbers

x	$f''(x)$
1	−6 (−)
3	6 (+)

From the sign chart, we see that the graph of f has an inflection point at $x = 2$. The graph of f is shown in the figure. (See Example 4 in Section 5-1.)

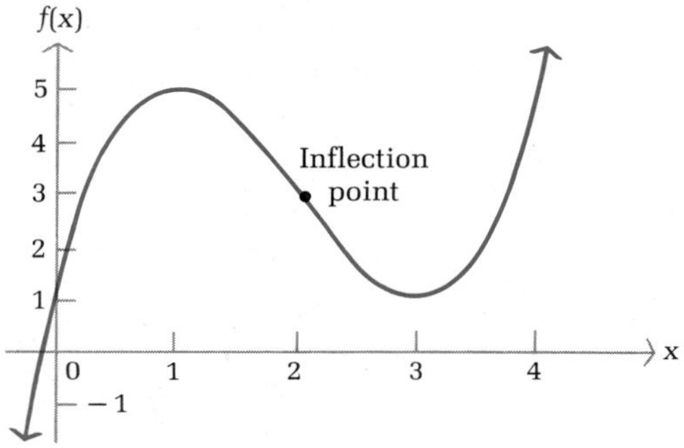

◆

PROBLEM 6 Find the inflection points of $f(x) = x^3 - 9x^2 + 24x - 10$. (See Problem 4 in Section 5-1 for the graph of f.) ◆

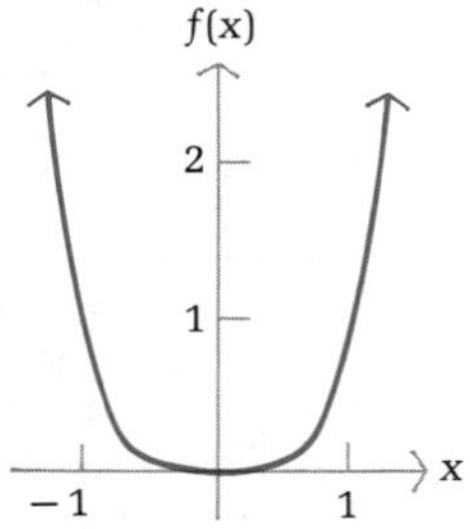

FIGURE 9
$f(x) = x^4$

It is important to remember that the values of x where $f''(x) = 0$ [or $f''(x)$ does not exist] are only candidates for inflection points. The second derivative must change sign at $x = c$ in order for the graph of f to have an inflection point at $x = c$. For example, consider

$$f(x) = x^4 \qquad f'(x) = 4x^3 \qquad f''(x) = 12x^2$$

The second derivative is 0 at $x = 0$, but $f''(x) > 0$ for all other values of x. Since $f''(x)$ does not change sign at $x = 0$, the graph of f does not have an inflection point at $x = 0$, as illustrated in Figure 9.

◆ SECOND-DERIVATIVE TEST

Now we want to see how the second derivative can be used to find local extrema. Suppose f is a function satisfying $f'(c) = 0$ and $f''(c) > 0$. First, note that if $f''(c) > 0$, then it follows from the properties of limits* that $f''(x) > 0$ in some

* Actually, we are assuming that $f''(x)$ is continuous in an interval containing c. It is very unlikely that we will encounter a function for which $f''(c)$ exists, but $f''(x)$ is not continuous in an interval containing c.

interval (m, n) containing c. Thus, the graph of f must be concave upward in this interval. But this implies that $f'(x)$ is increasing in this interval. Since $f'(c) = 0$, $f'(x)$ must change from negative to positive at $x = c$ and $f(c)$ is a local minimum (see Figure 10). Reasoning in the same fashion, we conclude that if $f'(c) = 0$ and $f''(c) < 0$, then $f(c)$ is a local maximum. Of course, it is possible that both $f'(c) = 0$ and $f''(c) = 0$. In this case, the second derivative cannot be used to determine the shape of the graph around $x = c$; $f(c)$ may be a local minimum, a local maximum, or neither.

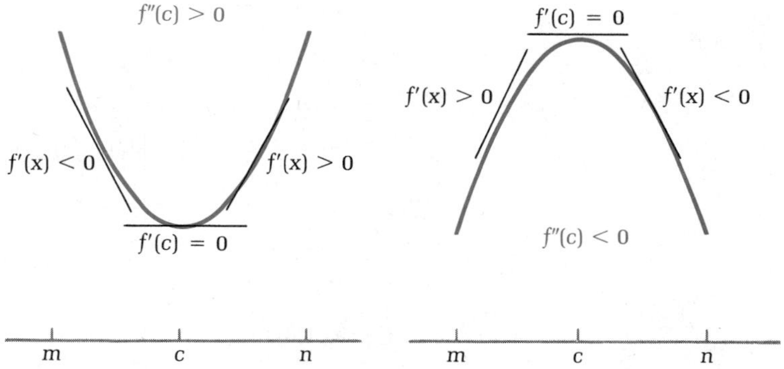

(A) $f'(c) = 0$ and $f''(c) > 0$ implies $f(c)$ is a local minimum.

(B) $f'(c) = 0$ and $f''(c) < 0$ implies $f(c)$ is a local maximum.

FIGURE 10
The second derivative and local extrema

The sign of the second derivative thus provides a simple test for identifying local maxima and minima. This test is most useful when we do not want to draw the graph of the function. If we are interested in drawing the graph and have already constructed the sign chart for $f'(x)$, then the first-derivative test can be used to identify the local extrema.

Second-Derivative Test for Local Maxima and Minima

Let c be a critical value for $f(x)$.

$f'(c)$	$f''(c)$	GRAPH OF f IS	$f(c)$	EXAMPLE
0	+	Concave upward	Local minimum	
0	−	Concave downward	Local maximum	
0	0	?	Test fails	

The first-derivative test must be used whenever $f''(c) = 0$ or $f''(c)$ does not exist.

◆ EXAMPLE 7 Find the local maxima and minima for each function. Use the second-derivative test when it applies.

(A) $f(x) = x^3 - 6x^2 + 9x + 1$ (B) $f(x) = \frac{1}{6}x^6 - 4x^5 + 25x^4$

Solutions (A) Take first and second derivatives and find critical values:

$$f(x) = x^3 - 6x^2 + 9x + 1$$
$$f'(x) = 3x^2 - 12x + 9 = 3(x - 1)(x - 3)$$
$$f''(x) = 6x - 12 = 6(x - 2)$$

Critical values are $x = 1$ and $x = 3$.

$f''(1) = -6 < 0$ f has a local maximum at $x = 1$.

$f''(3) = 6 > 0$ f has a local minimum at $x = 3$.

(B)
$$f(x) = \frac{1}{6}x^6 - 4x^5 + 25x^4$$
$$f'(x) = x^5 - 20x^4 + 100x^3 = x^3(x - 10)^2$$
$$f''(x) = 5x^4 - 80x^3 + 300x^2$$

Critical values are $x = 0$ and $x = 10$.

$f''(0) = 0$, $f''(10) = 0$: The second-derivative test fails at both critical values, so the first-derivative test must be used.

Sign chart for $f'(x) = x^3(x - 10)^2$ (partition numbers are 0 and 10):

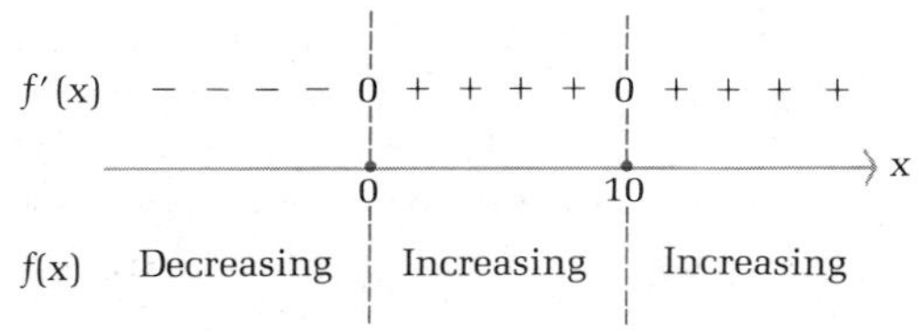

Test Numbers

x	$f'(x)$	
-1	-121	(−)
1	81	(+)
11	1,331	(+)

From the sign chart, we see that $f(x)$ has a local minimum at $x = 0$ and does not have a local extremum at $x = 10$. ◆

PROBLEM 7 Find the local maxima and minima for each function. Use the second-derivative test when it applies.

(A) $f(x) = x^3 - 9x^2 + 24x - 10$ (B) $f(x) = 10x^6 - 24x^5 + 15x^4$ ◆

A common error is to assume that $f''(c) = 0$ implies that $f(c)$ is not a local extreme point. As Example 7B illustrates, if $f''(c) = 0$, then $f(c)$ may or may not be a local extreme point. **The first-derivative test *must* be used whenever $f''(c) = 0$ or $f''(c)$ does not exist.**

◆ APPLICATION

◆ EXAMPLE 8

Maximum Rate of Change

Using past records, a company estimates that it will sell $N(x)$ units of a product after spending $x thousand on advertising, as given by

$$N(x) = 2{,}000 - 2x^3 + 60x^2 - 450x \qquad 5 \le x \le 15$$

When is the rate of change of sales per unit (thousand dollars) change in advertising increasing? Decreasing? What is the maximum rate of change? Graph N and N' on the same coordinate system and interpret.

Solution

The rate of change of sales per unit (thousand dollars) change in advertising expenditure is

$$N'(x) = -6x^2 + 120x - 450 = -6(x - 5)(x - 15)$$

To determine when this rate is increasing and decreasing, we find $N''(x)$, the derivative of $N'(x)$:

$$N''(x) = -12x + 120 = 12(10 - x)$$

The information obtained by analyzing the signs of $N'(x)$ and $N''(x)$ is summarized in the table (sign charts are omitted).

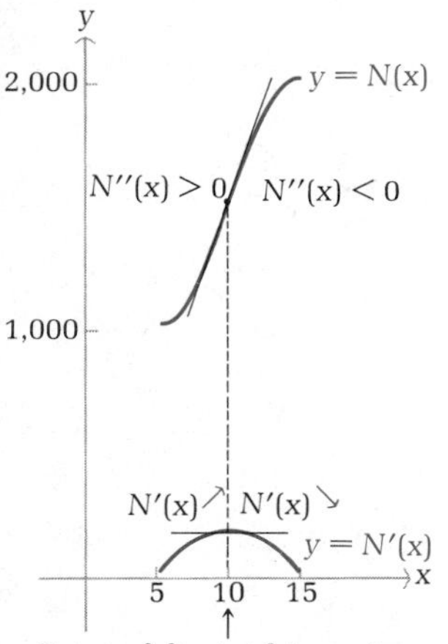

Point of diminishing returns

x	$N''(x)$	$N'(x)$	$N'(x)$	$N(x)$
$5 < x < 10$	+	+	Increasing	Increasing, concave upward
$x = 10$	0	+	Local maximum	Inflection point
$10 < x < 15$	−	+	Decreasing	Increasing, concave downward

Thus, we see that $N'(x)$, the rate of change of sales, is increasing on (5, 10) and decreasing on (10, 15). Both N and N' are graphed in the figure in the margin. An examination of the graph of $N'(x)$ shows that the maximum rate of change is $N'(10) = 150$. Notice that $N'(x)$ has a local maximum and $N(x)$ has an inflection point at $x = 10$. This value of x is referred to as the **point of diminishing returns,** since the rate of change of sales begins to decrease at this point. ◆

PROBLEM 8

Repeat Example 8 for $N(x) = 5{,}000 - x^3 + 60x^2 - 900x$, $10 \le x \le 30$. ◆

Answers to Matched Problems

5. Concave upward on $(-\infty, 0)$; concave downward on $(0, \infty)$

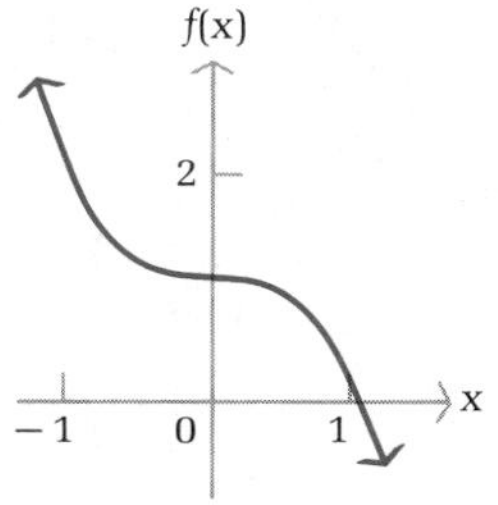

6. Inflection point at $x = 3$
7. (A) $f(2)$ is a local maximum; $f(4)$ is a local minimum
   (B) $f(0)$ is a local minimum; no local extremum at $x = 1$
8. $N'(x)$ is increasing on $(10, 20)$, decreasing on $(20, 30)$; maximum rate of change is $N'(20) = 300$; $x = 20$ is point of diminishing returns

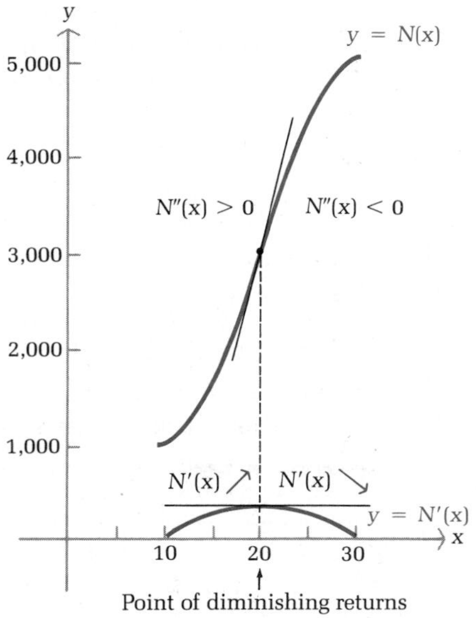

EXERCISE 5-2

A *Problems 1–4 refer to the following graph of $y = f(x)$:*

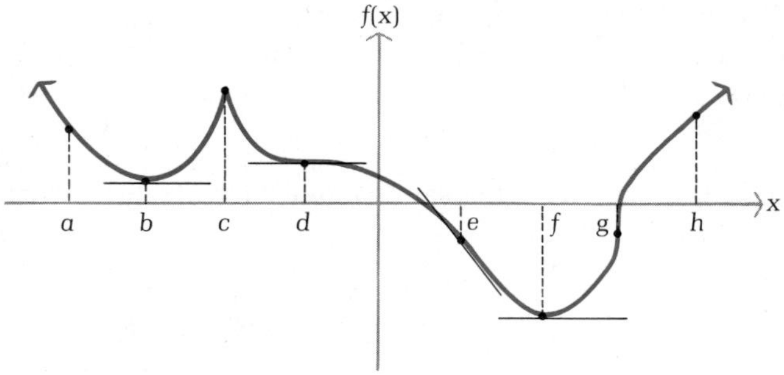

1. Identify intervals over which the graph of f is concave upward.
2. Identify intervals over which the graph of f is concave downward.
3. Identify inflection points.
4. Identify local extrema.

Find the indicated derivative for each function.

5. $f''(x)$ for $f(x) = x^3 - 2x^2 - 1$
6. $g''(x)$ for $g(x) = x^4 - 3x^2 + 5$
7. d^2y/dx^2 for $y = 2x^5 - 3$
8. d^2y/dx^2 for $y = 3x^4 - 7x$
9. $D_x^2(1 - 2x + x^3)$
10. $D_x^2(3x^2 - x^3)$
11. y'' for $y = (x^2 - 1)^3$
12. y'' for $y = (x^2 + 4)^4$
13. $f''(x)$ for $f(x) = 3x^{-1} + 2x^{-2} + 5$
14. $f''(x)$ for $f(x) = x^2 - x^{1/3}$

B *Find all local maxima and minima using the second-derivative test whenever it applies (do not graph). If the second-derivative test fails, use the first-derivative test.*

15. $f(x) = 2x^2 - 8x + 6$
16. $f(x) = 6x - x^2 + 4$
17. $f(x) = 2x^3 - 3x^2 - 12x - 5$
18. $f(x) = 2x^3 + 3x^2 - 12x - 1$
19. $f(x) = 3 - x^3 + 3x^2 - 3x$
20. $f(x) = x^3 + 6x^2 + 12x + 2$
21. $f(x) = x^4 - 8x^2 + 10$
22. $f(x) = x^4 - 18x^2 + 50$
23. $f(x) = x^6 + 3x^4 + 2$
24. $f(x) = 4 - x^6 - 6x^4$
25. $f(x) = x + \dfrac{16}{x}$
26. $f(x) = x + \dfrac{25}{x}$

Find the intervals where the graph of f is concave upward, the intervals where the graph is concave downward, and the inflection points.

27. $f(x) = x^2 - 4x + 5$
28. $f(x) = 9 + 3x - 4x^2$
29. $f(x) = x^3 - 18x^2 + 10x - 11$
30. $f(x) = x^3 + 24x^2 + 15x - 12$
31. $f(x) = x^4 - 24x^2 + 10x - 5$
32. $f(x) = x^4 + 6x^2 + 9x + 11$
33. $f(x) = -x^4 + 4x^3 + 3x + 7$
34. $f(x) = -x^4 - 2x^3 + 12x^2 + 15$

Find local maxima, local minima, and inflection points. Sketch the graph of each function. Include tangent lines at each local extreme point and inflection point.

35. $f(x) = x^3 - 6x^2 + 16$
36. $f(x) = x^3 - 9x^2 + 15x + 10$
37. $f(x) = x^3 + x + 2$
38. $f(x) = 1 - 3x - x^3$
39. $f(x) = (2 - x)^3 + 1$
40. $f(x) = (1 + x)^3 - 1$
41. $f(x) = x^3 - 12x$
42. $f(x) = 27x - x^3$

C *Find the inflection points. Do not graph.*

43. $f(x) = \dfrac{1}{x^2 + 12}$
44. $f(x) = \dfrac{x^2}{x^2 + 12}$
45. $f(x) = \dfrac{x}{x^2 + 12}$
46. $f(x) = \dfrac{x^3}{x^2 + 12}$

APPLICATIONS

Business & Economics

47. Revenue. The marketing research department for a computer company used a large city to test market their new product. They found that the relationship between price p (dollars per unit) and the demand x (units per

week) was given approximately by

$$p = 1{,}296 - 0.12x^2 \qquad 0 < x < 80$$

Thus, the weekly revenue can be approximated by

$$R(x) = xp = 1{,}296x - 0.12x^3 \qquad 0 < x < 80$$

(A) Find the local extrema for the revenue function.
(B) Over which intervals is the graph of the revenue function concave upward? Concave downward?

48. *Profit.* Suppose the cost equation for the company in Problem 47 is

$$C(x) = 830 + 396x$$

(A) Find the local extrema for the profit function.
(B) Over which intervals is the graph of the profit function concave upward? Concave downward?

49. *Advertising.* A company estimates that it will sell $N(x)$ units of a product after spending $x thousand on advertising, as given by

$$N(x) = -3x^3 + 225x^2 - 3{,}600x + 17{,}000 \qquad 10 \leq x \leq 40$$

(A) When is the rate of change of sales $N'(x)$ increasing? Decreasing?
(B) Find the inflection points for the graph of N.
(C) Graph N and N' on the same coordinate system.
(D) What is the maximum rate of change of sales?

50. *Advertising.* A company estimates that it will sell $N(x)$ units of a product after spending $x thousand on advertising, as given by

$$N(x) = -2x^3 + 90x^2 - 750x + 2{,}000 \qquad 5 \leq x \leq 25$$

(A) When is the rate of change of sales $N'(x)$ increasing? Decreasing?
(B) Find the inflection points for the graph of N.
(C) Graph N and N' on the same coordinate system.
(D) What is the maximum rate of change of sales?

Life Sciences

51. *Population growth—bacteria.* A drug that stimulates reproduction is introduced into a colony of bacteria. After t minutes, the number of bacteria is given approximately by

$$N(t) = 1{,}000 + 30t^2 - t^3 \qquad 0 \leq t \leq 20$$

(A) When is the rate of growth $N'(t)$ increasing? Decreasing?
(B) Find the inflection points for the graph of N.
(C) Sketch the graphs of N and N' on the same coordinate system.
(D) What is the maximum rate of growth?

52. *Drug sensitivity.* One hour after x milligrams of a particular drug are given to a person, the change in body temperature $T(x)$, in degrees Fahrenheit, is given by

$$T(x) = x^2\left(1 - \frac{x}{9}\right) \qquad 0 \leq x \leq 6$$

The rate at which $T(x)$ changes with respect to the size of the dosage x, $T'(x)$, is called the *sensitivity* of the body to the dosage.

(A) When is $T'(x)$ increasing? Decreasing?
(B) Where does the graph of T have inflection points?
(C) Sketch the graphs of T and T′ on the same coordinate system.
(D) What is the maximum value of $T'(x)$?

Social Sciences

53. *Learning.* The time T (in minutes) it takes a person to learn a list of length n is

$$T(n) = \tfrac{2}{25}n^3 - \tfrac{6}{5}n^2 + 6n \qquad n \geq 0$$

(A) When is the rate of change of T with respect to the length of the list increasing? Decreasing?
(B) Where does the graph of T have inflection points? Graph T and T′ on the same coordinate system.
(C) What is the minimum value of $T'(n)$?

SECTION 5-3 Curve Sketching Techniques: Unified and Extended

- ASYMPTOTES
- GRAPHING STRATEGY
- USING THE STRATEGY
- APPLICATION

In this section we will apply, in a systematic way, all the graphing concepts discussed in Sections 5-1 and 5-2. Before outlining a graphing strategy and considering the graphs of specific functions, we need to discuss in more detail the concept of *asymptotes*, which we introduced in Section 4-2.

◆ ASYMPTOTES

To refresh your memory concerning horizontal and vertical asymptotes, refer to Figure 11 on the next page. The lines $y = 2$ and $y = -2$ are horizontal asymptotes, since $\lim_{x\to\infty} f(x) = 2$ and $\lim_{x\to-\infty} f(x) = -2$. The line $x = -4$ is a vertical asymptote, since $\lim_{x\to-4} f(x) = \infty$. The line $x = 4$ is also a vertical asymptote, since $\lim_{x\to4^-} f(x) = -\infty$ and $\lim_{x\to4^+} f(x) = \infty$.

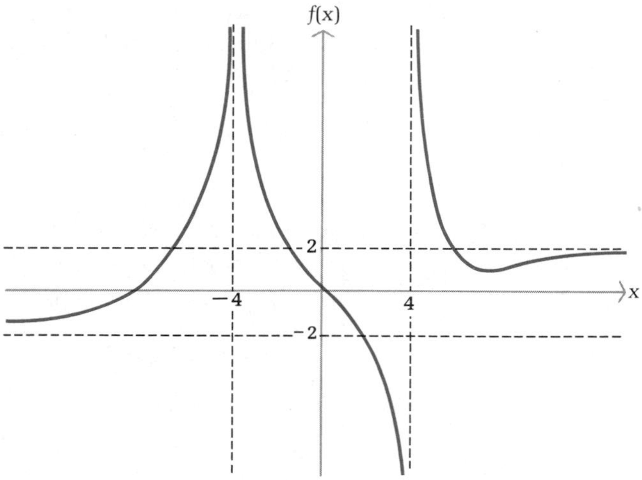

FIGURE 11

We restate the definitions of horizontal and vertical asymptotes in the box for convenient reference.

Horizontal and Vertical Asymptotes

A line $y = b$ is a **horizontal asymptote** for the graph of $y = f(x)$ if

$$\lim_{x \to -\infty} f(x) = b \qquad \text{or} \qquad \lim_{x \to \infty} f(x) = b$$

A line $x = a$ is a **vertical asymptote** for the graph of $y = f(x)$ if

$$\lim_{x \to a^-} f(x) = \infty \text{ (or } -\infty), \lim_{x \to a^+} f(x) = \infty \text{ (or } -\infty), \text{ or } \lim_{x \to a} f(x) = \infty \text{ (or } -\infty)$$

We are interested in locating any horizontal and vertical asymptotes as an aid to graphing a function. Consequently, we are interested in locating horizontal and vertical asymptotes from an equation that defines a function rather than from its graph. Consider the rational function given by

$$f(x) = \frac{2x - 3}{x - 1}$$

To locate horizontal asymptotes, we compute $\lim_{x \to -\infty} f(x)$ and $\lim_{x \to \infty} f(x)$ as follows:

$$\lim_{x \to -\infty} \frac{2x - 3}{x - 1} = \lim_{x \to -\infty} \frac{x\left(2 - \frac{3}{x}\right)}{x\left(1 - \frac{1}{x}\right)} = 2 \qquad \lim_{x \to \infty} \frac{2x - 3}{x - 1} = \lim_{x \to \infty} \frac{x\left(2 - \frac{3}{x}\right)}{x\left(1 - \frac{1}{x}\right)} = 2$$

Thus, $y = 2$ is the only horizontal asymptote, and it could have been obtained using either limit. In general, **a rational function has at most one horizontal asymptote.** (Notice that this implies that the function graphed in Figure 11 is not a rational function, since it has two horizontal asymptotes.)

Since the given function f is continuous for all values of x except those that make the denominator 0, the only candidate for a vertical asymptote is $x = 1$. How can we tell if $x = 1$ is a vertical asymptote? We could try to compute the limits in the definition of a vertical asymptote above—a difficult task in many problems. But Theorem 3, which we state without proof, eliminates the need to compute any of the limits in the definition, and it applies to f, as well as a wide variety of other functions.

THEOREM 3

On Vertical Asymptotes

Let $f(x) = N(x)/D(x)$, where both N and D are continuous at $x = c$. If at $x = c$ the denominator $D(x)$ is 0 and the numerator $N(x)$ is not 0, then the line $x = c$ is a vertical asymptote for the graph of f.

[*Note:* Since a rational function is the ratio of two polynomial functions, and polynomial functions are continuous for all real numbers, this theorem includes rational functions as a special case.]

Returning to the example above, $x = 1$ is a vertical asymptote for the graph of $f(x) = (2x - 3)/(x - 1)$, since the denominator is 0 at $x = 1$ and the numerator is not.

◆ EXAMPLE 9

Find horizontal and vertical asymptotes, if they exist, for:

(A) $f(x) = \dfrac{6x + 5}{2x - 4}$ (B) $f(x) = \dfrac{x}{x^2 + 1}$ (C) $f(x) = \dfrac{x^2 - 4}{x}$

Solutions

(A) *Horizontal asymptotes:*

$$\lim_{x \to \infty} \frac{6x + 5}{2x - 4} = \lim_{x \to \infty} \frac{x\left(6 + \frac{5}{x}\right)}{x\left(2 - \frac{4}{x}\right)} = \frac{6}{2} = 3$$

The line $y = 3$ is a horizontal asymptote. [Since a rational function cannot have more than one horizontal asymptote, there is no need to compute $\lim_{x \to -\infty} f(x)$.]

Vertical asymptotes:

$$f(x) = \frac{6x + 5}{2x - 4} = \frac{6x + 5}{2(x - 2)}$$

Using Theorem 3, we search for values of x that make the denominator 0 without making the numerator 0 at the same time. The denominator is 0 for $x = 2$, and we see that the line $x = 2$ is a vertical asymptote, since the numerator is not 0 for $x = 2$.

(B) *Horizontal asymptotes:*

$$\lim_{x\to\infty} \frac{x}{x^2 + 1} = \lim_{x\to\infty} \frac{x}{x^2\left(1 + \frac{1}{x^2}\right)} = \lim_{x\to\infty} \left[\frac{1}{x} \cdot \frac{1}{\left(1 + \frac{1}{x^2}\right)}\right] = 0$$

The line $y = 0$ (the x axis) is a horizontal asymptote.

Vertical asymptotes: Since the denominator, $x^2 + 1$, is never 0, there are no vertical asymptotes.

(C) *Horizontal asymptotes:*

$$\lim_{x\to\infty} \frac{x^2 - 4}{x} = \lim_{x\to\infty} \frac{x^2\left(1 - \frac{4}{x^2}\right)}{x} = \lim_{x\to\infty} x\left(1 - \frac{4}{x^2}\right) = \infty$$

There are no horizontal asymptotes.

Vertical asymptotes: The line $x = 0$ (the y axis) is a vertical asymptote, since for $x = 0$, the denominator is 0 and the numerator is not 0. ◆

PROBLEM 9 Find horizontal and vertical asymptotes, if they exist, for:

(A) $f(x) = \frac{3x + 5}{x + 2}$ (B) $f(x) = \frac{x + 1}{x^2}$ (C) $f(x) = \frac{x^3}{x^2 + 4}$ ◆

◆ GRAPHING STRATEGY

We now have powerful tools to determine the shape of a graph of a function, even before we plot any points. We can accurately sketch the graphs of many functions using these tools and point-by-point plotting as needed (often, very little point-by-point plotting is necessary). We organize these tools in the graphing strategy summarized in the box.

A Graphing Strategy for $y = f(x)$

Omit any of the following steps if procedures involved appear to be too difficult or impossible (what may seem too difficult now, will become less so with a little practice).

Step 1. *Use $f(x)$:*

(A) Find the domain of f. [The domain of f is the set of all real numbers x that produce real values for $f(x)$.]

(B) Find intercepts. [The y intercept is $f(0)$, if it exists; the x intercepts are the solutions to $f(x) = 0$, if they exist.]

(C) Find asymptotes. [Find any horizontal asymptotes by calculating $\lim_{x\to\pm\infty} f(x)$. Find any vertical asymptotes by using Theorem 3.]

Step 2. *Use $f'(x)$:* Find any critical values for $f(x)$ and any partition numbers for $f'(x)$. [Remember, every critical value for $f(x)$ is also a partition number for $f'(x)$, but some partition numbers for $f'(x)$ may not be critical values for $f(x)$.] Construct a sign chart for $f'(x)$, determine the intervals where $f(x)$ is increasing and decreasing, and find local maxima and minima.

Step 3. *Use $f''(x)$:* Construct a sign chart for $f''(x)$, determine where the graph of f is concave upward and concave downward, and find any inflection points.

Step 4. *Sketch the graph of f:* Draw asymptotes and locate intercepts, local maxima and minima, and inflection points. Sketch in what you know from steps 1–3. In regions of uncertainty, use point-by-point plotting to complete the graph.

◆ USING THE STRATEGY

Some examples will illustrate the use of the graphing strategy.

◆ EXAMPLE 10 Graph $f(x) = x^4 - 2x^3$ using the graphing strategy.

Solution Step 1. *Use $f(x)$:* $f(x) = x^4 - 2x^3$

(A) Domain: All real x

(B) Intercepts:

y intercept: $f(0) = 0$

x intercepts:
$$\begin{aligned} f(x) &= 0 \\ x^4 - 2x^3 &= 0 \\ x^3(x-2) &= 0 \\ x &= 0, 2 \end{aligned}$$

(C) Asymptotes: Since f is a polynomial, there are no horizontal or vertical asymptotes.

Step 2. *Use $f'(x)$:* $f'(x) = 4x^3 - 6x^2 = 4x^2(x - \frac{3}{2})$

Critical values: 0 and $\frac{3}{2}$

Partition numbers: 0 and $\frac{3}{2}$

Sign chart for $f'(x)$:

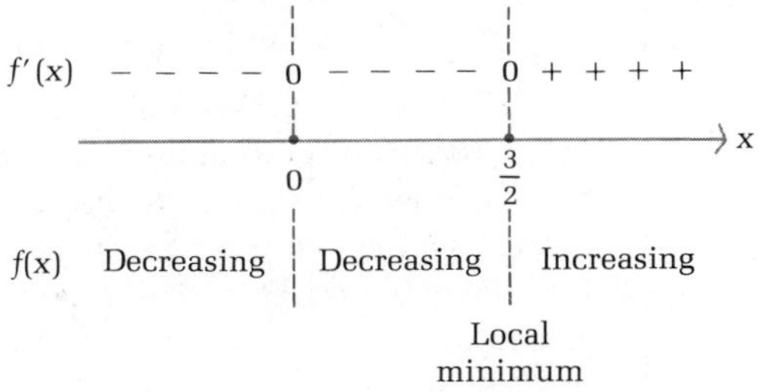

Test Numbers

x	$f'(x)$	
−1	−10	(−)
1	−2	(−)
2	8	(+)

Thus, $f(x)$ is decreasing on $(-\infty, \frac{3}{2})$, increasing on $(\frac{3}{2}, \infty)$, and has a local minimum at $x = \frac{3}{2}$.

Step 3. *Use $f''(x)$:* $f''(x) = 12x^2 - 12x = 12x(x - 1)$

Partition numbers for $f''(x)$: 0 and 1

Sign chart for $f''(x)$:

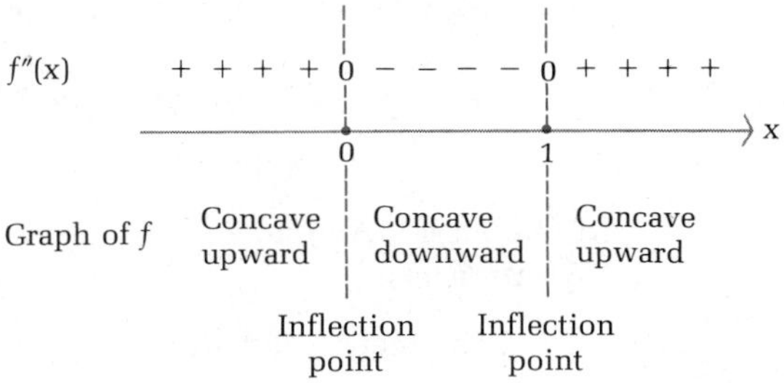

Test Numbers

x	$f''(x)$	
−1	24	(+)
$\frac{1}{2}$	−3	(−)
2	24	(+)

Thus, the graph of f is concave upward on $(-\infty, 0)$ and $(1, \infty)$, concave downward on $(0, 1)$, and has inflection points at $x = 0$ and $x = 1$.

Step 4. *Sketch the graph of f:*

x	$f(x)$
0	0
1	-1
$\frac{3}{2}$	$-\frac{27}{16}$
2	0

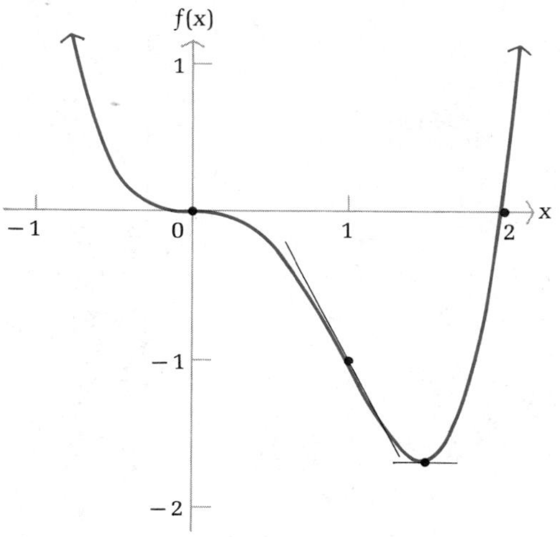

PROBLEM 10 Graph $f(x) = x^4 + 4x^3$ using the graphing strategy.

EXAMPLE 11 Graph $f(x) = \dfrac{x - 1}{x - 2}$ using the graphing strategy.

Solution *Step 1.* *Use f(x):* $f(x) = \dfrac{x - 1}{x - 2}$

(A) Domain: All real x, except $x = 2$

(B) Intercepts:

y intercept: $f(0) = \dfrac{0 - 1}{0 - 2} = \dfrac{1}{2}$

x intercepts: Since a fraction is 0 when its numerator is 0 and the denominator is not 0, the x intercept is $x = 1$.

(C) Asymptotes:

Horizontal asymptote: $\lim_{x\to\infty} \dfrac{x - 1}{x - 2} = \lim_{x\to\infty} \dfrac{x\left(1 - \dfrac{1}{x}\right)}{x\left(1 - \dfrac{2}{x}\right)} = 1$

Thus, the line $y = 1$ is a horizontal asymptote.

Vertical asymptote: The denominator is 0 for $x = 2$, and the numerator is not 0 for this value. Therefore, the line $x = 2$ is a vertical asymptote.

Step 2. *Use $f'(x)$:* $f'(x) = \dfrac{(x-2)(1)-(x-1)(1)}{(x-2)^2} = \dfrac{-1}{(x-2)^2}$

Critical values: None

Partition number: $x = 2$

Sign chart for $f'(x)$:

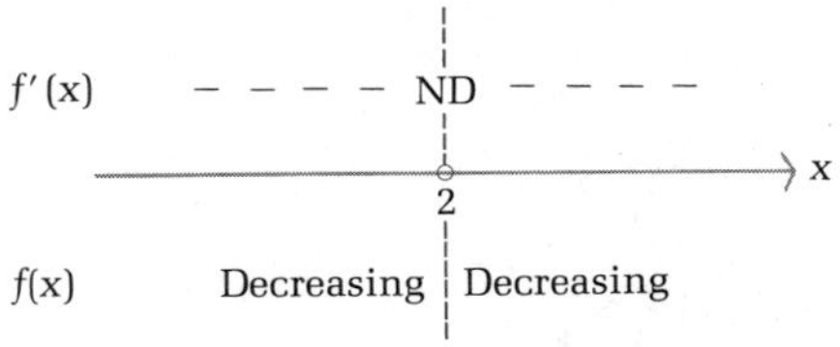

Test Numbers

x	$f'(x)$
1	-1 (–)
3	-1 (–)

Thus, $f(x)$ is decreasing on $(-\infty, 2)$ and $(2, \infty)$. There are no local extrema.

Step 3. *Use $f''(x)$:* $f''(x) = \dfrac{2}{(x-2)^3}$

Partition number for $f''(x)$: 2

Sign chart for $f''(x)$:

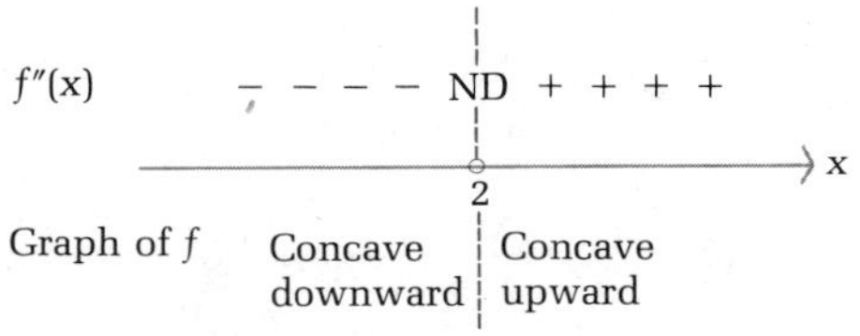

Test Numbers

x	$f''(x)$
1	-2 (–)
3	2 (+)

Thus, the graph of f is concave downward on $(-\infty, 2)$ and concave upward on $(2, \infty)$. Since $f(2)$ is not defined, there is no inflection point at $x = 2$, even though $f''(x)$ changes sign at $x = 2$.

Step 4. *Sketch a graph of f:* Insert intercepts and asymptotes, and plot a few additional points (for functions with asymptotes, plotting additional points is often helpful). Then sketch the graph.

x	$f(x)$
-2	$\frac{3}{4}$
0	$\frac{1}{2}$
1	0
$\frac{3}{2}$	-1
$\frac{5}{2}$	3
3	2
4	$\frac{3}{2}$

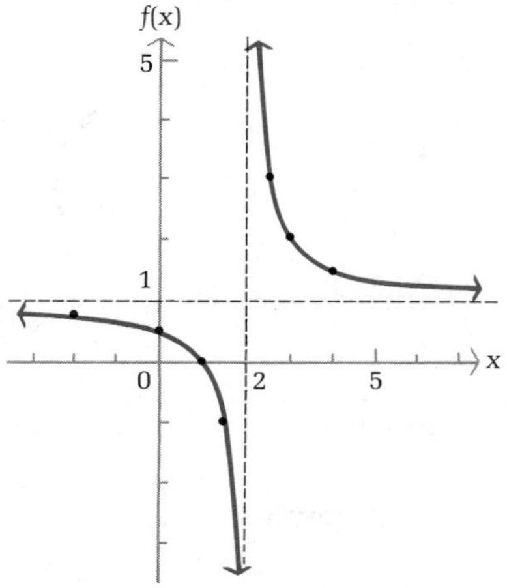

◆

PROBLEM 11 Graph $f(x) = \dfrac{2x}{1-x}$ using the graphing strategy. ◆

◆ APPLICATION

◆ EXAMPLE 12 *Average Cost*

Given the cost function $C(x) = 5{,}000 + \frac{1}{2}x^2$, where x is the number of units produced, graph the average cost function and the marginal cost function on the same set of coordinate axes.

Solution The average cost function is

$$\overline{C}(x) = \frac{C(x)}{x} = \frac{5{,}000}{x} + \frac{x}{2} = \frac{10{,}000 + x^2}{2x}$$

Graph $\overline{C}$ using the graphing strategy.

Step 1. Use $\overline{C}(x)$:

(A) Domain: Since negative values of x do not make sense and $\overline{C}(0)$ is not defined, the domain is the set of all positive real numbers.
(B) Intercepts: None
(C) Asymptotes:

Horizontal asymptote:

$$\lim_{x\to\infty} \frac{10{,}000 + x^2}{2x} = \lim_{x\to\infty} \frac{x^2\left(\dfrac{10{,}000}{x^2} + 1\right)}{2x} = \lim_{x\to\infty} \frac{x}{2}\left(\frac{10{,}000}{x^2} + 1\right) = \infty$$

Thus, there is no horizontal asymptote.

Vertical asymptote: The line $x = 0$ (the y axis) is a vertical asymptote, since the denominator is 0 and the numerator is not 0 for $x = 0$.

Oblique asymptotes: Some graphs have asymptotes that are neither vertical nor horizontal. These are called **oblique asymptotes.** If we look at

$$\overline{C}(x) = \frac{5{,}000}{x} + \frac{x}{2}$$

we can see that for x near and to the right of 0, $\overline{C}(x)$ is approximated by $5{,}000/x$. On the other hand, as x increases without bound, $\overline{C}(x)$ approaches $x/2$; that is,

$$\lim_{x\to\infty} \left[\overline{C}(x) - \frac{x}{2}\right] = \lim_{x\to\infty} \frac{5{,}000}{x} = 0$$

This implies that the graph of $y = \overline{C}(x)$ approaches the line $y = x/2$ as x approaches ∞. This line is an oblique asymptote for the graph of $y = \overline{C}(x)$.

Step 2. *Use $\overline{C}'(x)$:* $\overline{C}'(x) = -\dfrac{5,000}{x^2} + \dfrac{1}{2} = \dfrac{x^2 - 10,000}{2x^2}$

Critical value: 100

Partition numbers: 0 and 100

Sign chart for $\overline{C}'(x)$:

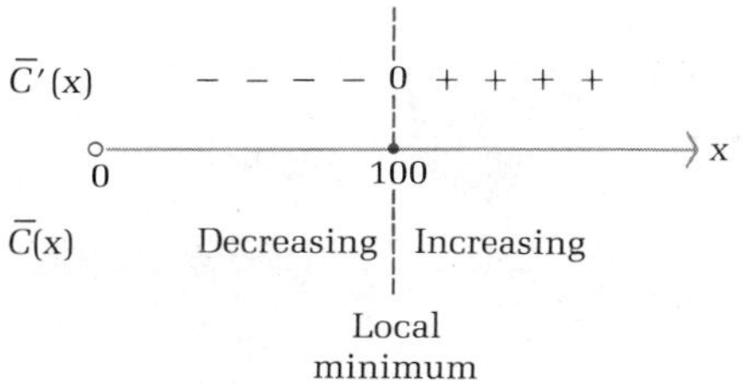

Test Numbers

x	$\overline{C}'(x)$	
50	−1.5	(−)
125	0.18	(+)

Thus, $\overline{C}(x)$ is decreasing on (0, 100), increasing on (100, ∞), and has a local minimum at $x = 100$.

Step 3. *Use $\overline{C}''(x)$:* $\overline{C}''(x) = \dfrac{10,000}{x^3}$

$\overline{C}''(x)$ is positive for all positive x; therefore, the graph of $y = \overline{C}(x)$ is concave upward on (0, ∞).

Step 4. *Sketch the graph of $\overline{C}$:* The graph of $\overline{C}$ is shown in the figure. The marginal cost function is $C'(x) = x$. The graph of this linear function is also shown in the figure.

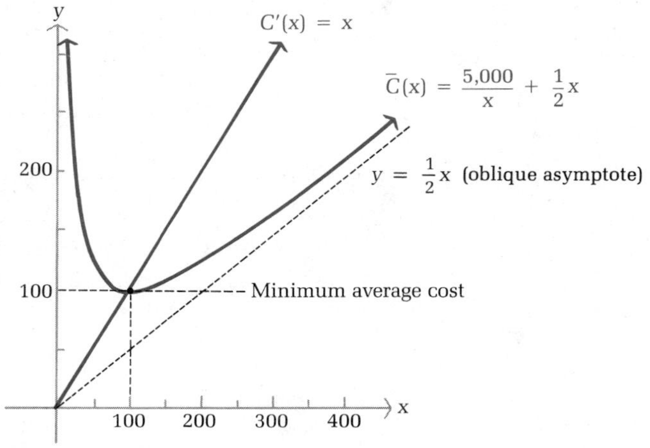

◆

The graph in Example 12 illustrates an important principle in economics.:

The minimum average cost occurs when the average cost is equal to the marginal cost.

PROBLEM 12 Given the cost function $C(x) = 1{,}600 + \frac{1}{4}x^2$, where x is the number of units produced:

(A) Graph the average cost function (using the graphing strategy) and the marginal cost function on the same set of coordinate axes. Include any oblique asymptotes.

(B) Find the minimum average cost. ◆

Answers to Matched Problems

9. (A) Horizontal asymptote: $y = 3$; vertical asymptote: $x = -2$

(B) Horizontal asymptote: $y = 0$ (x axis); vertical asymptote: $x = 0$ (y axis)

(C) No horizontal or vertical asymptotes

10. Domain: $(-\infty, \infty)$

y intercept: $f(0) = 0$; x intercepts: $-4, 0$

Asymptotes: No horizontal or vertical asymptotes

Decreasing on $(-\infty, -3)$; increasing on $(-3, \infty)$; local minimum at $x = -3$

Concave upward on $(-\infty, -2)$ and $(0, \infty)$; concave downward on $(-2, 0)$

Inflection points at $x = -2$ and $x = 0$

x	f(x)
−4	0
−3	−27
−2	−16
0	0

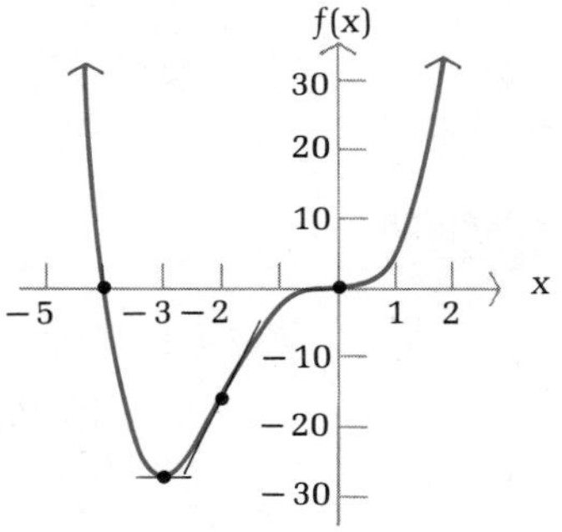

11. Domain: All real x, except $x = 1$

y intercept: $f(0) = 0$;

x intercept: 0

Horizontal asymptote: $y = -2$;

vertical asymptote: $x = 1$

Increasing on $(-\infty, 1)$ and $(1, \infty)$
Concave upward on $(-\infty, 1)$; concave downward on $(1, \infty)$

x	f(x)
−1	−1
0	0
$\frac{1}{2}$	2
$\frac{3}{2}$	−6
2	−4
5	$-\frac{5}{2}$

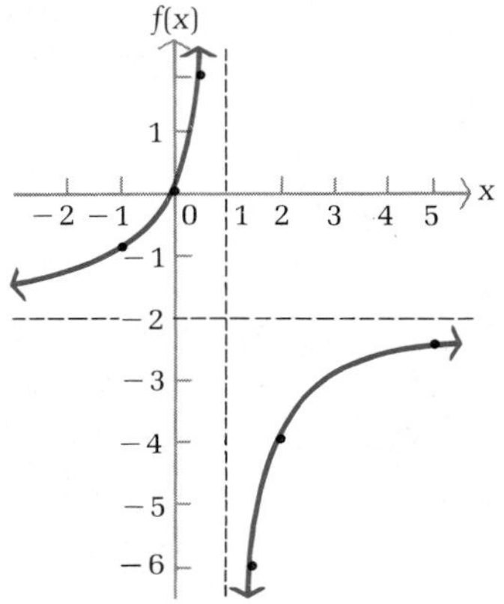

12. (A)

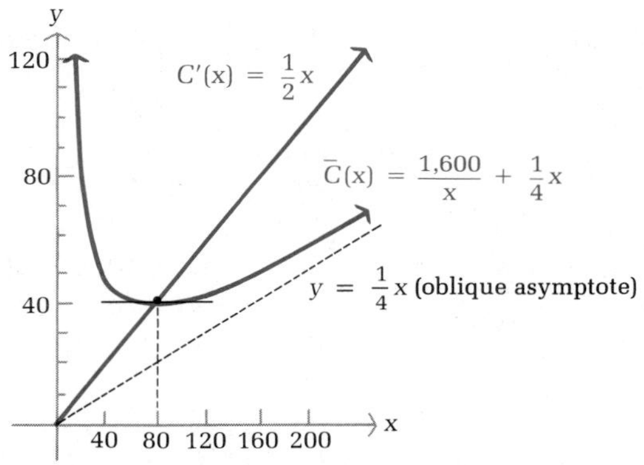

(B) Minimum average cost is 40 at $x = 80$.

EXERCISE 5-3

A *Problems 1–10 refer to the graph of $y = f(x)$ at the top of the next page:*

1. Identify the intervals over which $f(x)$ is increasing.
2. Identify the intervals over which $f(x)$ is decreasing.
3. Identify the points where $f(x)$ has a local maximum.
4. Identify the points where $f(x)$ has a local minimum.
5. Identify the intervals over which the graph of f is concave upward.
6. Identify the intervals over which the graph of f is concave downward.
7. Identify the inflection points.
8. Identify the horizontal asymptotes.
9. Identify the vertical asymptotes.
10. Identify the x and y intercepts.

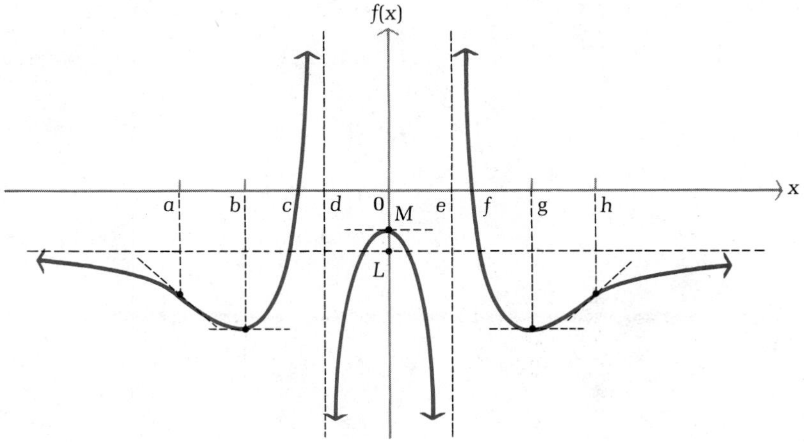

B *Find any horizontal and vertical asymptotes.*

11. $f(x) = \dfrac{2x}{x+2}$
12. $f(x) = \dfrac{3x+2}{x-4}$
13. $f(x) = \dfrac{x^2+1}{x^2-1}$
14. $f(x) = \dfrac{x^2-1}{x^2+2}$
15. $f(x) = \dfrac{x^3}{x^2+6}$
16. $f(x) = \dfrac{x}{x^2-4}$
17. $f(x) = \dfrac{x}{x^2+4}$
18. $f(x) = \dfrac{x^2+9}{x}$
19. $f(x) = \dfrac{x^2}{x-3}$
20. $f(x) = \dfrac{x+5}{x^2}$

Sketch a graph of $y = f(x)$ using the graphing strategy.

21. $f(x) = x^2 - 6x + 5$
22. $f(x) = 3 + 2x - x^2$
23. $f(x) = x^3 - 6x^2$
24. $f(x) = 3x^2 - x^3$
25. $f(x) = (x+4)(x-2)^2$
26. $f(x) = (2-x)(x+1)^2$
27. $f(x) = 8x^3 - 2x^4$
28. $f(x) = x^4 - 4x^3$
29. $f(x) = \dfrac{x+3}{x-3}$
30. $f(x) = \dfrac{2x-4}{x+2}$
31. $f(x) = \dfrac{x}{x-2}$
32. $f(x) = \dfrac{2+x}{3-x}$

C *In Problems 33 and 34, show that the line $y = x$ is an oblique asymptote for the graph of $y = f(x)$, and then use the graphing strategy to sketch a graph of $y = f(x)$.*

33. $f(x) = x + \dfrac{1}{x}$
34. $f(x) = x - \dfrac{1}{x}$

Sketch a graph of $y = f(x)$ using the graphing strategy.

35. $f(x) = x^3 - x$
36. $f(x) = x^3 + x$
37. $f(x) = (x^2+3)(9-x^2)$
38. $f(x) = (x^2+3)(x^2-1)$

39. $f(x) = (x^2 - 4)^2$

40. $f(x) = (x^2 - 1)(x^2 - 5)$

41. $f(x) = 2x^6 - 3x^5$

42. $f(x) = 3x^5 - 5x^4$

43. $f(x) = \dfrac{x}{x^2 - 4}$

44. $f(x) = \dfrac{1}{x^2 - 4}$

45. $f(x) = \dfrac{1}{1 + x^2}$

46. $f(x) = \dfrac{x^2}{1 + x^2}$

APPLICATIONS

Business & Economics

47. *Revenue.* The marketing research department for a computer company used a large city to test market their new product. They found that the relationship between price p (dollars per unit) and the demand x (units per week) was given approximately by

$$p = 1{,}296 - 0.12x^2 \qquad 0 < x < 80$$

Thus, the weekly revenue can be approximated by

$$R(x) = xp = 1{,}296x - 0.12x^3 \qquad 0 < x < 80$$

Graph the revenue function R.

48. *Profit.* Suppose the cost equation for the company in Problem 47 is

$$C(x) = 830 + 396x$$

(A) Write an equation for the profit $P(x)$.
(B) Graph the profit function P.

49. *Pollution.* In Silicon Valley (California), a number of computer-related manufacturing firms were found to be contaminating underground water supplies with toxic chemicals stored in leaking underground containers. A water quality control agency ordered the companies to take immediate corrective action and to contribute to a monetary pool for testing and cleanup of the underground contamination. Suppose the required monetary pool (in millions of dollars) for the testing and cleanup is estimated to be given by

$$P(x) = \frac{2x}{1 - x} \qquad 0 \leq x < 1$$

where x is the percentage (expressed as a decimal fraction) of the total contaminant removed.

(A) Where is $P(x)$ increasing? Decreasing?
(B) Where is the graph of P concave upward? Downward?
(C) Find any horizontal and vertical asymptotes.
(D) Find the x and y intercepts.
(E) Sketch a graph of P.

50. *Employee training.* A company producing computer components has established that on the average a new employee can assemble $N(t)$ components per day after t days of on-the-job training, as given by

$$N(t) = \frac{100t}{t+9} \qquad t \geq 0$$

(A) Where is $N(t)$ increasing? Decreasing?
(B) Where is the graph of N concave upward? Downward?
(C) Find any horizontal and vertical asymptotes.
(D) Find the intercepts.
(E) Sketch a graph of N.

51. *Replacement time.* An office copier has an initial price of $3,200. A maintenance/service contract costs $300 for the first year and increases $100 per year thereafter. It can be shown that the total cost of the copier after n years is given by

$$C(n) = 3{,}200 + 250n + 50n^2$$

(A) Write an expression for the average cost per year, $\overline{C}(n)$, for n years.
(B) Graph the average cost function found in part A.
(C) When is the average cost per year minimum? (This is frequently referred to as the **replacement time** for this piece of equipment.)

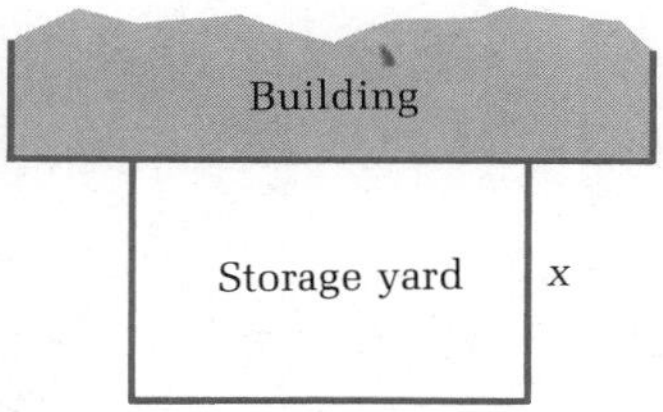

52. *Construction costs.* The management of a manufacturing plant wishes to add a fenced-in rectangular storage yard of 20,000 square feet, using the plant building as one side of the yard (see the figure in the margin). If x is the distance from the building to the fence parallel to the building, then show that the length of the fence required for the yard is given by

$$L(x) = 2x + \frac{20{,}000}{x} \qquad x > 0$$

(A) Graph L.
(B) What are the dimensions of the rectangle requiring the least amount of fencing?

53. *Average and marginal costs.* The cost of producing x units of a certain product is given by

$$C(x) = 1{,}000 + 5x + \tfrac{1}{10}x^2$$

(A) Sketch the graphs of the average cost function and the marginal cost function on the same set of coordinate axes. Include any oblique asymptotes.
(B) Find the minimum average cost.

54. Repeat Problem 53 for $C(x) = 500 + 2x + \frac{1}{5}x^2$.

Life Sciences

55. *Medicine.* A drug is injected into the bloodstream of a patient through her right arm. The concentration of the drug in the bloodstream of the left arm

t hours after the injection is given by

$$C(t) = \frac{0.14t}{t^2 + 1}$$

Graph C.

56. *Physiology.* In a study on the speed of muscle contraction in frogs under various loads, researchers W. O. Fems and J. Marsh found that the speed of contraction decreases with increasing loads. More precisely, they found that the relationship between speed of contraction S (in centimeters per second) and load w (in grams) is given approximately by

$$S(w) = \frac{26 + 0.06w}{w} \qquad w \geq 5$$

Graph S.

Social Sciences

57. *Psychology—retention.* An experiment on retention is conducted in a psychology class. Each student in the class is given 1 day to memorize the same list of 30 special characters. The lists are turned in at the end of the day, and for each succeeding day for 30 days each student is asked to turn in a list of as many of the symbols as can be recalled. Averages are taken, and it is found that

$$N(t) = \frac{5t + 20}{t} \qquad t \geq 1$$

provides a good approximation of the average number of symbols, $N(t)$, retained after t days. Graph N.

SECTION 5-4 Optimization; Absolute Maxima and Minima

- ◆ ABSOLUTE MAXIMA AND MINIMA
- ◆ APPLICATIONS

We are now ready to consider one of the most important applications of the derivative, namely, the use of derivatives to find the *absolute maximum* or *minimum* value of a function. As we mentioned earlier, an economist may be interested in the price or production level of a commodity that will bring a maximum profit; a doctor may be interested in the time it takes for a drug to reach its maximum concentration in the bloodstream after an injection; and a city planner might be interested in the location of heavy industry in a city to produce minimum pollution in residential and business areas. Before we launch an attack on problems of this type, which are called *optimization* problems, we have to say a few words about the procedures needed to find absolute maximum and absolute minimum values of functions. We have most of the tools we need from the previous sections.

◆ ABSOLUTE MAXIMA AND MINIMA

First, what do we mean by *absolute maximum* and *absolute minimum*? We say that $f(c)$ is an **absolute maximum** of f if

$$f(c) \geq f(x)$$

for all x in the domain of f. Similarly, $f(c)$ is called an **absolute minimum** of f if

$$f(c) \leq f(x)$$

for all x in the domain of f. Figure 12 illustrates several typical examples.

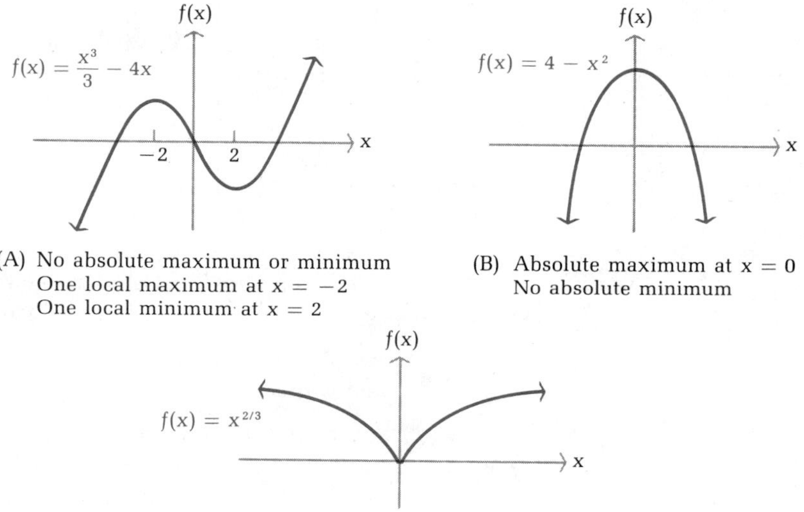

(A) No absolute maximum or minimum
One local maximum at $x = -2$
One local minimum at $x = 2$

(B) Absolute maximum at $x = 0$
No absolute minimum

(C) Absolute minimum at $x = 0$
No absolute maximum

FIGURE 12

In many practical problems, the domain of a function is restricted because of practical or physical considerations. If the domain is restricted to some closed interval, as is often the case, then Theorem 4 can be proved.

THEOREM 4

> A function f continuous on a closed interval $[a, b]$ (see Section 4-1) assumes both an absolute maximum and an absolute minimum on that interval.

It is important to understand that the absolute maximum and minimum depend on both the function f and the interval $[a, b]$. Figure 13 (at the top of the next page) illustrates four cases.

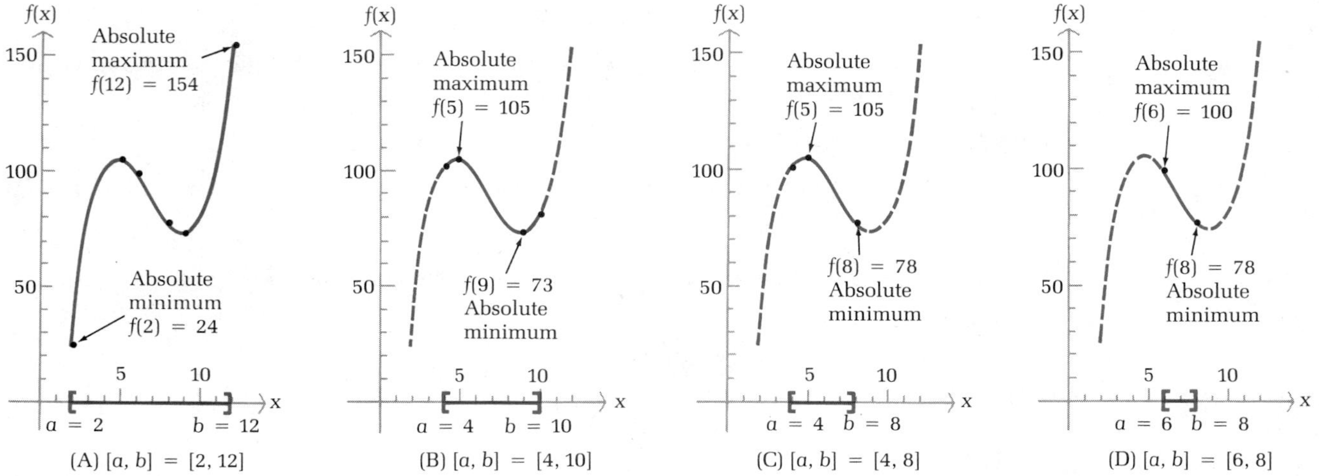

FIGURE 13
Absolute extrema for $f(x) = x^3 - 21x^2 + 135x - 170$ for various closed intervals

In all four cases illustrated in Figure 13, the absolute maximum and absolute minimum both occur at a critical value or an end point. In general:

Absolute extrema (if they exist) must always occur at critical values or at end points.

Thus, to find the absolute maximum or minimum value of a continuous function on a closed interval, we simply identify the end points and the critical values in the interval, evaluate each, and then choose the largest and smallest values out of this group.

Steps in Finding Absolute Maximum and Minimum Values of a Continuous Function f

Step 1. Check to make certain that f is continuous over $[a, b]$.

Step 2. Find the critical values in the interval (a, b).

Step 3. Evaluate f at the end points a and b and at the critical values found in step 2.

Step 4. The absolute maximum $f(x)$ on $[a, b]$ is the largest of the values found in step 3.

Step 5. The absolute minimum $f(x)$ on $[a, b]$ is the smallest of the values found in step 3.

◆ EXAMPLE 13 Find the absolute maximum and absolute minimum values of

$$f(x) = x^3 + 3x^2 - 9x - 7$$

on each of the following intervals:

(A) $[-6, 4]$ (B) $[-4, 2]$ (C) $[-2, 2]$

Solutions (A) The function is continuous for all values of x.

$$f'(x) = 3x^2 + 6x - 9 = 3(x - 1)(x + 3)$$

Thus, $x = -3$ and $x = 1$ are critical values in the interval $(-6, 4)$. Evaluate f at the end points and critical values, -6, -3, 1, and 4, and choose the maximum and minimum from these:

$f(-6) = -61$ Absolute minimum
$f(-3) = 20$
$f(1) = -12$
$f(4) = 69$ Absolute maximum

(B) Interval: $[-4, 2]$

x	f(x)	
−4	13	
−3	20	Absolute maximum
1	−12	Absolute minimum
2	−5	

(C) Interval: $[-2, 2]$

x	f(x)	
−2	15	Absolute maximum
1	−12	Absolute minimum
2	−5	

The critical value $x = -3$ is not included in this table, because it is not in the interval $[-2, 2]$. ◆

PROBLEM 13 Find the absolute maximum and absolute minimum values of

$$f(x) = x^3 - 12x$$

on each of the following intervals:

(A) $[-5, 5]$ (B) $[-3, 3]$ (C) $[-3, 1]$ ◆

Now, suppose we want to find the absolute maximum or minimum value of a function that is continuous on an interval that is not closed. Since Theorem 4 no longer applies, we cannot be certain that the absolute maximum or minimum value exists. Figure 14 (at the top of the next page) illustrates several ways that functions can fail to have absolute extrema.

In general, the best procedure to follow when the interval is not a closed interval (that is, not of the form $[a, b]$) is to sketch the graph of the function. However, one special case that occurs frequently in applications can be ana-

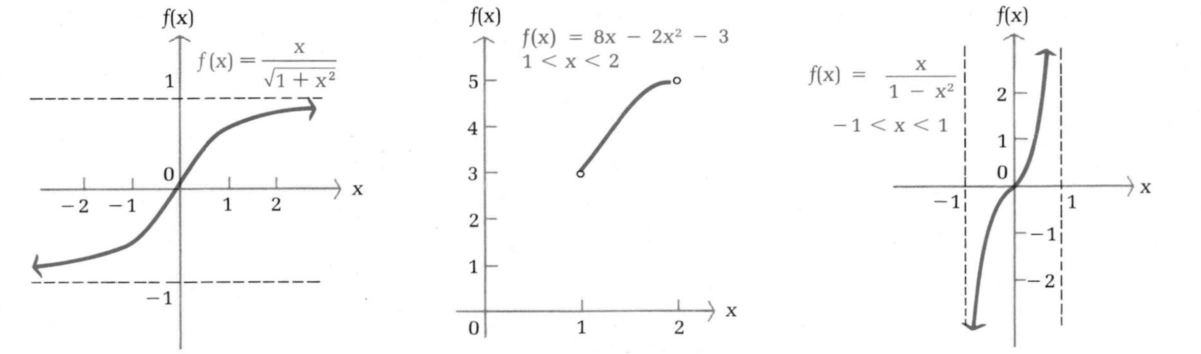

(A) No absolute extrema on $(-\infty, \infty)$: $-1 < f(x) < 1$ for all x [$f(x) \neq 1$ or -1 for any x]

(B) No absolute extrema on $(1, 2)$: $3 < f(x) < 5$ for $x \in (1, 2)$ [$f(x) \neq 3$ or 5 for any $x \in (1, 2)$]

(C) No absolute extrema on $(-1, 1)$: Graph has vertical asymptotes at $x = -1$ and $x = 1$.

FIGURE 14
Functions with no absolute extrema

lyzed without drawing a graph. It often happens that f is continuous on an interval I and has only one critical value c in the interval I (here, I can be any type of interval—open, closed, or half-closed). If this is the case and if $f''(c)$ exists, then we have the second-derivative test for absolute extrema given in the box below.

Second-Derivative Test for Absolute Maximum and Minimum When *f* Is Continuous on an Interval *I* and *c* Is the Only Critical Value in *I*

$f'(c)$	$f''(c)$	$f(c)$	EXAMPLE
0	+	Absolute minimum	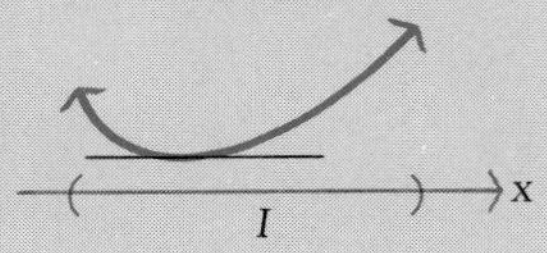
0	−	Absolute maximum	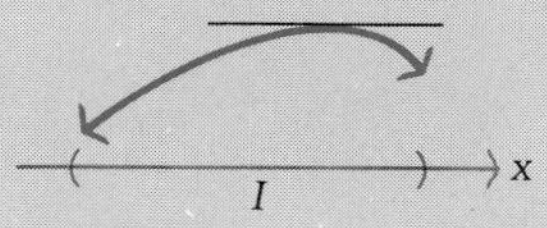
0	0	Test fails	

◆ EXAMPLE 14 Find the absolute minimum value of $f(x) = x + \frac{4}{x}$ on the interval $(0, \infty)$.

Solution

$$f'(x) = 1 - \frac{4}{x^2} = \frac{x^2 - 4}{x^2} = \frac{(x-2)(x+2)}{x^2} \qquad f''(x) = \frac{8}{x^3}$$

The only critical value in the interval $(0, \infty)$ is $x = 2$. Since $f''(2) = 1 > 0$, $f(2) = 4$ is the absolute minimum value of f on $(0, \infty)$. ◆

PROBLEM 14 Find the absolute maximum value of $f(x) = 12 - x - \frac{9}{x}$ on the interval $(0, \infty)$. ◆

◆ APPLICATIONS

Now we want to solve some applied problems that involve absolute extrema. Before beginning, we outline in the next box the steps to follow in solving this type of problem. The first step is the most difficult one. The techniques used to solve optimization problems are best illustrated through a series of examples.

A Strategy for Solving Applied Optimization Problems

Step 1. Introduce variables and a function f, including the domain I of f, and then construct a mathematical model of the form

Maximize (or minimize) $f(x)$ on the interval I

Step 2. Find the absolute maximum (or minimum) value of $f(x)$ on the interval I and the value(s) of x where this occurs.

Step 3. Use the solution to the mathematical model to answer the questions asked in the problem.

◆ EXAMPLE 15 A company manufactures and sells x transistor radios per week. If the weekly cost and demand equations are

Cost–Demand

$$C(x) = 5{,}000 + 2x$$

$$p = 10 - \frac{x}{1{,}000} \qquad 0 \leq x \leq 8{,}000$$

find for each week:

(A) The maximum revenue
(B) The maximum profit, the production level that will realize the maximum profit, and the price that the company should charge for each radio to realize the maximum profit

Solutions (A) The revenue received for selling x radios at \$p per radio is

$$R(x) = xp$$
$$= x\left(10 - \frac{x}{1{,}000}\right)$$
$$= 10x - \frac{x^2}{1{,}000}$$

Thus, the mathematical model is

$$\text{Maximize} \quad R(x) = 10x - \frac{x^2}{1{,}000} \qquad 0 \leq x \leq 8{,}000$$

$$R'(x) = 10 - \frac{x}{500}$$

$$10 - \frac{x}{500} = 0$$

$$x = 5{,}000 \qquad \text{Only critical value}$$

Use the second-derivative test for absolute extrema:

$$R''(x) = -\frac{1}{500} < 0 \qquad \text{for all } x$$

Thus, the maximum revenue is

$$\text{Max } R(x) = R(5{,}000) = \$25{,}000$$

(B) Profit = Revenue − Cost

$$P(x) = R(x) - C(x)$$
$$= 10x - \frac{x^2}{1{,}000} - 5{,}000 - 2x$$
$$= 8x - \frac{x^2}{1{,}000} - 5{,}000$$

The mathematical model is

$$\text{Maximize} \quad P(x) = 8x - \frac{x^2}{1{,}000} - 5{,}000 \qquad 0 \leq x \leq 8{,}000$$

$$P'(x) = 8 - \frac{x}{500}$$

$$8 - \frac{x}{500} = 0$$

$$x = 4{,}000$$

$$P''(x) = -\frac{1}{500} < 0 \qquad \text{for all } x$$

Since $x = 4{,}000$ is the only critical value and $P''(x) < 0$,

$$\text{Max } P(x) = P(4{,}000) = \$11{,}000$$

Using the price–demand equation with $x = 4{,}000$, we find

$$p = 10 - \frac{4{,}000}{1{,}000} = \$6$$

Thus, a maximum profit of \$11,000 per week is realized when 4,000 radios are produced weekly and sold for \$6 each. Notice that this is not the same level of production that produces the maximum revenue. ◆

All the results in Example 15 are illustrated in Figure 15. We also note that profit is maximum when

$$P'(x) = R'(x) - C'(x) = 0$$

that is, when the marginal revenue is equal to the marginal cost (the rate of increase in revenue is the same as the rate of increase in cost at the 4,000 output level—notice that the slopes of the two curves are the same at this point).

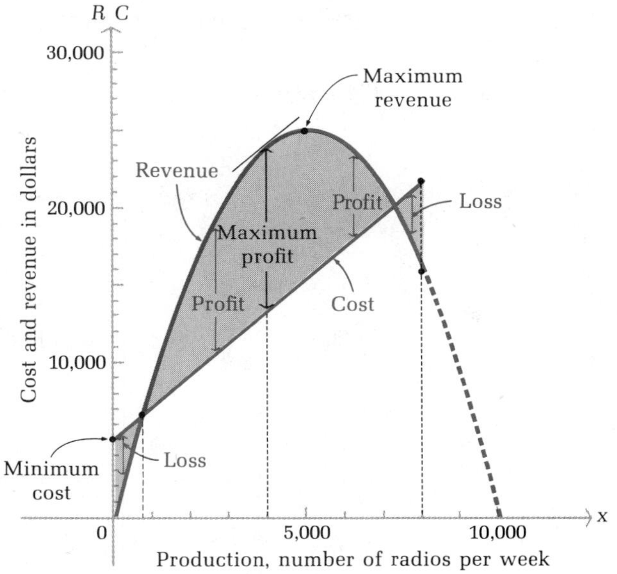

FIGURE 15

PROBLEM 15 Repeat Example 15 for

$$C(x) = 90{,}000 + 30x$$

$$p = 300 - \frac{x}{30} \qquad 0 \leq x \leq 9{,}000$$ ◆

◆ EXAMPLE 16 *Profit*

In Example 15 the government has decided to tax the company $2 for each radio produced. Taking into account this additional cost, how many radios should the company manufacture each week in order to maximize its weekly profit? What is the maximum weekly profit? How much should it charge for the radios to realize the maximum weekly profit?

Solution The tax of $2 per unit changes the company's cost equation:

$$\begin{aligned} C(x) &= \text{Original cost} + \text{Tax} \\ &= 5{,}000 + 2x + 2x \\ &= 5{,}000 + 4x \end{aligned}$$

The new profit function is

$$\begin{aligned} P(x) &= R(x) - C(x) \\ &= 10x - \frac{x^2}{1{,}000} - 5{,}000 - 4x \\ &= 6x - \frac{x^2}{1{,}000} - 5{,}000 \end{aligned}$$

Thus, we must solve the following:

$$\text{Maximize} \quad P(x) = 6x - \frac{x^2}{1{,}000} - 5{,}000 \qquad 0 \leq x \leq 8{,}000$$

$$P'(x) = 6 - \frac{x}{500}$$

$$6 - \frac{x}{500} = 0$$

$$x = 3{,}000$$

$$P''(x) = -\frac{1}{500} < 0 \qquad \text{for all } x$$

$$\text{Max } P(x) = P(3{,}000) = \$4{,}000$$

Using the price–demand equation with $x = 3{,}000$, we find

$$p = 10 - \frac{3{,}000}{1{,}000} = \$7$$

Thus, the company's maximum profit is $4,000 when 3,000 radios are produced and sold weekly at a price of $7.

Even though the tax caused the company's cost to increase by $2 per radio, the price that the company should charge to maximize its profit increases by only $1. The company must absorb the other $1 with a resulting decrease of $7,000 in maximum profit. ◆

PROBLEM 16 Repeat Example 16 if

$$C(x) = 90{,}000 + 30x$$

$$p = 300 - \frac{x}{30} \qquad 0 \leq x \leq 9{,}000$$

and the government decides to tax the company \$20 for each unit produced. Compare the results with the results in Problem 15B. ◆

◆ EXAMPLE 17

Maximize Yield

A walnut grower estimates from past records that if 20 trees are planted per acre, each tree will average 60 pounds of nuts per year. If for each additional tree planted per acre (up to 15) the average yield per tree drops 2 pounds, how many trees should be planted to maximize the yield per acre? What is the maximum yield?

Solution Let x be the number of additional trees planted per acre. Then

$$20 + x = \text{Total number of trees per acre}$$

$$60 - 2x = \text{Yield per tree}$$

$$\text{Yield per acre} = (\text{Total number of trees per acre})(\text{Yield per tree})$$

$$Y(x) = (20 + x)(60 - 2x)$$

$$= 1{,}200 + 20x - 2x^2 \qquad 0 \leq x \leq 15$$

Thus, we must solve the following:

$$\text{Maximize} \quad Y(x) = 1{,}200 + 20x - 2x^2 \qquad 0 \leq x \leq 15$$

$$Y'(x) = 20 - 4x$$

$$20 - 4x = 0$$

$$x = 5$$

$$Y''(x) = -4 < 0 \qquad \text{for all } x$$

Hence,

$$\text{Max } Y(x) = Y(5) = 1{,}250 \text{ pounds per acre}$$

Thus, a maximum yield of 1,250 pounds of nuts per acre is realized if 25 trees are planted per acre. ◆

PROBLEM 17 Repeat Example 17 starting with 30 trees per acre and a reduction of 1 pound per tree for each additional tree planted. ◆

◆ EXAMPLE 18

Maximize Area

A farmer wants to construct a rectangular pen next to a barn 60 feet long, using all of the barn as part of one side of the pen. Find the dimensions of the pen with the largest area that the farmer can build if:

(A) 160 feet of fencing material is available
(B) 250 feet of fencing material is available

Solutions

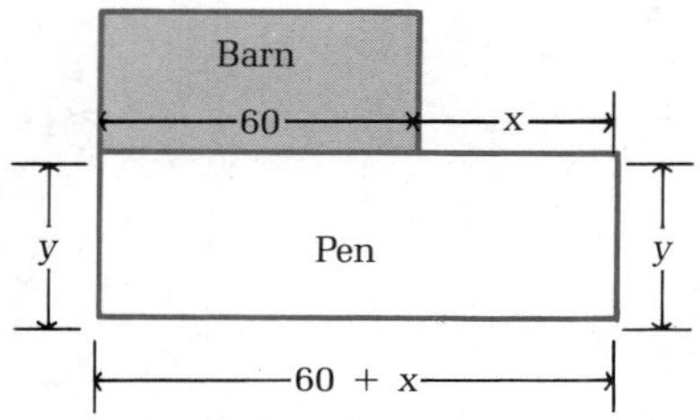

(A) We begin by constructing and labeling the figure in the margin. The area of the pen is

$$A = (x + 60)y$$

Before we can maximize the area, we must determine a relationship between x and y in order to express A as a function of one variable. In this case, x and y are related to the total amount of available fencing material:

$$x + y + 60 + x + y = 160$$
$$2x + 2y = 100$$
$$y = 50 - x$$

Thus,

$$A(x) = (x + 60)(50 - x)$$

Now we need to determine the permissible values of x; that is, the domain of the function A. Since the farmer wants to use all of the barn as part of one side of the pen, x cannot be negative. Since y is the other dimension of the pen, y cannot be negative. Thus,

$$y = 50 - x \geq 0$$
$$50 \geq x$$

The domain of A is [0, 50]. Thus, we must solve the following:

Maximize $A(x) = (x + 60)(50 - x) \qquad 0 \leq x \leq 50$

$$A(x) = 3{,}000 - 10x - x^2$$
$$A'(x) = -10 - 2x$$
$$-10 - 2x = 0$$
$$x = -5$$

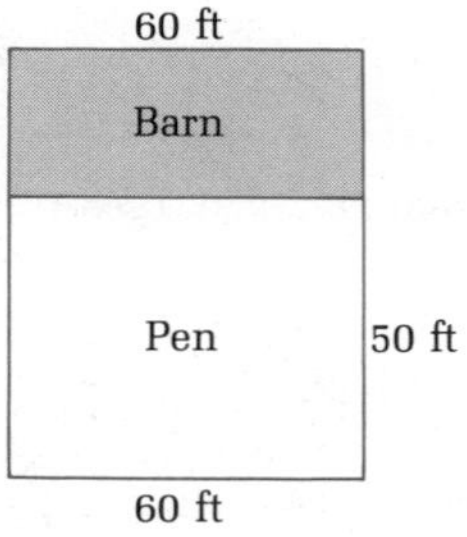

Since $x = -5$ is not in the interval [0, 50], there are no critical values in the interval. A(x) is continuous on [0, 50], so the absolute maximum must occur at one of the end points.

$A(0) = 3{,}000$ Maximum area

$A(50) = 0$

If $x = 0$, then $y = 50$. Thus, the dimensions of the pen with largest area are 60 feet by 50 feet.

(B) If 250 feet of fencing material is available, then

$$x + y + 60 + x + y = 250$$
$$2x + 2y = 190$$
$$y = 95 - x$$

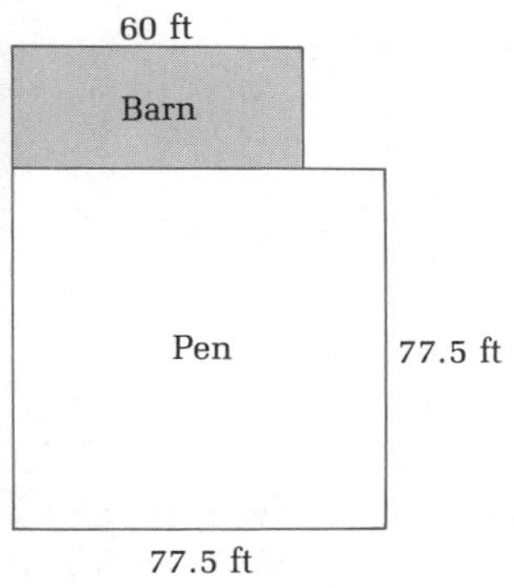

The model becomes

$$\text{Maximize} \quad A(x) = (x + 60)(95 - x) \qquad 0 \leq x \leq 95$$

$$A(x) = 5{,}700 + 35x - x^2$$

$$A'(x) = 35 - 2x$$

$$35 - 2x = 0$$

$$x = \tfrac{35}{2} = 17.5 \qquad \text{The only critical value}$$

$$A''(x) = -2 < 0 \qquad \text{for all } x$$

$$\text{Max } A(x) = A(17.5) = 6{,}006.25$$

$$y = 95 - 17.5 = 77.5$$

This time, the dimensions of the pen with the largest area are 77.5 feet by 77.5 feet. ◆

PROBLEM 18 Repeat Example 18 if the barn is 80 feet long. ◆

◆ EXAMPLE 19 *Inventory Control*

A record company anticipates that there will be a demand for 20,000 copies of a certain album during the following year. It costs the company $0.50 to store a record for 1 year. Each time it must press additional records, it costs $200 to set up the equipment. How many records should the company press during each production run in order to minimize its total storage and set-up costs?

Solution

This type of problem is called an **inventory control problem.** One of the basic assumptions made in such problems is that the demand is uniform. For example, if there are 250 working days in a year, then the daily demand would be $20{,}000/250 = 800$ records. The company could decide to produce all 20,000 records at the beginning of the year. This would certainly minimize the set-up costs, but would result in very large storage costs. At the other extreme, it could produce 800 records each day. This would minimize the storage costs, but would result in very large set-up costs. Somewhere between these two extremes is the optimal solution that will minimize the total storage and set-up costs. Let

x = Number of records pressed during each production run

y = Number of production runs

It is easy to see that the total set-up cost for the year is $200y$, but what is the total storage cost? If the demand is uniform, then the number of records in storage between production runs will decrease from x to 0, and the average number in storage each day is $x/2$. This result is illustrated in the figure on page 302.

Since it costs $0.50 to store a record for 1 year, the total storage cost is $0.5(x/2) = 0.25x$ and the total cost is

$$\text{Total cost} = \text{Set-up cost} + \text{Storage cost}$$

$$C = 200y + 0.25x$$

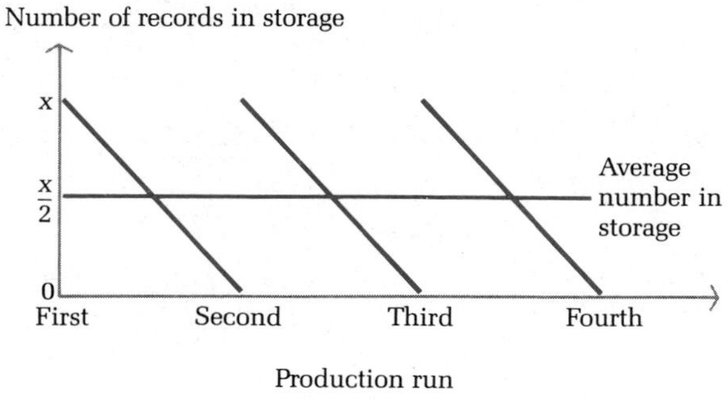

In order to write the total cost C as a function of one variable, we must find a relationship between x and y. If the company produces x records in each of y production runs, then the total number of records produced is xy. Thus,

$$xy = 20{,}000$$
$$y = \frac{20{,}000}{x}$$

Certainly, x must be at least 1 and cannot exceed 20,000. Thus, we must solve the following:

$$\text{Minimize} \quad C(x) = 200\left(\frac{20{,}000}{x}\right) + 0.25x \qquad 1 \leq x \leq 20{,}000$$

$$C(x) = \frac{4{,}000{,}000}{x} + 0.25x$$

$$C'(x) = -\frac{4{,}000{,}000}{x^2} + 0.25$$

$$-\frac{4{,}000{,}000}{x^2} + 0.25 = 0$$

$$x^2 = \frac{4{,}000{,}000}{0.25}$$

$$x^2 = 16{,}000{,}000$$

$$x = 4{,}000 \qquad -4{,}000 \text{ is not a critical value since } 1 \leq x \leq 20{,}000$$

$$C''(x) = \frac{8{,}000{,}000}{x^3} > 0 \qquad \text{for } x \in (1, 20{,}000)$$

Thus,

$$\text{Min } C(x) = C(4{,}000) = 2{,}000$$

$$y = \frac{20{,}000}{4{,}000} = 5$$

The company will minimize its total cost by pressing 4,000 records five times during the year. ◆

PROBLEM 19 Repeat Example 19 if it costs \$250 to set up a production run and \$0.40 to store a record for 1 year. ◆

Answers to Matched Problems

13. (A) Absolute maximum: $f(5) = 65$; absolute minimum: $f(-5) = -65$
    (B) Absolute maximum: $f(-2) = 16$; absolute minimum: $f(2) = -16$
    (C) Absolute maximum: $f(-2) = 16$; absolute minimum: $f(1) = -11$
14. $f(3) = 6$
15. (A) Max $R(x) = R(4{,}500) = \$675{,}000$
    (B) Max $P(x) = P(4{,}050) = \$456{,}750$; $p = \$165$
16. Max $P(x) = P(3{,}750) = \$378{,}750$; $p = \$175$; price increases \$10, profit decreases \$78,000
17. Max $Y(x) = Y(15) = 2{,}025$ lb/acre
18. (A) 80 ft by 40 ft (B) 82.5 ft by 82.5 ft
19. Press 5,000 records four times during the year

EXERCISE 5-4

A *Find the absolute maximum and minimum, if either exists, for each function.*

1. $f(x) = x^2 - 4x + 5$
2. $f(x) = x^2 + 6x + 7$
3. $f(x) = 10 + 8x - x^2$
4. $f(x) = 6 - 8x - x^2$
5. $f(x) = 1 - x^3$
6. $f(x) = 1 - x^4$

B *Find the indicated extremum of each function.*

7. Absolute maximum value of $f(x) = 24 - 2x - \dfrac{8}{x}$, $x > 0$
8. Absolute minimum value of $f(x) = 3x + \dfrac{27}{x}$, $x > 0$
9. Absolute minimum value of $f(x) = 5 + 3x + \dfrac{12}{x^2}$, $x > 0$
10. Absolute maximum value of $f(x) = 10 - 2x - \dfrac{27}{x^2}$, $x > 0$

Find the absolute maximum and minimum, if either exists, for each function on the indicated intervals.

11. $f(x) = x^3 - 6x^2 + 9x - 6$
    (A) $[-1, 5]$ (B) $[-1, 3]$ (C) $[2, 5]$
12. $f(x) = 2x^3 - 3x^2 - 12x + 24$
    (A) $[-3, 4]$ (B) $[-2, 3]$ (C) $[-2, 1]$
13. $f(x) = (x - 1)(x - 5)^3 + 1$
    (A) $[0, 3]$ (B) $[1, 7]$ (C) $[3, 6]$

14. $f(x) = x^4 - 8x^2 + 16$
(A) $[-1, 3]$ (B) $[0, 2]$ (C) $[-3, 4]$

Preliminary word problems:

C

15. How would you divide a 10 inch line so that the product of the two lengths is a maximum?
16. What quantity should be added to 5 and subtracted from 5 in order to produce the maximum product of the results?
17. Find two numbers whose difference is 30 and whose product is a minimum.
18. Find two positive numbers whose sum is 60 and whose product is a maximum.
19. Find the dimensions of a rectangle with perimeter 100 centimeters that has maximum area. Find the maximum area.
20. Find the dimensions of a rectangle of area 225 square centimeters that has the least perimeter. What is the perimeter?

APPLICATIONS

Business & Economics

21. *Average costs.* If the average manufacturing cost (in dollars) per pair of sunglasses is given by

$$\overline{C}(x) = x^2 - 6x + 12 \qquad 0 \leq x \leq 6$$

where x is the number (in thousands) of pairs manufactured, how many pairs of glasses should be manufactured to minimize the average cost per pair? What is the minimum average cost per pair?

22. *Maximum revenue and profit.* A company manufactures and sells x television sets per month. The monthly cost and demand equations are

$$C(x) = 72{,}000 + 60x$$

$$p = 200 - \frac{x}{30} \qquad 0 \leq x \leq 6{,}000$$

(A) Find the maximum revenue.
(B) Find the maximum profit, the production level that will realize the maximum profit, and the price the company should charge for each television set.
(C) If the government decides to tax the company \$5 for each set it produces, how many sets should the company manufacture each month in order to maximize its profit? What is the maximum profit? What should the company charge for each set?

23. *Car rental.* A car rental agency rents 200 cars per day at a rate of \$30 per day. For each \$1 increase in rate, 5 fewer cars are rented. At what rate

should the cars be rented to produce the maximum income? What is the maximum income?

24. *Rental income.* A 300 room hotel in Las Vegas is filled to capacity every night at \$80 a room. For each \$1 increase in rent, 3 fewer rooms are rented. If each rented room costs \$10 to service per day, how much should the management charge for each room to maximize gross profit? What is the maximum gross profit?

25. *Agriculture.* A commercial cherry grower estimates from past records that if 30 trees are planted per acre, each tree will yield an average of 50 pounds of cherries per season. If for each additional tree planted per acre (up to 20), the average yield per tree is reduced by 1 pound, how many trees should be planted per acre to obtain the maximum yield per acre? What is the maximum yield?

26. *Agriculture.* A commercial pear grower must decide on the optimum time to have fruit picked and sold. If the pears are picked now, they will bring 30¢ per pound, with each tree yielding an average of 60 pounds of salable pears. If the average yield per tree increases 6 pounds per tree per week for the next 4 weeks, but the price drops 2¢ per pound per week, when should the pears be picked to realize the maximum return per tree? What is the maximum return?

27. *Manufacturing.* A candy box is to be made out of a piece of cardboard that measures 8 by 12 inches. Squares of equal size will be cut out of each corner, and then the ends and sides will be folded up to form a rectangular box. What size square should be cut from each corner to obtain a maximum volume?

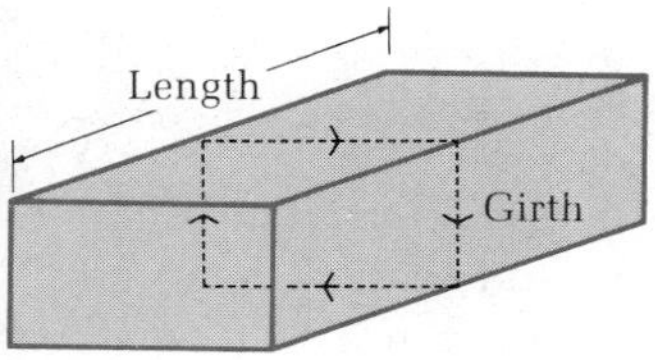

28. *Packaging.* A parcel delivery service will deliver a package only if the length plus girth (distance around) does not exceed 108 inches.
    (A) Find the dimensions of a rectangular box with square ends that satisfies the delivery service's restriction and has maximum volume. What is the maximum volume?
    (B) Find the dimensions (radius and height) of a cylindrical container that meets the delivery service's requirement and has maximum volume. What is the maximum volume?

29. *Construction costs.* A fence is to be built to enclose a rectangular area of 800 square feet. The fence along three sides is to be made of material that costs \$6 per foot. The material for the fourth side costs \$18 per foot. Find the dimensions of the rectangle that will allow the most economical fence to be built.

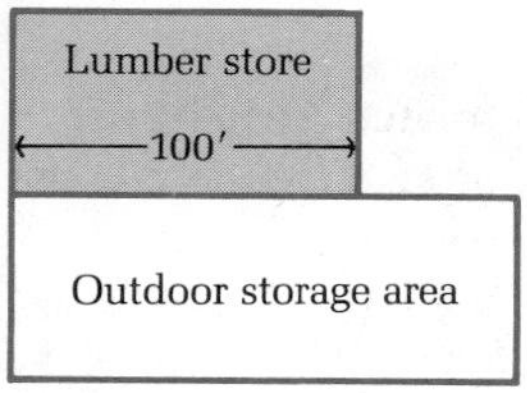

30. *Construction costs.* The owner of a retail lumber store wants to construct a fence to enclose an outdoor storage area adjacent to the store, as indicated in the figure in the margin. Find the dimensions that will enclose the largest area if:
    (A) 240 feet of fencing material is used.
    (B) 400 feet of fencing material is used.

31. *Inventory control.* A publishing company sells 50,000 copies of a certain book each year. It costs the company \$1.00 to store a book for 1 year. Each time it must print additional copies, it costs the company \$1,000 to set up the presses. How many books should the company produce during each printing in order to minimize its total storage and set-up costs?

32. *Operational costs.* The cost per hour for fuel to run a train is $v^2/4$ dollars, where v is the speed of the train in miles per hour. (Note that the cost goes up as the square of the speed.) Other costs, including labor, are \$300 per hour. How fast should the train travel on a 360 mile trip to minimize the total cost for the trip?

33. *Construction costs.* A freshwater pipeline is to be run from a source on the edge of a lake to a small resort community on an island 5 miles off-shore, as indicated in the figure.

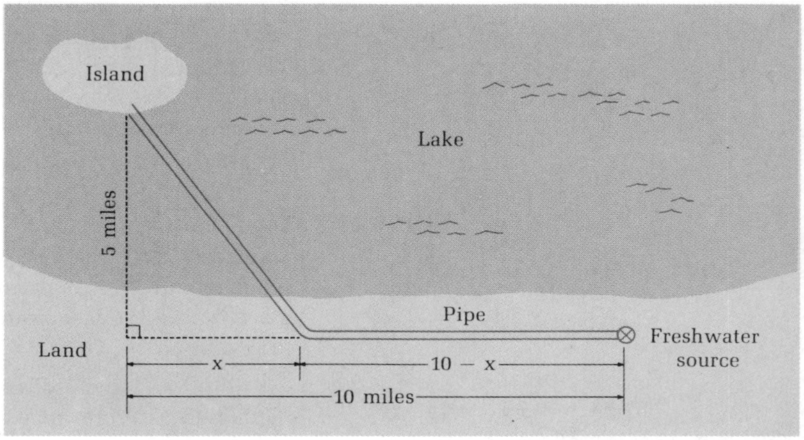

(A) If it costs 1.4 times as much to lay the pipe in the lake as it does on land, what should x be (in miles) to minimize the total cost of the project?
(B) If it costs only 1.1 times as much to lay the pipe in the lake as it does on land, what should x be to minimize the total cost of the project? [*Note:* Compare with Problem 38.]

34. *Manufacturing costs.* A manufacturer wants to produce cans that will hold 12 ounces (approximately 22 cubic inches) in the form of a right circular cylinder. Find the dimensions (radius of an end and height) of the can that will use the smallest amount of material. Assume the circular ends are cut out of squares, with the corner portions wasted, and the sides are made from rectangles, with no waste.

Life Sciences

35. *Bacteria control.* A recreational swimming lake is treated periodically to control harmful bacteria growth. Suppose t days after a treatment, the concentration of bacteria per cubic centimeter is given by

$$C(t) = 30t^2 - 240t + 500 \qquad 0 \leq t \leq 8$$

How many days after a treatment will the concentration be minimal? What is the minimum concentration?

36. *Drug concentration.* The concentration $C(t)$, in milligrams per cubic centimeter, of a particular drug in a patient's bloodstream is given by

$$C(t) = \frac{0.16t}{t^2 + 4t + 4}$$

where t is the number of hours after the drug is taken. How many hours after the drug is given will the concentration be maximum? What is the maximum concentration?

37. *Laboratory management.* A laboratory uses 500 white mice each year for experimental purposes. It costs \$4.00 to feed a mouse for 1 year. Each time mice are ordered from a supplier, there is a service charge of \$10 for processing the order. How many mice should be ordered each time in order to minimize the total cost of feeding the mice and of placing the orders for the mice?

38. *Bird flights.* Some birds tend to avoid flights over large bodies of water during daylight hours. (It is speculated that more energy is required to fly over water than land, because air generally rises over land and falls over water during the day.) Suppose an adult bird with this tendency is taken from its nesting area on the edge of a large lake to an island 5 miles off-shore and is then released (see the accompanying figure).

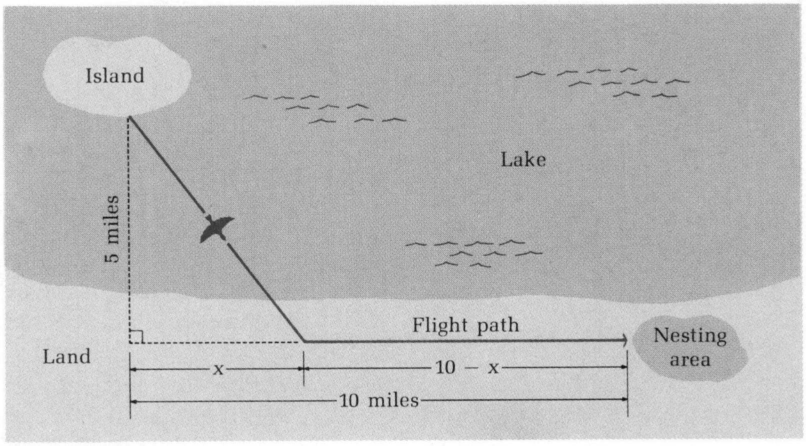

(A) If it takes 1.4 times as much energy to fly over water as land, how far up-shore (x, in miles) should the bird head in order to minimize the total energy expended in returning to the nesting area?

(B) It it takes only 1.1 times as much energy to fly over water as land, how far up-shore should the bird head in order to minimize the total energy expended in returning to the nesting area? [*Note:* Compare with Problem 33.]

39. *Botany.* If it is known from past experiments that the height (in feet) of a given plant after t months is given approximately by

$$H(t) = 4t^{1/2} - 2t \qquad 0 \leq t \leq 2$$

how long, on the average, will it take a plant to reach its maximum height? What is the maximum height?

40. *Pollution.* Two heavy industrial areas are located 10 miles apart, as indicated in the figure. If the concentration of particulate matter (in parts per million) decreases as the reciprocal of the square of the distance from the source, and area A_1 emits eight times the particulate matter as A_2, then the concentration of particulate matter at any point between the two areas is given by

$$C(x) = \frac{8k}{x^2} + \frac{k}{(10 - x)^2} \qquad 0.5 \leq x \leq 9.5, \quad k > 0$$

How far from A_1 will the concentration of particulate matter be at a minimum?

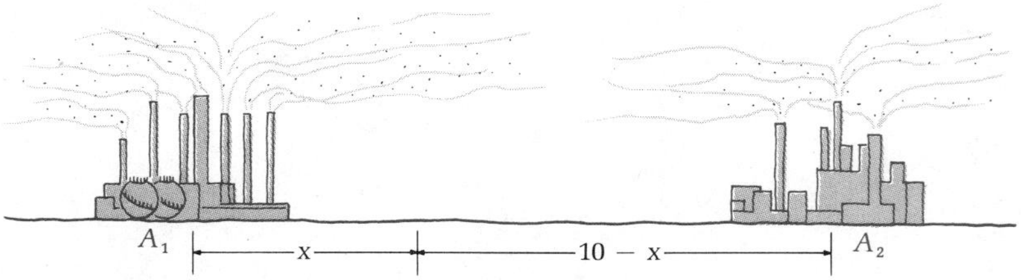

Social Sciences

41. *Politics.* In a newly incorporated city, it is estimated that the voting population (in thousands) will increase according to

$$N(t) = 30 + 12t^2 - t^3 \qquad 0 \leq t \leq 8$$

where t is time in years. When will the rate of increase be most rapid?

42. *Learning.* A large grocery chain found that, on the average, a checker can memorize P% of a given price list in x continuous hours, as given approximately by

$$P(x) = 96x - 24x^2 \qquad 0 \leq x \leq 3$$

How long should a checker plan to take to memorize the maximum percentage? What is the maximum?

SECTION 5-5 Increments and Differentials

- ◆ INCREMENTS
- ◆ DIFFERENTIALS
- ◆ APPROXIMATIONS USING DIFFERENTIALS

We now introduce the concepts of *increments* and *differentials*. Increments are useful in their own right, and they also provide an alternate notation for defining the derivative. Differentials are often easier to compute than increments, and can be used to approximate increments. Differential notation will be used extensively in Chapters 7 and 8 (where we discuss integral calculus).

◆ INCREMENTS

Given $y = f(x)$, we are sometimes interested in how y is affected by a small change in x. For example, if y represents profit and x represents price per item, we might be interested in the change in profit resulting from a small change in price.

Given

$$y = f(x) = x^3$$

if x changes from 2 to 2.1, then y will change from $y = f(2) = 2^3 = 8$ to $y = f(2.1) = 2.1^3 = 9.261$. Mathematically, the change in x and the corresponding change in y, called *increments in x and y*, respectively, are denoted by Δx and Δy (read "delta x" and "delta y," since Δ is the Greek letter delta). In terms of the given example, we write

$$\Delta x = 2.1 - 2 = 0.1 \qquad \text{Change in } x$$

$$\begin{aligned}\Delta y &= f(2.1) - f(2)\\ &= 2.1^3 - 2^3\\ &= 9.261 - 8 = 1.261 \qquad \text{Corresponding change in } y\end{aligned}$$

It is important to remember that Δx and Δy do not represent products—each is a variable with replacements from the real numbers. The general definitions are given in the following box:

Increments

For $y = f(x)$,

$$\Delta x = x_2 - x_1 \qquad \Delta y = y_2 - y_1$$
$$x_2 = x_1 + \Delta x \qquad = f(x_2) - f(x_1)$$
$$= f(x_1 + \Delta x) - f(x_1)$$

Δy represents the change in y corresponding to a Δx change in x.
Δx can be either positive or negative.

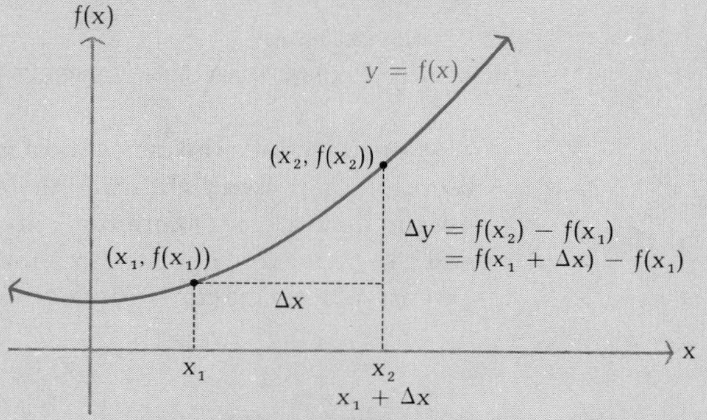

[*Note:* Δy depends on the function f, the input x, and the increment Δx.]

◆ EXAMPLE 20 Given the function: $y = f(x) = \dfrac{x^2}{2}$

(A) Find Δx, Δy, and $\Delta y/\Delta x$ for $x_1 = 1$ and $x_2 = 2$.

(B) Find $\dfrac{f(x_1 + \Delta x) - f(x_1)}{\Delta x}$ for $x_1 = 1$ and $\Delta x = 2$.

Solutions (A) $\Delta x = x_2 - x_1 = 2 - 1 = 1$

$$\Delta y = f(x_2) - f(x_1)$$
$$= f(2) - f(1) = \frac{4}{2} - \frac{1}{2} = \frac{3}{2}$$
$$\frac{\Delta y}{\Delta x} = \frac{f(x_2) - f(x_1)}{x_2 - x_1} = \frac{\frac{3}{2}}{1} = \frac{3}{2}$$

(B) $$\frac{f(x_1 + \Delta x) - f(x_1)}{\Delta x} = \frac{f(1 + 2) - f(1)}{2}$$
$$= \frac{f(3) - f(1)}{2} = \frac{\frac{9}{2} - \frac{1}{2}}{2} = \frac{4}{2} = 2$$ ◆

PROBLEM 20 Given the function: $y = f(x) = x^2 + 1$

(A) Find Δx, Δy, and $\Delta y/\Delta x$ for $x_1 = 2$ and $x_2 = 3$.

(B) Find $\dfrac{f(x_1 + \Delta x) - f(x_1)}{\Delta x}$ for $x_1 = 1$ and $\Delta x = 2$. ◆

In Example 20, we observe another notation for the familiar difference quotient

$$\frac{f(x + h) - f(x)}{h} \tag{1}$$

It is common to refer to h, the change in x, as Δx. Then, the difference quotient (1) takes on the form

$$\frac{f(x + \Delta x) - f(x)}{\Delta x} \quad \text{or} \quad \frac{\Delta y}{\Delta x} \qquad \Delta y = f(x + \Delta x) - f(x)$$

and the derivative is defined by

$$f'(x) = \lim_{\Delta x \to 0} \frac{f(x + \Delta x) - f(x)}{\Delta x}$$

or

$$f'(x) = \lim_{\Delta x \to 0} \frac{\Delta y}{\Delta x} \tag{2}$$

if the limits exists.

◆ DIFFERENTIALS

Assume that the limit in equation (2) exists. Then, for small Δx, the difference quotient $\Delta y/\Delta x$ provides a good approximation for $f'(x)$. Also, $f'(x)$ provides a good approximation for $\Delta y/\Delta x$. We indicate the latter by writing

$$\frac{\Delta y}{\Delta x} \approx f'(x) \qquad \Delta x \text{ is small but} \neq 0 \tag{3}$$

Multiplying both sides of (3) by Δx gives us

$$\Delta y \approx f'(x)\, \Delta x \qquad \Delta x \text{ is small but} \neq 0 \tag{4}$$

From equation (4) we see that $f'(x)\, \Delta x$ provides a good approximation for Δy when Δx is small.

Because of the practical and theoretical importance of $f'(x)\, \Delta x$, we give it the special name **differential** and represent it by special symbols, ***dy*** or ***df***:

$$dy = f'(x)\, \Delta x \qquad \text{or} \qquad df = f'(x)\, \Delta x$$

For example,

$$d(2x^3) = (2x^3)' \Delta x = 6x^2 \Delta x$$
$$d(x) = (x)' \Delta x = 1 \Delta x = \Delta x$$

In the second example, we usually drop the parentheses in $d(x)$ and simply write

$$dx = \Delta x$$

In summary, we have the following:

Differentials

If $y = f(x)$ defines a differentiable function, then the **differential *dy*, or *df*,** is defined as the product of $f'(x)$ and dx, where $dx = \Delta x$. Symbolically,

$$dy = f'(x)\, dx \qquad \text{or} \qquad df = f'(x)\, dx$$

where

$$dx = \Delta x$$

[*Note:* The differential dy (or df) is actually a function involving two independent variables, x and dx; a change in either one or both will affect dy (or df).]

◆ EXAMPLE 21 Find dy for $f(x) = x^2 + 3x$. Evaluate dy for:

(A) $x = 2$ and $dx = 0.1$ (B) $x = 3$ and $dx = 0.1$ (C) $x = 1$ and $dx = 0.02$

Solution

$$dy = f'(x)\, dx$$
$$= (2x + 3)\, dx$$

(A) When $x = 2$ and $dx = 0.1$,

$$dy = [2(2) + 3]0.1 = 0.7$$

(B) When $x = 3$ and $dx = 0.1$,

$$dy = [2(3) + 3]0.1 = 0.9$$

(C) When $x = 1$ and $dx = 0.02$,

$$dy = [2(1) + 3]0.02 = 0.1$$

◆

PROBLEM 21 Find dy for $f(x) = \sqrt{x} + 3$. Evaluate dy for:

(A) $x = 4$ and $dx = 0.1$ (B) $x = 9$ and $dx = 0.12$ (C) $x = 1$ and $dx = 0.01$

◆

We now have two interpretations of the symbol dy/dx. Referring to the function $y = f(x) = x^2 + 3x$ in Example 21 with $x = 2$ and $dx = 0.1$, we have

$$\frac{dy}{dx} = f'(2) = 7 \qquad \text{Derivative}$$

and

$$\frac{dy}{dx} = \frac{0.7}{0.1} = 7 \qquad \text{Ratio of differentials}$$

◆ APPROXIMATIONS USING DIFFERENTIALS

Earlier, we noted that for small Δx,

$$\frac{\Delta y}{\Delta x} \approx f'(x) \qquad \text{and} \qquad \Delta y \approx f'(x)\,\Delta x$$

Also, since

$$dy = f'(x)\,dx$$

it follows that

$$\boldsymbol{\Delta y \approx dy}$$

and dy can be used to approximate Δy.

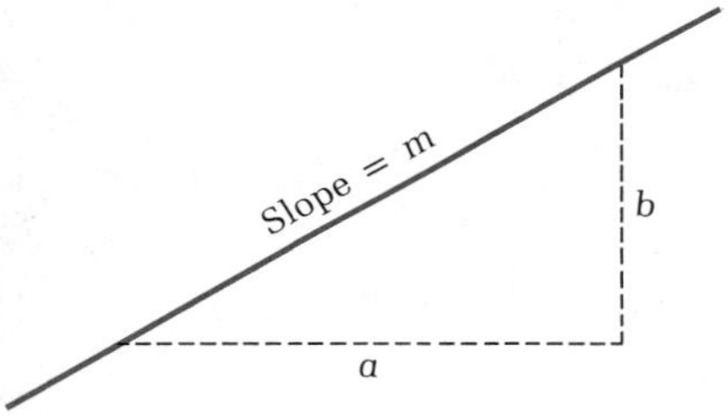

$m = \frac{b}{a}$ or $b = ma$

FIGURE 16

To interpret this result geometrically, we need to recall a basic property of slope. The vertical change in a line is equal to the product of the slope and the horizontal change, as shown in Figure 16.

Now consider the line tangent to the graph of $y = f(x)$, as shown in Figure 17. Since $f'(x)$ is the slope of the tangent line and dx is the horizontal change in the tangent line, it follows that the vertical change in the tangent line is given by $dy = f'(x)\,dx$, as indicated in Figure 17.

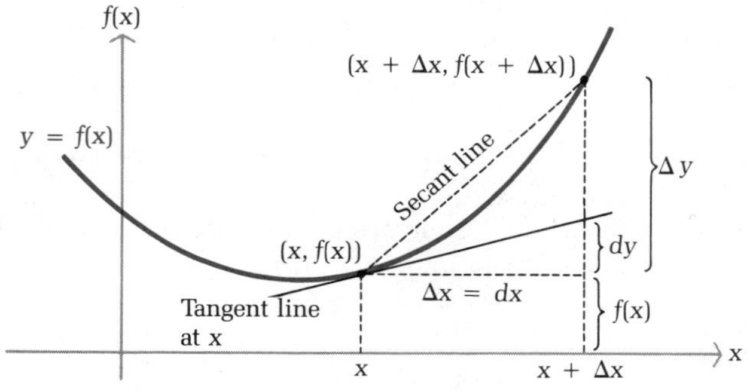

FIGURE 17

Why do we want to use dy to approximate Δy? Because, as we mentioned earlier, dy often is much easier to compute than Δy.

◆ EXAMPLE 22

Find Δy and dy for $f(x) = 6x - x^2$ when $x = 2$ and $\Delta x = dx = 0.1$.

Solution

$$\begin{aligned} \Delta y &= f(x + \Delta x) - f(x) \\ &= f(2.1) - f(2) \\ &= [6(2.1) - (2.1)^2] - [6(2) - 2^2] \\ &= 8.19 - 8 = 0.19 \end{aligned}$$

$$\begin{aligned} dy &= f'(x)\,dx \\ &= (6 - 2x)\,dx \\ &= [6 - 2(2)](0.1) \\ &= 0.2 \end{aligned}$$

Notice that dy and Δy differ by only 0.01 in this case. ◆

PROBLEM 22

Repeat Example 22 for $x = 4$ and $\Delta x = dx = -0.2$. ◆

◆ EXAMPLE 23

Cost–Revenue

A company manufactures and sells x transistor radios per week. If the weekly cost and revenue equations are

$$C(x) = 5{,}000 + 2x \qquad R(x) = 10x - \frac{x^2}{1{,}000} \qquad 0 \leq x \leq 8{,}000$$

find the approximate changes in revenue and profit if production is increased from 2,000 to 2,010 units per week.

Solution

We will approximate ΔR and ΔP with dR and dP, respectively, using $x = 2{,}000$ and $dx = 2{,}010 - 2{,}000 = 10$.

$$R(x) = 10x - \frac{x^2}{1{,}000}$$

$$\begin{aligned} dR &= R'(x)\,dx \\ &= \left(10 - \frac{x}{500}\right) dx \\ &= \left(10 - \frac{2{,}000}{500}\right) 10 \\ &= \$60 \text{ per week} \end{aligned}$$

$$\begin{aligned} P(x) &= R(x) - C(x) = 10x - \frac{x^2}{1{,}000} - 5{,}000 - 2x \\ &= 8x - \frac{x^2}{1{,}000} - 5{,}000 \end{aligned}$$

$$\begin{aligned} dP &= P'(x)\,dx \\ &= \left(8 - \frac{x}{500}\right) dx \\ &= \left(8 - \frac{2{,}000}{500}\right) 10 \\ &= \$40 \text{ per week} \end{aligned}$$

◆

PROBLEM 23

Repeat Example 23 with production increasing from 6,000 to 6,010. ◆

Comparing the results in Example 23 and Problem 23, we see that an increase in production results in a revenue and profit increase at the 2,000 production level, but a revenue and profit loss at the 6,000 production level.

Before closing, we should mention that even though differentials can be used to approximate certain quantities, the error can be substantial in certain cases. Generally, the smaller Δx is, the better the approximation $\Delta y \approx dy$.

Answers to Matched Problems

20. (A) $\Delta x = 1$, $\Delta y = 5$, $\Delta y/\Delta x = 5$ (B) 4

21. $dy = \frac{1}{2\sqrt{x}}\,dx$ (A) 0.025 (B) 0.02 (C) 0.005

22. $\Delta y = 0.36$, $dy = 0.4$

23. $dR = -\$20/\text{wk}$; $dP = -\$40/\text{wk}$

EXERCISE 5-5

A

In Problems 1–6, find the indicated quantities for $y = f(x) = 3x^2$.

1. Δx, Δy, and $\frac{\Delta y}{\Delta x}$; given $x_1 = 1$ and $x_2 = 4$

2. Δx, Δy, and $\frac{\Delta y}{\Delta x}$; given $x_1 = 2$ and $x_2 = 5$

3. $\frac{f(x_1 + \Delta x) - f(x_1)}{\Delta x}$; given $x_1 = 1$ and $\Delta x = 2$

4. $\frac{f(x_1 + \Delta x) - f(x_1)}{\Delta x}$; given $x_1 = 2$ and $\Delta x = 1$

5. $\frac{\Delta y}{\Delta x}$; given $x_1 = 1$ and $x_2 = 3$

6. $\frac{\Delta y}{\Delta x}$; given $x_1 = 2$ and $x_2 = 3$

In Problems 7–12, find dy for each function

7. $y = 30 + 12x^2 - x^3$

8. $y = 200x - \frac{x^2}{30}$

9. $y = x^2\left(1 - \frac{x}{9}\right)$

10. $y = x^3(60 - x)$

11. $y = \frac{590}{\sqrt{x}}$

12. $y = 52\sqrt{x}$

B

In Problems 13 and 14, find the indicated quantities for $y = f(x) = 3x^2$.

13. (A) $\frac{f(2 + \Delta x) - f(2)}{\Delta x}$ (simplify)

    (B) What does the quantity in part A approach as Δx approaches 0?

14. (A) $\frac{f(3 + \Delta x) - f(3)}{\Delta x}$ (simplify)

    (B) What does the quantity in part A approach as Δx approaches 0?

In Problems 15–18, find dy for each function.

15. $y = (2x + 1)^3$

16. $y = (3x + 5)^5$

17. $y = \frac{x}{x^2 + 9}$

18. $y = \frac{x^2}{(x + 1)^2}$

In Problems 19–22, evaluate dy and Δy for each function for the indicated values.

19. $y = f(x) = x^2 - 3x + 2; \quad x = 5, \quad dx = \Delta x = 0.2$

20. $y = f(x) = 30 + 12x^2 - x^3; \quad x = 2, \quad dx = \Delta x = 0.1$

21. $y = f(x) = 75\left(1 - \frac{2}{x}\right); \quad x = 5, \quad dx = \Delta x = -0.5$

22. $y = f(x) = 100\left(x - \frac{4}{x^2}\right); \quad x = 2, \quad dx = \Delta x = -0.1$

23. A cube with sides 10 inches long is covered with a coat of fiberglass 0.2 inch thick. Use differentials to estimate the volume of the fiberglass shell.

24. A sphere with a radius of 5 centimeters is coated with ice 0.1 centimeter thick. Use differentials to estimate the volume of the ice. [Recall that $V = \frac{4}{3}\pi r^3$, $\pi \approx 3.14$.]

C

25. Find dy if $y = \sqrt[3]{3x^2 - 2x + 1}$.

26. Find dy if $y = (2x^2 - 4)\sqrt{x + 2}$.

27. Find dy and Δy for $y = 52\sqrt{x}$, $x = 4$, and $\Delta x = dx = 0.3$.

28. Find dy and Δy for $y = 590/\sqrt{x}$, $x = 64$, and $\Delta x = dx = 1$.

APPLICATIONS

Use differential approximations in the following problems.

Business & Economics

29. *Advertising.* Using past records, it is estimated that a company will sell N units of a product after spending \$x thousand in advertising, as given by

$$N = 60x - x^2 \qquad 5 \leq x \leq 30$$

Approximately what increase in sales will result by increasing the advertising budget from \$10,000 to \$11,000? From \$20,000 to \$21,000?

30. *Price–demand.* Suppose, in a grocery chain, the daily demand (in pounds) for chocolate candy at \$x per pound is given by

$$D = 1{,}000 - 40x^2 \qquad 1 \leq x \leq 5$$

If the price is increased from \$3.00 per pound to \$3.20 per pound, what is the approximate change in demand?

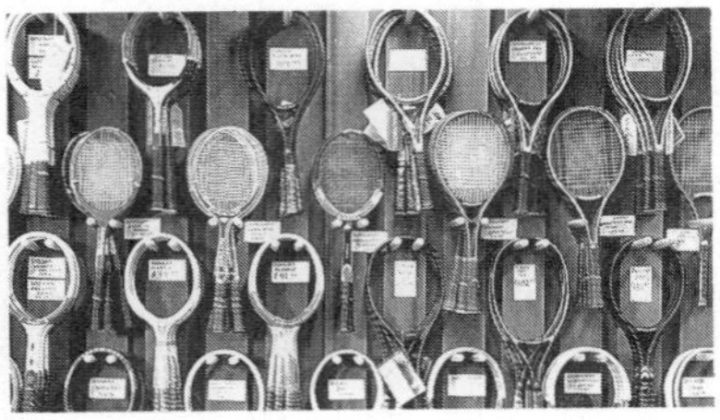

31. *Average cost.* For a company that manufactures tennis rackets, the average cost per racket, $\overline{C}$, is found to be

$$\overline{C} = x^2 - 20x + 110 \qquad 6 \leq x \leq 14$$

where x is the number of rackets produced per hour. What will the approximate change in cost per racket be if production is increased from 7 per hour to 8 per hour? From 12 per hour to 13 per hour?

32. *Revenue and profit.* A company manufactures and sells x televisions per month. If the cost and revenue equations are

$$C(x) = 72{,}000 + 60x \qquad R(x) = 200x - \frac{x^2}{30} \qquad 0 \leq x \leq 6{,}000$$

what will the approximate changes in revenue and profit be if production is increased from 1,500 to 1,501? From 4,500 to 4,501?

Life Sciences

33. *Pulse rate.* The average pulse rate y (in beats per minute) of a healthy person x inches tall is given approximately by

$$y = \frac{590}{\sqrt{x}} \qquad 30 \leq x \leq 75$$

Approximately how will the pulse rate change for a height change from 36 to 37 inches? From 64 to 65 inches?

34. *Measurement.* An egg of a particular bird is very nearly spherical. If the radius to the inside of the shell is 5 millimeters and the radius to the outside of the shell is 5.3 millimeters, approximately what is the volume of the shell? [Remember that $V = \frac{4}{3}\pi r^3$, $\pi \approx 3.14$.]

35. *Medicine.* A drug is given to a patient to dilate her arteries. If the radius of an artery is increased from 2 to 2.1 millimeters, approximately how much is the cross-sectional area increased? [Assume the cross-section of the artery is circular; $A = \pi r^2$, $\pi \approx 3.14$.]

36. *Drug sensitivity.* One hour after x milligrams of a particular drug are given to a person, the change in body temperature T (in degrees Fahrenheit) is given by

$$T = x^2\left(1 - \frac{x}{9}\right) \qquad 0 \leq x \leq 6$$

Approximate the changes in body temperature produced by the following changes in drug dosages:

(A) From 2 to 2.1 milligrams (B) From 3 to 3.1 milligrams
(C) From 4 to 4.1 milligrams

Social Sciences

37. *Learning.* A particular person learning to type has an achievement record given approximately by

$$N = 75\left(1 - \frac{2}{t}\right) \qquad 3 \leq t \leq 20$$

where N is the number of words per minute typed after t weeks of practice. What is the approximate improvement from 5 to 5.5 weeks of practice?

38. *Learning.* If a person learns y items in x hours, as given approximately by

$$y = 52\sqrt{x} \qquad 0 \leq x \leq 9$$

what is the approximate increase in the number of items learned when x changes from 1 to 1.1 hours? From 4 to 4.1 hours?

39. *Politics.* In a newly incorporated city, it is estimated that the voting population (in thousands) will increase according to

$$N(t) = 30 + 12t^2 - t^3 \qquad 0 \leq t \leq 8$$

where t is time in years. Find the approximate change in votes for the following time changes:

(A) From 1 to 1.1 years (B) From 4 to 4.1 years
(C) From 7 to 7.1 years

SECTION 5-6 Chapter Review

Important Terms and Symbols

5-1 *First Derivative and Graphs.* Solving inequalities using continuity properties; sign chart; partition numbers; increasing and decreasing functions; rising and falling graphs; local extremum; local maximum; local minimum; critical values; first-derivative test for local extrema

5-2 *Second Derivative and Graphs.* Concave upward; concave downward; second derivative; concavity and the second derivative; inflection point; second-derivative test for local maxima and minima

$$f''(x); \quad \frac{d^2y}{dx^2}; \quad y''; \quad D_x^2 f(x)$$

5-3 *Curve Sketching Techniques: Unified and Extended.* Horizontal asymptote; vertical asymptote; graphing strategy: use $f(x)$ to find the domain, intercepts, horizontal and vertical asymptotes; use $f'(x)$ to find increasing and decreasing regions, and local extrema; use $f''(x)$ to find concave upward and downward regions, and inflection points; oblique asymptote

5-4 *Optimization; Absolute Maxima and Minima.* Absolute maximum; absolute minimum; absolute extrema of a function continuous on a closed interval; second-derivative test for absolute maximum and minimum; optimization problems

5-5 *Increments and Differentials.* Increments; differentials; approximations using differentials

$$\Delta x; \quad \Delta y; \quad \frac{\Delta y}{\Delta x}; \quad \lim_{\Delta x \to 0} \frac{\Delta y}{\Delta x}; \quad dy; \quad df; \quad dx$$

EXERCISE 5-6 Chapter Review

Work through all the problems in this chapter review and check your answers in the back of the book. (Answers to all review problems are there.) Where weaknesses show up, review appropriate sections in the text.

A *Problems 1–8 refer to the following graph of $y = f(x)$:*

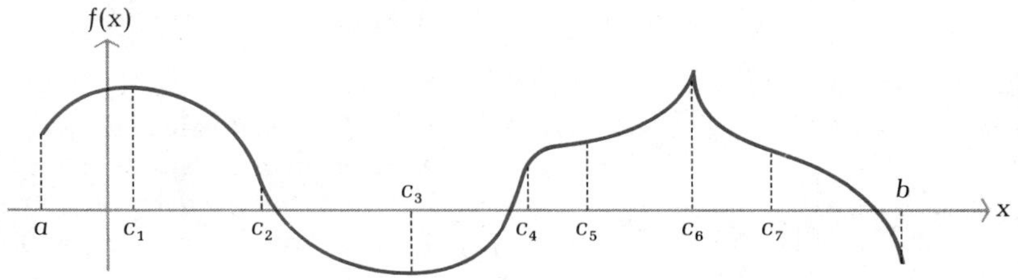

Identify the points or intervals on the x axis that produce the indicated behavior.

1. $f(x)$ is increasing
2. $f'(x) < 0$
3. Graph of f is concave downward
4. Local minima
5. Absolute maxima
6. $f'(x)$ appears to be 0
7. $f'(x)$ does not exist
8. Inflection points
9. Find $f''(x)$ for $f(x) = x^4 + 5x^3$.
10. Find y'' for $y = 3x + 4/x$.
11. Find dy for $y = f(x) = x^3 + 4x$.
12. Find dy for $y = f(x) = (3x^2 - 7)^3$.

B *In Problems 13 and 14, solve each inequality.*

13. $x^2 - x < 12$
14. $\frac{x - 5}{x^2 + 3x} > 0$

Problems 15–18 refer to the function: $f(x) = x^3 - 18x^2 + 81x$

15. Using $f(x)$:
    (A) Determine the domain of f.
    (B) Find any intercepts for the graph of f.
    (C) Find any horizontal or vertical asymptotes for the graph of f.
16. Using $f'(x)$:
    (A) Find critical values for $f(x)$.
    (B) Find partition numbers for $f'(x)$.
    (C) Find intervals over which $f(x)$ is increasing; decreasing.
    (D) Find any local maxima and minima.
17. Using $f'(x)$:
    (A) Find intervals over which the graph of f is concave upward; concave downward.
    (B) Find any inflection points.

18. Use the results of Problems 15–17 to graph f.

Problems 19–22 refer to the function: $y = f(x) = \dfrac{3x}{x+2}$

19. Using $f(x)$:
    (A) Determine the domain of f.
    (B) Find any intercepts for the graph of f.
    (C) Find any horizontal or vertical asymptotes for the graph of f.

20. Using $f'(x)$:
    (A) Find critical values for $f(x)$.
    (B) Find partition numbers for $f'(x)$.
    (C) Find intervals over which $f(x)$ is increasing; decreasing.
    (D) Find any local maxima and minima.

21. Using $f''(x)$:
    (A) Find intervals over which the graph of f is concave upward; concave downward.
    (B) Find any inflection points.

22. Use the results of Problems 19–21 to graph f.

23. Find the absolute maximum and absolute minimum for

$$y = f(x) = x^3 - 12x + 12 \qquad -3 \leq x \leq 5$$

24. Find the absolute minimum for

$$y = f(x) = x^2 + \frac{16}{x^2} \qquad x > 0$$

Find horizontal and vertical asymptotes (if they exist) in Problems 25 and 26.

25. $f(x) = \dfrac{x}{x^2+9}$

26. $f(x) = \dfrac{x^3}{x^2-9}$

27. Find dy and Δy for $f(x) = x^3 - 2x + 1$, $x = 5$, and $\Delta x = dx = 0.1$.

C

28. Find the absolute maximum for $f'(x)$ if

$$f(x) = 6x^2 - x^3 + 8$$

Graph f and f' on the same coordinate system.

29. Find two positive numbers whose product is 400 and whose sum is a minimum. What is the minimum sum?

30. Sketch the graph of $f(x) = (x-1)^3(x+3)$ using the graphing strategy discussed in Section 5-3.

31. Find dy and Δy for $y = (2/\sqrt{x}) + 8$, $x = 16$, and $\Delta x = dx = 0.2$.

Business & Economics

32. *Profit.* The profit for a company manufacturing and selling x units per month is given by

$$P(x) = 150x - \frac{x^2}{40} - 50{,}000 \qquad 0 \leq x \leq 5{,}000$$

What production level will produce the maximum profit? What is the maximum profit?

33. *Average cost.* The total cost of producing x units per month is given by

$$C(x) = 4{,}000 + 10x + \tfrac{1}{10}x^2$$

Find the minimum average cost. Graph the average cost and the marginal cost functions on the same coordinate system. Include any oblique asymptotes.

34. *Rental income.* A 200 room hotel in Fresno is filled to capacity every night at a rate of $40 per room. For each $1 increase in the nightly rate, 4 fewer rooms are rented. If each rented room costs $8 a day to service, how much should the management charge per room in order to maximize gross profit? What is the maximum gross profit?

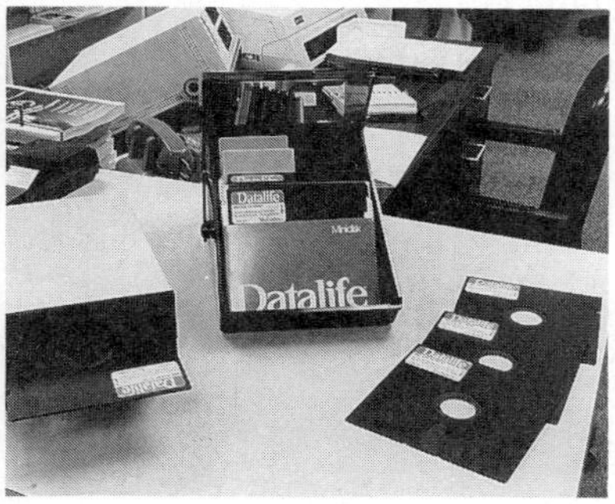

35. *Inventory control.* A computer store sells 7,200 boxes of floppy disks annually. It costs the store $0.20 to store a box of disks for 1 year. Each time it reorders disks, the store must pay a $5.00 service charge for processing the order. How many times during the year should the store order disks in order to minimize the total storage and reorder costs?

36. *Rate of change of revenue.* A company is manufacturing a new video game and can sell all it manufactures. The revenue (in dollars) is given by

$$R = 36x - \frac{x^2}{20}$$

where the production output in 1 day is x games. Use dR to approximate the change in revenue if production is increased from 250 to 260 games per day.

Life Sciences

37. *Bacteria control.* If t days after a treatment, the bacteria count per cubic centimeter in a body of water is given by

$$C(t) = 20t^2 - 120t + 800 \qquad 0 \leq t \leq 9$$

in how many days will the count be a minimum?

Social Sciences

38. *Politics.* In a new suburb, it is estimated that the number of registered voters will grow according to

$$N = 10 + 6t^2 - t^3 \qquad 0 \leq t \leq 5$$

where t is time in years and N is in thousands. When will the rate of increase be maximum?

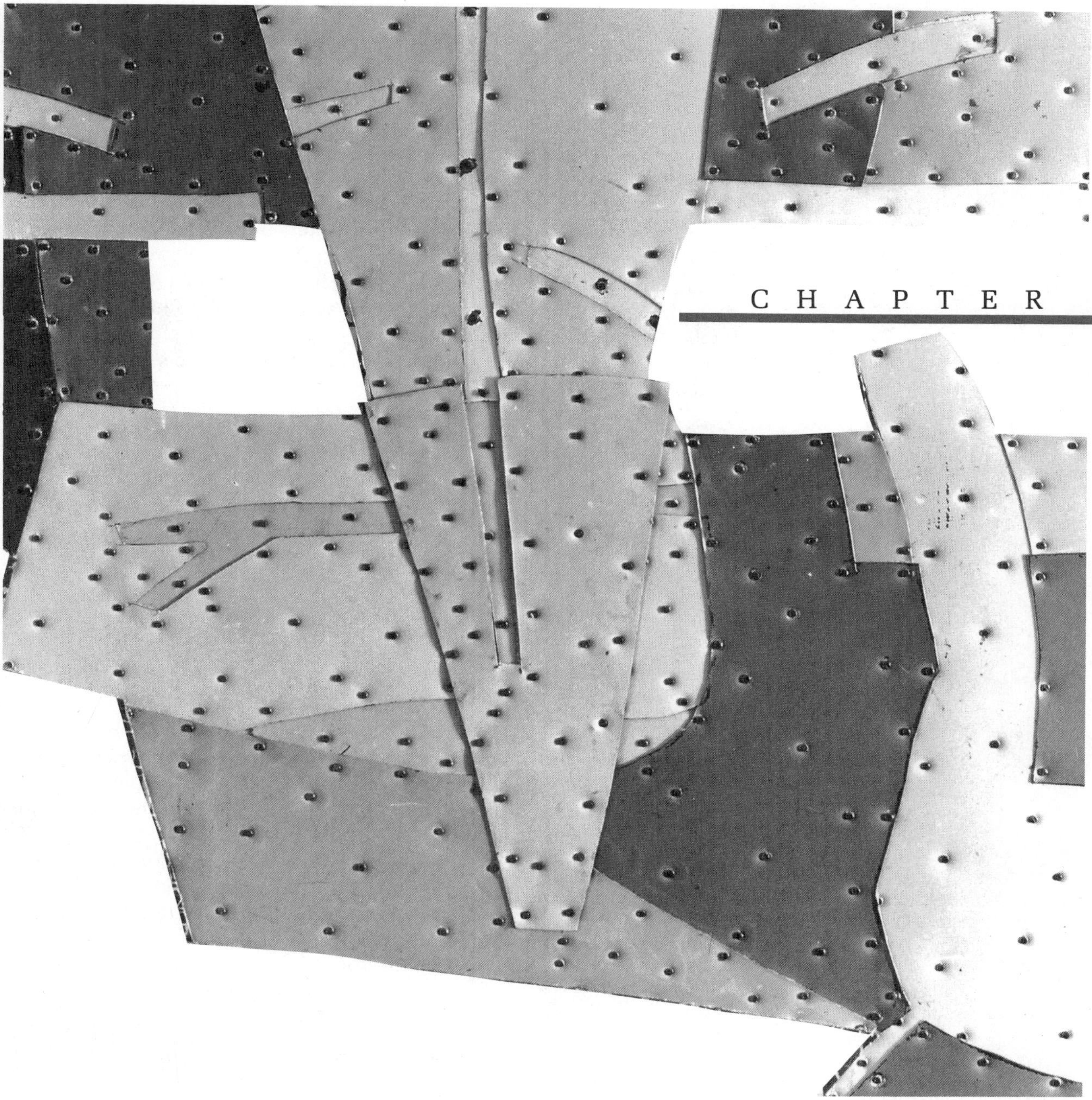

CHAPTER 6

Additional Derivative Topics

CHAPTER 6

Contents

We now know how to differentiate algebraic functions—that is, functions that can be defined using the algebraic operations of addition, subtraction, multiplication, division, powers, and roots. In this chapter, we will discuss differentiation of forms that involve the exponential and logarithmic functions. You will probably find it helpful to review some of the important properties of these functions given in Chapter 3 before proceeding further. We begin with a discussion of the irrational number e.

SECTION 6-1

The Constant e and Continuous Compound Interest

- THE CONSTANT e
- CONTINUOUS COMPOUND INTEREST

In Chapter 3, both the exponential function with base e and continuous compound interest were introduced informally. Now, with limit concepts at our disposal, we can give precise definitions of e and continuous compound interest.

THE CONSTANT e

The special irrational number e is a particularly suitable base for both exponential and logarithmic functions. The reasons for choosing this number as a base will become clear as we develop differentiation formulas for the exponential function e^x and the natural logarithmic function $\ln x$.

In precalculus treatments (Chapter 3), the number e is informally defined as an irrational number that can be approximated by the expression $[1 + (1/n)]^n$ by taking n sufficiently large. Now we will use the limit concept to formally define e as either of the following two limits:

The Number e

$$e = \lim_{n\to\infty}\left(1 + \frac{1}{n}\right)^n \quad \text{or, alternately,} \quad e = \lim_{s\to 0}(1 + s)^{1/s}$$

$$e = 2.718\ 281\ 828\ 459\ldots$$

We will use both these limit forms. [*Note:* If $s = 1/n$, then as $n \to \infty$, $s \to 0$.]

The proof that the indicated limits exist and represent an irrational number between 2 and 3 is not easy and is omitted here. Many people reason (incorrectly) that the limits are 1, since "$(1 + s)$ approaches 1 as $s \to 0$, and 1 to any power is 1." A little experimentation with a calculator can convince you otherwise. Consider the table of values for s and $f(s) = (1 + s)^{1/s}$ and the graph shown in Figure 1 for s close to 0.

s approaches 0 from the left → 0 ← s approaches 0 from the right

s	−0.5	−0.2	−0.1	−0.01 → 0 ← 0.01	0.1	0.2	0.5
$(1 + s)^{1/s}$	4.0000	3.0518	2.8680	2.7320 → e ← 2.7048	2.5937	2.4883	2.2500

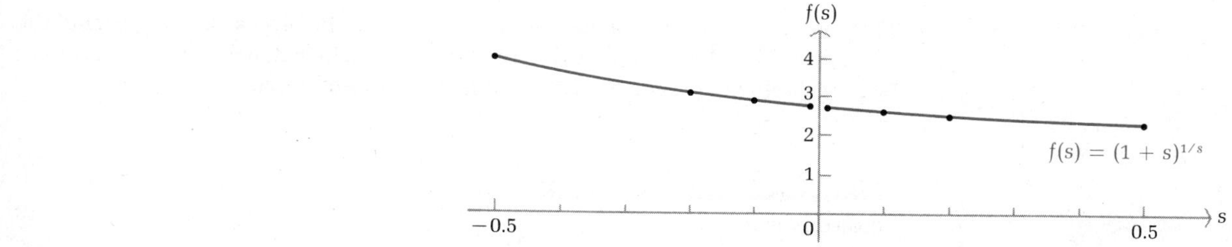

FIGURE 1

Compute some of the table values with a calculator yourself, and also try several values of s even closer to 0. Note that the function is discontinuous at $s = 0$.

Exactly who discovered e is still being debated. It is named after the great mathematician Leonard Euler (1707–1783), who computed e to twenty-three decimal places using $[1 + (1/n)]^n$.

◆ CONTINUOUS COMPOUND INTEREST

Now we will see how e appears quite naturally in the important application of compound interest. Let us start with simple interest, move on to compound interest, and then on to continuous compound interest.

If a principal P is borrowed at an annual rate of r,* then after t years at simple interest the borrower will owe the lender an amount A given by

$$A = P + Prt = P(1 + rt) \qquad \text{Simple interest} \tag{1}$$

* If r is the interest rate written as a decimal, then 100r% is the rate using %. For example, if $r = 0.12$, then we have $100r\% = 100(0.12)\% = 12\%$. The expressions 0.12 and 12% are therefore equivalent. Unless stated otherwise, all formulas in this book use r as a decimal.

On the other hand, if interest is compounded n times a year, then the borrower will owe the lender an amount A given by

$$A = P\left(1 + \frac{r}{n}\right)^{nt} \qquad \text{Compound interest} \tag{2}$$

where r/n is the interest rate per compounding period and nt is the number of compounding periods. Suppose P, r, and t in (2) are held fixed and n is increased. Will the amount A increase without bound, or will it tend to approach some limiting value?

Let us perform a calculator experiment before we attack the general limit problem. If $P = \$100$, $r = 0.06$, and $t = 2$ years, then

$$A = 100\left(1 + \frac{0.06}{n}\right)^{2n}$$

We compute A for several values of n in Table 1. The biggest gain appears in the first step; then the gains slow down as n increases. In fact, it appears that A might be tending to approach \$112.75 as n gets larger and larger.

TABLE 1

COMPOUNDING FREQUENCY	n	$A = 100\left(1 + \frac{0.06}{n}\right)^{2n}$
Annually	1	\$112.3600
Semiannually	2	112.5509
Quarterly	4	112.6493
Weekly	52	112.7419
Daily	365	112.7486
Hourly	8,760	112.7496

Now we turn back to the general problem for a moment. Keeping P, r, and t fixed in equation (2), we compute the following limit and observe an interesting and useful result.

$$\lim_{n\to\infty} P\left(1 + \frac{r}{n}\right)^{nt} = P \lim_{n\to\infty}\left(1 + \frac{r}{n}\right)^{(n/r)rt}$$

Insert r/r in the exponent and let $s = r/n$. Note that $n \to \infty$ implies $s \to 0$.

$$= P \lim_{s\to 0}[(1 + s)^{1/s}]^{rt}$$

Use the limit property given in the footnote below.*

$$= P[\lim_{s\to 0}(1 + s)^{1/s}]^{rt}$$

$\lim_{s\to 0}(1 + s)^{1/s} = e$

$$= Pe^{rt}$$

* The following new limit property is used: If $\lim_{x\to c} f(x)$ exists, then $\lim_{x\to c}[f(x)]^p = [\lim_{x\to c} f(x)]^p$, provided the last expression names a real number.

The resulting formula is called the **continuous compound interest formula,** a very important and widely used formula in business and economics.

Continuous Compound Interest

$$A = Pe^{rt}$$

where

P = Principal

r = Annual nominal interest rate compounded continuously

t = Time in years

A = Amount at time t

◆ EXAMPLE 1 If \$100 is invested at an annual nominal rate of 6% compounded continuously, what amount will be in the account after 2 years?

Solution

$$\begin{aligned} A &= Pe^{rt} \\ &= 100e^{(0.06)(2)} \qquad \text{6\% is equivalent to } r = 0.06. \\ &\approx \$112.7497 \end{aligned}$$

(Compare this result with the values calculated in Table 1.) ◆

PROBLEM 1 What amount (to the nearest cent) will an account have after 5 years if \$100 is invested at an annual nominal rate of 8% compounded annually? Semiannually? Continuously? ◆

◆ EXAMPLE 2 If \$100 is invested at 12% compounded continuously,* graph the amount in the account relative to time for a period of 10 years.

Solution We want to graph

$$A = 100e^{0.12t} \qquad 0 \le t \le 10$$

* Following common usage, we will often write the form "at 12% compounded continuously," understanding that this means "at an annual nominal rate of 12% compounded continuously."

We construct a table of values using a calculator, graph the points from the table, and join the points with a smooth curve.

t	A
0	100
1	113
2	127
3	143
4	162
5	182
6	205
7	232
8	261
9	294
10	332

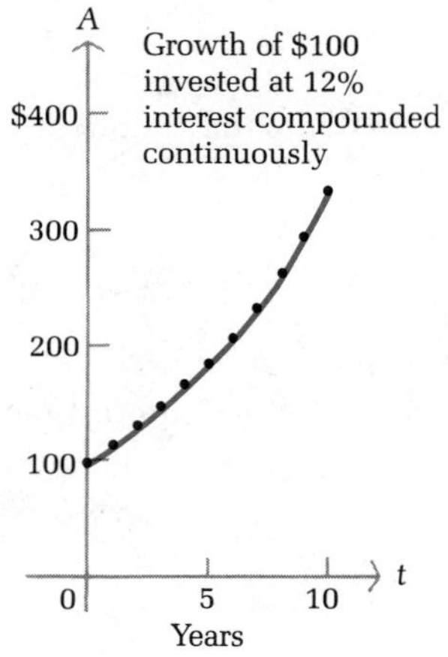

◆

PROBLEM 2 If $5,000 is invested at 20% compounded continuously, graph the amount in the account relative to time for a period of 10 years. ◆

◆ EXAMPLE 3 How long will it take an investment of $5,000 to grow to $8,000 if it is invested at 12% compounded continuously?

Solution Starting with the continuous compound interest formula $A = Pe^{rt}$, we must solve for t:

$$A = Pe^{rt}$$

$$8{,}000 = 5{,}000e^{0.12t}$$ Divide both sides by 5,000 and reverse the equation.

$$e^{0.12t} = 1.6$$ Take the natural logarithm of both sides—recall that $\log_b b^x = x$.

$$\ln e^{0.12t} = \ln 1.6$$

$$0.12t = \ln 1.6$$

$$t = \frac{\ln 1.6}{0.12}$$

$$t \approx 3.92 \text{ years}$$

◆

PROBLEM 3 How long will it take an investment of $10,000 to grow to $15,000 if it is invested at 9% compounded continuously? ◆

◆ EXAMPLE 4 How long will it take money to double if it is invested at 18% compounded continuously?

Solution Starting with the continuous compound interest formula $A = Pe^{rt}$, we must solve for t given $A = 2P$ and $r = 0.18$:

$$2P = Pe^{0.18t} \quad \text{Divide both sides by } P \text{ and reverse the equation.}$$
$$e^{0.18t} = 2 \quad \text{Take the natural logarithm of both sides.}$$
$$\ln e^{0.18t} = \ln 2$$
$$0.18t = \ln 2$$
$$t = \frac{\ln 2}{0.18}$$
$$t \approx 3.85 \text{ years}$$

◆

PROBLEM 4 How long will it take money to triple if it is invested at 12% compounded continuously? ◆

Answers to Matched Problems

1. \$146.93; \$148.02; \$149.18
2. $A = 5{,}000e^{0.2t}$

t	A
0	5,000
1	6,107
2	7,459
3	9,111
4	11,128
5	13,591
6	16,601
7	20,276
8	24,765
9	30,248
10	36,945

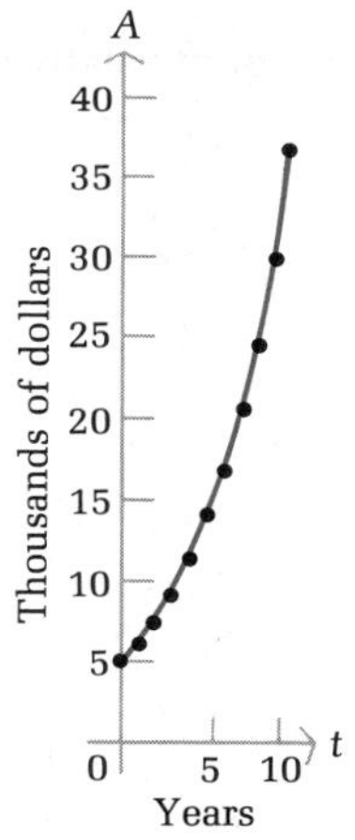

3. 4.51 yr
4. 9.16 yr

EXERCISE 6-1

A *Use a calculator or table to evaluate A to the nearest cent in Problems 1 and 2.*

1. $A = \$1{,}000e^{0.1t}$ for $t = 2, 5,$ and 8
2. $A = \$5{,}000e^{0.08t}$ for $t = 1, 4,$ and 10

B *In Problems 3–8, solve for t or r to two decimal places.*

3. $2 = e^{0.06t}$ **4.** $2 = e^{0.03t}$ **5.** $3 = e^{0.1t}$

6. $3 = e^{0.25t}$ **7.** $2 = e^{5r}$ **8.** $3 = e^{10r}$

C *In Problems 9 and 10, complete each table to five decimal places using a calculator.*

9.

n	$[1 + (1/n)]^n$
10	2.593 74
100	
1,000	
10,000	
100,000	
1,000,000	
10,000,000	
↓	↓
∞	$e = 2.718\ 281\ 828\ 459\ldots$

10.

s	$(1 + s)^{1/s}$
0.01	2.704 81
−0.01	
0.001	
−0.001	
0.000 1	
−0.000 1	
0.000 01	
−0.000 01	
↓	↓
0	$e = 2.718\ 281\ 828\ 459\ldots$

APPLICATIONS

Business & Economics

11. *Continuous compound interest.* If $20,000 is invested at an annual nominal rate of 12% compounded continuously, how much will it be worth in 8.5 years?

12. *Continuous compound interest.* Assume $1 had been invested at an annual nominal rate of 4% compounded continuously, at the time of the birth of Christ. What would be the value of the account in solid gold Earths in the year 2000? Assume that the Earth weighs approximately 2.11×10^{26} ounces and that gold will be worth $1,000 an ounce in the year 2000. What would be the value of the account in dollars at simple interest?

13. *Present value.* A note will pay $20,000 at maturity 10 years from now. How much should you be willing to pay for the note now if money is worth 7% compounded continuously?

14. *Present value.* A note will pay $50,000 at maturity 5 years from now. How much should you be willing to pay for the note now if money is worth 8% compounded continuously?

15. *Continuous compound interest.* An investor bought stock for $20,000. Four years later, the stock was sold for $30,000. If interest is compounded continuously, what annual nominal rate of interest did the original $20,000 investment earn?

16. *Continuous compound interest.* A family paid $40,000 cash for a house. Ten years later, they sold the house for $100,000. If interest is compounded continuously, what annual nominal rate of interest did the original $40,000 investment earn?

17. *Present value.* Solving $A = Pe^{rt}$ for P, we obtain

$$P = Ae^{-rt}$$

which is the present value of the amount A due in t years if money earns interest at an annual nominal rate r compounded continuously.

(A) Graph $P = 10{,}000e^{-0.08t}$, $0 \leq t \leq 50$.

(B) $\lim_{t\to\infty} 10{,}000e^{-0.08t} = ?$ (Guess, using part A.)

[*Conclusion:* The longer the duration of time until the amount A is due, the smaller its present value, as we would expect.]

18. *Present value.* Referring to Problem 17, in how many years will the $10,000 have to be due in order for its present value to be $5,000?

19. *Doubling time.* How long will it take money to double if it is invested at 25% compounded continuously?

20. *Doubling time.* How long will it take money to double if it is invested at 5% compounded continuously?

21. *Doubling rate.* At what nominal rate compounded continuously must money be invested to double in 5 years?

22. *Doubling rate.* At what nominal rate compounded continuously must money be invested to double in 3 years?

23. *Doubling time.* It is instructive to look at doubling times for money invested at various nominal rates of interest compounded continuously. Show that doubling time t at an annual rate r compounded continuously is given by

$$t = \frac{\ln 2}{r}$$

24. *Doubling time.* Graph the doubling-time equation from Problem 23 for $0 < r < 1.00$. Identify vertical and horizontal asymptotes.

Life Sciences

25. *World population.* A mathematical model for world population growth over short periods of time is given by

$$P = P_0 e^{rt}$$

where

P_0 = Population at time $t = 0$

r = Continuous compound rate of growth

t = Time in years

P = Population at time t

How long will it take the world population to double if it continues to grow at its current continuous compound rate of 2% per year?

26. *World population.* Repeat Problem 25 under the assumption that the world population is growing at a continuous compound rate of 1% per year.

27. *Population growth.* Some underdeveloped nations have population doubling times of 20 years. At what continuous compound rate is the population growing? (Use the population growth model in Problem 25.)

28. *Population growth.* Some developed nations have population doubling times of 120 years. At what continuous compound rate is the population growing? (Use the population growth model in Problem 25.)

29. *Radioactive decay.* A mathematical model for the decay of radioactive substances is given by

$$Q = Q_0 e^{rt}$$

where

Q_0 = Amount of the substance at time $t = 0$

r = Continuous compound rate of decay

t = Time in years

Q = Amount of the substance at time t

If the continuous compound rate of decay of radium per year is $r = -0.000\ 433\ 2$, how long will it take an amount of radium to decay to half the original amount? (This period of time is the half-life of the substance.)

30. *Radioactive decay.* The continuous compound rate of decay of carbon-14 per year is $r = -0.000\ 123\ 8$. How long will it take an amount of carbon-14 to decay to half the original amount? (Use the radioactive decay model in Problem 29.)

31. *Radioactive decay.* A cesium isotope has a half-life of 30 years. What is the continuous compound rate of decay? (Use the radioactive decay model in Problem 29.)

32. *Radioactive decay.* A strontium isotope has a half-life of 90 years. What is the continuous compound rate of decay? (Use the radioactive decay model in Problem 29.)

Social Sciences

33. *World population.* If the world population is now 5 billion (5×10^9) people and if it continues to grow at a continuous compound rate of 2% per year, how long will it be before there is only 1 square yard of land per person? (The Earth has approximately 1.68×10^{14} square yards of land.)

SECTION 6-2 Derivatives of Logarithmic and Exponential Functions

- DERIVATIVE FORMULAS FOR ln x AND e^x
- GRAPHING TECHNIQUES
- APPLICATION

In this section, we discuss derivative formulas for ln x and e^x. Out of all the possible choices for bases for the logarithmic and exponential functions, $\log_b x$ and b^x, it turns out (as we will see in this and the next section) that the simplest derivative formulas occur when the base b is chosen to be e.

◆ DERIVATIVE FORMULAS FOR ln x AND e^x

We are now ready to derive a formula for the derivative of

$$f(x) = \ln x = \log_e x \qquad x > 0$$

using the definition of the derivative

$$f'(x) = \lim_{h \to 0} \frac{f(x+h) - f(x)}{h}$$

and the two-step process discussed in Section 4-3.

Step 1. Simplify the difference quotient first:

$$\frac{f(x+h) - f(x)}{h} = \frac{\ln(x+h) - \ln x}{h}$$

$$= \frac{1}{h}[\ln(x+h) - \ln x] \qquad \text{Use } \ln A - \ln B = \ln \frac{A}{B}.$$

$$= \frac{1}{h} \ln \frac{x+h}{x} \qquad \text{Multiply by } 1 = x/x \text{ to change form.}$$

$$= \frac{x}{x} \cdot \frac{1}{h} \ln \frac{x+h}{x}$$

$$= \frac{1}{x}\left[\frac{x}{h} \ln\left(1 + \frac{h}{x}\right)\right] \qquad \text{Use } p \ln A = \ln A^p.$$

$$= \frac{1}{x} \ln\left(1 + \frac{h}{x}\right)^{x/h}$$

Step 2. Find the limit. Let $s = h/x$. For x fixed, if $h \to 0$, then $s \to 0$. Thus,

$$D_x \ln x = \lim_{h\to 0} \frac{f(x+h) - f(x)}{h}$$

$$= \lim_{h\to 0}\left[\frac{1}{x}\ln\left(1+\frac{h}{x}\right)^{x/h}\right] \qquad \text{Let } s = h/x. \text{ Note that } h \to 0 \text{ implies } s \to 0.$$

$$= \frac{1}{x}\lim_{s\to 0}[\ln(1+s)^{1/s}] \qquad \text{Use the new limit property given in the footnote below.*}$$

$$= \frac{1}{x}\ln[\lim_{s\to 0}(1+s)^{1/s}] \qquad \text{Use the definition of } e.$$

$$= \frac{1}{x}\ln e \qquad \ln e = \log_e e = 1$$

$$= \frac{1}{x}$$

Thus,

$$\boldsymbol{D_x \ln x = \frac{1}{x}}$$

In the next section, we will show that, in general,

$$D_x \log_b x = \frac{1}{\ln b}\left(\frac{1}{x}\right)$$

which is a somewhat more complicated result than the above—unless $b = e$.

We now apply the two-step process to the exponential function $f(x) = e^x$. In the process, we will use (without proof) the fact that

$$\lim_{h\to 0}\left(\frac{e^h - 1}{h}\right) = 1$$

[Try computing $(e^h - 1)/h$ for values of h closer and closer to 0 and on either side of 0 to convince yourself of the reasonableness of this limit.]

Step 1. Simplify the difference quotient first:

$$\frac{f(x+h) - f(x)}{h} = \frac{e^{x+h} - e^x}{h} \qquad \text{Use } e^{a+b} = e^a e^b.$$

$$= \frac{e^x e^h - e^x}{h} \qquad \text{Factor out } e^x.$$

$$= e^x\left(\frac{e^h - 1}{h}\right)$$

---

* The following new limit property is used: If $\lim_{x\to c} f(x)$ exists and is positive, then $\lim_{x\to c}[\ln f(x)] = \ln[\lim_{x\to c} f(x)]$.

Step 2. Compute the limit of the result in step 1:

$$D_x e^x = \lim_{h\to 0} \frac{f(x+h) - f(x)}{h}$$

$$= \lim_{h\to 0} e^x\left(\frac{e^h - 1}{h}\right)$$

$$= e^x \lim_{h\to 0} \left(\frac{e^h - 1}{h}\right) \qquad \text{Use the assumed limit given above.}$$

$$= e^x \cdot 1 = e^x$$

Thus,

$$\boldsymbol{D_x e^x = e^x}$$

In the next section, we will show that

$$D_x b^x = b^x \ln b$$

which is, again, a somewhat more complicated result than the above—unless $b = e$.

The two results just obtained explain why e^x is so widely used that it is sometimes referred to as *the* exponential function. These two new and important derivative formulas are restated in the box for reference.

Derivatives of the Natural Logarithmic and Exponential Functions

$$D_x \ln x = \frac{1}{x} \qquad D_x e^x = e^x$$

These new derivative formulas can be combined with the rules of differentiation discussed in Chapter 4 to differentiate a wide variety of functions.

◆ EXAMPLE 5 Find $f'(x)$ for:

(A) $f(x) = 2e^x + 3 \ln x$ (B) $f(x) = \dfrac{e^x}{x^3}$

(C) $f(x) = (\ln x)^4$ (D) $f(x) = \ln x^4$

Solutions

(A) $f'(x) = \boxed{2\, D_x e^x + 3\, D_x \ln x}$

$$= 2e^x + 3\left(\frac{1}{x}\right) = 2e^x + \frac{3}{x}$$

(B) $f'(x) = \boxed{\dfrac{x^3\, D_x e^x - e^x\, D_x x^3}{(x^3)^2}}$ Quotient rule

$$= \frac{x^3 e^x - e^x 3x^2}{x^6} = \frac{x^2 e^x (x-3)}{x^6} = \frac{e^x (x-3)}{x^4}$$

(C) $D_x(\ln x)^4 = 4(\ln x)^3 D_x \ln x$ Power rule for functions

$$= 4(\ln x)^3\left(\frac{1}{x}\right) = \frac{4(\ln x)^3}{x}$$

(D) $D_x \ln x^4 = D_x(4 \ln x)$ Property of logarithms

$$= 4\left(\frac{1}{x}\right) = \frac{4}{x}$$

◆

PROBLEM 5 Find $f'(x)$ for:

(A) $f(x) = 4 \ln x - 5e^x$ (B) $f(x) = x^2e^x$

(C) $f(x) = \ln x^3$ (D) $f(x) = (\ln x)^3$

◆

Common Error

$D_x e^x \neq xe^{x-1}$ $D_x e^x = e^x$

The power rule cannot be used to differentiate the exponential function. The power rule applies to exponential forms x^n where the exponent is a constant and the base is a variable. In the exponential form e^x, the base is a constant and the exponent is a variable.

◆ GRAPHING TECHNIQUES

Using the techniques discussed in Chapter 5, we can use first and second derivatives to gain useful information about the graphs of $y = \ln x$ and $y = e^x$. Using the derivative formulas given above, we can construct Table 2.

TABLE 2

ln x		e^x	
$y = \ln x$	$x > 0$	$y = e^x$	$-\infty < x < \infty$
$y' = 1/x > 0$	$x > 0$	$y' = e^x > 0$	$-\infty < x < \infty$
$y'' = -1/x^2 < 0$	$x > 0$	$y'' = e^x > 0$	$-\infty < x < \infty$

From the table, we can see that both functions are increasing throughout their respective domains, the graph of $y = \ln x$ is always concave downward, and the graph of $y = e^x$ is always concave upward. It can be shown that the y axis is a vertical asymptote for the graph of $y = \ln x$ ($\lim_{x\to 0^+} \ln x = -\infty$), and the x axis is a horizontal asymptote for the graph of $y = e^x$ ($\lim_{x\to -\infty} e^x = 0$). Both equations are graphed in Figure 2.

Notice that if we fold the page along the dashed line $y = x$, the two graphs match exactly (see Section 3-3). Also notice that both graphs are unbounded as $x \to \infty$. Comparing each graph with the graph of $y = x$ (the dashed line), we

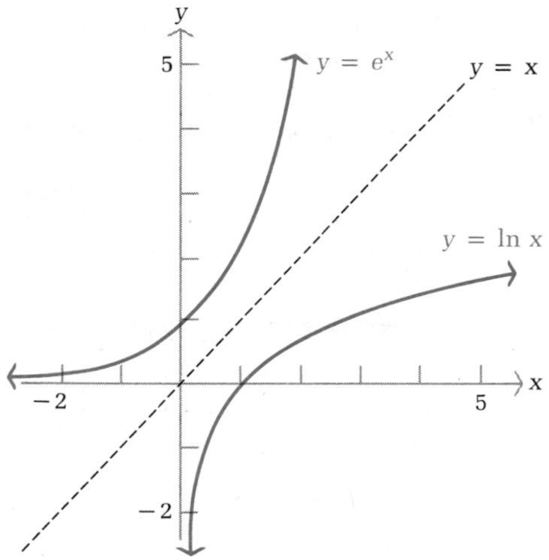

FIGURE 2
e^x is continuous on $(-\infty, \infty)$
ln x is continuous on $(0, \infty)$

conclude that e^x grows more rapidly than x and ln x grows more slowly than x. In fact, the following limits can be established:

$$\lim_{x \to \infty} \frac{x^p}{e^x} = 0, \quad p > 0 \qquad \text{and} \qquad \lim_{x \to \infty} \frac{\ln x}{x^p} = 0, \quad p > 0$$

These limits indicate that e^x grows more rapidly than any positive power of x, and ln x grows more slowly than any positive power of x.

Now we will apply graphing techniques to a slightly more complicated function.

◆ EXAMPLE 6 Sketch the graph of $f(x) = xe^x$ using the graphing strategy discussed in Section 5-3.

Solution Step 1. Use $f(x)$: $f(x) = xe^x$

(A) Domain: All real numbers

(B) Intercepts:
y intercept: $f(0) = 0$
x intercept: $xe^x = 0$ for $x = 0$ only, since $e^x > 0$ for all x (see Figure 2).

(C) Asymptotes:
Vertical asymptotes: None
Horizontal asymptotes: We have not developed limit techniques for functions of this type to determine the behavior of $f(x)$ as $x \to -\infty$ and $x \to \infty$.

However, the following tables of values suggest the nature of the graph of f as $x \to -\infty$ and $x \to \infty$:

x	1	5	10	$\to\infty$
$f(x)$	2.72	742.07	220,264.66	$\to\infty$

x	-1	-5	-10	$\to -\infty$
$f(x)$	-0.37	-0.03	$-0.000\ 45$	$\to 0$

Step 2. Use $f'(x)$:

$$f'(x) = x\, D_x\, e^x + e^x\, D_x\, x$$
$$= xe^x + e^x = e^x(x + 1)$$

Critical value: -1
Partition number: -1
Sign chart for $f'(x)$:

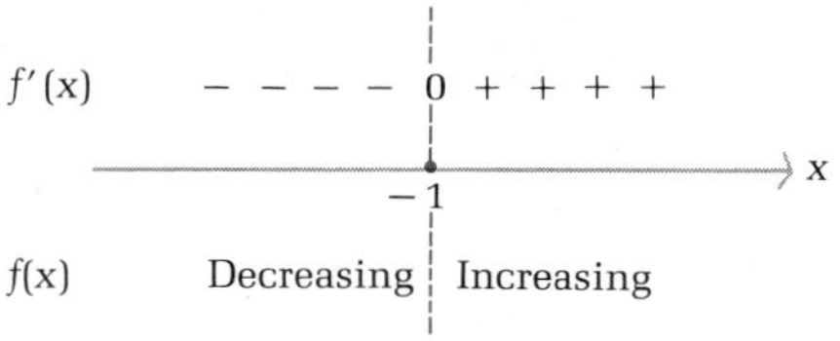

Test Numbers

x	$f'(x)$	
-2	$-e^{-2}$	(−)
0	1	(+)

Thus, $f(x)$ decreases on $(-\infty, -1)$, has a local minimum at $x = -1$, and increases on $(-1, \infty)$. [Since $e^x > 0$ for all x, we do not have to evaluate e^{-2} to conclude that $-e^{-2} < 0$ when using the test number -2.]

Step 3. Use $f''(x)$:

$$f''(x = e^x\, D_x(x + 1) + (x + 1)\, D_x\, e^x$$
$$= e^x + (x + 1)e^x = e^x(x + 2)$$

Sign chart for $f''(x)$ (partition number is -2):

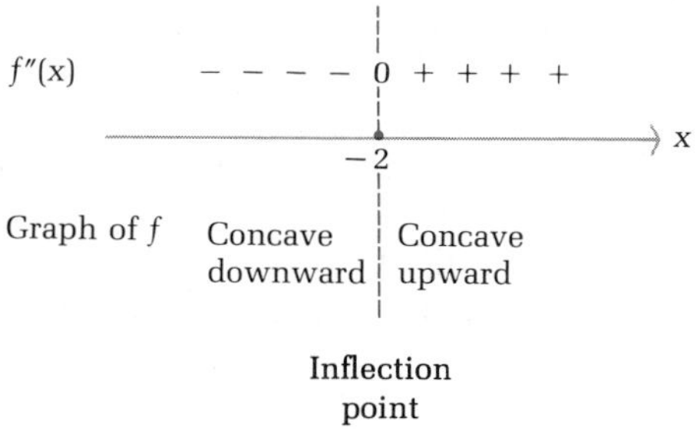

Test Numbers

x	$f''(x)$	
-3	$-e^{-3}$	(−)
-1	e^{-1}	(+)

Thus, the graph of f is concave downward on $(-\infty, -2)$, has an inflection point at $x = -2$, and is concave upward on $(-2, \infty)$.

Step 4. Sketch the graph of f using the information from steps 1–3:

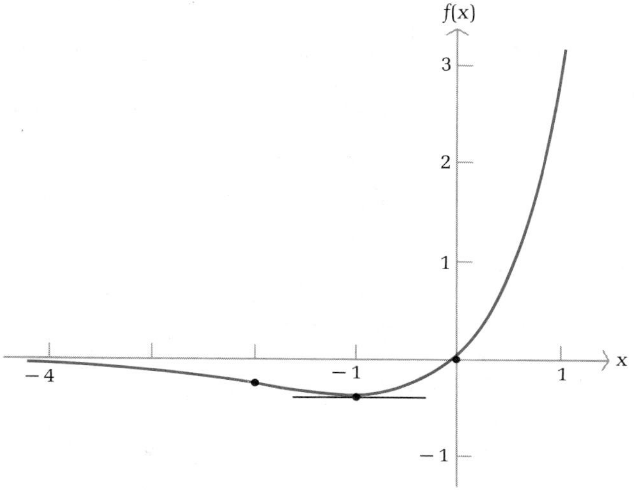

x	f(x)
−2	−0.27
−1	−0.37
0	0

◆

PROBLEM 6 Sketch the graph of $f(x) = x \ln x$. ◆

◆ APPLICATION

◆ EXAMPLE 7

Maximum profit

The market research department of a chain of pet stores test marketed their aquarium pumps (as well as other items) in several of their stores in a test city. They found that the weekly demand for aquarium pumps is given approximately by

$$p = 12 - 2 \ln x \qquad 0 < x < 90$$

where x is the number of pumps sold each week and $\$p$ is the price of one pump. If each pump costs the chain \$3, how should it be priced in order to maximize the weekly profit?

Solution

Although we want to find the price that maximizes the weekly profit, it will be easier to first find the number of pumps that will maximize the weekly profit. The revenue equation is

$$R(x) = xp = 12x - 2x \ln x$$

The cost equation is

$$C(x) = 3x$$

and the profit equation is

$$\begin{aligned} P(x) &= R(x) - C(x) \\ &= 12x - 2x \ln x - 3x \\ &= 9x - 2x \ln x \end{aligned}$$

Thus, we must solve the following:

$$\text{Maximize} \quad P(x) = 9x - 2x \ln x \qquad 0 < x < 90$$

$$\begin{aligned} P'(x) &= 9 - 2x\left(\frac{1}{x}\right) - 2 \ln x \\ &= 7 - 2 \ln x = 0 \\ 2 \ln x &= 7 \\ \ln x &= 3.5 \\ x &= e^{3.5} \end{aligned}$$

$$P''(x) = -2\left(\frac{1}{x}\right) = -\frac{2}{x}$$

Since $x = e^{3.5}$ is the only critical value and $P''(e^{3.5}) < 0$, the maximum weekly profit occurs when $x = e^{3.5} \approx 33$ and $p = 12 - 2 \ln e^{3.5} = \$5$. ◆

PROBLEM 7 Repeat Example 7 if each pump costs the chain \$3.50. ◆

Answers to Matched Problems

5. (A) $(4/x) - 5e^x$ (B) $xe^x(x + 2)$ (C) $3/x$ (D) $3(\ln x)^2/x$
6. Domain: $(0, \infty)$
   y intercept: None [$f(0)$ is not defined]
   x intercept: $x = 1$
   Increasing on (e^{-1}, ∞)
   Decreasing on $(0, e^{-1})$
   Local minimum at $x = e^{-1} \approx 0.368$
   Concave upward on $(0, \infty)$

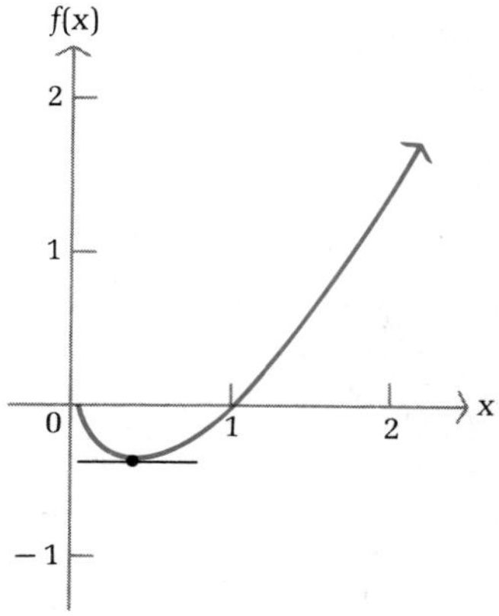

x	5	10	100	$\to\infty$
$f(x)$	8.05	23.03	460.52	$\to\infty$

x	0.1	0.01	0.001	0.000 1	$\to 0$
$f(x)$	−0.23	−0.046	−0.006 9	−0.000 92	$\to 0$

7. Maximum profit occurs for $x = e^{3.25} \approx 26$ and $p = \$5.50$.

EXERCISE 6-2

A *Find f′(x).*

1. $f(x) = 6e^x - 7 \ln x$
2. $f(x) = 4e^x + 5 \ln x$
3. $f(x) = 2x^e + 3e^x$
4. $f(x) = 4e^x - ex^e$
5. $f(x) = \ln x^5$
6. $f(x) = (\ln x)^5$
7. $f(x) = (\ln x)^2$
8. $f(x) = \ln x^2$

B

9. $f(x) = x^4 \ln x$
10. $f(x) = x^3 \ln x$
11. $f(x) = x^3 e^x$
12. $f(x) = x^4 e^x$
13. $f(x) = \dfrac{e^x}{x^2 + 9}$
14. $f(x) = \dfrac{e^x}{x^2 + 4}$
15. $f(x) = \dfrac{\ln x}{x^4}$
16. $f(x) = \dfrac{\ln x}{x^3}$
17. $f(x) = (x + 2)^3 \ln x$
18. $f(x) = (x - 1)^2 \ln x$
19. $f(x) = (x + 1)^3 e^x$
20. $f(x) = (x - 2)^3 e^x$
21. $f(x) = \dfrac{x^2 + 1}{e^x}$
22. $f(x) = \dfrac{x + 1}{e^x}$
23. $f(x) = x(\ln x)^3$
24. $f(x) = x(\ln x)^2$
25. $f(x) = (4 - 5e^x)^3$
26. $f(x) = (5 - \ln x)^4$
27. $f(x) = \sqrt{1 + \ln x}$
28. $f(x) = \sqrt{1 + e^x}$
29. $f(x) = xe^x - e^x$
30. $f(x) = x \ln x - x$
31. $f(x) = 2x^2 \ln x - x^2$
32. $f(x) = x^2 e^x - 2xe^x + 2e^x$

Find the equation of the line tangent to the graph of y = f(x) at the indicated value of x.

33. $f(x) = e^x; \quad x = 1$
34. $f(x) = e^x; \quad x = 2$
35. $f(x) = \ln x; \quad x = e$
36. $f(x) = \ln x; \quad x = 1$

C *Find the indicated extremum of each function for $x > 0$.*

37. Absolute maximum value of: $f(x) = 4x - x \ln x$
38. Absolute minimum value of: $f(x) = x \ln x - 3x$
39. Absolute minimum value of: $f(x) = \dfrac{e^x}{x}$
40. Absolute maximum value of: $f(x) = \dfrac{x^2}{e^x}$
41. Absolute maximum value of: $f(x) = \dfrac{1 + 2 \ln x}{x}$
42. Absolute minimum value of: $f(x) = \dfrac{1 - 5 \ln x}{x}$

Sketch the graph of $y = f(x)$.

43. $f(x) = 1 - e^x$
44. $f(x) = 1 - \ln x$
45. $f(x) = x - \ln x$
46. $f(x) = e^x - x$
47. $f(x) = (3 - x)e^x$
48. $f(x) = (x - 2)e^x$
49. $f(x) = x^2 \ln x$
50. $f(x) = \dfrac{\ln x}{x}$

APPLICATIONS

Business & Economics

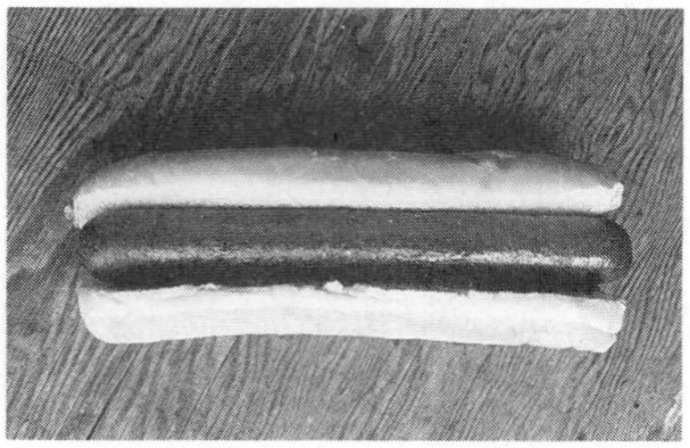

51. *Maximum profit.* A national food service runs food concessions for sporting events throughout the country. Their marketing research department chose a particular football stadium to test market a new jumbo hot dog. It was found that the demand for the new hot dog is given approximately by

$$p = 5 - \ln x \qquad 5 \le x \le 50$$

where x is the number of hot dogs (in thousands) that can be sold during one game at a price of \$$p$. If the concessionaire pays \$1 for each hot dog, how should the hot dogs be priced to maximize the profit per game?

52. *Maximum profit.* On a national tour of a rock band, the demand for T-shirts is given by

$$p = 15 - 4 \ln x \qquad 1 \le x \le 40$$

where x is the number of T-shirts (in thousands) that can be sold during a single concert at a price of \$$p$. If the shirts cost the band \$5 each, how should they be priced in order to maximize the profit per concert?

53. *Minimum average cost.* The cost of producing x units of a product is given by

$$C(x) = 600 + 100x - 100 \ln x \qquad x \ge 1$$

Find the minimum average cost.

54. *Minimum average cost.* The cost of producing x units of a product is given by

$$C(x) = 1{,}000 + 200x - 200 \ln x \qquad x \ge 1$$

Find the minimum average cost.

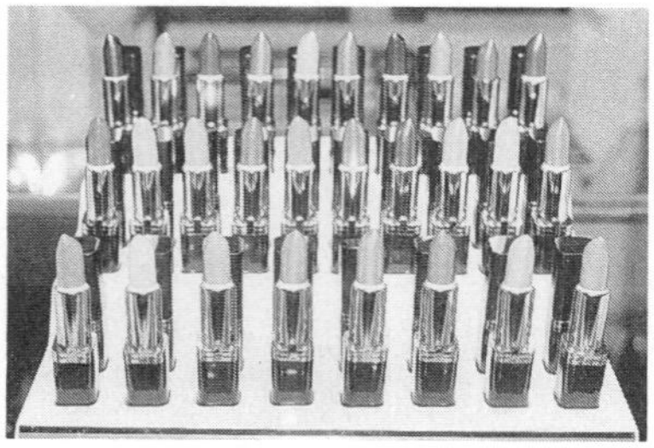

55. *Maximizing revenue.* A cosmetic company is planning the introduction and promotion of a new lipstick line. The marketing research department, after test marketing the new line in a carefully selected large city, found that the demand in that city is given approximately by

$$p = 10e^{-x} \qquad 0 \le x \le 2$$

where x thousand lipsticks were sold per week at a price of p dollars each.

(A) At what price will the weekly revenue $R(x) = xp$ be maximum? What is the maximum weekly revenue in the test city?

(B) Graph R for $0 \le x \le 2$.

56. *Maximizing revenue.* Repeat Problem 55 using the demand equation $p = 12e^{-x}$, $0 \leq x \leq 2$.

Life Sciences

57. *Blood pressure.* An experiment was set up to find a relationship between weight and systolic blood pressure in normal children. Using hospital records for 5,000 normal children, it was found that the systolic blood pressure was given approximately by

$$P(x) = 17.5(1 + \ln x) \qquad 10 \leq x \leq 100$$

where $P(x)$ is measured in millimeters of mercury and x is measured in pounds. What is the rate of change of blood pressure with respect to weight at the 40 pound weight level? At the 90 pound weight level?

58. *Blood pressure.* Graph the systolic blood pressure equation in Problem 57.

59. *Drug concentration.* The concentration of a drug in the bloodstream t hours after injection is given approximately by

$$C(t) = 4.35e^{-t} \qquad 0 \leq t \leq 5$$

where $C(t)$ is concentration in milligrams per milliliter.

(A) What is the rate of change of concentration after 1 hour? After 4 hours?
(B) Graph C.

60. *Water pollution.* The use of iodine crystals is a popular way of making small quantities of nonpotable water safe to drink. Crystals placed in a 1 ounce bottle of water will dissolve until the solution is saturated. After saturation, half of this solution is poured into a quart container of nonpotable water, and after about an hour, the water is usually safe to drink. The half empty 1 ounce bottle is then refilled to be used again in the same way. Suppose the concentration of iodine in the 1 ounce bottle t minutes after the crystals are introduced can be approximated by

$$C(t) = 250(1 - e^{-t}) \qquad t \geq 0$$

where $C(t)$ is the concentration of iodine in micrograms per milliliter.

(A) What is the rate of change of the concentration after 1 minute? After 4 minutes?
(B) Graph C for $0 \leq t \leq 5$.

Social Sciences

61. *Psychology—stimulus/response.* In psychology, the Weber–Fechner law for stimulus response is

$$R = k \ln\left(\frac{S}{S_0}\right)$$

where R is the response, S is the stimulus, and S_0 is the lowest level of stimulus that can be detected. Find dR/dS.

62. *Psychology—learning.* A mathematical model for the average of a group of people learning to type is given by

$$N(t) = 10 + 6 \ln t \qquad t \geqslant 1$$

where $N(t)$ is the number of words per minute typed after t hours of instruction and practice (2 hours per day, 5 days per week). What is the rate of learning after 10 hours of instruction and practice? After 100 hours?

SECTION 6-3 Chain Rule: General Form

- COMPOSITE FUNCTIONS
- CHAIN RULE
- GENERALIZED DERIVATIVE RULES
- OTHER LOGARITHMIC AND EXPONENTIAL FUNCTIONS

In Section 4-6, we introduced the power form of the chain rule:

$$D_x[u(x)]^n = n[u(x)]^{n-1}\, u'(x)$$

For example,

$$D_x(x^2 - 3)^5 = 5(x^2 - 3)^4\, D_x(x^2 - 3) = 10x(x^2 - 3)^4$$

This general power rule is a special case of one of the most important derivative rules of all—the *chain rule*—which will enable us to determine the derivatives of some fairly complicated functions in terms of derivatives of more elementary functions.

Suppose you were asked to find the derivative of

$$h(x) = \ln(2x + 1) \qquad \text{or} \qquad m(x) = e^{3x^2-1}$$

We have formulas for computing derivatives of $\ln x$ and e^x, and polynomial functions in general, but not in the indicated combinations. The chain rule is used to compute derivatives of functions that are *compositions* of more elementary functions whose derivatives are known. We start the section with a brief review of *composite functions*.

◆ COMPOSITE FUNCTIONS

Let us look at function h more closely:

$$h(x) = \ln(2x + 1)$$

The function h is a combination of the natural logarithm function and a linear function. To see this more clearly, let

$$y = f(u) = \ln u \qquad \text{and} \qquad u = g(x) = 2x + 1$$

Then we can express y as a function of x as follows:

$$y = f(u) = f[g(x)] = \ln(2x + 1) = h(x)$$

The function h is said to be the *composite* of the two simpler functions f and g. (Loosely speaking, we can think of h as a function of a function.) In general, we have the following:

Composite Functions

A function h is a **composite** of functions f and g if

$$h(x) = f[g(x)]$$

The domain of h is the set of all numbers x such that x is in the domain of g and g(x) is in the domain of f.

◆ EXAMPLE 8 Let $f(u) = e^u$, $g(x) = 3x^2 + 1$, and $m(v) = v^{3/2}$. Find:

(A) $f[g(x)]$ (B) $g[f(u)]$ (C) $m[g(x)]$

Solutions

(A) $f[g(x)] = e^{g(x)} = e^{3x^2+1}$
(B) $g[f(u)] = 3[f(u)]^2 + 1 = 3(e^u)^2 + 1 = 3e^{2u} + 1$
(C) $m[g(x)] = [g(x)]^{3/2} = (3x^2 + 1)^{3/2}$ ◆

PROBLEM 8 Let $f(u) = \ln u$, $g(x) = 2x^3 + 4$, and $m(v) = v^{-5}$. Find:

(A) $f[g(x)]$ (B) $g[f(u)]$ (C) $m[g(x)]$ ◆

◆ EXAMPLE 9 Write each function as a composition of the natural logarithm or exponential function and a polynomial.

(A) $y = \ln(x^3 - 2x^2 + 1)$ (B) $y = e^{x^2+4}$

Solutions

(A) Let

$$y = f(u) = \ln u$$
$$u = g(x) = x^3 - 2x^2 + 1$$

Check: $y = f[g(x)] = \ln[g(x)] = \ln(x^3 - 2x^2 + 1)$

(B) Let

$$y = f(u) = e^u$$
$$u = g(x) = x^2 + 4$$

Check: $y = f[g(x)] = e^{g(x)} = e^{x^2+4}$ ◆

PROBLEM 9 Repeat Example 9 for:

(A) $y = e^{2x^3+7}$ (B) $y = \ln(x^4 + 10)$ ◆

◆ CHAIN RULE

The word "chain" in the name *chain rule* comes from the fact that a function formed by composition (such as those in Example 8) involves a chain of functions—that is, a function of a function. The *chain rule* will enable us to compute the derivative of a composite function in terms of the derivatives of the functions making up the composition.

Suppose

$$y = h(x) = f[g(x)]$$

is a composite of f and g, where

$$y = f(u) \qquad \text{and} \qquad u = g(x)$$

We would like to express the derivative dy/dx in terms of the derivatives of f and g. From the definition of a derivative in increment form (see Section 5-5), we have

$$\begin{aligned}\frac{dy}{dx} &= \lim_{\Delta x \to 0} \frac{h(x + \Delta x) - h(x)}{\Delta x} \\ &= \lim_{\Delta x \to 0} \frac{\Delta y}{\Delta x} \qquad (1)\end{aligned}$$

Noting that

$$\frac{\Delta y}{\Delta x} = \frac{\Delta y}{\Delta u}\frac{\Delta u}{\Delta x} \qquad (2)$$

we substitute (2) into (1) to obtain

$$\frac{dy}{dx} = \lim_{\Delta x \to 0} \frac{\Delta y}{\Delta u}\frac{\Delta u}{\Delta x}$$

and reason that $\Delta u \to 0$ as $\Delta x \to 0$ so that

$$\begin{aligned}\frac{dy}{dx} &= \left(\lim_{\Delta u \to 0} \frac{\Delta y}{\Delta u}\right)\left(\lim_{\Delta x \to 0} \frac{\Delta u}{\Delta x}\right) \\ &= \frac{dy}{du}\frac{du}{dx}\end{aligned}$$

The result is correct under rather general conditions, and is called the *chain rule*, but our "derivation" is superficial, because it ignores a number of hidden problems. Since a formal proof of the chain rule is beyond the scope of this book, we simply state it as follows:

Chain Rule

If $y = f(u)$ and $u = g(x)$, define the composite function

$$y = h(x) = f[g(x)]$$

Then

$$\frac{dy}{dx} = \frac{dy}{du}\frac{du}{dx} \qquad \text{provided } \frac{dy}{du} \text{ and } \frac{du}{dx} \text{ exist}$$

◆ EXAMPLE 10 Find dy/dx, given:

(A) $y = \ln(x^2 - 4x + 2)$ (B) $y = e^{2x^3+5}$ (C) $y = (3x^2 + 1)^{3/2}$

Solutions (A) Let $y = \ln u$ and $u = x^2 - 4x + 2$. Then

$$\frac{dy}{dx} = \frac{dy}{du}\frac{du}{dx} = \frac{1}{u}(2x - 4)^{*}$$

$$= \frac{1}{x^2 - 4x + 2}(2x - 4) \qquad \text{Since } u = x^2 - 4x + 2$$

$$= \frac{2x - 4}{x^2 - 4x + 2}$$

(B) Let $y = e^u$ and $u = 2x^3 + 5$. Then

$$\frac{dy}{dx} = \frac{dy}{du}\frac{du}{dx} = e^u(6x^2)$$

$$= 6x^2e^{2x^3+5} \qquad \text{Since } u = 2x^3 + 5$$

* After some experience with the chain rule, the steps in the dashed boxes are usually done mentally.

(C) We have two methods:

Method 1. Chain rule—general form: Let $y = u^{3/2}$ and $u = 3x^2 + 1$. Then

$$\frac{dy}{dx} = \frac{dy}{du}\frac{du}{dx}$$
$$= \tfrac{3}{2}u^{1/2}(6x)$$
$$= \tfrac{3}{2}(3x^2 + 1)^{1/2}(6x) \qquad \text{Since } u = 3x^2 + 1$$
$$= 9x(3x^2 + 1)^{1/2} \quad \text{or} \quad 9x\sqrt{3x^2 + 1}$$

Method 2. Chain rule—power form (general power rule):

$$D_x(3x^2 + 1)^{3/2} = \tfrac{3}{2}(3x^2 + 1)^{1/2}\, D_x(3x^2 + 1) \qquad D_x[u(x)]^n = n[u(x)]^{n-1}\, D_x\, u(x)$$
$$= \tfrac{3}{2}(3x^2 + 1)^{1/2}(6x)$$
$$= 9x(3x^2 + 1)^{1/2} \quad \text{or} \quad 9x\sqrt{3x^2 + 1}$$

◆

The general power rule can be derived using the chain rule as follows: Given $y = [u(x)]^n$, let $y = v^n$ and $v = u(x)$. Then

$$\frac{dy}{dx} = \frac{dy}{dv}\frac{dv}{dx}$$
$$= nv^{n-1}\, D_x\, u(x)$$
$$= n[u(x)]^{n-1}\, D_x\, u(x) \qquad \text{Since } v = u(x)$$

PROBLEM 10 Find dy/dx, given:

(A) $y = e^{3x^4+6}$ (B) $y = \ln(x^2 + 9x + 4)$ (C) $y = (2x^3 + 4)^{-5}$ (Use two methods.)

◆

The chain rule can be extended to compositions of three or more functions. For example, if $y = f(w)$, $w = g(u)$, and $u = h(x)$, then

$$\frac{dy}{dx} = \frac{dy}{dw}\frac{dw}{du}\frac{du}{dx}$$

◆ EXAMPLE 11 For $y = h(x) = e^{1+(\ln x)^2}$, find dy/dx.

Solution Note that h is of the form $y = e^w$, where $w = 1 + u^2$ and $u = \ln x$. Thus,

$$\frac{dy}{dx} = \frac{dy}{dw}\frac{dw}{du}\frac{du}{dx}$$
$$= e^w(2u)\left(\frac{1}{x}\right)$$
$$= e^{1+u^2}(2u)\left(\frac{1}{x}\right) \qquad \text{Since } w = 1 + u^2$$
$$= e^{1+(\ln x)^2}(2\ln x)\left(\frac{1}{x}\right) \qquad \text{Since } u = \ln x$$
$$= \frac{2}{x}(\ln x)e^{1+(\ln x)^2}$$

◆

PROBLEM 11 For $y = h(x) = [\ln(1 + e^x)]^3$, find dy/dx. ◆

◆ GENERALIZED DERIVATIVE RULES

In practice, it is not necessary to introduce additional variables when using the chain rule, as we did in Examples 10 and 11. Instead, the chain rule can be used to extend the derivative rules for specific functions to general derivative rules for compositions. This is what we did above when we showed that the general power rule is a consequence of the chain rule. The same technique can be applied to functions of the form $y = e^{f(x)}$ and $y = \ln[f(x)]$ (see Problems 59 and 60 at the end of this section). The results are summarized in the following box:

General Derivative Rules

$$D_x[f(x)]^n = n[f(x)]^{n-1}f'(x) \qquad (3)$$

$$D_x \ln[f(x)] = \frac{1}{f(x)} f'(x) \qquad (4)$$

$$D_x e^{f(x)} = e^{f(x)}f'(x) \qquad (5)$$

For power, natural logarithm, or exponential forms, we can either use the chain rule discussed earlier or these special differentiation formulas based on the chain rule. Use whichever is easier for you. In Example 12, we will use the general derivative rules.

◆ EXAMPLE 12

(A) $D_x e^{2x} = e^{2x} D_x 2x$ Using (5)

$= e^{2x}(2) = 2e^{2x}$

(B) $D_x \ln(x^2 + 9) = \dfrac{1}{x^2 + 9} D_x(x^2 + 9)$ Using (4)

$= \dfrac{1}{x^2 + 9} 2x = \dfrac{2x}{x^2 + 9}$

(C) $D_x(1 + e^{x^2})^3 = 3(1 + e^{x^2})^2 D_x(1 + e^{x^2})$ Using (3)

$= 3(1 + e^{x^2})^2 e^{x^2} D_x x^2$ Using (5).

$= 3(1 + e^{x^2})^2 e^{x^2}(2x)$

$= 6xe^{x^2}(1 + e^{x^2})^2$ ◆

PROBLEM 12 Find:

(A) $D_x \ln(x^3 + 2x)$ (B) $D_x e^{3x^2+2}$ (C) $D_x(2 + e^{-x^2})^4$ ◆

◆ OTHER LOGARITHMIC AND EXPONENTIAL FUNCTIONS

In most applications involving logarithmic or exponential functions, the number e is the preferred base. However, there are situations where it is convenient to use a base other than e. Derivatives of $y = \log_b x$ and $y = b^x$ can be obtained by expressing these functions in terms of the natural logarithmic and exponential functions. We begin by finding a relationship between $\log_b x$ and $\ln x$ for any base b, $b > 0$ and $b \neq 1$.

$$y = \log_b x \qquad \text{Change to exponential form.}$$
$$b^y = x \qquad \text{Take the natural logarithm of both sides.}$$
$$\ln b^y = \ln x \qquad \text{Recall that } \ln b^y = y \ln b.$$
$$y \ln b = \ln x \qquad \text{Solve for } y.$$
$$y = \frac{1}{\ln b} \ln x$$

Thus,

$$\log_b x = \frac{1}{\ln b} \ln x \qquad \text{Change-of-base formula*} \qquad (6)$$

Differentiating both sides of (6), we have

$$\boldsymbol{D_x \log_b x = \frac{1}{\ln b} D_x \ln x = \frac{1}{\ln b}\left(\frac{1}{x}\right)}$$

◆ EXAMPLE 13 Find $f'(x)$ for:

(A) $f(x) = \log_2 x$ (B) $f(x) = \log(1 + x^3)$

Solutions (A) $f(x) = \log_2 x = \dfrac{1}{\ln 2} \ln x$ Using (6)

$$f'(x) = \frac{1}{\ln 2}\left(\frac{1}{x}\right)$$

(B) $f(x) = \log(1 + x^3)$ Recall that $\log r = \log_{10} r$.

$$= \frac{1}{\ln 10} \ln(1 + x^3) \qquad \text{Using (6)}$$

$$f'(x) = \frac{1}{\ln 10}\left(\frac{1}{1 + x^3} 3x^2\right) = \frac{1}{\ln 10}\left(\frac{3x^2}{1 + x^3}\right)$$ ◆

PROBLEM 13 Find $f'(x)$ for:

(A) $f(x) = \log x$ (B) $f(x) = \log_3(x + x^2)$ ◆

---

* Equation (6) is a special case of the **general change-of-base formula** for logarithms (which can be derived in the same way): $\log_b x = (\log_a x)/(\log_a b)$.

Now we want to find a relationship between b^x and e^x for any base b, $b > 0$ and $b \neq 1$.

$$\begin{aligned} y &= b^x && \text{Take the natural logarithm of both sides.} \\ \ln y &= \ln b^x \\ &= x \ln b && \text{If } \ln A = B \text{, then } A = e^B. \\ y &= e^{x \ln b} \end{aligned}$$

Thus,

$$b^x = e^{x \ln b} \qquad (7)$$

Differentiating both sides of (7), we have

$$D_x\, b^x = e^{x \ln b} \ln b = b^x \ln b$$

◆ EXAMPLE 14 Find $f'(x)$ for:

(A) $f(x) = 2^x$ (B) $f(x) = 10^{x^5+x}$

Solutions

(A) $f(x) = 2^x = e^{x \ln 2}$ Using (7)
$f'(x) = e^{x \ln 2} \ln 2 = 2^x \ln 2$

(B) $f(x) = 10^{x^5+x} = e^{(x^5+x)\ln 10}$ Using (7)
$f'(x) = e^{(x^5+x)\ln 10}(5x^4 + 1) \ln 10$
$= 10^{x^5+x}(5x^4 + 1) \ln 10$ ◆

PROBLEM 14 Find $f'(x)$ for:

(A) $f(x) = 5^x$ (B) $f(x) = 4^{x^2+3x}$ ◆

Answers to Matched Problems

8. (A) $\ln(2x^3 + 4)$ (B) $2(\ln u)^3 + 4$ (C) $(2x^3 + 4)^{-5}$
9. (A) $y = f(u) = e^u$; $u = g(x) = 2x^3 + 7$
   (B) $y = f(u) = \ln u$; $u = g(x) = x^4 + 10$
10. (A) $12x^3e^{3x^4+6}$ (B) $\dfrac{2x + 9}{x^2 + 9x + 4}$ (C) $-30x^2(2x^3 + 4)^{-6}$
11. $\dfrac{3e^x[\ln(1 + e^x)]^2}{1 + e^x}$
12. (A) $\dfrac{3x^2 + 2}{x^3 + 2x}$ (B) $6xe^{3x^2+2}$ (C) $-8xe^{-x^2}(2 + e^{-x^2})^3$
13. (A) $\dfrac{1}{\ln 10}\left(\dfrac{1}{x}\right)$ (B) $\dfrac{1}{\ln 3}\left(\dfrac{1 + 2x}{x + x^2}\right)$
14. (A) $5^x \ln 5$ (B) $4^{x^2+3x}(2x + 3) \ln 4$

EXERCISE 6-3

A *Write each composite function in the form $y = f(u)$ and $u = g(x)$.*

1. $y = (2x + 5)^3$
2. $y = (3x - 7)^5$
3. $y = \ln(2x^2 + 7)$
4. $y = \ln(x^2 - 2x + 5)$
5. $y = e^{x^2-2}$
6. $y = e^{3x^3+5x}$

Express y in terms of x. Use the chain rule to find dy/dx, and then express dy/dx in terms of x.

7. $y = u^2; \quad u = 2 + e^x$
8. $y = u^3; \quad u = 3 - \ln x$
9. $y = e^u; \quad u = 2 - x^4$
10. $y = e^u; \quad u = x^6 + 5x^2$
11. $y = \ln u; \quad u = 4x^5 - 7$
12. $y = \ln u; \quad u = 2 + 3x^4$

Find each derivative.

13. $D_x \ln(x - 3)$
14. $D_w \ln(w + 100)$
15. $D_t \ln(3 - 2t)$
16. $D_y \ln(4 - 5y)$
17. $D_x\, 3e^{2x}$
18. $D_y\, 2e^{3y}$
19. $D_t\, 2e^{-4t}$
20. $D_r\, 6e^{-3r}$

B

21. $D_x\, 100e^{-0.03x}$
22. $D_t\, 1{,}000e^{0.06t}$
23. $D_x \ln(x + 1)^4$
24. $D_x \ln(x + 1)^{-3}$
25. $D_x(2e^{2x} - 3e^x + 5)$
26. $D_t(1 + e^{-t} - e^{-2t})$
27. $D_x\, e^{3x^2-2x}$
28. $D_x\, e^{x^3-3x^2+1}$
29. $D_t \ln(t^2 + 3t)$
30. $D_x \ln(x^3 - 3x^2)$
31. $D_x \ln(x^2 + 1)^{1/2}$
32. $D_x \ln(x^4 + 5)^{3/2}$
33. $D_t[\ln(t^2 + 1)]^4$
34. $D_w\, [\ln(w^3 - 1)]^2$
35. $D_x(e^{2x} - 1)^4$
36. $D_x(e^{x^2} + 3)^5$
37. $D_x \dfrac{e^{2x}}{x^2 + 1}$
38. $D_x \dfrac{e^{x+1}}{x + 1}$
39. $D_x(x^2 + 1)e^{-x}$
40. $D_x(1 - x)e^{2x}$
41. $D_x(e^{-x} \ln x)$
42. $D_x \dfrac{\ln x}{e^x + 1}$
43. $D_x \dfrac{1}{\ln(1 + x^2)}$
44. $D_x \dfrac{1}{\ln(1 - x^3)}$
45. $D_x \sqrt[3]{\ln(1 - x^2)}$
46. $D_t \sqrt[5]{\ln(1 - t^5)}$

C *Sketch the graph of $y = f(x)$.*

47. $f(x) = 1 - e^{-x}$
48. $f(x) = 2 - 3e^{-2x}$
49. $f(x) = \ln(1 - x)$
50. $f(x) = \ln(2x + 4)$
51. $f(x) = e^{-(1/2)x^2}$
52. $f(x) = \ln(x^2 + 4)$

Express y in terms of x. Use the chain rule to find dy/dx, and express dy/dx in terms of x.

53. $y = 1 + w^2; \quad w = \ln u; \quad u = 2 + e^x$
54. $y = \ln w; \quad w = 1 + e^u; \quad u = x^2$

Find each derivative.

55. $D_x \log_2(3x^2 - 1)$
56. $D_x \log(x^3 - 1)$
57. $D_x\, 10^{x^2+x}$
58. $D_x\, 8^{1-2x^2}$
59. Use the chain rule to derive the formula: $D_x \ln[f(x)] = \dfrac{1}{f(x)} f'(x)$
60. Use the chain rule to derive the formula: $D_x\, e^{f(x)} = e^{f(x)} f'(x)$

Business & Economics

61. *Maximum revenue.* Suppose the price–demand equation for x units of a commodity is determined from empirical data to be

$$p = 100e^{-0.05x}$$

where x units are sold per day at a price of \$p each. Find the production level and price that maximize revenue. What is the maximum revenue?

62. *Maximum revenue.* Repeat Problem 61 using the price–demand equation

$$p = 10e^{-0.04x}$$

63. *Salvage value.* The salvage value S (in dollars) of a company airplane after t years is estimated to be given by

$$S(t) = 300{,}000e^{-0.1t}$$

What is the rate of depreciation (in dollars per year) after 1 year? 5 years? 10 years?

64. *Resale value.* The resale value R (in dollars) of a company car after t years is estimated to be given by

$$R(t) = 20{,}000e^{-0.15t}$$

What is the rate of depreciation (in dollars per year) after 1 year? 2 years? 3 years?

65. *Promotion and maximum profit.* A recording company has produced a new compact disk featuring a very popular recording group. Before launching a national sales campaign, the marketing research department chose to test market the disk in a bellwether city. Their interest is in determining the length of a sales campaign that will maximize total profits. From empirical data, the research department estimates that the proportion of a target group of 50,000 persons buying the disk after t days of television promotion is given by $1 - e^{-0.03t}$. If \$4 is received for each disk sold, then the total revenue after t days of promotion will be approximated by

$$R(t) = (4)(50{,}000)(1 - e^{-0.03t}) \qquad t \geq 0$$

Television promotion costs are

$$C(t) = 4{,}000 + 3{,}000t \qquad t \geq 0$$

(A) How many days of television promotion should be used to maximize total profit? What is the maximum total profit? What percentage of the target market will have purchased the disk when the maximum profit is reached?
(B) Graph the profit function.

66. *Promotion and maximum profit.* Repeat Problem 65 using the revenue equation

$$R(t) = (3)(60{,}000)(1 - e^{-0.04t})$$

Life Sciences

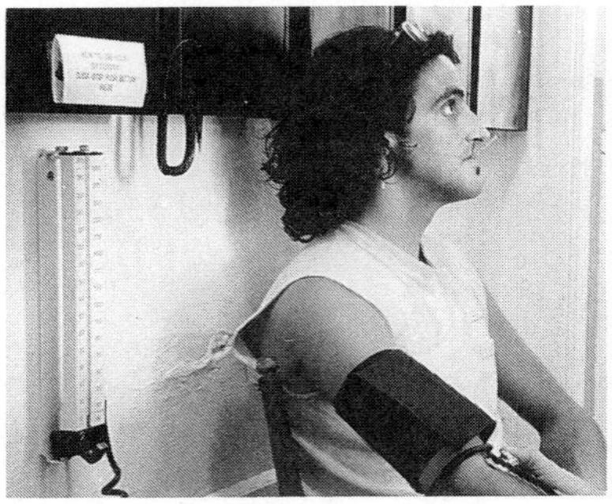

67. *Blood pressure and age.* A research group using hospital records developed the following approximate mathematical model relating systolic blood pressure and age:

$$P(x) = 40 + 25 \ln(x + 1) \qquad 0 \leq x \leq 65$$

where $P(x)$ is pressure measured in millimeters of mercury and x is age in years. What is the rate of change of pressure at the end of 10 years? At the end of 30 years? At the end of 60 years?

68. *Biology.* A yeast culture at room temperature (68°F) is placed in a refrigerator maintaining a constant temperature of 38°F. After t hours, the temperature T of the culture is given approximately by

$$T = 30e^{-0.58t} + 38 \qquad t \geq 0$$

What is the rate of change of temperature of the culture at the end of 1 hour? At the end of 4 hours?

69. *Bacterial growth.* A single cholera bacterium divides every 0.5 hour to produce two complete cholera bacteria. If we start with a colony of 5,000 bacteria, then after t hours there will be

$$A(t) = 5{,}000 \cdot 2^{2t}$$

bacteria. Find $A'(t)$, $A'(1)$, and $A'(5)$, and interpret the results.

70. *Bacterial growth.* Repeat Problem 69 for a starting colony of 1,000 bacteria where a single bacterium divides every 0.25 hour.

Social Sciences

71. *Sociology.* Daniel Lowenthal, a sociologist at Columbia University, made a 5 year study on the sale of popular records relative to their position in the top 20. He found that the average number of sales $N(n)$ of the nth ranking record was given approximately by

$$N(n) = N_1 e^{-0.09(n-1)} \qquad 1 \leq n \leq 20$$

where N_1 was the number of sales of the number one record on the list at a given time. Graph N for $N_1 = 1{,}000{,}000$ records.

72. *Political science.* Thomas W. Casstevens, a political scientist at Oakland University, has studied legislative turnover. He (with others) found that the number $N(t)$ of continuously serving members of an elected legislative body remaining t years after an election is given approximately by a function of the form

$$N(t) = N_0 e^{-ct}$$

In particular, for the 1965 election for the U.S. House of Representatives, it was found that

$$N(t) = 434e^{-0.0866t}$$

What is the rate of change after 2 years? After 10 years?

SECTION 6-4 Implicit Differentiation

- SPECIAL FUNCTION NOTATION
- IMPLICIT DIFFERENTIATION

SPECIAL FUNCTION NOTATION

The equation

$$y = 2 - 3x^2 \tag{1}$$

defines a function f with y as a dependent variable and x as an independent variable. Using function notation, we would write

$$y = f(x) \qquad \text{or} \qquad f(x) = 2 - 3x^2$$

In order to reduce to a minimum the number of symbols involved in a discussion, we will often write equation (1) in the form

$$y = 2 - 3x^2 = y(x)$$

where y is *both* a dependent variable and a function symbol. This is a convenient notation and no harm is done as long as one is aware of the double role of y. Other examples are

$$x = 2t^2 - 3t + 1 = x(t)$$
$$z = \sqrt{u^2 - 3u} = z(u)$$
$$r = \frac{1}{(s^2 - 3s)^{2/3}} = r(s)$$

This type of notation will simplify much of the discussion and work that follows.

Until now we have considered functions involving only one independent variable. There is no reason to stop there. The concept can be generalized to functions involving two or more independent variables, and this will be done in detail in Chapter 9. For now, we will "borrow" the notation for a function involving two independent variables. For example,

$$F(x, y) = x^2 - 2xy + 3y^2 - 5$$

specifies a function F involving two independent variables.

◆ IMPLICIT DIFFERENTIATION

Consider the equation

$$3x^2 + y - 2 = 0 \tag{2}$$

and the equation obtained by solving (2) for y in terms of x,

$$y = 2 - 3x^2 \tag{3}$$

Both equations define the same function using x as the independent variable and y as the dependent variable. For (3), we can write

$$y = f(x)$$

where

$$f(x) = 2 - 3x^2 \tag{4}$$

and we have an **explicit** (clearly stated) rule that enables us to determine y for each value of x. On the other hand, the y in equation (2) is the same y as in equation (3), and equation (2) **implicitly** gives (implies though does not plainly express) y as a function of x. Thus, we say that equations (3) and (4) define the function f explicitly, and equation (2) defines f implicitly.

The direct use of an equation that defines a function implicitly to find the derivative of the dependent variable with respect to the independent variable is called **implicit differentiation.** Let us differentiate (2) implicitly and (3) directly, and compare results.

Starting with

$$3x^2 + y - 2 = 0$$

we think of y as a function of x—that is, $y = y(x)$—and write

$$3x^2 + y(x) - 2 = 0$$

Then we differentiate both sides with respect to x:

$$D_x[3x^2 + y(x) - 2] = D_x\, 0$$
$$D_x\, 3x^2 + D_x\, y(x) - D_x\, 2 = 0$$
$$6x + y' - 0 = 0$$

Since y is a function of x, but is not explicitly given, we simply write $D_x\, y(x) = y'$ to indicate its derivative.

Now, we solve for y':

$$y' = -6x$$

Note that we get the same result if we start with equation (3) and differentiate directly:

$$y = 2 - 3x^2$$
$$y' = -6x$$

Why are we interested in implicit differentiation? In general, why do we not solve for y in terms of x and differentiate directly? The answer is that there are many equations of the form

$$F(x, y) = 0 \tag{5}$$

that are either difficult or impossible to solve for y explicitly in terms of x (try it for $x^2y^5 - 3xy + 5 = 0$ or for $e^y - y = 3x$, for example). But it can be shown that, under fairly general conditions on F, equation (5) will define one or more functions where y is a dependent variable and x is an independent variable. To find y' under these conditions, we differentiate (5) implicitly.

◆ EXAMPLE 15 Given

$$F(x, y) = x^2 + y^2 - 25 = 0 \tag{6}$$

find y' and the slope of the graph at $x = 3$.

Solution We start with the graph of $x^2 + y^2 - 25 = 0$ (a circle, as shown in the margin) so that we can interpret our results geometrically. From the graph, it is clear that equation (6) does not define a function. But with a suitable restriction on the variables, equation (6) can define two or more functions. For example, the upper half and the lower half of the circle each define a function. A point on each half-circle that corresponds to $x = 3$ is found by substituting $x = 3$ into (6) and solving for y:

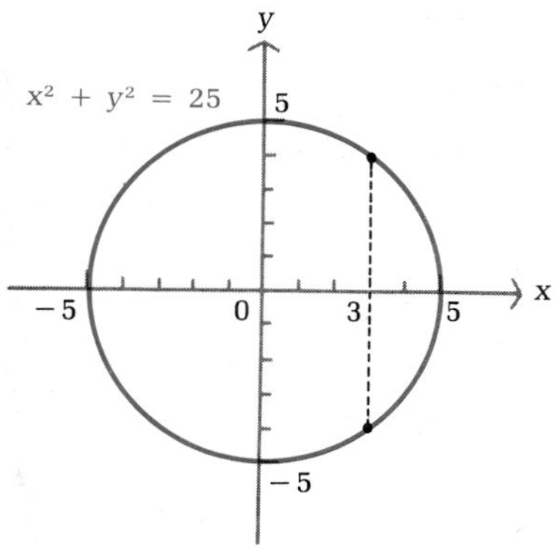

$$x^2 + y^2 - 25 = 0$$
$$(3)^2 + y^2 = 25$$
$$y^2 = 16$$
$$y = \pm 4$$

Thus, the point (3, 4) is on the upper half-circle and the point (3, −4) is on the lower half-circle. We will use these results in a moment. We now differentiate (6) implicitly, treating y as a function of x; that is, $y = y(x)$:

$$x^2 + y^2 - 25 = 0$$
$$x^2 + [y(x)]^2 - 25 = 0$$
$$D_x\{x^2 + [y(x)]^2 - 25\} = D_x\, 0$$
$$D_x\, x^2 + D_x[y(x)]^2 - D_x\, 25 = 0 \qquad \text{Use the chain rule.}$$
$$2x + 2[y(x)]^{2-1}y'(x) - 0 = 0$$
$$2x + 2yy' = 0 \qquad \text{Solve for } y' \text{ in terms of } x \text{ and } y.$$
$$y' = -\frac{2x}{2y}$$
$$y' = -\frac{x}{y} \qquad \text{Leave the answer in terms of } x \text{ and } y.$$

We have found y' without first solving $x^2 + y^2 - 25 = 0$ for y in terms of x. And by leaving y' in terms of x and y, we can use $y' = -x/y$ to find y' for *any* point on the graph of $x^2 + y^2 - 25 = 0$ (except where $y = 0$). In particular, for $x = 3$, we found that $(3, 4)$ and $(3, -4)$ are on the graph; thus, the slope of the graph at $(3, 4)$ is

$$y'|_{(3,4)} = -\tfrac{3}{4} \qquad \text{The slope of the graph at (3, 4)}$$

and the slope at $(3, -4)$ is

$$y'|_{(3,-4)} = -\tfrac{3}{-4} = \tfrac{3}{4} \qquad \text{The slope of the graph at (3, −4)}$$

The symbol

$$y'|_{(a,b)}$$

is used to indicate that we are evaluating y' at $x = a$ and $y = b$.

The results are interpreted geometrically on the original graph as follows:

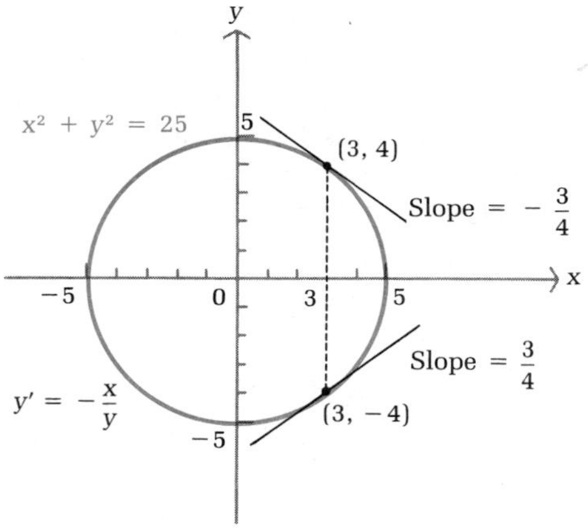

◆

In Example 15, the fact that y' is given in terms of both x and y is not a great disadvantage. We have only to make certain that when we want **to evaluate y' for a particular value of x and y, say, (x_0, y_0), the ordered pair must satisfy the original equation.**

PROBLEM 15 Graph $x^2 + y^2 - 169 = 0$, find y' by implicit differentiation, and find the slope of the graph when $x = 5$. ◆

◆ EXAMPLE 16 Find the equation(s) of the tangent line(s) to the graph of

$$y - xy^2 + x^2 + 1 = 0 \tag{7}$$

at the point(s) where $x = 1$.

Solution We first find y when $x = 1$:

$$y - xy^2 + x^2 + 1 = 0$$
$$y - (1)y^2 + (1)^2 + 1 = 0$$
$$y - y^2 + 2 = 0$$
$$y^2 - y - 2 = 0$$
$$(y - 2)(y + 1) = 0$$
$$y = -1, 2$$

Thus, there are two points on the graph of (7) where $x = 1$; namely, $(1, -1)$ and $(1, 2)$. We next find the slope of the graph at these two points by differentiating (7) implicitly:

$$y - xy^2 + x^2 + 1 = 0$$
$$D_x y - D_x xy^2 + D_x x^2 + D_x 1 = D_x 0$$
$$y' - (x \cdot 2yy' + y^2) + 2x = 0$$
$$y' - 2xyy' - y^2 + 2x = 0$$
$$y' - 2xyy' = y^2 - 2x$$
$$(1 - 2xy)y' = y^2 - 2x$$
$$y' = \frac{y^2 - 2x}{1 - 2xy}$$

Use the product rule and the chain rule for $D_x xy^2$.

Solve for y' by getting all terms involving y' on one side.

Now, find the slope at each point:

$$y'|_{(1,-1)} = \frac{(-1)^2 - 2(1)}{1 - 2(1)(-1)} = \frac{1 - 2}{1 + 2} = \frac{-1}{3} = -\frac{1}{3}$$
$$y'|_{(1,2)} = \frac{(2)^2 - 2(1)}{1 - 2(1)(2)} = \frac{4 - 2}{1 - 4} = \frac{2}{-3} = -\frac{2}{3}$$

Equation of the tangent line at $(1, -1)$:

$$y - y_1 = m(x - x_1)$$
$$y + 1 = -\tfrac{1}{3}(x - 1)$$
$$y + 1 = -\tfrac{1}{3}x + \tfrac{1}{3}$$
$$y = -\tfrac{1}{3}x - \tfrac{2}{3}$$

Equation of the tangent line at $(1, 2)$:

$$y - y_1 = m(x - x_1)$$
$$y - 2 = -\tfrac{2}{3}(x - 1)$$
$$y - 2 = -\tfrac{2}{3}x + \tfrac{2}{3}$$
$$y = -\tfrac{2}{3}x + \tfrac{8}{3}$$

◆

PROBLEM 16 Repeat Example 16 for $x^2 + y^2 - xy - 7 = 0$ at $x = 1$. ◆

◆ EXAMPLE 17 Find x' for $x = x(t)$ defined implicitly by

$$t \ln x = xe^t - 1$$

and evaluate x' at $(t, x) = (0, 1)$.

Solution It is important to remember that x is the dependent variable and t is the independent variable. Therefore, we differentiate both sides of the equation with respect to t (using product and chain rules where appropriate), and then solve for x':

$$t \ln x = xe^t - 1 \qquad \text{Differentiate implicitly with respect to } t.$$

$$D_t(t \ln x) = D_t(xe^t) - D_t\, 1$$

$$t\frac{x'}{x} + \ln x = xe^t + x'e^t \qquad \text{Clear fractions.}$$

$$x\left(t\frac{x'}{x}\right) + x \ln x = x \cdot xe^t + xe^t x' \qquad x \neq 0$$

$$tx' + x \ln x = x^2e^t + xe^tx' \qquad \text{Solve for } x'.$$

$$tx' - xe^tx' = x^2e^t - x \ln x \qquad \text{Factor out } x'.$$

$$(t - xe^t)x' = x^2e^t - x \ln x$$

$$x' = \frac{x^2e^t - x \ln x}{t - xe^t}$$

Now, we evaluate x' at (t, x) = (0, 1), as requested:

$$x'|_{(0,1)} = \frac{(1)^2e^0 - 1 \ln 1}{0 - 1e^0}$$

$$= \frac{1}{-1} = -1$$

◆

PROBLEM 17 Find x' for x = x(t) defined implicitly by

$$1 + x \ln t = te^x$$

and evaluate x' at (t, x) = (1, 0). ◆

Answers to Matched Problems

15. $y' = -x/y$; when $x = 5$, $y = \pm 12$, thus, $y'|_{(5,12)} = -\frac{5}{12}$ and $y'|_{(5,-12)} = \frac{5}{12}$
16. $y' = \dfrac{y - 2x}{2y - x}$; $y = \frac{4}{5}x - \frac{14}{5}$, $y = \frac{1}{5}x + \frac{14}{5}$
17. $x' = \dfrac{te^x - x}{t \ln t - t^2e^x}$; $x'|_{(1,0)} = -1$

EXERCISE 6-4

In Problems 1–18, find y' without solving for y in terms of x (use implicit differentiation). Evaluate y' at the indicated point.

A

1. $y - 3x^2 + 5 = 0$; (1, −2)
2. $3x^4 + y - 2 = 0$; (1, −1)
3. $y^2 - 3x^2 + 8 = 0$; (2, 2)
4. $3y^2 + 2x^3 - 14 = 0$; (1, 2)
5. $y^2 + y - x = 0$; (2, 1)
6. $2y^3 + y^2 - x = 0$; (3, 1)

B

7. $xy - 6 = 0$; $(2, 3)$
8. $3xy - 2x - 2 = 0$; $(2, 1)$
9. $2xy + y + 2 = 0$; $(-1, 2)$
10. $2y + xy - 1 = 0$; $(-1, 1)$
11. $x^2y - 3x^2 - 4 = 0$; $(2, 4)$
12. $2x^3y - x^3 + 5 = 0$; $(-1, 3)$
13. $e^y = x^2 + y^2$; $(1, 0)$
14. $x^2 - y = 4e^y$; $(2, 0)$
15. $x^3 - y = \ln y$; $(1, 1)$
16. $\ln y = 2y^2 - x$; $(2, 1)$
17. $x \ln y + 2y = 2x^3$; $(1, 1)$
18. $xe^y - y = x^2 - 2$; $(2, 0)$

In Problems 19 and 20, find x′ for x = x(t) defined implicitly by the given equation. Evaluate x′ at the indicated point.

19. $x^2 - t^2x + t^3 + 11 = 0$; $(-2, 1)$
20. $x^3 - tx^2 - 4 = 0$; $(-3, -2)$

Find the equation(s) of the tangent line(s) to the graphs of the indicated equations at the point(s) with abscissas as indicated.

21. $xy - x - 4 = 0$; $x = 2$
22. $3x + xy + 1 = 0$; $x = -1$
23. $y^2 - xy - 6 = 0$; $x = 1$
24. $xy^2 - y - 2 = 0$; $x = 1$

C

Find y′ and the slope of the tangent line to the graph of each equation at the indicated point.

25. $(1 + y)^3 + y = x + 7$; $(2, 1)$
26. $(y - 3)^4 - x = y$; $(-3, 4)$
27. $(x - 2y)^3 = 2y^2 - 3$; $(1, 1)$
28. $(2x - y)^4 - y^3 = 8$; $(-1, -2)$
29. $\sqrt{7 + y^2} - x^3 + 4 = 0$; $(2, 3)$
30. $6\sqrt{y^3 + 1} - 2x^{3/2} - 2 = 0$; $(4, 2)$
31. $\ln(xy) = y^2 - 1$; $(1, 1)$
32. $e^{xy} - 2x = y + 1$; $(0, 0)$

APPLICATIONS

Business & Economics

For the demand equations in Problems 33–36, find the rate of change of p with respect to x by differentiating implicitly (x is the number of items that can be sold at a price of \$p.)

33. $x = p^2 - 2p + 1{,}000$
34. $x = p^3 - 3p^2 + 200$
35. $x = \sqrt{10{,}000 - p^2}$
36. $x = \sqrt[3]{1{,}500 - p^3}$

Life Sciences

37. Biophysics. In biophysics, the equation

$$(L + m)(V + n) = k$$

is called the *fundamental equation of muscle contraction*, where m, n, and k are constants, and V is the velocity of the shortening of muscle fibers for a muscle subjected to a load of L. Find dL/dV using implicit differentiation.

38. Biophysics. In Problem 37, find dV/dL using implicit differentiation.

SECTION 6-5 Related Rates

We start this discussion of *related rates* with an example, out of which will come a list of helpful suggestions for solving related rates problems in general.

◆ EXAMPLE 18

A 26 foot ladder is placed against a wall. If the top of the ladder is sliding down the wall at 2 feet per second, at what rate is the bottom of the ladder moving away from the wall when the bottom of the ladder is 10 feet away from the wall?

Solution

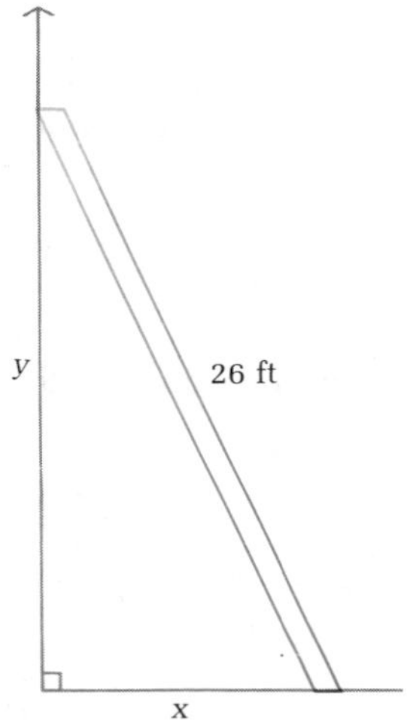

Many people reason that since the ladder is of constant length, the bottom of the ladder will move away from the wall at the same rate that the top of the ladder is moving down the wall. This is not the case, as we will see.

At any moment in time, let x be the distance of the bottom of the ladder from the wall, and let y be the distance of the top of the ladder on the wall (see the figure in the margin). Both x and y are changing with respect to time and can be thought of as functions of time; that is, $x = x(t)$ and $y = y(t)$. Furthermore, x and y are related by the Pythagorean relationship:

$$x^2 + y^2 = 26^2 \tag{1}$$

Differentiating (1) implicitly with respect to time t, and using the chain rule where appropriate, we obtain

$$2x\frac{dx}{dt} + 2y\frac{dy}{dt} = 0 \tag{2}$$

The rates dx/dt and dy/dt are related by equation (2); hence, this type of problem is referred to as a **related rates problem.**

Now our problem is to find dx/dt when $x = 10$ feet, given that $dy/dt = -2$ (y is decreasing at a constant rate of 2 feet per second). We have all the quantities we need in equation (2) to solve for dx/dt, except y. When $x = 10$, y can be found using (1):

$$10^2 + y^2 = 26^2$$
$$y = \sqrt{26^2 - 10^2} = 24 \text{ feet}$$

Substitute $dy/dt = -2$, $x = 10$, and $y = 24$ into (2); then solve for dx/dt:

$$2(10)\frac{dx}{dt} + 2(24)(-2) = 0$$
$$\frac{dx}{dt} = \frac{-2(24)(-2)}{2(10)} = 4.8 \text{ feet per second}$$

Thus, the bottom of the ladder is moving away from the wall at a rate of 4.8 feet per second. ◆

PROBLEM 18

A 26 foot ladder is placed against a wall. If the bottom of the ladder is moving away from the wall at 3 feet per second, at what rate is the top moving down when the top of the ladder is 24 feet up the wall? ◆

Suggestions for Solving Related Rates Problems

Step 1. Sketch a figure if helpful.

Step 2. Identify all relevant variables, including those whose rates are given and those whose rates are to be found.

Step 3. Express all given rates and rates to be found as derivatives.

Step 4. Find an equation connecting the variables in step 2.

Step 5. Implicitly differentiate the equation found in step 4, using the chain rule where appropriate, and substitute in all given values.

Step 6. Solve for the derivative that will give the unknown rate.

◆ EXAMPLE 19 Suppose two motor boats leave from the same point at the same time. If one travels north at 15 miles per hour and the other travels east at 20 miles per hour, how fast will the distance between them be changing after 2 hours?

Solution First, draw a picture:

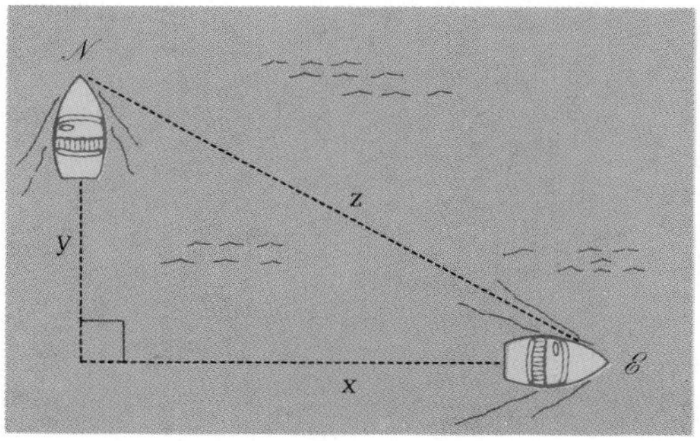

All variables, x, y, and z, are changing with time. Hence, they can be thought of as functions of time; $x = x(t)$, $y = y(t)$, and $z = z(t)$, given implicitly. It now makes sense to take derivatives of each variable with respect to time. From the Pythagorean theorem,

$$z^2 = x^2 + y^2 \tag{3}$$

We also know that

$$\frac{dx}{dt} = 20 \text{ miles per hour} \qquad \text{and} \qquad \frac{dy}{dt} = 15 \text{ miles per hour}$$

We would like to find dz/dt at the end of 2 hours; that is, when $x = 40$ miles and $y = 30$ miles. To do this, we differentiate both sides of (3) with respect to t and solve for dz/dt:

$$2z \frac{dz}{dt} = 2x \frac{dx}{dt} + 2y \frac{dy}{dt} \tag{4}$$

We have everything we need except z. When $x = 40$ and $y = 30$, we find z from (3) to be 50. Substituting the known quantities into (4), we obtain

$$2(50) \frac{dz}{dt} = 2(40)(20) + 2(30)(15)$$
$$\frac{dz}{dt} = 25 \text{ miles per hour}$$

Thus, the boats will be separating at a rate of 25 miles per hour. ◆

PROBLEM 19 Repeat Example 19 for the situation at the end of 3 hours. ◆

◆ EXAMPLE 20 Suppose a point is moving on the graph of $x^2 + y^2 = 25$. When the point is at $(-3, 4)$, its x coordinate is increasing at the rate of 0.4 unit per second. How fast is the y coordinate changing at that moment?

Solution Since both x and y are changing with respect to time, we can think of each as a function of time:

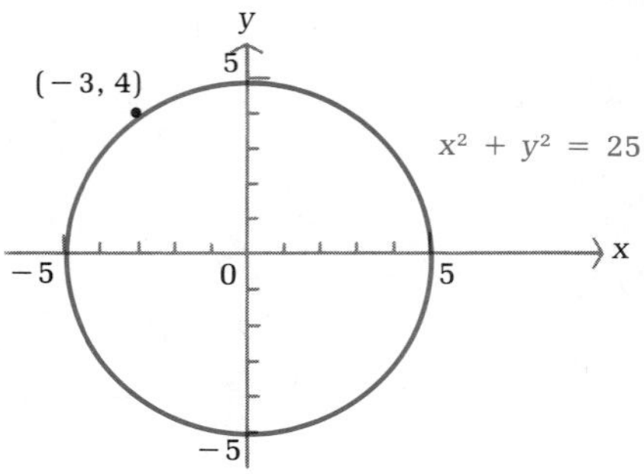

$$x = x(t) \qquad \text{and} \qquad y = y(t)$$

but restricted so that

$$x^2 + y^2 = 25 \tag{5}$$

Our problem is now to find dy/dt, given $x = -3$, $y = 4$, and $dx/dt = 0.4$. Implicitly differentiating both sides of (5) with respect to t, we have

$$x^2 + y^2 = 25$$
$$2x \frac{dx}{dt} + 2y \frac{dy}{dt} = 0 \qquad \text{Divide both sides by 2.}$$
$$x \frac{dx}{dt} + y \frac{dy}{dt} = 0 \qquad \text{Substitute } x = -3,\ y = 4, \text{ and } dx/dt = 0.4, \text{ and solve for } dy/dt.$$
$$(-3)(0.4) + 4 \frac{dy}{dt} = 0$$
$$\frac{dy}{dt} = 0.3 \text{ unit per second}$$

◆

PROBLEM 20 A point is moving on the graph of $y^3 = x^2$. When the point is at $(-8, 4)$, its y coordinate is decreasing at 2 units per second. How fast is the x coordinate changing at that moment? ◆

◆ EXAMPLE 21 Suppose that for a company manufacturing transistor radios, the cost, revenue, and profit equations are given by

$$C = 5{,}000 + 2x \qquad \text{Cost equation}$$

$$R = 10x - \frac{x^2}{1{,}000} \qquad \text{Revenue equation}$$

$$P = R - C \qquad \text{Profit equation}$$

where the production output in 1 week is x radios. If production is increasing at the rate of 500 radios per week when production is 2,000 radios, find the rate of increase in:

(A) Cost (B) Revenue (C) Profit

Solutions If production x is a function of time (it must be, since it is changing with respect to time), then C, R, and P must also be functions of time. These functions are implicitly (rather than explicitly) given. Letting t represent time in weeks, we differentiate both sides of each of the three equations above with respect to t, and then substitute $x = 2{,}000$ and $dx/dt = 500$ to find the desired rates.

(A)

$$C = 5{,}000 + 2x \qquad \text{Think: } C = C(t) \text{ and } x = x(t).$$

$$\frac{dC}{dt} = \frac{d}{dt}(5{,}000) + \frac{d}{dt}(2x) \qquad \text{Differentiate both sides with respect to } t.$$

$$\frac{dC}{dt} = 0 + 2\frac{dx}{dt} = 2\frac{dx}{dt}$$

Since $dx/dt = 500$ when $x = 2{,}000$,

$$\frac{dC}{dt} = 2(500) = \$1{,}000 \text{ per week}$$

Cost is increasing at a rate of \$1,000 per week.

(B)

$$R = 10x - \frac{x^2}{1{,}000}$$

$$\frac{dR}{dt} = \frac{d}{dt}(10x) - \frac{d}{dt}\frac{x^2}{1{,}000}$$

$$\frac{dR}{dt} = 10\frac{dx}{dt} - \frac{x}{500}\frac{dx}{dt}$$

$$\frac{dR}{dt} = \left(10 - \frac{x}{500}\right)\frac{dx}{dt}$$

Since $dx/dt = 500$ when $x = 2{,}000$,

$$\frac{dR}{dt} = \left(10 - \frac{2{,}000}{500}\right)(500) = \$3{,}000 \text{ per week}$$

Revenue is increasing at a rate of \$3,000 per week.

(C) $P = R - C$

$$\frac{dP}{dt} = \frac{dR}{dt} - \frac{dC}{dt}$$

$= \$3,000 - \$1,000$ Results from parts A and B

$= \$2,000$ per week

Profit is increasing at a rate of \$2,000 per week. ◆

PROBLEM 21 Repeat Example 21 for a production level of 6,000 radios per week. ◆

Answers to Matched Problems

18. $dy/dt = -1.25$ ft/sec 19. $dz/dt = 25$ mi/hr
20. $dx/dt = 6$ units/sec
21. (A) $dC/dt = \$1,000/\text{wk}$ (B) $dR/dt = -\$1,000/\text{wk}$
(C) $dP/dt = -\$2,000/\text{wk}$

EXERCISE 6-5

A

In Problems 1–6, assume $x = x(t)$ and $y = y(t)$. Find the indicated rate, given the other information.

1. $y = 2x^2 - 1$; $dy/dt = ?$, $dx/dt = 2$ when $x = 30$
2. $y = 2x^{1/2} + 3$; $dy/dt = ?$, $dx/dt = 8$ when $x = 4$
3. $x^2 + y^2 = 25$; $dy/dt = ?$, $dx/dt = -3$ when $x = 3$ and $y = 4$
4. $y^2 + x = 11$; $dx/dt = ?$, $dy/dt = -2$ when $x = 2$ and $y = 3$
5. $x^2 + xy + 2 = 0$; $dy/dt = ?$, $dx/dt = -1$ when $x = 2$ and $y = -3$
6. $y^2 + xy - 3x = -3$; $dx/dt = ?$, $dy/dt = -2$ when $x = 1$ and $y = 0$

B

7. A point is moving on the graph of $xy = 36$. When the point is at (4, 9), its x coordinate is increasing at 4 units per second. How fast is the y coordinate changing at that moment?
8. A point is moving on the graph of $4x^2 + 9y^2 = 36$. When the point is at (3, 0), its y coordinate is decreasing at 2 units per second. How fast is its x coordinate changing at that moment?
9. A boat is being pulled toward a dock as indicated in the accompanying figure. If the rope is being pulled in at 3 feet per second, how fast is the distance between the dock and the boat decreasing when it is 30 feet from the dock?

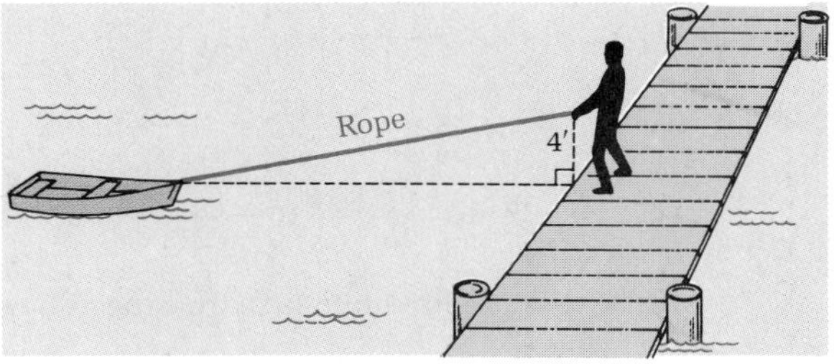

10. Refer to Problem 9. Suppose the distance between the boat and the dock is decreasing at 3.05 feet per second. How fast is the rope being pulled in when the boat is 10 feet from the dock?

11. A rock is thrown into a still pond and causes a circular ripple. If the radius of the ripple is increasing at 2 feet per second, how fast is the area changing when the radius is 10 feet? [Use $A = \pi R^2$, $\pi \approx 3.14$.]

12. Refer to Problem 11. How fast is the circumference of a circular ripple changing when the radius is 10 feet? [Use $C = 2\pi R$, $\pi \approx 3.14$.]

13. The radius of a spherical balloon is increasing at the rate of 3 centimeters per minute. How fast is the volume changing when the radius is 10 centimeters? [Use $V = \frac{4}{3}\pi R^3$, $\pi \approx 3.14$.]

14. Refer to Problem 13. How fast is the surface area of the sphere increasing? [Use $S = 4\pi R^2$, $\pi \approx 3.14$.]

15. Boyle's law for enclosed gases states that if the volume is kept constant, then the pressure P and temperature T are related by the equation

$$\frac{P}{T} = k$$

where k is a constant. If the temperature is increasing at 3 degrees per hour, what is the rate of change of pressure when the temperature is 250° (Kelvin) and the pressure is 500 pounds per square inch?

16. Boyle's law for enclosed gases states that if the temperature is kept constant, then the pressure P and volume V of the gas are related by the equation

$$VP = k$$

where k is a constant. If the volume is decreasing by 5 cubic inches per second, what is the rate of change of the pressure when the volume is 1,000 cubic inches and the pressure is 40 pounds per square inch?

17. A 10 foot ladder is placed against a vertical wall. Suppose the bottom slides away from the wall at a constant rate of 3 feet per second. How fast is the top sliding down the wall (negative rate) when the bottom is 6 feet from the wall? [*Hint:* Use the Pythagorean theorem: $a^2 + b^2 = c^2$, where c is the length of the hypotenuse of a right triangle and a and b are the lengths of the two shorter sides.]

18. A weather balloon is rising vertically at the rate of 5 meters per second. An observer is standing on the ground 300 meters from the point where the balloon was released. At what rate is the distance between the observer and the balloon changing when the balloon is 400 meters high?

C

19. A street light is on top of a 20 foot pole. A person who is 5 feet tall walks away from the pole at the rate of 5 feet per second. At what rate is the tip of the person's shadow moving away from the pole when he is 20 feet from the pole?

20. Refer to Problem 19. At what rate is the person's shadow growing when he is 20 feet from the pole?

Business & Economics

21. *Cost, revenue, and profit rates.* Suppose that for a company manufacturing calculators, the cost, revenue, and profit equations are given by

$$C = 90{,}000 + 30x \qquad R = 300x - \frac{x^2}{30} \qquad P = R - C$$

where the production output in 1 week is x calculators. If production is increasing at a rate of 500 calculators per week when production output is 6,000 calculators, find the rate of increase (decrease) in:

(A) Cost (B) Revenue (C) Profit

22. *Cost, revenue, and profit rates.* Repeat Problem 21 for

$$C = 72{,}000 + 60x \qquad R = 200x - \frac{x^2}{30} \qquad P = R - C$$

where production is increasing at a rate of 500 calculators per week at a production level of 1,500 calculators.

Life Sciences

23. *Pollution.* An oil tanker aground on a reef is leaking oil that forms a circular oil slick about 0.1 foot thick (see the figure). To estimate the rate dV/dt (in cubic feet per minute) at which the oil is leaking from the tanker, it was found that the radius of the slick was increasing at 0.32 foot per minute ($dR/dt = 0.32$) when the radius R was 500 feet. Find dV/dt, using $\pi \approx 3.14$.

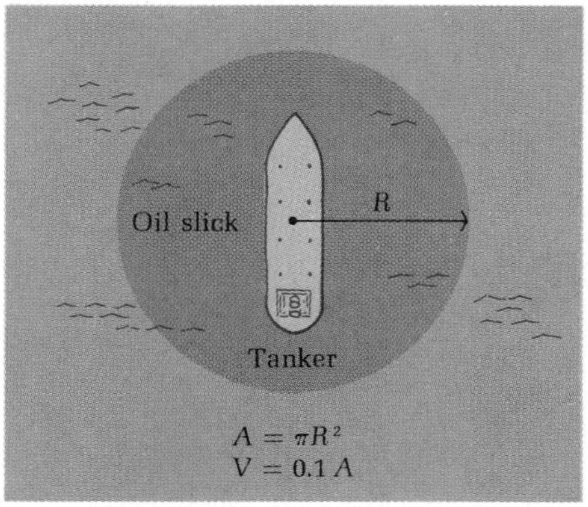

Social Sciences

24. *Learning.* A person who is new on an assembly line performs an operation in T minutes after x performances of the operation, as given by

$$T = 6\left(1 + \frac{1}{\sqrt{x}}\right)$$

If $dx/dt = 6$ operations per hour, where t is time in hours, find dT/dt after 36 performances of the operation.

SECTION 6-6 Chapter Review

Important Terms and Symbols

6-1 *The Constant e and Continuous Compound Interest.* Definition of e; continuous compound interest

6-2 *Derivatives of Logarithmic and Exponential Functions.* Derivative formulas for the natural logarithmic and exponential functions; graph properties of $y = \ln x$ and $y = e^x$

6-3 *Chain Rule: General Form.* Composite functions; chain rule; general derivative formulas; derivative formulas for $y = \log_b x$ and $y = b^x$

6-4 *Implicit Differentiation.* Special function notation; function explicitly defined; function implicitly defined; implicit differentiation

$$y = f(x); \quad y = y(x); \quad F(x, y) = 0; \quad y'|_{(a, b)}$$

6-5 *Related Rates.* Related rates

$$x = x(t); \quad y = y(t)$$

Additional Rules of Differentiation

$$D_x \ln x = \frac{1}{x} \qquad D_x e^x = e^x$$

$$D_x \ln[f(x)] = \frac{1}{f(x)} f'(x) \qquad D_x e^{f(x)} = e^{f(x)} f'(x)$$

$$D_x \log_b x = D_x \frac{1}{\ln b} \ln x = \frac{1}{\ln b}\left(\frac{1}{x}\right)$$

$$D_x b^x = D_x e^{x \ln b} = e^{x \ln b} \ln b = b^x \ln b$$

$$D_x[f(x)]^n = n[f(x)]^{n-1} f'(x)$$

$$\frac{dy}{dx} = \frac{dy}{du}\frac{du}{dx}, \quad \frac{dy}{dx} = \frac{dy}{dw}\frac{dw}{du}\frac{du}{dx}, \quad \text{and so on}$$

EXERCISE 6-6 Chapter Review

Work through all the problems in this chapter review and check your answers in the back of the book. (Answers to all review problems are there.) Where weaknesses show up, review appropriate sections in the text.

A

1. Use a calculator to evaluate $A = 2{,}000e^{0.09t}$ to the nearest cent for $t = 5$, 10, and 20.

Find the indicated derivatives in Problems 2–4.

2. $D_x(2 \ln x + 3e^x)$
3. $D_x e^{2x-3}$
4. y' for $y = \ln(2x + 7)$
5. Let $y = \ln u$ and $u = 3 + e^x$.
   (A) Express y in terms of x.
   (B) Use the chain rule to find dy/dx, and then express dy/dx in terms of x.

6. Find y' for $y = y(x)$ defined implicitly by $2y^2 - 3x^3 - 5 = 0$, and evaluate at $(x, y) = (1, 2)$.

7. For $y = 3x^2 - 5$, where $x = x(t)$ and $y = y(t)$, find dy/dt if $dx/dt = 3$ when $x = 12$.

B

8. Graph $y = 100e^{-0.1x}$.

Find the indicated derivatives in Problems 9–14.

9. $D_z[(\ln z)^7 + \ln z^7]$

10. $D_x(x^6 \ln x)$

11. $D_x \dfrac{e^x}{x^6}$

12. y' for $y = \ln(2x^3 - 3x)$

13. $f'(x)$ for $f(x) = e^{x^3 - x^2}$

14. dy/dx for $y = e^{-2x} \ln 5x$

15. Find the equation of the line tangent to the graph of $y = f(x) = 1 + e^{-x}$ at $x = 0$. At $x = -1$.

16. Find y' for $y = y(x)$ defined implicitly by $x^2 - 3xy + 4y^2 = 23$, and find the slope of the graph at $(-1, 2)$.

17. Find x' for $x = x(t)$ defined implicitly by $x^3 - 2t^2x + 8 = 0$, and evaluate at $(t, x) = (-2, 2)$.

18. Find y' for $y = y(x)$ defined implicitly by $x - y^2 = e^y$, and evaluate at $(1, 0)$.

19. Find y' for $y = y(x)$ defined implicitly by $\ln y = x^2 - y^2$, and evaluate at $(1, 1)$.

20. A point is moving on the graph of $y^2 - 4x^2 = 12$ so that its x coordinate is decreasing at 2 units per second when $(x, y) = (1, 4)$. Find the rate of change of the y coordinate.

21. A 17 foot ladder is placed against a wall. If the foot of the ladder is pushed toward the wall at 0.5 foot per second, how fast is the top rising when the foot of the ladder is 8 feet from the wall?

22. Water from a water heater is leaking onto a floor. A circular pool is created whose area is increasing at the rate of 24 square inches per minute. How fast is the radius R of the pool increasing when the radius is 12 inches? $[A = \pi R^2]$

C

In Problems 23 and 24, find the absolute maximum value of $f(x)$ for $x > 0$.

23. $f(x) = 11x - 2x \ln x$

24. $f(x) = 10xe^{-2x}$

Sketch the graph of $y = f(x)$ in Problems 25 and 26.

25. $f(x) = 5 - 5e^{-x}$

26. $f(x) = x^3 \ln x$

27. Let $y = w^3$, $w = \ln u$, and $u = 4 - e^x$.
    (A) Express y in terms of x.
    (B) Use the chain rule to find dy/dx, and then express dy/dx in terms of x.

Find the indicated derivatives in Problems 28–30.

28. y' for $y = 5^{x^2 - 1}$

29. $D_x \log_5(x^2 - x)$

30. $D_x \sqrt{\ln(x^2 + x)}$

31. Find y' for $y = y(x)$ defined implicitly by $e^{xy} = x^2 + y + 1$, and evaluate at $(0, 0)$.

Business & Economics

32. *Doubling time.* How long will it take money to double if it is invested at 5% interest compounded:

(A) Annually? (B) Continuously?

33. *Continuous compound interest.* If \$100 is invested at 10% interest compounded continuously, the amount (in dollars) at the end of t years is given by

$$A = 100e^{0.1t}$$

Find $A'(t)$, $A'(1)$, and $A'(10)$.

34. *Marginal analysis.* If the price–demand equation for x units of a commodity is

$$p(x) = 1{,}000e^{-0.02x}$$

find the marginal revenue equation.

35. *Maximum revenue.* For the price–demand equation in Problem 34, find the production level and price per unit that produces the maximum revenue. What is the maximum revenue?

36. *Maximum revenue.* Graph the revenue function from Problems 34 and 35 for $0 \leq x \leq 100$.

37. *Minimum average cost.* The cost of producing x units of a product is given by

$$C(x) = 200 + 50x - 50 \ln x \qquad x \geq 1$$

Find the minimum average cost.

38. *Demand equation.* Given the demand equation

$$x = \sqrt{5{,}000 - 2p^3}$$

find the rate of change of p with respect to x by implicit differentiation (x is the number of items that can be sold at a price of \$$p$ per item).

39. *Rate of change of revenue.* A company is manufacturing a new video game and can sell all it manufactures. The revenue (in dollars) is given by

$$R = 36x - \frac{x^2}{20}$$

where the production output in 1 day is x games. If production is increasing at 10 games per day when production is 250 games per day, find the rate of increase in revenue.

Life Sciences

40. *Drug concentration.* The concentration of a drug in the bloodstream t hours after injection is given approximately by

$$C(t) = 5e^{-0.3t}$$

where $C(t)$ is concentration in milligrams per milliliter. What is the rate of change of concentration after 1 hour? After 5 hours?

41. *Wound healing.* A circular wound on an arm is healing at the rate of 45 square millimeters per day (the area of wound is decreasing at this rate). How fast is the radius R of the wound decreasing when $R = 15$ millimeters? $[A = \pi R^2]$

Social Sciences

42. *Psychology—learning.* In a computer assembly plant, a new employee, on the average, is able to assemble

$$N(t) = 10(1 - e^{-0.4t})$$

units after t days of on-the-job training.

(A) What is the rate of learning after 1 day? After 5 days?
(B) Graph N for $0 \leq t \leq 10$.

43. *Learning.* A new worker on the production line performs an operation in T minutes after x performances of the operation, as given by

$$T = 2\left(1 + \frac{1}{x^{3/2}}\right)$$

If, after performing the operation 9 times, the rate of improvement is $dx/dt = 3$ operations per hour, find the rate of improvement in time dT/dt in performing each operation.

CHAPTER 7

Integration

Contents

CHAPTER 7

The last three chapters dealt with differential calculus. We now begin the development of the second main part of calculus, called *integral calculus*. Two types of integrals will be introduced, the *indefinite integral* and the *definite integral*. Each is quite different from the other, but both are intimately related to differentiation.

SECTION 7-1 Antiderivatives and Indefinite Integrals

- ANTIDERIVATIVES
- INDEFINITE INTEGRALS
- INDEFINITE INTEGRALS INVOLVING ALGEBRAIC FUNCTIONS
- INDEFINITE INTEGRALS INVOLVING EXPONENTIAL AND LOGARITHMIC FUNCTIONS
- APPLICATIONS

◆ ANTIDERIVATIVES

Many operations in mathematics have reverses—compare addition and subtraction, multiplication and division, and powers and roots. The function $f(x) = \frac{1}{3}x^3$ has the derivative $f'(x) = x^2$. Reversing this process is referred to as **antidifferentiation.** Thus,

$$\frac{x^3}{3} \quad \text{is an antiderivative of} \quad x^2$$

since

$$D_x\left(\frac{x^3}{3}\right) = x^2$$

In general, we say that $F(x)$ is an **antiderivative** of $f(x)$ if

$$F'(x) = f(x)$$

Note that

$$D_x\left(\frac{x^3}{3} + 2\right) = x^2 \qquad D_x\left(\frac{x^3}{3} - \pi\right) = x^2 \qquad D_x\left(\frac{x^3}{3} + \sqrt{5}\right) = x^2$$

Hence,

$$\frac{x^3}{3} + 2 \qquad \frac{x^3}{3} - \pi \qquad \frac{x^3}{3} + \sqrt{5}$$

are also antiderivatives of x^2, since each has x^2 as a derivative. In fact, it appears that

$$\frac{x^3}{3} + C$$

for any real number C, is an antiderivative of x^2, since

$$D_x\left(\frac{x^3}{3} + C\right) = x^2$$

Thus, antidifferentiation of a given function does not lead to a unique function, but to a whole set of functions.

Does the expression

$$\frac{x^3}{3} + C$$

with C any real number, include all antiderivatives of x^2? Theorem 1 (which we state without proof) indicates that the answer is yes.

THEOREM 1

If F and G are differentiable functions on the interval (a, b) and $F'(x) = G'(x)$, then $F(x) = G(x) + k$ for some constant k.

◆ INDEFINITE INTEGRALS

In words, Theorem 1 states that **if the derivatives of two functions are equal, then the functions differ by at most a constant.** We use the symbol

$$\int f(x)\,dx$$

called the **indefinite integral,** to represent the set of all antiderivatives of $f(x)$, and we write

$$\int f(x)\,dx = F(x) + C \qquad \text{if} \qquad F'(x) = f(x)$$

The symbol $\int$ is called an **integral sign,** and the function $f(x)$ is called the **integrand.** The symbol dx indicates that the antidifferentiation is performed with respect to the variable x. (We will have more to say about the symbol dx

later.) The arbitrary constant C is called the **constant of integration.** Referring to the preceding discussion, we can write

$$\int x^2\,dx = \frac{x^3}{3} + C \qquad \text{since} \qquad D_x\left(\frac{x^3}{3} + C\right) = x^2$$

Of course, variables other than x can be used in indefinite integrals. Thus, we also can write

$$\int t^2\,dt = \frac{t^3}{3} + C \qquad \text{since} \qquad D_t\left(\frac{t^3}{3} + C\right) = t^2$$

or

$$\int u^2\,du = \frac{u^3}{3} + C \qquad \text{since} \qquad D_u\left(\frac{u^3}{3} + C\right) = u^2$$

The fact that indefinite integration and differentiation are reverse operations, except for the addition of the constant of integration, can be expressed symbolically as

$$D_x\left[\int f(x)\,dx\right] = f(x)$$ The derivative of the indefinite integral of $f(x)$ is $f(x)$.

and

$$\int F'(x)\,dx = F(x) + C$$ The indefinite integral of the derivative of $F(x)$ is $F(x) + C$.

◆ INDEFINITE INTEGRALS INVOLVING ALGEBRAIC FUNCTIONS

Just as with differentiation, we can develop formulas and special properties that will enable us to find indefinite integrals of many frequently encountered functions. To start, we list some formulas that can be established using the definitions of antiderivative and indefinite integral, and the properties of derivatives considered in Chapter 4.

Indefinite Integral Formulas and Properties

For k and C constants:

1. $\int k\,dx = kx + C$
2. $\int x^n\,dx = \dfrac{x^{n+1}}{n+1} + C \qquad n \neq -1$
3. $\int kf(x)\,dx = k\int f(x)\,dx$
4. $\int [f(x) \pm g(x)]\,dx = \int f(x)\,dx \pm \int g(x)\,dx$

We will establish formula 2 and property 3 here (the others may be shown to be true in a similar manner). To establish formula 2, we simply differentiate the right side to obtain the integrand on the left side. Thus,

$$D_x\left(\frac{x^{n+1}}{n+1}+C\right)=\frac{(n+1)x^n}{n+1}+0=x^n \qquad n\neq -1$$

(Notice that formula 2 cannot be used when $n=-1$; that is, when the integrand is x^{-1} or $1/x$. The indefinite integral of $x^{-1}=1/x$ will be considered later in this section.)

To establish property 3, let F be a function such that $F'(x)=f(x)$. Then

$$k\int f(x)\,dx = k\int F'(x)\,dx = k[F(x)+C_1] = kF(x)+kC_1$$

and since $(kF(x))'=kF'(x)=kf(x)$, we have

$$\int kf(x)\,dx=\int kF'(x)\,dx=kF(x)+C_2$$

But $kF(x)+kC_1$ and $kF(x)+C_2$ describe the same set of functions, since C_1 and C_2 are arbitrary real numbers. Thus, property 3 is established.

It is important to remember that property 3 states that **a constant factor can be moved across an integral sign; a variable factor cannot be moved across an integral sign:**

Constant Factor	Variable Factor
$\int 5x^{1/2}\,dx = 5\int x^{1/2}\,dx$	$\int xx^{1/2}\,dx \neq x\int x^{1/2}\,dx$

Now let us put the formulas and properties to use.

◆ EXAMPLE 1

(A) $\int 5\,dx = 5x+C$

(B) $\int x^4\,dx=\frac{x^{4+1}}{4+1}+C=\frac{x^5}{5}+C$

(C) $\int 5t^7\,dt = 5\int t^7\,dt = 5\frac{t^8}{8}+C=\frac{5}{8}t^8+C$

(D) $\int (4x^3+2x-1)\,dx = \int 4x^3\,dx+\int 2x\,dx-\int dx$

$= 4\int x^3\,dx + 2\int x\,dx - \int dx$

$= \frac{4x^4}{4}+\frac{2x^2}{2}-x+C$

$= x^4+x^2-x+C$

Property 4 can be extended to the sum and difference of an arbitrary number of functions.

(E) $\int \frac{3\,dx}{x^2} = \int 3x^{-2}\,dx = \frac{3x^{-2+1}}{-2+1} + C = -3x^{-1} + C$

(F) $\int 5\sqrt[3]{u^2}\,du = 5\int u^{2/3}\,du = 5\,\frac{u^{(2/3)+1}}{\frac{2}{3}+1} + C$

$$= 5\,\frac{u^{5/3}}{\frac{5}{3}} + C = 3u^{5/3} + C$$

◆

To check any of the results in Example 1, we differentiate the final result to obtain the integrand in the original indefinite integral. When you evaluate an indefinite integral, do not forget to include the arbitrary constant C.

PROBLEM 1 Find each of the following:

(A) $\int dx$ (B) $\int 3t^4\,dt$ (C) $\int (2x^5 - 3x^2 + 1)\,dx$

(D) $\int 4\sqrt[5]{w^3}\,dw$ (E) $\int \left(2x^{2/3} - \frac{3}{x^4}\right) dx$

◆

◆ EXAMPLE 2 (A) $\int \frac{x^3 - 3}{x^2}\,dx = \int \left(\frac{x^3}{x^2} - \frac{3}{x^2}\right) dx$

$$= \int (x - 3x^{-2})\,dx$$

$$= \int x\,dx - 3\int x^{-2}\,dx$$

$$= \frac{x^{1+1}}{1+1} - 3\,\frac{x^{-2+1}}{-2+1} + C$$

$$= \tfrac{1}{2}x^2 + 3x^{-1} + C$$

(B) $\int \left(\frac{2}{\sqrt[3]{x}} - 6\sqrt{x}\right) dx = \int (2x^{-1/3} - 6x^{1/2})\,dx$

$$= 2\int x^{-1/3}\,dx - 6\int x^{1/2}\,dx$$

$$= 2\,\frac{x^{(-1/3)+1}}{-\frac{1}{3}+1} - 6\,\frac{x^{(1/2)+1}}{\frac{1}{2}+1} + C$$

$$= 2\,\frac{x^{2/3}}{\frac{2}{3}} - 6\,\frac{x^{3/2}}{\frac{3}{2}} + C$$

$$= 3x^{2/3} - 4x^{3/2} + C$$

◆

PROBLEM 2 Find each indefinite integral.

(A) $\int \frac{x^4 - 8x^3}{x^2}\,dx$ (B) $\int \left(8\sqrt[3]{x} - \frac{6}{\sqrt{x}}\right) dx$

◆

◆ INDEFINITE INTEGRALS INVOLVING EXPONENTIAL AND LOGARITHMIC FUNCTIONS

We now give indefinite integral formulas for e^x and $1/x$. (Recall that the form $x^{-1} = 1/x$ is not covered by formula 2, given earlier.)

Indefinite Integral Formulas

5. $\int e^x\, dx = e^x + C$ 6. $\int \frac{1}{x}\, dx = \ln|x| + C \qquad x \neq 0$

Formula 5 follows immediately from the derivative formula for the exponential function discussed in the last chapter. Because of the absolute value, formula 6 does not follow directly from the derivative formula for the natural logarithm function. Let us show that

$$D_x \ln|x| = \frac{1}{x} \qquad x \neq 0$$

We consider two cases, $x > 0$ and $x < 0$:

Case 1. $x > 0$:

$$\begin{aligned} D_x \ln|x| &= D_x \ln x && \text{Since } |x| = x \text{ for } x > 0 \\ &= \frac{1}{x} \end{aligned}$$

Case 2. $x < 0$:

$$\begin{aligned} D_x \ln|x| &= D_x \ln(-x) && \text{Since } |x| = -x \text{ for } x < 0 \\ &= \frac{1}{-x} D_x(-x) \\ &= \frac{-1}{-x} = \frac{1}{x} \end{aligned}$$

Thus,

$$D_x \ln|x| = \frac{1}{x} \qquad x \neq 0$$

and hence,

$$\int \frac{1}{x}\, dx = \ln|x| + C \qquad x \neq 0$$

What about the indefinite integral of $\ln x$? We postpone a discussion of $\int \ln x\, dx$ until Section 8-3, where we will be able to find it using a technique called *integration by parts.*

◆ EXAMPLE 3 $\int \left(2e^x + \frac{3}{x}\right) dx = 2\int e^x\,dx + 3\int \frac{1}{x}\,dx$

$= 2e^x + 3\ln|x| + C$ ◆

PROBLEM 3 Find: $\int \left(\frac{5}{x} - 4e^x\right) dx$ ◆

◆ APPLICATIONS

Let us now consider some applications of the indefinite integral to see why we are interested in finding antiderivatives of functions.

◆ EXAMPLE 4 *Curves* Find the equation of the curve that passes through (2, 5) if its slope is given by $dy/dx = 2x$ at any point x.

Solution We are interested in finding a function $y = f(x)$ such that

$$\frac{dy}{dx} = 2x \tag{1}$$

and

$$y = 5 \qquad \text{when} \qquad x = 2 \tag{2}$$

If $dy/dx = 2x$, then

$$y = \int 2x\,dx$$
$$= x^2 + C \tag{3}$$

Since $y = 5$ when $x = 2$, we determine the *particular value of C* so that

$$5 = 2^2 + C$$

Thus, $C = 1$, and

$$y = x^2 + 1$$

is the *particular antiderivative* out of all those possible from (3) that satisfies both (1) and (2). See Figure 1.

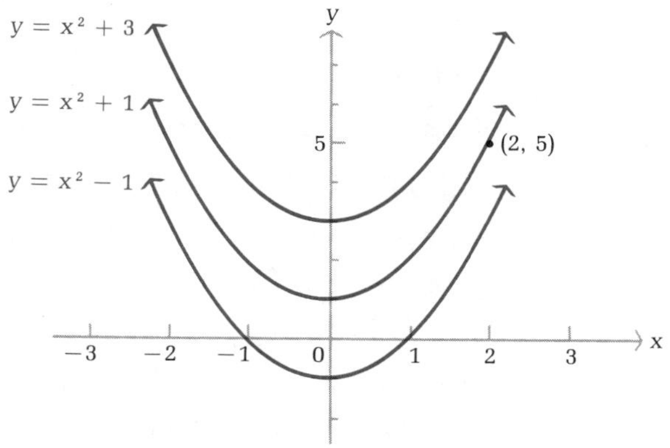

FIGURE 1
$y = x^2 + C$

PROBLEM 4 Find the equation of the curve that passes through (2, 6) if the slope of the curve at any point x is given by $dy/dx = 3x^2$.

In certain situations, it is easier to determine the rate at which something happens than how much of it has happened in a given length of time (for example, population growth rates, business growth rates, rate of healing of a wound, rates of learning or forgetting). If a rate function (derivative) is given and we know the value of the dependent variable for a given value of the independent variable, then—if the rate function is not too complicated—we can often find the original function by integration.

EXAMPLE 5 *Cost Function*

If the marginal cost of producing x units is given by

$$C'(x) = 0.3x^2 + 2x$$

and the fixed cost is \$2,000, find the cost function $C(x)$ and the cost of producing 20 units.

Solution Recall that marginal cost is the derivative of the cost function and that fixed cost is cost at a 0 production level. Thus, the mathematical problem is to find $C(x)$ given

$$C'(x) = 0.3x^2 + 2x \qquad C(0) = 2{,}000$$

We now find the indefinite integral of $0.3x^2 + 2x$ and determine the arbitrary integration constant using $C(0) = 2{,}000$:

$$C'(x) = 0.3x^2 + 2x$$

$$C(x) = \int (0.3x^2 + 2x)\, dx$$

$$= 0.1x^3 + x^2 + K$$ Since C represents the cost, we use K for the constant of integration.

But

$$C(0) = (0.1)0^3 + 0^2 + K = 2{,}000$$

Thus, $K = 2{,}000$, and the particular cost function is

$$C(x) = 0.1x^3 + x^2 + 2{,}000$$

We now find $C(20)$, the cost of producing 20 units:

$$\begin{aligned} C(20) &= (0.1)20^3 + 20^2 + 2{,}000 \\ &= \$3{,}200 \end{aligned}$$

◆

PROBLEM 5 Find the revenue function $R(x)$ when the marginal revenue is

$$R'(x) = 400 - 0.4x$$

and no revenue results at a 0 production level. What is the revenue at a production level of 1,000 units? ◆

◆ EXAMPLE 6 *Advertising*

An FM radio station is launching an aggressive advertising campaign in order to increase the number of daily listeners. The station currently has 27,000 daily listeners, and management expects the number of daily listeners, $S(t)$, to grow at the rate of

$$S'(t) = 60t^{1/2}$$

listeners per day, where t is the number of days since the campaign began. How long should the campaign last if the station wants the number of daily listeners to grow to 41,000?

Solution We must solve the equation $S(t) = 41{,}000$ for t, given that

$$S'(t) = 60t^{1/2} \qquad \text{and} \qquad S(0) = 27{,}000$$

First, we use integration to find $S(t)$:

$$\begin{aligned} S(t) &= \int 60t^{1/2}\, dt \\ &= 60\frac{t^{3/2}}{\frac{3}{2}} + C \\ &= 40t^{3/2} + C \end{aligned}$$

Since

$$S(0) = 40(0)^{3/2} + C = 27{,}000$$

we have $C = 27{,}000$, and

$$S(t) = 40t^{3/2} + 27{,}000$$

Now we solve the equation $S(t) = 41{,}000$ for t:

$$
\begin{aligned}
40t^{3/2} + 27{,}000 &= 41{,}000 \\
40t^{3/2} &= 14{,}000 \\
t^{3/2} &= 350 \\
t &= 350^{2/3} \qquad \text{Use a calculator.} \\
&= 49.664\ 419\ldots
\end{aligned}
$$

Thus, the advertising campaign should last approximately 50 days. ◆

PROBLEM 6

The current monthly circulation of the magazine *Computing News* is 640,000 copies. Due to competition from a new magazine in the same field, the monthly circulation of *Computing News*, $C(t)$, is expected to decrease at the rate of

$$C'(t) = -6{,}000t^{1/3}$$

copies per month, where t is the time in months since the new magazine began publication. How long will it take for the circulation of *Computing News* to decrease to 460,000 copies per month? ◆

Common Errors

1. $\displaystyle\int e^x\,dx \neq \frac{e^{x+1}}{x+1} + C$

   The power rule only applies to power functions of the form x^n where the exponent n is a real constant not equal to -1 and the base x is the variable. The function e^x is an exponential function with variable exponent x and constant base e. The correct form for this problem is

   $$\int e^x\,dx = e^x + C$$

2. $\displaystyle\int x(x^2 + 2)\,dx \neq \frac{x^2}{2}\left(\frac{x^3}{3} + 2x\right) + C$

   The integral of a product is not equal to the product of the integrals. The correct form for this problem is

   $$\int x(x^2 + 2)\,dx = \int (x^3 + 2x)\,dx = \frac{x^4}{4} + x^2 + C$$

Answers to Matched Problems

1. (A) $x + C$ (B) $\frac{3}{5}t^5 + C$ (C) $(x^6/3) - x^3 + x + C$ (D) $\frac{5}{2}w^{8/5} + C$
   (E) $\frac{6}{5}x^{5/3} + x^{-3} + C$
2. (A) $\frac{1}{3}x^3 - 4x^2 + C$ (B) $6x^{4/3} - 12x^{1/2} + C$ 3. $5\ln|x| - 4e^x + C$
4. $y = x^3 - 2$ 5. $R(x) = 400x - 0.2x^2$; $R(1{,}000) = \$200{,}000$
6. $t = (40)^{3/4} \approx 16$ months

EXERCISE 7-1

A *Find each indefinite integral. (Check by differentiating.)*

1. $\int 7\,dx$ **2.** $\int \pi\,dx$ **3.** $\int x^6\,dx$

4. $\int x^3\,dx$ **5.** $\int 8t^3\,dt$ **6.** $\int 10t^4\,dt$

7. $\int (2u + 1)\,du$ **8.** $\int (1 - 2u)\,du$ **9.** $\int (3x^2 + 2x - 5)\,dx$

10. $\int (2 + 4x - 6x^2)\,dx$ **11.** $\int (s^4 - 8s^5)\,ds$ **12.** $\int (t^5 + 6t^3)\,dt$

13. $\int 3e^t\,dt$ **14.** $\int 2e^t\,dt$ **15.** $\int 2z^{-1}\,dz$

16. $\int \frac{3}{s}\,ds$

Find all the antiderivatives for each derivative.

17. $\frac{dy}{dx} = 200x^4$ **18.** $\frac{dx}{dt} = 42t^5$ **19.** $\frac{dP}{dx} = 24 - 6x$

20. $\frac{dy}{dx} = 3x^2 - 4x^3$ **21.** $\frac{dy}{du} = 2u^5 - 3u^2 - 1$ **22.** $\frac{dA}{dt} = 3 - 12t^3 - 9t^5$

23. $\frac{dy}{dx} = e^x + 3$ **24.** $\frac{dy}{dx} = x - e^x$ **25.** $\frac{dx}{dt} = 5t^{-1} + 1$

26. $\frac{du}{dv} = \frac{4}{v} + \frac{v}{4}$

B *Find each indefinite integral. (Check by differentiation.)*

27. $\int 6x^{1/2}\,dx$ **28.** $\int 8t^{1/3}\,dt$

29. $\int 8x^{-3}\,dx$ **30.** $\int 12u^{-4}\,du$

31. $\int \frac{du}{\sqrt{u}}$ **32.** $\int \frac{dt}{\sqrt[3]{t}}$

33. $\int \frac{dx}{4x^3}$ **34.** $\int \frac{6\,dm}{m^2}$

35. $\int \frac{du}{2u^5}$ **36.** $\int \frac{dy}{3y^4}$

37. $\int \left(3x^2 - \frac{2}{x^2}\right) dx$ **38.** $\int \left(4x^3 + \frac{2}{x^3}\right) dx$

39. $\int \left(10x^4 - \frac{8}{x^5} - 2\right) dx$

40. $\int \left(\frac{6}{x^4} - \frac{2}{x^3} + 1\right) dx$

41. $\int \left(3\sqrt{x} + \frac{2}{\sqrt{x}}\right) dx$

42. $\int \left(\frac{2}{\sqrt[3]{x}} - \sqrt[3]{x^2}\right) dx$

43. $\int \left(\sqrt[3]{x^2} - \frac{4}{x^3}\right) dx$

44. $\int \left(\frac{12}{x^5} - \frac{1}{\sqrt[3]{x^2}}\right) dx$

45. $\int \frac{e^x - 3x}{4}\, dx$

46. $\int \frac{e^x - 3x^2}{2}\, dx$

47. $\int (2z^{-3} + z^{-2} + z^{-1})\, dz$

48. $\int (3x^{-2} - x^{-1})\, dx$

for each derivative

In Problems 49–58, find the particular antiderivative ~~of each derivative~~ that satisfies the given condition.

49. $\frac{dy}{dx} = 2x - 3; \quad y(0) = 5$

50. $\frac{dy}{dx} = 5 - 4x; \quad y(0) = 20$

51. $C'(x) = 6x^2 - 4x; \quad C(0) = 3{,}000$

52. $R'(x) = 600 - 0.6x; \quad R(0) = 0$

53. $\frac{dx}{dt} = \frac{20}{\sqrt{t}}; \quad x(1) = 40$

54. $\frac{dR}{dt} = \frac{100}{t^2}; \quad R(1) = 400$

55. $\frac{dy}{dx} = 2x^{-2} + 3x^{-1} - 1; \quad y(1) = 0$

56. $\frac{dy}{dx} = 3x^{-1} + x^{-2}; \quad y(1) = 1$

57. $\frac{dx}{dt} = 4e^t - 2; \quad x(0) = 1$

58. $\frac{dy}{dt} = 5e^t - 4; \quad y(0) = -1$

59. Find the equation of the curve that passes through (2, 3) if its slope is given by

$$\frac{dy}{dx} = 4x - 3$$

for each x.

60. Find the equation of the curve that passes through (1, 3) if its slope is given by

$$\frac{dy}{dx} = 12x^2 - 12x$$

for each x.

C

Find each indefinite integral.

61. $\int \frac{2x^4 - x}{x^3}\, dx$

62. $\int \frac{x^{-1} - x^4}{x^2}\, dx$

63. $\int \frac{x^5 - 2x}{x^4}\, dx$

64. $\int \frac{1 - 3x^4}{x^2}\, dx$

65. $\int \frac{x^2e^x - 2x}{x^2}\, dx$

66. $\int \frac{1 - xe^x}{x}\, dx$

For each derivative, find an antiderivative that satisfies the given condition.

67. $\dfrac{dM}{dt} = \dfrac{t^2 - 1}{t^2}$; $M(4) = 5$

68. $\dfrac{dR}{dx} = \dfrac{1 - x^4}{x^3}$; $R(1) = 4$

69. $\dfrac{dy}{dx} = \dfrac{5x + 2}{\sqrt[3]{x}}$; $y(1) = 0$

70. $\dfrac{dx}{dt} = \dfrac{\sqrt{t^3} - t}{\sqrt{t^3}}$; $x(9) = 4$

71. $p'(x) = -\dfrac{10}{x^2}$; $p(1) = 20$

72. $p'(x) = \dfrac{10}{x^3}$; $p(1) = 15$

APPLICATIONS

Business & Economics

73. *Profit function.* If the marginal profit for producing x units is given by

$$P'(x) = 50 - 0.04x \qquad P(0) = 0$$

where $P(x)$ is the profit in dollars, find the profit function P and the profit on 100 units of production.

74. *Natural resources.* The world demand for wood is increasing. In 1975, the demand was 12.6 billion cubic feet, and the rate of increase in demand is given approximately by

$$d'(t) = 0.009t$$

where t is time in years after 1975 (data from the U.S. Department of Agriculture and Forest Service). Noting that $d(0) = 12.6$, find $d(t)$. Also find $d(25)$, the demand in the year 2000.

75. *Revenue function.* The marginal revenue from the sale of x digital sports watches is given by

$$R'(x) = 100 - \tfrac{1}{5}x \qquad R(0) = 0$$

where $R(x)$ is the revenue in dollars. Find the revenue function and the price–demand equation. What is the price when the demand is 700 units?

76. *Cost function.* The marginal average cost for producing x digital sports watches is given by

$$\overline{C}'(x) = -\frac{1{,}000}{x^2} \qquad \overline{C}(100) = 25$$

where $\overline{C}(x)$ is the average cost in dollars. Find the average cost function and the cost function. What are the fixed costs?

77. *Sales analysis.* The monthly sales of a particular personal computer are expected to decline at the rate of

$$S'(t) = -25t^{2/3}$$

computers per month, where t is time in months and $S(t)$ is the number of computers sold each month. The company plans to stop manufacturing this computer when the monthly sales reach 800 computers. If the monthly sales now ($t = 0$) are 2,000 computers, find $S(t)$. How long will the company continue to manufacture this computer?

78. *Sales analysis.* The rate of change of the monthly sales of a new home video game cartridge is given by

$$S'(t) = 500t^{1/4} \qquad S(0) = 0$$

where t is the number of months since the game was released and $S(t)$ is the number of cartridges sold each month. Find $S(t)$. When will the monthly sales reach 20,000 cartridges?

79. *Labor costs and learning.* A defense contractor is starting production on a new missile control system. On the basis of data collected while assembling the first 16 control systems, the production manager obtained the following function describing the rate of labor use:

$$g(x) = 2{,}400x^{-1/2}$$

where $g(x)$ is the number of labor-hours required to assemble the xth unit of the control system. For example, after assembling 16 units, the rate of assembly is 600 labor-hours per unit, and after assembling 25 units, the rate of assembly is 480 labor-hours per unit. The more units assembled, the more efficient the process because of learning. If 19,200 labor-hours are required to assemble the first 16 units, how many labor-hours, $L(x)$, will be required to assemble the first x units? The first 25 units?

80. *Labor costs and learning.* If the rate of labor use in Problem 79 is

$$g(x) = 2{,}000x^{-1/3}$$

and if the first 8 control units require 12,000 labor-hours, how many labor-hours, $L(x)$, will be required for the first x control units? The first 27 control units?

Life Sciences

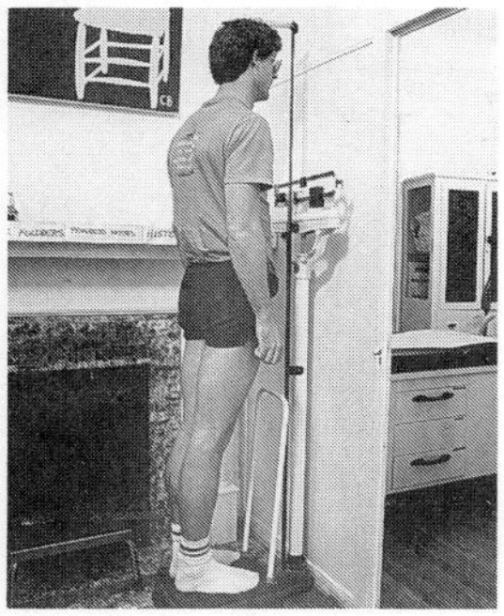

81. *Weight–height.* For an average person, the rate of change of weight W (in pounds) with respect to height h (in inches) is given approximately by

$$\frac{dW}{dh} = 0.0015h^2$$

Find $W(h)$ if $W(60) = 108$ pounds. Also find the weight for a person who is 5 feet 10 inches tall.

82. *Wound healing.* If the area A of a healing wound changes at a rate given approximately by

$$\frac{dA}{dt} = -4t^{-3} \qquad 1 \leq t \leq 10$$

where t is time in days and $A(1) = 2$ square centimeters, what will the area of the wound be in 10 days?

Social Sciences

83. *Urban growth.* The rate of growth of the population, $N(t)$, of a newly incorporated city t years after incorporation is estimated to be

$$\frac{dN}{dt} = 400 + 600\sqrt{t} \qquad 0 \leqslant t \leqslant 9$$

If the population was 5,000 at the time of incorporation, find the population 9 years later.

84. *Learning.* A beginning high school language class was chosen for an experiment in learning. Using a list of 50 words, the experiment involved measuring the rate of vocabulary memorization at different times during a continuous 5 hour study session. It was found that the average rate of learning for the whole class was inversely proportional to the time spent studying and was given approximately by

$$V'(t) = \frac{15}{t} \qquad 1 \leqslant t \leqslant 5$$

If the average number of words memorized after 1 hour of study was 15 words, what was the average number of words learned after t hours of study for $1 \leqslant t \leqslant 5$? After 4 hours of study? (Round answer to the nearest whole number.)

SECTION 7-2 Integration by Substitution

- ◆ INTEGRATION BY SUBSTITUTION
- ◆ SUBSTITUTION TECHNIQUES
- ◆ APPLICATION

◆ INTEGRATION BY SUBSTITUTION

In Sections 4-6 and 6-3, we saw that the chain rule was a very important tool for finding derivatives. Now we will see that the chain rule also has applications to integration. To begin, consider the indefinite integral

$$\int \frac{2x}{x^2 + 5}\, dx \tag{1}$$

This integral cannot be evaluated directly by any of the techniques discussed in Section 7-1, but it can be simplified by making a *substitution.* If we let

$$u = x^2 + 5$$

then the differential of u is (see Section 5-5)

$$du = 2x\, dx$$

If we rewrite the integrand in (1) so that 2x is grouped with dx, then we can express the integrand entirely in terms of u and du:

$$\int \frac{2x}{x^2+5}\,dx = \int \frac{1}{x^2+5}\,2x\,dx \qquad \text{Substitute } u = x^2+5 \text{ and } du = 2x\,dx.$$

$$= \int \frac{1}{u}\,du$$

This last integral looks just like the integral in formula 6 from Section 7-1, except that u is not an independent variable; it is a function of x. For now, let us assume that formula 6 applies in this situation. Then we have

$$\int \frac{2x}{x^2+5}\,dx = \int \frac{1}{u}\,du \qquad \text{Use formula 6.}$$

$$= \ln|u| + C \qquad \text{Substitute } u = x^2+5.$$

$$= \ln(x^2+5) + C \qquad \text{Absolute value signs can be omitted since } x^2+5>0 \text{ for all } x.$$

We use the chain rule to check this result:

$$D_x \ln(x^2+5) = \frac{1}{x^2+5}\,D_x(x^2+5) = \frac{1}{x^2+5}\,2x = \frac{2x}{x^2+5}$$

which is the integrand in (1). Thus, our assumption concerning formula 6 was valid for this problem. Is this always the case? Yes, it is, as we will now verify. Since the chain rule implies that

$$D_x \ln|f(x)| = \frac{1}{f(x)}\,f'(x)$$

it follows that

$$\int \frac{1}{f(x)}\,f'(x)\,dx = \ln|f(x)| + C$$

or, substituting u for $f(x)$ and du for $f'(x)\,dx$,

$$\int \frac{1}{u}\,du = \ln|u| + C$$

Similarly, all the integration formulas from Section 7-1 are valid if the variable of integration is a function u and du is its differential. We restate these formulas in the box on the next page for convenient reference.

General Indefinite Integral Formulas

For k and C constants:

1. $\int k\,du = ku + C$

2. $\int u^n\,du = \frac{u^{n+1}}{n+1} + C \qquad n \neq -1$

3. $\int kf(u)\,du = k\int f(u)\,du$

4. $\int [f(u) \pm g(u)]\,du = \int f(u)\,du \pm \int g(u)\,du$

5. $\int e^u\,du = e^u + C$

6. $\int \frac{1}{u}\,du = \ln|u| + C$

These formulas are valid if u is an independent variable or if u is a function of another variable and du is its differential with respect to that variable.

The substitution method for evaluating certain indefinite integrals is outlined in the following box:

Integration by Substitution

Step 1. Select a substitution that appears to simplify the integrand. In particular, try to select u so that du is a factor in the integrand.

Step 2. Express the integrand entirely in terms of u and du, completely eliminating the original variable and its differential.

Step 3. Evaluate the new integral, if possible.

Step 4. Express the antiderivative found in step 3 in terms of the original variable.

◆ EXAMPLE 7 Use a substitution to find the following:

(A) $\int (3x+4)^6 3\,dx$ (B) $\int e^{t^2} 2t\,dt$

Solutions (A) If we let $u = 3x + 4$, then $du = 3\,dx$, and

$$\int (3x + 4)^6 3\,dx = \int u^6\,du \qquad \text{Use formula 2.}$$

$$= \frac{u^7}{7} + C$$

$$= \frac{(3x + 4)^7}{7} + C \qquad \text{Since } u = 3x + 4$$

Check: $D_x \frac{(3x + 4)^7}{7} = \frac{7(3x + 4)^6}{7} D_x(3x + 4) = (3x + 4)^6 3$

(B) If we let $u = t^2$, then $du = 2t\,dt$, and

$$\int e^{t^2} 2t\,dt = \int e^u\,du \qquad \text{Use formula 5.}$$

$$= e^u + C$$

$$= e^{t^2} + C \qquad \text{Since } u = t^2$$

Check: $D_t\, e^{t^2} = e^{t^2} D_t\, t^2 = e^{t^2} 2t$ ◆

PROBLEM 7 Use a substitution to find each indefinite integral.

(A) $\int (2x^3 - 3)^4 6x^2\,dx$ (B) $\int e^{5w} 5\,dw$ ◆

Integration by substitution is an effective procedure for some indefinite integrals, but not all. Substitution is not helpful for $\int e^{x^2}\,dx$ or $\int (\ln x)\,dx$, for example.

◆ SUBSTITUTION TECHNIQUES

In order to use the substitution method, **the integrand must be expressed entirely in terms of *u* and *du*.** In some cases, the integrand will have to be modified before making a substitution and using one of the integration formulas. Example 8 illustrates this process.

◆ EXAMPLE 8 Integrate: (A) $\int \frac{1}{4x + 7}\,dx$ (B) $\int te^{-t^2}\,dt$ (C) $\int 4x^2\sqrt{x^3 + 5}\,dx$

Solutions (A) If $u = 4x + 7$, then $du = 4\,dx$. We are missing a factor of 4 in the integrand to match formula 6 exactly. Recalling that a constant factor can be moved across an integral sign, we proceed as follows:

$$\int \frac{1}{4x + 7}\,dx = \int \frac{1}{4x + 7}\frac{4}{4}\,dx$$

$$= \frac{1}{4}\int \frac{1}{4x + 7} 4\,dx \qquad \text{Substitute } u = 4x + 7 \text{ and } du = 4\,dx.$$

$$\int \frac{1}{4x+7}\,dx = \frac{1}{4}\int \frac{1}{u}\,du \qquad \text{Use formula 6.}$$

$$= \tfrac{1}{4}\ln|u| + C$$

$$= \tfrac{1}{4}\ln|4x+7| + C \qquad \text{Since } u = 4x+7$$

Check: $D_x \frac{1}{4}\ln|4x+7| = \frac{1}{4}\frac{1}{4x+7}D_x(4x+7) = \frac{1}{4}\frac{1}{4x+7}4 = \frac{1}{4x+7}$

(B) If $u = -t^2$, then $du = -2t\,dt$. Proceed as in part A:

$$\int te^{-t^2}\,dt = \int e^{-t^2}\frac{-2}{-2}t\,dt$$

$$= -\frac{1}{2}\int e^{-t^2}(-2t)\,dt \qquad \text{Substitute } u = -t^2 \text{ and } du = -2t\,dt.$$

$$= -\frac{1}{2}\int e^u\,du \qquad \text{Use formula 5.}$$

$$= -\tfrac{1}{2}e^u + C$$

$$= -\tfrac{1}{2}e^{-t^2} + C \qquad \text{Since } u = -t^2$$

Check: $D_t(-\frac{1}{2}e^{-t^2}) = -\frac{1}{2}e^{-t^2}D_t(-t^2) = -\frac{1}{2}e^{-t^2}(-2t) = te^{-t^2}$

(C) $\int 4x^2\sqrt{x^3+5}\,dx = 4\int \sqrt{x^3+5}(x^2)\,dx$ — Move the 4 across the integral sign and proceed as before.

$$= 4\int \sqrt{x^3+5}\,\frac{3}{3}(x^2)\,dx$$

$$= \frac{4}{3}\int \sqrt{x^3+5}(3x^2)\,dx \qquad \text{Substitute } u = x^3+5 \text{ and } du = 3x^2\,dx.$$

$$= \frac{4}{3}\int \sqrt{u}\,du$$

$$= \frac{4}{3}\int u^{1/2}\,du \qquad \text{Use formula 2.}$$

$$= \frac{4}{3}\frac{u^{3/2}}{\frac{3}{2}} + C$$

$$= \tfrac{8}{9}u^{3/2} + C$$

$$= \tfrac{8}{9}(x^3+5)^{3/2} + C \qquad \text{Since } u = x^3+5$$

Check: $D_x[\frac{8}{9}(x^3+5)^{3/2}] = \frac{4}{3}(x^3+5)^{1/2}D_x(x^3+5)$
$= \frac{4}{3}(x^3+5)^{1/2}3x^2 = 4x^2\sqrt{x^3+5}$ ◆

PROBLEM 8 Integrate:

(A) $\int e^{-3x}\,dx$ (B) $\int \frac{x}{x^2-9}\,dx$ (C) $\int 5t^2(t^3+4)^{-2}\,dt$ ◆

Even if it is not possible to find a substitution that makes an integrand match one of the integration formulas exactly, a substitution may sufficiently simplify the integrand so that other techniques can be used.

◆ EXAMPLE 9 Find: $\int \frac{x}{\sqrt{x+2}}\,dx$

Solution Proceeding as before, if we let $u = x + 2$, then $du = dx$ and

$$\int \frac{x}{\sqrt{x+2}}\,dx = \int \frac{x}{\sqrt{u}}\,du$$

Notice that this substitution is not yet complete, because we have not expressed the integrand entirely in terms of u and *du*. As we noted earlier, only a constant factor can be moved across an integral sign, so we cannot move x outside the integral sign (as much as we would like to). Instead, we must return to the original substitution, solve for x in terms of u, and use the resulting equation to complete the substitution:

$u = x + 2$ Solve for x in terms of u.

$u - 2 = x$ Substitute this expression for x.

Thus,

$$\begin{aligned}
\int \frac{x}{\sqrt{x+2}}\,dx &= \int \frac{u-2}{\sqrt{u}}\,du && \text{Simplify the integrand.}\\
&= \int \frac{u-2}{u^{1/2}}\,du\\
&= \int (u^{1/2} - 2u^{-1/2})\,du\\
&= \boxed{\int u^{1/2}\,du - 2\int u^{-1/2}\,du}\\
&= \frac{u^{3/2}}{\frac{3}{2}} - 2\,\frac{u^{1/2}}{\frac{1}{2}} + C\\
&= \tfrac{2}{3}(x+2)^{3/2} - 4(x+2)^{1/2} + C && \text{Since } u = x + 2
\end{aligned}$$

Check: $$\begin{aligned}
D_x[\tfrac{2}{3}(x+2)^{3/2} - 4(x+2)^{1/2}] &= (x+2)^{1/2} - 2(x+2)^{-1/2}\\
&= \frac{x+2}{(x+2)^{1/2}} - \frac{2}{(x+2)^{1/2}}\\
&= \frac{x}{(x+2)^{1/2}}
\end{aligned}$$ ◆

PROBLEM 9 Find: $\int x\sqrt{x+1}\,dx$ ◆

◆ APPLICATION

◆ EXAMPLE 10

Price–Demand

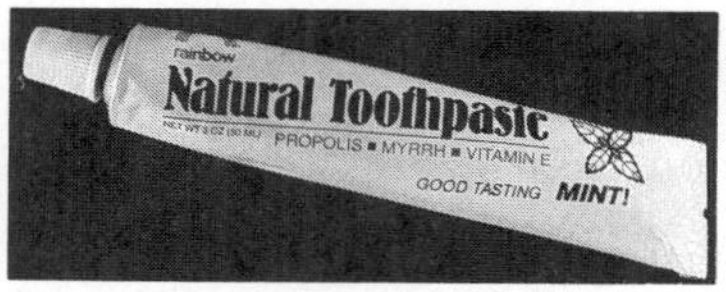

The market research department for a supermarket chain has determined that for one store the marginal price $p'(x)$ at x tubes per week for a certain brand of toothpaste is given by

$$p'(x) = -0.015e^{-0.01x}$$

Find the price–demand equation if the weekly demand is 50 when the price of a tube is \$2.35. Find the weekly demand when the price of a tube is \$1.89.

Solution

$$\begin{aligned} p(x) &= \int -0.015e^{-0.01x}\,dx \\ &= -0.015\int e^{-0.01x}\,dx \\ &= -0.015\int e^{-0.01x}\frac{-0.01}{-0.01}\,dx \\ &= \frac{-0.015}{-0.01}\int e^{-0.01x}(-0.01)\,dx && \text{Substitute } u = -0.01x \text{ and } du = -0.01\,dx. \\ &= 1.5\int e^{u}\,du \\ &= 1.5e^{u} + C \\ &= 1.5e^{-0.01x} + C && \text{Since } u = -0.01x \end{aligned}$$

We find C by noting that

$$\begin{aligned} p(50) = 1.5e^{-0.01(50)} + C &= \$2.35 \\ C &= \$2.35 - 1.5e^{-0.5} && \text{Use a calculator.} \\ C &= \$2.35 - 0.91 \\ C &= \$1.44 \end{aligned}$$

Thus,

$$p(x) = 1.5e^{-0.01x} + 1.44$$

To find the demand when the price is \$1.89, we solve $p(x) = \$1.89$ for x:

$$\begin{aligned} 1.5e^{-0.01x} + 1.44 &= 1.89 \\ 1.5e^{-0.01x} &= 0.45 \\ e^{-0.01x} &= 0.3 \\ -0.01x &= \ln 0.3 \\ x &= -100 \ln 0.3 \approx 120 \text{ tubes} \end{aligned}$$

◆

PROBLEM 10 The marginal price $p'(x)$ at a supply level of x tubes per week for a certain brand of toothpaste is given by

$$p'(x) = 0.001e^{0.01x}$$

Find the price–supply equation if the supplier is willing to supply 100 tubes per week at a price of \$1.65 each. How many tubes would the supplier be willing to supply at a price of \$1.98 each? ◆

Integrals of the form encountered in Example 10 occur so frequently that it is worthwhile to state a general formula for integrals of this type:

$$\int e^{au}\,du = \frac{1}{a}e^{au} + C \qquad \text{where } a \text{ is a constant, } a \neq 0$$

Common Errors

1. $$\int (x^2+3)^2\,dx = \int (x^2+3)^2 \frac{2x}{2x}\,dx$$
$$\neq \frac{1}{2x}\int (x^2+3)^2 2x\,dx$$

Remember: A variable factor cannot be moved across an integral sign! The correct procedure for this problem is

$$\int (x^2+3)^2\,dx = \int (x^4+6x^2+9)\,dx$$
$$= \frac{x^5}{5} + 2x^3 + 9x + C$$

2. $$\int \frac{1}{10x+3}\,dx = \int \frac{1}{u}\,dx \qquad u = 10x+3$$
$$\neq \ln|u| + C$$

Remember: An integral must be expressed entirely in terms of *u* and *du* before applying integration formulas 1–6. The correct procedure for this problem is

$$\int \frac{1}{10x+3}\,dx = \frac{1}{10}\int \frac{1}{10x+3}10\,dx \qquad u = 10x+3,\ du = 10\,dx$$
$$= \frac{1}{10}\int \frac{1}{u}\,du$$
$$= \tfrac{1}{10}\ln|u| + C$$
$$= \tfrac{1}{10}\ln|10x+3| + C$$

Answers to Matched Problems

7. (A) $\frac{1}{5}(2x^3-3)^5 + C$ (B) $e^{5w} + C$
8. (A) $-\frac{1}{3}e^{-3x} + C$ (B) $\frac{1}{2}\ln|x^2-9| + C$ (C) $-\frac{5}{3}(t^3+4)^{-1} + C$
9. $\frac{2}{5}(x+1)^{5/2} - \frac{2}{3}(x+1)^{3/2} + C$ 10. $p(x) = 0.1e^{0.01x} + 1.38$; 179 tubes

EXERCISE 7-2

A *Find each indefinite integral, and check the result by differentiating.*

1. $\int (x^2-4)^5 2x\,dx$ **2.** $\int (x^3+1)^4 3x^2\,dx$ **3.** $\int e^{4x}4\,dx$

4. $\int e^{-3x}(-3)\,dx$ **5.** $\int \frac{1}{2t+3}2\,dt$ **6.** $\int \frac{1}{5t-7}5\,dt$

B **7.** $\int (3x-2)^7\,dx$ **8.** $\int (5x+3)^9\,dx$ **9.** $\int (x^2+3)^7 x\,dx$

10. $\int (x^3-5)^4 x^2\,dx$ **11.** $\int 10e^{-0.5t}\,dt$ **12.** $\int 4e^{0.01t}\,dt$

13. $\int \frac{1}{10x+7}\,dx$ **14.** $\int \frac{1}{100-3x}\,dx$ **15.** $\int xe^{2x^2}\,dx$

16. $\int x^2e^{4x^3}\,dx$ **17.** $\int \frac{x^2}{x^3+4}\,dx$ **18.** $\int \frac{x}{x^2-2}\,dx$

19. $\int \frac{t}{(3t^2+1)^4}\,dt$ **20.** $\int \frac{t^2}{(t^3-2)^5}\,dt$ **21.** $\int \frac{x^2}{(4-x^3)^2}\,dx$

22. $\int \frac{x}{(5-2x^2)^5}\,dx$ **23.** $\int x\sqrt{x+4}\,dx$ **24.** $\int x\sqrt{x-9}\,dx$

25. $\int \frac{x}{\sqrt{x-3}}\,dx$ **26.** $\int \frac{x}{\sqrt{x+5}}\,dx$ **27.** $\int x(x-4)^9\,dx$

28. $\int x(x+6)^8\,dx$ **29.** $\int e^{2x}(1+e^{2x})^3\,dx$ **30.** $\int e^{-x}(1-e^{-x})^4\,dx$

31. $\int \frac{1+x}{4+2x+x^2}\,dx$ **32.** $\int \frac{x^2-1}{x^3-3x+7}\,dx$ **33.** $\int (2x+1)e^{x^2+x+1}\,dx$

34. $\int (x^2+2x)e^{x^3+3x^2}\,dx$ **35.** $\int (e^x-2x)^3(e^x-2)\,dx$

36. $\int (x^2-e^x)^4(2x-e^x)\,dx$ **37.** $\int \frac{x^3+x}{(x^4+2x^2+1)^4}\,dx$

38. $\int \frac{x^2-1}{(x^3-3x+7)^2}\,dx$

C **39.** $\int x\sqrt{3x^2+7}\,dx$ **40.** $\int x^2\sqrt{2x^3+1}\,dx$ **41.** $\int x(x^3+2)^2\,dx$

42. $\int x(x^2+2)^2\,dx$ **43.** $\int x^2(x^3+2)^2\,dx$ **44.** $\int (x^2+2)^2\,dx$

45. $\int \frac{x^3}{\sqrt{2x^4+3}}\,dx$ **46.** $\int \frac{x^2}{\sqrt{4x^3-1}}\,dx$ **47.** $\int \frac{(\ln x)^3}{x}\,dx$

48. $\int \frac{e^x}{1+e^x}\,dx$ **49.** $\int \frac{1}{x^2}e^{-1/x}\,dx$ **50.** $\int \frac{1}{x\ln x}\,dx$

Find the antiderivative of each derivative.

51. $\frac{dx}{dt} = 7t^2(t^3 + 5)^6$ **52.** $\frac{dm}{dn} = 10n(n^2 - 8)^7$ **53.** $\frac{dy}{dt} = \frac{3t}{\sqrt{t^2 - 4}}$

54. $\frac{dy}{dx} = \frac{5x^2}{(x^3 - 7)^4}$ **55.** $\frac{dp}{dx} = \frac{e^x + e^{-x}}{(e^x - e^{-x})^2}$ **56.** $\frac{dm}{dt} = \frac{\ln(t - 5)}{t - 5}$

Use substitution techniques to derive the integration formulas in Problems 57 and 58. Then check your work by differentiation.

57. $\int e^{au}\,du = \frac{1}{a}e^{au} + C, \quad a \neq 0$

58. $\int \frac{1}{au + b}\,du = \frac{1}{a}\ln|au + b| + C, \quad a \neq 0$

APPLICATIONS

Business & Economics

59. *Price–demand equation.* The marginal price for a weekly demand of x bottles of baby shampoo in a drug store is given by

$$p'(x) = \frac{-6{,}000}{(3x + 50)^2}$$

Find the price–demand equation if the weekly demand is 150 when the price of a bottle of shampoo is $4. What is the weekly demand when the price is $2.50?

60. *Price–supply equation.* The marginal price at a supply level of x bottles of baby shampoo per week is given by

$$p'(x) = \frac{300}{(3x + 25)^2}$$

Find the price–supply equation if the distributor of the shampoo is willing to supply 75 bottles a week at a price of $1.60 per bottle. How many bottles would the supplier be willing to supply at a price of $1.75 per bottle?

61. *Cost function.* The weekly marginal cost of producing x pairs of tennis shoes is given by

$$C'(x) = 12 + \frac{500}{x + 1}$$

where C(x) is cost in dollars. If the fixed costs are $2,000 per week, find the cost function. What is the average cost per pair of shoes if 1,000 pairs of shoes are produced each week?

62. *Revenue function.* The weekly marginal revenue from the sale of x pairs of tennis shoes is given by

$$R'(x) = 40 - 0.02x + \frac{200}{x + 1} \qquad R(0) = 0$$

where $R(x)$ is revenue in dollars. Find the revenue function. Find the revenue from the sale of 1,000 pairs of shoes.

63. *Marketing.* An automobile company is ready to introduce a new line of cars with a national sales campaign. After test marketing the line in a carefully selected city, the marketing research department estimates that sales (in millions of dollars) will increase at the monthly rate of

$$S'(t) = 10 - 10e^{-0.1t} \qquad 0 \leq t \leq 24$$

t months after the national campaign has started. What will be the total sales, $S(t)$, t months after the beginning of the national campaign if we assume 0 sales at the beginning of the campaign? What are the estimated total sales for the first 12 months of the campaign?

64. *Marketing.* Repeat Problem 63 if the monthly rate of increase in sales is found to be approximated by

$$S'(t) = 20 - 20e^{-0.05t} \qquad 0 \leq t \leq 24$$

65. *Oil production.* Using data from the first 3 years of production as well as geological studies, the management of an oil company estimates that oil will be pumped from a producing field at a rate given by

$$R(t) = \frac{100}{t+1} + 5 \qquad 0 \leq t \leq 20$$

where $R(t)$ is the rate of production (in thousands of barrels per year) t years after pumping begins. How many barrels of oil, $Q(t)$, will the field produce the first t years if $Q(0) = 0$? How many barrels will be produced the first 9 years?

66. *Oil production.* In Problem 65, if the rate is found to be

$$R(t) = \frac{120t}{t^2+1} + 3 \qquad 0 \leq t \leq 20$$

how many barrels of oil, $Q(t)$, will the field produce the first t years if $Q(0) = 0$? How many barrels will be produced the first 5 years?

Life Sciences

67. *Biology.* A yeast culture is growing at the rate of $W'(t) = 0.2e^{0.1t}$ grams per hour. If the starting culture weighs 2 grams, what will be the weight of the culture, $W(t)$, after t hours? After 8 hours?

68. *Medicine.* The rate of healing for a skin wound (in square centimeters per day) is approximated by $A'(t) = -0.9e^{-0.1t}$. If the initial wound has an area of 9 square centimeters, what will its area, $A(t)$, be after t days? After 5 days?

69. *Pollution.* A contaminated lake is treated with a bactericide. The rate of decrease in harmful bacteria t days after the treatment is given by

$$\frac{dN}{dt} = -\frac{2{,}000t}{1+t^2} \qquad 0 \leq t \leq 10$$

where $N(t)$ is the number of bacteria per milliliter of water. If the initial count was 5,000 bacteria per milliliter, find $N(t)$ and then find the bacteria count after 10 days.

70. *Pollution.* An oil tanker aground on a reef is losing oil and producing an oil slick that is radiating outward at a rate given approximately by

$$\frac{dR}{dt} = \frac{60}{\sqrt{t + 9}} \qquad t \geq 0$$

where R is the radius (in feet) of the circular slick after t minutes. Find the radius of the slick after 16 minutes if the radius is 0 when $t = 0$.

Social Sciences

71. *Learning.* In a particular business college, it was found that an average student enrolled in an advanced typing class progressed at a rate of $N'(t) = 6e^{-0.1t}$ words per minute per week, t weeks after enrolling in a 15 week course. If at the beginning of the course a student could type 40 words per minute, how many words per minute, $N(t)$, would the student be expected to type t weeks into the course? After completing the course?

72. *Learning.* In the same business college, it was also found that an average student enrolled in a beginning shorthand class progressed at a rate of $N'(t) = 12e^{-0.06t}$ words per minute per week, t weeks after enrolling in a 15 week course. If at the beginning of the course a student could take dictation in shorthand at 0 words per minute, how many words per minute, $N(t)$, would the student be expected to handle t weeks into the course? After completing the course?

73. *College enrollment.* The projected rate of increase in enrollment in a new college is estimated by

$$\frac{dE}{dt} = 5{,}000(t + 1)^{-3/2} \qquad t \geq 0$$

where $E(t)$ is the projected enrollment in t years. If enrollment is 2,000 when $t = 0$, find the projected enrollment 15 years from now.

SECTION 7-3 Differential Equations — Growth and Decay

- DIFFERENTIAL EQUATIONS
- CONTINUOUS COMPOUND INTEREST REVISITED
- EXPONENTIAL GROWTH LAW
- POPULATION GROWTH; RADIOACTIVE DECAY; LEARNING
- A COMPARISON OF EXPONENTIAL GROWTH PHENOMENA

◆ DIFFERENTIAL EQUATIONS

In the previous section, we considered equations of the form

$$\frac{dy}{dx} = 6x^2 - 4x \qquad p'(x) = -400e^{-0.04x}$$

These are examples of *differential equations*. In general, an equation is a **differential equation** if it involves an unknown function (often denoted by y) and one or more of its derivatives. Other examples of differential equations are

$$\frac{dy}{dx} = ky \qquad y'' - xy' + x^2 = 5$$

Finding solutions to different types of differential equations (functions that satisfy the equation) is the subject matter for whole books and courses on this topic. Here, we will consider only a few very special but very important types of equations that have immediate and significant applications. We start by considering the problem of continuous compound interest from another point of view, which will enable us to generalize the concept and apply the results to problems from a number of different fields.

◆ CONTINUOUS COMPOUND INTEREST REVISITED

Let P be the initial amount of money deposited in an account, and let A be the amount in the account at any time t. Instead of assuming that the money in the account earns a particular rate of interest, suppose we say that the rate of growth of the amount of money in the account at any time t is proportional to the amount present at that time. Since dA/dt is the rate of growth of A with respect to t, we have

$$\frac{dA}{dt} = rA \qquad A(0) = P \qquad A, P > 0 \tag{1}$$

where r is an appropriate constant. We would like to find a function $A = A(t)$ that satisfies these conditions. Multiplying both sides of equation (1) by $1/A$, we obtain

$$\frac{1}{A}\frac{dA}{dt} = r$$

Now we integrate each side with respect to t:

$$\int \frac{1}{A}\frac{dA}{dt}\,dt = \int r\,dt \qquad \frac{dA}{dt}\,dt = A'(t)\,dt = dA$$

$$\int \frac{1}{A}\,dA = \int r\,dt \qquad \text{Use formula 6 to evaluate the left side.}$$

$$\ln|A| = rt + C \qquad |A| = A \text{ since } A > 0$$

$$\ln A = rt + C$$

We convert this last equation into the equivalent exponential form

$$A = e^{rt+C} \qquad \text{Definition of logarithmic function: } y = \ln x \text{ if and only if } x = e^y$$

$$= e^C e^{rt} \qquad \text{Property of exponents: } b^m b^n = b^{m+n}$$

Since $A(0) = P$, we evaluate $A(t) = e^C e^{rt}$ at $t = 0$ and set it equal to P:

$$A(0) = e^C e^0 = e^C = P$$

Hence, $e^C = P$, and we can rewrite $A = e^C e^{rt}$ in the form

$$A = Pe^{rt}$$

This is the same continuous compound interest formula obtained in Section 6-1, where the principal P is invested at an annual nominal rate of r compounded continuously for t years.

◆ EXPONENTIAL GROWTH LAW

In general, if the rate of change with respect to time of a quantity Q is proportional to the amount present and $Q(0) = Q_0$, then proceeding in exactly the same way as above, we obtain the following:

Exponential Growth Law

If $\dfrac{dQ}{dt} = rQ$ and $Q(0) = Q_0$, then $Q = Q_0 e^{rt}$,

where

Q_0 = Amount at $t = 0$

r = Continuous compound growth rate (expressed as a decimal)

t = Time

Q = Quantity at time t

The constant r in the exponential growth law is sometimes called the **growth constant,** or the **growth rate.** This last term can be misleading, since the rate of growth of Q with respect to time is dQ/dt, not r. Notice that if $r < 0$, then $dQ/dt < 0$ and Q is decreasing. This type of growth is called **exponential decay.**

Once we know that the rate of growth of something is proportional to the amount present, then we know it has exponential growth and we can use the results summarized in the box without having to solve the differential equation each time. The exponential growth law applies not only to money invested at interest compounded continuously, but also to many other types of problems—population growth, radioactive decay, natural resource depletion, and so on.

◆ POPULATION GROWTH; RADIOACTIVE DECAY; LEARNING

The world population is growing at an ever-increasing rate, as illustrated in Figure 2 at the top of the next page. **Population growth** over certain periods of time often can be approximated by the exponential growth law described above.

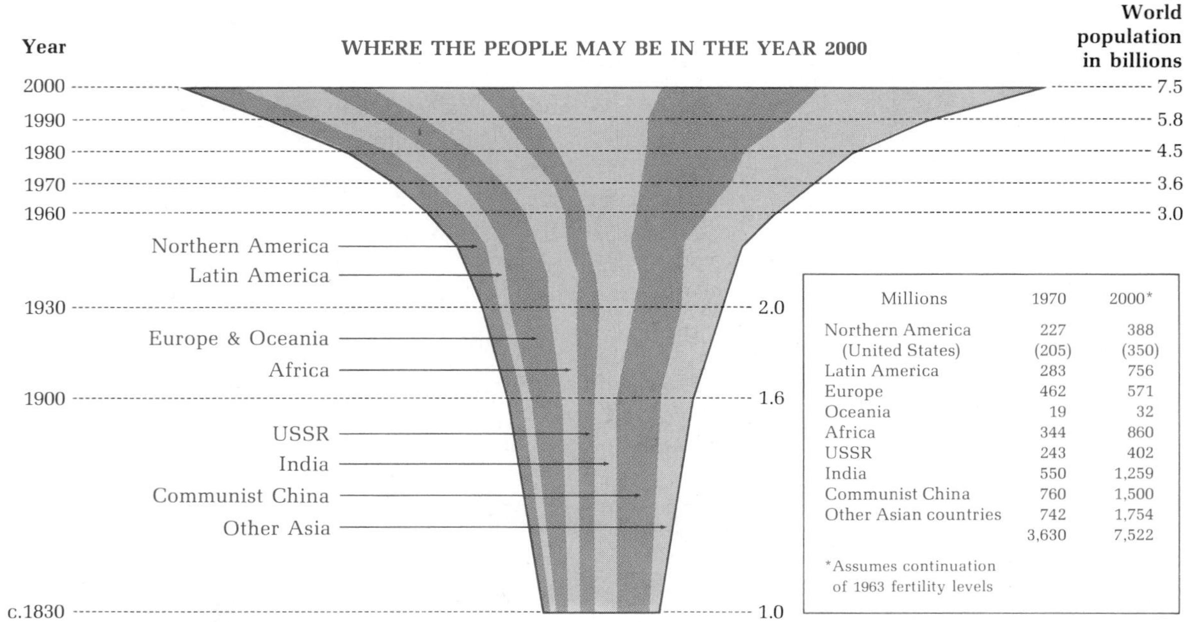

Millions	1970	2000*
Northern America	227	388
(United States)	(205)	(350)
Latin America	283	756
Europe	462	571
Oceania	19	32
Africa	344	860
USSR	243	402
India	550	1,259
Communist China	760	1,500
Other Asian countries	742	1,754
	3,630	7,522

*Assumes continuation of 1963 fertility levels

FIGURE 2
The population explosion
Source: U.S. State Department

◆ EXAMPLE 11

Population Growth

India had a population of 800 million people in 1987 ($t = 0$) and a growth rate of 3% per year (which we will assume is compounded continuously). If P is the population (in millions) t years after 1987 and the same growth rate continues, then

$$\frac{dP}{dt} = 0.03P \qquad P(0) = 800$$

Thus, using the exponential growth law, we obtain

$$P = 800e^{0.03t}$$

With this result, we can estimate the population of India in 2007 ($t = 20$) to be

$$P(20) = 800e^{0.03(20)}$$
$$\approx 1{,}458 \text{ million people}$$

◆

PROBLEM 11

Assuming the same continuous compound growth rate as in Example 11, what will India's population be in the year 2012? ◆

◆ **EXAMPLE 12** *Population Growth*

If the exponential growth law applies to ~~Russia's~~ the Soviet Union's population growth, at what continuous compound growth rate will the population double over the next 100 years?

Solution The problem is to find r, given $P = 2P_0$ and $t = 100$:

$$\begin{aligned} P &= P_0e^{rt} \\ 2P_0 &= P_0e^{100r} \\ 2 &= e^{100r} \\ 100r &= \ln 2 \\ r &= \frac{\ln 2}{100} \\ &\approx 0.0069 \quad \text{or} \quad 0.69\% \end{aligned}$$

Take the natural logarithm of both sides and reverse the equation. ◆

PROBLEM 12 If the exponential growth law applies to population growth in Mexico, find the doubling time of the population if it continues to grow at 3.2% per year compounded continuously. ◆

We now turn to another type of exponential growth—**radioactive decay.** In 1946, Willard Libby (who later received a Nobel prize in chemistry) found that as long as a plant or animal is alive, radioactive carbon-14 is maintained at a constant level in its tissues. Once the plant or animal is dead, however, the radioactive carbon-14 diminishes by radioactive decay at a rate proportional to the amount present. Thus,

$$\frac{dQ}{dt} = rQ \qquad Q(0) = Q_0$$

and we have another example of the exponential growth law. The continuous compound rate of decay for radioactive carbon-14 has been found to be 0.000 123 8; thus, $r = -0.000\ 123\ 8$, since decay implies a negative continuous compound growth rate.

◆ **EXAMPLE 13** *Archaeology*

A piece of human bone was found at an archaeological site in Africa. If 10% of the original amount of radioactive carbon-14 was present, estimate the age of the bone.

Solution Using the exponential growth law for

$$\frac{dQ}{dt} = -0.000\ 123\ 8Q \qquad Q(0) = Q_0$$

we find that

$$Q = Q_0 e^{-0.0001238t}$$

and our problem is to find t so that $Q = 0.1Q_0$ (since the amount of carbon-14 present now is 10% of the amount present, Q_0, at the death of the person). Thus,

$$\begin{aligned} 0.1Q_0 &= Q_0 e^{-0.0001238t} \\ 0.1 &= e^{-0.0001238t} \\ \ln 0.1 &= \ln e^{-0.0001238t} \\ t &= \frac{\ln 0.1}{-0.0001238} \approx 18{,}600 \text{ years} \end{aligned}$$

◆

PROBLEM 13 Estimate the age of the bone in Example 13 if 50% of the original amount of carbon-14 is present. ◆

In learning certain skills such as typing and swimming, a mathematical model often used is one that assumes there is a maximum skill attainable, say, M, and the rate of improving is proportional to the difference between that achieved, y, and that attainable, M. Mathematically,

$$\frac{dy}{dt} = k(M - y) \qquad y(0) = 0$$

We solve this type of problem using the same technique that was used to obtain the exponential growth law. First, multiply both sides of the first equation by $1/(M - y)$ to obtain

$$\frac{1}{M - y}\frac{dy}{dt} = k$$

and then integrate each side with respect to t:

$$\begin{aligned} \int \frac{1}{M - y}\frac{dy}{dt}\,dt &= \int k\,dt \\ -\int \frac{1}{M - y}\left(-\frac{dy}{dt}\right)dt &= \int k\,dt && \text{Substitute } u = M - y \text{ and } du = -dy = -\frac{dy}{dt}\,dt. \\ -\int \frac{1}{u}\,du &= \int k\,dt \\ -\ln|u| &= kt + C && \text{Substitute } M - y = u. \\ -\ln(M - y) &= kt + C && \text{Absolute value signs are not required.} \\ \ln(M - y) &= -kt - C && \text{(Why?)} \end{aligned}$$

Change this last equation to equivalent exponential form:

$$M - y = e^{-kt-C}$$
$$M - y = e^{-C}e^{-kt}$$
$$y = M - e^{-C}e^{-kt}$$

Now, $y(0) = 0$; hence,

$$y(0) = M - e^{-C}e^{0} = 0$$

Solving for e^{-C}, we obtain

$$e^{-C} = M$$

and our final solution is

$$y = M - Me^{-kt} = M(1 - e^{-kt})$$

◆ EXAMPLE 14 *Learning*

For a particular person who is learning to swim, it is found that the distance y (in feet) the person is able to swim in 1 minute after t hours of practice is given approximately by

$$y = 50(1 - e^{-0.04t})$$

What is the rate of improvement after 10 hours of practice?

Solution

$$y = 50 - 50e^{-0.04t}$$
$$y'(t) = 2e^{-0.04t}$$
$$y'(10) = 2e^{-0.04(10)} \approx 1.34 \text{ feet per hour of practice}$$ ◆

PROBLEM 14 In Example 14, what is the rate of improvement after 50 hours of practice? ◆

◆ A COMPARISON OF EXPONENTIAL GROWTH PHENOMENA

The graphs and equations given in Table 1 (on the next page) compare several widely used growth models. These are divided basically into two groups: unlimited growth and limited growth. Following each equation and graph is a short (and necessarily incomplete) list of areas in which the models are used. This only touches on a subject that has been extensively developed and which you are likely to encounter in greater depth in the future.

TABLE 1

Exponential Growth

DESCRIPTION	MODEL	SOLUTION	GRAPH	USES
Unlimited growth: Rate of growth is proportional to the amount present	$\frac{dy}{dt} = ky$ $k, t > 0$ $y(0) = c$	$y = ce^{kt}$	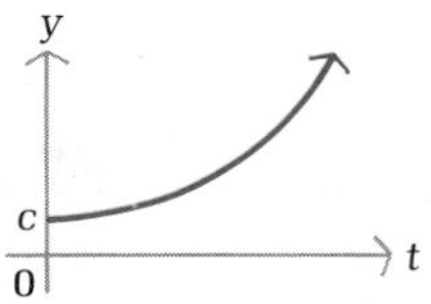	• Short-term population growth (people, bacteria, etc.) • Growth of money at continuous compound interest • Price–supply curves • Depletion of natural resources
Exponential decay: Rate of growth is proportional to the amount present	$\frac{dy}{dt} = -ky$ $k, t > 0$ $y(0) = c$	$y = ce^{-kt}$	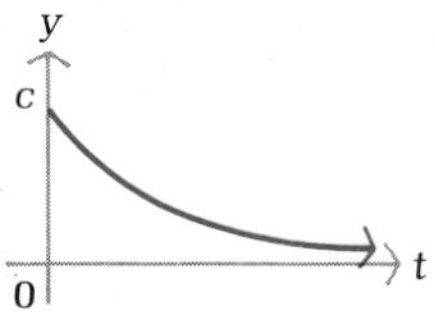	• Radioactive decay • Light absorption in water • Price–demand curves • Atmospheric pressure (t is altitude)
Limited growth: Rate of growth is proportional to the difference between the amount present and a fixed limit	$\frac{dy}{dt} = k(M - y)$ $k, t > 0$ $y(0) = 0$	$y = M(1 - e^{-kt})$	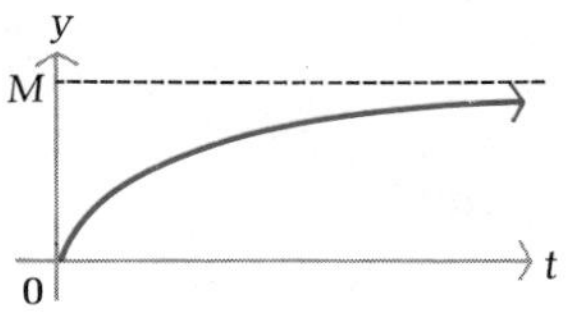	• Sales fads (e.g., skateboards) • Depreciation of equipment • Company growth • Learning
Logistic growth: Rate of growth is proportional to the amount present and to the difference between the amount present and a fixed limit	$\frac{dy}{dt} = ky(M - y)$ $k, t > 0$ $y(0) = \frac{M}{1 + c}$	$y = \frac{M}{1 + ce^{-kMt}}$	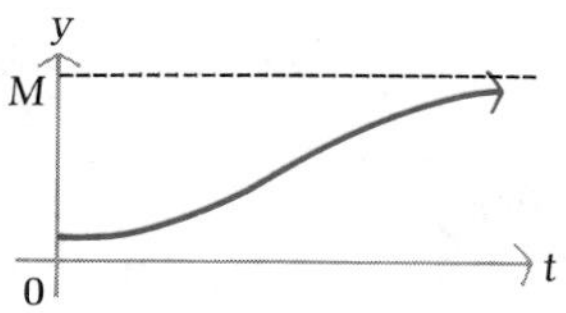	• Long-term population growth • Epidemics • Sales of new products • Rumor spread • Company growth

Answers to Matched Problems

11. 1,694 million people
12. Approx. 22 yr
13. Approx. 5,600 yr
14. Approx. 0.27 ft/hr

EXERCISE 7-3

APPLICATIONS

Business & Economics

1. *Continuous compound interest.* Find the amount A in an account after t years if

$$\frac{dA}{dt} = 0.08A \qquad \text{and} \qquad A(0) = 1{,}000$$

2. *Continuous compound interest.* Find the amount A in an account after t years if

$$\frac{dA}{dt} = 0.12A \qquad \text{and} \qquad A(0) = 5{,}250$$

3. *Continuous compound interest.* Find the amount A in an account after t years if

$$\frac{dA}{dt} = rA \qquad A(0) = 8{,}000 \qquad A(2) = 9{,}020$$

4. *Continuous compound interest.* Find the amount A in an account after t years if

$$\frac{dA}{dt} = rA \qquad A(0) = 5{,}000 \qquad A(5) = 7{,}460$$

5. *Price–demand.* If the marginal price dp/dx at x units of demand per week is proportional to the price p, and if at a price of \$100 there is no weekly demand [$p(0) = 100$] and at a price of \$77.88 there is a weekly demand of 5 units [$p(5) = 77.88$], find the price–demand equation.

6. *Price–supply.* If the marginal price dp/dx at x units of supply per day is proportional to the price p, and if at a price of \$10 there is no daily supply [$p(0) = 10$] and at a price of \$12.84 there is a daily supply of 50 units [$p(50) = 12.84$], find the price–supply equation.

7. *Advertising.* A company is trying to expose a new product to as many people as possible through television advertising. Suppose the rate of exposure to new people is proportional to the number of those who have not seen the product out of L possible viewers. If no one is aware of the product at the start of the campaign and after 10 days 40% of L are aware of the product, solve

$$\frac{dN}{dt} = k(L - N) \qquad N(0) = 0 \qquad N(10) = 0.4L$$

for $N = N(t)$, the number of people who are aware of the product after t days of advertising.

8. *Advertising.* Repeat Problem 7 for

$$\frac{dN}{dt} = k(L - N) \qquad N(0) = 0 \qquad N(10) = 0.1L$$

Life Sciences

9. *Ecology.* For relatively clear bodies of water, light intensity is reduced according to

$$\frac{dI}{dx} = -kI \qquad I(0) = I_0$$

where I is the intensity of light at x feet below the surface. For the Sargasso Sea off the West Indies, $k = 0.009\ 42$. Find I in terms of x, and find the depth at which the light is reduced to half of that at the surface.

10. *Blood pressure.* It can be shown under certain assumptions that blood pressure P in the largest artery in the human body (the aorta) changes between beats with respect to time t according to

$$\frac{dP}{dt} = -aP \qquad P(0) = P_0$$

where a is a constant. Find $P = P(t)$ that satisfies both conditions.

11. *Drug concentrations.* A single injection of a drug is administered to a patient. The amount Q in the body then decreases at a rate proportional to the amount present, and for this particular drug the rate is 4% per hour. Thus,

$$\frac{dQ}{dt} = -0.04Q \qquad Q(0) = Q_0$$

where t is time in hours. If the initial injection is 3 milliliters $[Q(0) = 3]$, find $Q = Q(t)$ that satisfies both conditions. How many milliliters of the drug are still in the body after 10 hours?

12. *Simple epidemic.* A community of 1,000 individuals is assumed to be homogeneously mixed. One individual who has just returned from another community has influenza. Assume the home community has not had influenza shots and all are susceptible. One mathematical model for an influenza epidemic assumes that influenza tends to spread at a rate in direct proportion to the number who have it, N, and to the number who have not contracted it, in this case, $1{,}000 - N$. Mathematically,

$$\frac{dN}{dt} = kN(1{,}000 - N) \qquad N(0) = 1$$

where N is the number of people who have contracted influenza after t days. For $k = 0.0004$, it can be shown that $N(t)$ is given by

$$N(t) = \frac{1{,}000}{1 + 999e^{-0.4t}}$$

See Table 1 (logistic growth) for the characteristic graph.

(A) How many people have contracted influenza after 10 days? After 20 days?

(B) How many days will it take until half the community has contracted influenza?

(C) Find $\lim_{t\to\infty} N(t)$.

Reuters/Bettmann Newsphotos

13. *Nuclear accident.* One of the dangerous radioactive isotopes detected after the Chernobyl nuclear accident in 1986 was cesium-137. If 93.3% of the cesium-137 emitted during the accident is still present 3 years later, find the continuous compound rate of decay of this isotope.

14. *Insecticides.* Many countries have banned the use of the insecticide DDT because of its long-term adverse effects. Five years after a particular country stopped using DDT, the amount of DDT in the ecosystem had declined to 75% of the amount present at the time of the ban. Find the continuous compound rate of decay of DDT.

Social Sciences

15. *Archaeology.* A skull from an ancient tomb was discovered and was found to have 5% of the original amount of radioactive carbon-14 present. Estimate the age of the skull. (See Example 13.)

16. *Learning.* For a particular person learning to type, it was found that the number of words per minute, N, the person was able to type after t hours of practice was given approximately by

$$N = 100(1 - e^{-0.02t})$$

See Table 1 (limited growth) for a characteristic graph. What is the rate of improvement after 10 hours of practice? After 40 hours of practice?

17. *Small group analysis.* In a study on small group dynamics, sociologists Stephan and Mischler found that, when the members of a discussion group of 10 were ranked according to the number of times each participated, the number of times $N(k)$ the kth-ranked person participated was given approximately by

$$N(k) = N_1 e^{-0.11(k-1)} \qquad 1 \leq k \leq 10$$

where N_1 is the number of times the 1st-ranked person participated in the discussion. If, in a particular discussion group of 10 people, $N_1 = 180$, estimate how many times the 6th-ranked person participated. The 10th-ranked person.

18. *Perception.* One of the oldest laws in mathematical psychology is the Weber–Fechner law (discovered in the middle of the nineteenth century). It concerns a person's sensed perception of various strengths of stimulation involving weights, sound, light, shock, taste, and so on. One form of the law states that the rate of change of sensed sensation S with respect to stimulus R is inversely proportional to the strength of the stimulus R. Thus,

$$\frac{dS}{dR} = \frac{k}{R}$$

where k is a constant. If we let R_0 be the threshold level at which the stimulus R can be detected (the least amount of sound, light, weight, and so on, that can be detected), then it is appropriate to write

$$S(R_0) = 0$$

Find a function S in terms of R that satisfies the above conditions.

19. *Rumor spread.* A group of 400 parents, relatives, and friends are waiting anxiously at Kennedy Airport for a student charter flight to return after a

year in Europe. It is stormy and the plane is late. A particular parent thought he had heard that the plane's radio had gone out and related this news to some friends, who in turn passed it on to others, and so on. Sociologists have studied rumor propagation and have found that a rumor tends to spread at a rate in direct proportion to the number who have heard it, x, and to the number who have not, $P - x$, where P is the total population. Mathematically, for our case, $P = 400$ and

$$\frac{dx}{dt} = 0.001x(400 - x) \qquad x(0) = 1$$

where t is time (in minutes). From this, it can be shown that

$$x(t) = \frac{400}{1 + 399e^{-0.4t}}$$

See Table 1 (logistic growth) for a characteristic graph.

(A) How many people have heard the rumor after 5 minutes? 20 minutes?
(B) Find $\lim_{t\to\infty} x(t)$.

SECTION 7-4 Definite Integrals

- DEFINITE INTEGRALS
- PROPERTIES
- DEFINITE INTEGRALS AND SUBSTITUTION
- APPLICATIONS

◆ DEFINITE INTEGRALS

We start this discussion with a simple example, out of which will evolve a new integral form, called the *definite integral*. Our approach in this section will be intuitive and informal; these concepts will be made more precise in the next chapter.

Suppose a manufacturing company's marginal cost equation for a particular product is given by

$$C'(x) = 2 - 0.2x \qquad 0 \leq x \leq 8$$

where the marginal cost is in thousands of dollars and production is x units per day. What is the total change in cost per day going from a production level of 2 units per day to 6 units per day? If $C = C(x)$ is the cost function, then

$$\begin{pmatrix}\text{Total net change in cost}\\ \text{between } x = 2 \text{ and } x = 6\end{pmatrix} = C(6) - C(2) = C(x)\Big|_2^6 \tag{1}$$

The special symbol $C(x)|_2^6$ is a convenient way of representing the center expression, and it will prove useful to us later.

To evaluate (1), we need to find the antiderivative of $C'(x)$; that is,

$$C(x) = \int (2 - 0.2x)\, dx = 2x - 0.1x^2 + K \tag{2}$$

Thus, we are within a constant of knowing the original cost function. However, we do not need to know the constant K to solve the original problem (1). We compute $C(6) - C(2)$ for $C(x)$ found in (2):

$$\begin{aligned} C(6) - C(2) &= [2(6) - 0.1(6)^2 + K] - [2(2) - 0.1(2)^2 + K] \\ &= 12 - 3.6 + K - 4 + 0.4 - K \\ &= \$4.8 \text{ thousand per day increase in costs for a production} \\ &\qquad \text{increase from 2 to 6 units per day} \end{aligned}$$

The unknown constant K canceled out! Thus, we conclude that any antiderivative of $C'(x) = 2 - 0.2x$ will do, since antiderivatives of a given function can differ by at most a constant (see Section 7-1). Thus, we really do not have to find the constant in the original cost function to solve the problem.

Since $C(x)$ is an antiderivative of $C'(x)$, the above discussion suggests the following notation:

$$C(6) - C(2) = C(x)\Big|_2^6 = \int_2^6 C'(x)\, dx \tag{3}$$

The integral form on the right in (3) is called a *definite integral*—it represents the number found by evaluating an antiderivative of the integrand at 6 and 2 and taking the difference as indicated.

Definite Integral

The **definite integral** of a continuous function f over an interval from $x = a$ to $x = b$ is the net change of an antiderivative of f over the interval. Symbolically, if $F(x)$ is an antiderivative of $f(x)$, then

$$\int_a^b f(x)\, dx = F(x)\Big|_a^b = F(b) - F(a) \qquad \text{where} \quad F'(x) = f(x)$$

INTEGRAND: $f(x)$ UPPER LIMIT: b LOWER LIMIT: a

The numbers a and b are referred to as the **limits of integration.** Note that the lower limit a can be larger than the upper limit b, as long as f is continuous on an interval containing both a and b.

In the next chapter, we will formally define a definite integral as a limit of a special sum. Then the relationship in the box above turns out to be the most important theorem in calculus—the *fundamental theorem of calculus.* Our intent in this and the next section is to give you some intuitive experience with the definite integral concept and its use. You will then be better able to understand a formal definition and to appreciate the significance of the fundamental theorem.

◆ EXAMPLE 15 Evaluate: $\int_{-1}^{2} (3x^2 - 2x)\, dx$

Solution We choose the simplest antiderivative of the integrand $3x^2 - 2x$, namely, $x^3 - x^2$, since any antiderivative will do:

$$\int_{-1}^{2} (3x^2 - 2x)\, dx = (x^3 - x^2)\Big|_{-1}^{2}$$

$$= (2^3 - 2^2) - [(-1)^3 - (-1)^2]$$

$$= 4 - (-2) = 6$$

Be careful of sign errors here. ◆

PROBLEM 15 Evaluate: $\int_{-2}^{2} (2x - 1)\, dx$ ◆

On Definite and Indefinite Integrals

Do not confuse a definite integral with an indefinite integral. The definite integral $\int_a^b f(x)\, dx$ is a real number; the indefinite integral $\int f(x)\, dx$ is a whole set of functions—all the antiderivatives of $f(x)$.

◆ PROPERTIES

In the next box we state several useful properties of the definite integral. Note that properties 3 and 4 parallel properties 3 and 4 given earlier for the indefinite integral.

Definite Integral Properties

1. $\int_a^a f(x)\, dx = 0$
2. $\int_a^b f(x)\, dx = -\int_b^a f(x)\, dx$
3. $\int_a^b kf(x)\, dx = k\int_a^b f(x)\, dx$ k a constant
4. $\int_a^b [f(x) \pm g(x)]\, dx = \int_a^b f(x)\, dx \pm \int_a^b g(x)\, dx$
5. $\int_a^b f(x)\, dx = \int_a^c f(x)\, dx + \int_c^b f(x)\, dx$

These properties are justified as follows: If $F'(x) = f(x)$, then:

1. $\int_a^a f(x)\,dx = F(x)\Big|_a^a = F(a) - F(a) = 0$

2. $\int_a^b f(x)\,dx = F(x)\Big|_a^b = F(b) - F(a) = -[F(a) - F(b)] = -\int_b^a f(x)\,dx$

3. $\int_a^b kf(x)\,dx = kF(x)\Big|_a^b = kF(b) - kF(a) = k[F(b) - F(a)] = k\int_a^b f(x)\,dx$

And so on.

◆ EXAMPLE 16 Evaluate: $\int_1^2 \left(2x + 3e^x - \frac{4}{x}\right) dx$

Solution

$$\begin{aligned}\int_1^2 \left(2x + 3e^x - \frac{4}{x}\right) dx &= 2\int_1^2 x\,dx + 3\int_1^2 e^x\,dx - 4\int_1^2 \frac{1}{x}\,dx \\ &= 2\,\frac{x^2}{2}\Big|_1^2 + 3e^x\Big|_1^2 - 4\ln|x|\Big|_1^2 \\ &= (2^2 - 1^2) + (3e^2 - 3e^1) - (4\ln 2 - 4\ln 1) \\ &= 3 + 3e^2 - 3e - 4\ln 2 \approx 14.24\end{aligned}$$ ◆

PROBLEM 16 Evaluate: $\int_1^3 \left(4x - 2e^x + \frac{5}{x}\right) dx$ ◆

◆ DEFINITE INTEGRALS AND SUBSTITUTION

The evaluation of a definite integral is a two-step process: First, find an antiderivative, and then find the net change in that antiderivative. If substitution techniques are required to find the antiderivative, there are two different ways to proceed. The next example illustrates both methods.

◆ EXAMPLE 17 Evaluate: $\int_0^5 \frac{x}{x^2 + 10}\,dx$

Solution We will solve this problem using substitution in two different ways:

Method 1. Use substitution in an indefinite integral to find an antiderivative as a function of x; then evaluate the definite integral:

$$\begin{aligned}\int \frac{x}{x^2 + 10}\,dx &= \tfrac{1}{2}\int \frac{1}{x^2 + 10}\,2x\,dx && \text{Substitute } u = x^2 + 10 \text{ and } du = 2x\,dx. \\ &= \tfrac{1}{2}\int \frac{1}{u}\,du \\ &= \tfrac{1}{2}\ln|u| + C \\ &= \tfrac{1}{2}\ln(x^2 + 10) + C && \text{Since } u = x^2 + 10 > 0\end{aligned}$$

We choose $C = 0$ and use the antiderivative $\frac{1}{2}\ln(x^2 + 10)$ to evaluate the definite integral:

$$\int_0^5 \frac{x}{x^2+10}\,dx = \frac{1}{2}\ln(x^2+10)\Big|_0^5$$
$$= \tfrac{1}{2}\ln 35 - \tfrac{1}{2}\ln 10 \approx 0.626$$

Method 2. Substitute directly in the definite integral, changing both the variable of integration and the limits of integration: In the definite integral

$$\int_0^5 \frac{x}{x^2+10}\,dx$$

the upper limit is $x = 5$ and the lower limit is $x = 0$. When we make the substitution $u = x^2 + 10$ in this definite integral, we must change the limits of integration to the corresponding values of u:

$x = 5$	implies	$u = 5^2 + 10 = 35$	New upper limit
$x = 0$	implies	$u = 0^2 + 10 = 10$	New lower limit

Thus, we have

$$\int_0^5 \frac{x}{x^2+10}\,dx = \frac{1}{2}\int_0^5 \frac{1}{x^2+10}\,2x\,dx$$
$$= \frac{1}{2}\int_{10}^{35} \frac{1}{u}\,du$$
$$= \frac{1}{2}\left(\ln|u|\Big|_{10}^{35}\right)$$
$$= \tfrac{1}{2}(\ln 35 - \ln 10) \approx 0.626$$

◆

PROBLEM 17 Use both methods described in Example 17 to evaluate

$$\int_0^1 \frac{1}{2x+4}\,dx$$

◆

◆ EXAMPLE 18 Use method 2 described in Example 17 to evaluate

$$\int_{-4}^{1} \sqrt{5-t}\,dt$$

Solution If $u = 5 - t$, then $du = -dt$, and

$t = 1$	implies	$u = 5 - 1 = 4$	New upper limit
$t = -4$	implies	$u = 5 - (-4) = 9$	New lower limit

Notice that the lower limit for u is larger than the upper limit. Be careful not to reverse these two values when substituting in the definite integral.

$$\begin{aligned}
\int_{-4}^{1} \sqrt{5-t}\,dt &= -\int_{-4}^{1} \sqrt{5-t}\,(-dt) \\
&= -\int_{9}^{4} \sqrt{u}\,du \\
&= -\int_{9}^{4} u^{1/2}\,du \\
&= -\left(\frac{u^{3/2}}{\frac{3}{2}}\bigg|_{9}^{4}\right) \\
&= -[\tfrac{2}{3}(4)^{3/2} - \tfrac{2}{3}(9)^{3/2}] \\
&= -[\tfrac{16}{3} - \tfrac{54}{3}] = \tfrac{38}{3} \approx 12.667
\end{aligned}$$

◆

PROBLEM 18 Use method 2 described in Example 17 to evaluate

$$\int_{2}^{5} \frac{1}{\sqrt{6-t}}\,dt$$

◆

◆ APPLICATIONS

In Section 7-1, we used the indefinite integral of the rate of change (derivative) of a function to find the original function. Now we can use the definite integral of the derivative of a function to find the total net change in the function. That is, the **total change** in a function $F(x)$ as x changes from a to b is

$$\int_{a}^{b} F'(x)\,dx = F(x)\bigg|_{a}^{b} = F(b) - F(a)$$

◆ EXAMPLE 19 A company manufactures x television sets per month. The monthly marginal profit (in dollars) is given by

Change in Profit

$$P'(x) = 165 - 0.1x \qquad 0 \leq x \leq 4{,}000$$

The company is currently manufacturing 1,500 sets per month, but is planning to increase production. Find the total change in the monthly profit if monthly production is increased to:

(A) 1,600 sets (B) 2,000 sets

Solutions

(A)
$$\begin{aligned}
P(1{,}600) - P(1{,}500) &= \int_{1{,}500}^{1{,}600} (165 - 0.1x)\,dx \\
&= (165x - 0.05x^2)\bigg|_{1{,}500}^{1{,}600} \\
&= [165(1{,}600) - 0.05(1{,}600)^2] \\
&\quad - [165(1{,}500) - 0.05(1{,}500)^2] \\
&= 136{,}000 - 135{,}000 \\
&= 1{,}000
\end{aligned}$$

Thus, increasing monthly production from 1,500 units to 1,600 units will increase the monthly profit by only \$1,000.

(B) $$P(2{,}000) - P(1{,}500) = \int_{1{,}500}^{2{,}000} (165 - 0.1x)\,dx$$

$$= (165x - 0.05x^2)\Big|_{1{,}500}^{2{,}000}$$

$$= [165(2{,}000) - 0.05(2{,}000)^2] - [165(1{,}500) - 0.05(1{,}500)^2]$$
$$= 130{,}000 - 135{,}000 = -5{,}000$$

Thus, increasing monthly production from 1,500 units to 2,000 units will decrease the monthly profit by \$5,000. ◆

PROBLEM 19 Repeat Example 19 if

$$P'(x) = 300 - 0.2x \qquad 0 \leq x \leq 3{,}000$$

and monthly production is increased from 1,400 sets to:

(A) 1,500 sets (B) 1,700 sets ◆

◆ EXAMPLE 20 *Useful Life*

An amusement company maintains records for each video game it installs in an arcade. Suppose that $C(t)$ and $R(t)$ represent the total accumulated costs and revenues (in thousands of dollars), respectively, t years after a particular game has been installed and that

$$C'(t) = 2 \qquad R'(t) = 9e^{-0.5t}$$

The value of t for which $C'(t) = R'(t)$ is called the **useful life** of the game.

(A) Find the useful life of the game to the nearest year.
(B) Find the total profit accumulated during the useful life of the game.

Solutions

(A)
$$R'(t) = C'(t)$$
$$9e^{-0.5t} = 2$$
$$e^{-0.5t} = \tfrac{2}{9} \qquad \text{Convert to equivalent logarithmic form.}$$
$$-0.5t = \ln \tfrac{2}{9}$$
$$t = -2 \ln \tfrac{2}{9} \approx 3 \text{ years}$$

Thus, the game has a useful life of 3 years.

(B) The total profit accumulated during the useful life of the game is

$$P(3) - P(0) = \int_0^3 P'(t)\,dt$$

$$= \int_0^3 [R'(t) - C'(t)]\,dt$$

$$= \int_0^3 (9e^{-0.5t} - 2)\,dt$$

$$= \left(\frac{9}{-0.5} e^{-0.5t} - 2t \right) \Big|_0^3 \qquad \textit{Recall: } \int e^{ax}\,dx = \frac{1}{a} e^{ax} + C$$

$$= (-18e^{-0.5t} - 2t) \Big|_0^3$$

$$= (-18e^{-1.5} - 6) - (-18e^0 - 0)$$

$$= 12 - 18e^{-1.5} \approx 7.984 \quad \text{or} \quad \$7{,}984$$

◆

PROBLEM 20 Repeat Example 20 if $C'(t) = 1$ and $R'(t) = 7.5e^{-0.5t}$. ◆

Common Errors

1. $\int_0^2 e^x\,dx = e^x \Big|_0^2 \neq e^2$

Do not forget to evaluate the antiderivative at both the upper and lower limits of integration, and do not assume that the antiderivative is 0 just because the lower limit is 0. The correct procedure for this integral is

$$\int_0^2 e^x\,dx = e^x \Big|_0^2 = e^2 - e^0 = e^2 - 1$$

2. $\int_2^5 \frac{1}{2x+3}\,dx \neq \frac{1}{2} \int_2^5 \frac{1}{u}\,du \qquad u = 2x + 3,\ du = 2\,dx$

If a substitution is made in a definite integral, the limits of integration also must be changed. The new limits are determined by the particular substitution used in the integral. The correct procedure for this integral is

$$\int_2^5 \frac{1}{2x+3}\,dx = \frac{1}{2} \int_7^{13} \frac{1}{u}\,du \qquad \begin{array}{l} x = 5 \text{ implies } u = 2(5) + 3 = 13 \\ x = 2 \text{ implies } u = 2(2) + 3 = 7 \end{array}$$

$$= \frac{1}{2} \left(\ln|u| \Big|_7^{13} \right)$$

$$= \tfrac{1}{2}(\ln 13 - \ln 7)$$

Answers to Matched Problems

15. -4 16. $16 + 2e - 2e^3 + 5 \ln 3 \approx -13.241$
17. $\frac{1}{2}(\ln 6 - \ln 4) \approx 0.203$ 18. 2
19. (A) \$1,000 (B) −\$3,000
20. (A) $-2 \ln \frac{2}{15} \approx 4$ yr (B) $11 - 15e^{-2} \approx 8.970$ or \$8,970

EXERCISE 7-4

Evaluate.

A

1. $\int_2^3 2x\,dx$
2. $\int_1^2 3x^2\,dx$
3. $\int_3^4 5\,dx$
4. $\int_{12}^{20} dx$
5. $\int_1^3 (2x-3)\,dx$
6. $\int_1^3 (6x+5)\,dx$
7. $\int_0^4 (3x^2-4)\,dx$
8. $\int_0^2 (6x^2-2x)\,dx$
9. $\int_{-3}^4 (4-x^2)\,dx$
10. $\int_{-1}^2 (x^2-4x)\,dx$
11. $\int_0^1 24x^{11}\,dx$
12. $\int_0^2 30x^5\,dx$
13. $\int_0^1 e^{2x}\,dx$
14. $\int_{-1}^1 e^{5x}\,dx$
15. $\int_1^{3.5} 2x^{-1}\,dx$
16. $\int_1^2 \frac{dx}{x}$

B

17. $\int_1^2 (2x^{-2}-3)\,dx$
18. $\int_1^2 (5-16x^{-3})\,dx$
19. $\int_1^4 3\sqrt{x}\,dx$
20. $\int_4^{25} \frac{2}{\sqrt{x}}\,dx$
21. $\int_2^3 12(x^2-4)^5 x\,dx$
22. $\int_0^1 32(x^2+1)^7 x\,dx$
23. $\int_3^9 \frac{1}{x-1}\,dx$
24. $\int_2^8 \frac{1}{x+1}\,dx$
25. $\int_{-5}^{10} e^{-0.05x}\,dx$
26. $\int_{-10}^{25} e^{-0.01x}\,dx$
27. $\int_{-6}^{0} \sqrt{4-2x}\,dx$
28. $\int_{-4}^{2} \frac{1}{\sqrt{8-2x}}\,dx$
29. $\int_{-1}^{7} \frac{x}{\sqrt{x+2}}\,dx$
30. $\int_0^3 x\sqrt{x+1}\,dx$
31. $\int_0^1 (e^{2x}-2x)^2(e^{2x}-1)\,dx$
32. $\int_0^1 \frac{2e^{4x}-3}{e^{2x}}\,dx$
33. $\int_{-2}^{-1} (x^{-1}+2x)\,dx$
34. $\int_{-3}^{-1} (-3x^{-2}+x^{-1})\,dx$

C

35. $\int_2^3 x\sqrt{2x^2-3}\,dx$
36. $\int_0^1 x\sqrt{3x^2+2}\,dx$
37. $\int_0^1 \frac{x-1}{x^2-2x+3}\,dx$
38. $\int_1^2 \frac{x+1}{2x^2+4x+4}\,dx$
39. $\int_{-1}^1 \frac{e^{-x}-e^x}{(e^{-x}+e^x)^2}\,dx$
40. $\int_6^7 \frac{\ln(t-5)}{t-5}\,dt$

APPLICATIONS

Business & Economics

41. *Salvage value.* A new piece of industrial equipment will depreciate in value rapidly at first, then less rapidly as time goes on. Suppose the rate (in dollars per year) at which the book value of a new milling machine changes

is given approximately by

$$V'(t) = f(t) = 500(t - 12) \qquad 0 \le t \le 10$$

where $V(t)$ is the value of the machine after t years. What is the total loss in value of the machine in the first 5 years? In the second 5 years? Set up appropriate integrals and solve.

42. *Maintenance costs.* Maintenance costs for an apartment house generally increase as the building gets older. From past records, a managerial service determines that the rate of increase in maintenance costs (in dollars per year) for a particular apartment complex is given approximately by

$$M'(x) = f(x) = 90x^2 + 5{,}000$$

where x is the age of the apartment complex in years and $M(x)$ is the total (accumulated) cost of maintenance for x years. Write a definite integral that will give the total maintenance costs from 2 to 7 years after the apartment complex was built, and evaluate it.

43. *Useful life.* The total accumulated costs $C(t)$ and revenues $R(t)$ (in thousands of dollars), respectively, for a coin-operated photocopying machine satisfy

$$C'(t) = \tfrac{1}{11}t \qquad \text{and} \qquad R'(t) = 5te^{-t^2}$$

where t is time in years. Find the useful life of the machine to the nearest year. What is the total profit accumulated during the useful life of the machine?

44. *Useful life.* The total accumulated costs $C(t)$ and revenues $R(t)$ (in thousands of dollars), respectively, for a coal mine satisfy

$$C'(t) = 3 \qquad \text{and} \qquad R'(t) = 15e^{-0.1t}$$

where t is the number of years the mine has been in operation. Find the useful life of the mine to the nearest year. What is the total profit accumulated during the useful life of the mine?

45. *Labor costs and learning.* A defense contractor is starting production on a new missile control system. On the basis of data collected while assembling the first 16 control systems, the production manager obtained the following function for rate of labor use:

$$g(x) = 2{,}400x^{-1/2}$$

where $g(x)$ is the number of labor-hours required to assemble the xth unit of a control system. Approximately how many labor-hours will be required to assemble the 17th through the 25th control units? [*Hint:* Let $a = 16$ and $b = 25$.]

46. *Labor costs and learning.* If the rate of labor use in Problem 45 is

$$g(x) = 2{,}000x^{-1/3}$$

approximately how many labor-hours will be required to assemble the 9th through the 27th control units? [*Hint:* Let $a = 8$ and $b = 27$.]

47. *Oil production.* Using data from the first 3 years of production as well as geological studies, the management of an oil company estimates that oil will be pumped from a producing field at a rate given by

$$R(t) = \frac{100}{t+1} + 5 \qquad 0 \le t \le 20$$

where $R(t)$ is the rate of production (in thousands of barrels per year) t years after pumping begins. Approximately how many barrels of oil will the field produce during the first 10 years of production? From the end of the 10th year to the end of the 20th year of production?

48. *Oil production.* In Problem 47, if the rate is found to be

$$R(t) = \frac{120t}{t^2+1} + 3 \qquad 0 \le t \le 20$$

approximately how many barrels of oil will the field produce during the first 5 years of production? The second 5 years of production?

49. *Marketing.* An automobile company is ready to introduce a new line of cars with a national sales campaign. After test marketing the line in a carefully selected city, the marketing research department estimates that sales (in millions of dollars) will increase at the monthly rate of

$$S'(t) = 10 - 10e^{-0.1t} \qquad 0 \le t \le 24$$

t months after the national campaign is started. What will be the approximate total sales during the first 12 months of the campaign? The second 12 months of the campaign?

50. *Marketing.* Repeat Problem 49 if the monthly rate of increase in sales is found to be approximated by

$$S'(t) = 20 - 20e^{-0.05t} \qquad 0 \le t \le 24$$

Life Sciences

51. *Natural resource depletion.* The instantaneous rate of change of demand for wood in the United States since 1970 ($t = 0$) in billions of cubic feet per year is estimated to be given by

$$Q'(t) = 12 + 0.006t^2 \qquad 0 \le t \le 50$$

where $Q(t)$ is the total amount of wood consumed t years after 1970. How many billions of cubic feet of wood will have been consumed from 1980 to 1990?

52. *Natural resource depletion.* Repeat Problem 51 for the time interval from 1990 to 2000.

53. *Biology.* A yeast culture weighing 2 grams is removed from a refrigerator unit and is expected to grow at the rate of $W'(t) = 0.2e^{0.1t}$ grams per hour at a higher controlled temperature. How much will the weight of the culture increase during the first 8 hours of growth? How much will the weight of the culture increase from the end of the 8th hour to the end of the 16th hour of growth?

54. *Medicine.* The rate of healing for a skin wound (in square centimeters per day) is approximated by $A'(t) = -0.9e^{-0.1t}$. The initial wound has an area of 9 square centimeters. How much will the area change during the first 5 days? The second 5 days?

Social Sciences

55. *Learning.* In a particular business college, it was found that an average student enrolled in an advanced typing class progressed at a rate of $N'(t) = 6e^{-0.1t}$ words per minute per week, t weeks after enrolling in a 15 week course. At the beginning of the course, an average student could type 40 words per minute. How much improvement would be expected during the first 5 weeks of the course? The second 5 weeks of the course? The last 5 weeks of the course?

56. *Learning.* In the same business college, it was also found that an average student enrolled in a beginning shorthand class progressed at a rate of $N'(t) = 12e^{-0.06t}$ words per minute per week, t weeks after enrolling in a 15 week course. At the beginning of the course, none of the students could take any dictation by shorthand. How much improvement would be expected during the first 5 weeks of the course? The second 5 weeks of the course? The last 5 weeks of the course?

SECTION 7-5 Area and the Definite Integral

- AREA UNDER A CURVE
- AREA BETWEEN A CURVE AND THE x AXIS
- AREA BETWEEN TWO CURVES
- APPLICATION: DISTRIBUTION OF INCOME

AREA UNDER A CURVE

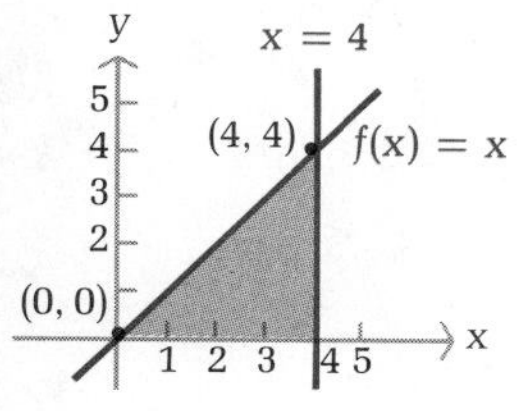

FIGURE 3

Consider the graph of $f(x) = x$ from $x = 0$ to $x = 4$ (Fig. 3). We can easily compute the area of the triangle bounded by $f(x) = x$, the x axis ($y = 0$), and the line $x = 4$, using the formula for the area of a triangle:

$$A = \frac{bh}{2} = \frac{4 \cdot 4}{2} = 8$$

Now, let us integrate $f(x) = x$ from $x = 0$ to $x = 4$:

$$\int_0^4 x\,dx = \frac{x^2}{2}\bigg|_0^4 = \frac{(4)^2}{2} - \frac{(0)^2}{2} = 8$$

We get the same result! It turns out that this is not a coincidence. In general, we can prove the following:

Area under a Curve

If f is continuous and $f(x) \geq 0$ over the interval $[a, b]$, then the area bounded by $y = f(x)$, the x axis ($y = 0$), and the vertical lines $x = a$ and $x = b$ is given exactly by

$$A = \int_a^b f(x)\,dx$$

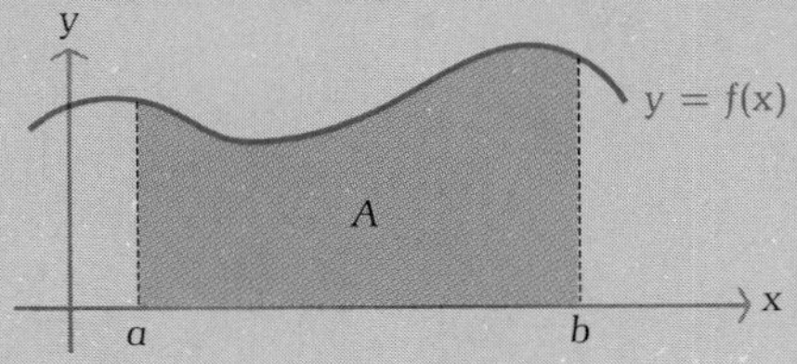

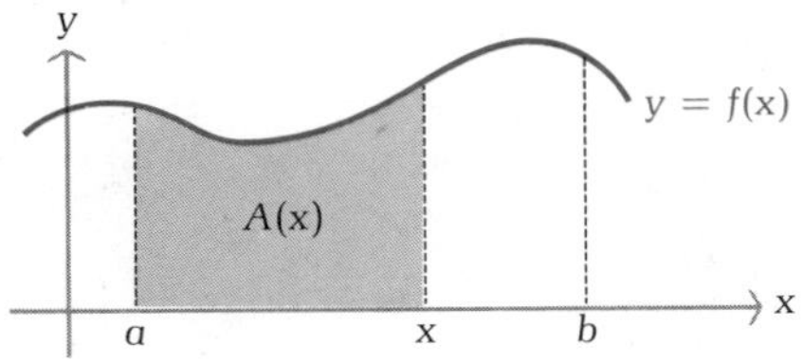

FIGURE 4
$A(x) =$ Area from a to x

Let us see why the definite integral gives us the area exactly. Let $A(x)$ be the area under the graph of $y = f(x)$ from a to x, as indicated in Figure 4. If we can show that $A(x)$ is an antiderivative of $f(x)$, then we can write

$$\int_a^b f(x)\,dx = A(x)\Big|_a^b = A(b) - A(a)$$

$$= \begin{pmatrix}\text{Area from}\\ x = a \text{ to } x = b\end{pmatrix} - \begin{pmatrix}\text{Area from}\\ x = a \text{ to } x = a\end{pmatrix}$$

$$= A - 0 = A$$

To show that $A(x)$ is an antiderivative of $f(x)$—that is, $A'(x) = f(x)$—we use the definition of a derivative (Section 4-3) and write

$$A'(x) = \lim_{h \to 0} \frac{A(x + h) - A(x)}{h}$$

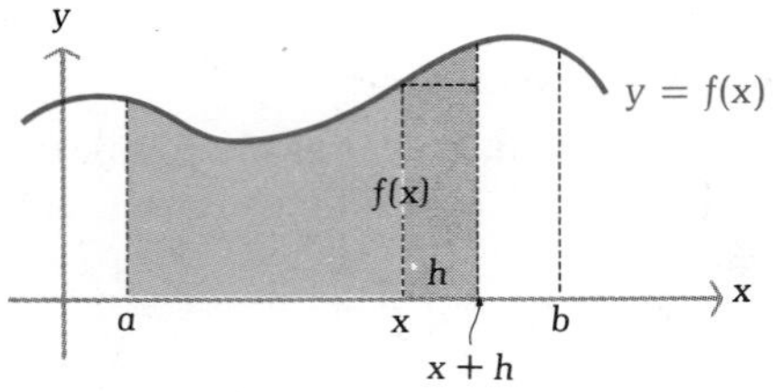

FIGURE 5

Geometrically, $A(x + h) - A(x)$ is the area from x to $x + h$ (see Fig. 5). This area is given approximately by the area of the rectangle $h \cdot f(x)$, and the smaller h is, the better the approximation. Using

$$A(x + h) - A(x) \approx h \cdot f(x)$$

and dividing both sides by h, we obtain

$$\frac{A(x + h) - A(x)}{h} \approx f(x)$$

Now, if we let $h \to 0$, then the left side has $A'(x)$ as a limit, which is equal to the right side. Hence,

$$A'(x) = f(x)$$

that is, $A(x)$ is an antiderivative of $f(x)$. Thus,

$$\int_a^b f(x)\,dx = A(x)\Big|_a^b = A(b) - A(a) = A - 0 = A$$

This is a remarkable result: The area under the graph of $y = f(x)$, $f(x) \geq 0$, can be obtained simply by evaluating the antiderivative of $f(x)$ at the end points of the interval $[a, b]$. We have now solved, at least in part, the third basic problem of calculus stated at the beginning of Chapter 4.

◆ EXAMPLE 21 Find the area bounded by $f(x) = 6x - x^2$ and $y = 0$ for $1 \leq x \leq 4$.

Solution We sketch a graph of the region first. (The solution of every area problem should begin with a sketch.) Then,

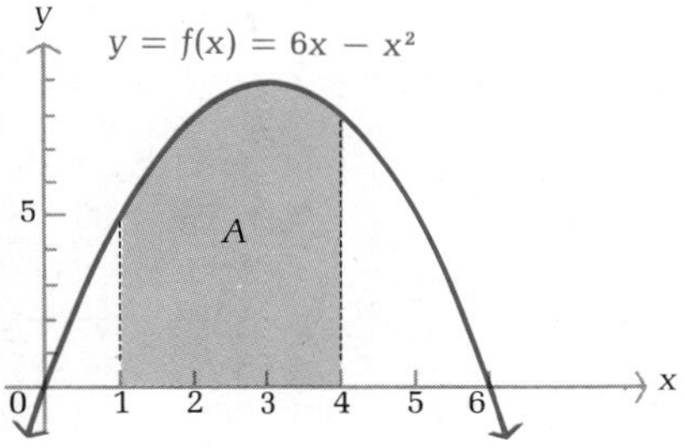

$$A = \int_1^4 (6x - x^2)\,dx = \left(3x^2 - \frac{x^3}{3}\right)\Big|_1^4$$

$$= \left[3(4)^2 - \frac{(4)^3}{3}\right] - \left[3(1)^2 - \frac{(1)^3}{3}\right]$$

$$= 48 - \tfrac{64}{3} - 3 + \tfrac{1}{3}$$

$$= 48 - 21 - 3 = 24$$ ◆

PROBLEM 21 Find the area bounded by $f(x) = x^2 + 1$ and $y = 0$ for $-1 \leq x \leq 3$. ◆

◆ AREA BETWEEN A CURVE AND THE X AXIS

The condition $f(x) \geq 0$ is essential to the relationship between an area under a graph and the definite integral. How can we find the area between the graph of f and the x axis if $f(x) \leq 0$ on $[a, b]$ or if $f(x)$ is both positive and negative on $[a, b]$? To begin, suppose $f(x) \leq 0$ and A is the area between the graph of f and the x axis for $a \leq x \leq b$, as illustrated in Figure 6A.

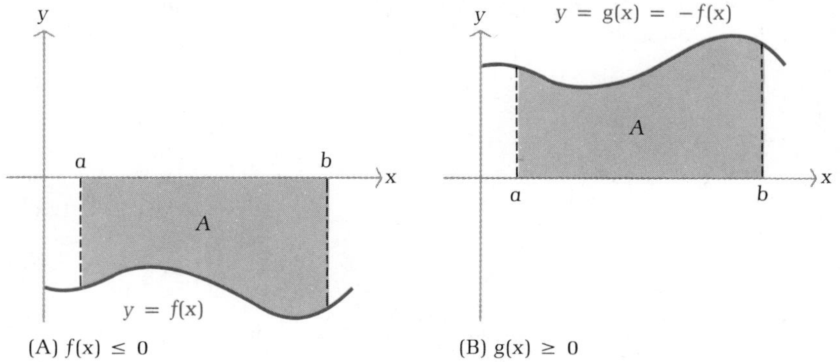

FIGURE 6

If we let $g(x) = -f(x)$, then A is also the area between the graph of g and the x axis for $a \leq x \leq b$ (see Fig. 6B). Since $g(x) \geq 0$ for $a \leq x \leq b$, we can use the definite integral of g to find A:

$$A = \int_a^b g(x)\,dx = \int_a^b [-f(x)]\,dx$$

Thus, **the area between the graph of a negative function and the x axis is equal to the definite integral of the negative of the function.** Finally, if $f(x)$ is positive for some values of x and negative for others, the area between the graph of f and the x axis can be obtained by dividing $[a, b]$ into intervals over which f is always positive or always negative, finding the area over each interval, and then summing these areas.

◆ EXAMPLE 22

Find the area between the graph of $f(x) = x^2 - 2x$ and the x axis over the indicated intervals:

(A) $[1, 2]$ (B) $[-1, 1]$

Solutions

We begin by sketching the graph of f, as shown in the margin.

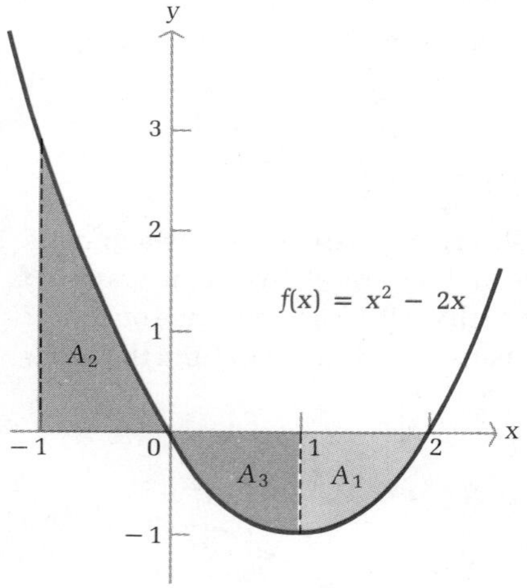

(A) From the graph, we see that $f(x) \leq 0$ for $1 \leq x \leq 2$, so we integrate $-f(x)$:

$$\begin{aligned} A_1 &= \int_1^2 [-f(x)]\,dx \\ &= \int_1^2 (2x - x^2)\,dx \\ &= \left(x^2 - \frac{x^3}{3}\right)\Big|_1^2 \\ &= \left[(2)^2 - \frac{(2)^3}{3}\right] - \left[(1)^2 - \frac{(1)^3}{3}\right] \\ &= 4 - \tfrac{8}{3} - 1 + \tfrac{1}{3} = \tfrac{2}{3} \end{aligned}$$

(B) Since the graph shows that $f(x) \geq 0$ on $[-1, 0]$ and $f(x) \leq 0$ on $[0, 1]$, the computation of this area will require two integrals:

$$\begin{aligned} A &= A_2 + A_3 \\ &= \int_{-1}^0 f(x)\,dx + \int_0^1 [-f(x)]\,dx \\ &= \int_{-1}^0 (x^2 - 2x)\,dx + \int_0^1 (2x - x^2)\,dx \\ &= \left(\frac{x^3}{3} - x^2\right)\Big|_{-1}^0 + \left(x^2 - \frac{x^3}{3}\right)\Big|_0^1 \\ &= \tfrac{4}{3} + \tfrac{2}{3} = 2 \end{aligned}$$

◆

PROBLEM 22 Find the area between the graph of $f(x) = x^2 - 9$ and the x axis over the indicated intervals:

(A) [0, 2] (B) [2, 4] ◆

◆ AREA BETWEEN TWO CURVES

Consider the area bounded by $y = f(x)$ and $y = g(x)$, where $f(x) \geq g(x) \geq 0$, for $a \leq x \leq b$, as indicated in Figure 7.

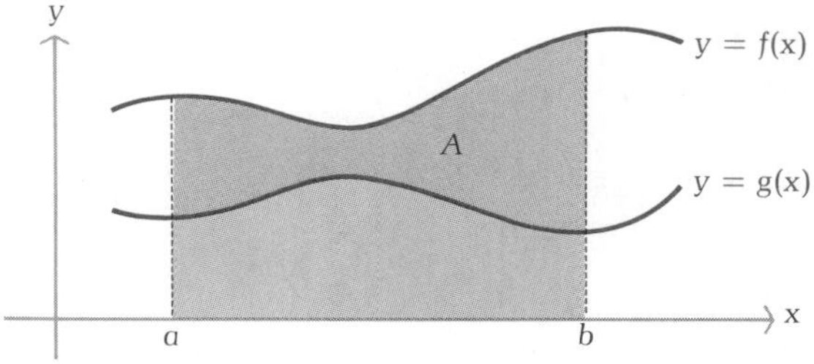

FIGURE 7

$$\begin{pmatrix}\text{Area } A \text{ between}\\ f(x) \text{ and } g(x)\end{pmatrix} = \begin{pmatrix}\text{Area}\\ \text{under } f(x)\end{pmatrix} - \begin{pmatrix}\text{Area}\\ \text{under } g(x)\end{pmatrix}$$ Areas are from $x = a$ to $x = b$ above the x axis.

$$= \int_a^b f(x)\,dx - \int_a^b g(x)\,dx$$ Use definite integral property 4 (Section 7-4).

$$= \int_a^b [f(x) - g(x)]\,dx$$

It can be shown that the above result does not require $f(x)$ or $g(x)$ to remain positive over the interval $[a, b]$. A more general result is stated in the box:

Area between Two Curves

If f and g are continuous and $f(x) \geq g(x)$ over the interval $[a, b]$, then the area bounded by $y = f(x)$ and $y = g(x)$ for $a \leq x \leq b$ is given exactly by

$$A = \int_a^b [f(x) - g(x)]\,dx$$

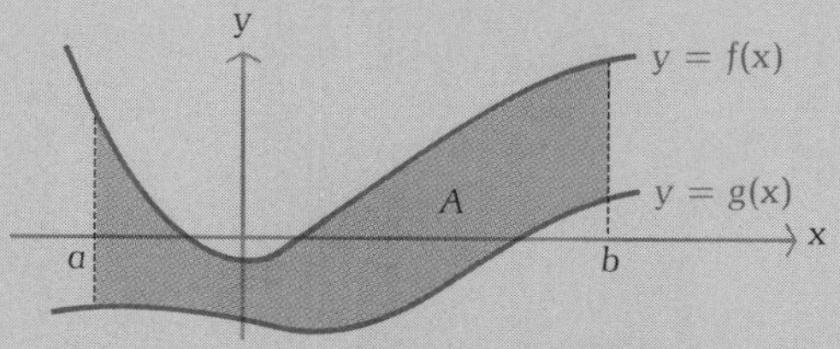

◆ EXAMPLE 23 Find the area bounded by $f(x) = \frac{1}{2}x + 3$, $g(x) = -x^2 + 1$, $x = -2$, and $x = 1$.

Solution We first sketch the area, then set up and evaluate an appropriate definite integral:

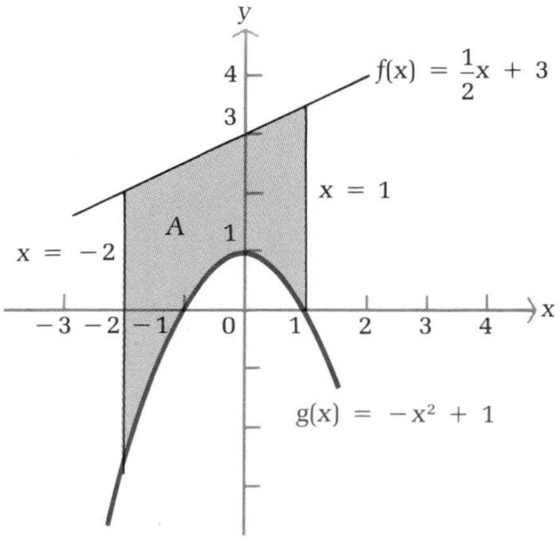

We observe from the graph that $f(x) \geqslant g(x)$ for $-2 \leqslant x \leqslant 1$, so

$$A = \int_{-2}^{1} [f(x) - g(x)]\, dx = \int_{-2}^{1} \left[\left(\frac{x}{2} + 3\right) - (-x^2 + 1)\right] dx$$

$$= \int_{-2}^{1} \left(x^2 + \frac{x}{2} + 2\right) dx$$

$$= \left(\frac{x^3}{3} + \frac{x^2}{4} + 2x\right)\Big|_{-2}^{1} = \left(\frac{1}{3} + \frac{1}{4} + 2\right) - \left(\frac{-8}{3} + \frac{4}{4} - 4\right) = \frac{33}{4}$$ ◆

PROBLEM 23 Find the area bounded by $f(x) = x^2 - 1$, $g(x) = -\frac{1}{2}x - 3$, $x = -1$, and $x = 2$. ◆

◆ EXAMPLE 24 Find the area bounded by $f(x) = 5 - x^2$ and $g(x) = 2 - 2x$.

Solution First, graph f and g on the same coordinate system, as shown in the margin. Since the statement of the problem does not include any limits on the values of x, we must determine the appropriate values from the graph. The graph of f is a parabola and the graph of g is a line, as shown in the figure. The area bounded by these two graphs extends from the intersection point on the left to the intersection point on the right. To find these intersection points, we solve the equation $f(x) = g(x)$ for x:

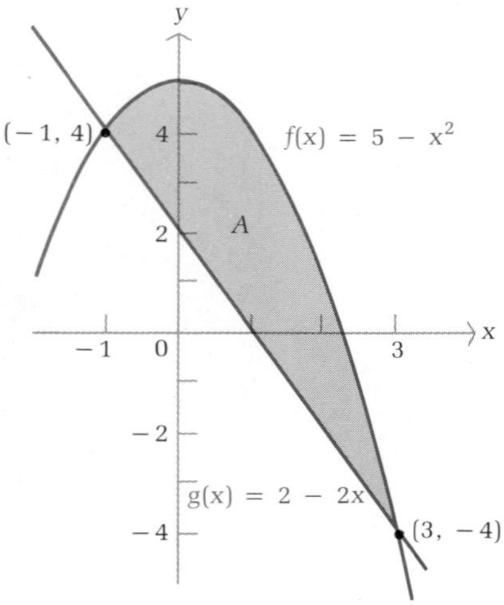

$$f(x) = g(x)$$
$$5 - x^2 = 2 - 2x$$
$$x^2 - 2x - 3 = 0$$
$$x = -1, 3$$

(Note that the area between the graphs for $x < -1$ is unbounded on the left, and the area between the graphs for $x > 3$ is unbounded on the right.) The figure shows that $f(x) \geq g(x)$ over the interval $[-1, 3]$, so we have

$$A = \int_{-1}^{3} [f(x) - g(x)]\, dx = \int_{-1}^{3} [5 - x^2 - (2 - 2x)]\, dx$$

$$= \int_{-1}^{3} (3 + 2x - x^2)\, dx$$

$$= \left(3x + x^2 - \frac{x^3}{3}\right)\Big|_{-1}^{3}$$

$$= \left[3(3) + (3)^2 - \frac{(3)^3}{3}\right] - \left[3(-1) + (-1)^2 - \frac{(-1)^3}{3}\right]$$

$$= 9 + 9 - 9 + 3 - 1 - \tfrac{1}{3} = \tfrac{32}{3}$$ ◆

PROBLEM 24 Find the area bounded by $f(x) = 6 - x^2$ and $g(x) = x$. ◆

◆ EXAMPLE 25 Find the area bounded by $f(x) = x^2 - x$ and $g(x) = 2x$ for $-2 \leq x \leq 3$.

Solution The graphs of f and g are shown in the figure. Examining the graph, we see that $f(x) \geq g(x)$ on the interval $[-2, 0]$, but $g(x) \geq f(x)$ on the interval $[0, 3]$. Thus, two integrals are required to compute this area:

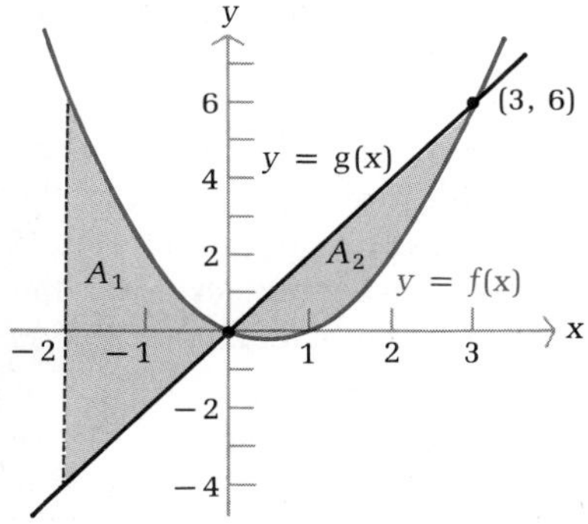

$$A_1 = \int_{-2}^{0} [f(x) - g(x)]\, dx \qquad f(x) \geq g(x) \text{ on } [-2, 0]$$

$$= \int_{-2}^{0} [x^2 - x - 2x]\, dx$$

$$= \int_{-2}^{0} (x^2 - 3x)\, dx$$

$$= \left(\frac{x^3}{3} - \frac{3}{2}x^2\right)\Big|_{-2}^{0}$$

$$= (0) - \left[\frac{(-2)^3}{3} - \frac{3}{2}(-2)^2\right]$$

$$= \tfrac{8}{3} + 6 = \tfrac{26}{3}$$

$$A_2 = \int_{0}^{3} [g(x) - f(x)]\, dx \qquad g(x) \geq f(x) \text{ on } [0, 3]$$

$$= \int_{0}^{3} [2x - (x^2 - x)]\, dx$$

$$= \int_{0}^{3} (3x - x^2)\, dx$$

$$= \left(\frac{3}{2}x^2 - \frac{x^3}{3}\right)\Big|_{0}^{3}$$

$$= \left[\frac{3}{2}(3)^2 - \frac{(3)^3}{3}\right] - (0)$$

$$= \tfrac{27}{2} - 9 = \tfrac{9}{2}$$

The total area area between the two graphs is

$$A = A_1 + A_2 = \frac{26}{3} + \frac{9}{2} = \frac{79}{6}$$ ◆

PROBLEM 25 Find the area bounded by $f(x) = 2x^2$ and $g(x) = 4 - 2x$ for $-2 \leq x \leq 2$. ◆

◆ APPLICATION: DISTRIBUTION OF INCOME

TABLE 2

Income Distribution in the United States (1983)

INCOME LEVEL	x	y
Under $10,000	0.16	0.03
Under $15,000	0.28	0.08
Under $25,000	0.51	0.24
Under $35,000	0.71	0.44
Under $50,000	0.88	0.69
Under $75,000	0.97	0.88

Source: U.S. Bureau of the Census

Economists often use a graph called a **Lorenz curve** to provide a graphical description of the distribution of income among various groups of people. For example, the distribution of personal income in the United States given in Table 2 can be represented by the Lorenz curve shown in Figure 8. In both Table 2 and Figure 8, **x represents the cumulative percentage of families at or below a given income level** and **y represents the cumulative percentage of total personal income received by all these families.** For example, the point (0.51, 0.24) on the Lorenz curve in Figure 8 indicates that the bottom 51% of families (those with incomes under $25,000) received 24% of the total income, the point (0.71, 0.44) indicates that the bottom 71% of families received 44% of the total income, and so on.

Absolute equality of income would occur if every family received the same income. That is, 20% of the families receive 20% of the income, 40% of the families receive 40% of the income, and so on. This distribution is represented by the graph of $y = x$ in Figure 8. The area between the Lorenz curve and the line $y = x$ can be used to measure how much the distribution of income differs from absolute equality.

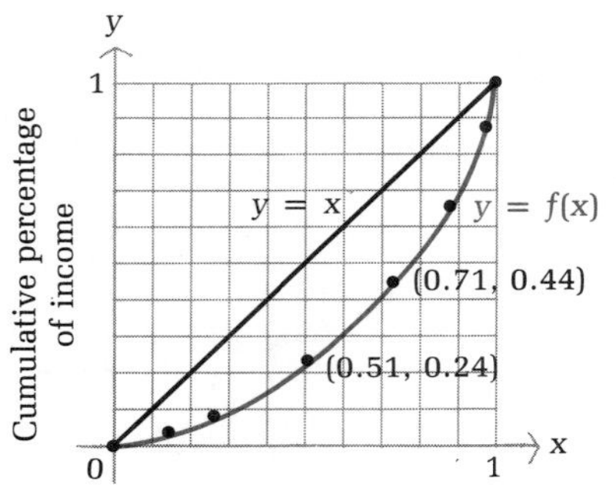

FIGURE 8
Lorenz curve, $y = f(x)$

More precisely, the **coefficient of inequality** of income distribution is defined to be the ratio of the area between the line $y = x$ and the Lorenz curve to the area between the line $y = x$ and the x axis. If we are given a function f whose graph is a Lorenz curve, then the area between the line $y = x$ and the Lorenz curve is $\int_0^1 [x - f(x)]\,dx$. Since the area between the graph of the line $y = x$ and the x axis from $x = 0$ to $x = 1$ is $\frac{1}{2}$, it follows that the coefficient of inequality is given by

$$\frac{\text{Area between } y = x \text{ and } y = f(x)}{\text{Area between } y = x \text{ and x axis}} = \frac{\int_0^1 [x - f(x)]\,dx}{\frac{1}{2}} = 2\int_0^1 [x - f(x)]\,dx$$

Coefficient of Inequality

If $y = f(x)$ is the equation of a Lorenz curve, then

$$\text{Coefficient of inequality} = 2\int_0^1 [x - f(x)]\,dx$$

The coefficient of inequality is always a number between 0 and 1. If the coefficient is near 0, then the income distribution is close to absolute equality. The closer the coefficient is to 1, the greater the inequality (or disparity) of the distribution. The coefficient of inequality can be used to compare income distributions at various points in time, between different groups of people, before and after taxes are paid, between different countries, and so on.

◆ EXAMPLE 26

Distribution of Income

The Lorenz curve for the distribution of income in a certain country in 1990 is given by $f(x) = x^{2.6}$. Economists predict that the Lorenz curve for the country in the year 2010 will be given by $g(x) = x^{1.8}$. Find the coefficient of inequality for each curve, and interpret the results.

Solution

The Lorenz curves are shown in the figure below:

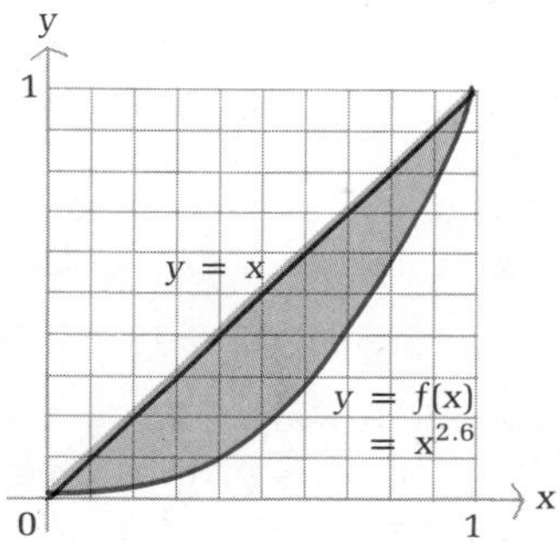

(A) Lorenz curve for 1990

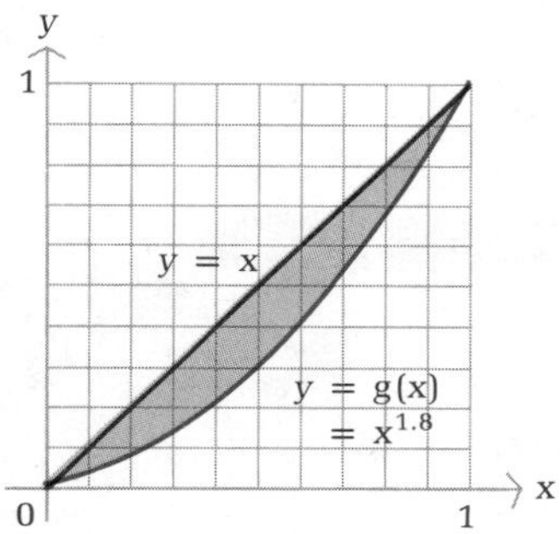

(B) Projected Lorenz curve for 2010

The coefficient of inequality in 1990 is (see figure A)

$$2\int_0^1 [x - f(x)]\,dx = 2\int_0^1 [x - x^{2.6}]\,dx$$

$$= 2\left(\frac{1}{2}x^2 - \frac{1}{3.6}x^{3.6}\right)\Big|_0^1$$

$$= 2\left(\frac{1}{2} - \frac{1}{3.6}\right) \approx 0.444$$

The projected coefficient of inequality in 2010 is (see figure B)

$$2\int_0^1 [x - g(x)]\,dx = 2\int_0^1 [x - x^{1.8}]\,dx$$

$$= 2\left(\frac{1}{2}x^2 - \frac{1}{2.8}x^{2.8}\right)\Big|_0^1$$

$$= 2\left(\frac{1}{2} - \frac{1}{2.8}\right) \approx 0.286$$

If this projection is correct, the coefficient of inequality will decrease, and income will be more equally distributed in the year 2010 than in 1990. ◆

PROBLEM 26 Repeat Example 26 if the projected Lorenz curve in the year 2010 is given by $g(x) = x^{3.8}$. ◆

Answers to Matched Problems

21. $A = \int_{-1}^{3} (x^2 + 1)\,dx = \frac{40}{3}$
22. (A) $A = \int_{0}^{2} (9 - x^2)\,dx = \frac{46}{3}$
    (B) $A = \int_{2}^{3} (9 - x^2)\,dx + \int_{3}^{4} (x^2 - 9)\,dx = 6$
23. $A = \int_{-1}^{2} \left[(x^2 - 1) - \left(-\frac{x}{2} - 3\right)\right] dx = \frac{39}{4}$
24. $A = \int_{-3}^{2} [(6 - x^2) - x]\,dx = \frac{125}{6}$
25. $A = \int_{-2}^{1} [(4 - 2x) - 2x^2]\,dx + \int_{1}^{2} [2x^2 - (4 - 2x)]\,dx = \frac{38}{3}$
26. Coefficient of inequality ≈ 0.583; income will be less equally distributed in 2010.

EXERCISE 7-5

Find the area bounded by the graphs of the indicated equations.

A

1. $y = 2x + 4, \quad y = 0, \quad 1 \le x \le 3$
2. $y = -2x + 6, \quad y = 0, \quad 0 \le x \le 2$
3. $y = 3x^2 \quad y = 0, \quad 1 \le x \le 2$
4. $y = 4x^3, \quad y = 0, \quad 1 \le x \le 2$
5. $y = x^2 + 2, \quad y = 0, \quad -1 \le x \le 0$
6. $y = 3x^2 + 1, \quad y = 0, \quad -2 \le x \le 0$
7. $y = 4 - x^2, \quad y = 0, \quad -1 \le x \le 2$
8. $y = 12 - 3x^2, \quad y = 0, \quad -2 \le x \le 1$
9. $y = e^x, \quad y = 0, \quad -1 \le x \le 2$
10. $y = e^{-x}, \quad y = 0, \quad -2 \le x \le 1$
11. $y = 1/t, \quad y = 0, \quad 0.5 \le t \le 1$
12. $y = 1/t, \quad y = 0, \quad 0.1 \le t \le 1$

B

13. $y = 12, \quad y = -2x + 8, \quad -1 \le x \le 2$
14. $y = 3, \quad y = 2x + 6, \quad -1 \le x \le 2$
15. $y = 3x^2, \quad y = 12$
16. $y = x^2, \quad y = 9$
17. $y = 4 - x^2, \quad y = -5$
18. $y = x^2 - 1, \quad y = 3$
19. $y = x^2 + 1, \quad y = 2x - 2, \quad -1 \le x \le 2$
20. $y = x^2 - 1, \quad y = x - 2, \quad -2 \le x \le 1$
21. $y = -x, \quad y = 0, \quad -2 \le x \le 1$
22. $y = -x + 1, \quad y = 0, \quad -1 \le x \le 2$
23. $y = e^{0.5x}, \quad y = -1/x, \quad 1 \le x \le 2$
24. $y = 1/x, \quad y = -e^x, \quad 0.5 \le x \le 1$

C

25. $y = x^2 - 4, \quad y = 0, \quad 0 \leq x \leq 3$
26. $y = 4\sqrt[3]{x}, \quad y = 0, \quad -1 \leq x \leq 8$
27. $y = 10 - 2x, \quad y = 4 + 2x, \quad 0 \leq x \leq 4$
28. $y = 3x \quad y = x + 5, \quad 0 \leq x \leq 5$
29. $y = 5x - x^2, \quad y = x + 3, \quad 0 \leq x \leq 5$
30. $y = x^2 - 6x + 9, \quad y = 5 - x, \quad 0 \leq x \leq 5$
31. $y = x^2 + 2x + 3, \quad y = 2x + 4$
32. $y = 8 + 4x - x^2, \quad y = x^2 - 2x$
33. $y = x^2 - 4x - 10, \quad y = 14 - 2x - x^2$
34. $y = 6 + 6x - x^2, \quad y = 13 - 2x$
35. $y = x^3, \quad y = 4x$
36. $y = x^3 + 1, \quad y = x + 1$
37. $y = x^3 - 3x^2 - 9x + 12, \quad y = x + 12$
38. $y = x^3 - 6x^2 + 9x, \quad y = x$

APPLICATIONS

Business & Economics

Photo courtesy of Shell Oil Company

39. *Oil production.* Using data from the first 3 years of production as well as geological studies, the management of an oil company estimates that oil will be pumped from a producing field at a rate given by

$$R(t) = \frac{100}{t + 10} + 10 \qquad 0 \leq t \leq 15$$

where $R(t)$ is the rate of production (in thousands of barrels per year) t years after pumping begins. Find the area between the graph of R and the t axis over the interval [5, 10] and interpret the results.

40. *Oil production.* In Problem 39, if the rate is found to be

$$R(t) = \frac{100t}{t^2 + 25} + 4 \qquad 0 \leq t \leq 25$$

find the area between the graph of R and the t axis over the interval [5, 15] and interpret the results.

41. *Useful life.* An amusement company maintains records for each video game it installs in an arcade. Suppose that $C(t)$ and $R(t)$ represent the total accumulated costs and revenues (in thousands of dollars), respectively, t years after a particular game has been installed. If

$$C'(t) = 2 \qquad \text{and} \qquad R'(t) = 9e^{-0.3t}$$

find the area between the graphs of C' and R' over the interval on the t axis from 0 to the useful life of the game and interpret the results.

42. *Useful life.* Repeat Problem 41 if

$$C'(t) = 2t \qquad \text{and} \qquad R'(t) = 5te^{-0.1t^2}$$

43. *Income distribution.* As part of a study of the effects of World War II on the economy of the United States, an economist used data from the U.S. Bureau of the Census to produce the following Lorenz curves for distribution of income in the United States in 1935 and in 1947:

$f(x) = x^{2.4}$	Lorenz curve for 1935
$g(x) = x^{1.6}$	Lorenz curve for 1947

Find the coefficient of inequality for each Lorenz curve and interpret the results.

44. *Income distribution.* Using data from the U.S. Bureau of the Census, an economist produced the following Lorenz curves for distribution of income in the United States in 1962 and in 1972:

$f(x) = \frac{3}{10}x + \frac{7}{10}x^2$	Lorenz curve for 1962
$g(x) = \frac{1}{2}x + \frac{1}{2}x^2$	Lorenz curve for 1972

Find the coefficient of inequality for each Lorenz curve and interpret the results.

45. *Distribution of wealth.* Lorenz curves also can be used to provide a relative measure of the distribution of the total assets of a country. Using data in a report by the U.S. Congressional Joint Economic Committee, an economist produced the following Lorenz curves for the distribution of total assets in the United States in 1963 and in 1983:

$f(x) = x^{10}$	Lorenz curve for 1963
$g(x) = x^{12}$	Lorenz curve for 1983

Find the coefficient of inequality for each Lorenz curve and interpret the results.

46. *Income distribution.* The government of a small country is planning sweeping changes in the tax structure in order to provide a more equitable distribution of income. The Lorenz curves for the current income distribution and for the projected income distribution after enactment of the tax changes are given below. Find the coefficient of inequality for each Lorenz curve. Will the proposed changes provide a more equitable income distribution?

$f(x) = x^{2.3}$	Current Lorenz curve
$g(x) = 0.4x + 0.6x^2$	Projected Lorenz curve after changes in tax laws

Life Sciences

47. *Biology.* A yeast culture is growing at a rate of $W'(t) = 0.3e^{0.1t}$ grams per hour. Find the area between the graph of W' and the t axis over the interval $[0, 10]$ and interpret the results.

48. *Natural resource depletion.* The instantaneous rate of change of the demand for lumber in the United States since 1970 ($t = 0$) in billions of cubic feet per year is estimated to be given by

$$Q'(t) = 12 + 0.006t^2 \qquad 0 \le t \le 50$$

Find the area between the graph of Q' and the t axis over the interval [15, 20] and interpret the results.

Social Sciences

49. *Learning.* A beginning high school language class was chosen for an experiment on learning. Using a list of 50 words, the experiment involved measuring the rate of vocabulary memorization at different times during a continuous 5 hour study session. It was found that the average rate of learning for the whole class was inversely proportional to the time spent studying and was given approximately by

$$V'(t) = \frac{15}{t} \qquad 1 \le t \le 5$$

Find the area between the graph of V' and the t axis over the interval [2, 4] and interpret the results.

SECTION 7-6 Chapter Review

Important Terms and Symbols

7-1 *Antiderivatives and Indefinite Integrals.* Antiderivative; indefinite integral; integral sign; integrand; constant of integration

$$\int f(x)\, dx$$

7-2 *Integration by Substitution.* General integral formulas; method of substitution

7-3 *Differential Equations—Growth and Decay.* Differential equation; continuous compound interest; exponential growth law; population growth; radioactive decay

$$\frac{dQ}{dt} = rQ; \quad Q = Q_0e^{rt}$$

7-4 *Definite Integrals.* Definite integral; integrand; upper limit; lower limit; limits of integration; properties of definite integrals; substitution methods; total change; useful life

$$\int_a^b f(x)\, dx$$

7-5 *Area and the Definite Integral.* Area under a curve; area between a curve and the x axis; area between two curves; distribution of income; Lorenz curve; coefficient of inequality

Integration Formulas

$$\int k\,dx = kx + C$$

$$\int kf(x)\,dx = k\int f(x)\,dx$$

$$\int [f(x) \pm g(x)]\,dx = \int f(x)\,dx \pm \int g(x)\,dx$$

$$\int u^n\,du = \frac{u^{n+1}}{n+1} + C \qquad n \neq -1$$

$$\int e^u\,du = e^u + C \qquad \text{and} \qquad \int e^{au}\,du = \frac{1}{a}e^{au} + C \qquad a \neq 0$$

$$\int \frac{1}{u}\,du = \ln|u| + C \qquad u \neq 0$$

EXERCISE 7-6

Chapter Review

Work through all the problems in this chapter review and check your answers in the back of the book. (Answers to all review problems are there.) Where weaknesses show up, review appropriate sections in the text.

A *Find each integral in Problems 1–6.*

1. $\int (3t^2 - 2t)\,dt$ **2.** $\int_2^5 (2x - 3)\,dx$ **3.** $\int (3t^{-2} - 3)\,dt$

4. $\int_1^4 x\,dx$ **5.** $\int e^{-0.5x}\,dx$ **6.** $\int_1^5 \frac{2}{u}\,du$

7. Find a function $y = f(x)$ that satisfies both conditions:

$$\frac{dy}{dx} = 3x^2 - 2 \qquad f(0) = 4$$

8. Find the area bounded by the graphs of $y = 3x^2 + 1$ and $y = 0, -1 \leq x \leq 2$.

B *Find each integral in Problems 9–14.*

9. $\int \sqrt[3]{6x - 5}\,dx$ **10.** $\int_0^1 10(2x - 1)^4\,dx$ **11.** $\int \left(\frac{2}{x^2} - 2xe^{x^2}\right) dx$

12. $\int_0^4 x\sqrt{x^2 + 4}\,dx$ **13.** $\int (e^{-2x} + x^{-1})\,dx$ **14.** $\int_0^{10} 10e^{-0.02x}\,dx$

15. Find a function $y = f(x)$ that satisfies both conditions:

$$\frac{dy}{dx} = 3x^{-1} - x^{-2} \qquad f(1) = 5$$

16. Find the equation of the curve that passes through (2, 10) if its slope is given by

$$\frac{dy}{dx} = 6x + 1$$

for each x.

C

17. Find the area bounded by the graphs of $y = x^2 - 4$ and $y = 0$, $-2 \le x \le 4$.

Find each integral in Problems 18–27.

18. $\int_0^3 \frac{x}{1 + x^2}\,dx$

19. $\int_0^3 \frac{x}{(1 + x^2)^2}\,dx$

20. $\int x^3(2x^4 + 5)^5\,dx$

21. $\int \frac{e^{-x}}{e^{-x} + 3}\,dx$

22. $\int \frac{e^x}{(e^x + 2)^2}\,dx$

23. $\int \frac{(\ln x)^2}{x}\,dx$

24. $\int x(x^3 - 1)^2\,dx$

25. $\int \frac{x}{\sqrt{6 - x}}\,dx$

26. $\int_0^7 x\sqrt{16 - x}\,dx$

27. $\int_{-1}^{1} x(x + 1)^4\,dx$

28. Find a function $y = f(x)$ that satisfies both conditions:

$$\frac{dy}{dx} = 9x^2e^{x^3} \qquad f(0) = 2$$

29. Solve the differential equation:

$$\frac{dN}{dt} = 0.06N \qquad N(0) = 800 \qquad N > 0$$

30. Find the area bounded by the graphs of $y = 6 - x^2$ and $y = x^2 - 2$, $0 \le x \le 3$.

APPLICATIONS

Business & Economics

31. *Profit function.* If the marginal profit for producing x units per day is given by

$$P'(x) = 100 - 0.02x \qquad P(0) = 0$$

where P(x) is the profit in dollars, find the profit function P and the profit on 10 units of production per day.

32. *Resource depletion.* An oil well starts out producing oil at the rate of 60,000 barrels of oil per year, but the production rate is expected to decrease by

4,000 barrels per year. Thus, if $P(t)$ is the total production (in thousands of barrels) in t years, then

$$P'(t) = f(t) = 60 - 4t \qquad 0 \leq t \leq 15$$

Write a definite integral that will give the total production after 15 years of operation, and evaluate it.

33. *Profit and production.* The weekly marginal profit for an output of x units is given approximately by

$$P'(x) = 150 - \frac{x}{10} \qquad 0 \leq x \leq 40$$

What is the total change in profit for a production change from 10 units per week to 40 units? Set up a definite integral and evaluate it.

34. *Useful life.* The total accumulated costs $C(t)$ and revenues $R(t)$ in thousands of dollars), respectively, for a coal mine satisfy

$$C'(t) = 3 \qquad \text{and} \qquad R'(t) = 20e^{-0.1t}$$

where t is the number of years the mine has been in operation. Find the useful life of the mine to the nearest year. What is the total profit accumulated during the useful life of the mine?

35. *Marketing.* The market research department for an automobile company estimates that the sales (in millions of dollars) of a new automobile will increase at the monthly rate of

$$S'(t) = 4e^{-0.08t} \qquad 0 \leq t \leq 24$$

t months after the introduction of the automobile. What will be the total sales $S(t)$, t months after the automobile is introduced if we assume that there were 0 sales at the time the automobile entered the marketplace? What are the estimated total sales during the first 12 months after the introduction of the automobile? How long will it take for the total sales to reach $40 million?

36. *Income distribution.* An economist produced the following Lorenz curves for the current income distribution and the projected income distribution 10 years from now in a certain country:

$$f(x) = \tfrac{1}{10}x + \tfrac{9}{10}x^2 \qquad \text{Current Lorenz curve}$$

$$g(x) = x^{1.5} \qquad \text{Projected Lorenz curve}$$

Find the coefficient of inequality for each Lorenz curve and interpret the results.

Life Sciences

37. *Wound healing.* The area of a small, healing surface wound changes at a rate given approximately by

$$\frac{dA}{dt} = -5t^{-2} \qquad 1 \leq t \leq 5$$

where t is time in days and $A(1) = 5$ square centimeters. What will the area of the wound be in 5 days?

38. *Pollution.* An environmental protection agency estimates that the rate of seepage of toxic chemicals from a waste dump (in gallons per year) is given by

$$R(t) = \frac{1{,}000}{(1+t)^2}$$

where t is time in years since the discovery of the seepage. Find the total amount of toxic chemicals that seep from the dump during the first 4 years after the seepage is discovered.

39. *Population.* The population of the United States in 1980 was approximately 226 million. The continuous compound growth rate for the decade from 1970 to 1980 was 1.1% (data from the 1980 census).

(A) If the population continued to grow at the same continuous compound growth rate, what is the 1990 population?
(B) How long will it take the population to double at this continuous compound growth rate?

Social Sciences

40. *Archaeology.* The continuous compound rate of decay for carbon-14 is $r = -0.000\ 123\ 8$. A piece of animal bone found at an archaeological site contains 4% of the original amount of carbon-14. Estimate the age of the bone.

41. *Learning.* In a particular business college, it was found that an average student enrolled in a typing class progressed at a rate of $N'(t) = 7e^{-0.1t}$ words per minute t weeks after enrolling in a 15 week course. If at the beginning of the course a student could type 25 words per minute, how many words per minute, $N(t)$, would the student be expected to type t weeks into the course? After completing the course?

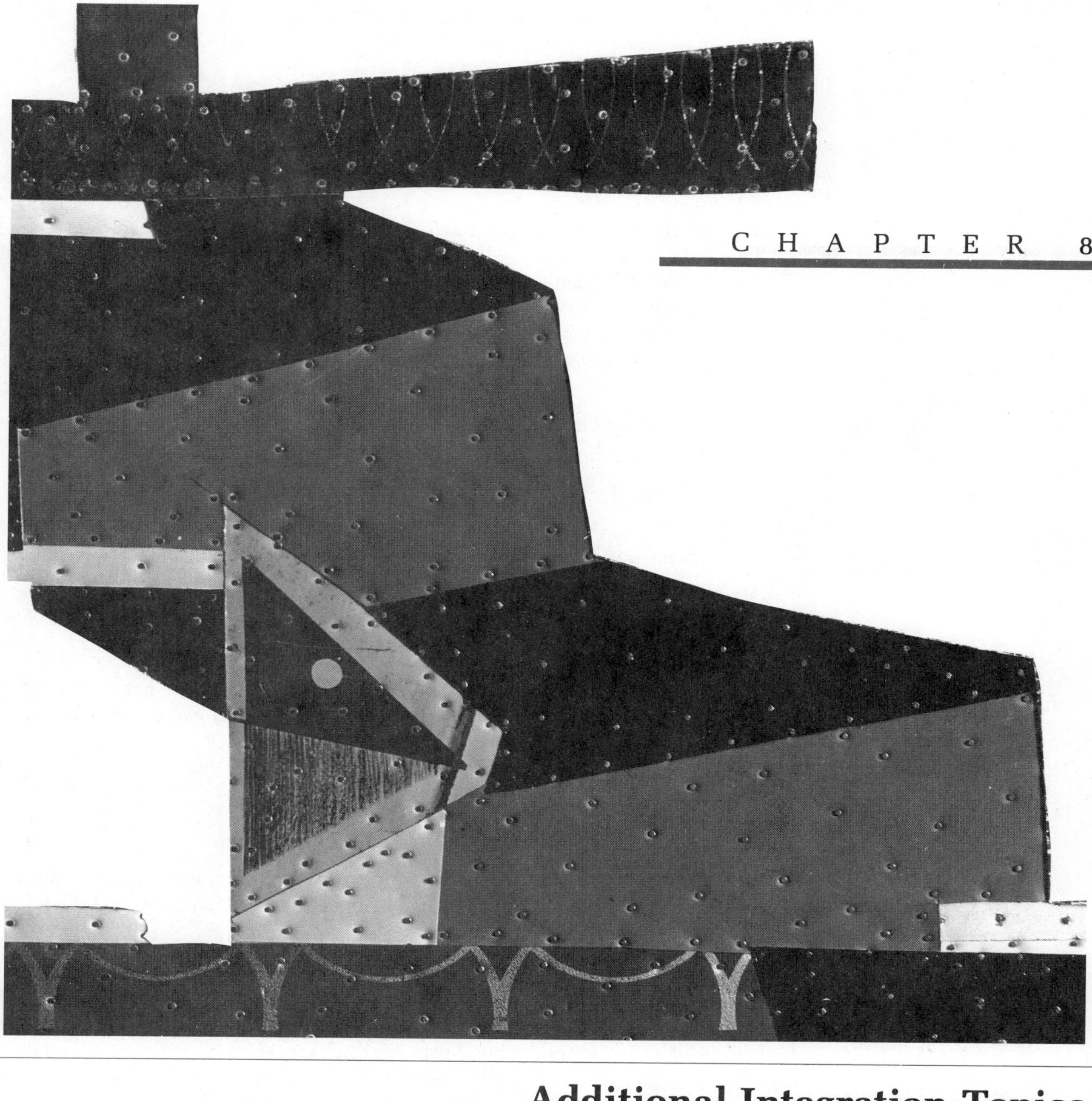

CHAPTER 8

Additional Integration Topics

CHAPTER 8

Contents

SECTION 8-1

Definite Integral as a Limit of a Sum

- RECTANGLE RULE FOR APPROXIMATING DEFINITE INTEGRALS
- DEFINITE INTEGRAL AS A LIMIT OF A SUM
- RECOGNIZING A DEFINITE INTEGRAL
- AVERAGE VALUE OF A CONTINUOUS FUNCTION

Up to this point, in order to evaluate a definite integral

$$\int_a^b f(x)\,dx$$

we need to find an antiderivative of the function f so that we can write

$$\int_a^b f(x)\,dx = F(x)\Big|_a^b = F(b) - F(a) \qquad F'(x) = f(x)$$

But suppose we cannot find an antiderivative of f (it may not even exist in a convenient or closed form). For example, how would you evaluate the following?

$$\int_2^8 \sqrt{x^3+1}\,dx \qquad \int_1^5 \left(\frac{x}{x+1}\right)^3 dx \qquad \int_0^5 e^{-x^2}\,dx$$

We now introduce the *rectangle rule* for approximating definite integrals, and out of this discussion will evolve a new way of looking at definite integrals.

◆ RECTANGLE RULE FOR APPROXIMATING DEFINITE INTEGRALS

In Section 7-5, we saw that any definite integral of a positive continuous function f over an interval $[a, b]$ always can be interpreted as the area bounded by $y = f(x)$, $y = 0$, $x = a$, and $x = b$ (see Fig. 1). What we need is a way of approximating such areas, given $y = f(x)$ and an interval $[a, b]$.

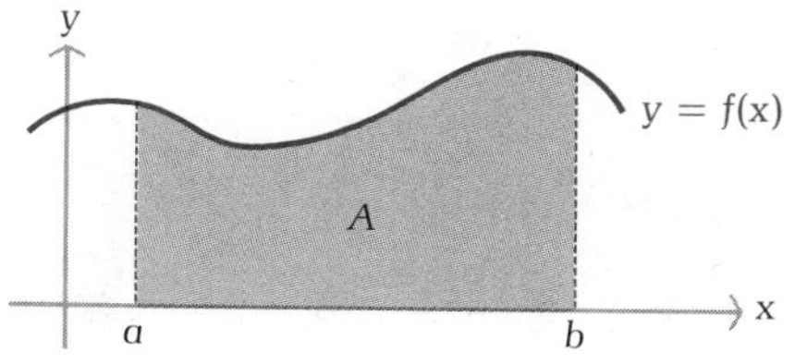

FIGURE 1

$A = \int_a^b f(x)\, dx$

Let us start with a concrete example and generalize from the experience. We begin with a simple definite integral we can evaluate exactly:

$$\int_1^5 (x^2 + 3)\, dx = \left(\frac{x^3}{3} + 3x\right)\Big|_1^5$$
$$= \left[\frac{(5)^3}{3} + 3(5)\right] - \left[\frac{(1)^3}{3} + 3(1)\right]$$
$$= (\tfrac{125}{3} + 15) - (\tfrac{1}{3} + 3)$$
$$= \tfrac{160}{3} = 53\tfrac{1}{3}$$

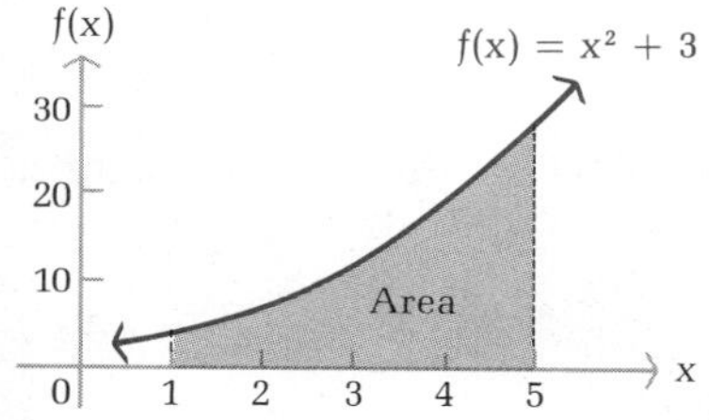

FIGURE 2

This integral represents the area bounded by $y = x^2 + 3$ and $y = 0$, $1 \leq x \leq 5$, as indicated in Figure 2.

Since areas of rectangles are easy to compute, we cover the area in Figure 2 with rectangles (see Fig. 3) so that the top of each rectangle has a point in common with the graph of $y = f(x)$. As our first approximation, we divide the interval [1, 5] into two equal subintervals, each with length $(b - a)/2 = (5 - 1)/2 = 2$, and use the midpoint of each subinterval to compute the altitude of the rectangle sitting on top of that subinterval (see Fig. 3):

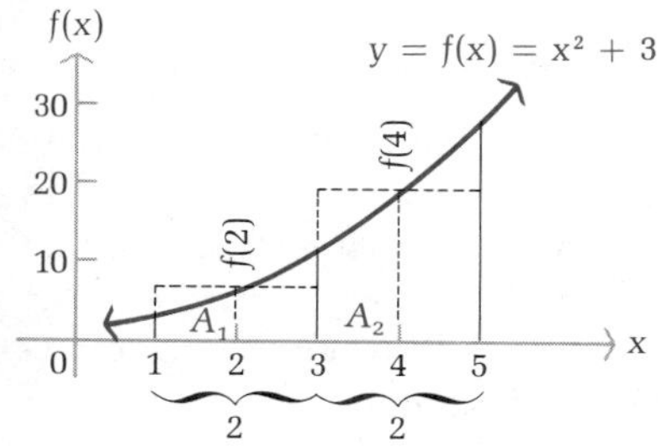

FIGURE 3

$$\int_1^5 (x^2 + 3)\, dx \approx A_1 + A_2$$
$$= f(2) \cdot 2 + f(4) \cdot 2$$
$$= 2[f(2) + f(4)]$$
$$= 2(7 + 19) = 52$$

This approximation is less than 3% off the exact area we found above ($53\frac{1}{3}$).

Now let us divide the interval [1, 5] into four equal subintervals, each of length $(b - a)/4 = (5 - 1)/4 = 1$, and use the midpoint* of each subinterval to

---

* We actually do not need to choose the midpoint of each subinterval; any point from each subinterval will do. The midpoint is often a convenient point to choose, because then the rectangle tops are usually above part of the graph and below part of the graph. This tends to cancel some of the error that occurs.

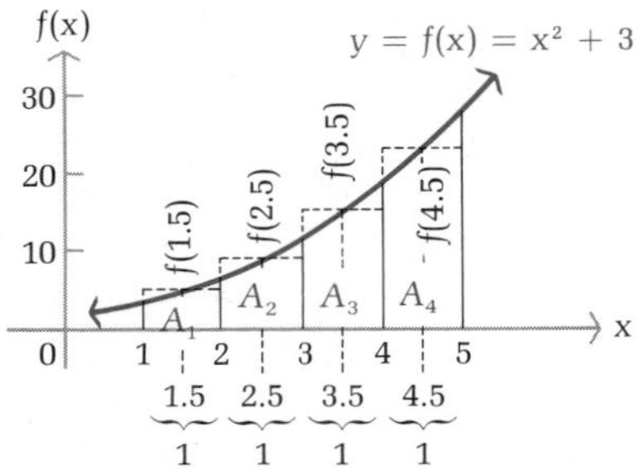

FIGURE 4

compute the altitude of the rectangle corresponding to that subinterval (see Fig. 4):

$$\int_1^5 (x^3 + 3)\,dx \approx A_1 + A_2 + A_3 + A_4$$

$$= f(1.5) \cdot 1 + f(2.5) \cdot 1 + f(3.5) \cdot 1 + f(4.5) \cdot 1$$
$$= f(1.5) + f(2.5) + f(3.5) + f(4.5)$$
$$= 5.25 + 9.25 + 15.25 + 23.25 = 53$$

Now we are less than 1% off the exact area ($53\frac{1}{3}$).

We would expect the approximations to continue to improve as we use more and more rectangles with smaller and smaller bases. We now state the **rectangle rule** for approximating definite integrals of a continuous function f over the interval from $x = a$ to $x = b$.

Rectangle Rule

Divide the interval from $x = a$ to $x = b$ into n equal subintervals of length $\Delta x = (b - a)/n$. Let c_k be any point in the kth subinterval. Then

$$\int_a^b f(x)\,dx \approx f(c_1)\Delta x + f(c_2)\Delta x + \cdots + f(c_n)\Delta x$$

$$= \Delta x[f(c_1) + f(c_2) + \cdots + f(c_n)]$$

The rectangle rule is valid for any continuous function f. If f is positive on $[a, b]$, then each term $f(c_k)\Delta x$ in the approximating sum can be interpreted as the area of a rectangle, and the sum of these terms approximates the area under the graph of f. If f is not positive on $[a, b]$, then $f(c_k)\Delta x$ no longer represents the area of a rectangle (area is always positive), but it is still true that the sum of these terms approximates the definite integral of f from a to b.

◆ EXAMPLE 1

Use the rectangle rule to approximate

$$\int_0^2 \frac{12x - 12}{x + 4}\,dx$$

with $n = 4$ and c_k the midpoint of each subinterval. Compute the approximation to three decimal places.

Solution

Step 1. Find Δx, the length of each subinterval:

$$\Delta x = \frac{b - a}{n} = \frac{2 - 0}{4} = 0.5$$

Step 2. Use the midpoint of each interval for c_k:

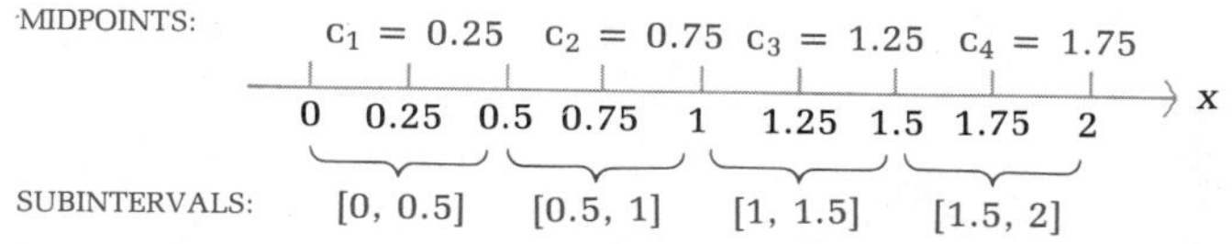

Step 3. Use the rectangle rule with $n = 4$ (see the figure below):

$$\begin{aligned}\int_0^2 \frac{12x - 12}{x + 4}\,dx &\approx f(c_1)\Delta x + f(c_2)\Delta x + f(c_3)\Delta x + f(c_4)\Delta x \\ &= [f(c_1) + f(c_2) + f(c_3) + f(c_4)]\Delta x \\ &= [f(0.25) + f(0.75) + f(1.25) + f(1.75)](0.5) \\ &= [(-2.117\ 647) + (-0.631\ 579) + (0.571\ 429) + (1.565\ 217)](0.5) \\ &= (-0.612\ 58)(0.5) = -0.306 \qquad \text{To three decimal places}\end{aligned}$$

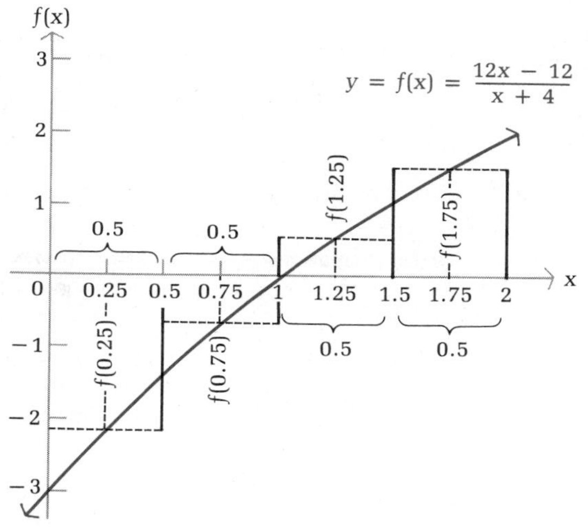

◆

PROBLEM 1 Use the rectangle rule to approximate

$$\int_{-2}^{2} \frac{4x}{x + 6}\,dx$$

with $n = 4$ and c_k the midpoint of each subinterval. Compute the approximation to three decimal places. ◆

One important application of the rectangle rule is the approximation of definite integrals involving **tabular functions**—that is, functions defined by tables rather than by formulas. The next example illustrates this approach.

◆ EXAMPLE 2 *Real Estate*

A developer is interested in estimating the area of the irregularly shaped property shown in Figure 5A (on the next page). A surveyor used the straight horizontal road at the bottom of the property as the x axis and measured the vertical

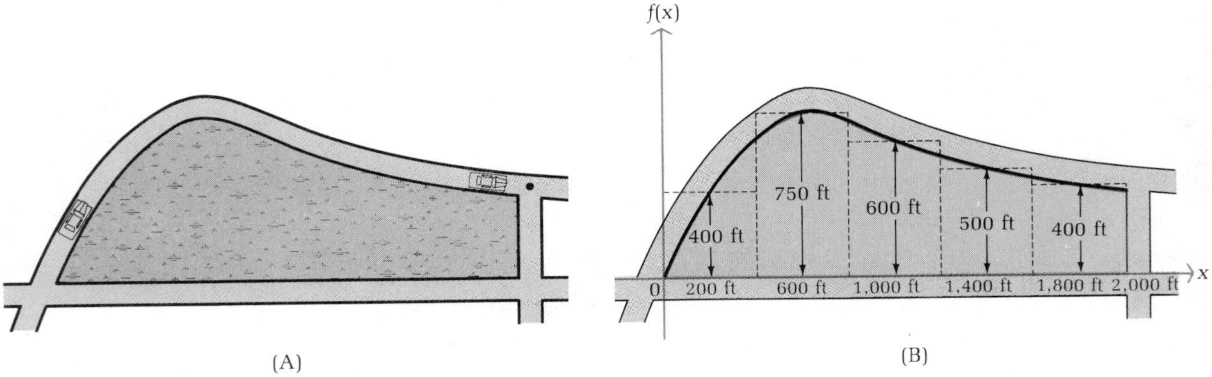

FIGURE 5

distance across the property at 400 foot intervals, starting at 200 (see Fig. 5B). These distances can be viewed as the values of the continuous function f whose graph forms the top of the property. Use the rectangle rule to approximate the area of the property.

Solution List the values of the function f in a table:

x	200	600	1,000	1,400	1,800
f(x)	400	750	600	500	400

The area of the property is given by

$$A = \int_0^{2,000} f(x)\, dx$$

Using the rectangle rule with $n = 5$ and $\Delta x = 400$, and the values of f in the table, we have

$$\begin{aligned} A &\approx \Delta x[f(200) + f(600) + f(1,000) + f(1,400) + f(1,800)] \\ &= 400[400 + 750 + 600 + 500 + 400] \\ &= 1,060,000 \text{ square feet} \end{aligned}$$

◆

PROBLEM 2 To obtain a more accurate approximation of the area of the property shown in Figure 5A, the surveyor measured the vertical distances at 200 foot intervals, starting at 100. The results are listed in the table. Use these values and the rectangle rule with $n = 10$ and $\Delta x = 200$ to approximate the area of the property.

x	100	300	500	700	900	1,100	1,300	1,500	1,700	1,900
f(x)	225	500	700	725	650	575	525	475	425	375

◆

◆ DEFINITE INTEGRAL AS A LIMIT OF A SUM

In using the rectangle rule to approximate a definite integral, one might expect

$$\lim_{\Delta x \to 0} [f(c_1)\Delta x + f(c_2)\Delta x + \cdots + f(c_n)\Delta x] = \int_a^b f(x)\, dx$$

This idea motivates the formal definition of a definite integral that we referred to in Section 7-4.

Definition of a Definite Integral

Let f be a continuous function defined on the closed interval $[a, b]$, and let

1. $a = x_0 < x_1 < \cdots < x_{n-1} < x_n = b$
2. $\Delta x_k = x_k - x_{k-1}$ for $k = 1, 2, \ldots, n$
3. $\Delta x_k \to 0$ as $n \to \infty$
4. $x_{k-1} \leq c_k \leq x_k$ for $k = 1, 2, \ldots, n$

Then

$$\int_a^b f(x)\, dx = \lim_{n \to \infty} [f(c_1)\Delta x_1 + f(c_2)\Delta x_2 + \cdots + f(c_n)\Delta x_n]$$

is called a **definite integral.**

In the definition of a definite integral, we divide the closed interval $[a, b]$ into n subintervals of arbitrary lengths in such a way that the length of each subinterval $\Delta x_k = x_k - x_{k-1}$ tends to 0 as n increases without bound. From each of the n subintervals we then select a point c_k and form the sum

$$f(c_1)\Delta x_1 + f(c_2)\Delta x_2 + \cdots + f(c_n)\Delta x_n$$

which is called a **Riemann sum** [named after the celebrated German mathematician Georg Riemann (1826–1866)].

Under the conditions stated in the definition, it can be shown that the limit of the Riemann sum always exists, and it is a real number. The limit is independent of the nature of the subdivisions of $[a, b]$ as long as condition 3 holds, and it is independent of the choice of c_k as long as condition 4 holds. In particular, when using Riemann sums to approximate definite integrals, we can always choose subintervals of equal length and c_k the midpoint of each subinterval, as we did in the previous examples in this section.

In a more formal treatment of the subject, we would then prove the remarkable **fundamental theorem of calculus,** stated below. This theorem shows that the limit in the definition of a definite integral can be determined exactly by evaluating an antiderivative of $f(x)$, if it exists, at the end points of the interval $[a, b]$ and taking the difference.

Fundamental Theorem of Calculus

Under the conditions stated in the definition of a definite integral:

DEFINITION:

$$\int_a^b f(x)\,dx = \lim_{n\to\infty} [f(c_1)\Delta x_1 + f(c_2)\Delta x_2 + \cdots + f(c_n)\Delta x_n]$$

THEOREM:

$$= F(b) - F(a) \qquad \text{where } F'(x) = f(x)$$

Now we are free to evaluate a definite integral by using the fundamental theorem if an antiderivative of $f(x)$ can be found; otherwise, we can approximate it using the formal definition in the form of the rectangle rule.

◆ RECOGNIZING A DEFINITE INTEGRAL

Recall that the derivative of a function f was defined in Section 4-3 by

$$f'(x) = \lim_{h\to 0} \frac{f(x+h) - f(x)}{h}$$

This form is generally not easy to compute directly, but is easy to recognize in certain practical problems (slope, instantaneous velocity, rates of change, and so on). Once it is recognized that we are dealing with a derivative, we then proceed to try to compute it using derivative formulas and rules.

Similarly, evaluating a definite integral using the definition

$$\int_a^b f(x)\,dx = \lim_{n\to\infty}[f(c_1)\Delta x_1 + f(c_2)\Delta x_2 + \cdots + f(c_n)\Delta x_n] \tag{1}$$

is generally not easy; but the form on the right occurs naturally in many practical problems. We can use the fundamental theorem to evaluate the integral (once it is recognized) if an antiderivative can be found; otherwise, we will approximate it using the rectangle rule. We will now illustrate these points by finding the *average value* of a continuous function.

◆ AVERAGE VALUE OF A CONTINUOUS FUNCTION

Suppose the temperature F (in degrees Fahrenheit) in the middle of a small shallow lake from 8 AM ($t = 0$) to 6 PM ($t = 10$) during the month of May is given approximately as shown in Figure 6.

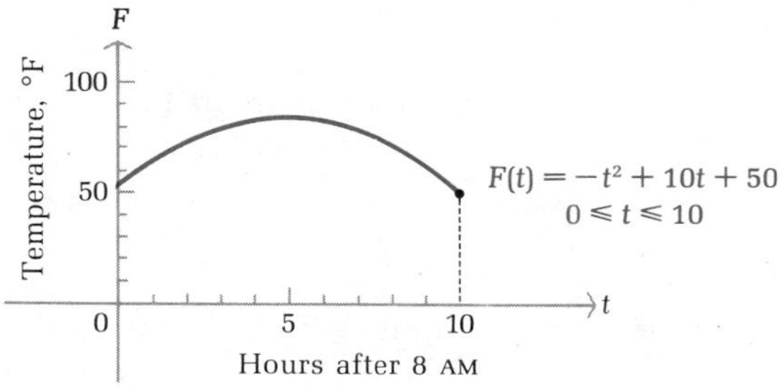

FIGURE 6

How can we compute the average temperature from 8 AM to 6 PM? We know that the average of a finite number of values

$$a_1, a_2, \ldots, a_n$$

is given by

$$\text{Average} = \frac{a_1 + a_2 + \cdots + a_n}{n}$$

But how can we handle a continuous function with infinitely many values? It would seem reasonable to divide the time interval [0, 10] into n equal subintervals, compute the temperature at a point in each subinterval, and then use the average of these values as an approximation of the average value of the continuous function $F = F(t)$ over [0, 10]. We would expect the approximations to improve as n increases. In fact, we would be inclined to define the limit of the average of n values as $n \to \infty$ as the *average value* of F over [0, 10], if the limit exists. This is exactly what we will do:

$$\begin{pmatrix}\text{Average temperature}\\ \text{for } n \text{ values}\end{pmatrix} = \frac{1}{n}[F(t_1) + F(t_2) + \cdots + F(t_n)] \tag{2}$$

where t_k is a point in the kth subinterval. We will call the limit of (2) as $n \to \infty$ the *average temperature over the time interval* [0, 10].

Form (2) looks sort of like form (1), but we are missing the Δt_k. We take care of this by multiplying (2) by $(b - a)/(b - a)$, which will change the form of (2) without changing its value:

$$\begin{aligned}\frac{b-a}{b-a} \cdot \frac{1}{n}[F(t_1) + F(t_2) + \cdots + F(t_n)] &= \frac{1}{b-a} \cdot \frac{b-a}{n}[F(t_1) + F(t_2) + \cdots + F(t_n)]\\ &= \frac{1}{b-a}\left[F(t_1)\frac{b-a}{n} + F(t_2)\frac{b-a}{n} + \cdots + F(t_n)\frac{b-a}{n}\right]\\ &= \frac{1}{b-a}[F(t_1)\Delta t + F(t_2)\Delta t + \cdots + F(t_n)\Delta t]\end{aligned}$$

Thus,

$$\begin{pmatrix}\text{Average temperature}\\ \text{over } [a, b] = [0, 10]\end{pmatrix} = \lim_{n\to\infty}\left\{\frac{1}{b-a}[F(t_1)\Delta t + F(t_2)\Delta t + \cdots + F(t_n)\Delta t]\right\}$$

$$= \frac{1}{b-a}\{\lim_{n\to\infty}[F(t_1)\Delta t + F(t_2)\Delta t + \cdots + F(t_n)\Delta t]\}$$

Now the limit inside the braces is of form (1)—that is, a definite integral. Thus,

$$\begin{pmatrix}\text{Average temperature}\\ \text{over } [a, b] = [0, 10]\end{pmatrix} = \frac{1}{b-a}\int_a^b F(t)\,dt$$

$$= \frac{1}{10-0}\int_0^{10}(-t^2 + 10t + 50)\,dt$$

We now evaluate the definite integral using the fundamental theorem.

$$= \frac{1}{10}\left(-\frac{t^3}{3} + 5t^2 + 50t\right)\Big|_0^{10}$$

$$= \tfrac{200}{3} \approx 67°\text{F}$$

In general, proceeding as above for an arbitrary continuous function f over an interval $[a, b]$, we obtain the following general formula:

Average Value of a Continuous Function f over $[a, b]$

$$\frac{1}{b-a}\int_a^b f(x)\,dx$$

◆ EXAMPLE 3 Find the average value of $f(x) = x - 3x^2$ over the interval $[-1, 2]$.

Solution

$$\frac{1}{b-a}\int_a^b f(x)\,dx = \frac{1}{2-(-1)}\int_{-1}^{2}(x - 3x^2)\,dx$$

$$= \frac{1}{3}\left(\frac{x^2}{2} - x^3\right)\Big|_{-1}^{2} = -\frac{5}{2}$$ ◆

PROBLEM 3 Find the average value of $g(t) = 6t^2 - 2t$ over the interval $[-2, 3]$. ◆

◆ EXAMPLE 4 *Average Price*

Given the demand function

$$p = D(x) = 100e^{-0.05x}$$

find the average price (in dollars) over the demand interval [40, 60].

Solution

$$\text{Average price} = \frac{1}{b-a}\int_a^b D(x)\,dx$$

$$= \frac{1}{60-40}\int_{40}^{60} 100e^{-0.05x}\,dx$$

$$= \frac{100}{20}\int_{40}^{60} e^{-0.05x}\,dx \qquad \text{Use } \int e^{ax}\,dx = \frac{1}{a}e^{ax},\ a \neq 0.$$

$$= -\frac{5}{0.05}e^{-0.05x}\Big|_{40}^{60}$$

$$= 100(e^{-2} - e^{-3}) \approx \$8.55$$

◆

PROBLEM 4 Given the supply equation

$$p = S(x) = 10e^{0.05x}$$

find the average price (in dollars) over the supply interval [20, 30]. ◆

◆ EXAMPLE 5

Advertising

A metropolitan newspaper currently has a daily circulation of 50,000 papers (weekdays and Sunday). The management of the paper decides to initiate an aggressive advertising campaign to increase circulation. Suppose that the daily circulation (in thousands of papers) t days after the beginning of the campaign is given by

$$S(t) = 100 - 50e^{-0.01t}$$

What is the average daily circulation during the first 30 days of the campaign? During the second 30 days of the campaign?

Solution

$$\begin{pmatrix}\text{Average daily circulation}\\ \text{over } [a, b] = [0, 30]\end{pmatrix} = \frac{1}{b-a}\int_a^b S(t)\,dt$$

$$= \frac{1}{30}\int_0^{30} (100 - 50e^{-0.01t})\,dt$$

$$= \frac{1}{30}(100t + 5{,}000e^{-0.01t})\Big|_0^{30}$$

$$= \frac{1}{30}(3{,}000 + 5{,}000e^{-0.3} - 5{,}000)$$

$$\approx 56.8 \quad \text{or} \quad 56{,}800 \text{ papers}$$

$$\begin{pmatrix}\text{Average daily circulation}\\ \text{over } [a, b] = [30, 60]\end{pmatrix} = \frac{1}{60-30}\int_{30}^{60} (100 - 50e^{-0.01t})\,dt$$

$$= \frac{1}{30}(100t + 5{,}000e^{-0.01t})\Big|_{30}^{60}$$

$$= \frac{1}{30}(6{,}000 + 5{,}000e^{-0.6} - 3{,}000 - 5{,}000e^{-0.3})$$

$$\approx 68.0 \quad \text{or} \quad 68{,}000 \text{ papers}$$

◆

PROBLEM 5 Refer to Example 5. Satisfied with the increase in circulation, management decides to terminate the advertising campaign. Suppose that the daily circulation (in thousands of papers) t days after the end of the advertising campaign is given by

$$S(t) = 65 + 8e^{-0.02t}$$

What is the average daily circulation during the first 30 days after the end of the campaign? During the second 30 days after the end of the campaign? ◆

Answers to Matched Problems

1. -0.589 2. 1,035,000 ft^2 3. 13 4. $35.27
5. $\dfrac{2{,}350 - 400e^{-0.6}}{30} \approx 71.0$ or 71,000 papers;
$\dfrac{1{,}950 - 400e^{-1.2} + 400e^{-0.6}}{30} \approx 68.3$ or 68,300 papers

EXERCISE 8-1

For Problems 1–8:

(A) Use the rectangle rule to approximate each definite integral for the indicated number of subintervals n. Choose c_k as the midpoint of each subinterval.
(B) Evaluate each integral exactly using an antiderivative.

A

1. $\int_1^5 3x^2\,dx;\quad n = 2$

2. $\int_2^6 x^2\,dx;\quad n = 2$

3. $\int_1^5 3x^2\,dx;\quad n = 4$

4. $\int_2^6 x^2\,dx;\quad n = 4$

B

5. $\int_0^4 (4 - x^2)\,dx;\quad n = 2$

6. $\int_0^4 (3x^2 - 12)\,dx;\quad n = 2$

7. $\int_0^4 (4 - x^2)\,dx;\quad n = 4$

8. $\int_0^4 (3x^2 - 12)\,dx;\quad n = 4$

In Problems 9–12, use the rectangle rule with $n = 4$ and the values given for f in each table to approximate the indicated definite integral. Choose c_k as the midpoint of each interval.

9. $\int_0^8 f(x)\,dx$

x	1	3	5	7
f(x)	4.5	3.2	2.4	1.6

10. $\int_1^9 f(x)\,dx$

x	2	4	6	8
f(x)	3.2	4.5	7.9	9.4

11. $\int_1^5 f(x)\,dx$

x	1.5	2.5	3.5	4.5
f(x)	12.5	16.7	15.4	10.7

12. $\int_0^4 f(x)\,dx$

x	0.5	1.5	2.5	3.5
f(x)	9.4	14.7	11.5	6.4

Find the average value of each function over the indicated interval.

13. $f(x) = 500 - 50x$; $[0, 10]$
14. $g(x) = 2x + 7$; $[0, 5]$
15. $f(t) = 3t^2 - 2t$; $[-1, 2]$
16. $g(t) = 4t - 3t^2$; $[-2, 2]$
17. $f(x) = \sqrt[3]{x}$; $[1, 8]$
18. $g(x) = \sqrt{x + 1}$; $[3, 8]$
19. $f(x) = 4e^{-0.2x}$; $[0, 10]$
20. $f(x) = 64e^{0.08x}$; $[0, 10]$

Use the rectangle rule to approximate (to three decimal places) each quantity in Problems 21–24. Use $n = 4$ and choose c_k as the midpoint of each subinterval.

21. The average value of $f(x) = (x + 1)/(x^2 + 1)$ for $[-1, 1]$
22. The average value of $f(x) = x/(x + 1)$ for $[0, 4]$
23. The area under the graph of $f(x) = \ln(1 + x^3)$ from $x = 0$ to $x = 2$
24. The area under the graph of $f(x) = 1/(2 + x^3)$ from $x = -1$ to $x = 1$

C

In Problems 25–28, use the rectangle rule to approximate (to three decimal places) each definite integral for the indicated number of subintervals n. Choose c_k as the midpoint of each subinterval.

25. $\int_0^1 e^{-x^2}\,dx$; $n = 5$
26. $\int_0^1 e^{x^2}\,dx$; $n = 5$
27. $\int_0^1 e^{-x^2}\,dx$; $n = 10$
28. $\int_0^1 e^{x^2}\,dx$; $n = 10$
29. Find the average value of $f'(x)$ over the interval $[a, b]$ for any differentiable function f.
30. Show that the average value of $f(x) = Ax + B$ over the interval $[a, b]$ is

$$f\left(\frac{a + b}{2}\right)$$

APPLICATIONS

Business & Economics

31. *Inventory.* A store orders 600 units of a product every 3 months. If the product is steadily depleted to 0 by the end of each 3 months, the inventory on hand, I, at any time t during the year is illustrated as follows:

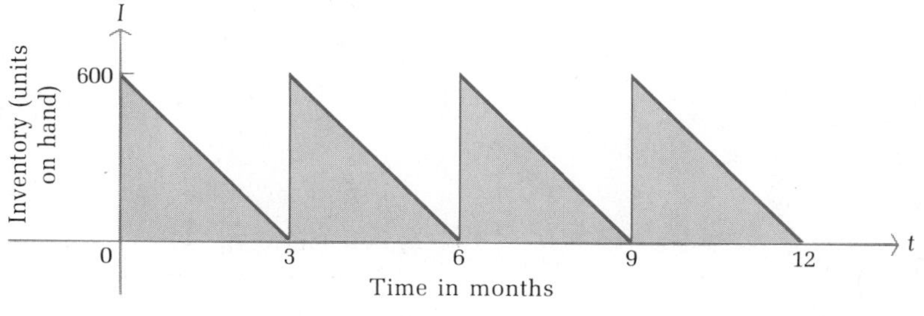

(A) Write an inventory function (assume it is continuous) for the first 3 months. [The graph is a straight line joining (0, 600) and (3, 0).]
(B) What is the average number of units on hand for a 3 month period?

32. Repeat Problem 31 with an order of 1,200 units every 4 months.

33. *Cash reserves.* Suppose cash reserves (in thousands of dollars) are approximated by

$$C(x) = 1 + 12x - x^2 \qquad 0 \leq x \leq 12$$

where x is the number of months after the first of the year. What is the average cash reserve for the first quarter?

34. Repeat Problem 33 for the second quarter.

35. *Average cost.* The total cost (in dollars) of manufacturing x auto body frames is

$$C(x) = 60{,}000 + 300x$$

(A) Find the average cost per unit if 500 frames are produced. [*Hint:* Recall that $\overline{C}(x)$ is the average cost per unit.]
(B) Find the average value of the cost function over the interval [0, 500].

36. *Average cost.* The total cost (in dollars) of printing x dictionaries is $C(x) = 20{,}000 + 10x$.

(A) Find the average cost per unit if 1,000 dictionaries are produced.
(B) Find the average value of the cost function over the interval [0, 1,000].

37. *Continuous compound interest.* If $100 is deposited in an account that earns interest at an annual nominal rate of 8% compounded continuously, find the amount in the account after 5 years and the average amount in the account during this 5 year period. [See Section 6-1 for a discussion of continuous compound interest.]

38. *Continuous compound interest.* If $500 is deposited in an account that earns interest at an annual nominal rate of 12% compounded continuously, find the amount in the account after 10 years and the average amount in the account during this 10 year period.

39. *Supply function.* Given the supply function

$$p = S(x) = 10(e^{0.02x} - 1)$$

find the average price (in dollars) over the supply interval [20, 30].

40. *Demand function.* Given the demand function

$$p = D(x) = \frac{1{,}000}{x}$$

find the average price (in dollars) over the demand interval [400, 600].

41. *Advertising.* The number of hamburgers (in thousands) sold each day by a chain of restaurants t days after the beginning of an advertising campaign is

given by

$$S(t) = 20 - 10e^{-0.1t}$$

What is the average number of hamburgers sold each day during the first week of the advertising campaign? During the second week of the campaign?

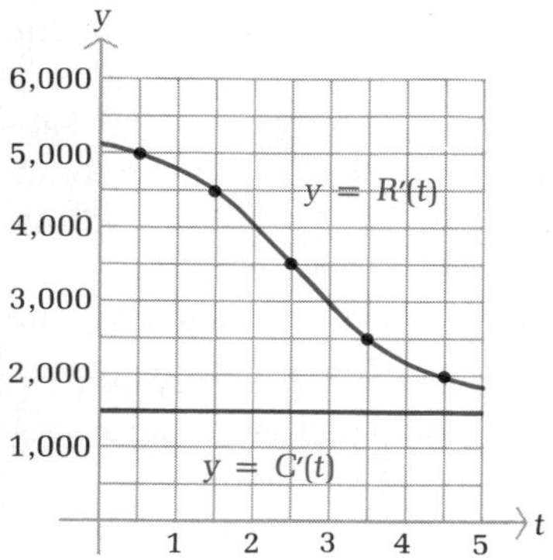

42. *Advertising.* The number of hamburgers (in thousands) sold each day by a chain of restaurants t days after the end of an advertising campaign is given by

$$S(t) = 10 + 8e^{-0.2t}$$

What is the average number of hamburgers sold each day during the first week after the end of the advertising campaign? During the second week after the end of the campaign?

43. *Profit.* Let $R(t)$ and $C(t)$ represent the total accumulated revenues and costs (in dollars), respectively, for a producing oil well, where t is time in years. The graphs of the derivatives of R and C over a 5 year period are shown in the figure in the margin. Use the rectangle rule with $n = 5$ and c_k as the midpoint of each subinterval to approximate the total accumulated profits from the well over this 5 year period. Estimate necessary function values from the graphs.

44. *Revenue.* Use the figure in Problem 43 and the rectangle rule with $n = 5$ and c_k as the midpoint of each subinterval to approximate the average annual revenue from the oil well.

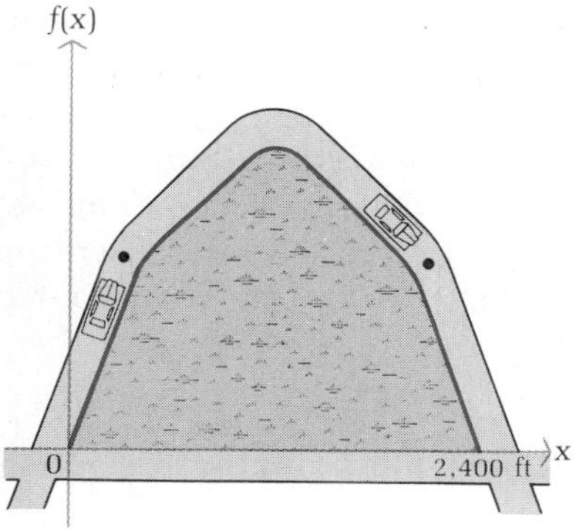

45. *Real estate.* A surveyor produced the table below by measuring the vertical distance (in feet) across a piece of real estate at 600 foot intervals, starting at 300 (see the figure in the margin). Use these values and the rectangle rule to estimate the area of the property.

x	300	900	1,500	2,100
f(x)	900	1,700	1,700	900

46. *Real estate.* Repeat Problem 45 for the following table of measurements:

x	200	600	1,000	1,400	1,800	2,200
f(x)	600	1,400	1,800	1,800	1,400	600

Life Sciences

47. *Temperature.* If the temperature $C(t)$ in an aquarium is made to change according to

$$C(t) = t^3 - 2t + 10 \qquad 0 \leq t \leq 2$$

(in degrees Celsius) over a 2 hour period, what is the average temperature over this period?

48. *Medicine.* A drug is injected into the bloodstream of a patient through her right arm. The concentration of the drug in the bloodstream of the left arm

t hours after the injection is given by

$$C(t) = \frac{0.14t}{t^2 + 1}$$

What is the average concentration of the drug in the bloodstream of the left arm during the first hour after the injection? During the first 2 hours after the injection?

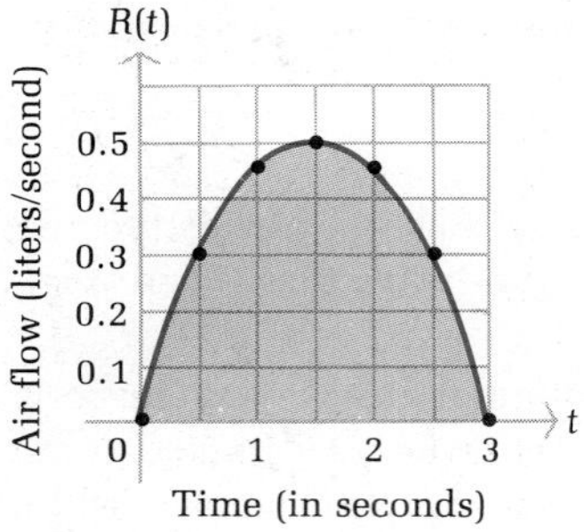

49. *Medicine—respiration.* Physiologists use a machine called a pneumotachograph to produce a graph of the rate of flow $R(t)$ of air into the lungs (inspiration) and out of the lungs (expiration). The figure in the margin gives the graph of the inspiration phase of the breathing cycle for an individual at rest. The area under this graph represents the total volume of air inhaled during the inspiration phase. Use the rectangle rule with $n = 3$ and c_k the midpoint of each subinterval to approximate the area under the graph. Estimate the necessary function values from the graph.

50. *Medicine—respiration.* Use the result obtained in Problem 49 to approximate the average volume of air in the lungs during the inspiration phase.

Social Sciences

51. *Politics.* Public awareness of a Congressional candidate before and after a successful campaign was approximated by

$$P(t) = \frac{8.4t}{t^2 + 49} + 0.1 \qquad 0 \leq t \leq 24$$

where t is time in months after the campaign started and $P(t)$ is the fraction of people in the Congressional district who could recall the candidate's name. What is the average fraction of people who could recall the candidate's name during the first 7 months after the campaign began? During the first 2 years after the campaign began?

52. *Population composition.* Because of various factors (such as birth rate expansion, then contraction; family flights from urban areas; and so on), the number of children in a large city was found to increase and then decrease rather drastically. If the number of children over a 6 year period was found to be given approximately by

$$N(t) = -\tfrac{1}{4}t^2 + t + 4 \qquad 0 \leq t \leq 6$$

what was the average number of children in the city over the 6 year time period? [Assume $N = N(t)$ is continuous.]

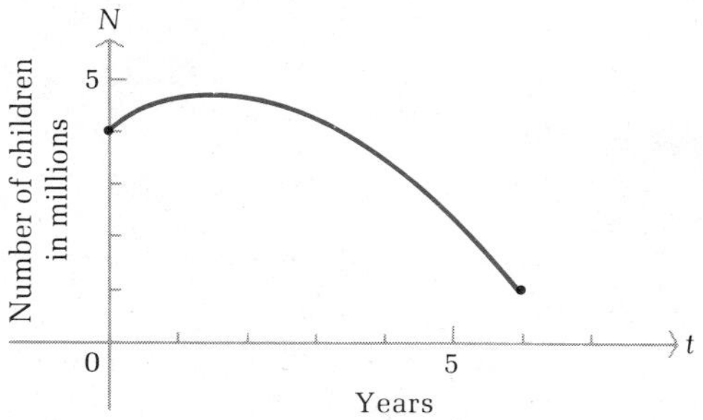

SECTION 8-2 Applications in Business and Economics

- CONTINUOUS INCOME STREAM
- FUTURE VALUE OF A CONTINUOUS INCOME STREAM
- CONSUMERS' AND PRODUCERS' SURPLUS

◆ CONTINUOUS INCOME STREAM

We start this discussion with an example.

◆ EXAMPLE 6 The rate of change of the income produced by a vending machine (in dollars per year) is given by

$$f(t) = 5{,}000e^{0.04t}$$

where t is time in years since the installation of the machine. Find the total income produced by the machine during the first 5 years of operation.

Solution Since we have been given the rate of change of income, we can find the total income by using a definite integral:

$$\begin{aligned}\text{Total income} &= \int_0^5 5{,}000e^{0.04t}\,dt\\ &= 125{,}000e^{0.04t}\Big|_0^5\\ &= 125{,}000e^{0.04(5)} - 125{,}000e^{0.04(0)}\\ &= 152{,}675 - 125{,}000\\ &= \$27{,}675 \qquad \text{Rounded to the nearest dollar}\end{aligned}$$

Thus, the vending machine produces a total income of \$27,675 during the first 5 years of operation. ◆

PROBLEM 6 Refer to Example 6. Find the total income produced during the second 5 years of operation. ◆

In reality, income from a vending machine is not received as a single payment at the end of the 5 year period. Instead, the income is collected on a regular basis, perhaps daily or weekly. In problems of this type, it is convenient to assume that income is actually received in a **continuous stream;** that is, we assume that income is a continuous function of time. The rate of change of income is called the **rate of flow** of the continuous income stream. In general, we have the following:

Total Income for a Continuous Income Stream

If $f(t)$ is the rate of flow of a continuous income stream, then the **total income** produced during the time period from $t = a$ to $t = b$ is

$$\text{Total income} = \int_a^b f(t)\,dt$$

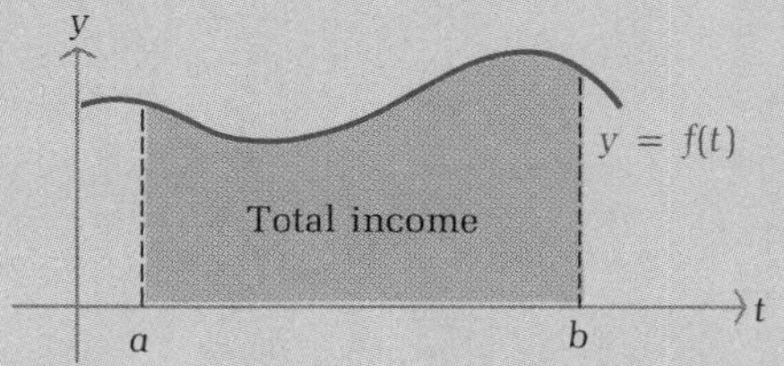

◆ FUTURE VALUE OF A CONTINUOUS INCOME STREAM

In Section 6-1, we discussed the continuous compound interest formula

$$A = Pe^{rt}$$

where P is the principal (or present value), A is the amount (or future value), r is the annual rate of continuous compounding (expressed as a decimal), and t is time in years. For example, if money is worth 12% compounded continuously, then the future value of a $10,000 investment in 5 years is (to the nearest dollar)

$$A = 10{,}000e^{0.12(5)} = \$18{,}221$$

Now we want to apply the future value concept to the income produced by a continuous income stream. Suppose $f(t)$ is the rate of flow of a continuous income stream, and the income produced by this continuous income stream is invested as soon as it is received at a rate r, compounded continuously. We already know how to find the total income produced after T years, but how can we find the total of the income produced and the interest earned by this income? Since the income is received in a continuous flow, we cannot just use the formula $A = Pe^{rt}$. This formula is valid only for a single deposit P, not for a continuous flow of income. Instead, we use a Riemann sum approach that will allow us to apply the formula $A = Pe^{rt}$ repeatedly. To begin, we divide the time interval $[0, T]$ into n equal subintervals of length Δt and choose an arbitrary point c_k in each subinterval, as illustrated in Figure 7.

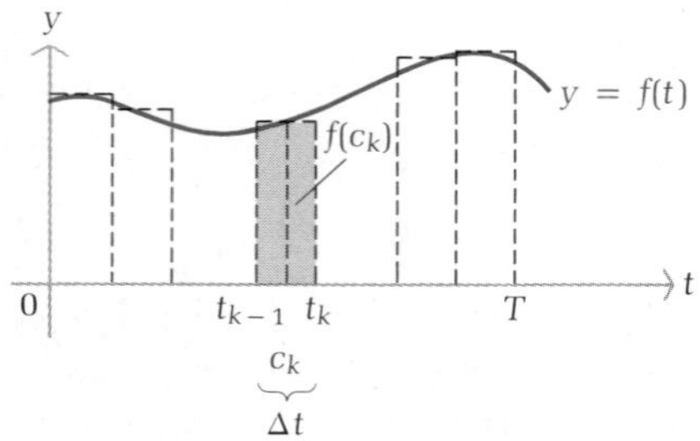

FIGURE 7

The total income produced during the time period from $t = t_{k-1}$ to $t = t_k$ is equal to the area under the graph of $f(t)$ over this subinterval and is approximately equal to $f(c_k)\Delta t$, the area of the shaded rectangle in Figure 7. The income received during this time period will earn interest for approximately $T - c_k$ years. Thus, using the future value formula $A = Pe^{rt}$ with $P = f(c_k)\Delta t$ and $t = T - c_k$, the future value of the income produced during the time period from $t = t_{k-1}$ to $t = t_k$ is approximately equal to

$$f(c_k)\Delta te^{(T-c_k)r}$$

The total of these approximate future values over n subintervals is then

$$f(c_1)\Delta te^{(T-c_1)r} + f(c_2)\Delta te^{(T-c_2)r} + \cdots + f(c_n)\Delta te^{(T-c_n)r}$$

This has the form of a Riemann sum, and the limit of this sum is a definite integral. (See the definition of definite integral in Section 8-1.) Thus, the *future value*, *FV*, of the income produced by the continuous income stream is given by

$$FV = \int_0^T f(t)e^{r(T-t)}\,dt$$

Since r and T are constants, we also can write

$$FV = \int_0^T f(t)e^{rT}e^{-rt}\,dt = e^{rT}\int_0^T f(t)e^{-rt}\,dt \qquad (1)$$

This last form is preferable, since the integral is usually easier to evaluate than the intermediate form in (1).

Future Value of a Continuous Income Stream

If $f(t)$ is the rate of flow of a continuous income stream, $0 \le t \le T$, and if the income is continuously invested at a rate r compounded continuously, then the **future value, *FV*,** at the end of T years is given by

$$FV = \int_0^T f(t)e^{r(T-t)}dt = e^{rT}\int_0^T f(t)e^{-rt}\,dt$$

The future value of a continuous income stream is the total value of all money produced by the continuous income stream (income and interest) at the end of T years.

◆ EXAMPLE 7 Let $f(t) = 5{,}000e^{0.04t}$ be the rate of flow of the income produced by the vending machine in Example 6. Find the future value of this income stream at 12% compounded continuously for 5 years, and find the total interest earned.

Solution Using the formula

$$FV = e^{rT}\int_0^T f(t)e^{-rt}\,dt$$

with $r = 0.12$, $T = 5$, and $f(t) = 5{,}000e^{0.04t}$, we have

$$FV = e^{0.12(5)} \int_0^5 5{,}000e^{0.04t}e^{-0.12t}\,dt$$

$$= 5{,}000e^{0.6} \int_0^5 e^{-0.08t}\,dt$$

$$= 5{,}000e^{0.6} \left(\frac{e^{-0.08t}}{-0.08}\right)\Bigg|_0^5$$

$$= 5{,}000e^{0.6}(-12.5e^{-0.4} + 12.5)$$

$$= \$37{,}545 \qquad \text{Rounded to the nearest dollar}$$

In Example 6, we saw that the total income produced by this vending machine over a 5 year period was \$27,675. Since the future value is the total of the income produced and the interest earned, the difference between the future value and income is interest. Thus,

$$\$37{,}545 - \$27{,}675 = \$9{,}870$$

is the interest earned by the income produced by the vending machine during the 5 year period. ◆

PROBLEM 7 Repeat Example 7 if the interest rate is 9% compounded continuously. ◆

◆ CONSUMERS' AND PRODUCERS' SURPLUS

Let $p = D(x)$ be the price–demand equation for a product, where x is the number of units of the product that consumers will purchase at a price of \$$p$ per unit. Notice that we are expressing price $p = D(x)$ as a function of demand x (see Section 4-7). Suppose $\bar{p}$ is the current price and $\bar{x}$ is the number of units that can be sold at that price. The price–demand curve in Figure 8 shows that if the price is higher than $\bar{p}$, then the demand x is less than $\bar{x}$, but some consumers are still

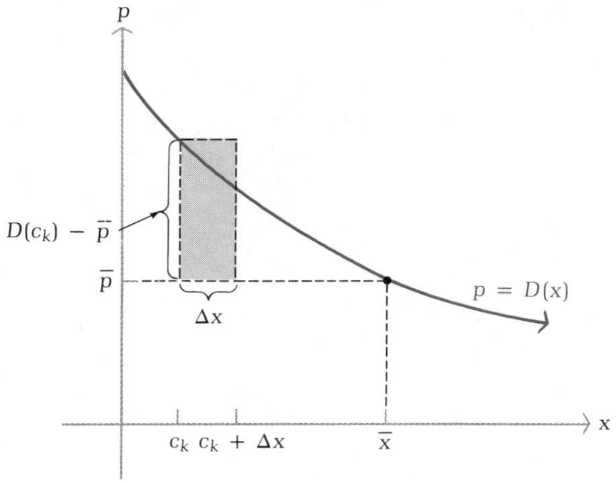

FIGURE 8

willing to pay the higher price. The consumers who were willing to pay more than $\bar{p}$ have saved money. We want to determine the total amount saved by all the consumers who were willing to pay a higher price for this product.

To do this, consider the interval $[c_k, c_k + \Delta x]$, where $c_k + \Delta x < \bar{x}$. If the price remained constant over this interval, then the savings on each unit would be the difference between $D(c_k)$, the price consumers are willing to pay, and $\bar{p}$, the price they actually pay. The total savings to consumers over this interval is approximately equal to

$$[D(c_k) - \bar{p}]\Delta x$$

which is the area of the shaded rectangle shown in Figure 8. If we divide the interval $[0, \bar{x}]$ into n equal subintervals, then the total savings to consumers is approximately equal to

$$[D(c_1) - \bar{p}]\Delta x + [D(c_2) - \bar{p}]\Delta x + \cdots + [D(c_n) - \bar{p}]\Delta x$$

which we recognize as a Riemann sum for the integral

$$\int_0^{\bar{x}} [D(x) - \bar{p}]\, dx$$

Thus, we define the *consumers' surplus* to be this integral.

Consumers' Surplus

If $(\bar{x}, \bar{p})$ is a point on the graph of the price–demand equation $p = D(x)$ for a particular product, then the **consumers' surplus, *CS*,** at a price level of $\bar{p}$ is

$$CS = \int_0^{\bar{x}} [D(x) - \bar{p}]\, dx$$

which is the area between $p = \bar{p}$ and $p = D(x)$ from $x = 0$ to $x = \bar{x}$.

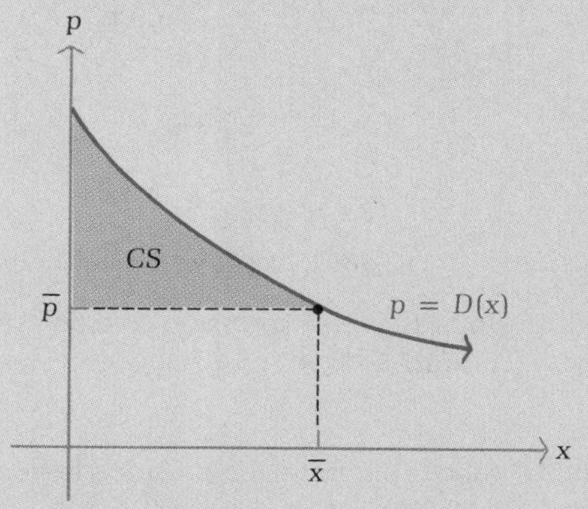

The consumers' surplus represents the total savings to consumers who are willing to pay a higher price for the product.

◆ EXAMPLE 8 Find the consumers' surplus at a price level of \$8 for the price–demand equation

$$p = D(x) = 20 - \tfrac{1}{20}x$$

Solution *Step 1.* Find $\overline{x}$, the demand when the price is $\overline{p} = 8$:

$$\overline{p} = 20 - \tfrac{1}{20}\overline{x}$$
$$8 = 20 - \tfrac{1}{20}\overline{x}$$
$$\tfrac{1}{20}\overline{x} = 12$$
$$\overline{x} = 240$$

Step 2. Sketch a graph:

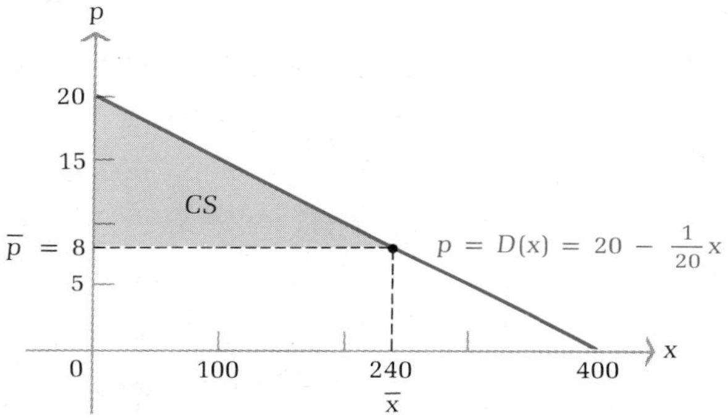

Step 3. Find the consumers' surplus (the shaded area in the graph):

$$CS = \int_0^{\overline{x}} [D(x) - \overline{p}]\,dx$$
$$= \int_0^{240} (20 - \tfrac{1}{20}x - 8)\,dx$$
$$= \int_0^{240} (12 - \tfrac{1}{20}x)\,dx$$
$$= (12x - \tfrac{1}{40}x^2)\Big|_0^{240}$$
$$= 2{,}880 - 1{,}440 = \$1{,}440$$

Thus, the total savings to consumers who are willing to pay a higher price for the product is \$1,440. ◆

PROBLEM 8 Repeat Example 8 for a price level of \$4. ◆

If $p = S(x)$ is the price–supply equation for a product, $\overline{p}$ is the current price, and $\overline{x}$ is the current supply, then some suppliers are still willing to supply some units at a lower price. The additional money that these suppliers gain from the higher price is called the *producers' surplus* and can be expressed in terms of a definite integral (proceeding as we did for the consumers' surplus).

Producers' Surplus

If $(\bar{x}, \bar{p})$ is a point on the graph of the price–supply equation $p = S(x)$, then the **producers' surplus, *PS*,** at a price level of $\bar{p}$ is

$$PS = \int_0^{\bar{x}} [\bar{p} - S(x)]\, dx$$

which is the area between $p = \bar{p}$ and $p = S(x)$ from $x = 0$ to $x = \bar{x}$.

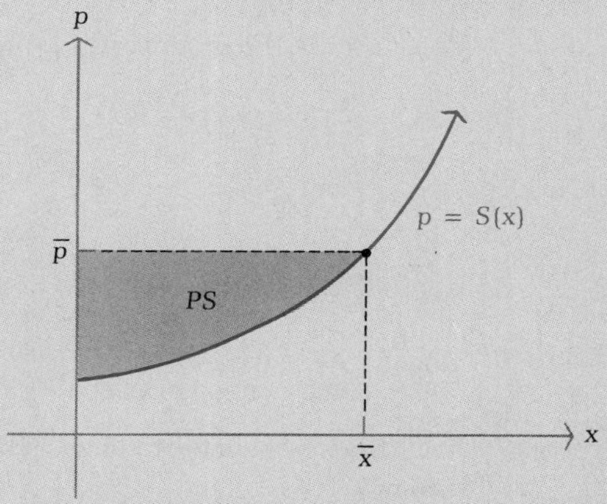

The producers' surplus represents the total gain to producers who are willing to supply units at a lower price.

◆ EXAMPLE 9 Find the producers' surplus at a price level of \$20 for the price–supply equation

$$p = S(x) = 2 + \tfrac{1}{5{,}000}x^2$$

Solution *Step 1.* Find $\bar{x}$, the supply when the price is $\bar{p} = 20$:

$$\begin{aligned} \bar{p} &= 2 + \tfrac{1}{5{,}000}\bar{x}^2 \\ 20 &= 2 + \tfrac{1}{5{,}000}\bar{x}^2 \\ \tfrac{1}{5{,}000}\bar{x}^2 &= 18 \\ \bar{x}^2 &= 90{,}000 \\ \bar{x} &= 300 \quad \text{There is only one solution since } \bar{x} \geq 0. \end{aligned}$$

Step 2. Sketch a graph:

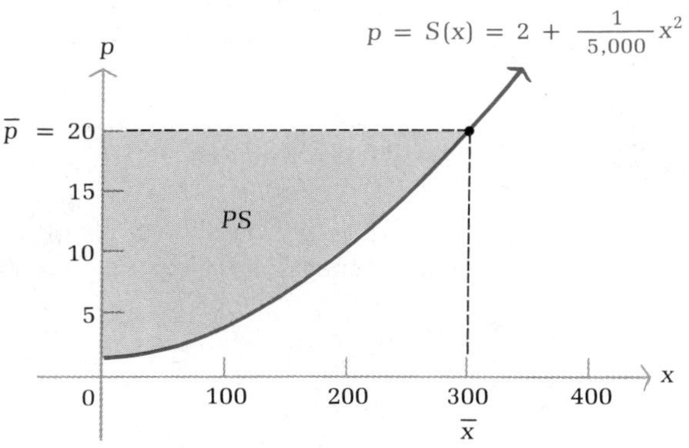

Step 3. Find the producers' surplus (the shaded area in the graph):

$$PS = \int_0^{\overline{x}} [\overline{p} - S(x)]\, dx$$

$$= \int_0^{300} [20 - (2 + \tfrac{1}{5{,}000}x^2)]\, dx$$

$$= \int_0^{300} (18 - \tfrac{1}{5{,}000}x^2)\, dx$$

$$= (18x - \tfrac{1}{15{,}000}x^3)\Big|_0^{300}$$

$$= 5{,}400 - 1{,}800 = \$3{,}600$$

Thus, the total gain to producers who are willing to supply units at a lower price is \$3,600. ◆

PROBLEM 9 Repeat Example 9 for a price level of \$4. ◆

In a free competitive market, the price of a product is determined by the relationship between supply and demand. If $p = D(x)$ and $p = S(x)$ are the price–demand and price–supply equations, respectively, for a product and if $(\overline{x}, \overline{p})$ is the point of intersection of these equations, then $\overline{p}$ is called the **equilibrium price** and $\overline{x}$ is called the **equilibrium quantity.** If the price stabilizes at the equilibrium price $\overline{p}$, then this is the price level that will determine both the consumers' surplus and the producers' surplus.

◆ EXAMPLE 10 Find the equilibrium price and then find the consumers' surplus and producers' surplus at the equilibrium price level if

$$p = D(x) = 20 - \tfrac{1}{20}x \qquad \text{and} \qquad p = S(x) = 2 + \tfrac{1}{5{,}000}x^2$$

Solution *Step 1.* Find the equilibrium point. Set $D(x)$ equal to $S(x)$ and solve:

$$D(x) = S(x)$$
$$20 - \tfrac{1}{20}x = 2 + \tfrac{1}{5,000}x^2$$
$$\tfrac{1}{5,000}x^2 + \tfrac{1}{20}x - 18 = 0$$
$$x^2 + 250x - 90,000 = 0$$
$$x = 200, -450$$

Since x cannot be negative, the only solution is $x = 200$. The equilibrium price can be determined by using $D(x)$ or $S(x)$. We will use both to check our work:

$$\bar{p} = D(200) = 20 - \tfrac{1}{20}(200) = 10 \qquad \bar{p} = S(200) = 2 + \tfrac{1}{5,000}(200)^2 = 10$$

Thus, the equilibrium price is $\bar{p} = 10$, and the equilibrium quantity is $\bar{x} = 200$.

Step 2. Sketch a graph:

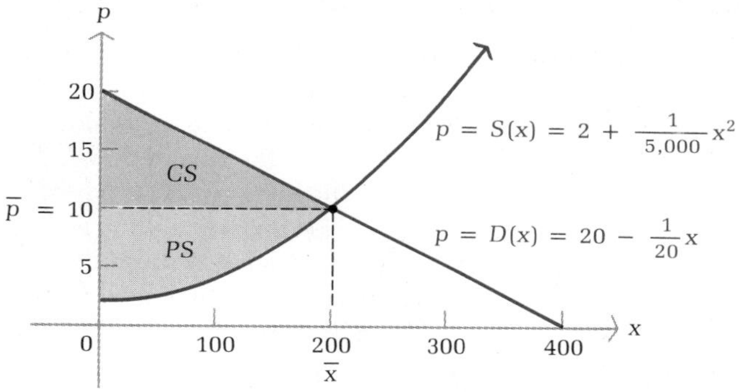

Step 3. Find the consumers' surplus:

$$CS = \int_0^{\bar{x}} [D(x) - \bar{p}]\,dx$$
$$= \int_0^{200} (20 - \tfrac{1}{20}x - 10)\,dx$$
$$= \int_0^{200} (10 - \tfrac{1}{20}x)\,dx$$
$$= (10x - \tfrac{1}{40}x^2)\Big|_0^{200}$$
$$= 2,000 - 1,000 = \$1,000$$

Step 4. Find the producers' surplus:

$$PS = \int_0^{\bar{x}} [\bar{p} - S(x)]\,dx$$
$$= \int_0^{200} [10 - (2 + \tfrac{1}{5,000}x^2)]\,dx$$

$$PS = \int_0^{200} (8 - \tfrac{1}{5,000}x^2)\, dx$$

$$= (8x - \tfrac{1}{15,000}x^3)\Big|_0^{200}$$

$$= 1,600 - \tfrac{1,600}{3} \approx \$1,067 \qquad \text{Rounded to the nearest dollar}$$

◆

PROBLEM 10 Repeat Example 10 for

$$p = D(x) = 25 - \tfrac{1}{1,000}x^2 \qquad \text{and} \qquad p = S(x) = 5 + \tfrac{1}{10}x$$

◆

Answers to Matched Problems

6. \$33,803 7. FV = \$34,691; interest = \$7,016 8. \$2,560
9. \$133 10. $\overline{p} = 15$; CS = \$667; PS = \$500

EXERCISE 8-2

APPLICATIONS

Business & Economics

1. Find the total income produced by a continuous income stream in the first 5 years if the rate of flow is $f(t) = 2,500$.
2. Find the total income produced by a continuous income stream in the first 10 years if the rate of flow is $f(t) = 3,000$.
3. Find the total income produced by a continuous income stream in the first 3 years if the rate of flow is $f(t) = 400e^{0.05t}$.
4. Find the total income produced by a continuous income stream in the first 2 years if the rate of flow is $f(t) = 600e^{0.06t}$.
5. Find the future value at 8% interest compounded continuously for 10 years for the continuous income stream with rate of flow $f(t) = 800$.
6. Find the future value at 6% interest compounded continuously for 8 years for the continuous income stream with rate of flow $f(t) = 1,200$.
7. Find the future value at 10% interest compounded continuously for 4 years for the continuous income stream with rate of flow $f(t) = 1,500e^{-0.02t}$.
8. Find the future value at 7% interest compounded continuously for 6 years for the continuous income stream with rate of flow $f(t) = 2,000e^{0.06t}$.
9. Find the interest earned at 12% compounded continuously for 3 years by a continuous income stream with rate of flow $f(t) = 3,000$.
10. Find the interest earned at 14% compounded continuously for 2 years by a continuous income stream with rate of flow $f(t) = 7,000$.
11. Find the consumers' surplus at a price level of $\overline{p}$ = \$150 for the price–demand equation

$$p = D(x) = 400 - \tfrac{1}{20}x$$

12. Find the consumers' surplus at a price level of $\overline{p}$ = \$120 for the price–demand equation

$$p = D(x) = 200 - \tfrac{1}{50}x$$

13. Find the producers' surplus at a price level of $\bar{p} = \$65$ for the price–supply equation

$$p = S(x) = 10 + \tfrac{1}{10}x + \tfrac{1}{3,600}x^2$$

14. Find the producers' surplus at a price level of $\bar{p} = \$55$ for the price–supply equation

$$p = S(x) = 15 + \tfrac{1}{10}x + \tfrac{3}{1,000}x^2$$

15. Find the consumers' surplus and the producers' surplus at the equilibrium price level for

$$p = D(x) = 50 - \tfrac{1}{10}x \quad \text{and} \quad p = S(x) = 11 + \tfrac{1}{20}x$$

16. Find the consumers' surplus and the producers' surplus at the equilibrium price level for

$$p = D(x) = 20 - \tfrac{1}{600}x^2 \quad \text{and} \quad p = S(x) = 2 + \tfrac{1}{300}x^2$$

17. Find the consumers' surplus and the producers' surplus at the equilibrium price level (rounded to the nearest dollar) for

$$p = D(x) = 80e^{-0.001x} \quad \text{and} \quad p = S(x) = 30e^{0.001x}$$

18. Find the consumers' surplus and the producers' surplus at the equilibrium price level (rounded to the nearest dollar) for

$$p = D(x) = 185e^{-0.005x} \quad \text{and} \quad p = S(x) = 25e^{0.005x}$$

19. Suppose that in a monopolistic market, the production level for a certain product is always chosen so that the profit is maximized. The price–demand equation for this product is

$$p = 100 - \tfrac{1}{10}x$$

and the cost equation is

$$C(x) = 5,000 + 60x$$

Find the consumers' surplus at the production level that maximizes profit.

20. Suppose that in a certain market, the production level for a product is always chosen so that the revenue is maximized. The price–demand equation for this product is

$$p = 100 - \tfrac{1}{10}x$$

Find the consumers' surplus at the production level that maximizes revenue.

21. An investor is presented with a choice of two investments, an established clothing store and a new computer store. Each choice requires the same initial investment and each produces a continuous income stream at 10% compounded continuously. The rate of flow of income from the clothing store is $f(t) = 12,000$, and the rate of flow of income from the computer store is expected to be $g(t) = 10{,}000e^{0.05t}$. Compare the future values of these investments to determine which is the better choice over the next 5 years.

22. Refer to Problem 21. Which investment is the better choice over the next 10 years?

23. An investor has $10,000 to invest in either a bond that matures in 5 years or a business that will produce a continuous stream of income over the next 5 years with rate of flow $f(t) = 2{,}000$. If both the bond and the continuous income stream earn 8% compounded continuously, which is the better investment?

24. Refer to Problem 23. Which is the better investment if the rate of flow of the income from the business is $f(t) = 3{,}000$?

25. A business is planning to purchase a piece of equipment that will produce a continuous stream of income for 8 years with rate of flow $f(t) = 9{,}000$. If the continuous income stream earns 12% compounded continuously, what single deposit into an account earning the same interest rate will produce the same future value as the continuous income stream? (This deposit is called the **present value** of the continuous income stream.)

26. Refer to Problem 25. Find the present value of a continuous income stream at 8% compounded continuously for 12 years if the rate of flow is $f(t) = 1{,}000^{0.03t}$.

27. Find the future value at a rate r compounded continuously for T years for a continuous income stream with rate of flow $f(t) = k$, where k is a constant.

28. Find the future value at a rate r compounded continuously for T years for a continuous income stream with rate of flow $f(t) = ke^{ct}$, where c and k are constants, $c \neq r$.

SECTION 8-3 Integration by Parts

In Section 7-1, we said we would return later to the indefinite integral

$$\int \ln x \, dx$$

since none of the integration techniques considered up to that time could be used to find an antiderivative for ln x. We will now develop a very useful technique, called *integration by parts,* that will enable us to find not only the above integral, but also many others, including integrals such as

$$\int x \ln x \, dx \qquad \text{and} \qquad \int xe^x \, dx$$

The technique of integration by parts is also used to derive many integration formulas that are tabulated in mathematical handbooks.

The method of integration by parts is based on the product formula for derivatives. If f and g are differentiable functions, then

$$D_x[f(x)g(x)] = f(x)g'(x) + g(x)f'(x)$$

which can be written in the equivalent form

$$f(x)g'(x) = D_x[f(x)g(x)] - g(x)f'(x)$$

Integrating both sides, we obtain

$$\int f(x)g'(x)\,dx = \int D_x[f(x)g(x)]\,dx - \int g(x)f'(x)\,dx$$

The first integral to the right of the equal sign is $f(x)g(x) + C$. (Why?) We will leave out the constant of integration for now, since we can add it after integrating the second integral to the right of the equal sign. So we have

$$\boldsymbol{\int f(x)g'(x)\,dx = f(x)g(x) - \int g(x)f'(x)\,dx}$$

This equation can be transformed into a more convenient form by letting $u = f(x)$ and $v = g(x)$; then $du = f'(x)\,dx$ and $dv = g'(x)\,dx$. Making these substitutions, we obtain the **integration by parts formula:**

Integration by Parts Formula

$$\int u\,dv = uv - \int v\,du$$

This formula can be very useful when the integral on the left is difficult to integrate using standard formulas. If u and dv are chosen with care, then the integral on the right side may be easier to integrate than the one on the left. Several examples will demonstrate the use of the formula.

◆ EXAMPLE 11 Find $\int xe^x\,dx$ using integration by parts.

Solution First, write the integration by parts formula:

$$\int u\,dv = uv - \int v\,du$$

Then, try to identify u and dv in $\int u\,dv$ (this is the key step) so that when $\int u\,dv$ is written in the form $uv - \int v\,du$, the new integral will be easier to evaluate.

Suppose we choose

$$u = x \qquad \text{and} \qquad dv = e^x\,dx \qquad u\,dv = xe^x\,dx$$

Then

$$du = (D_x\, x)\, dx = dx \qquad \text{and} \qquad v = \int e^x\, dx = e^x$$

Any constant may be added to v (we always choose 0 for simplicity). The general arbitrary constant of integration will be added at the end of the process.

Using the chosen u and dv and the corresponding du and v in the integration by parts formula, we obtain

$$\int u\, dv = uv - \int v\, du$$

$$\int xe^x\, dx = xe^x - \int e^x\, dx \qquad \text{This new integral is easy to evaluate.}$$

$$= xe^x - e^x + C$$

Check $\quad D_x(xe^x - e^x + C) = xe^x + e^x - e^x = xe^x$ ◆

PROBLEM 11 Find: $\int xe^{2x}\, dx$ ◆

◆ EXAMPLE 12 Find: $\int x \ln x\, dx$

Solution We write the integration by parts formula,

$$\int u\, dv = uv - \int v\, du$$

and choose

$$u = \ln x \qquad \text{and} \qquad dv = x\, dx \qquad u\, dv = (\ln x)x\, dx = x \ln x\, dx$$

Then,

$$du = (D_x \ln x)\, dx = \frac{1}{x}\, dx \qquad \text{and} \qquad v = \int x\, dx = \frac{x^2}{2}$$

and we have

$$\int x \ln x\, dx = (\ln x)\left(\frac{x^2}{2}\right) - \int \left(\frac{x^2}{2}\right)\left(\frac{1}{x}\, dx\right)$$

$$= \frac{x^2}{2} \ln x - \int \frac{x}{2}\, dx \qquad \text{This integral is easy to evaluate.}$$

$$= \frac{x^2}{2} \ln x - \frac{x^2}{4} + C$$

Check $\quad D_x\left(\frac{x^2}{2} \ln x - \frac{x^2}{4} + C\right) = x \ln x + \frac{x^2}{2}\left(\frac{1}{x}\right) - \frac{x}{2} = x \ln x$ ◆

PROBLEM 12 Find: $\int x \ln 2x\,dx$ ◆

Before considering additional examples, let us examine the process of selecting u and dv more closely. In the integral $\int xe^x\,dx$ (Example 11), suppose we had chosen $u = e^x$ and $dv = x\,dx$. Then $du = e^x\,dx$, $v = \frac{1}{2}x^2$, and the integration by parts formula gives

$$\int xe^x\,dx = e^x(\tfrac{1}{2}x^2) - \int \tfrac{1}{2}x^2e^x\,dx$$

$$= \tfrac{1}{2}x^2e^x - \tfrac{1}{2}\int x^2e^x\,dx$$

For this choice of u and dv, the new integral $\int v\,du$ is more complicated than the original integral $\int u\,dv$. This does not mean there is an error in these calculations or in the integration by parts formula. It simply means that this choice of u and dv does not change the original problem into one we can solve. When this happens, we must look for a different choice of u and dv. In some problems, it is possible that no choice will work. These observations and some guidelines for selecting u and dv are summarized below.

Integration by Parts: Selection of *u* and *dv*

1. The product $u\,dv$ must equal the original integrand.
2. It must be possible to integrate dv (preferably by using standard formulas or simple substitutions).
3. The new integral $\int v\,du$, should not be any more involved than the original integral $\int u\,dv$.
4. For integrals involving x^pe^{ax}, try

   $u = x^p$ and $dv = e^{ax}\,dx$

5. For integrals involving $x^p(\ln x)^q$, try

   $u = (\ln x)^q$ and $dv = x^p\,dx$

In some cases, repeated use of the integration by parts formula will lead to the evaluation of the original integral. The next example provides an illustration of such a case.

◆ EXAMPLE 13 Find: $\int x^2e^{-x}\,dx$

Solution Following suggestion 4 in the box, we choose

$u = x^2 \qquad dv = e^{-x}\,dx$

Then,

$$du = 2x\,dx \qquad v = -e^{-x}$$

and

$$\int x^2e^{-x}\,dx = x^2(-e^{-x}) - \int (-e^{-x})2x\,dx$$

$$= -x^2e^{-x} + 2\int xe^{-x}\,dx \tag{1}$$

The new integral is not one we can evaluate by standard formulas, but it is simpler than the original integral. Applying the integration by parts formula to it will produce an even simpler integral. For the integral $\int xe^{-x}\,dx$, we choose

$$u = x \qquad dv = e^{-x}\,dx$$

Then,

$$du = dx \qquad v = -e^{-x}$$

and

$$\int xe^{-x}\,dx = x(-e^{-x}) - \int (-e^{-x})\,dx$$

$$= -xe^{-x} + \int e^{-x}\,dx$$

$$= -xe^{-x} - e^{-x} + C \tag{2}$$

Substituting (2) into (1), we have

$$\int x^2e^{-x}\,dx = -x^2e^{-x} + 2(-xe^{-x} - e^{-x}) + C$$

$$= -x^2e^{-x} - 2xe^{-x} - 2e^{-x} + C$$ ◆

PROBLEM 13 Find: $\int x^2e^{2x}\,dx$ ◆

◆ EXAMPLE 14 Find: $\int_1^e \ln x\,dx$

Solution First, we will find $\int \ln x\,dx$, and then return to the definite integral. Following suggestion 5 in the box (with $p = 0$), we choose

$$u = \ln x \qquad dv = dx$$

Then,

$$du = \frac{1}{x}\,dx \qquad v = x$$

Hence,

$$\int \ln x\, dx = (\ln x)(x) - \int (x)\frac{1}{x}\, dx$$
$$= x \ln x - x + C$$

This is the important result we mentioned at the beginning of this section.
Now, we have

$$\int_1^e \ln x\, dx = (x \ln x - x)\Big|_1^e$$
$$= (e \ln e - e) - (1 \ln 1 - 1)$$
$$= (e - e) - (0 - 1)$$
$$= 1$$

◆

PROBLEM 14 Find: $\int_1^2 \ln 3x\, dx$ ◆

Answers to Matched Problems

11. $\frac{x}{2}e^{2x} - \frac{1}{4}e^{2x} + C$ 12. $\frac{x^2}{2}\ln 2x - \frac{x^2}{4} + C$

13. $\frac{x^2}{2}e^{2x} - \frac{x}{2}e^{2x} + \frac{1}{4}e^{2x} + C$ 14. $2 \ln 6 - \ln 3 - 1 \approx 1.4849$

EXERCISE 8-3

A *Integrate using the method of integration by parts. Assume $x > 0$ whenever the natural logarithm function is involved.*

1. $\int xe^{3x}\, dx$
2. $\int xe^{4x}\, dx$
3. $\int x^2 \ln x\, dx$
4. $\int x^3 \ln x\, dx$

B *Problems 5–18 are mixed—some require integration by parts and others can be solved using techniques we have considered earlier. Integrate as indicated, assuming $x > 0$ whenever the natural logarithm function is involved.*

5. $\int xe^{-x}\, dx$
6. $\int (x - 1)e^{-x}\, dx$
7. $\int xe^{x^2}\, dx$
8. $\int xe^{-x^2}\, dx$
9. $\int_0^1 (x - 3)e^x\, dx$
10. $\int_0^2 (x + 5)e^x\, dx$
11. $\int_1^3 \ln 2x\, dx$
12. $\int_2^3 \ln 7x\, dx$
13. $\int \frac{2x}{x^2 + 1}\, dx$
14. $\int \frac{x^2}{x^3 + 5}\, dx$
15. $\int \frac{\ln x}{x}\, dx$
16. $\int \frac{e^x}{e^x + 1}\, dx$
17. $\int \sqrt{x} \ln x\, dx$
18. $\int \frac{\ln x}{\sqrt{x}}\, dx$

C Problems 19–36 are mixed—some may require use of the integration by parts formula along with techniques we have considered earlier; others may require repeated use of the integration by parts formula. Assume $g(x) > 0$ whenever $\ln g(x)$ is involved.

19. $\int x^2 e^x \, dx$

20. $\int x^3 e^x \, dx$

21. $\int x e^{ax} \, dx, \quad a \neq 0$

22. $\int \ln ax \, dx, \quad a > 0$

23. $\int_1^e \frac{\ln x}{x^2} \, dx$

24. $\int_1^2 x^3 e^{x^2} \, dx$

25. $\int_0^2 \ln(x+4) \, dx$

26. $\int_0^2 \ln(4-x) \, dx$

27. $\int x e^{x-2} \, dx$

28. $\int x e^{x+1} \, dx$

29. $\int x \ln(1+x^2) \, dx$

30. $\int x \ln(1+x) \, dx$

31. $\int e^x \ln(1+e^x) \, dx$

32. $\int \frac{\ln(1+\sqrt{x})}{\sqrt{x}} \, dx$

33. $\int (\ln x)^2 \, dx$

34. $\int x(\ln x)^2 \, dx$

35. $\int (\ln x)^3 \, dx$

36. $\int x(\ln x)^3 \, dx$

Find the area bounded by the graphs of the indicated equations on the given intervals.

37. $y = x - \ln x$ and $y = 0, \quad 1 \leq x \leq e$
38. $y = 1 - \ln x$ and $y = 0, \quad 1 \leq x \leq 4$
39. $y = (x-2)e^x$ and $y = 0, \quad 0 \leq x \leq 3$
40. $y = (3-x)e^x$ and $y = 0, \quad 0 \leq x \leq 3$

APPLICATIONS

Business & Economics

41. *Profit.* If the rate of change of profit (in millions of dollars per year) is given by

$$P'(t) = 2t - te^{-t}$$

where t is time in years and the profit at time 0 is 0, find $P = P(t)$.

42. *Production.* An oil field is estimated to produce oil at a rate of $R(t)$ thousand barrels per month t months from now, as given by

$$R(t) = 10te^{-0.1t}$$

Estimate the total production in the first year of operation by use of an appropriate definite integral.

43. *Continuous income stream.* Find the future value at 8% compounded continuously for 5 years for the continuous income stream with rate of flow

$$f(t) = 1{,}000 - 200t$$

44. *Continuous income stream.* Find the interest earned at 10% compounded continuously for 4 years for the continuous income stream with rate of flow

$$f(t) = 1{,}000 - 250t$$

45. *Income distribution.* Find the coefficient of inequality for the Lorenz curve with equation

$$y = xe^{x-1}$$

46. *Producers' surplus.* Find the producers' surplus at a price level of $\overline{p} = \$26$ for the price–supply equation

$$p = S(x) = 5\ \ln(x + 1)$$

Life Sciences

47. *Pollution.* The concentration of particulate matter (in parts per million) t hours after a factory ceases operation for the day is given by

$$C(t) = \frac{20\ \ln(t + 1)}{(t + 1)^2}$$

Find the average concentration for the time period from $t = 0$ to $t = 5$.

48. *Medicine.* After a person takes a pill, the drug contained in the pill is assimilated into the bloodstream. The rate of assimilation t minutes after taking the pill is

$$R(t) = te^{-0.2t}$$

Find the total amount of the drug that is assimilated into the bloodstream during the first 10 minutes after the pill is taken.

Social Sciences

49. *Politics.* The number of voters (in thousands) in a certain city is given by

$$N(t) = 20 + 4t - 5te^{-0.1t}$$

where t is time in years. Find the average number of voters during the time period from $t = 0$ to $t = 5$.

SECTION 8-4 Integration Using Tables

- ◆ INTRODUCTION
- ◆ USING A TABLE OF INTEGRALS
- ◆ SUBSTITUTION AND INTEGRAL TABLES
- ◆ REDUCTION FORMULAS
- ◆ APPLICATION

◆ INTRODUCTION

A **table of integrals** is a list of integration formulas used to evaluate integrals. People who frequently evaluate complex integrals may refer to tables that contain hundreds of formulas. Tables of this type are included in mathematical handbooks available in most college book stores. Table II of Appendix B contains a short list of integral formulas illustrating the types found in more extensive tables. Some of these formulas can be derived using the integration techniques discussed earlier, while others require techniques we have not considered. However, it is possible to verify each formula by differentiating the right side.

◆ USING A TABLE OF INTEGRALS

The formulas in Table II (and in larger integral tables) are organized by categories, such as "*Integrals Involving* $a + bu$," "*Integrals Involving* $\sqrt{u^2 - a^2}$," and so on. The variable u is the variable of integration. All other symbols represent constants. To use a table to evaluate an integral, you must first find the category that most closely agrees with the form of the integrand and then find a formula in that category that can be made to match the integrand exactly by assigning values to the constants in the formula. The following examples illustrate this process.

◆ EXAMPLE 15 Use Table II to find: $\displaystyle\int \frac{x}{(5 + 2x)(4 + 3x)}\,dx$

Solution Since the integrand

$$f(x) = \frac{x}{(5 + 2x)(4 + 3x)}$$

is a rational function involving terms of the form $a + bu$ and $c + du$, we examine formulas 15–20 in Table II to see if any of the integrands in these formulas can be made to match $f(x)$ exactly. Comparing the integrand in formula 16 with $f(x)$, we see that this integrand will match $f(x)$ if we let $a = 5$, $b = 2$, $c = 4$, and $d = 3$. Letting $u = x$ and substituting for a, b, c, and d in formula 16, we have

$$\int \frac{u}{(a + bu)(c + du)}\,du = \frac{1}{ad - bc}\left(\frac{a}{b}\ln|a + bu| - \frac{c}{d}\ln|c + du|\right) \qquad \text{Formula 16}$$

$$\int \frac{x}{\underset{a}{(5} + \underset{b}{2}x)(\underset{c}{4} + \underset{d}{3}x)}\,dx = \frac{1}{5 \cdot 3 - 2 \cdot 4}\left(\frac{5}{2}\ln|5 + 2x| - \frac{4}{3}\ln|4 + 3x|\right) + C$$

$$a \cdot d - b \cdot c = 5 \cdot 3 - 2 \cdot 4 = 7$$

$$= \tfrac{5}{14}\ln|5 + 2x| - \tfrac{4}{21}\ln|4 + 3x| + C$$

Notice that the constant of integration C is not included in any of the formulas in Table II. However, you must still include C in all antiderivatives. ◆

PROBLEM 15 Use Table II to find: $\displaystyle\int \frac{1}{(5 + 3x)^2(1 + x)}\,dx$ ◆

◆ EXAMPLE 16 Evaluate: $\int_3^4 \frac{1}{x\sqrt{25-x^2}}\,dx$

Solution First, we use Table II to find

$$\int \frac{1}{x\sqrt{25-x^2}}\,dx$$

Since the integrand involves the expression $\sqrt{25-x^2}$, we examine formulas 29–31 and select formula 29 with $a^2 = 25$ and $a = 5$:

$$\int \frac{1}{u\sqrt{a^2-u^2}}\,du = -\frac{1}{a}\ln\left|\frac{a+\sqrt{a^2-u^2}}{u}\right| \qquad \text{Formula 29}$$

$$\int \frac{1}{x\sqrt{25-x^2}}\,dx = -\frac{1}{5}\ln\left|\frac{5+\sqrt{25-x^2}}{x}\right| + C$$

Thus,

$$\int_3^4 \frac{1}{x\sqrt{25-x^2}}\,dx = -\frac{1}{5}\ln\left|\frac{5+\sqrt{25-x^2}}{x}\right|\Bigg|_3^4$$

$$= -\frac{1}{5}\ln\left|\frac{5+3}{4}\right| + \frac{1}{5}\ln\left|\frac{5+4}{3}\right|$$

$$= -\tfrac{1}{5}\ln 2 + \tfrac{1}{5}\ln 3 = \tfrac{1}{5}\ln 1.5 \approx 0.0811$$ ◆

PROBLEM 16 Evaluate: $\int_6^8 \frac{1}{x^2\sqrt{100-x^2}}\,dx$ ◆

◆ SUBSTITUTION AND INTEGRAL TABLES

As Examples 15 and 16 illustrate, if the integral we want to evaluate can be made to match one in the table exactly, then evaluating the indefinite integral consists of simply substituting the correct values of the constants into the formula. What happens if we cannot match an integral with one of the formulas in the table? In many cases, a substitution will change the given integral into one that corresponds to a table entry. The following examples illustrate several frequently used substitutions.

◆ EXAMPLE 17 Find: $\int \frac{x^2}{\sqrt{16x^2-25}}\,dx$

Solution In order to relate this integral to one of the formulas involving $\sqrt{u^2-a^2}$ (formulas 40–45), we observe that if $u = 4x$, then

$$u^2 = 16x^2 \qquad \text{and} \qquad \sqrt{16x^2-25} = \sqrt{u^2-25}$$

Thus, we will use the substitution $u = 4x$ to change this integral into one that appears in the table:

$$\int \frac{x^2}{\sqrt{16x^2 - 25}}\,dx = \frac{1}{4}\int \frac{\frac{1}{16}u^2}{\sqrt{u^2 - 25}}\,du$$

$$= \frac{1}{64}\int \frac{u^2}{\sqrt{u^2 - 25}}\,du$$

Substitution:
$u = 4x$, $du = 4\,dx$
$x = \frac{1}{4}u$

This last integral can be evaluated by using formula 44 with $a = 5$:

$$\int \frac{u^2}{\sqrt{u^2 - a^2}}\,du = \frac{1}{2}\left(u\sqrt{u^2 - a^2} + a^2 \ln|u + \sqrt{u^2 - a^2}|\right) \qquad \text{Formula 44}$$

$$\int \frac{x^2}{\sqrt{16x^2 - 25}}\,dx = \frac{1}{64}\int \frac{u^2}{\sqrt{u^2 - 25}}\,du \qquad \text{Use formula 44 with } a = 5.$$

$$= \frac{1}{128}\left(u\sqrt{u^2 - 25} + 25 \ln|u + \sqrt{u^2 - 25}|\right) + C \qquad \text{Substitute } u = 4x.$$

$$= \frac{1}{128}\left(4x\sqrt{16x^2 - 25} + 25 \ln|4x + \sqrt{16x^2 - 25}|\right) + C$$ ◆

PROBLEM 17 Find: $\int \sqrt{9x^2 - 16}\,dx$ ◆

◆ EXAMPLE 18 Find: $\int \frac{x}{\sqrt{x^4 + 1}}\,dx$

Solution None of the formulas in the table involve fourth powers; however, if we let $u = x^2$, then

$$\sqrt{x^4 + 1} = \sqrt{u^2 + 1}$$

and this form does appear in formulas 32–39. Thus, we substitute $u = x^2$:

$$\int \frac{1}{\sqrt{x^4 + 1}}\,x\,dx = \frac{1}{2}\int \frac{1}{\sqrt{u^2 + 1}}\,du$$

Substitution:
$u = x^2$, $du = 2x\,dx$

We recognize the last integral as formula 36 with $a = 1$:

$$\int \frac{1}{\sqrt{u^2 + a^2}}\,du = \ln|u + \sqrt{u^2 + a^2}| \qquad \text{Formula 36}$$

$$\int \frac{x}{\sqrt{x^4 + 1}}\,dx = \frac{1}{2}\int \frac{1}{\sqrt{u^2 + 1}}\,du \qquad \text{Use formula 36 with } a = 1.$$

$$= \tfrac{1}{2}\ln|u + \sqrt{u^2 + 1}| + C \qquad \text{Substitute } u = x^2.$$

$$= \tfrac{1}{2}\ln|x^2 + \sqrt{x^4 + 1}| + C$$ ◆

PROBLEM 18 Find: $\int x\sqrt{x^4+1}\,dx$ ◆

◆ REDUCTION FORMULAS

◆ EXAMPLE 19 Use Table II to find: $\int x^2e^{3x}\,dx$

Solution Since the integrand involves the function e^{3x}, we examine formulas 46–48 and conclude that formula 47 can be used for this problem. Letting $u = x$, $n = 2$, and $a = 3$ in formula 47, we have

$$\int u^n e^{au}\,du = \frac{u^n e^{au}}{a} - \frac{n}{a}\int u^{n-1}e^{au}\,du \qquad \text{Formula 47}$$

$$\int x^2e^{3x}\,dx = \frac{x^2e^{3x}}{3} - \frac{2}{3}\int xe^{3x}\,dx$$

Notice that the expression on the right still contains an integral, but the exponent of x has been reduced by 1. Formulas of this type are called **reduction formulas** and are designed to be applied repeatedly until an integral that can be evaluated is obtained. Applying formula 47 to $\int xe^{3x}\,dx$ with $n = 1$, we have

$$\int x^2e^{3x}\,dx = \frac{x^2e^{3x}}{3} - \frac{2}{3}\left(\frac{xe^{3x}}{3} - \frac{1}{3}\int e^{3x}\,dx\right)$$

$$= \frac{x^2e^{3x}}{3} - \frac{2xe^{3x}}{9} + \frac{2}{9}\int e^{3x}\,dx$$

This last expression contains an integral that is easy to evaluate:

$$\int e^{3x}\,dx = \tfrac{1}{3}e^{3x}$$

After making a final substitution and adding a constant of integration, we have

$$\int x^2e^{3x}\,dx = \frac{x^2e^{3x}}{3} - \frac{2xe^{3x}}{9} + \frac{2}{27}e^{3x} + C$$ ◆

PROBLEM 19 Use Table II to find: $\int (\ln x)^2\,dx$ ◆

◆ APPLICATION

◆ EXAMPLE 20 Find the producers' surplus at a price level of $20 for the price–supply equation

Producers' Surplus

$$p = S(x) = \frac{5x}{500 - x}$$

Solution *Step 1.* Find $\bar{x}$, the supply when the price is $\bar{p} = 20$:

$$\bar{p} = \frac{5\bar{x}}{500 - \bar{x}}$$

$$20 = \frac{5\bar{x}}{500 - \bar{x}}$$

$$10{,}000 - 20\bar{x} = 5\bar{x}$$

$$10{,}000 = 25\bar{x}$$

$$\bar{x} = 400$$

Step 2. Sketch a graph, as shown in the margin.

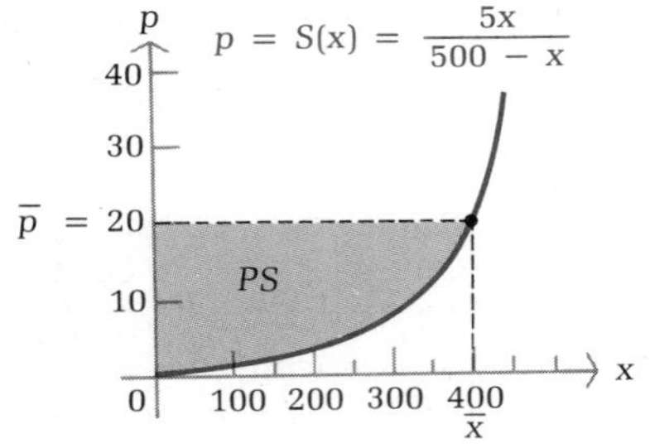

Step 3. Find the producers' surplus (the shaded area in the graph):

$$PS = \int_0^{\bar{x}} [\bar{p} - S(x)]\, dx$$

$$= \int_0^{400} \left(20 - \frac{5x}{500 - x}\right) dx$$

$$= \int_0^{400} \frac{10{,}000 - 25x}{500 - x}\, dx$$

Use formula 20 with $a = 10{,}000$, $b = -25$, $c = 500$, and $d = -1$:

$$\int \frac{a + bu}{c + du}\, du = \frac{bu}{d} + \frac{ad - bc}{d^2} \ln|c + du| \qquad \text{Formula 20}$$

$$PS = (25x + 2{,}500 \ln|500 - x|)\Big|_0^{400}$$

$$= 10{,}000 + 2{,}500 \ln|100| - 2{,}500 \ln|500|$$

$$\approx \$5{,}976$$

◆

PROBLEM 20 Find the consumers' surplus at a price level of \$10 for the price–demand equation

$$p = D(x) = \frac{20x - 8{,}000}{x - 500}$$

◆

Answers to Matched Problems

15. $\frac{1}{2}\left(\frac{1}{5 + 3x}\right) + \frac{1}{4} \ln\left|\frac{1 + x}{5 + 3x}\right| + C$ 16. $\frac{7}{1{,}200} \approx 0.0058$
17. $\frac{1}{6}(3x\sqrt{9x^2 - 16} - 16 \ln|3x + \sqrt{9x^2 - 16}|) + C$
18. $\frac{1}{4}(x^2\sqrt{x^4 + 1} + \ln|x^2 + \sqrt{x^4 + 1}|) + C$
19. $x(\ln x)^2 - 2x \ln x + 2x + C$
20. $3{,}000 + 2{,}000 \ln 200 - 2{,}000 \ln 500 \approx \$1{,}167$

EXERCISE 8-4

A *Use Table II to find each indefinite integral.*

1. $\int \frac{1}{x(1+x)}\,dx$
2. $\int \frac{1}{x^2(1+x)}\,dx$
3. $\int \frac{1}{(3+x)^2(5+2x)}\,dx$
4. $\int \frac{x}{(5+2x)^2(2+x)}\,dx$
5. $\int \frac{x}{\sqrt{16+x}}\,dx$
6. $\int \frac{1}{x\sqrt{16+x}}\,dx$
7. $\int \frac{1}{x\sqrt{x^2+4}}\,dx$
8. $\int \frac{1}{x^2\sqrt{x^2-16}}\,dx$
9. $\int x^2 \ln x\,dx$
10. $\int x^3 \ln x\,dx$

Evaluate each definite integral. Use Table II to find the antiderivative.

11. $\int_1^3 \frac{x^2}{3+x}\,dx$
12. $\int_2^6 \frac{x}{(6+x)^2}\,dx$
13. $\int_0^7 \frac{1}{(3+x)(1+x)}\,dx$
14. $\int_0^7 \frac{x}{(3+x)(1+x)}\,dx$
15. $\int_0^4 \frac{1}{\sqrt{x^2+9}}\,dx$
16. $\int_4^5 \sqrt{x^2-16}\,dx$

B *Use substitution techniques and Table II to find each indefinite integral.*

17. $\int \frac{\sqrt{4x^2+1}}{x^2}\,dx$
18. $\int x^2\sqrt{9x^2-1}\,dx$
19. $\int \frac{x}{\sqrt{x^4-16}}\,dx$
20. $\int x\sqrt{x^4-16}\,dx$
21. $\int x^2\sqrt{x^6+4}\,dx$
22. $\int \frac{x^2}{\sqrt{x^6+4}}\,dx$
23. $\int \frac{1}{x^3\sqrt{4-x^4}}\,dx$
24. $\int \frac{\sqrt{x^4+4}}{x}\,dx$
25. $\int \frac{e^x}{(2+e^x)(3+4e^x)}\,dx$
26. $\int \frac{e^x}{(4+e^x)^2(2+e^x)}\,dx$
27. $\int \frac{\ln x}{x\sqrt{4+\ln x}}\,dx$
28. $\int \frac{1}{x \ln x\sqrt{4+\ln x}}\,dx$

C *Use Table II to find each indefinite integral.*

29. $\int x^2e^{5x}\,dx$
30. $\int x^2e^{-4x}\,dx$
31. $\int x^3e^{-x}\,dx$
32. $\int x^3e^{2x}\,dx$
33. $\int (\ln x)^3\,dx$
34. $\int (\ln x)^4\,dx$

Problems 35–42 are mixed—some require the use of Table II and others can be solved using techniques we considered earlier.

35. $\int_3^5 x\sqrt{x^2-9}\,dx$ **36.** $\int_3^5 x^2\sqrt{x^2-9}\,dx$ **37.** $\int_2^4 \frac{1}{x^2-1}\,dx$

38. $\int_2^4 \frac{x}{(x^2-1)^2}\,dx$ **39.** $\int \frac{x+1}{x^2+2x}\,dx$ **40.** $\int \frac{x+1}{x^2+x}\,dx$

41. $\int \frac{x+1}{x^2+3x}\,dx$ **42.** $\int \frac{x^2+1}{x^2+3x}\,dx$

APPLICATIONS

Use Table II to evaluate all integrals involved in the solutions of Problems 43–54.

Business & Economics

43. *Consumers' surplus.* Find the consumers' surplus at a price level of $\bar{p} = \$15$ for the price–demand equation

$$p = D(x) = \frac{7{,}500 - 30x}{300 - x}$$

44. *Producers' surplus.* Find the producers' surplus at a price level of $\bar{p} = \$20$ for the price–supply equation

$$p = S(x) = \frac{10x}{300 - x}$$

45. *Continuous income stream.* Find the future value at 10% compounded continuously for 10 years for the continuous income stream with rate of flow $f(t) = 50t^2$.

46. *Continuous income stream.* Find the interest earned at 8% compounded continuously for 5 years for the continuous income stream with rate of flow $f(t) = 200t$.

47. *Income distribution.* Find the coefficient of inequality for the Lorenz curve with equation

$$y = \tfrac{1}{2}x\sqrt{1+3x}$$

48. *Income distribution.* Find the coefficient of inequality for the Lorenz curve with equation

$$y = \tfrac{1}{2}x^2\sqrt{1+3x}$$

49. *Marketing.* After test marketing a new high-fiber cereal, the market research department of a major food producer estimates that monthly sales (in millions of dollars) will grow at the monthly rate of

$$S'(t) = \frac{t^2}{(1+t)^2}$$

t months after the cereal is introduced. If we assume 0 sales at the time the cereal is first introduced, find the total sales, $S(t)$, t months after the cereal is introduced. Find the total sales during the first 2 years this cereal is on the market.

50. *Average price.* At a discount department store, the price–demand equation for premium motor oil is given by

$$p = p(x) = \frac{50}{\sqrt{100 + 6x}}$$

where x is the number of cans of oil that can be sold at a price of \$$p$. Find the average price over the demand interval [50, 250].

Life Sciences

51. *Pollution.* An oil tanker aground on a reef is losing oil and producing an oil slick that is radiating outward at a rate given approximately by

$$\frac{dR}{dt} = \frac{100}{\sqrt{t^2 + 9}} \qquad t \geqslant 0$$

where R is the radius (in feet) of the circular slick after t minutes. Find the radius of the slick after 4 minutes if the radius is 0 when $t = 0$.

52. *Pollution.* The concentration of particulate matter (in parts per million) during a 24 hour period is given approximately by

$$C(t) = t\sqrt{24 - t} \qquad 0 \leqslant t \leqslant 24$$

where t is time in hours. Find the average concentration during the time period from $t = 0$ to $t = 24$.

Social Sciences

53. *Learning.* A person learns N items at a rate given approximately by

$$N'(t) = \frac{60}{\sqrt{t^2 + 25}} \qquad t \geqslant 0$$

where t is the number of hours of continuous study. Determine the total number of items learned in the first 12 hours of continuous study.

54. *Politics.* The number of voters (in thousands) in a metropolitan area is given approximately by

$$f(t) = \frac{500}{2 + 3e^{-t}} \qquad t \geqslant 0$$

where t is time in years. Find the average number of voters during the time period from $t = 0$ to $t = 10$.

SECTION 8-5 Chapter Review

Important Terms and Symbols

8-1 *Definite Integral as a Limit of a Sum.* Rectangle rule; tabular function; definite integral (as a limit of a Riemann sum); fundamental theorem of calculus; average value of a continuous function

8-2 *Applications in Business and Economics.* Continuous income stream; rate of flow; total income; future value; consumers' surplus; producers' surplus; equilibrium price; equilibrium quantity

8-3 *Integration by Parts.*

$$\int u\,dv = uv - \int v\,du$$

8-4 *Integration Using Tables.* Table of integrals; substitution and integral tables; reduction formulas

EXERCISE 8-5

Chapter Review

Work through all the problems in this chapter review and check your answers in the back of the book. (Answers to all review problems are there.) Where weaknesses show up, review appropriate sections in the text.

A *Evaluate each integral.*

1. $\int xe^{4x}\,dx$
2. $\int x \ln x\,dx$
3. $\int \frac{1}{x(1+x)^2}\,dx$
4. $\int \frac{1}{x^2\sqrt{1+x}}\,dx$
5. Approximate $\int_1^5 (x^2+1)\,dx$ using the rectangle rule with $n = 2$ and c_k the midpoint of each subinterval.
6. Use the table of values below and the rectangle rule with $n = 4$ and c_k the midpoint of each subinterval to approximate $\int_1^{17} f(x)\,dx$.

x	3	7	11	15
f(x)	1.2	3.4	2.6	0.5

7. Find the average value of $f(x) = 6x^2 + 2x$ over the interval $[-1, 2]$.

B *Evaluate each integral.*

8. $\int_0^1 xe^x\,dx$
9. $\int_0^3 \frac{x^2}{\sqrt{x^2+16}}\,dx$
10. $\int \sqrt{9x^2-49}\,dx$
11. $\int te^{-0.5t}\,dt$
12. $\int x^2 \ln x\,dx$
13. $\int \frac{1}{1+2e^x}\,dx$
14. Find the area bounded by the graphs of $y = \ln x$ and $y = 0$ for $e \le x \le 2e$.
15. Approximate $\int_0^1 e^{2x^2}\,dx$ to three decimal places using $n = 5$ and c_k the midpoint of each subinterval.
16. Find the average value of $f(x) = 3x^{1/2}$ over the interval $[1, 9]$.

C *Evaluate each integral.*

17. $\int \frac{(\ln x)^2}{x}\,dx$

18. $\int x(\ln x)^2\,dx$

19. $\int \frac{x}{\sqrt{x^2 - 36}}\,dx$

20. $\int \frac{x}{\sqrt{x^4 - 36}}\,dx$

21. $\int_0^4 x\ln(10 - x)\,dx$

22. $\int (\ln x)^2\,dx$

APPLICATIONS

Business & Economics

23. *Inventory.* Suppose the inventory of a certain item t months after the first of the year is given approximately by

$$I(t) = 10 + 36t - 3t^2 \qquad 0 \leq t \leq 12$$

What is the average inventory for the second quarter of the year?

24. *Supply function.* Given the supply function

$$p = S(x) = 8(e^{0.05x} - 1)$$

find the average price (in dollars) over the supply interval [40, 50].

25. *Continuous income stream.* The rate of flow of a continuous income stream for a 5 year period is given by

$$f(t) = 2{,}500e^{0.05t} \qquad 0 \leq t \leq 5$$

(A) Find the total income over this 5 year period.
(B) Find the future value at the end of this 5 year period at 15% compounded continuously.
(C) Find the interest earned during this 5 year period.

26. *Consumers' and producers' surplus.* Given the price–demand and price–supply equations

$$p = D(x) = 70 - \tfrac{1}{5}x \qquad \text{and} \qquad p = S(x) = 13 + \tfrac{3}{2{,}500}x^2$$

(A) Find the consumers' surplus at a price level of $\overline{p} = \$50$.
(B) Find the producers' surplus at a price level of $\overline{p} = \$25$.
(C) Find the equilibrium price, and then find the consumers' surplus and producers' surplus at the equilibrium price level.

27. *Production.* An oil field is estimated to produce oil at the rate of $R(t)$ thousand barrels per month t months from now, as given by

$$R(t) = 25te^{-0.05t}$$

How much oil is produced during the first 2 years of operation?

Life Sciences

28. *Drug assimilation.* The rate at which the body eliminates a drug (in milliliters per hour) is given by

$$R(t) = \frac{60t}{(t + 1)^2(t + 2)}$$

where t is the number of hours since the drug was administered. How much of the drug is eliminated in the first hour after it was administered?

29. *Pollution.* In an industrial area, the concentration $C(t)$ of particulate matter (in parts per million) during a 12 hour period is given in the figure. Use the rectangle rule with $n = 6$ and c_k the midpoint of each subinterval to approximate the average concentration during this 12 hour period. Estimate the necessary function values from the graph.

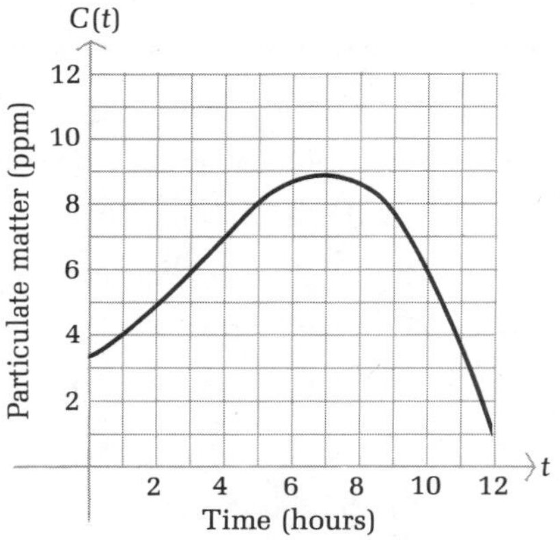

Social Sciences

30. *Politics.* The rate of change of the voting population of a city, $N'(t)$, with respect to time t in years is estimated to be

$$N'(t) = \frac{25}{\sqrt{t^2 + 1}}$$

where $N(t)$ is in thousands. If $N(0)$ is the current voting population, how much will this population increase during the next 3 years?

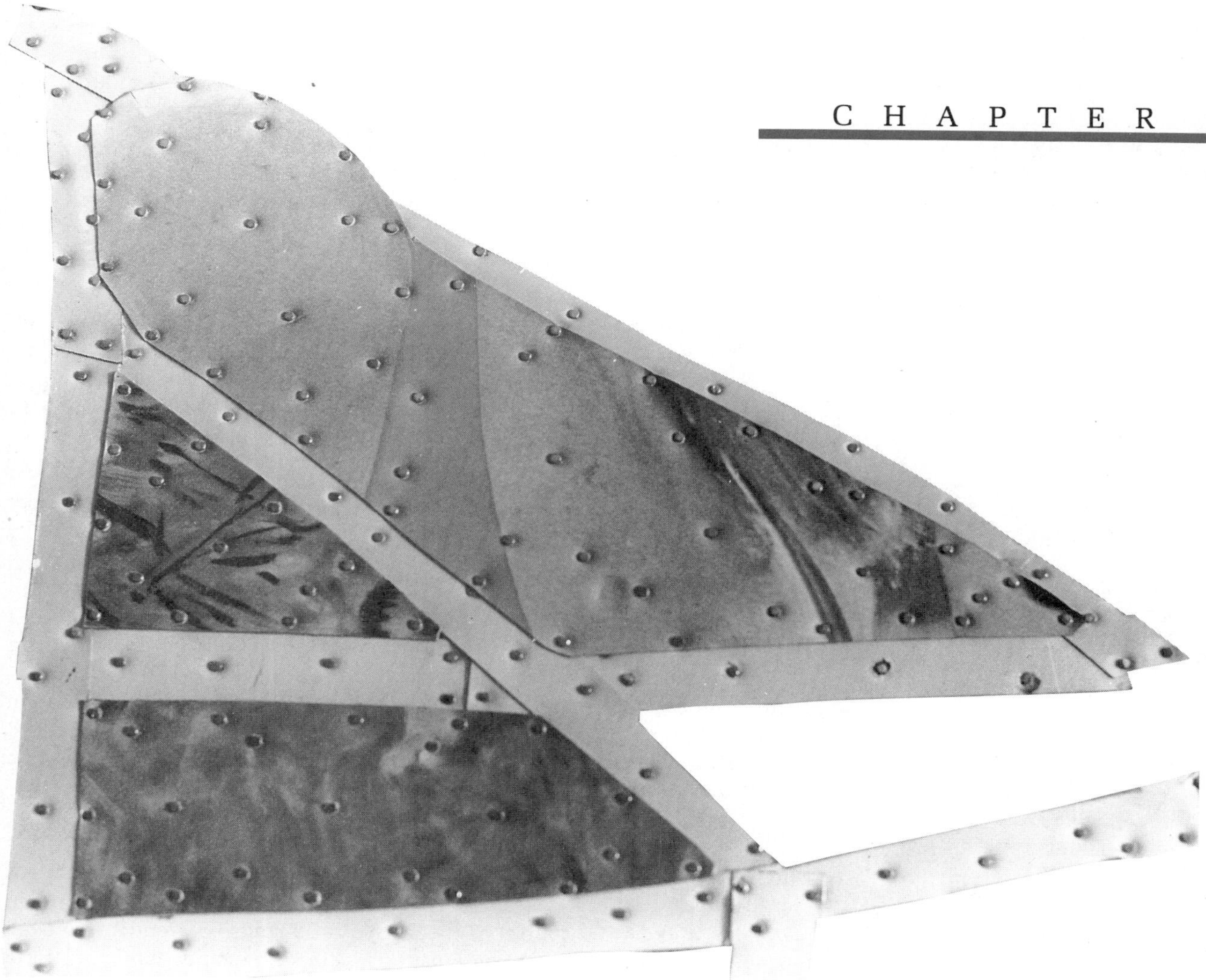

CHAPTER 9

Multivariable Calculus

CHAPTER 9

Contents

SECTION 9-1 Functions of Several Variables

- FUNCTIONS OF TWO OR MORE INDEPENDENT VARIABLES
- EXAMPLES OF FUNCTIONS OF SEVERAL VARIABLES
- THREE-DIMENSIONAL COORDINATE SYSTEMS

◆ FUNCTIONS OF TWO OR MORE INDEPENDENT VARIABLES

In Section 2-4, we introduced the concept of a function with one independent variable. Now we will broaden the concept to include functions with more than one independent variable. We start with an example.

A small manufacturing company produces a standard type of surfboard and no other products. If fixed costs are \$500 per week and variable costs are \$70 per board produced, then the weekly cost function is given by

$$C(x) = 500 + 70x \tag{1}$$

where x is the number of boards produced per week. The cost function is a function of a single independent variable x. For each value of x from the domain of C there exists exactly one value of C(x) in the range of C.

Now, suppose the company decides to add a high-performance competition board to its line. If the fixed costs for the competition board are \$200 per week and the variable costs are \$100 per board, then the cost function (1) must be modified to

$$C(x, y) = 700 + 70x + 100y \tag{2}$$

where C(x, y) is the cost for weekly output of x standard boards and y competition boards. Equation (2) is an example of a function with two independent variables, x and y. Of course, as the company expands its product line even further, its weekly cost function must be modified to include more and more independent variables, one for each new product produced.

In general, an equation of the form

$$z = f(x, y)$$

describes a **function of two independent variables** if for each permissible ordered pair (x, y), there is one and only one value of z determined by $f(x, y)$. The variables x and y are **independent variables,** and the variable z is a **dependent variable.** The set of all ordered pairs of permissible values of x and y is the **domain** of the function, and the set of all corresponding values $f(x, y)$ is the **range** of the function. Unless otherwise stated, we will assume that the domain of a function specified by an equation of the form $z = f(x, y)$ is the set of all ordered pairs of real numbers (x, y) such that $f(x, y)$ is also a real number. It should be noted, however, that certain conditions in practical problems often lead to further restrictions of the domain of a function.

We can similarly define functions of three independent variables, $w = f(x, y, z)$; of four independent variables, $u = f(w, x, y, z)$; and so on. In this chapter, we will primarily concern ourselves with functions of two independent variables.

◆ EXAMPLE 1 For the cost function $C(x, y) = 700 + 70x + 100y$ described earlier, find $C(10, 5)$.

Solution

$$C(10, 5) = 700 + 70(10) + 100(5) = \$1{,}900$$ ◆

PROBLEM 1 Find $C(20, 10)$ for the cost function in Example 1. ◆

◆ EXAMPLE 2 For $f(x, y, z) = 2x^2 - 3xy + 3z + 1$, find $f(3, 0, -1)$.

Solution

$$f(3, 0, -1) = 2(3)^2 - 3(3)(0) + 3(-1) + 1 = 18 - 0 - 3 + 1 = 16$$ ◆

PROBLEM 2 Find $f(-2, 2, 3)$ for f in Example 2. ◆

◆ EXAMPLE 3 *Revenue, Cost, and Profit Functions*

The surfboard company discussed at the beginning of this section has determined that the demand equations for the two types of boards they produce are given by

$$p = 210 - 4x + y$$
$$q = 300 + x - 12y$$

where p is the price of the standard board, q is the price of the competition board, x is the weekly demand for standard boards, and y is the weekly demand for competition boards.

(A) Find the weekly revenue function $R(x, y)$ and evaluate $R(20, 10)$.
(B) If the weekly cost function is

$$C(x, y) = 700 + 70x + 100y$$

find the weekly profit function $P(x, y)$ and evaluate $P(20, 10)$.

Solutions (A) $\text{Revenue} = \begin{pmatrix}\text{Demand for}\\ \text{standard}\\ \text{boards}\end{pmatrix} \times \begin{pmatrix}\text{Price of a}\\ \text{standard}\\ \text{board}\end{pmatrix} + \begin{pmatrix}\text{Demand for}\\ \text{competition}\\ \text{boards}\end{pmatrix} \times \begin{pmatrix}\text{Price of a}\\ \text{competition}\\ \text{board}\end{pmatrix}$

$$\begin{aligned} R(x, y) &= xp + yq \\ &= x(210 - 4x + y) + y(300 + x - 12y) \\ &= 210x + 300y - 4x^2 + 2xy - 12y^2 \\ R(20, 10) &= 210(20) + 300(10) - 4(20)^2 + 2(20)(10) - 12(10)^2 \\ &= \$4{,}800 \end{aligned}$$

(B)

$$\begin{aligned} \text{Profit} &= \text{Revenue} - \text{Cost} \\ P(x, y) &= R(x, y) - C(x, y) \\ &= 210x + 300y - 4x^2 + 2xy - 12y^2 - 700 - 70x - 100y \\ &= 140x + 200y - 4x^2 + 2xy - 12y^2 - 700 \\ P(20, 10) &= 140(20) + 200(10) - 4(20)^2 + 2(20)(10) - 12(10)^2 - 700 \\ &= \$1{,}700 \end{aligned}$$

◆

PROBLEM 3 Repeat Example 3 if the demand and cost equations are given by

$$\begin{aligned} p &= 220 - 6x + y \\ q &= 300 + 3x - 10y \\ C(x, y) &= 40x + 80y + 1{,}000 \end{aligned}$$

◆

◆ EXAMPLES OF FUNCTIONS OF SEVERAL VARIABLES

A number of concepts we have already considered can be thought of in terms of functions of two or more variables. We list a few of these below.

Area of a rectangle $A(x, y) = xy$

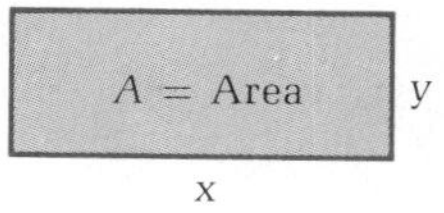

Volume of a box $V(x, y, z) = xyz$

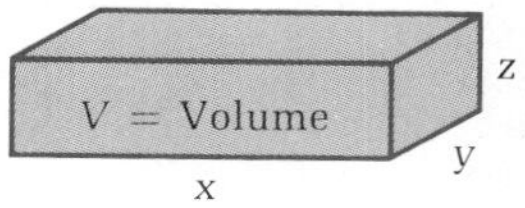

Volume of a right circular cylinder $V(r, h) = \pi r^2 h$

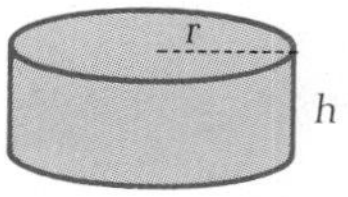

Simple interest	$A(P, r, t) = P(1 + rt)$	A = Amount P = Principal r = Annual rate t = Time in years
Compound interest	$A(P, r, t, n) = P\left(1 + \frac{r}{n}\right)^{nt}$	A = Amount P = Principal r = Annual rate t = Time in years n = Compound periods per year
IQ	$Q(M, C) = \frac{M}{C}(100)$	Q = IQ = Intelligence quotient M = MA = Mental age C = CA = Chronological age
Resistance for blood flow in a vessel (Poiseuille's law)	$R(L, r) = k\frac{L}{r^4}$	R = Resistance L = Length of vessel r = Radius of vessel k = Constant

◆ EXAMPLE 4

Package Design

A company uses a box with a square base and an open top for one of its products (see the figure). If x is the length (in inches) of each side of the base and y is the height (in inches), find the total amount of material $M(x, y)$ required to construct one of these boxes, and evaluate $M(5, 10)$.

Solution

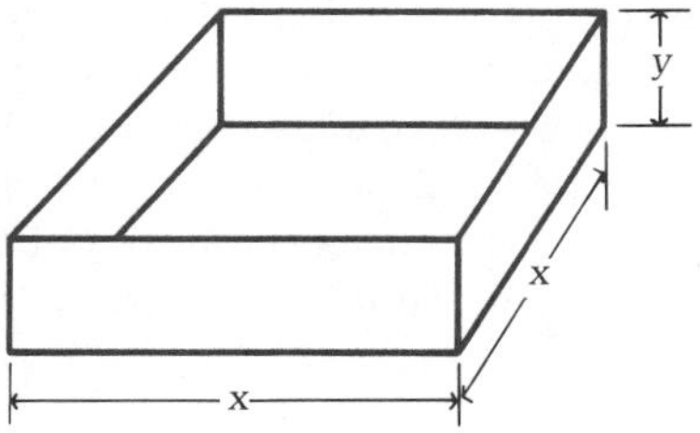

$$\begin{aligned}
\text{Area of base} &= x^2 \\
\text{Area of one side} &= xy \\
\text{Total material} &= (\text{Area of base}) + 4(\text{Area of one side}) \\
M(x, y) &= x^2 + 4xy \\
M(5, 10) &= (5)^2 + 4(5)(10) \\
&= 225 \text{ square inches}
\end{aligned}$$

◆

PROBLEM 4

For the box in Example 4, find the volume $V(x, y)$, and evaluate $V(5, 10)$. ◆

The next example concerns the **Cobb–Douglas production function,**

$$f(x, y) = kx^m y^n$$

where k, m, and n are positive constants with $m + n = 1$. Economists use this function to describe the number of units $f(x, y)$ produced from the utilization of x units of labor and y units of capital (for equipment such as tools, machinery, buildings, and so on). Cobb–Douglas production functions are also used to describe the productivity of a single industry, of a group of industries producing the same product, or even of an entire country.

◆ EXAMPLE 5 *Productivity*

The productivity of a steel manufacturing company is given approximately by the function

$$f(x, y) = 10x^{0.2}y^{0.8}$$

with the utilization of x units of labor and y units of capital. If the company uses 3,000 units of labor and 1,000 units of capital, how many units of steel will be produced?

Solution

The number of units of steel produced is given by

$$f(3{,}000, 1{,}000) = 10(3{,}000)^{0.2}(1{,}000)^{0.8} \quad \text{Use a calculator.}$$
$$\approx 12{,}457 \text{ units}$$

◆

PROBLEM 5

Refer to Example 5. Find the steel production if the company uses 1,000 units of labor and 2,000 units of capital. ◆

◆ THREE-DIMENSIONAL COORDINATE SYSTEMS

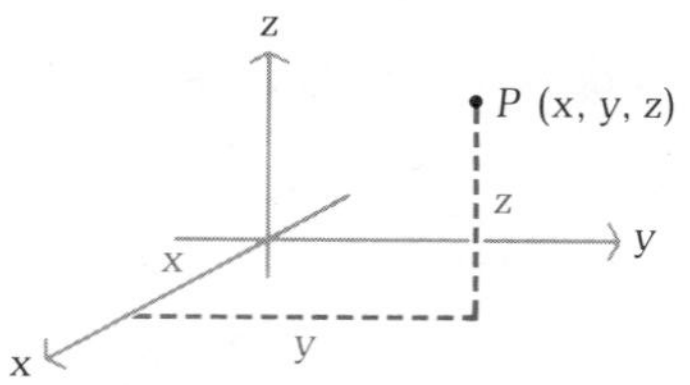

FIGURE 1
Rectangular coordinate system

We now take a brief look at some graphs of functions of two independent variables. Since functions of the form $z = f(x, y)$ involve two independent variables, x and y, and one dependent variable, z, we need a *three-dimensional coordinate system* for their graphs. A **three-dimensional coordinate system** is formed by three mutually perpendicular number lines intersecting at their origins. To represent a three-dimensional coordinate system on a piece of paper, we draw two of the number lines (usually the y and z axes) in the plane of the paper and draw the third (the x axis) as a diagonal line which we visualize as projecting out from the paper (see Fig. 1). In such a system, every ordered **triplet of numbers (x, y, z)** can be associated with a unique point, and conversely.

◆ EXAMPLE 6

Locate (−3, 5, 2) in a rectangular coordinate system.

Solution

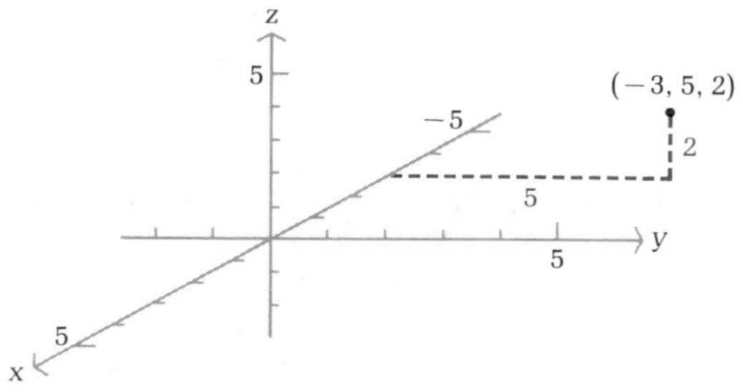

◆

PROBLEM 6 Find the coordinates of the corners A, C, G, and D of the rectangular box shown in the figure.

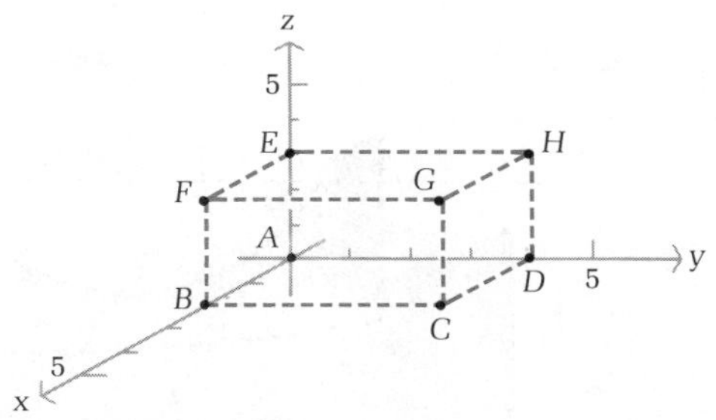

◆

What does the graph of $z = x^2 + y^2$ look like? If we let $x = 0$ and graph $z = 0^2 + y^2 = y^2$ in the yz plane, we obtain a parabola; if we let $y = 0$ and graph $z = x^2 + 0^2 = x^2$ in the xz plane, we obtain another parabola. It can be shown that the graph of $z = x^2 + y^2$ is either one of these parabolas rotated around the z axis (see Fig. 2). This cup-shaped figure is a *surface* and is called a **paraboloid.**

In general, the graph of any function of the form $z = f(x, y)$ is called a **surface.** The graph of such a function is the graph of all ordered triplets of numbers (x, y, z) that satisfy the equation. Graphing functions of two independent variables is often a very difficult task, and the general process will not be dealt with in this book. We present only a few simple graphs to suggest extensions of earlier geometric interpretations of the derivative and local maxima and minima to functions of two variables. Note that $z = f(x, y) = x^2 + y^2$ appears (see Fig. 2) to have a local minimum at $(x, y) = (0, 0)$. Figure 3 shows a local maximum at $(x, y) = (0, 0)$.

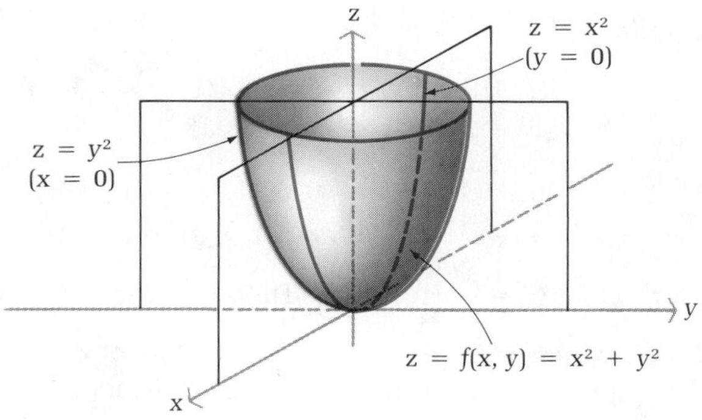

FIGURE 2
Paraboloid

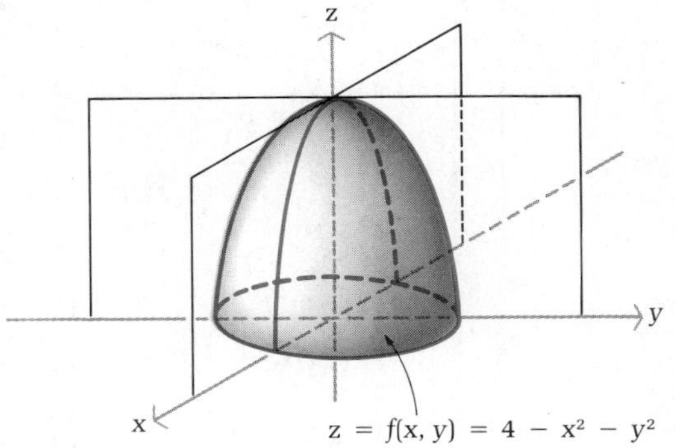

FIGURE 3
Local maximum: $f(0, 0) = 4$

Figure 4 shows a point at $(x, y) = (0, 0)$, called a **saddle point,** which is neither a local minimum nor a local maximum. More will be said about local maxima and minima in Section 9-3.

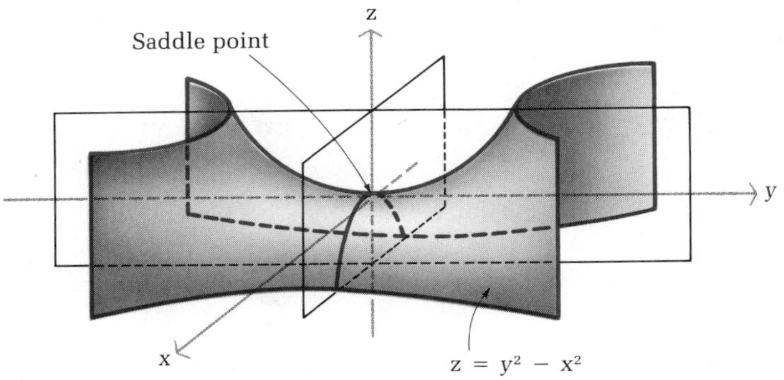

FIGURE 4
Saddle point at (0, 0, 0)

Answers to Matched Problems

1. \$3,100 2. 30
3. (A) $R(x, y) = 220x + 300y - 6x^2 + 4xy - 10y^2$; $R(20, 10) = \$4,800$
   (B) $P(x, y) = 180x + 220y - 6x^2 + 4xy - 10y^2 - 1,000$; $P(20, 10) = \$2,200$
4. $V(x, y) = x^2y$; $V(5, 10) = 250$ in.^{3} 5. 17,411 units
6. $A(0, 0, 0)$; $C(2, 4, 0)$; $G(2, 4, 3)$; $D(0, 4, 0)$

EXERCISE 9-1

A

For the functions

$$f(x, y) = 10 + 2x - 3y \quad \text{and} \quad g(x, y) = x^2 - 3y^2$$

find each of the following:

1. $f(0, 0)$ **2.** $f(2, 1)$ **3.** $f(-3, 1)$ **4.** $f(2, -7)$
5. $g(0, 0)$ **6.** $g(0, -1)$ **7.** $g(2, -1)$ **8.** $g(-1, 2)$

B

Find each of the following:

9. $A(2, 3)$ for $A(x, y) = xy$ **10.** $V(2, 4, 3)$ for $V(x, y, z) = xyz$

11. $Q(12, 8)$ for $Q(M, C) = \frac{M}{C}(100)$ **12.** $T(50, 17)$ for $T(V, x) = \frac{33V}{x + 33}$

13. $V(2, 4)$ for $V(r, h) = \pi r^2 h$ **14.** $S(4, 2)$ for $S(x, y) = 5x^2y^3$

15. $R(1, 2)$ for $R(x, y) = -5x^2 + 6xy - 4y^2 + 200x + 300y$

16. $P(2, 2)$ for $P(x, y) = -x^2 + 2xy - 2y^2 - 4x + 12y + 5$

17. $R(6, 0.5)$ for $R(L, r) = 0.002 \frac{L}{r^4}$

18. $L(2{,}000, 50)$ for $L(w, v) = (1.25 \times 10^{-5})wv^2$
19. $A(100, 0.06, 3)$ for $A(P, r, t) = P + Prt$
20. $A(10, 0.04, 3, 2)$ for $A(P, r, t, n) = P\left(1 + \frac{r}{n}\right)^{tn}$
21. $A(100, 0.08, 10)$ for $A(P, r, t) = Pe^{rt}$
22. $A(1{,}000, 0.06, 8)$ for $A(P, r, t) = Pe^{rt}$

C

23. For the function $f(x, y) = x^2 + 2y^2$, find: $\frac{f(x + h, y) - f(x, y)}{h}$
24. For the function $f(x, y) = x^2 + 2y^2$, find: $\frac{f(x, y + k) - f(x, y)}{k}$
25. For the function $f(x, y) = 2xy^2$, find: $\frac{f(x + h, y) - f(x, y)}{h}$
26. For the function $f(x, y) = 2xy^2$, find: $\frac{f(x, y + k) - f(x, y)}{k}$
27. Find the coordinates of E and F in the figure for Problem 6 in the text.
28. Find the coordinates of B and H in the figure for Problem 6 in the text.

APPLICATIONS

Business & Economics

29. *Cost function.* A small manufacturing company produces two models of a surfboard: a standard model and a competition model. If the standard model is produced at a variable cost of \$70 each, the competition model at a variable cost of \$100 each, and the total fixed costs per month are \$2,000, then the monthly cost function is given by

$$C(x, y) = 2{,}000 + 70x + 100y$$

where x and y are the numbers of standard and competition models produced per month, respectively. Find $C(20, 10)$, $C(50, 5)$, and $C(30, 30)$.

30. *Advertising and sales.* A company spends \$$x$ thousand per week on newspaper advertising and \$$y$ thousand per week on television advertising. Its weekly sales are found to be given by

$$S(x, y) = 5x^2y^3$$

Find $S(3, 2)$ and $S(2, 3)$.

31. *Revenue function.* A supermarket sells two brands of coffee: brand A at \$$p$ per pound and brand B at \$$q$ per pound. The daily demand equations for brands A and B are, respectively,

$$x = 200 - 5p + 4q$$
$$y = 300 + 2p - 4q$$

(both in pounds). Find the daily revenue function $R(p, q)$. Evaluate $R(2, 3)$ and $R(3, 2)$.

32. *Revenue, cost, and profit functions.* A company manufactures ten-speed and three-speed bicycles. The weekly demand and cost equations are

$$p = 230 - 9x + y$$
$$q = 130 + x - 4y$$
$$C(x, y) = 200 + 80x + 30y$$

where \$$p$ is the price of a ten-speed bicycle, \$$q$ is the price of a three-speed bicycle, x is the weekly demand for ten-speed bicycles, y is the weekly demand for three-speed bicycles, and $C(x, y)$ is the cost function. Find the weekly revenue function $R(x, y)$ and the weekly profit function $P(x, y)$. Evaluate $R(10, 15)$ and $P(10, 15)$.

33. *Productivity.* The Cobb–Douglas production function for a petroleum company is given by

$$f(x, y) = 20x^{0.4}y^{0.6}$$

where x is the utilization of labor and y is the utilization of capital. If the company uses 1,250 units of labor and 1,700 units of capital, how many units of petroleum will be produced?

34. *Productivity.* The petroleum company in Problem 33 is taken over by another company that decides to double both the units of labor and the units of capital utilized in the production of petroleum. Use the Cobb–Douglas production function given in Problem 33 to find the amount of petroleum that will be produced by this increased utilization of labor and capital. What is the effect on productivity of doubling both the units of labor and the units of capital?

35. *Future value.* At the end of each year, \$2,000 is invested into an IRA earning 9% compounded annually. How much will be in the account at the end of 30 years? Use the annuity formula

$$F(P, i, n) = P\frac{(1 + i)^n - 1}{i}$$

where

P = Periodic payment	n = Number of payments (periods)
i = Rate per period	$F = FV$ = Future value

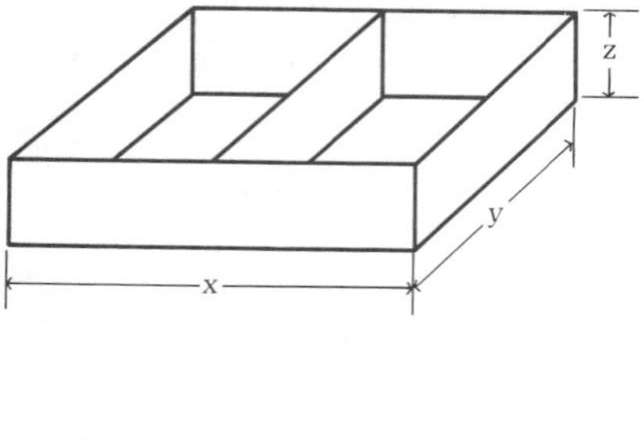

36. *Package design.* The packaging department in a company has been asked to design a rectangular box with no top and a partition down the middle (see the figure in the margin). If x, y, and z are the dimensions (in inches), find the total amount of material $M(x, y, z)$ used in constructing one of these boxes, and evaluate $M(10, 12, 6)$.

Life Sciences

37. *Marine biology.* In using scuba diving gear, a marine biologist estimates the time of a dive according to the equation

$$T(V, x) = \frac{33V}{x + 33}$$

where

$T =$ Time of dive in minutes

$V =$ Volume of air, at sea level pressure, compressed into tanks

$x =$ Depth of dive in feet

Find $T(70, 47)$ and $T(60, 27)$.

38. *Blood flow.* Poiseuille's law states that the resistance, R, for blood flowing in a blood vessel varies directly as the length of the vessel, L, and inversely as the fourth power of its radius, r. Stated as an equation,

$$R(L, r) = k\frac{L}{r^4} \qquad k \text{ a constant}$$

Find $R(8, 1)$ and $R(4, 0.2)$.

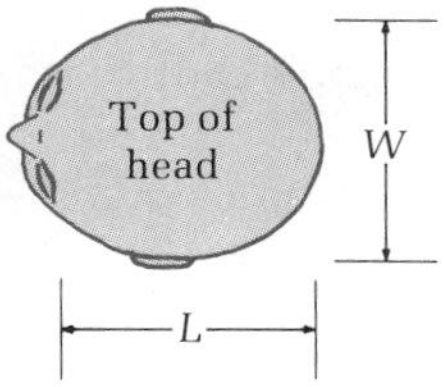

39. *Physical anthropology.* Anthropologists, in their study of race and human genetic groupings, often use an index called the *cephalic index*. The cephalic index, C, varies directly as the width, W, of the head, and inversely as the length, L, of the head (both viewed from the top). In terms of an equation,

$$C(W, L) = 100\frac{W}{L}$$

where

$W =$ Width in inches $\qquad L =$ Length in inches

Find $C(6, 8)$ and $C(8.1, 9)$.

Social Sciences

40. *Safety research.* Under ideal conditions, if a person driving a car slams on the brakes and skids to a stop, the length of the skid marks (in feet) is given by the formula

$$L(w, v) = kwv^2$$

where

$k =$ Constant $\qquad w =$ Weight of car in pounds

$v =$ Speed of car in miles per hour

For $k = 0.000\ 013\ 3$, find $L(2{,}000, 40)$ and $L(3{,}000, 60)$.

41. *Psychology.* The intelligence quotient (IQ) is defined to be the ratio of mental age (MA), as determined by certain tests, and chronological age (CA), multiplied by 100. Stated as an equation,

$$Q(M, C) = \frac{M}{C} \cdot 100$$

where

$$Q = \text{IQ} \qquad M = \text{MA} \qquad C = \text{CA}$$

Find $Q(12, 10)$ and $Q(10, 12)$.

SECTION 9-2 Partial Derivatives

- PARTIAL DERIVATIVES
- SECOND-ORDER PARTIAL DERIVATIVES

PARTIAL DERIVATIVES

We know how to differentiate many kinds of functions of one independent variable and how to interpret the results. What about functions with two or more independent variables? Let us return to the surfboard example considered at the beginning of the chapter.

For the company producing only the standard board, the cost function was

$$C(x) = 500 + 70x$$

Differentiating with respect to x, we obtain the marginal cost function

$$C'(x) = 70$$

Since the marginal cost is constant, \$70 is the change in cost for a 1 unit increase in production at any output level.

For the company producing two types of boards, a standard model and a competition model, the cost function was

$$C(x, y) = 700 + 70x + 100y$$

Now suppose we differentiate with respect to x, holding y fixed, and denote this by $C_x(x, y)$; or suppose we differentiate with respect to y, holding x fixed, and denote this by $C_y(x, y)$. Differentiating in this way, we obtain

$$C_x(x, y) = 70 \qquad C_y(x, y) = 100$$

Each of these is called a **partial derivative,** and, in this example, each represents marginal cost. The first is the change in cost due to a 1 unit increase in production of the standard board with the production of the competition model held fixed. The second is the change in cost due to a 1 unit increase in production of the competition board with the production of the standard board held fixed.

In general, if $z = f(x, y)$, then the **partial derivative of f with respect to x,** denoted by $\partial z/\partial x$, f_x, or $f_x(x, y)$, is defined by

$$\frac{\partial z}{\partial x} = \lim_{h\to 0} \frac{f(x+h, y) - f(x, y)}{h}$$

provided the limit exists. We recognize this as the ordinary derivative of f with respect to x, holding y constant. Thus, we are able to continue to use all the derivative rules and properties discussed in Chapters 4–6 for partial derivatives.

Similarly, the **partial derivative of f with respect to y,** denoted by $\partial z/\partial y$, f_y, or $f_y(x, y)$, is defined by

$$\frac{\partial z}{\partial y} = \lim_{k\to 0} \frac{f(x, y+k) - f(x, y)}{k}$$

which is the ordinary derivative with respect to y, holding x constant.

Parallel definitions and interpretations hold for functions with three or more independent variables.

◆ EXAMPLE 7 For $z = f(x, y) = 2x^2 - 3x^2y + 5y + 1$, find:

(A) $\partial z/\partial x$ (B) $f_x(2, 3)$

Solutions (A) $z = 2x^2 - 3x^2y + 5y + 1$

Differentiating with respect to x, holding y constant (that is, treating y as a constant), we obtain

$$\frac{\partial z}{\partial x} = 4x - 6xy$$

(B) $f(x, y) = 2x^2 - 3x^2y + 5y + 1$

First, differentiate with respect to x (part A) to obtain

$$f_x(x, y) = 4x - 6xy$$

Then evaluate at (2, 3):

$$f_x(2, 3) = 4(2) - 6(2)(3) = -28$$ ◆

PROBLEM 7 For f in Example 7, find:

(A) $\partial z/\partial y$ (B) $f_y(2, 3)$ ◆

◆ EXAMPLE 8 For $z = f(x, y) = e^{x^2+y^2}$, find:

(A) $\partial z/\partial x$ (B) $f_y(2, 1)$

Solutions (A) Using the chain rule [thinking of $z = e^u$, $u = u(x)$; y is held constant], we obtain

$$\frac{\partial z}{\partial x} = e^{x^2+y^2} \frac{\partial(x^2 + y^2)}{\partial x}$$
$$= 2xe^{x^2+y^2}$$

(B) $$f_y(x, y) = e^{x^2+y^2} \frac{\partial(x^2 + y^2)}{\partial y} = 2ye^{x^2+y^2}$$

$$f_y(2, 1) = 2(1)e^{(2)^2+(1)^2}$$
$$= 2e^5$$

◆

PROBLEM 8 For $z = f(x, y) = (x^2 + 2xy)^5$, find:

(A) $\partial z/\partial y$ (B) $f_x(1, 0)$

◆

◆ EXAMPLE 9 *Profit*

The profit function for the surfboard company in Example 3 in Section 9-1 was

$$P(x, y) = 140x + 200y - 4x^2 + 2xy - 12y^2 - 700$$

Find $P_x(15, 10)$ and $P_x(30, 10)$, and interpret the results.

Solution

$$P_x(x, y) = 140 - 8x + 2y$$
$$P_x(15, 10) = 140 - 8(15) + 2(10) = 40$$
$$P_x(30, 10) = 140 - 8(30) + 2(10) = -80$$

At a production level of 15 standard and 10 competition boards per week, increasing the production of standard boards by 1 unit and holding the production of competition boards fixed at 10 will increase profit by approximately \$40. At a production level of 30 standard and 10 competition boards per week, increasing the production of standard boards by 1 unit and holding the production of competition boards fixed at 10 will decrease profit by approximately \$80. ◆

PROBLEM 9 For the profit function in Example 9, find $P_y(25, 10)$ and $P_y(25, 15)$, and interpret the results.

◆

◆ EXAMPLE 10 *Productivity*

The productivity of a major computer manufacturer is given approximately by the Cobb–Douglas production function

$$f(x, y) = 15x^{0.4}y^{0.6}$$

with the utilization of x units of labor and y units of capital. The partial derivative $f_x(x, y)$ represents the rate of change of productivity with respect to labor and is called the **marginal productivity of labor.** The partial derivative $f_y(x, y)$ represents the rate of change of productivity with respect to capital and is called the **marginal productivity of capital.** If the company is currently utilizing 4,000 units of labor and 2,500 units of capital, find the marginal productivity of labor and the marginal productivity of capital. For the greatest increase in productivity, should the management of the company encourage increased use of labor or increased use of capital?

Solution

$$f_x(x, y) = 6x^{-0.6}y^{0.6}$$

$$f_x(4{,}000, 2{,}500) = 6(4{,}000)^{-0.6}(2{,}500)^{0.6} \approx 4.53 \qquad \text{Marginal productivity of labor}$$

$$f_y(x, y) = 9x^{0.4}y^{-0.4}$$

$$f_y(4{,}000, 2{,}500) = 9(4{,}000)^{0.4}(2{,}500)^{-0.4} \approx 10.86 \qquad \text{Marginal productivity of capital}$$

At the current level of utilization of 4,000 units of labor and 2,500 units of capital, each 1 unit increase in labor utilization (keeping capital utilization fixed at 2,500 units) will increase production by approximately 4.53 units and each 1 unit increase in capital utilization (keeping labor utilization fixed at 4,000 units) will increase production by approximately 10.86 units. Thus, the management of the company should encourage increased use of capital. ◆

PROBLEM 10 The productivity of an airplane manufacturing company is given approximately by the Cobb–Douglas production function

$$f(x, y) = 40x^{0.3}y^{0.7}$$

(A) Find $f_x(x, y)$ and $f_y(x, y)$.
(B) If the company is currently using 1,500 units of labor and 4,500 units of capital, find the marginal productivity of labor and the marginal productivity of capital.
(C) For the greatest increase in productivity, should the management of the company encourage increased use of labor or increased use of capital? ◆

Partial derivatives have simple geometric interpretations, as indicated in Figure 5. If we hold x fixed, say, $x = a$, then $f_y(a, y)$ is the slope of the curve obtained by intersecting the plane $x = a$ with the surface $z = f(x, y)$. A similar interpretation is given to $f_x(x, b)$.

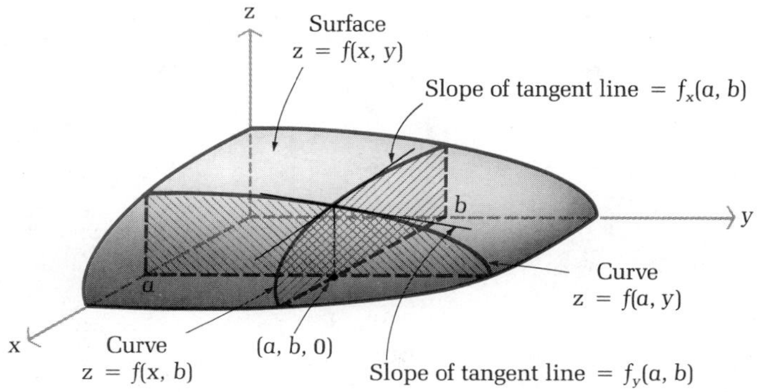

FIGURE 5

◆ SECOND-ORDER PARTIAL DERIVATIVES

The function

$$z = f(x, y) = x^4y^7$$

has two **first-order partial derivatives,**

$$\frac{\partial z}{\partial x} = f_x = f_x(x, y) = 4x^3y^7 \qquad \text{and} \qquad \frac{\partial z}{\partial y} = f_y = f_y(x, y) = 7x^4y^6$$

Each of these partial derivatives, in turn, has two partial derivatives which are called **second-order partial derivatives** of $z = f(x, y)$. Generalizing the various notations we have for first-order partial derivatives, the four second-order partial derivatives of $z = f(x, y) = x^4y^7$ are written as

Equivalent notations

$$f_{xx} = f_{xx}(x, y) = \frac{\partial^2 z}{\partial x^2} = \frac{\partial}{\partial x}\left(\frac{\partial z}{\partial x}\right) = \frac{\partial}{\partial x}(4x^3y^7) = 12x^2y^7$$

$$f_{xy} = f_{xy}(x, y) = \frac{\partial^2 z}{\partial y\,\partial x} = \frac{\partial}{\partial y}\left(\frac{\partial z}{\partial x}\right) = \frac{\partial}{\partial y}(4x^3y^7) = 28x^3y^6$$

$$f_{yx} = f_{yx}(x, y) = \frac{\partial^2 z}{\partial x\,\partial y} = \frac{\partial}{\partial x}\left(\frac{\partial z}{\partial y}\right) = \frac{\partial}{\partial x}(7x^4y^6) = 28x^3y^6$$

$$f_{yy} = f_{yy}(x, y) = \frac{\partial^2 z}{\partial y^2} = \frac{\partial}{\partial y}\left(\frac{\partial z}{\partial y}\right) = \frac{\partial}{\partial y}(7x^4y^6) = 42x^4y^5$$

In the mixed partial derivative $\partial^2 z/\partial y\,\partial x = f_{xy}$, we started with $z = f(x, y)$ and first differentiated with respect to x (holding y constant). Then we differentiated with respect to y (holding x constant). In the other mixed partial derivative, $\partial^2 z/\partial x\,\partial y = f_{yx}$, the order of differentiation was reversed; however, the final result was the same—that is, $f_{xy} = f_{yx}$. Although it is possible to find functions for which $f_{xy} \neq f_{yx}$, such functions rarely occur in applications involving partial derivatives. Thus, for all the functions in this text, we will assume that $f_{xy} = f_{yx}$.

In general, we have the following definitions:

Second-Order Partial Derivatives

If $z = f(x, y)$, then

$$f_{xx} = f_{xx}(x, y) = \frac{\partial^2 z}{\partial x^2} = \frac{\partial}{\partial x}\left(\frac{\partial z}{\partial x}\right)$$

$$f_{xy} = f_{xy}(x, y) = \frac{\partial^2 z}{\partial y\,\partial x} = \frac{\partial}{\partial y}\left(\frac{\partial z}{\partial x}\right)$$

$$f_{yx} = f_{yx}(x, y) = \frac{\partial^2 z}{\partial x\,\partial y} = \frac{\partial}{\partial x}\left(\frac{\partial z}{\partial y}\right)$$

$$f_{yy} = f_{yy}(x, y) = \frac{\partial^2 z}{\partial y^2} = \frac{\partial}{\partial y}\left(\frac{\partial z}{\partial y}\right)$$

◆ EXAMPLE 11 For $z = f(x, y) = 3x^2 - 2xy^3 + 1$, find:

(A) $\dfrac{\partial^2 z}{\partial x\, \partial y}, \dfrac{\partial^2 z}{\partial y\, \partial x}$ (B) $\dfrac{\partial^2 z}{\partial x^2}$ (C) $f_{yx}(2, 1)$

Solutions (A) First differentiate with respect to y and then with respect to x:

$$\frac{\partial z}{\partial y} = -6xy^2 \qquad \frac{\partial^2 z}{\partial x\, \partial y} = \frac{\partial}{\partial x}\left(\frac{\partial z}{\partial y}\right) = \frac{\partial}{\partial x}(-6xy^2) = -6y^2$$

First differentiate with respect to x and then with respect to y:

$$\frac{\partial z}{\partial x} = 6x - 2y^3 \qquad \frac{\partial^2 z}{\partial y\, \partial x} = \frac{\partial}{\partial y}\left(\frac{\partial z}{\partial x}\right) = \frac{\partial}{\partial y}(6x - 2y^3) = -6y^2$$

(B) Differentiate with respect to x twice:

$$\frac{\partial z}{\partial x} = 6x - 2y^3 \qquad \frac{\partial^2 z}{\partial x^2} = \frac{\partial}{\partial x}\left(\frac{\partial z}{\partial x}\right) = 6$$

(C) First find $f_{yx}(x, y)$; then evaluate at (2, 1). Again, remember that f_{yx} means to differentiate with respect to y first and then with respect to x. Thus,

$$f_y(x, y) = -6xy^2 \qquad f_{yx}(x, y) = -6y^2$$

and

$$f_{yx}(2, 1) = -6(1)^2 = -6$$ ◆

PROBLEM 11 For $z = f(x, y) = x^3y - 2y^4 + 3$, find:

(A) $\dfrac{\partial^2 z}{\partial y\, \partial x}$ (B) $\dfrac{\partial^2 z}{\partial y^2}$ (C) $f_{xy}(2, 3)$ (D) $f_{yx}(2, 3)$ ◆

Answers to Matched Problems

7. (A) $\partial z/\partial y = -3x^2 + 5$ (B) $f_y(2, 3) = -7$
8. (A) $10x(x^2 + 2xy)^4$ (B) 10
9. $P_y(25, 10) = 10$: at a production level of $x = 25$ and $y = 10$, increasing y by 1 unit and holding x fixed at 25 will increase profit by approx. \$10; $P_y(25, 15) = -110$: at a production level of $x = 25$ and $y = 15$, increasing y by 1 unit and holding x fixed at 25 will decrease profit by approx. \$110
10. (A) $f_x(x, y) = 12x^{-0.7}y^{0.7}$; $f_y(x, y) = 28x^{0.3}y^{-0.3}$
    (B) Marginal productivity of labor ≈ 25.89; marginal productivity of capital ≈ 20.14
    (C) Labor
11. (A) $3x^2$ (B) $-24y^2$ (C) 12 (D) 12

EXERCISE 9-2

A *For $z = f(x, y) = 10 + 3x + 2y$, find each of the following:*

1. $\partial z/\partial x$ 2. $\partial z/\partial y$ 3. $f_y(1, 2)$ 4. $f_x(1, 2)$

For $z = f(x, y) = 3x^2 - 2xy^2 + 1$, find each of the following:

5. $\partial z/\partial y$
6. $\partial z/\partial x$
7. $f_x(2, 3)$
8. $f_y(2, 3)$

For $S(x, y) = 5x^2y^3$, find each of the following:

9. $S_x(x, y)$
10. $S_y(x, y)$
11. $S_y(2, 1)$
12. $S_x(2, 1)$

B

For $C(x, y) = x^2 - 2xy + 2y^2 + 6x - 9y + 5$, find each of the following:

13. $C_x(x, y)$
14. $C_y(x, y)$
15. $C_x(2, 2)$
16. $C_y(2, 2)$
17. $C_{xy}(x, y)$
18. $C_{yx}(x, y)$
19. $C_{xx}(x, y)$
20. $C_{yy}(x, y)$

For $z = f(x, y) = e^{2x+3y}$, find each of the following:

21. $\dfrac{\partial z}{\partial x}$
22. $\dfrac{\partial z}{\partial y}$
23. $\dfrac{\partial^2 z}{\partial x\, \partial y}$
24. $\dfrac{\partial^2 z}{\partial y\, \partial x}$
25. $f_{xy}(1, 0)$
26. $f_{yx}(0, 1)$
27. $f_{xx}(0, 1)$
28. $f_{yy}(1, 0)$

Find $f_x(x, y)$ and $f_y(x, y)$ for each function f given by:

29. $f(x, y) = (x^2 - y^3)^3$
30. $f(x, y) = \sqrt{2x - y^2}$
31. $f(x, y) = (3x^2y - 1)^4$
32. $f(x, y) = (3 + 2xy^2)^3$
33. $f(x, y) = \ln(x^2 + y^2)$
34. $f(x, y) = \ln(2x - 3y)$
35. $f(x, y) = y^2e^{xy^2}$
36. $f(x, y) = x^3e^{x^2y}$
37. $f(x, y) = \dfrac{x^2 - y^2}{x^2 + y^2}$
38. $f(x, y) = \dfrac{2x^2y}{x^2 + y^2}$

Find $f_{xx}(x, y)$, $f_{xy}(x, y)$, $f_{yx}(x, y)$, and $f_{yy}(x, y)$ for each function f given by:

39. $f(x, y) = x^2y^2 + x^3 + y$
40. $f(x, y) = x^3y^3 + x + y^2$
41. $f(x, y) = \dfrac{x}{y} - \dfrac{y}{x}$
42. $f(x, y) = \dfrac{x^2}{y} - \dfrac{y^2}{x}$
43. $f(x, y) = xe^{xy}$
44. $f(x, y) = x \ln(xy)$

C

45. For

$$P(x, y) = -x^2 + 2xy - 2y^2 - 4x + 12y - 5$$

find values of x and y such that

$$P_x(x, y) = 0 \qquad \text{and} \qquad P_y(x, y) = 0$$

simultaneously.

46. For

$$C(x, y) = 2x^2 + 2xy + 3y^2 - 16x - 18y + 54$$

find values of x and y such that

$$C_x(x, y) = 0 \qquad \text{and} \qquad C_y(x, y) = 0$$

simultaneously.

In Problems 47 and 48, show that the function f satisfies $f_{xx}(x, y) + f_{yy}(x, y) = 0$.

47. $f(x, y) = \ln(x^2 + y^2)$

48. $f(x, y) = x^3 - 3xy^2$

49. For $f(x, y) = x^2 + 2y^2$, find:

(A) $\lim_{h \to 0} \frac{f(x + h, y) - f(x, y)}{h}$ (B) $\lim_{k \to 0} \frac{f(x, y + k) - f(x, y)}{k}$

50. For $f(x, y) = 2xy^2$, find:

(A) $\lim_{h \to 0} \frac{f(x + h, y) - f(x, y)}{h}$ (B) $\lim_{k \to 0} \frac{f(x, y + k) - f(x, y)}{k}$

APPLICATIONS

Business & Economics

51. *Profit function.* A firm produces two types of calculators, x of type A and y of type B each week. The weekly revenue and cost functions (in dollars) are

$$R(x, y) = 80x + 90y + 0.04xy - 0.05x^2 - 0.05y^2$$
$$C(x, y) = 8x + 6y + 20{,}000$$

Find $P_x(1{,}200, 1{,}800)$ and $P_y(1{,}200, 1{,}800)$, and interpret the results.

52. *Advertising and sales.* A company spends \$x per week on newspaper advertising and \$y per week on television advertising. Its weekly sales were found to be given by

$$S(x, y) = 10x^{0.4}y^{0.8}$$

Find $S_x(3{,}000, 2{,}000)$ and $S_y(3{,}000, 2{,}000)$, and interpret the results.

53. *Demand equations.* A supermarket sells two brands of coffee, brand A at \$p per pound and brand B at \$q per pound. The daily demand equations for brands A and B are, respectively,

$$x = 200 - 5p + 4q$$
$$y = 300 + 2p - 4q$$

Find $\partial x/\partial p$ and $\partial y/\partial p$, and interpret the results.

54. *Revenue and profit functions.* A company manufactures ten-speed and three-speed bicycles. The weekly demand and cost functions are

$$p = 230 - 9x + y$$
$$q = 130 + x - 4y$$
$$C(x, y) = 200 + 80x + 30y$$

where \$p is the price of a ten-speed bicycle, \$q is the price of a three-speed bicycle, x is the weekly demand for ten-speed bicycles, y is the weekly demand for three-speed bicycles, and C(x, y) is the cost function. Find $R_x(10, 5)$ and $P_x(10, 5)$, and interpret the results.

55. *Productivity.* The productivity of a certain third-world country is given approximately by the function

$$f(x, y) = 10x^{0.75}y^{0.25}$$

with the utilization of x units of labor and y units of capital.

(A) Find $f_x(x, y)$ and $f_y(x, y)$.
(B) If the country is now using 600 units of labor and 100 units of capital, find the marginal productivity of labor and the marginal productivity of capital.
(C) For the greatest increase in the country's productivity, should the government encourage increased use of labor or increased use of capital?

56. *Productivity.* The productivity of an automobile manufacturing company is given approximately by the function

$$f(x, y) = 50\sqrt{xy} = 50x^{0.5}y^{0.5}$$

with the utilization of x units of labor and y units of capital.

(A) Find $f_x(x, y)$ and $f_y(x, y)$.
(B) If the company is now using 250 units of labor and 125 units of capital, find the marginal productivity of labor and the marginal productivity of capital.
(C) For the greatest increase in the company's productivity, should the management encourage increased use of labor or increased use of capital?

Problems 57–60 refer to the following: If a decrease in demand for one product results in an increase in demand for another product, then the two products are said to be ***competitive,*** *or* ***substitute, products.*** *(Real whipping cream and imitation whipping cream are examples of competitive, or substitute, products.) If a decrease in demand for one product results in a decrease in demand for another product, then the two products are said to be* ***complementary products.*** *(Fishing boats and outboard motors are examples of complementary products.) Partial derivatives can be used to test whether two products are competitive, complementary, or neither. We start with demand functions for two products where the demand for either depends on the prices for both:*

$x = f(p, q)$ *Demand function for product A*

$y = g(p, q)$ *Demand function for product B*

The variables x and y represent the number of units demanded of products A and B, respectively, at a price p for 1 unit of product A and a price q for 1 unit of product B. Normally, if the price of A increases while the price of B is held constant, then the demand for A will decrease; that is, $f_p(p, q) < 0$. *Then, if A and B are competitive products, the demand for B will increase; that is,* $g_p(p, q) > 0$. *Similarly, if the price of B increases while the price of A is held constant, then the* *demand for B will decrease; that is, $g_q(p, q) < 0$. And if A and B are competitive products, then the demand for A will increase; that is, $f_q(p, q) > 0$. Reasoning similarly for complementary products, we arrive at the following test:*

Test for Competitive and Complementary Products

PARTIAL DERIVATIVES			PRODUCTS *A* AND *B*
$f_q(p, q) > 0$	and	$g_p(p, q) > 0$	Competitive (Substitute)
$f_q(p, q) < 0$	and	$g_p(p, q) < 0$	Complementary
$f_q(p, q) \geq 0$	and	$g_p(p, q) \leq 0$	Neither
$f_q(p, q) \leq 0$	and	$g_p(p, q) \geq 0$	Neither

Use this test in Problems 57–60 to determine whether the indicated products are competitive, complementary, or neither.

57. *Product demand.* The weekly demand equations for the sale of butter and margarine in a supermarket are

$$x = f(p, q) = 8{,}000 - 0.09p^2 + 0.08q^2 \quad \text{Butter}$$
$$y = g(p, q) = 15{,}000 + 0.04p^2 - 0.3q^2 \quad \text{Margarine}$$

58. *Product demand.* The daily demand equations for the sale of brand *A* coffee and brand *B* coffee in a supermarket are

$$x = f(p, q) = 200 - 5p + 4q \quad \text{Brand } A \text{ coffee}$$
$$y = g(p, q) = 300 + 2p - 4q \quad \text{Brand } B \text{ coffee}$$

59. *Product demand.* The monthly demand equations for the sale of skis and ski boots in a sporting goods store are

$$x = f(p, q) = 800 - 0.004p^2 - 0.003q^2 \quad \text{Skis}$$
$$y = g(p, q) = 600 - 0.003p^2 - 0.002q^2 \quad \text{Ski boots}$$

60. *Product demand.* The monthly demand equations for the sale of tennis rackets and tennis balls in a sporting goods store are

$$x = f(p, q) = 500 - 0.5p - q^2 \quad \text{Tennis rackets}$$
$$y = g(p, q) = 10{,}000 - 8p - 100q^2 \quad \text{Tennis balls (cans)}$$

Life Sciences

61. *Medicine.* The following empirical formula relates the surface area *A* (in square inches) of an average human body to its weight *w* (in pounds) and its height *h* (in inches):

$$A = f(w, h) = 15.64w^{0.425}h^{0.725}$$

Knowing the surface area of a human body is useful, for example, in studies pertaining to hypothermia (heat loss due to exposure).

(A) Find $f_w(w, h)$ and $f_h(w, h)$.
(B) For a 65 pound child who is 57 inches tall, find $f_w(65, 57)$ and $f_h(65, 57)$, and interpret the results.

62. *Blood flow.* Poiseuille's law states that the resistance, R, for blood flowing in a blood vessel varies directly as the length of the vessel, L, and inversely as the fourth power of its radius, r. Stated as an equation,

$$R(L, r) = k\frac{L}{r^4} \qquad k \text{ a constant}$$

Find $R_L(4, 0.2)$ and $R_r(4, 0.2)$, and interpret the results.

Social Sciences

63. *Physical anthropology.* Anthropologists, in their study of race and human genetic groupings, often use the cephalic index, C, which varies directly as the width, W, of the head, and inversely as the length, L, of the head (both viewed from the top). In terms of an equation,

$$C(W, L) = 100\frac{W}{L}$$

where

W = Width in inches $\qquad$ L = Length in inches

Find $C_W(6, 8)$ and $C_L(6, 8)$, and interpret the results.

64. *Safety research.* Under ideal conditions, if a person driving a car slams on the brakes and skids to a stop, the length of the skid marks (in feet) is given by the formula

$$L(w, v) = kwv^2$$

where

k = Constant $\qquad$ w = Weight of car in pounds

v = Speed of car in miles per hour

For $k = 0.000\ 013\ 3$, find $L_w(2{,}500, 60)$ and $L_v(2{,}500, 60)$, and interpret the results.

SECTION 9-3 Maxima and Minima

We are now ready to undertake a brief but useful analysis of local maxima and minima for functions of the type $z = f(x, y)$. Basically, we are going to extend the second-derivative test developed for functions of a single independent variable. To start, we assume that all second-order partial derivatives exist for the function f in some circular region in the xy plane. This guarantees that the surface $z = f(x, y)$ has no sharp points, breaks, or ruptures. In other words, we are dealing only with smooth surfaces with no edges (like the edge of a box); or breaks (like an earthquake fault); or sharp points (like the bottom point of a golf tee). See Figure 6.

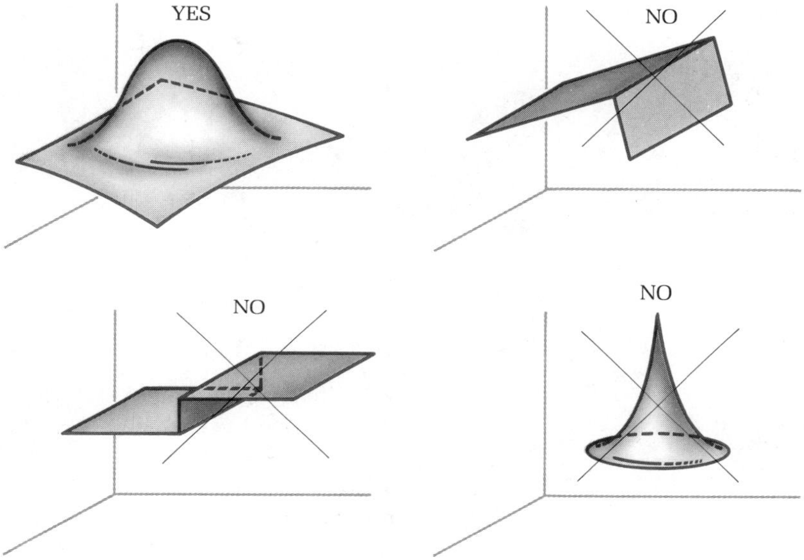

FIGURE 6

In addition, we will not concern ourselves with boundary points or absolute maxima–minima theory. In spite of these restrictions, the procedure we are now going to describe will help us solve a large number of useful problems.

What does it mean for $f(a, b)$ to be a local maximum or a local minimum? We say that **$f(a, b)$ is a local maximum** if there exists a circular region in the domain of f with (a, b) as the center, such that

$$f(a, b) \geq f(x, y)$$

for all (x, y) in the region. Similarly, we say that **$f(a, b)$ is a local minimum** if there exists a circular region in the domain of f with (a, b) as the center, such that

$$f(a, b) \leq f(x, y)$$

for all (x, y) in the region. Figure 7A illustrates a local maximum, Figure 7B a local minimum, and Figure 7C a **saddle point,** which is neither a local maximum nor a local minimum.

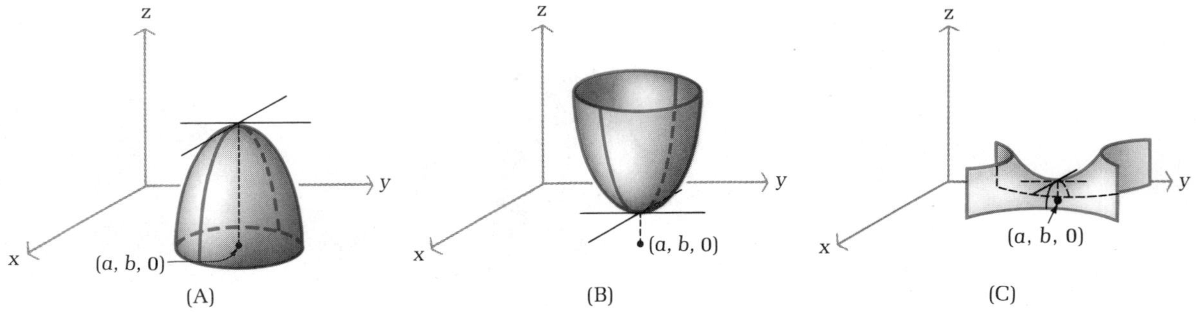

FIGURE 7

What happens to $f_x(a, b)$ and $f_y(a, b)$ if $f(a, b)$ is a local minimum or a local maximum and the partial derivatives of f exist in a circular region containing (a, b)? Figure 7 suggests that $f_x(a, b) = 0$ and $f_y(a, b) = 0$, since the tangent lines to the given curves are horizontal. Theorem 1 indicates that our intuitive reasoning is correct.

THEOREM 1

Let $f(a, b)$ be a local extremum (a local maximum or a local minimum) for the function f. If both f_x and f_y exist at (a, b), then

$$f_x(a, b) = 0 \quad \text{and} \quad f_y(a, b) = 0 \tag{1}$$

The converse of this theorem is false. That is, if $f_x(a, b) = 0$ and $f_y(a, b) = 0$, then $f(a, b)$ may or may not be a local extremum; for example, the point $(a, b, f(a, b))$ may be a saddle point (see Fig. 7C).

Theorem 1 gives us *necessary* (but not *sufficient*) conditions for $f(a, b)$ to be a local extremum. We thus find all points (a, b) such that $f_x(a, b) = 0$ and $f_y(a, b) = 0$ and test these further to determine whether $f(a, b)$ is a local extremum or a saddle point. Points (a, b) such that conditions (1) hold are called **critical points.** The next theorem, using second-derivative tests, gives us *sufficient* conditions for a critical point to produce a local extremum or a saddle point. (As was the case with Theorem 1, we state this theorem without proof.)

THEOREM 2

Second-Derivative Test for Local Extrema

Given:

1. $z = f(x, y)$
2. $f_x(a, b) = 0$ and $f_y(a, b) = 0$ [(a, b) is a critical point]
3. All second-order partial derivatives of f exist in some circular region containing (a, b) as a center
4. $A = f_{xx}(a, b)$, $B = f_{xy}(a, b)$, $C = f_{yy}(a, b)$

Then:

Case 1. If $AC - B^2 > 0$ and $A < 0$, then $f(a, b)$ is a local maximum.

Case 2. If $AC - B^2 > 0$ and $A > 0$, then $f(a, b)$ is a local minimum.

Case 3. If $AC - B^2 < 0$, then f has a saddle point at (a, b).

Case 4. If $AC - B^2 = 0$, the test fails.

To illustrate the use of Theorem 2, we will first find the local extrema for a very simple function whose solution is almost obvious: $z = f(x, y) = x^2 + y^2 + 2$. From the function f itself and its graph (Fig. 8), it is clear that a local minimum is found at $(0, 0)$. Let us see how Theorem 2 confirms this observation.

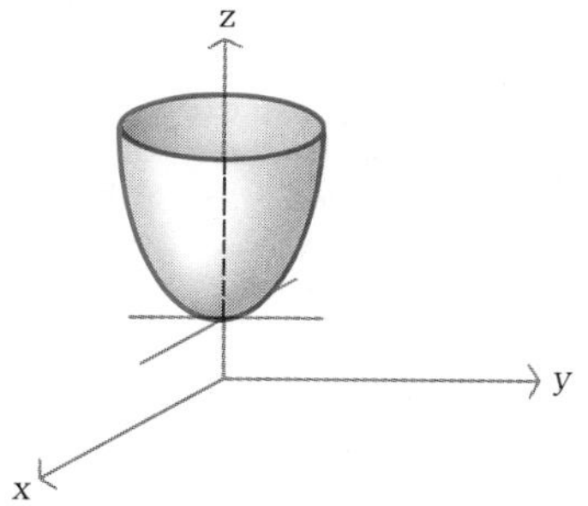

FIGURE 8

Step 1. Find critical points: Find (x, y) such that $f_x(x, y) = 0$ and $f_y(x, y) = 0$ simultaneously:

$$f_x(x, y) = 2x = 0 \qquad f_y(x, y) = 2y = 0$$
$$x = 0 \qquad\qquad y = 0$$

The only critical point is $(a, b) = (0, 0)$.

Step 2. Compute $A = f_{xx}(0, 0)$, $B = f_{xy}(0, 0)$, and $C = f_{yy}(0, 0)$:

$$f_{xx}(x, y) = 2 \quad \text{thus} \quad A = f_{xx}(0, 0) = 2$$
$$f_{xy}(x, y) = 0 \quad \text{thus} \quad B = f_{xy}(0, 0) = 0$$
$$f_{yy}(x, y) = 2 \quad \text{thus} \quad C = f_{yy}(0, 0) = 2$$

Step 3. Evaluate $AC - B^2$ and try to classify the critical point $(0, 0)$ using Theorem 2:

$$AC - B^2 = (2)(2) - (0)^2 = 4 > 0 \quad \text{and} \quad A = 2 > 0$$

Therefore, case 2 in Theorem 2 holds. That is, $f(0, 0) = 2$ is a local minimum.

We will now use Theorem 2 in the following examples to analyze extrema without the aid of graphs.

◆ EXAMPLE 12 Use Theorem 2 to find local extrema for: $f(x, y) = -x^2 - y^2 + 6x + 8y - 21$

Solution *Step 1.* Find critical points: Find (x, y) such that $f_x(x, y) = 0$ and $f_y(x, y) = 0$ simultaneously:

$$f_x(x, y) = -2x + 6 = 0 \qquad f_y(x, y) = -2y + 8 = 0$$
$$x = 3 \qquad y = 4$$

The only critical point is $(a, b) = (3, 4)$.

Step 2. Compute $A = f_{xx}(3, 4)$, $B = f_{xy}(3, 4)$, and $C = f_{yy}(3, 4)$:

$$f_{xx}(x, y) = -2 \quad \text{thus} \quad A = f_{xx}(3, 4) = -2$$
$$f_{xy}(x, y) = 0 \quad \text{thus} \quad B = f_{xy}(3, 4) = 0$$
$$f_{yy}(x, y) = -2 \quad \text{thus} \quad C = f_{yy}(3, 4) = -2$$

Step 3. Evaluate $AC - B^2$ and try to classify the critical point $(3, 4)$ using Theorem 2:

$$AC - B^2 = (-2)(-2) - (0)^2 = 4 > 0 \quad \text{and} \quad A = -2 < 0$$

Therefore, case 1 in Theorem 2 holds. That is, $f(3, 4) = 4$ is a local maximum. ◆

PROBLEM 12 Use Theorem 2 to find local extrema for: $f(x, y) = x^2 + y^2 - 10x - 2y + 36$ ◆

◆ EXAMPLE 13 Use Theorem 2 to find local extrema for: $f(x, y) = x^3 + y^3 - 6xy$

Solution *Step 1.* Find critical points for $f(x, y) = x^3 + y^3 - 6xy$:

$$f_x(x, y) = 3x^2 - 6y = 0 \qquad \text{Solve for } y.$$
$$6y = 3x^2$$
$$y = \tfrac{1}{2}x^2 \qquad (2)$$
$$f_y(x, y) = 3y^2 - 6x = 0$$
$$3y^2 = 6x \qquad \text{Use (2) to eliminate } y.$$
$$3(\tfrac{1}{2}x^2)^2 = 6x$$
$$\tfrac{3}{4}x^4 = 6x \qquad \text{Solve for } x.$$
$$3x^4 - 24x = 0$$
$$3x(x^3 - 8) = 0$$
$$x = 0 \quad \text{or} \quad x = 2$$
$$y = 0 \qquad y = \tfrac{1}{2}(2)^2 = 2$$

The critical points are $(0, 0)$ and $(2, 2)$. Since there are two critical points, steps 2 and 3 must be performed twice.

Test (0, 0) *Step 2.* Compute $A = f_{xx}(0, 0)$, $B = f_{xy}(0, 0)$, and $C = f_{yy}(0, 0)$:

$$f_{xx}(x, y) = 6x \quad \text{thus} \quad A = f_{xx}(0, 0) = 0$$
$$f_{xy}(x, y) = -6 \quad \text{thus} \quad B = f_{xy}(0, 0) = -6$$
$$f_{yy}(x, y) = 6y \quad \text{thus} \quad C = f_{yy}(0, 0) = 0$$

Step 3. Evaluate $AC - B^2$ and try to classify the critical point (0, 0) using Theorem 2:

$$AC - B^2 = (0)(0) - (-6)^2 = -36 < 0$$

Therefore, case 3 in Theorem 2 applies. That is, f has a saddle point at (0, 0).

Now we will consider the second critical point, (2, 2).

Test (2, 2) *Step 2.* Compute $A = f_{xx}(2, 2)$, $B = f_{xy}(2, 2)$, and $C = f_{yy}(2, 2)$:

$$f_{xx}(x, y) = 6x \quad \text{thus} \quad A = f_{xx}(2, 2) = 12$$
$$f_{xy}(x, y) = -6 \quad \text{thus} \quad B = f_{xy}(2, 2) = -6$$
$$f_{yy}(x, y) = 6y \quad \text{thus} \quad C = f_{yy}(2, 2) = 12$$

Step 3. Evaluate $AC - B^2$ and try to classify the critical point (2, 2) using Theorem 2:

$$AC - B^2 = (12)(12) - (-6)^2 = 108 > 0 \quad \text{and} \quad A = 12 > 0$$

Thus, case 2 in Theorem 2 applies, and $f(2, 2) = -8$ is a local minimum. ◆

PROBLEM 13 Use Theorem 2 to find local extrema for: $f(x, y) = x^3 + y^2 - 6xy$ ◆

◆ EXAMPLE 14 Suppose the surfboard company discussed earlier has developed the yearly profit equation

Profit

$$P(x, y) = -2x^2 + 2xy - y^2 + 10x - 4y + 107$$

where x is the number (in thousands) of standard surfboards produced per year, y is the number (in thousands) of competition surfboards produced per year, and P is profit (in thousands of dollars). How many of each type of board should be produced per year to realize a maximum profit? What is the maximum profit?

Solution *Step 1.* Find critical points:

$$P_x(x, y) = -4x + 2y + 10 = 0$$
$$P_y(x, y) = 2x - 2y - 4 = 0$$

Solving this system, we obtain (3, 1) as the only critical point.

Step 2. Compute $A = P_{xx}(3, 1)$, $B = P_{xy}(3, 1)$, and $C = P_{yy}(3, 1)$:

$$P_{xx}(x, y) = -4 \quad \text{thus} \quad A = P_{xx}(3, 1) = -4$$
$$P_{xy}(x, y) = 2 \quad \text{thus} \quad B = P_{xy}(3, 1) = 2$$
$$P_{yy}(x, y) = -2 \quad \text{thus} \quad C = P_{yy}(3, 1) = -2$$

Step 3. Evaluate $AC - B^2$ and try to classify the critical point (3, 1) using Theorem 2:

$$AC - B^2 = (-4)(-2) - (2)^2 = 8 - 4 = 4 > 0 \qquad \text{and} \qquad A = -4 < 0$$

Therefore, case 1 in Theorem 2 applies. That is, $P(3, 1) = \$120{,}000$ is a local maximum. This is obtained by producing 3,000 standard boards and 1,000 competition boards per year. ◆

PROBLEM 14

Repeat Example 14 with: $P(x, y) = -2x^2 + 4xy - 3y^2 + 4x - 2y + 77$ ◆

◆ EXAMPLE 15

Package Design

The packaging department in a company has been asked to design a rectangular box with no top and a partition down the middle. The box must have a volume of 48 cubic inches. Find the dimensions that will minimize the amount of material used to construct the box.

Solution

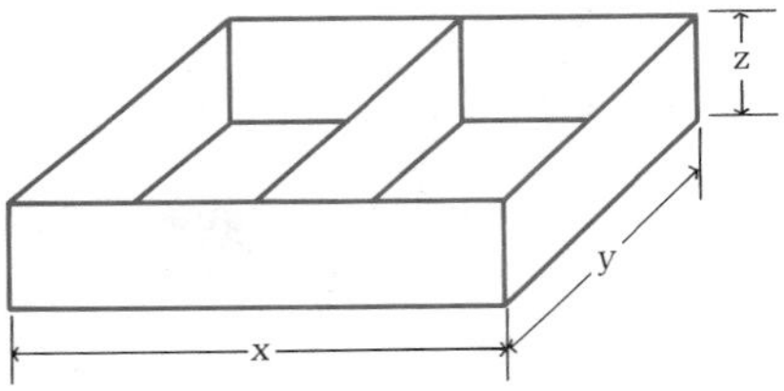

Refer to the figure in the margin. The amount of material used in constructing this box is

$$M = \underset{\text{Base}}{xy} + \underset{\text{Front, back}}{2xz} + \underset{\text{Sides, partition}}{3yz} \tag{3}$$

The volume of the box is

$$V = xyz = 48 \tag{4}$$

Since Theorem 2 applies only to functions with two independent variables, we must use (4) to eliminate one of the variables in (3):

$$\begin{aligned} M &= xy + 2xz + 3yz && \text{Substitute } z = 48/xy. \\ &= xy + 2x\left(\frac{48}{xy}\right) + 3y\left(\frac{48}{xy}\right) \\ &= xy + \frac{96}{y} + \frac{144}{x} \end{aligned}$$

Thus, we must find the minimum value of

$$M(x, y) = xy + \frac{96}{y} + \frac{144}{x} \qquad x > 0 \qquad \text{and} \qquad y > 0$$

Step 1. Find critical points:

$$M_x(x, y) = y - \frac{144}{x^2} = 0$$

$$y = \frac{144}{x^2} \tag{5}$$

$$M_y(x, y) = x - \frac{96}{y^2} = 0$$

$$x = \frac{96}{y^2}$$ Solve for y^2.

$$y^2 = \frac{96}{x}$$ Use (5) to eliminate y and solve for x.

$$\left(\frac{144}{x^2}\right)^2 = \frac{96}{x}$$

$$\frac{20{,}736}{x^4} = \frac{96}{x}$$ Multiply both sides by $x^4/96$ (recall, $x > 0$).

$$x^3 = \tfrac{20{,}736}{96} = 216$$

$$x = 6$$ Use (5) to find y.

$$y = \tfrac{144}{36} = 4$$

Thus, (6, 4) is the only critical point.

Step 2. Compute $A = M_{xx}(6, 4)$, $B = M_{xy}(6, 4)$, and $C = M_{yy}(6, 4)$:

$$M_{xx}(x, y) = \frac{288}{x^3} \quad \text{thus} \quad A = M_{xx}(6, 4) = \tfrac{288}{216} = \tfrac{4}{3}$$

$$M_{xy}(x, y) = 1 \quad \text{thus} \quad B = M_{xy}(6, 4) = 1$$

$$M_{yy}(x, y) = \frac{192}{y^3} \quad \text{thus} \quad C = M_{yy}(6, 4) = \tfrac{192}{64} = 3$$

Step 3. Evaluate $AC - B^2$ and try to classify the critical point (6, 4) using Theorem 2:

$$AC - B^2 = (\tfrac{4}{3})(3) - (1)^2 = 3 > 0 \qquad \text{and} \qquad A = \tfrac{4}{3} > 0$$

Therefore, case 2 in Theorem 2 applies; $M(x, y)$ has a local minimum at (6, 4). If $x = 6$ and $y = 4$, then

$$z = \frac{48}{xy} = \frac{48}{(6)(4)} = 2$$

Thus, the dimensions that will require the minimum amount of material are 6 inches by 4 inches by 2 inches. ◆

PROBLEM 15 If the box in Example 15 must have a volume of 384 cubic inches, find the dimensions that will require the least amount of material. ◆

Answers to Matched Problems

12. $f(5, 1) = 10$ is a local minimum
13. f has a saddle point at (0, 0); $f(6, 18) = -108$ is a local minimum
14. Local maximum for $x = 2$ and $y = 1$; $P(2, 1) = \$80{,}000$
15. 12 in. by 8 in. by 4 in.

EXERCISE 9-3

Find local extrema using Theorem 2.

A

1. $f(x, y) = 6 - x^2 - 4x - y^2$
2. $f(x, y) = 3 - x^2 - y^2 + 6y$
3. $f(x, y) = x^2 + y^2 + 2x - 6y + 14$
4. $f(x, y) = x^2 + y^2 - 4x + 6y + 23$

B

5. $f(x, y) = xy + 2x - 3y - 2$
6. $f(x, y) = x^2 - y^2 + 2x + 6y - 4$
7. $f(x, y) = -3x^2 + 2xy - 2y^2 + 14x + 2y + 10$
8. $f(x, y) = -x^2 + xy - 2y^2 + x + 10y - 5$
9. $f(x, y) = 2x^2 - 2xy + 3y^2 - 4x - 8y + 20$
10. $f(x, y) = 2x^2 - xy + y^2 - x - 5y + 8$

C

11. $f(x, y) = e^{xy}$
12. $f(x, y) = x^2y - xy^2$
13. $f(x, y) = x^3 + y^3 - 3xy$
14. $f(x, y) = 2y^3 - 6xy - x^2$
15. $f(x, y) = 2x^4 + y^2 - 12xy$
16. $f(x, y) = 16xy - x^4 - 2y^2$
17. $f(x, y) = x^3 - 3xy^2 + 6y^2$
18. $f(x, y) = 2x^2 - 2x^2y + 6y^3$

APPLICATIONS

Business & Economics

19. *Product mix for maximum profit.* A firm produces two types of calculators, x thousand of type A and y thousand of type B per year. If the revenue and cost equations for the year are (in millions of dollars)

$$R(x, y) = 2x + 3y$$
$$C(x, y) = x^2 - 2xy + 2y^2 + 6x - 9y + 5$$

determine how many of each type of calculator should be produced per year to maximize profit. What is the maximum profit?

20. *Automation–labor mix for minimum cost.* The annual labor and automated equipment cost (in millions of dollars) for a company's production of television sets is given by

$$C(x, y) = 2x^2 + 2xy + 3y^2 - 16x - 18y + 54$$

where x is the amount spent per year on labor and y is the amount spent per year on automated equipment (both in millions of dollars). Determine how much should be spent on each per year to minimize this cost. What is the minimum cost?

21. *Maximizing profit.* A department store sells two brands of inexpensive calculators. The store pays $6 for each brand A calculator and $8 for each brand B calculator. The research department has estimated the following weekly demand equations for these two competitive products:

$$x = 116 - 30p + 20q \qquad \text{Demand equation for brand } A$$
$$y = 144 + 16p - 24q \qquad \text{Demand equation for brand } B$$

where p is the selling price for brand A and q is the selling price for brand B.

(A) Determine the demands x and y when $p = \$10$ and $q = \$12$; when $p = \$11$ and $q = \$11$.

(B) How should the store price each calculator to maximize weekly profits? What is the maximum weekly profit? [*Hint:* $C = 6x + 8y$, $R = px + qy$, and $P = R - C$.]

22. *Maximizing profit.* A store sells two brands of color print film. The store pays \$2 for each roll of brand A film and \$3 for each roll of brand B film. A consulting firm has estimated the following daily demand equations for these two competitive products:

$$x = 75 - 40p + 25q \qquad \text{Demand equation for brand } A$$

$$y = 80 + 20p - 30q \qquad \text{Demand equation for brand } B$$

where p is the selling price for brand A and q is the selling price for brand B.

(A) Determine the demands x and y when $p = \$4$ and $q = \$5$; when $p = \$4$ and $q = \$4$.

(B) How should the store price each brand of film to maximize daily profits? What is the maximum daily profit? [*Hint:* $C = 2x + 3y$, $R = px + qy$, and $P = R - C$.]

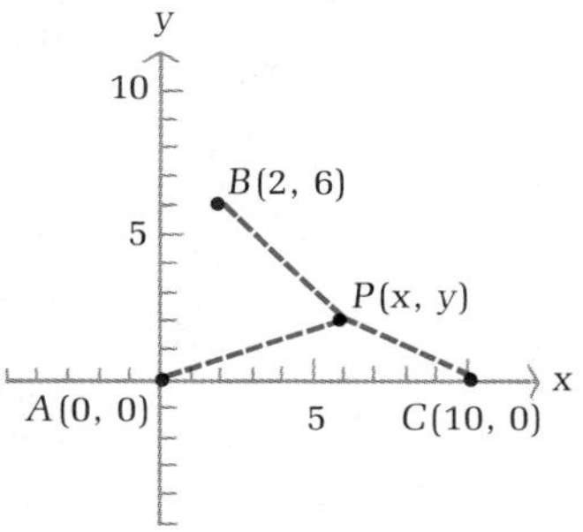

23. *Minimizing cost.* A satellite television reception station is to be located at $P(x, y)$ so that the sum of the squares of the distances from P to the three towns A, B, and C is minimum (see the figure). Find the coordinates of P. This location will minimize the cost of providing satellite cable television for all three towns.

24. *Minimizing cost.* Repeat Problem 23 replacing the coordinates of B with $B(6, 9)$ and the coordinates of C with $C(9, 0)$.

25. *Minimum material.* A rectangular box with no top and two parallel partitions (see the figure below) is to be made to hold a volume of 64 cubic inches. Find the dimensions that will require the least amount of material.

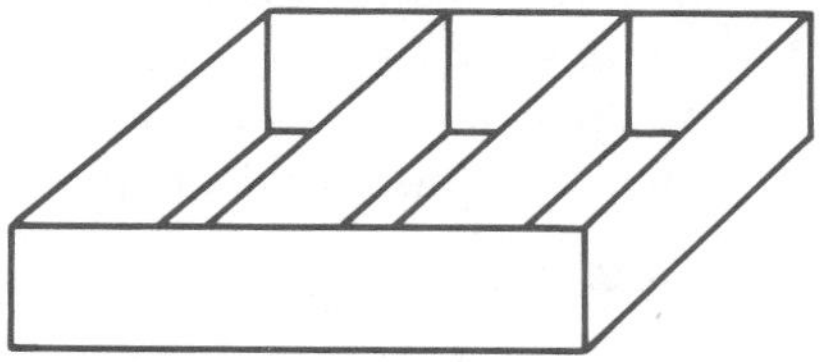

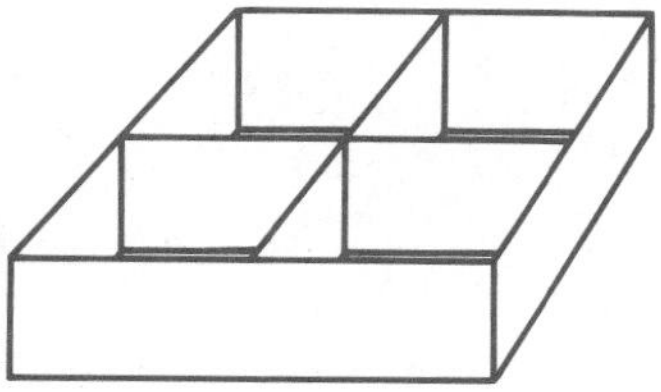

26. *Minimum material.* A rectangular box with no top and two intersecting partitions (see the figure in the margin) is to be made to hold a volume of 72 cubic inches. What should its dimensions be in order to use the least amount of material in its construction?

27. *Maximum volume.* A mailing service states that a rectangular package shall have the sum of the length and girth not to exceed 120 inches (see the figure). What are the dimensions of the largest (in volume) mailing carton that can be constructed meeting these restrictions?

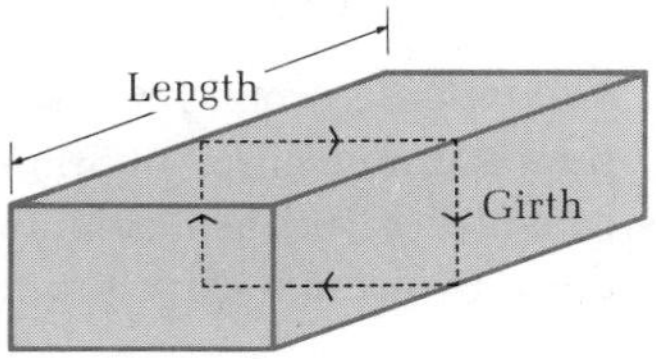

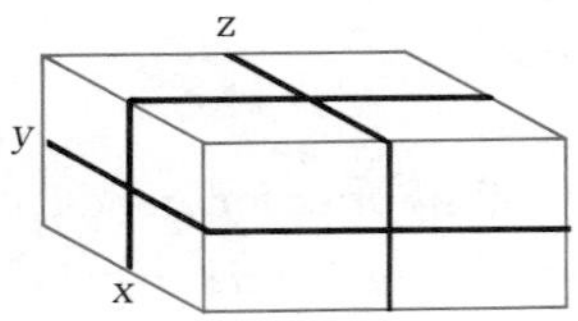

28. *Maximum shipping volume.* A shipping box is reinforced with steel bands in all three directions, as indicated in the figure in the margin. A total of 150 inches of steel tape are to be used, with 6 inches of waste because of a 2 inch overlap in each direction. Find the dimensions of the box with maximum volume that can be taped as indicated.

SECTION 9-4 Maxima and Minima Using Lagrange Multipliers

- FUNCTIONS OF TWO INDEPENDENT VARIABLES
- FUNCTIONS OF THREE INDEPENDENT VARIABLES

◆ FUNCTIONS OF TWO INDEPENDENT VARIABLES

We will now consider a particularly powerful method of solving a certain class of maxima–minima problems. The method is due to Joseph Louis Lagrange (1736–1813), an eminent eighteenth century French mathematician, and it is called the **method of Lagrange multipliers.** We introduce the method through an example; then we will formalize the discussion in the form of a theorem.

A rancher wants to construct two feeding pens of the same size along an existing fence (see Fig. 9). If the rancher has 720 feet of fencing materials available, how long should x and y be in order to obtain the maximum total area? What is the maximum area?

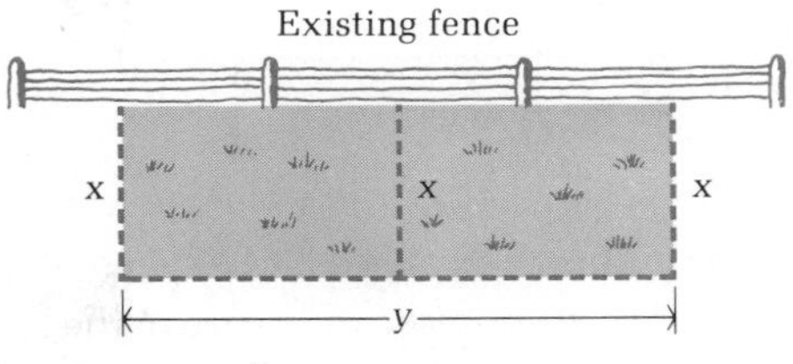

FIGURE 9

The total area is given by

$$f(x, y) = xy$$

which can be made as large as we like, providing there are no restrictions on x and y. But there are restrictions on x and y, since we have only 720 feet of fencing. That is, x and y must be chosen so that

$$3x + y = 720$$

This restriction on x and y, called a **constraint,** leads to the following maxima–minima problem:

$$\text{Maximize} \quad f(x, y) = xy \tag{1}$$

$$\text{Subject to} \quad 3x + y = 720 \quad \text{or} \quad 3x + y - 720 = 0 \tag{2}$$

This problem is a special case of a general class of problems of the form

$$\text{Maximize (or Minimize)} \quad z = f(x, y) \tag{3}$$

$$\text{Subject to} \quad g(x, y) = 0 \tag{4}$$

Of course, we could try to solve (4) for y in terms of x, or for x in terms of y, then substitute the result into (3), and use methods developed in Section 5-4 for functions of a single variable. But what if (4) were more complicated than (2), and solving for one variable in terms of the other was either very difficult or impossible? In the method of Lagrange multipliers, we will work with $g(x, y)$ directly and avoid having to solve (4) for one variable in terms of the other. In addition, the method generalizes to functions of arbitrarily many variables subject to one or more constraints.

Now, to the method. We form a new function F, using functions f and g in (3) and (4), as follows:

$$F(x, y, \lambda) = f(x, y) + \lambda g(x, y) \tag{5}$$

where λ (the Greek letter lambda) is called a **Lagrange multiplier.** Theorem 3 gives the basis for the method.

THEOREM 3

Any local maxima or minima of the function $z = f(x, y)$ subject to the constraint $g(x, y) = 0$ will be among those points (x_0, y_0) for which (x_0, y_0, λ_0) is a solution to the system

$$F_x(x, y, \lambda) = 0$$
$$F_y(x, y, \lambda) = 0$$
$$F_\lambda(x, y, \lambda) = 0$$

where $F(x, y, \lambda) = f(x, y) + \lambda g(x, y)$, provided all the partial derivatives exist.

We now solve the fence problem using the method of Lagrange multipliers.

Step 1. Formulate the problem in the form of equations (3) and (4):

Maximize $f(x, y) = xy$

Subject to $g(x, y) = 3x + y - 720 = 0$

Step 2. Form the function F, introducing the Lagrange multiplier λ:

$$\begin{aligned} F(x, y, \lambda) &= f(x, y) + \lambda g(x, y) \\ &= xy + \lambda(3x + y - 720) \end{aligned}$$

Step 3. Solve the system $F_x = 0$, $F_y = 0$, $F_\lambda = 0$. (The solutions are called **critical points** for F.)

$$F_x = y + 3\lambda = 0$$
$$F_y = x + \lambda = 0$$
$$F_\lambda = 3x + y - 720 = 0$$

From the first two equations, we see that

$$y = -3\lambda$$
$$x = -\lambda$$

Substitute these values for x and y into the third equation and solve for λ:

$$\begin{aligned} -3\lambda - 3\lambda &= 720 \\ -6\lambda &= 720 \\ \lambda &= -120 \end{aligned}$$

Thus,

$$y = -3(-120) = 360 \text{ feet}$$
$$x = -(-120) = 120 \text{ feet}$$

and $(x_0, y_0, \lambda_0) = (120, 360, -120)$ is the only critical point for F.

Step 4. According to Theorem 3, if the function $f(x, y)$, subject to the constraint $g(x, y) = 0$, has a local maximum or minimum, it must occur at $x = 120$, $y = 360$. Although it is possible to develop a test similar to Theorem 2 in Section 9-3 to determine the nature of this local extremum, we will not do so. [Note that Theorem 2 cannot be applied to $f(x, y)$ at (120, 360), since this point is not a critical point of the unconstrained function $f(x, y)$.] We will simply assume that the maximum value of $f(x, y)$ must occur for $x = 120$, $y = 360$. Thus,

$$\begin{aligned} \text{Max } f(x, y) &= f(120, 360) \\ &= (120)(360) = 43{,}200 \text{ square feet} \end{aligned}$$

The key steps in applying the method of Lagrange multipliers are listed in the following box:

Method of Lagrange Multipliers—Key Steps

Step 1. Formulate the problem in the form

Maximize (or Minimize) $z = f(x, y)$
Subject to $g(x, y) = 0$

Step 2. Form the function F:

$$F(x, y, \lambda) = f(x, y) + \lambda g(x, y)$$

Step 3. Find the critical points for F; that is, solve the system

$$F_x(x, y, \lambda) = 0$$
$$F_y(x, y, \lambda) = 0$$
$$F_\lambda(x, y, \lambda) = 0$$

Step 4. If (x_0, y_0, λ_0) is the only critical point of F, then we assume that (x_0, y_0) will always produce the solution to the problems we consider. If F has more than one critical point, then we evaluate $z = f(x, y)$ at (x_0, y_0) for each critical point (x_0, y_0, λ_0) of F. For the problems we consider, we assume that the largest of these values is the maximum value of $f(x, y)$, subject to the constraint $g(x, y) = 0$, and the smallest is the minimum value of $f(x, y)$, subject to the constraint $g(x, y) = 0$.

◆ EXAMPLE 16 Minimize $f(x, y) = x^2 + y^2$ subject to $x + y = 10$.

Solution

Step 1. Minimize $f(x, y) = x^2 + y^2$
Subject to $g(x, y) = x + y - 10 = 0$

Step 2. $F(x, y, \lambda) = x^2 + y^2 + \lambda(x + y - 10)$

Step 3. $F_x = 2x + \lambda = 0$
$F_y = 2y + \lambda = 0$
$F_\lambda = x + y - 10 = 0$

From the first two equations,

$$x = -\frac{\lambda}{2} \qquad y = -\frac{\lambda}{2}$$

Substituting these into the third equation, we obtain

$$-\frac{\lambda}{2} - \frac{\lambda}{2} = 10$$
$$-\lambda = 10$$
$$\lambda = -10$$

The only critical point is $(x_0, y_0, \lambda_0) = (5, 5, -10)$.

Step 4. Since (5, 5, −10) is the only critical point for F, we conclude that (see step 4 in the box)

$$\text{Min } f(x, y) = f(5, 5) = (5)^2 + (5)^2 = 50$$ ◆

PROBLEM 16 Maximize $f(x, y) = 25 - x^2 - y^2$ subject to $x + y = 4$. ◆

Figures 10 and 11 illustrate the results obtained in Example 16 and Problem 16, respectively.

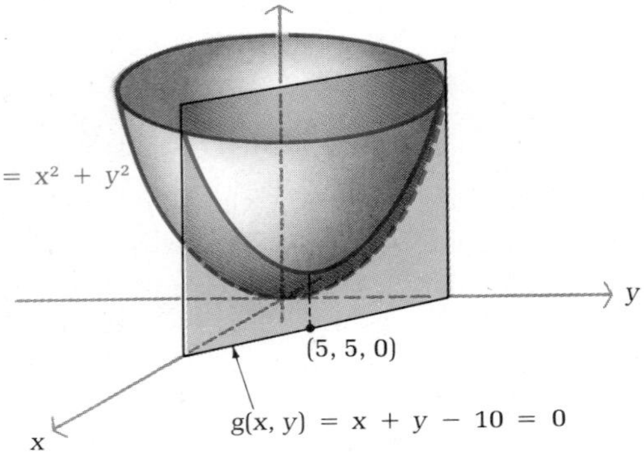

FIGURE 10

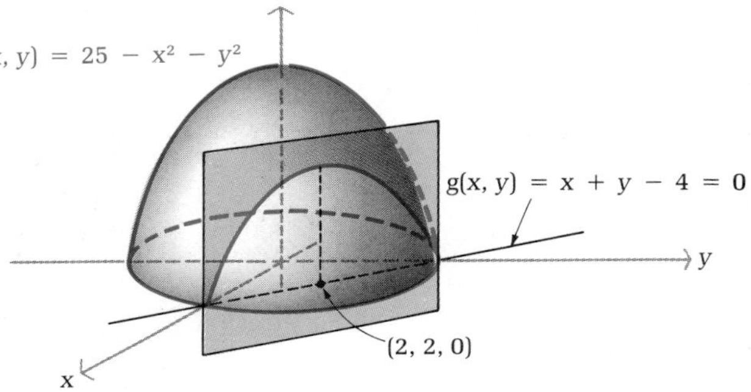

FIGURE 11

◆ EXAMPLE 17 *Productivity*

The Cobb–Douglas production function for a new product is given by

$$N(x, y) = 16x^{0.25}y^{0.75}$$

where x is the number of units of labor and y is the number of units of capital required to produce N(x, y) units of the product. Each unit of labor costs $50 and each unit of capital costs $100. If $500,000 has been budgeted for the production of this product, how should this amount be allocated between labor and capital in order to maximize production? What is the maximum number of units that can be produced?

Solution The total cost of using x units of labor and y units of capital is $50x + 100y$. Thus, the constraint imposed by the $500,000 budget is

$$50x + 100y = 500{,}000$$

Step 1. Maximize $N(x, y) = 16x^{0.25}y^{0.75}$
Subject to $g(x, y) = 50x + 100y - 500{,}000 = 0$

Step 2. $F(x, y, \lambda) = 16x^{0.25}y^{0.75} + \lambda(50x + 100y - 500{,}000)$

Step 3. $F_x = 4x^{-0.75}y^{0.75} + 50\lambda = 0$
$F_y = 12x^{0.25}y^{-0.25} + 100\lambda = 0$
$F_\lambda = 50x + 100y - 500{,}000 = 0$

From the first two equations,

$$\lambda = -\tfrac{2}{25}x^{-0.75}y^{0.75} \qquad \text{and} \qquad \lambda = -\tfrac{3}{25}x^{0.25}y^{-0.25}$$

Thus,

$$-\tfrac{2}{25}x^{-0.75}y^{0.75} = -\tfrac{3}{25}x^{0.25}y^{-0.25}$$ Multiply both sides by $x^{0.75}y^{0.25}$. (We can assume $x \neq 0$ and $y \neq 0$.)

$$-\tfrac{2}{25}y = -\tfrac{3}{25}x$$

$$y = \tfrac{3}{2}x$$

Now, substitute for y in the third equation and solve for x:

$$50x + 100(\tfrac{3}{2}x) - 500{,}000 = 0$$
$$200x = 500{,}000$$
$$x = 2{,}500$$

Thus,

$$y = \tfrac{3}{2}(2{,}500) = 3{,}750$$

and

$$\lambda = -\tfrac{2}{25}(2{,}500)^{-0.75}(3{,}750)^{0.75} \approx -0.1084$$

The only critical point of F is $(2{,}500, 3{,}750, -0.1084)$.

Step 4. Since F has only one critical point, we conclude that maximum productivity occurs when 2,500 units of labor and 3,750 units of capital are used (see step 4 in the box). Thus,

$$\begin{aligned}\text{Max } N(x, y) &= N(2{,}500, 3{,}750)\\ &= 16(2{,}500)^{0.25}(3{,}750)^{0.75}\\ &\approx 54{,}216 \text{ units}\end{aligned}$$

◆

The negative of the value of the Lagrange multiplier found in step 3 is called the **marginal productivity of money** and gives the approximate increase in production for each additional dollar spent on production. In Example 17, increasing the production budget from \$500,000 to \$600,000 would result in an approximate increase in production of

$$0.1084(100{,}000) = 10{,}840 \text{ units}$$

Note that simplifying the constraint equation

$$50x + 100y - 500{,}000 = 0$$

to

$$x + 2y - 10{,}000 = 0$$

before forming the function $F(x, y, \lambda)$ would make it difficult to interpret $-\lambda$ correctly. Thus, **in marginal productivity problems, the constraint equation should not be simplified.**

PROBLEM 17

The Cobb–Douglas production function for a new product is given by

$$N(x, y) = 20x^{0.5}y^{0.5}$$

where x is the number of units of labor and y is the number of units of capital required to produce $N(x, y)$ units of the product. Each unit of labor costs \$40 and each unit of capital costs \$120.

(A) If \$300,000 has been budgeted for the production of this product, how should this amount be allocated in order to maximize production? What is the maximum production?

(B) Find the marginal productivity of money in this case, and estimate the increase in production if an additional \$40,000 is budgeted for production.

◆

◆ FUNCTIONS OF THREE INDEPENDENT VARIABLES

We have indicated that the method of Lagrange multipliers can be extended to functions with arbitrarily many independent variables with one or more constraints. We now state a theorem for functions with three independent variables and one constraint, and consider an example that will demonstrate the advantage of the method of Lagrange multipliers over the method used in Section 9-3.

THEOREM 4

Any local maxima or minima of the function $w = f(x, y, z)$ subject to the constraint $g(x, y, z) = 0$ will be among the set of points (x_0, y_0, z_0) for which $(x_0, y_0, z_0, \lambda_0)$ is a solution to the system

$$F_x(x, y, z, \lambda) = 0$$
$$F_y(x, y, z, \lambda) = 0$$
$$F_z(x, y, z, \lambda) = 0$$
$$F_\lambda(x, y, z, \lambda) = 0$$

where $F(x, y, z, \lambda) = f(x, y, z) + \lambda g(x, y, z)$, provided all the partial derivatives exist.

◆ EXAMPLE 18

Package Design

A rectangular box with an open top and one partition is to be constructed from 162 square inches of cardboard. Find the dimensions that will result in a box with the largest possible volume.

Solution We must maximize

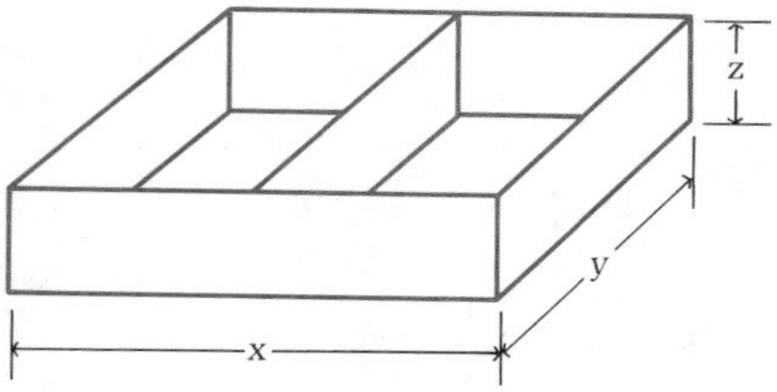

$$V(x, y, z) = xyz$$

subject to the constraint that the amount of material used is 162 square inches. Thus, x, y, and z must satisfy

$$xy + 2xz + 3yz = 162$$

Step 1. Maximize $V(x, y, z) = xyz$
Subject to $g(x, y, z) = xy + 2xz + 3yz - 162 = 0$

Step 2. $F(x, y, z, \lambda) = xyz + \lambda(xy + 2xz + 3yz - 162)$

Step 3.
$$F_x = yz + \lambda(y + 2z) = 0$$
$$F_y = xz + \lambda(x + 3z) = 0$$
$$F_z = xy + \lambda(2x + 3y) = 0$$
$$F_\lambda = xy + 2xz + 3yz - 162 = 0$$

From the first two equations, we can write

$$\lambda = \frac{-yz}{y + 2z} \qquad \lambda = \frac{-xz}{x + 3z}$$

Eliminating λ, we have

$$\frac{-yz}{y + 2z} = \frac{-xz}{x + 3z}$$
$$-xyz - 3yz^2 = -xyz - 2xz^2$$
$$3yz^2 = 2xz^2 \qquad \text{We can assume } z \neq 0.$$
$$3y = 2x$$
$$x = \tfrac{3}{2}y$$

From the second and third equations,

$$\lambda = \frac{-xz}{x + 3z} \qquad \lambda = \frac{-xy}{2x + 3y}$$

Eliminating λ, we have

$$\frac{-xz}{x + 3z} = \frac{-xy}{2x + 3y}$$
$$-2x^2z - 3xyz = -x^2y - 3xyz$$
$$2x^2z = x^2y \qquad \text{We can assume } x \neq 0.$$
$$2z = y$$
$$z = \tfrac{1}{2}y$$

Substituting $x = \frac{3}{2}y$ and $z = \frac{1}{2}y$ in the fourth equation, we have

$$(\tfrac{3}{2}y)y + 2(\tfrac{3}{2}y)(\tfrac{1}{2}y) + 3y(\tfrac{1}{2}y) - 162 = 0$$

$$\tfrac{3}{2}y^2 + \tfrac{3}{2}y^2 + \tfrac{3}{2}y^2 = 162$$

$$y^2 = 36 \qquad \text{We can assume } y > 0.$$

$$y = 6$$

$$x = \tfrac{3}{2}(6) = 9 \qquad \text{Using } x = \tfrac{3}{2}y$$

$$z = \tfrac{1}{2}(6) = 3 \qquad \text{Using } z = \tfrac{1}{2}y$$

and, finally,

$$\lambda = \frac{-(6)(3)}{6 + 2(3)} = -\frac{3}{2} \qquad \text{Using } \lambda = \frac{-yz}{y + 2z}$$

Thus, the only critical point of F with x, y, and z all positive is $(9, 6, 3, -\frac{3}{2})$.

Step 4. The box with maximum volume has dimensions 9 inches by 6 inches by 3 inches. ◆

PROBLEM 18 A box of the same type as described in Example 18 is to be constructed from 288 square inches of cardboard. Find the dimensions that will result in a box with the largest possible volume. ◆

Suppose we had decided to solve Example 18 by the method used in Section 9-3. First we would have to solve the material constraint for one of the variables, say, z:

$$z = \frac{162 - xy}{2x + 3y}$$

Then we would eliminate z in the volume function and maximize

$$V(x, y) = xy\,\frac{162 - xy}{2x + 3y}$$

Using the method of Lagrange multipliers allows us to avoid the formidable task of finding the partial derivatives of V.

Answers to Matched Problems

16. Max $f(x, y) = f(2, 2) = 17$ (see Fig. 11)
17. (A) 3,750 units of labor and 1,250 units of capital; Max $N(x, y) = N(3{,}750, 1{,}250) \approx 43{,}301$ units
    (B) Marginal productivity of money ≈ 0.1443; increase in production $\approx$ 5,774 units
18. 12 in. by 8 in. by 4 in.

EXERCISE 9-4

Use the method of Lagrange multipliers in the following problems:

A

1. Maximize $f(x, y) = 2xy$
   Subject to $x + y = 6$
2. Minimize $f(x, y) = 6xy$
   Subject to $y - x = 6$
3. Minimize $f(x, y) = x^2 + y^2$
   Subject to $3x + 4y = 25$
4. Maximize $f(x, y) = 25 - x^2 - y^2$
   Subject to $2x + y = 10$

B

5. Find the maximum and minimum of $f(x, y) = 2xy$ subject to $x^2 + y^2 = 18$.
6. Find the maximum and minimum of $f(x, y) = x^2 - y^2$ subject to $x^2 + y^2 = 25$.
7. Maximize the product of two numbers if their sum must be 10.
8. Minimize the product of two numbers if their difference must be 10.

C

9. Minimize $f(x, y, z) = x^2 + y^2 + z^2$
   Subject to $2x - y + 3z = -28$
10. Maximize $f(x, y, z) = xyz$
    Subject to $2x + y + 2z = 120$
11. Maximize and minimize $f(x, y, z) = x + y + z$
    Subject to $x^2 + y^2 + z^2 = 12$
12. Maximize and minimize $f(x, y, z) = 2x + 4y + 4z$
    Subject to $x^2 + y^2 + z^2 = 9$

APPLICATIONS

Business & Economics

13. *Budgeting for least cost.* A manufacturing company produces two models of a television set, x units of model *A* and y units of model *B* per week, at a cost (in dollars) of

$$C(x, y) = 6x^2 + 12y^2$$

If it is necessary (because of shipping considerations) that

$$x + y = 90$$

how many of each type of set should be manufactured per week to minimize cost? What is the minimum cost?

14. *Budgeting for maximum production.* A manufacturing firm has budgeted \$60,000 per month for labor and materials. If \$x thousand is spent on labor and \$y thousand is spent on materials, and if the monthly output (in units) is given by

$$N(x, y) = 4xy - 8x$$

how should the \$60,000 be allocated to labor and materials in order to maximize *N*? What is the maximum *N*?

15. *Productivity.* A consulting firm for a manufacturing company arrived at the following Cobb–Douglas production function for a particular product:

$$N(x, y) = 50x^{0.8}y^{0.2}$$

where x is the number of units of labor and y is the number of units of capital required to produce $N(x, y)$ units of the product. Each unit of labor costs \$40 and each unit of capital costs \$80.

(A) If \$400,000 is budgeted for production of the product, determine how this amount should be allocated to maximize production, and find the maximum production.

(B) Find the marginal productivity of money in this case, and estimate the increase in production if an additional \$50,000 is budgeted for the production of this product.

16. *Productivity.* The research department for a manufacturing company arrived at the following Cobb–Douglas production function for a particular product:

$$N(x, y) = 10x^{0.6}y^{0.4}$$

where x is the number of units of labor and y is the number of units of capital required to produce $N(x, y)$ units of the product. Each unit of labor costs \$30 and each unit of capital costs \$60.

(A) If \$300,000 is budgeted for production of the product, determine how this amount should be allocated to maximize production, and find the maximum production.

(B) Find the marginal productivity of money in this case, and estimate the increase in production if an additional \$80,000 is budgeted for the production of this product.

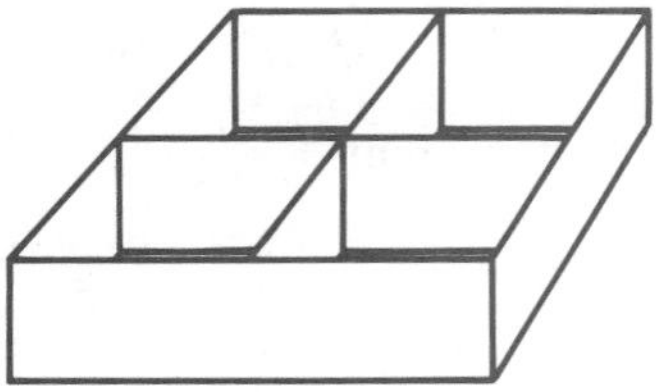

17. *Maximum volume.* A rectangular box with no top and two intersecting partitions is to be constructed from 192 square inches of cardboard (see the figure in the margin). Find the dimensions that will maximize the volume.

18. *Maximum volume.* A mailing service states that a rectangular package shall have the sum of the length and girth not to exceed 120 inches (see the figure). What are the dimensions of the largest (in volume) mailing carton that can be constructed meeting these restrictions?

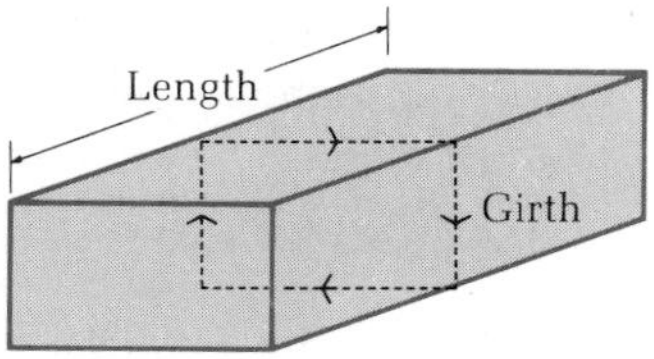

Life Sciences

19. *Agriculture.* Three pens of the same size are to be built along an existing fence (see the figure). If 400 feet of fencing are available, what length should x and y be to produce the maximum total area? What is the maximum area?

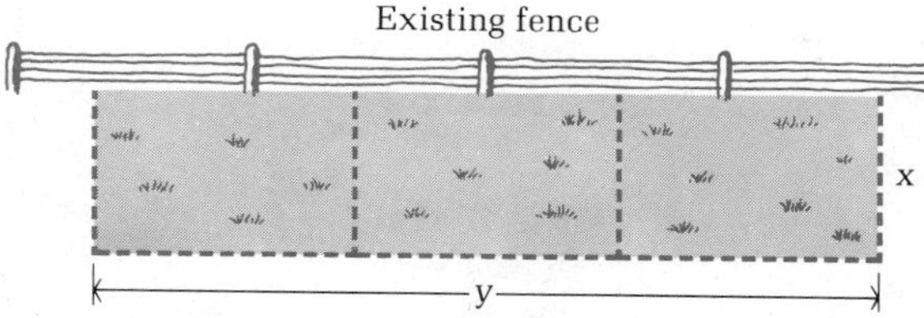

20. *Diet and minimum cost.* A group of guinea pigs is to receive 25,600 calories per week. Two available foods produce $200xy$ calories for a mixture of x kilograms of type M food and y kilograms of type N food. If type M costs \$1 per kilogram and type N costs \$2 per kilogram, how much of each type of food should be used to minimize weekly food costs? What is the minimum cost? [*Note:* $x \geq 0$, $y \geq 0$]

SECTION 9-5 Method of Least Squares

- LEAST SQUARES APPROXIMATION
- APPLICATIONS

◆ LEAST SQUARES APPROXIMATION

In this section we will use the optimization techniques discussed in Section 9-3 to find the equation of a line that is a "best" approximation to a set of points in a rectangular coordinate system. This very popular method is known as **least squares approximation,** or **linear regression.** Let us begin by considering a specific case.

A manufacturer wants to approximate the cost function for a product. The value of the cost function has been determined for certain levels of production, as listed in Table 1. Although these points do not all lie on a line (see Fig. 12, page 528), they are very close to being linear. The manufacturer would like to approximate the cost function by a linear function; that is, determine values m and d so that the line

$$y = mx + d$$

is, in some sense, the "best" approximation to the cost function.

TABLE 1

NUMBER OF UNITS x, *in hundreds*	COST y, *in thousands of dollars*
2	4
5	6
6	7
9	8

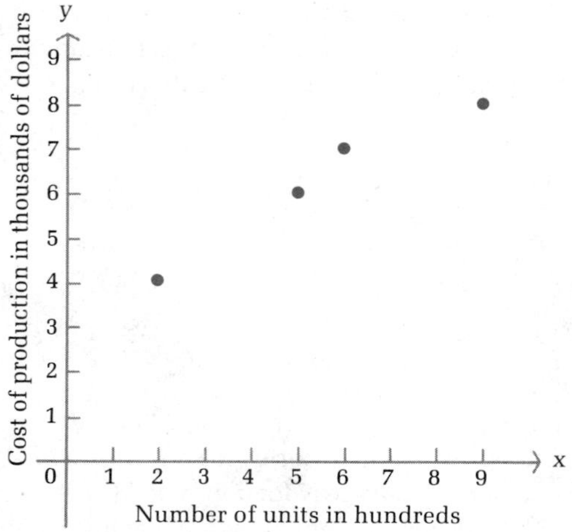

FIGURE 12

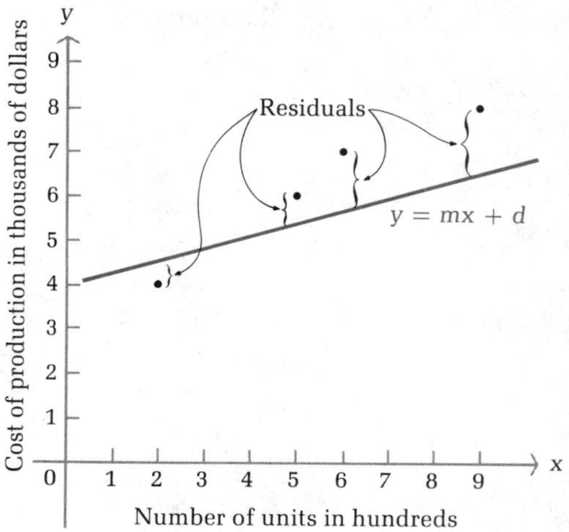

FIGURE 13

What do we mean by "best"? Since the line $y = mx + d$ will not go through all four points, it is reasonable to examine the differences between the y coordinates of the points listed in the table and the y coordinates of the corresponding points on the line. Each of these differences is called the **residual** at that point (see Fig. 13). For example, at $x = 2$, the point from Table 1 is (2, 4) and the point on the line is $(2, 2m + d)$, so the residual is

$$4 - (2m + d) = 4 - 2m - d$$

All the residuals are listed in Table 2.

TABLE 2

x	y	$mx + d$	*Residual*
2	4	$2m + d$	$4 - 2m - d$
5	6	$5m + d$	$6 - 5m - d$
6	7	$6m + d$	$7 - 6m - d$
9	8	$9m + d$	$8 - 9m - d$

Our criterion for the "best" approximation is the following: Determine the values of m and d that *minimize the sum of the squares* of the residuals. The resulting line is called the **least squares line,** or the **regression line.** To this end, we minimize

$$F(m, d) = (4 - 2m - d)^2 + (6 - 5m - d)^2 + (7 - 6m - d)^2 + (8 - 9m - d)^2$$

Step 1. Find critical points:

$$\begin{aligned} F_m(m, d) &= 2(4 - 2m - d)(-2) + 2(6 - 5m - d)(-5) \\ &\quad + 2(7 - 6m - d)(-6) + 2(8 - 9m - d)(-9) \\ &= -304 + 292m + 44d = 0 \end{aligned}$$

$$\begin{aligned} F_d(m, d) &= 2(4 - 2m - d)(-1) + 2(6 - 5m - d)(-1) \\ &\quad + 2(7 - 6m - d)(-1) + 2(8 - 9m - d)(-1) \\ &= -50 + 44m + 8d = 0 \end{aligned}$$

Solving the system

$$\begin{aligned} -304 + 292m + 44d &= 0 \\ -50 + 44m + 8d &= 0 \end{aligned}$$

we obtain $(m, d) = (0.58, 3.06)$ as the only critical point.

Step 2. Compute $A = F_{mm}(m, d)$, $B = F_{md}(m, d)$, and $C = F_{dd}(m, d)$:

$$F_{mm}(m, d) = 292 \quad \text{thus} \quad A = F_{mm}(0.58, 3.06) = 292$$
$$F_{md}(m, d) = 44 \quad \text{thus} \quad B = F_{md}(0.58, 3.06) = 44$$
$$F_{dd}(m, d) = 8 \quad \text{thus} \quad C = F_{dd}(0.58, 3.06) = 8$$

Step 3. Evaluate $AC - B^2$ and try to classify the critical point (m, d) using Theorem 2 in Section 9-3:

$$AC - B^2 = (292)(8) - (44)^2 = 400 > 0 \quad \text{and} \quad A = 292 > 0$$

Therefore, case 2 in Theorem 2 applies, and $F(m, d)$ has a local minimum at the critical point (0.58, 3.06).

Thus, the least squares line for the given data is

$y = 0.58x + 3.06$ Least squares line

The sum of the squares of the residuals is minimized for this choice of m and d (see Fig. 14).

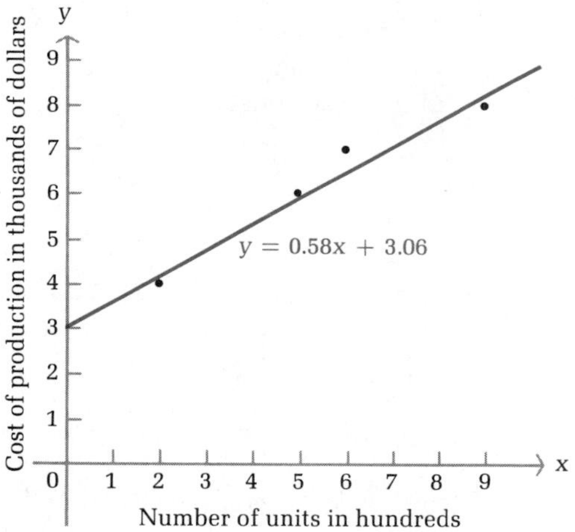

FIGURE 14

This linear function can now be used by the manufacturer to estimate any of the quantities normally associated with the cost function—such as costs, marginal costs, average costs, and so on. For example, the cost of producing 2,000 units is approximately

$$y = (0.58)(20) + 3.06 = 14.66 \quad \text{or} \quad \$14,660$$

The marginal cost function is

$$\frac{dy}{dx} = 0.58$$

The average cost function is

$$\bar{y} = \frac{0.58x + 3.06}{x}$$

In general, if we are given a set of n points $(x_1, y_1), (x_2, y_2), \ldots, (x_n, y_n)$, then proceeding as we did with the points in Table 1, it can be shown that the coefficients m and d of the least squares line $y = mx + d$ must satisfy the following system of *normal equations:**

$$\begin{aligned} \left(\sum_{k=1}^{n} x_k\right)m + nd &= \sum_{k=1}^{n} y_k \\ \left(\sum_{k=1}^{n} x_k^2\right)m + \left(\sum_{k=1}^{n} x_k\right)d &= \sum_{k=1}^{n} x_k y_k \end{aligned} \qquad (1)$$

Solving system (1) for m and d produces the formulas given in the box.

Least Squares Approximation

For a set of n points $(x_1, y_1), (x_2, y_2), \ldots, (x_n, y_n)$, the coefficients m and d of the least squares line

$$y = mx + d$$

are the solutions of the system of **normal equations**

$$\begin{aligned} \left(\sum_{k=1}^{n} x_k\right)m + nd &= \sum_{k=1}^{n} y_k \\ \left(\sum_{k=1}^{n} x_k^2\right)m + \left(\sum_{k=1}^{n} x_k\right)d &= \sum_{k=1}^{n} x_k y_k \end{aligned} \qquad (1)$$

and are given by the formulas

$$m = \frac{n\left(\sum_{k=1}^{n} x_k y_k\right) - \left(\sum_{k=1}^{n} x_k\right)\left(\sum_{k=1}^{n} y_k\right)}{n\left(\sum_{k=1}^{n} x_k^2\right) - \left(\sum_{k=1}^{n} x_k\right)^2} \qquad (2)$$

and

$$d = \frac{\sum_{k=1}^{n} y_k - m\left(\sum_{k=1}^{n} x_k\right)}{n} \qquad (3)$$

* See Appendix A for a review of the summation notation used in system (1).

To find m and d, we can either solve system (1) directly or use formulas (2) and (3). If the formulas are used, note that the value of m is needed in formula 3; thus, the value of m always must be computed first.

◆ APPLICATIONS

◆ EXAMPLE 19 *Educational Testing*

Table 3 lists the midterm and final examination scores for 10 students in a calculus course.

TABLE 3

MIDTERM	FINAL	MIDTERM	FINAL
49	61	78	77
53	47	83	81
67	72	85	79
71	76	91	93
74	68	99	99

(A) Find the least squares line for the data given in the table.
(B) Use the least squares line to predict the final examination score for a student who scored 95 on the midterm examination.
(C) Graph the data and the least squares line on the same set of axes.

Solutions (A) Table 4 shows a convenient way to compute all the sums in the formulas for m and d.

TABLE 4

	x_k	y_k	$x_k y_k$	x_k^2
	49	61	2,989	2,401
	53	47	2,491	2,809
	67	72	4,824	4,489
	71	76	5,396	5,041
	74	68	5,032	5,476
	78	77	6,006	6,084
	83	81	6,723	6,889
	85	79	6,715	7,225
	91	93	8,463	8,281
	99	99	9,801	9,801
Totals	750	753	58,440	58,496

Thus,

$$\sum_{k=1}^{10} x_k = 750 \qquad \sum_{k=1}^{10} y_k = 753 \qquad \sum_{k=1}^{10} x_k y_k = 58{,}440 \qquad \sum_{k=1}^{10} x_k^2 = 58{,}496$$

and the normal equations are

$$750m + \quad 10d = 753$$
$$58{,}496m + 750d = 58{,}440$$

We can either solve this system directly or use formulas (2) and (3). We choose to use the formulas:

$$m = \frac{10(58{,}440) - (750)(753)}{10(58{,}496) - (750)^2} = \frac{19{,}650}{22{,}460} \approx 0.875$$
$$d = \frac{753 - 0.875(750)}{10} = 9.675$$

The least squares line is given (approximately) by

$$y = 0.875x + 9.675$$

(B) If $x = 95$, then the predicted score on the final exam is

$$y = 0.875(95) + 9.675$$
$$\approx 93 \qquad \text{Assuming the score must be an integer}$$

(C)

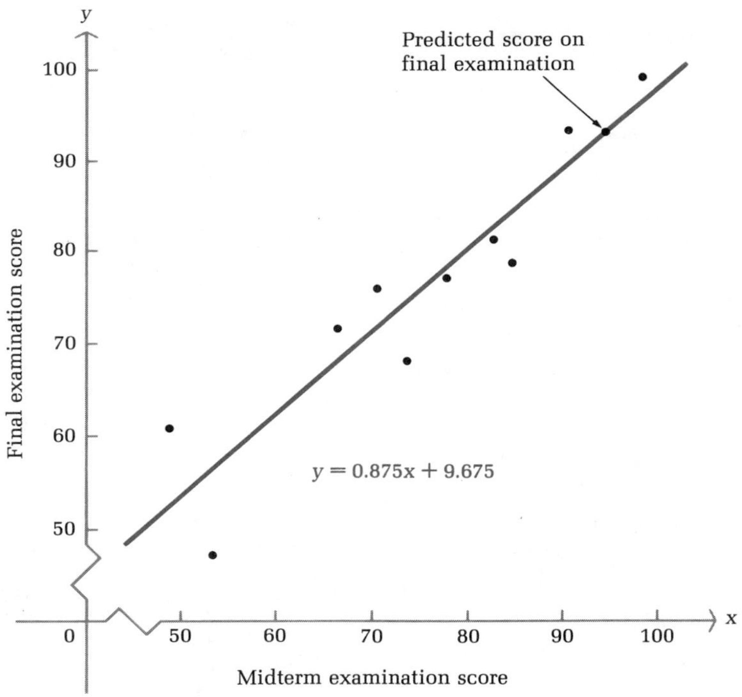

◆

PROBLEM 19 Repeat Example 19 for the scores listed in Table 5.

TABLE 5

MIDTERM	FINAL	MIDTERM	FINAL
54	50	84	80
60	66	88	95
75	80	89	85
76	68	97	94
78	71	99	86

◆

◆ EXAMPLE 20

Wool Production

Table 6 lists the annual production of wool throughout the world for the years 1970–1980. Use the data in the table to predict the worldwide wool production for 1981.

TABLE 6

World Wool Production

YEAR	MILLIONS OF POUNDS	YEAR	MILLIONS OF POUNDS
1970	6,107	1976	5,827
1971	5,972	1977	5,838
1972	5,560	1978	5,983
1973	5,474	1979	6,168
1974	5,769	1980	6,285
1975	5,911		

Solution

Solving this problem by hand is certainly possible, but would require considerable effort. Instead, we used a computer to perform the necessary computations. (The program we used can be found in the computer supplement for this text; see the Preface.) The computer output is listed in Table 7.

TABLE 7

INPUT TO PROGRAM

```
* - LEAST SQUARES TECHNIQUE - *
             DATA
PT.  X COORDINATE  Y COORDINATE
===  ============  ============
 1             0          6107
 2             1          5972
 3             2          5560
 4             3          5474
 5             4          5769
 6             5          5911
 7             6          5827
 8             7          5838
 9             8          5983
 10            9          6168
 11           10          6285
```

OUTPUT FROM PROGRAM

```
* - LEAST SQUARES TECHNIQUE - *
         POLYNOMIAL

Y = 33.9X + 5729.95

* - LEAST SQUARES TECHNIQUE - *
    EVALUATE POLYNOMIAL

      X          CALCULATED Y
===============================
     11               6102.85
```

Notice that we used x = 0 for 1970, x = 1 for 1971, and so on. Examining the computer output in Table 7, we see that the least squares line is

$$y = 33.9x + 5{,}729.95$$

and the estimated worldwide wool production in 1981 is 6,102.85 million pounds. ◆

PROBLEM 20 Use the least squares line in Example 20 to estimate the worldwide wool production in 1982. ◆

Answers to Matched Problems

19. (A) $y = 0.85x + 9.47$ (B) 90.2

(C)

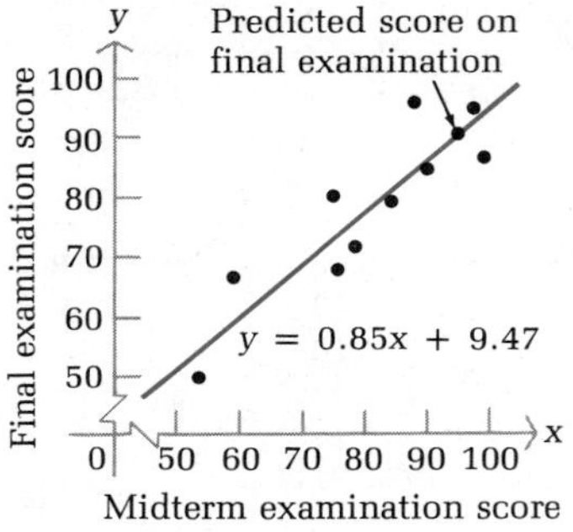

20. 6,136.75 million pounds

EXERCISE 9-5

A *Find the least squares line. Graph the data and the least squares line.*

1.

x	y
1	1
2	3
3	4
4	3

2.

x	y
1	−2
2	−1
3	3
4	5

3.

x	y
1	8
2	5
3	4
4	0

4.

x	y
1	20
2	14
3	11
4	3

5.

x	y
1	3
2	4
3	5
4	6

6.

x	y
1	2
2	3
3	3
4	2

B *Find the least squares line and use it to estimate y for the indicated value of x.*

7.

x	y
0	10
5	22
10	31
15	46
20	51

Estimate y when $x = 25$.

8.

x	y
−5	60
0	50
5	30
10	20
15	15

Estimate y when $x = 20$.

9.

x	y
−1	14
1	12
3	8
5	6
7	5

Estimate y when $x = 2$.

10.

x	y
2	−4
6	0
10	8
14	12
18	14

Estimate y when $x = 15$.

11.

x	y	x	y
0.5	25	9.5	12
2	22	11	11
3.5	21	12.5	8
5	21	14	5
6.5	18	15.5	1

Estimate y when $x = 8$.

12.

x	y	x	y
0	−15	12	11
2	−9	14	13
4	−7	16	19
6	−7	18	25
8	−1	20	33

Estimate y when $x = 10$.

C **13.** The method of least squares can be generalized to curves other than straight lines. To find the coefficients of the parabola

$$y = ax^2 + bx + c$$

that is the "best" fit for the points (1, 2), (2, 1), (3, 1), and (4, 3), minimize the sum of the squares of the residuals

$$F(a, b, c) = (a + b + c - 2)^2 + (4a + 2b + c - 1)^2 + (9a + 3b + c - 1)^2 + (16a + 4b + c - 3)^2$$

by solving the system

$$F_a(a, b, c) = 0 \qquad F_b(a, b, c) = 0 \qquad F_c(a, b, c) = 0$$

for a, b, and c. Graph the points and the parabola.

14. Repeat Problem 13 for the points (−1, −2), (0, 1), (1, 2), and (2, 0).

Problems 15 and 16 refer to the system of normal equations and the formulas for m and d stated on page 530.

15. Verify formulas (2) and (3) by solving the system of normal equations (1) for m and d.

16. If

$$\bar{x} = \frac{1}{n} \sum_{k=1}^{n} x_k \quad \text{and} \quad \bar{y} = \frac{1}{n} \sum_{k=1}^{n} y_k$$

are the averages of the x and y coordinates, respectively, show that the point $(\bar{x}, \bar{y})$ satisfies the equation of the least squares line $y = mx + d$.

APPLICATIONS

Business & Economics

17. *Income.* Data for the per capita personal income in the United States from 1980 to 1986 are given in the table below.

(A) Find the least squares line for the data using $x = 0$ for 1980.
(B) Use the least squares line to estimate the per capita income in 1990.

YEAR	INCOME Dollars	YEAR	INCOME Dollars
1980	9,910	1984	13,115
1981	10,940	1985	13,687
1982	11,470	1986	14,461
1983	12,093		

18. *Consumer price index.* Data for the consumer price index (CPI) in the United States (in 1967 dollars) from 1980 to 1985 are given in the table below.

(A) Find the least squares line for the data using $x = 0$ for 1980.
(B) Use the least squares line to estimate the consumer price index in 1990.

YEAR	CPI	YEAR	CPI
1980	246.8	1983	298.4
1981	272.4	1984	311.1
1982	289.1	1985	322.2

x	y
5.0	2.0
5.5	1.8
6.0	1.4
6.5	1.2
7.0	1.1

19. *Maximizing profit.* The market research department for a drug store chain chose two summer resort areas to test market a new sun screen lotion packaged in 4 ounce plastic bottles. After a summer of varying the selling price and recording the monthly demand, the research department arrived at the demand table given in the margin, where y is the number of bottles purchased per month (in thousands) at x dollars per bottle.

(A) Find a demand equation using the method of least squares.
(B) If each bottle of sun screen costs the drug store chain \$4, how should it be priced to achieve a maximum monthly profit? [*Hint:* Use the result of part A, with $C = 4y$, $R = xy$, and $P = R - C$.]

x	y
4.0	4.2
4.5	3.5
5.0	2.7
5.5	1.5
6.0	0.7

20. *Maximizing profit.* A market research consultant for a supermarket chain chose a large city to test market a new brand of mixed nuts packaged in 8 ounce cans. After a year of varying the selling price and recording the monthly demand, the consultant arrived at the demand table given in the margin, where y is the number of cans purchased per month (in thousands) at x dollars per can.

(A) Find a demand equation using the method of least squares.
(B) If each can of nuts costs the supermarket chain $3, how should it be priced to achieve a maximum monthly profit?

Life Sciences

WATER TEMPERATURE	PULSE RATE REDUCTION
50	15
55	13
60	10
65	6
70	2

21. *Medicine.* If a person dives into cold water, a neural reflex response automatically shuts off blood circulation to the skin and muscles and reduces the pulse rate. A medical research team conducted an experiment using a group of ten 2-year-olds. A child's face was placed momentarily in cold water, and the corresponding reduction in pulse rate was recorded. The data for the average reduction in heart rate for each temperature are summarized in the table in the margin.

(A) If T is water temperature (in degrees Fahrenheit) and P is pulse rate reduction (in beats per minute), use the method of least squares to find a linear equation relating T and P.
(B) Use the equation found in part A to find P when $T = 57$.

22. *Biology.* In biology there is an approximate rule, called the *bioclimatic rule for temperate climates*, that has been known for a couple of hundred years. This rule states that in spring and early summer, periodic phenomena such as blossoming of flowers, appearance of insects, and ripening of fruit usually come about 4 days later for each 500 feet of altitude. Stated as a formula,

$$d = 8h \qquad 0 \leq h \leq 4$$

h	d
0	0
1	7
2	18
3	28
4	33

where d is the change in days and h is the altitude (in thousands of feet). To test this rule, an experiment was set up to record the difference in blossoming time of the same type of apple tree at different altitudes. A summary of the results is given in the table in the margin.

(A) Use the method of least squares to find a linear equation relating h and d. Does the bioclimatic rule, $d = 8h$, appear to be approximately correct?
(B) How much longer will it take this type of apple tree to blossom at 3.5 thousand feet than at sea level? [Use the linear equation found in part A.]

Social Sciences

23. *Political science.* Association of economic class and party affiliation did not start with Roosevelt's New Deal; it goes back to the time of Andrew Jackson (1767–1845). Paul Lazarsfeld of Columbia University published an article

Political Affiliations, 1836

WARD	AVERAGE ASSESSED VALUE PER PERSON *(in $100)*	DEMOCRATIC VOTES *(%)*
12	1.7	51
3	2.1	49
1	2.3	53
5	2.4	36
2	3.6	65
11	3.7	35
10	4.7	29
4	6.2	40
6	7.1	34
9	7.4	29
8	8.7	20
7	11.9	23

in the November 1950 issue of *Scientific American* in which he discusses statistical investigations of the relationships between economic class and party affiliation. The data in the table are taken from this article.

(A) If A represents the average assessed value per person in a given ward in 1836 and D represents the percentage of people in that ward voting Democratic in 1836, use the method of least squares to find a linear equation relating A and D.

(B) If the average assessed value per person in a ward had been $300, what is the predicted percentage of people in that ward that would have voted Democratic?

24. *Education.* The table lists the high school grade-point averages (GPA's) of 10 students, along with their grade-point averages after one semester of college.

HIGH SCHOOL GPA	COLLEGE GPA	HIGH SCHOOL GPA	COLLEGE GPA
2.0	1.5	3.0	2.3
2.2	1.5	3.1	2.5
2.4	1.6	3.3	2.9
2.7	1.8	3.4	3.2
2.9	2.1	3.7	3.5

(A) Find the least squares line for the data.

(B) Estimate the college GPA for a student with a high school GPA of 3.5.

(C) Estimate the high school GPA necessary for a college GPA of 2.7.

SECTION 9-6 Double Integrals over Rectangular Regions

- ◆ INTRODUCTION
- ◆ DEFINITION OF THE DOUBLE INTEGRAL
- ◆ AVERAGE VALUE OVER RECTANGULAR REGIONS
- ◆ VOLUME AND DOUBLE INTEGRALS

◆ INTRODUCTION

We have generalized the concept of differentiation to functions with two or more independent variables. How can we do the same with integration, and how can we interpret the results? Let us first look at the operation of antidifferentiation. We can antidifferentiate a function of two or more variables with respect to one of the variables by treating all the other variables as though they were constants. Thus, this operation is the reverse operation of partial differentiation, just as ordinary antidifferentiation is the reverse operation of ordinary differentiation. We write $\int f(x, y)\,dx$ to indicate that we are to antidifferentiate $f(x, y)$ with respect to x, holding y fixed; we write $\int f(x, y)\,dy$ to indicate that we are to antidifferentiate $f(x, y)$ with respect to y, holding x fixed.

◆ EXAMPLE 21 Evaluate:

(A) $\int (6xy^2 + 3x^2)\,dy$ (B) $\int (6xy^2 + 3x^2)\,dx$

Solutions (A) Treating x as a constant and using the properties of antidifferentiation from Section 7-1, we have

$$\begin{aligned}\int (6xy^2 + 3x^2)\,dy &= \int 6xy^2\,dy + \int 3x^2\,dy\\ &= 6x\int y^2\,dy + 3x^2\int dy\\ &= 6x\left(\frac{y^3}{3}\right) + 3x^2(y) + C(x)\\ &= 2xy^3 + 3x^2y + C(x)\end{aligned}$$

The dy tells us we are looking for the antiderivative of $6xy^2 + 3x^2$ with respect to y only, holding x constant.

Notice that the constant of integration actually can be *any function of* x *alone*, since, for any such function

$$\frac{\partial}{\partial y} C(x) = 0$$

We can verify that our answer is correct by using partial differentiation:

$$\begin{aligned}\frac{\partial}{\partial y}[2xy^3 + 3x^2y + C(x)] &= 6xy^2 + 3x^2 + 0\\ &= 6xy^2 + 3x^2\end{aligned}$$

(B) Now we treat y as a constant:

$$\begin{aligned}\int (6xy^2 + 3x^2)\,dx &= \int 6xy^2\,dx + \int 3x^2\,dx\\ &= 6y^2\int x\,dx + 3\int x^2\,dx\\ &= 6y^2\left(\frac{x^2}{2}\right) + 3\left(\frac{x^3}{3}\right) + E(y)\\ &= 3x^2y^2 + x^3 + E(y)\end{aligned}$$

This time, the antiderivative contains an arbitrary function $E(y)$ of y alone.

Check: $$\begin{aligned}\frac{\partial}{\partial x}[3x^2y^2 + x^3 + E(y)] &= 6xy^2 + 3x^2 + 0\\ &= 6xy^2 + 3x^2\end{aligned}$$ ◆

PROBLEM 21 Evaluate:

(A) $\int (4xy + 12x^2y^3)\,dy$ (B) $\int (4xy + 12x^2y^3)\,dx$ ◆

Now that we have extended the concept of antidifferentiation to functions with two variables, we also can evaluate definite integrals of the form

$$\int_a^b f(x, y)\,dx \qquad \text{or} \qquad \int_c^d f(x, y)\,dy$$

◆ EXAMPLE 22 Evaluate, substituting the limits of integration in y if dy is used and in x if dx is used:

(A) $\int_0^2 (6xy^2 + 3x^2)\,dy$ (B) $\int_0^1 (6xy^2 + 3x^2)\,dx$

Solutions

(A) From Example 21A, we know that $\int (6xy^2 + 3x^2)\,dy = 2xy^3 + 3x^2y + C(x)$. According to the definition of the definite integral for a function of one variable, we can use any antiderivative to evaluate the definite integral. Thus, choosing $C(x) = 0$, we have

$$\begin{aligned} \int_0^2 (6xy^2 + 3x^2)\,dy &= (2xy^3 + 3x^2y)\Big|_{y=0}^{y=2} \\ &= [2x(2)^3 + 3x^2(2)] - [2x(0)^3 + 3x^2(0)] \\ &= 16x + 6x^2 \end{aligned}$$

(B) From Example 21B, we know that $\int (6xy^2 + 3x^2)\,dx = 3x^2y^2 + x^3 + E(y)$. Thus, choosing $E(y) = 0$, we have

$$\begin{aligned} \int_0^1 (6xy^2 + 3x^2)\,dx &= (3x^2y^2 + x^3)\Big|_{x=0}^{x=1} \\ &= [3y^2(1)^2 + (1)^3] - [3y^2(0)^2 + (0)^3] \\ &= 3y^2 + 1 \end{aligned}$$

◆

PROBLEM 22 Evaluate:

(A) $\int_0^1 (4xy + 12x^2y^3)\,dy$ (B) $\int_0^3 (4xy + 12x^2y^3)\,dx$ ◆

Notice that integrating and evaluating a definite integral with integrand $f(x, y)$ with respect to y produces a function of x alone (or a constant). Likewise, integrating and evaluating a definite integral with integrand $f(x, y)$ with respect to x produces a function of y alone (or a constant). Each of these results, involving at most one variable, can now be used as an integrand in a second definite integral.

◆ EXAMPLE 23 Evaluate:

$$\text{(A)} \int_0^1 \left[\int_0^2 (6xy^2 + 3x^2)\, dy \right] dx \qquad \text{(B)} \int_0^2 \left[\int_0^1 (6xy^2 + 3x^2)\, dx \right] dy$$

Solutions (A) Example 22A showed that

$$\int_0^2 (6xy^2 + 3x^2)\, dy = 16x + 6x^2$$

Thus,

$$\begin{aligned} \int_0^1 \left[\int_0^2 (6xy^2 + 3x^2)\, dy \right] dx &= \int_0^1 (16x + 6x^2)\, dx \\ &= (8x^2 + 2x^3)\Big|_{x=0}^{x=1} \\ &= [8(1)^2 + 2(1)^3] - [8(0)^2 + 2(0)^3] \\ &= 10 \end{aligned}$$

(B) Example 22B showed that

$$\int_0^1 (6xy^2 + 3x^2)\, dx = 3y^2 + 1$$

Thus,

$$\begin{aligned} \int_0^2 \left[\int_0^1 (6xy^2 + 3x^2)\, dx \right] dy &= \int_0^2 (3y^2 + 1)\, dy \\ &= (y^3 + y)\Big|_{y=0}^{y=2} \\ &= [(2)^3 + 2] - [(0)^3 + 0] \\ &= 10 \end{aligned}$$

◆

PROBLEM 23 Evaluate:

$$\text{(A)} \int_0^3 \left[\int_0^1 (4xy + 12x^2y^3)\, dy \right] dx \qquad \text{(B)} \int_0^1 \left[\int_0^3 (4xy + 12x^2y^3)\, dx \right] dy$$

◆

◆ DEFINITION OF THE DOUBLE INTEGRAL

Notice that the answers in Examples 23A and 23B are identical. This is not an accident. In fact, it is this property that enables us to define the *double integral*, as follows:

Double Integral

The **double integral** of a function $f(x, y)$ over a rectangle $R = \{(x, y) | a \leq x \leq b, \quad c \leq y \leq d\}$ is

$$\iint_R f(x, y)\,dA$$

$$= \int_a^b \left[\int_c^d f(x, y)\,dy\right] dx$$

$$= \int_c^d \left[\int_a^b f(x, y)\,dx\right] dy$$

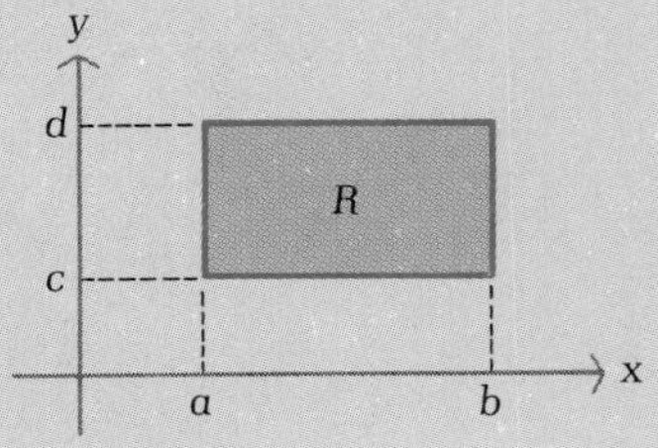

In the double integral $\iint_R f(x, y)\,dA$, $f(x, y)$ is called the **integrand** and R is called the **region of integration.** The expression dA indicates that this is an integral over a two-dimensional region. The integrals

$$\int_a^b \left[\int_c^d f(x, y)\,dy\right] dx \qquad \text{and} \qquad \int_c^d \left[\int_a^b f(x, y)\,dx\right] dy$$

are referred to as **iterated integrals** (the brackets are often omitted), and the order in which dx and dy are written indicates the order of integration. This is not the most general definition of the double integral over a rectangular region; however, it is equivalent to the general definition for all the functions we will consider.

◆ EXAMPLE 24 Evaluate:

$$\iint_R (x + y)\,dA \qquad \text{over} \qquad R = \{(x, y) | 1 \leq x \leq 3, \quad -1 \leq y \leq 2\}$$

Solution We can choose either order of iteration. As a check, we will evaluate the integral both ways:

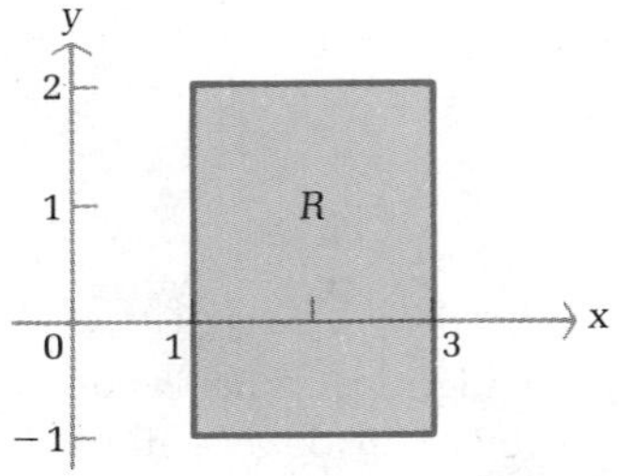

$$\iint_R (x + y)\,dA = \int_1^3 \int_{-1}^2 (x + y)\,dy\,dx$$

$$= \int_1^3 \left[\left(xy + \frac{y^2}{2}\right)\Big|_{y=-1}^{y=2}\right] dx$$

$$= \int_1^3 [(2x + 2) - (-x + \tfrac{1}{2})]\,dx$$

$$= \int_1^3 (3x + \tfrac{3}{2})\,dx$$

$$\int_1^3 (3x + \tfrac{3}{2})\,dx = (\tfrac{3}{2}x^2 + \tfrac{3}{2}x)\Big|_{x=1}^{x=3}$$

$$= (\tfrac{27}{2} + \tfrac{9}{2}) - (\tfrac{3}{2} + \tfrac{3}{2})$$
$$= 18 - 3 = 15$$

$$\iint_R (x + y)\,dA = \int_{-1}^{2} \int_1^3 (x + y)\,dx\,dy$$

$$= \int_{-1}^{2} \left[\left(\frac{x^2}{2} + xy\right)\Big|_{x=1}^{x=3}\right] dy$$

$$= \int_{-1}^{2} [(\tfrac{9}{2} + 3y) - (\tfrac{1}{2} + y)]\,dy$$

$$= \int_{-1}^{2} (4 + 2y)\,dy$$

$$= (4y + y^2)\Big|_{y=-1}^{y=2}$$

$$= (8 + 4) - (-4 + 1)$$
$$= 12 - (-3) = 15$$ ◆

PROBLEM 24 Evaluate both ways:

$$\iint_R (2x - y)\,dA \qquad \text{over} \qquad R = \{(x, y) | -1 \le x \le 5, \quad 2 \le y \le 4\}$$ ◆

◆ EXAMPLE 25 Evaluate:

$$\iint_R 2xe^{x^2+y}\,dA \qquad \text{over} \qquad R = \{(x, y) | 0 \le x \le 1, \quad -1 \le y \le 1\}$$

Solution

$$\iint_R 2xe^{x^2+y}\,dA = \int_{-1}^{1} \int_0^1 2xe^{x^2+y}\,dx\,dy$$

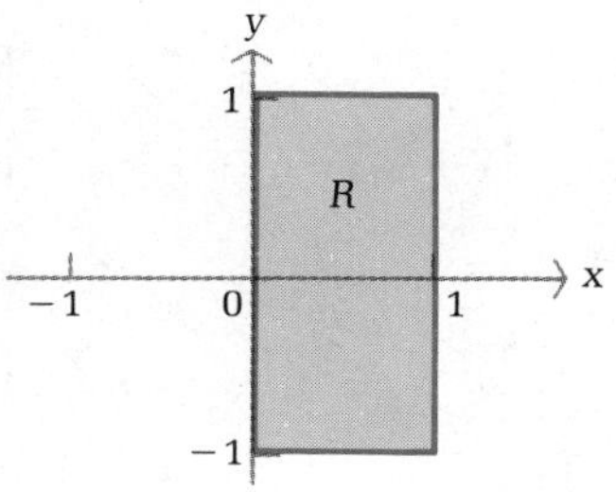

$$= \int_{-1}^{1} \left[(e^{x^2+y})\Big|_{x=0}^{x=1}\right] dy$$

$$= \int_{-1}^{1} (e^{1+y} - e^y)\,dy$$

$$= (e^{1+y} - e^y)\Big|_{y=-1}^{y=1}$$

$$= (e^2 - e) - (e^0 - e^{-1})$$
$$= e^2 - e - 1 + e^{-1}$$ ◆

PROBLEM 25 Evaluate:

$$\iint_R \frac{x}{y^2} e^{x/y}\,dA \qquad \text{over} \qquad R = \{(x, y) | 0 \le x \le 1, \quad 1 \le y \le 2\}$$ ◆

◆ AVERAGE VALUE OVER RECTANGULAR REGIONS

In Section 8-1, the average value of a function $f(x)$ over an interval $[a, b]$ was defined as

$$\frac{1}{b-a}\int_a^b f(x)\,dx$$

This definition is easily extended to functions of two variables over rectangular regions, as shown in the box. Notice that the denominator in the expression given in the box, $(b-a)(d-c)$, is simply the area of the rectangle R.

Average Value over Rectangular Regions

The **average value** of the function $f(x, y)$ over the rectangle $R = \{(x, y) | a \leq x \leq b, \quad c \leq y \leq d\}$ is

$$\frac{1}{(b-a)(d-c)}\iint_R f(x, y)\,dA$$

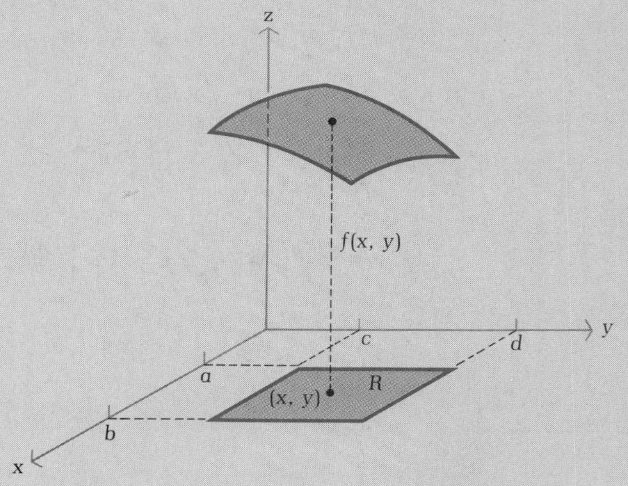

◆ EXAMPLE 26 Find the average value of $f(x, y) = 4 - \frac{1}{2}x - \frac{1}{2}y$ over the rectangle $R = \{(x, y) | 0 \leq x \leq 2, \; 0 \leq y \leq 2\}$.

Solution $$\frac{1}{(b-a)(d-c)}\iint_R f(x, y)\,dA = \frac{1}{(2-0)(2-0)}\iint_R \left(4 - \frac{1}{2}x - \frac{1}{2}y\right) dA$$

$$= \tfrac{1}{4}\int_0^2 \int_0^2 (4 - \tfrac{1}{2}x - \tfrac{1}{2}y)\,dy\,dx$$

$$= \tfrac{1}{4}\int_0^2 \left[(4y - \tfrac{1}{2}xy - \tfrac{1}{4}y^2)\Big|_{y=0}^{y=2}\right] dx$$

$$= \tfrac{1}{4}\int_0^2 (7 - x)\,dx$$

$$= \tfrac{1}{4}(7x - \tfrac{1}{2}x^2)\Big|_{x=0}^{x=2}$$

$$= \tfrac{1}{4}(12) = 3$$

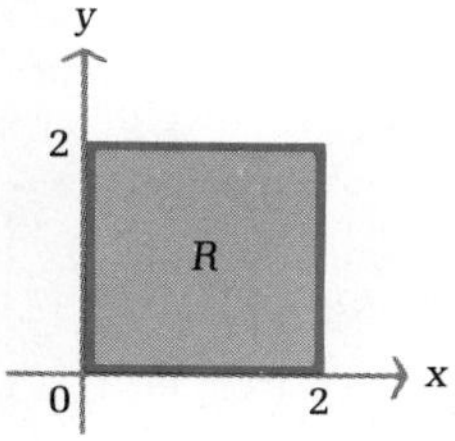

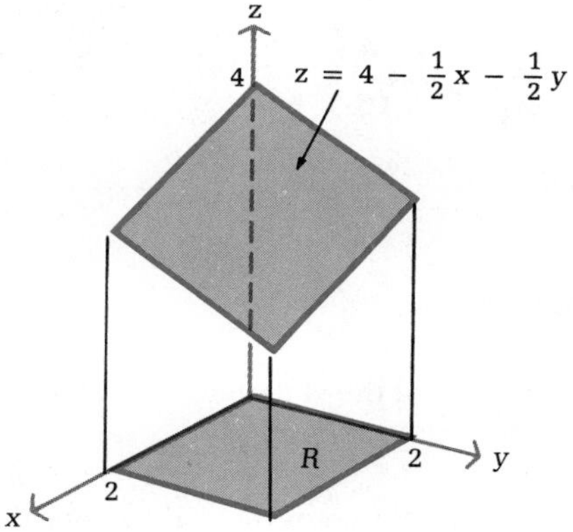

◆

PROBLEM 26 Find the average value of $f(x, y) = x + 2y$ over the rectangle $R = \{(x, y) | 0 \le x \le 2,\quad 0 \le y \le 1\}$. ◆

◆ VOLUME AND DOUBLE INTEGRALS

One application of the definite integral of a function with one variable is the calculation of areas, so it is not surprising that the definite integral of a function of two variables can be used to calculate volumes of solids.

Volume under a Surface

If $f(x, y) \geq 0$ over a rectangle $R = \{(x, y) | a \leq x \leq b, \quad c \leq y \leq d\}$, then the volume of the solid formed by graphing f over the rectangle R is given by

$$V = \iint_R f(x, y)\, dA$$

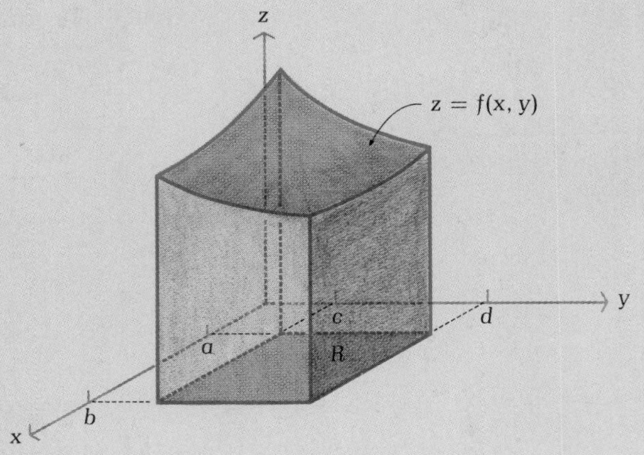

A proof of the statement in the box is left to a more advanced text.

◆ EXAMPLE 27 Find the volume of the solid under the graph of $f(x, y) = 1 + x^2 + y^2$ over the rectangle $R = \{(x, y) | 0 \leq x \leq 1, \quad 0 \leq y \leq 1\}$.

Solution

$$V = \iint_R (1 + x^2 + y^2)\, dA$$

$$= \int_0^1 \int_0^1 (1 + x^2 + y^2)\, dx\, dy$$

$$= \int_0^1 \left[(x + \tfrac{1}{3}x^3 + xy^2) \Big|_{x=0}^{x=1} \right] dy$$

$$= \int_0^1 (\tfrac{4}{3} + y^2)\, dy$$

$$= (\tfrac{4}{3}y + \tfrac{1}{3}y^3) \Big|_{y=0}^{y=1} = \tfrac{5}{3} \text{ cubic units}$$

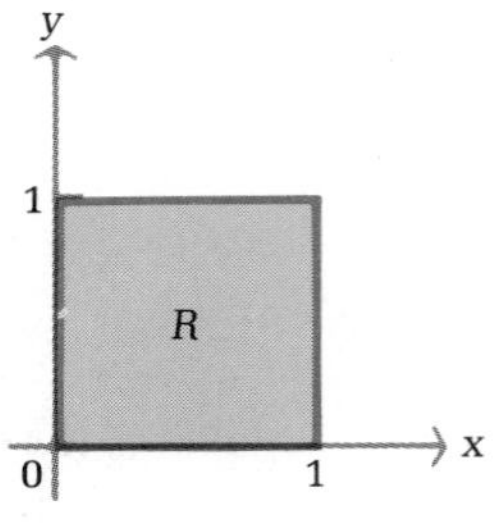

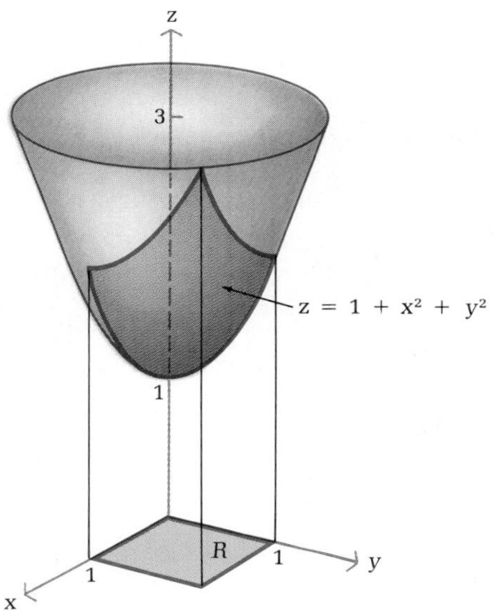

◆

PROBLEM 27 Find the volume of the solid under the graph of $f(x, y) = 1 + x + y$ over the rectangle $R = \{(x, y) | 0 \le x \le 1, \ 0 \le y \le 2\}$. ◆

Answers to Matched Problems

21. (A) $2xy^2 + 3x^2y^4 + C(x)$ (B) $2x^2y + 4x^3y^3 + E(y)$
22. (A) $2x + 3x^2$ (B) $18y + 108y^3$ 23. (A) 36 (B) 36
24. 12 25. $e - 2e^{1/2} + 1$ 26. 2 27. 5 cubic units

EXERCISE 9-6

A *Find each antiderivative. Then use the antiderivative to evaluate the definite integral.*

1. (A) $\int 12x^2y^3\,dy$ (B) $\int_0^1 12x^2y^3\,dy$

2. (A) $\int 12x^2y^3\,dx$ (B) $\int_{-1}^{2} 12x^2y^3\,dx$

3. (A) $\int (4x + 6y + 5)\,dx$ (B) $\int_{-2}^{3} (4x + 6y + 5)\,dx$

4. (A) $\int (4x + 6y + 5)\,dy$ (B) $\int_1^4 (4x + 6y + 5)\,dy$

5. (A) $\int \frac{x}{\sqrt{y + x^2}}\,dx$ (B) $\int_0^2 \frac{x}{\sqrt{y + x^2}}\,dx$

6. (A) $\int \frac{x}{\sqrt{y + x^2}}\,dy$ (B) $\int_1^5 \frac{x}{\sqrt{y + x^2}}\,dy$

B *Evaluate each iterated integral. (See the indicated problem for the evaluation of the inner integral.)*

7. $\int_{-1}^{2}\int_0^1 12x^2y^3\,dy\,dx$

(see Problem 1)

8. $\int_0^1\int_{-1}^{2} 12x^2y^3\,dx\,dy$

(see Problem 2)

9. $\int_1^4\int_{-2}^{3} (4x + 6y + 5)\,dx\,dy$

(see Problem 3)

10. $\int_{-2}^{3}\int_1^4 (4x + 6y + 5)\,dy\,dx$

(see Problem 4)

11. $\int_1^5\int_0^2 \frac{x}{\sqrt{y + x^2}}\,dx\,dy$

(see Problem 5)

12. $\int_0^2\int_1^5 \frac{x}{\sqrt{y + x^2}}\,dy\,dx$

(see Problem 6)

Use both orders of iteration to evaluate each double integral.

13. $\iint_R xy\,dA$; $R = \{(x, y)|0 \leq x \leq 2,\ \ 0 \leq y \leq 4\}$

14. $\iint_R \sqrt{xy}\,dA$; $R = \{(x, y)|1 \leq x \leq 4,\ \ 1 \leq y \leq 9\}$

15. $\iint_R (x + y)^5\,dA$; $R = \{(x, y)|-1 \leq x \leq 1,\ \ 1 \leq y \leq 2\}$

16. $\iint_R xe^y\,dA$; $R = \{(x, y)|-2 \leq x \leq 3,\ \ 0 \leq y \leq 2\}$

Find the average value of each function over the given rectangle.

17. $f(x, y) = (x + y)^2$; $R = \{(x, y)|1 \leq x \leq 5,\ \ -1 \leq y \leq 1\}$
18. $f(x, y) = x^2 + y^2$; $R = \{(x, y)|-1 \leq x \leq 2,\ \ 1 \leq y \leq 4\}$
19. $f(x, y) = x/y$; $R = \{(x, y)|1 \leq x \leq 4,\ \ 2 \leq y \leq 7\}$
20. $f(x, y) = x^2y^3$; $R = \{(x, y)|-1 \leq x \leq 1,\ \ 0 \leq y \leq 2\}$

Find the volume of the solid under the graph of each function over the given rectangle.

21. $f(x, y) = 2 - x^2 - y^2$; $R = \{(x, y)|0 \leq x \leq 1,\ \ 0 \leq y \leq 1\}$
22. $f(x, y) = 5 - x$; $R = \{(x, y)|0 \leq x \leq 5,\ \ 0 \leq y \leq 5\}$
23. $f(x, y) = 4 - y^2$; $R = \{(x, y)|0 \leq x \leq 2,\ \ 0 \leq y \leq 2\}$
24. $f(x, y) = e^{-x-y}$; $R = \{(x, y)|0 \leq x \leq 1,\ \ 0 \leq y \leq 1\}$

C *Evaluate each double integral. Select the order of integration carefully—each problem is easy to do one way and difficult the other.*

25. $\iint_R xe^{xy}\,dA; \quad R = \{(x, y)|0 \leq x \leq 1, \quad 1 \leq y \leq 2\}$

26. $\iint_R xye^{x^2y}\,dA; \quad R = \{(x, y)|0 \leq x \leq 1, \quad 1 \leq y \leq 2\}$

27. $\iint_R \frac{2y + 3xy^2}{1 + x^2}\,dA; \quad R = \{(x, y)|0 \leq x \leq 1, \quad -1 \leq y \leq 1\}$

28. $\iint_R \frac{2x + 2y}{1 + 4y + y^2}\,dA; \quad R = \{(x, y)|1 \leq x \leq 3, \quad 0 \leq y \leq 1\}$

APPLICATIONS

Business & Economics

29. *Multiplier principle.* Suppose Congress enacts a one-time-only 10% tax rebate that is expected to infuse \$y billion, $5 \leq y \leq 7$, into the economy. If every individual and corporation is expected to spend a proportion x, $0.6 \leq x \leq 0.8$, of each dollar received, then by the **multiplier principle** in economics, the total amount of spending S (in billions of dollars) generated by this tax rebate is given by

$$S(x, y) = \frac{y}{1 - x}$$

What is the average total amount of spending for the indicated ranges of the values of x and y? Set up a double integral and evaluate.

30. *Multiplier principle.* Repeat Problem 29 if $6 \leq y \leq 10$ and $0.7 \leq x \leq 0.9$.

31. *Cobb–Douglas production function.* If an industry invests x thousand labor-hours, $10 \leq x \leq 20$, and \$y million, $1 \leq y \leq 2$, in the production of N thousand units of a certain item, then N is given by

$$N(x, y) = x^{0.75}y^{0.25}$$

What is the average number of units produced for the indicated ranges of x and y? Set up a double integral and evaluate.

32. *Cobb–Douglas production function.* Repeat Problem 31 for

$$N(x, y) = x^{0.5}y^{0.5}$$

where $10 \leq x \leq 30$ and $1 \leq y \leq 3$.

Life Sciences

33. *Population distribution.* In order to study the population distribution of a certain species of insects, a biologist has constructed an artificial habitat in the shape of a rectangle 16 feet long and 12 feet wide. The only food available to the insects in this habitat is located at its center. The biologist has determined that the concentration C of insects per square foot at a point d units from the food supply (see the figure) is given approximately by

$$C = 10 - \tfrac{1}{10}d^2$$

What is the average concentration of insects throughout the habitat? Express C as a function of x and y, set up a double integral, and evaluate.

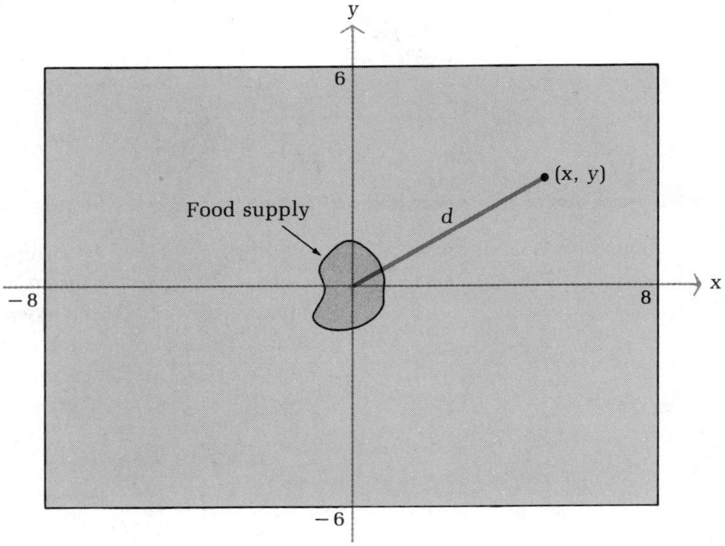

34. *Population distribution.* Repeat Problem 33 for a square habitat that measures 12 feet on each side, where the insect concentration is given by

$$C = 8 - \tfrac{1}{10}d^2$$

35. *Pollution.* A heavy industrial plant located in the center of a small town emits particulate matter into the atmosphere. Suppose the concentration of particulate matter (in parts per million) at a point d miles from the plant (see the figure at the top of the next page) is given by

$$C = 100 - 15d^2$$

If the boundaries of the town form a rectangle 4 miles long and 2 miles wide, what is the average concentration of particulate matter throughout the city? Express C as a function of x and y, set up a double integral, and evaluate.

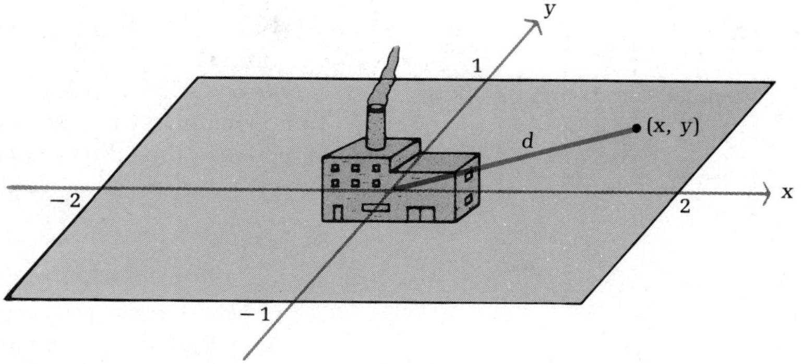

36. *Pollution.* Repeat Problem 35 if the boundaries of the town form a rectangle 8 miles long and 4 miles wide and the concentration of particulate matter is given by

$$C = 100 - 3d^2$$

Social Sciences

37. *Safety research.* Under ideal conditions, if a person driving a car slams on the brakes and skids to a stop, the length of the skid marks (in feet) is given by the formula

$$L = 0.000\ 013\ 3xy^2$$

where x is the weight of the car (in pounds) and y is the speed of the car (in miles per hour). What is the average length of the skid marks for cars weighing between 2,000 and 3,000 pounds and traveling at speeds between 50 and 60 miles per hour? Set up a double integral and evaluate.

38. *Safety research.* Repeat Problem 37 for cars weighing between 2,000 and 2,500 pounds and traveling at speeds between 40 and 50 miles per hour.

39. *Psychology.* The intelligence quotient Q for an individual with mental age x and chronological age y is given by

$$Q(x, y) = 100\,\frac{x}{y}$$

In a group of sixth graders, the mental age varies between 8 and 16 years and the chronological age varies between 10 and 12 years. What is the average intelligence quotient for this group? Set up a double integral and evaluate.

40. *Psychology.* Repeat Problem 39 for a group with mental ages between 6 and 14 years and chronological ages between 8 and 10 years.

SECTION 9-7 Chapter Review

Important Terms and Symbols

9-1 *Functions of Several Variables.* Functions of two independent variables; functions of several independent variables; Cobb–Douglas production function; three-dimensional coordinate system; triplet of numbers (x, y, z); surface; paraboloid; saddle point

$$z = f(x, y); \quad w = f(x, y, z)$$

9-2 *Partial Derivatives.* Partial derivative of f with respect to x; partial derivative of f with respect to y; marginal productivity of labor; marginal productivity of capital; second-order partial derivatives

$$\frac{\partial z}{\partial x}; \quad \frac{\partial z}{\partial y}; \quad f_x(x, y); \quad f_y(x, y); \quad \frac{\partial^2 z}{\partial x^2} = f_{xx}(x, y); \quad \frac{\partial^2 z}{\partial x\, \partial y} = f_{yx}(x, y);$$

$$\frac{\partial^2 z}{\partial y\, \partial x} = f_{xy}(x, y); \quad \frac{\partial^2 z}{\partial y^2} = f_{yy}(x, y)$$

9-3 *Maxima and Minima.* Local maximum; local minimum; saddle point; critical point; second-derivative test

9-4 *Maxima and Minima Using Lagrange Multipliers.* Constraint; Lagrange multiplier; critical point; method of Lagrange multipliers for functions of two variables; marginal productivity of money; method of Lagrange multipliers for functions of three variables

9-5 *Method of Least Squares.* Least squares approximation; linear regression; residual; least squares line; regression line; normal equations

9-6 *Double Integrals over Rectangular Regions.* Double integral; integrand; region of integration; iterated integral; average value over rectangular region; volume under a surface

$$\iint_R f(x, y)\, dA = \int_a^b \left[\int_c^d f(x, y)\, dy \right] dx = \int_c^d \left[\int_a^b f(x, y)\, dx \right] dy$$

$$\frac{1}{(b - a)(d - c)} \iint_R f(x, y)\, dA$$

EXERCISE 9-7 Chapter Review

Work through all the problems in this chapter review and check your answers in the back of the book. (Answers to all review problems are there.) Where weaknesses show up, review appropriate sections in the text.

A

1. For $f(x, y) = 2{,}000 + 40x + 70y$, find $f(5, 10)$, $f_x(x, y)$, and $f_y(x, y)$.
2. For $z = x^3y^2$, find $\partial^2 z/\partial x^2$ and $\partial^2 z/\partial x\, \partial y$.

3. Evaluate: $\int (6xy^2 + 4y)\,dy$ 4. Evaluate: $\int (6xy^2 + 4y)\,dx$

5. Evaluate: $\int_0^1 \int_0^1 4xy\,dy\,dx$

B

6. For $f(x, y) = 3x^2 - 2xy + y^2 - 2x + 3y - 7$, find $f(2, 3)$, $f_y(x, y)$, and $f_y(2, 3)$.
7. For $f(x, y) = -4x^2 + 4xy - 3y^2 + 4x + 10y + 81$, find

$$[f_{xx}(2, 3)][f_{yy}(2, 3)] - [f_{xy}(2, 3)]^2$$

8. If $f(x, y) = x + 3y$ and $g(x, y) = x^2 + y^2 - 10$, find the critical points of $F(x, y, \lambda) = f(x, y) + \lambda g(x, y)$.
9. Use the least squares line for the data in the table to estimate y when $x = 10$.

x	y
2	12
4	10
6	7
8	3

10. For $R = \{(x, y) \mid -1 \le x \le 1,\ \ 1 \le y \le 2\}$, evaluate the following in two ways:

$$\iint_R (4x + 6y)\,dA$$

C

11. For $f(x, y) = e^{x^2+2y}$, find f_x, f_y, and f_{xy}.
12. For $f(x, y) = (x^2 + y^2)^5$, find f_x and f_{xy}.
13. Find all critical points and test for extrema for

$$f(x, y) = x^3 - 12x + y^2 - 6y$$

14. Use Lagrange multipliers to maximize $f(x, y) = xy$ subject to $2x + 3y = 24$.
15. Use Lagrange multipliers to minimize $f(x, y, z) = x^2 + y^2 + z^2$ subject to $2x + y + 2z = 9$.
16. Find the least squares line for the data in the table in the margin.

x	y	x	y
10	50	60	80
20	45	70	85
30	50	80	90
40	55	90	90
50	65	100	110

17. Find the average value of $f(x, y) = x^{2/3}y^{1/3}$ over the rectangle

$$R = \{(x, y) \mid -8 \le x \le 8,\ \ 0 \le y \le 27\}$$

18. Find the volume of the solid under the graph of $z = 3x^2 + 3y^2$ over the rectangle

$$R = \{(x, y) \mid 0 \le x \le 1,\ \ -1 \le y \le 1\}$$

APPLICATIONS

Business & Economics

19. *Maximizing profit.* A company produces x units of product A and y units of product B (both in hundreds per month). The monthly profit equation (in thousands of dollars) is found to be

$$P(x, y) = -4x^2 + 4xy - 3y^2 + 4x + 10y + 81$$

(A) Find $P_x(1,3)$ and interpret the results.
(B) How many of each product should be produced each month to maximize profit? What is the maximum profit?

20. *Minimizing material.* A rectangular box with no top and six compartments (see the figure) is to have a volume of 96 cubic inches. Find the dimensions that will require the least amount of material.

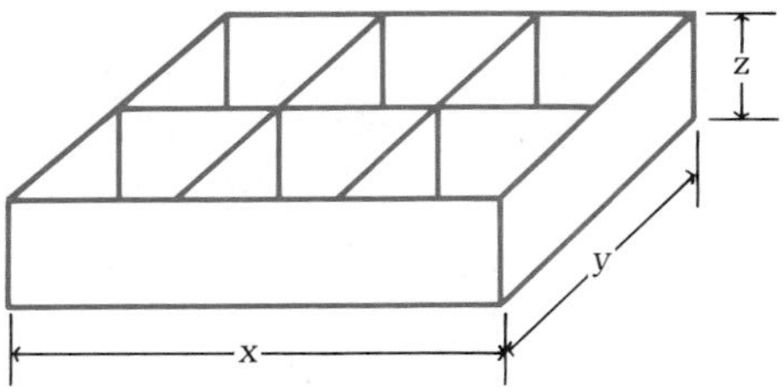

YEAR	PROFIT
1	2
2	2.5
3	3.1
4	4.2
5	4.3

21. *Profit.* A company's annual profits (in millions of dollars) over a 5 year period are given in the table. Use the least squares line to estimate the profit for the sixth year.

22. *Productivity.* The Cobb–Douglas production function for a product is

$$N(x, y) = 10x^{0.8}y^{0.2}$$

where x is the number of units of labor and y is the number of units of capital required to produce N units of the product.

(A) Find the marginal productivity of labor and the marginal productivity of capital at $x = 40$ and $y = 50$. For the greatest increase in productivity, should management encourage increased use of labor or increased use of capital?

(B) If each unit of labor costs $100, each unit of capital costs $50, and $10,000 is budgeted for production of this product, use the method of Lagrange multipliers to determine the allocations of labor and capital that will maximize the number of units produced and find the maximum production. Find the marginal productivity of money and approximate the increase in production that would result from an increase of $2,000 in the amount budgeted for production.

(C) If $50 \leq x \leq 100$ and $20 \leq y \leq 40$, find the average number of units produced. Set up a definite integral and evaluate.

Life Sciences

23. *Marine biology.* The function used for timing dives with scuba gear is

$$T(V, x) = \frac{33V}{x + 33}$$

where T is the time of the dive in minutes, V is the volume of air (at sea level pressure) compressed into tanks, and x is the depth of the dive in feet. Find $T_x(70, 17)$ and interpret the results.

24. *Pollution.* A heavy industrial plant located in the center of a small town emits particulate matter into the atmosphere. Suppose the concentration

of particulate matter (in parts per million) at a point d miles from the plant is given by

$$C = 100 - 24d^2$$

If the boundaries of the town form a square 4 miles long and 4 miles wide, what is the average concentration of particulate matter throughout the city? Express C as a function of x and y, set up a double integral, and evaluate.

Social Sciences

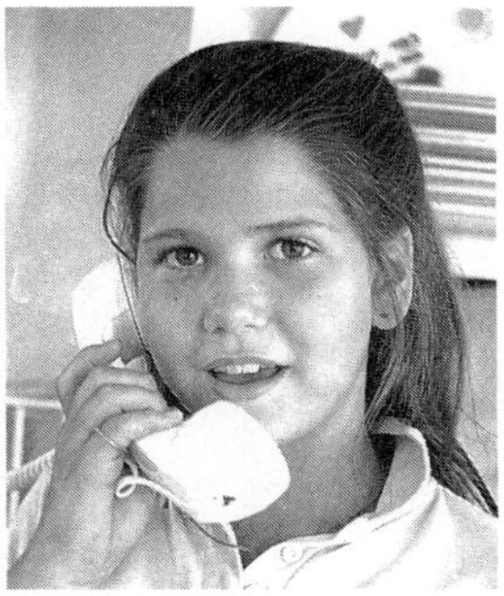

25. *Sociology.* Joseph Cavanaugh, a sociologist, found that the number of long-distance telephone calls, n, between two cities in a given period of time varied (approximately) jointly as the populations P_1 and P_2 of the two cities, and varied inversely as the distance, d, between the two cities. In terms of an equation for a time period of 1 week,

$$n(P_1, P_2, d) = 0.001 \frac{P_1 P_2}{d}$$

Find $n(100{,}000, 50{,}000, 100)$.

26. *Education.* At the beginning of the semester, students in a foreign language course are given a proficiency exam. The same exam is given at the end of the semester. The results for 5 students are given in the table. Use the least squares line to estimate the score on the second exam for a student who scored 40 on the first exam.

FIRST EXAM	SECOND EXAM
30	60
50	75
60	80
70	85
90	90

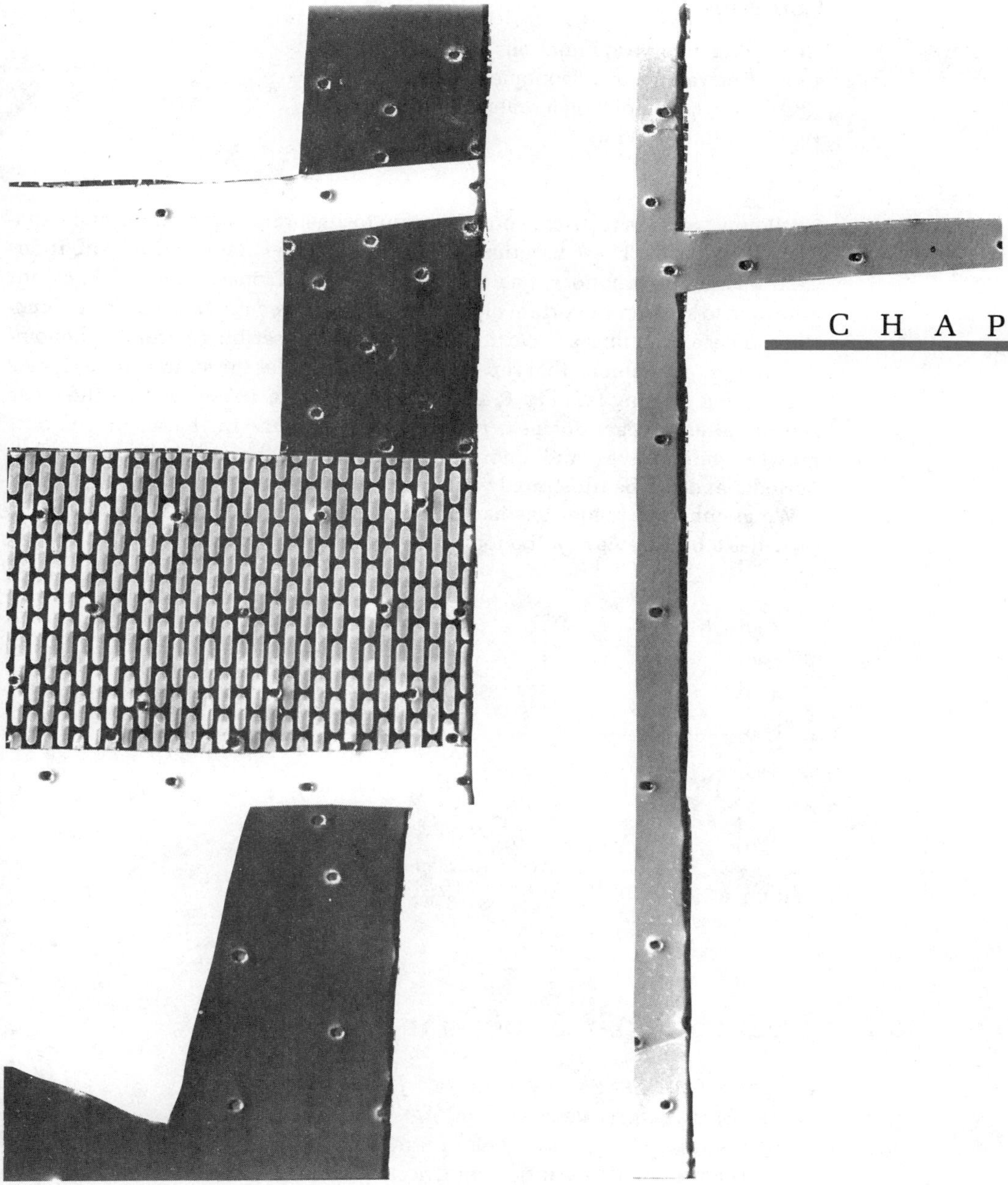

CHAPTER 10

Trigonometric Functions

Contents

CHAPTER 10

Until now we have restricted our attention to algebraic, logarithmic, and exponential functions. These functions were used to model many real-life situations from business, economics, and the life and social sciences. Now we turn our attention to another important class of functions, called the **trigonometric functions.** These functions are particularly useful in describing periodic phenomena; that is, phenomena that repeat in cycles. Consider the sunrise times for a 2 year period starting January 1, as pictured in Figure 1. We see that the cycle repeats after 1 year. Business cycles, blood pressure in the aorta, seasonal growth, water waves, and amounts of pollution in the atmosphere are often periodic and can be illustrated with similar types of graphs.

We assume the reader has had a course in trigonometry. The next section provides a brief review of the topics that are most important for our purposes.

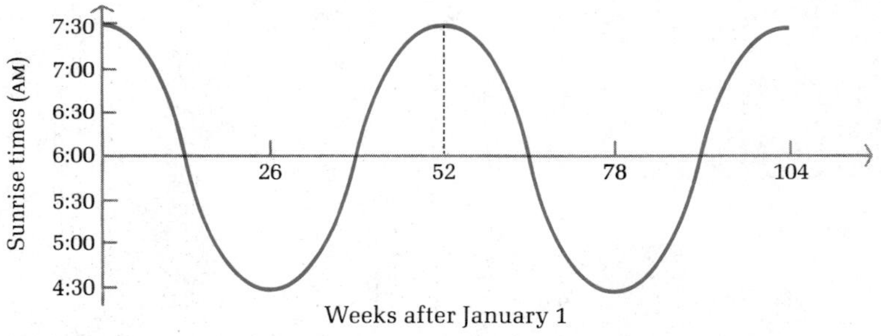

FIGURE 1
Sunrise times

SECTION 10-1 Trigonometric Functions Review

- ANGLES; DEGREE–RADIAN MEASURE
- TRIGONOMETRIC FUNCTIONS
- GRAPHS OF THE SINE AND COSINE FUNCTIONS
- FOUR OTHER TRIGONOMETRIC FUNCTIONS

◆ ANGLES; DEGREE–RADIAN MEASURE

We start our discussion of trigonometry with the concept of *angle*. A point P on a line divides the line into two parts. Either part together with the end point is

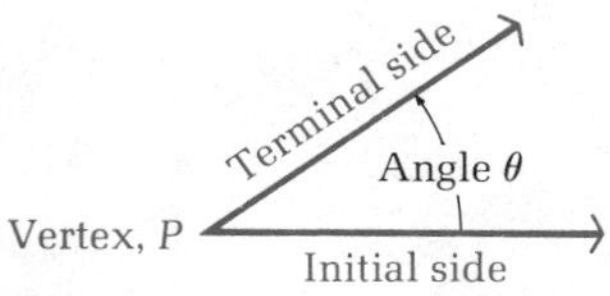

FIGURE 2
Angle θ

called a **half-line.** A geometric figure consisting of two half-lines with a common end point is called an **angle.** One of the half-lines is called the **initial side** and the other half-line is called the **terminal side.** The common end point is called the **vertex.** Figure 2 illustrates these concepts.

There are two widely used measures of angles—the *degree* and the *radian.* A central angle of a circle subtended by an arc $\frac{1}{360}$ of the circumference of the circle is said to have **degree measure 1,** written **1°** (see Fig. 3A). It follows that a central angle subtended by an arc $\frac{1}{4}$ the circumference has degree measure 90; $\frac{1}{2}$ the circumference has degree measure 180; and the whole circumference of a circle has degree measure 360.

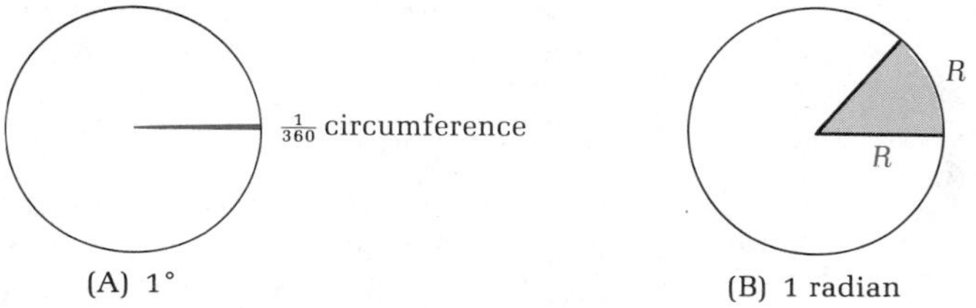

FIGURE 3
Degree and radian measure

The other measure of angles, which we will use extensively in the next two sections, is radian measure. A central angle subtended by an arc of length equal to the radius (R) of the circle is said to have **radian measure 1,** written **1 radian** (see Fig. 3B). In general, a central angle subtended by an arc of length s has radian measure determined as follows:

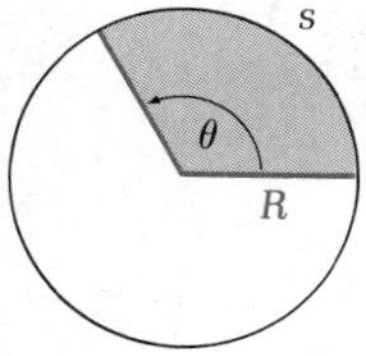

FIGURE 4

$$\theta^r = \text{Radian measure of } \theta = \frac{\text{Arc length}}{\text{Radius}} = \frac{s}{R}$$

See Figure 4. [*Note:* If $R = 1$, then $\theta^r = s$.]

What is the radian measure of an angle of 180°? A central angle of 180° is subtended by an arc of $\frac{1}{2}$ the circumference of a circle. Thus,

$$s = \frac{C}{2} = \frac{2\pi R}{2} = \pi R \qquad \text{and} \qquad \theta^r = \frac{s}{R} = \frac{\pi R}{R} = \pi \text{ radians}$$

The following proportion can be used to convert degree measure to radian measure and vice versa:

Degree–Radian Conversion Formula

$$\frac{\theta^\circ}{180^\circ} = \frac{\theta^r}{\pi \text{ radians}}$$

◆ EXAMPLE 1 Find the radian measure of 1°.

Solution

$$\frac{1°}{180°} = \frac{\theta^r}{\pi \text{ radians}}$$

$$\theta^r = \frac{\pi}{180} \text{ radian} \approx 0.0175 \text{ radian}$$ ◆

PROBLEM 1 Find the degree measure of 1 radian. ◆

A comparison of degree and radian measure for a few important angles is given in the following table:

RADIAN	0	$\pi/6$	$\pi/4$	$\pi/3$	$\pi/2$	π	2π
DEGREE	0	30°	45°	60°	90°	180°	360°

Before defining trigonometric functions, we will generalize the notion of angle defined above. Starting with a rectangular coordinate system and two half-lines coinciding with the nonnegative x axis, the initial side of the angle remains fixed and the terminal side rotates until it reaches its terminal position. When the terminal side is rotated counterclockwise, the angle formed is considered positive (see Figs. 5A and 5C). When it is rotated clockwise, the angle formed is considered negative (see Fig. 5B). Angles located in a coordinate system in this manner are said to be in a **standard position.**

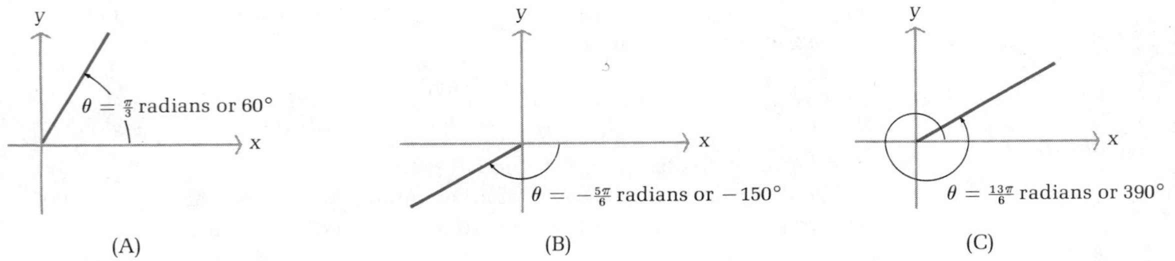

FIGURE 5
Generalized angles

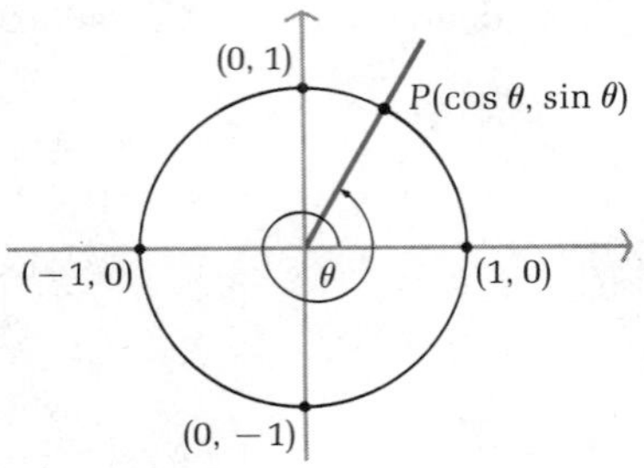

FIGURE 6

◆ TRIGONOMETRIC FUNCTIONS

Let us locate a unit circle (radius 1) in a coordinate system with center at the origin (Fig. 6). The terminal side of any angle in standard position will pass through this circle at some point P. The abscissa of this point P is called the **cosine of θ** (abbreviated **cos θ**), and the ordinate of the point is the **sine of θ** (abbreviated **sin θ**). Thus, the set of all ordered pairs of the form $(\theta, \cos\theta)$, and the set of all ordered pairs of the form $(\theta, \sin\theta)$ constitute, respectively, the **cosine** and **sine functions.** The **domain** of these two functions is the set of all angles, positive or negative, with measure either in degrees or radians. The **range** is a subset of the set of real numbers.

It is desirable, and necessary for our work in calculus, to define these two trigonometric functions in terms of real number domains. This is easily done as follows:

Sine and Cosine Functions with Real Number Domains

For any real number x,

$$\sin x = \sin(x \text{ radians}) \qquad \cos x = \cos(x \text{ radians})$$

◆ EXAMPLE 2 Referring to Figure 6, find:

(A) $\cos 90°$ (B) $\sin(-\pi/2 \text{ radians})$ (C) $\cos \pi$

Solutions

(A) The terminal side of an angle of degree measure 90 passes through (0, 1) on the unit circle. This point has abscissa 0. Thus,

$$\cos 90° = 0$$

(B) The terminal side of an angle of radian measure $-\pi/2$ ($-90°$) passes through $(0, -1)$ on the unit circle. This point has ordinate -1. Thus,

$$\sin\left(-\frac{\pi}{2} \text{ radians}\right) = -1$$

(C) $\cos \pi = \cos(\pi \text{ radians}) = -1$, since the terminal side of an angle of radian measure π (180°) passes through $(-1, 0)$ on the unit circle and this point has abscissa -1. ◆

PROBLEM 2 Referring to Figure 6, find:

(A) $\sin 180°$ (B) $\cos(2\pi \text{ radians})$ (C) $\sin(-\pi)$ ◆

To find the value of either the sine or the cosine function for any angle or any real number by direct use of the definition is not easy. Tables are available, but calculators with [sin] and [cos] keys are even more convenient. Calculators generally have degree and radian options, so we can use a calculator to evaluate these functions for most of the real numbers in which we might have an interest. The following table includes a few values produced by a calculator in the radian mode. The x value is entered, and then the [sin] or [cos] key is pressed to obtain the desired value in display.

x	1	−7	35.26	−105.9
sin x	0.8415	−0.6570	−0.6461	0.7920
cos x	0.5403	0.7539	−0.7632	0.6105

Exact values of the sine and cosine functions can be obtained for multiples of the special angles shown in the triangles in Figure 7, since these triangles can be used to find the coordinate of the intersection of the terminal side of each angle with the unit circle.

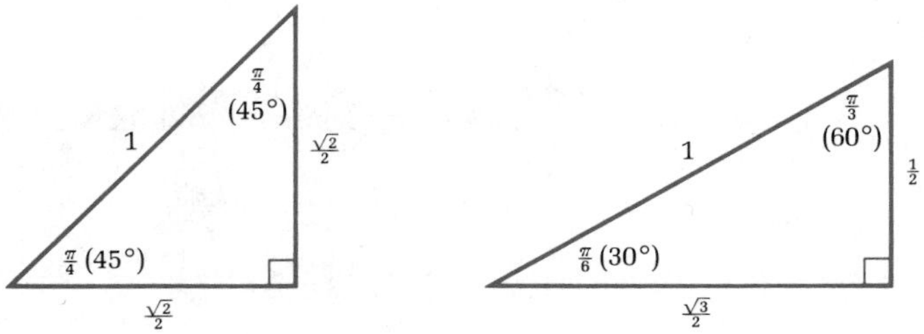

FIGURE 7

Remark

We now drop the word "radian" after $\pi/4$ and interpret $\pi/4$ as the radian measure of an angle or simply as a real number, depending on the context.

◆ EXAMPLE 3 Find the exact value of each of the following using Figure 7:

(A) $\cos\frac{\pi}{4}$ (B) $\sin\frac{\pi}{6}$ (C) $\sin\left(-\frac{\pi}{6}\right)$

Solutions (A) $\cos\frac{\pi}{4}=\frac{\sqrt{2}}{2}$ (B) $\sin\frac{\pi}{6}=\frac{1}{2}$

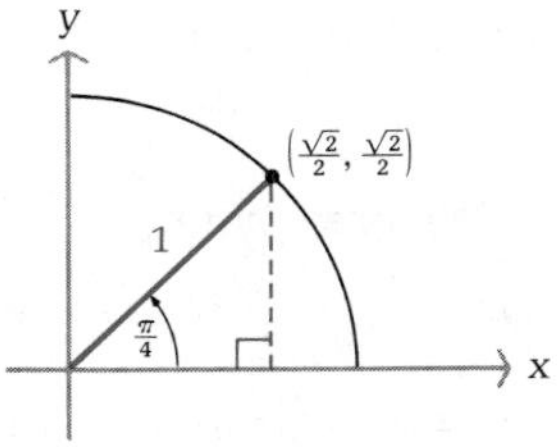

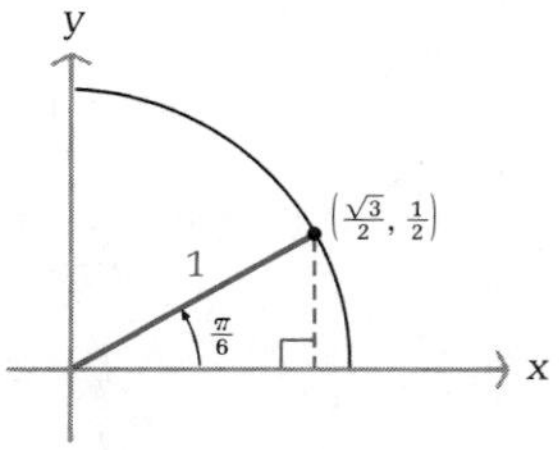

(C) $\sin\left(-\frac{\pi}{6}\right)=-\frac{1}{2}$

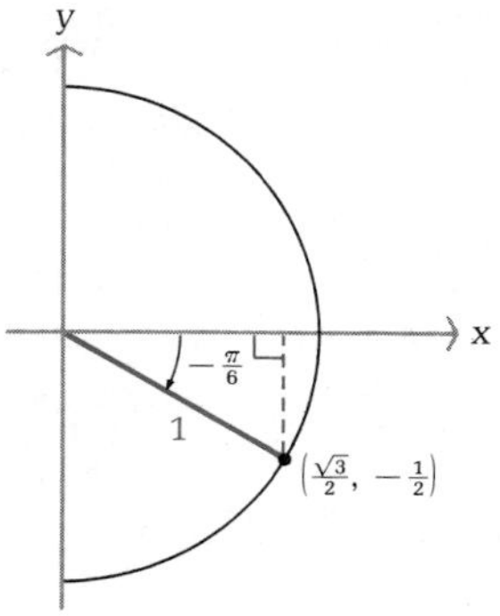

◆

PROBLEM 3 Find the exact value of each of the following using Figure 7:

(A) $\sin \frac{\pi}{4}$ (B) $\cos \frac{\pi}{3}$ (C) $\cos\left(-\frac{\pi}{3}\right)$ ◆

◆ GRAPHS OF THE SINE AND COSINE FUNCTIONS

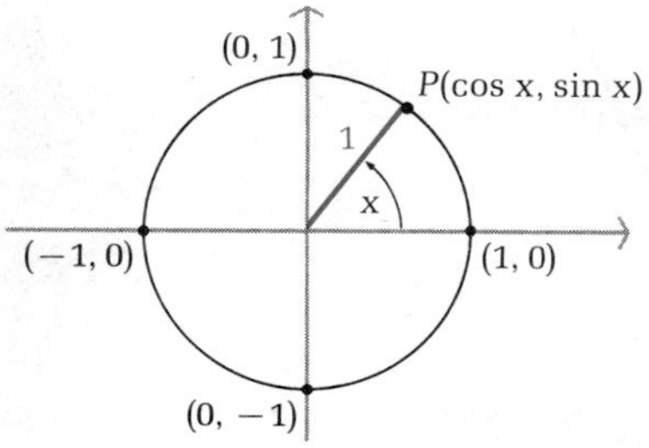

FIGURE 8

To graph $y = \sin x$ or $y = \cos x$ for x a real number, we could use a calculator to produce a table, and then plot the ordered pairs from the table in a coordinate system. However, we can speed up the process by returning to basic definitions. Referring to Figure 8, since cos x and sin x are the coordinates of a point on the unit circle, we see that

$$-1 \leq \sin x \leq 1 \qquad \text{and} \qquad -1 \leq \cos x \leq 1$$

for all real numbers x. Furthermore, as x increases and P moves around the unit circle in a counterclockwise (positive) direction, both sin x and cos x behave in uniform ways.

As x increases from	$y = \sin x$	$y = \cos x$
0 to $\pi/2$	Increases from 0 to 1	Decreases from 1 to 0
$\pi/2$ to π	Decreases from 1 to 0	Decreases from 0 to −1
π to $3\pi/2$	Decreases from 0 to −1	Increases from −1 to 0
$3\pi/2$ to 2π	Increases from −1 to 0	Increases from 0 to 1

Note that P has completed one revolution and is back at its starting place. If we let x continue to increase, then the second and third columns in the table will be repeated every 2π units. In general, it can be shown that

$$\sin(x + 2\pi) = \sin x \qquad \cos(x + 2\pi) = \cos x$$

for all real numbers x. Functions such that

$$f(x + p) = f(x)$$

for some positive constant p and all real numbers x for which the functions are defined are said to be **periodic.** The smallest such value of p is called the **period** of the function. Thus, both the sine and cosine functions are periodic (a very important property) with period 2π.

Putting all this information together, and, perhaps adding a few values obtained from a calculator or Figure 7, we obtain the graphs of the sine and cosine functions illustrated in Figure 9 (page 564). Notice that these curves are continuous. It can be shown that **the sine and cosine functions are continuous for all real numbers.**

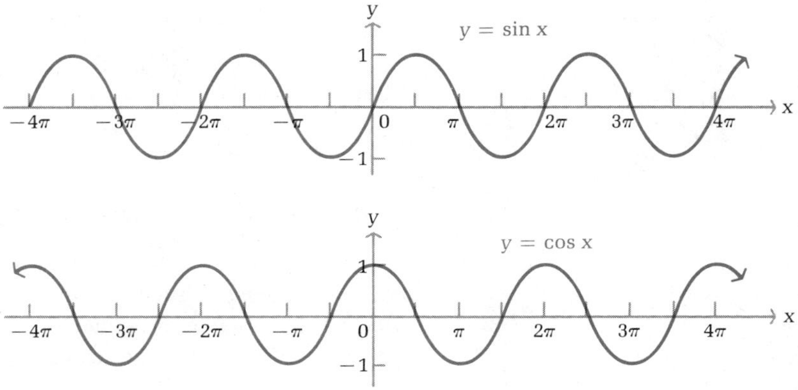

FIGURE 9

◆ FOUR OTHER TRIGONOMETRIC FUNCTIONS

The sine and cosine functions are only two of six trigonometric functions. They are, however, the most important of the six for many applications. We define the other four trigonometric functions below. Exercises involving these functions may be found in the exercise set that follows.

Four Other Trigonometric Functions

$$\tan x = \frac{\sin x}{\cos x} \qquad \cos x \neq 0 \qquad\qquad \sec x = \frac{1}{\cos x} \qquad \cos x \neq 0$$

$$\cot x = \frac{\cos x}{\sin x} \qquad \sin x \neq 0 \qquad\qquad \csc x = \frac{1}{\sin x} \qquad \sin x \neq 0$$

Answers to Matched Problems

1. $180/\pi \approx 57.3°$ 2. (A) 0 (B) 1 (C) 0
3. (A) $\sqrt{2}/2$ (B) $\frac{1}{2}$ (C) $\frac{1}{2}$

EXERCISE 10-1

A *Recall that 180° corresponds to π radians. Mentally convert each degree measure to radian measure in terms of π.*

1. 60° 2. 90° 3. 45° 4. 360° 5. 30° 6. 120°

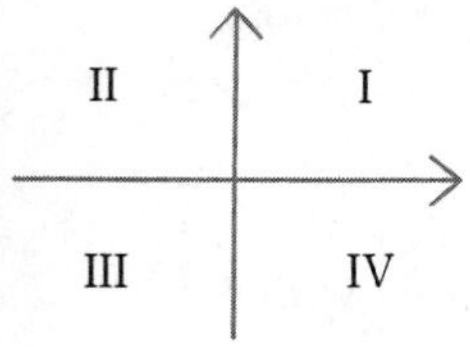

In Problems 7–12, indicate the quadrant in which the terminal side of each angle lies.

7. $150°$
8. $-190°$
9. $-\frac{\pi}{3}$ radians
10. $\frac{7\pi}{6}$ radians
11. $400°$
12. $-250°$

Use Figure 6 to find the exact value of each of the following:

13. $\cos 0°$
14. $\sin 90°$
15. $\sin \pi$
16. $\cos \frac{\pi}{2}$
17. $\cos(-\pi)$
18. $\sin \frac{3\pi}{2}$

B

Recall that π radians corresponds to 180°. Mentally convert each radian measure to degree measure.

19. $\frac{\pi}{3}$ radians
20. 2π radians
21. $\frac{\pi}{4}$ radian
22. $\frac{\pi}{2}$ radians
23. $\frac{\pi}{6}$ radian
24. $\frac{5\pi}{6}$ radians

Use Figure 7 to find the exact value of each of the following:

25. $\cos 30°$
26. $\sin(-45°)$
27. $\sin \frac{\pi}{6}$
28. $\sin\left(-\frac{\pi}{3}\right)$
29. $\cos \frac{5\pi}{6}$
30. $\cos(-120°)$

Use a calculator (set in radian mode) to find the value (to four decimal places) of each of the following:

31. $\sin 3$
32. $\cos 13$
33. $\cos 33.74$
34. $\sin 325.9$
35. $\sin(-43.06)$
36. $\cos(-502.3)$

C

Convert to radian measure.

37. $27°$
38. $18°$

Convert to degree measure.

39. $\frac{\pi}{12}$ radian
40. $\frac{\pi}{60}$ radian

Use Figure 7 to find the exact value of each of the following:

41. $\tan 45°$
42. $\cot 45°$
43. $\sec \frac{\pi}{3}$
44. $\csc \frac{\pi}{6}$
45. $\cot \frac{\pi}{3}$
46. $\tan \frac{\pi}{6}$

47. Refer to Figure 6 and use the Pythagorean theorem to show that

$$(\sin x)^2 + (\cos x)^2 = 1$$

for all x.

48. Use the results of Problem 47 and basic definitions to show that

(A) $(\tan x)^2 + 1 = (\sec x)^2$ (B) $1 + (\cot x)^2 = (\csc x)^2$

APPLICATIONS

Business & Economics

49. *Seasonal business cycle.* Suppose profits on the sale of swimming suits in a department store over a 2 year period are given approximately by

$$P(t) = 5 - 5\cos\frac{\pi t}{26} \qquad 0 \le t \le 104$$

where P is profit (in hundreds of dollars) for a week of sales t weeks after January 1. The graph of the profit function is shown in the figure.

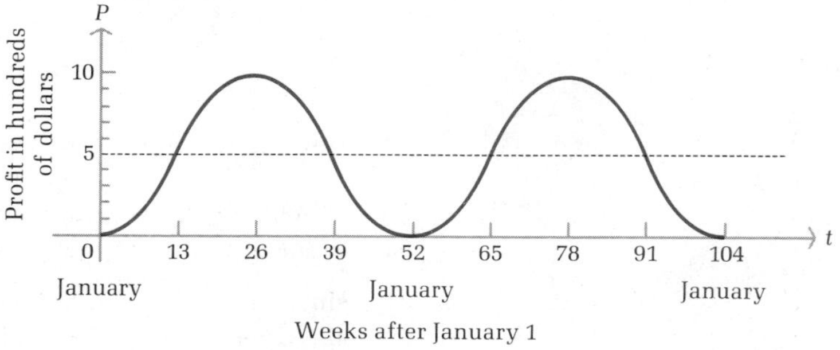

(A) Find the exact value of $P(13)$, $P(26)$, $P(39)$, and $P(52)$ by evaluating $P(t) = 5 - 5\cos(\pi t/26)$ without using a calculator.
(B) Use a calculator to find $P(30)$ and $P(100)$.

50. *Seasonal business cycle.* A soft drink company has revenues from sales over a 2 year period as given approximately by

$$R(t) = 4 - 3\cos\frac{\pi t}{6} \qquad 0 \le t \le 24$$

where $R(t)$ is revenue (in millions of dollars) for a month of sales t months after February 1. The graph of the revenue function is shown in the figure.

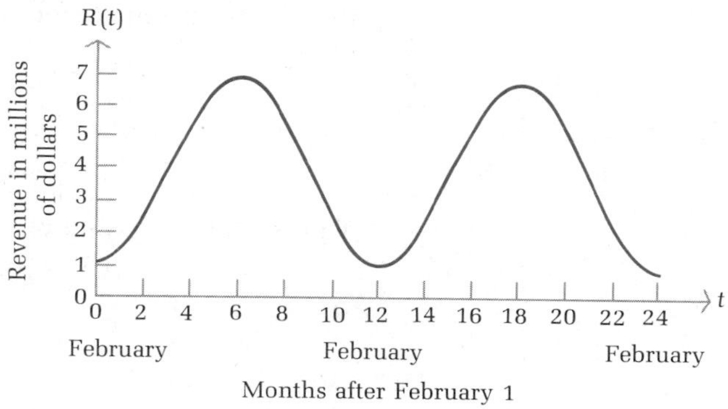

(A) Find the exact values of $R(0)$, $R(2)$, $R(3)$, and $R(18)$ without using a calculator.

(B) Use a calculator to find $R(5)$ and $R(23)$.

Life Sciences

51. *Physiology.* A normal seated adult breathes in and exhales about 0.8 liter of air every 4 seconds. The volume of air $V(t)$ in the lungs t seconds after exhaling is given approximately by

$$V(t) = 0.45 - 0.35 \cos \frac{\pi t}{2} \qquad 0 \leq t \leq 8$$

The graph for two complete respirations is shown in the figure.

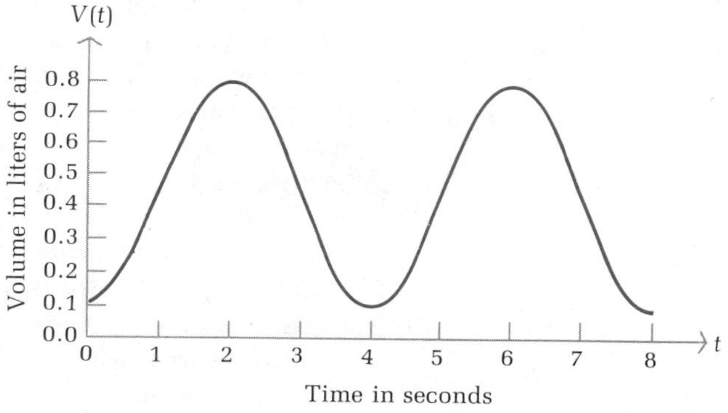

(A) Find the exact value of $V(0)$, $V(1)$, $V(2)$, $V(3)$, and $V(7)$ without using a calculator.

(B) Use a calculator to find $V(3.5)$ and $V(5.7)$.

52. *Pollution.* In a large city, the amount of sulfur dioxide pollutant released into the atmosphere due to the burning of coal and oil for heating purposes varies seasonally. Suppose the number of tons of pollutant released into the atmosphere during the nth week after January 1 is given approximately by

$$P(n) = 1 + \cos\frac{\pi n}{26} \qquad 0 \leq n \leq 104$$

The graph of the pollution function is shown in the figure.

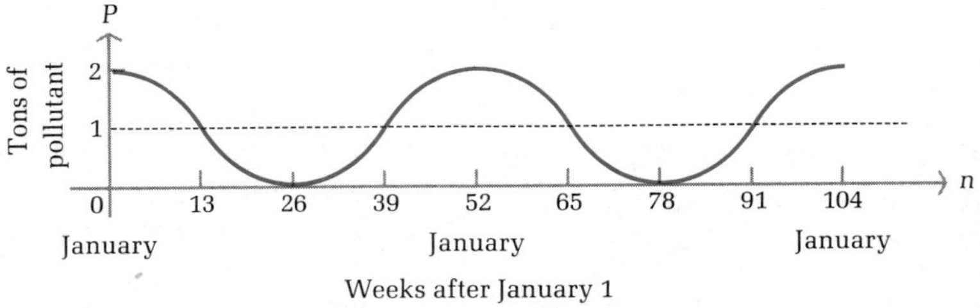

(A) Find the exact value of $P(0)$, $P(39)$, $P(52)$, and $P(65)$ by evaluating $P(n) = 1 + \cos(\pi n/26)$ without using a calculator.
(B) Use a calculator to find $P(10)$ and $P(95)$.

Social Sciences

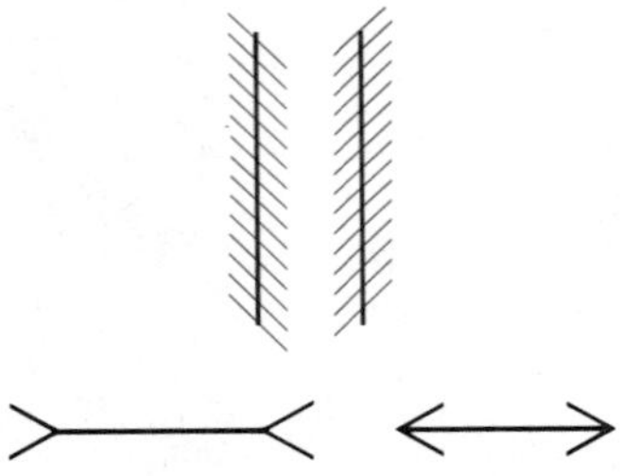

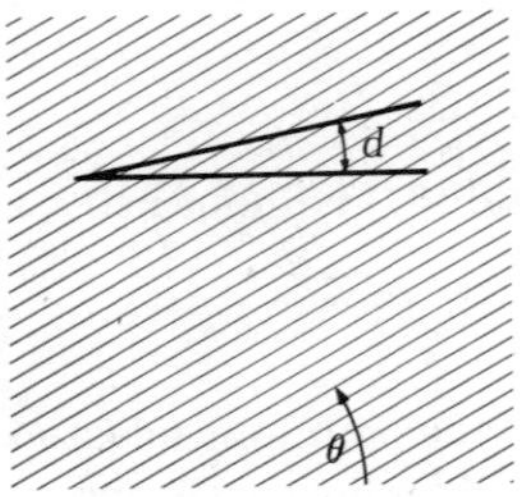

53. *Psychology—perception.* An important area of study in psychology is perception. Individuals perceive objects differently in different settings. Consider the well-known illusions shown in the top figure in the margin. Lines that appear parallel in one setting may appear to be curved in another (the two vertical lines are actually parallel). Lines of the same length may appear to be of different lengths in two different settings (the two horizontal lines are actually the same length). An interesting experiment in visual perception was conducted by psychologists Berliner and Berliner (*American Journal of Psychology*, 1952, 65:271–277). They reported that when subjects were presented with a large tilted field of parallel lines and were asked to estimate the position of a horizontal line in the field, most of the subjects were consistently off. They found that the difference in degrees, d, between the estimates and the actual horizontal could be approximated by the equation

$$d = a + b \sin 4\theta$$

where a and b are constants associated with a particular individual and θ is the angle of tilt of the visual field (in degrees). Suppose that for a given individual, $a = -2.1$ and $b = -4$. Find d if:

(A) $\theta = 30°$ (B) $\theta = 10°$

SECTION 10-2 Derivatives of Trigonometric Functions

- DERIVATIVE FORMULAS
- APPLICATION

DERIVATIVE FORMULAS

In this section, we will discuss derivative formulas for the sine and cosine functions. Once we have these formulas, we will automatically have integral formulas for the same functions, which we will discuss in the next section.

From the definition of derivative (Section 4-3),

$$D_x \sin x = \lim_{h\to 0} \frac{\sin(x+h) - \sin x}{h}$$

Using trigonometric identities and some special trigonometric limits, it can be shown that the limit on the right is $\cos x$. Similarly, it can be shown that $D_x \cos x = -\sin x$.

We now add the following important derivative formulas to our list of derivative formulas from preceding chapters.

Derivative Formulas

BASIC FORM:

$D_x \sin x = \cos x$

$D_x \cos x = -\sin x$

GENERALIZED FORM:

For $u = u(x)$:

$$D_x \sin u = \cos u \frac{du}{dx}$$

$$D_x \cos u = -\sin u \frac{du}{dx}$$

EXAMPLE 4

(A) $D_x \sin x^2 = (\cos x^2)\, D_x x^2 = (\cos x^2)2x = 2x \cos x^2$

(B) $D_x \cos(2x - 5) = -\sin(2x - 5)\, D_x(2x - 5) = -2 \sin(2x - 5)$

(C) $D_x(3x^2 - x) \cos x = (3x^2 - x)\, D_x \cos x + (\cos x)\, D_x(3x^2 - x)$
$= -(3x^2 - x) \sin x + (6x - 1) \cos x$
$= (x - 3x^2) \sin x + (6x - 1) \cos x$

PROBLEM 4 Find each of the following derivatives:

(A) $D_x \cos x^3$ (B) $D_x \sin(5 - 3x)$ (C) $D_x \dfrac{\sin x}{x}$

EXAMPLE 5 Find the slope of the graph of $f(x) = \sin x$ at $(\pi/2, 1)$, and sketch in the tangent line to the graph at this point.

Solution Slope at $(\pi/2, 1) = f'(\pi/2) = \cos(\pi/2) = 0$.

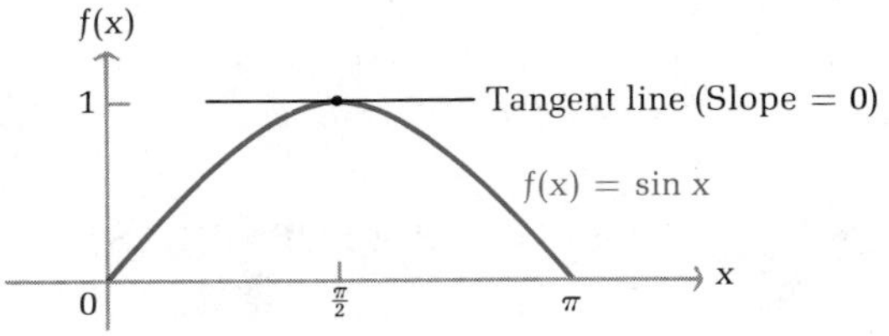

PROBLEM 5 Find the slope of the graph of $f(x) = \cos x$ at $(\pi/6, \sqrt{3}/2)$.

◆ EXAMPLE 6 Find: $D_x \sec x$

Solution

$$D_x \sec x = D_x \frac{1}{\cos x} \qquad \text{Since } \sec x = \frac{1}{\cos x}$$

$$= D_x(\cos x)^{-1}$$

$$= -(\cos x)^{-2} D_x \cos x$$

$$= -(\cos x)^{-2}(-\sin x)$$

$$= \frac{\sin x}{(\cos x)^2} = \left(\frac{\sin x}{\cos x}\right)\left(\frac{1}{\cos x}\right)$$

$$= \tan x \sec x \qquad \text{Since } \tan x = \frac{\sin x}{\cos x}$$

PROBLEM 6 Find: $D_x \csc x$

◆ APPLICATION

◆ EXAMPLE 7 A sporting goods store has revenues from the sale of ski jackets that are given approximately by

Revenue

$$R(t) = 1.55 + 1.45 \cos \frac{\pi t}{26} \qquad 0 \le t \le 104$$

where $R(t)$ is revenue (in thousands of dollars) for a week of sales t weeks after January 1.

(A) What is the rate of change of revenue t weeks after the first of the year?
(B) What is the rate of change of revenue 10 weeks after the first of the year? 26 weeks after the first of the year? 40 weeks after the first of the year?
(C) Find all local maxima and minima for $0 < t < 104$.
(D) Find the absolute maximum and minimum for $0 \le t \le 104$.

Solutions

(A) $R'(t) = -\frac{1.45\pi}{26} \sin \frac{\pi t}{26} \qquad 0 \le t \le 104$

(B) $R'(10) \approx -\$0.164$ thousand or $-\$164$ per week
$R'(26) = \$0$ per week
$R'(40) \approx \$0.174$ thousand or $\$174$ per week

(C) Find the critical points:

$$R'(t) = -\frac{1.45\pi}{26}\sin\frac{\pi t}{26} = 0 \qquad 0 < t < 104$$

$$\sin\frac{\pi t}{26} = 0$$

$$\frac{\pi t}{26} = \pi,\ 2\pi,\ 3\pi \qquad \textit{Note:}\ \ 0 < t < 104 \text{ implies } 0 < \frac{\pi t}{26} < 4\pi.$$

$$t = 26,\ 52,\ 78$$

Use the second-derivative test to get the results shown in the table:

$$R''(t) = -\frac{1.45\pi^2}{26^2}\cos\frac{\pi t}{26}$$

t	$R''(t)$	GRAPH OF R
26	+	Local minimum
52	−	Local maximum
78	+	Local minimum

(D) Evaluate $R(t)$ at end points $t = 0$ and $t = 104$ and at the critical points found in part C:

t	$R(t)$	
0	\$3,000	Absolute maximum
26	\$100	Absolute minimum
52	\$3,000	Absolute maximum
78	\$100	Absolute minimum
104	\$3,000	Absolute maximum

The results above can be visualized as shown in the graph of R in the figure.

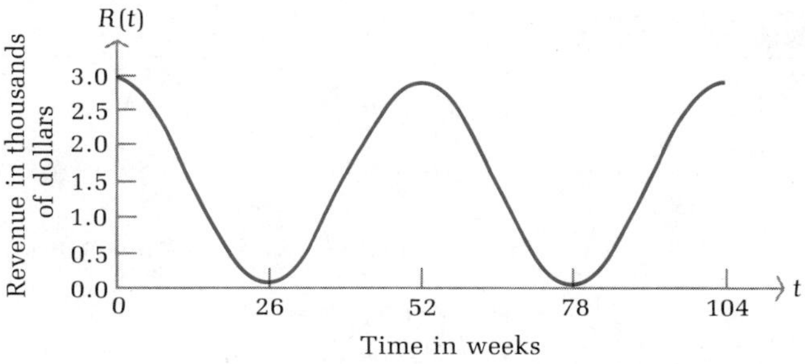

◆

PROBLEM 7 Suppose that in Example 7 revenues from the sale of ski jackets are given approximately by

$$R(t) = 6.2 + 5.8\cos\frac{\pi t}{6} \qquad 0 \leq t \leq 24$$

where $R(t)$ is revenue (in thousands of dollars) for a month of sales t months after January 1.

(A) What is the rate of change of revenue t months after the first of the year?
(B) What is the rate of change of revenue 2 months after the first of the year? 12 months after the first of the year? 23 months after the first of the year?
(C) Find all local maxima and minima for $0 < t < 24$.
(D) Find the absolute maximum and minimum for $0 \le t \le 24$. ◆

Answers to Matched Problems

4. (A) $-3x^2 \sin x^3$ (B) $-3 \cos(5 - 3x)$ (C) $\dfrac{x \cos x - \sin x}{x^2}$

5. $-\frac{1}{2}$ 6. $-\cot x \csc x$

7. (A) $R'(t) = -\dfrac{5.8\pi}{6} \sin \dfrac{\pi t}{6}$, $0 < t < 24$

(B) $R'(2) \approx -\$2.630$ thousand or $-\$2,630$/month; $R'(12) = \$0$/month; $R'(23) \approx \$1.518$ thousand or $\$1,518$/month

(C) Local minima at $t = 6$ and $t = 18$; local maximum at $t = 12$

(D)

	t	$R(t)$	
End point	0	\$12,000	Absolute maximum
	6	\$400	Absolute minimum
	12	\$12,000	Absolute maximum
	18	\$400	Absolute minimum
End point	24	\$12,000	Absolute maximum

EXERCISE 10-2

Find the following derivatives:

A

1. $D_t \cos t$ 2. $D_w \sin w$ 3. $D_x \sin x^3$ 4. $D_x \cos(x^2 - 1)$

B

5. $D_t\, t \sin t$ 6. $D_u\, u \cos u$ 7. $D_x \sin x \cos x$ 8. $D_x \dfrac{\sin x}{\cos x}$

9. $D_x(\sin x)^5$ 10. $D_x(\cos x)^8$ 11. $D_x \sqrt{\sin x}$ 12. $D_x \sqrt{\cos x}$

13. $D_x \cos \sqrt{x}$ 14. $D_x \sin \sqrt{x}$

15. Find the slope of the graph of $f(x) = \sin x$ at $x = \pi/6$.
16. Find the slope of the graph of $f(x) = \cos x$ at $x = \pi/4$.

C

Find the following derivatives:

17. $D_x \tan x$ 18. $D_x \cot x$ 19. $D_x \sin \sqrt{x^2 - 1}$ 20. $D_x \cos \sqrt{x^4 - 1}$

In Problems 21 and 22, find $f''(x)$.

21. $f(x) = e^x \sin x$ 22. $f(x) = e^x \cos x$

Business & Economics

23. *Profit.* Suppose profits on the sale of swimming suits in a department store are given approximately by

$$P(t) = 5 - 5\cos\frac{\pi t}{26} \qquad 0 \leq t \leq 104$$

where $P(t)$ is profit (in hundreds of dollars) for a week of sales t weeks after January 1.

(A) What is the rate of change of profit t weeks after the first of the year?
(B) What is the rate of change of profit 8 weeks after the first of the year? 26 weeks after the first of the year? 50 weeks after the first of the year?
(C) Find all local maxima and minima for $0 < t < 104$.
(D) Find the absolute maximum and minimum for $0 \leq t \leq 104$.

24. *Revenue.* A soft drink company has revenues from sales over a 2 year period as given approximately by

$$R(t) = 4 - 3\cos\frac{\pi t}{6} \qquad 0 \leq t \leq 24$$

where $R(t)$ is revenue (in millions of dollars) for a month of sales t months after February 1.

(A) What is the rate of change of revenue t months after February 1?
(B) What is the rate of change of revenue 1 month after February 1? 6 months after February 1? 11 months after February 1?
(C) Find all local maxima and minima for $0 < t < 24$.
(D) Find the absolute maximum and minimum for $0 \leq t \leq 24$.

Life Sciences

25. *Physiology.* A normal seated adult breathes in and exhales about 0.8 liter of air every 4 seconds. The volume of air $V(t)$ in the lungs t seconds after exhaling is given approximately by

$$V(t) = 0.45 - 0.35\cos\frac{\pi t}{2} \qquad 0 \leq t \leq 8$$

(A) What is the rate of flow of air t seconds after exhaling?
(B) What is the rate of flow of air 3 seconds after exhaling? 4 seconds after exhaling? 5 seconds after exhaling?
(C) Find all local maxima and minima for $0 < t < 8$.
(D) Find the absolute maximum and minimum for $0 \leq t \leq 8$.

26. *Pollution.* In a large city, the amount of sulfur dioxide pollutant released into the atmosphere due to the burning of coal and oil for heating purposes varies seasonally. Suppose the number of tons of pollutant released into the atmosphere during the nth week after January 1 is given approximately by

$$P(n) = 1 + \cos \frac{\pi n}{26} \qquad 0 \leq n \leq 104$$

(A) What is the rate of change of pollutant n weeks after the first of the year?
(B) What is the rate of change of pollutant 13 weeks after the first of the year? 26 weeks after the first of the year? 30 weeks after the first of the year?
(C) Find all local maxima and minima for $0 < t < 104$.
(D) Find the absolute maximum and minimum for $0 \leq t \leq 104$.

SECTION 10-3 Integration of Trigonometric Functions

- ◆ INTEGRAL FORMULAS
- ◆ APPLICATION

◆ INTEGRAL FORMULAS

Now that we know the derivative formulas

$$D_x \sin x = \cos x \qquad \text{and} \qquad D_x \cos x = -\sin x$$

from the definition of the indefinite integral of a function (Section 7-1), we automatically have the two integral formulas

$$\int \cos x\, dx = \sin x + C \qquad \text{and} \qquad \int \sin x\, dx = -\cos x + C$$

◆ EXAMPLE 8 Find the area under the sine curve $y = \sin x$ from 0 to π.

Solution

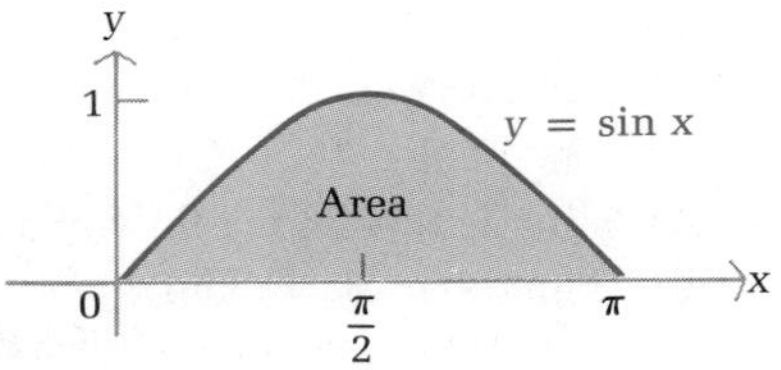

$$\text{Area} = \int_0^\pi \sin x\, dx = -\cos x \Big|_0^\pi$$
$$= (-\cos \pi) - (-\cos 0)$$
$$= [-(-1)] - [-(1)] = 2$$ ◆

PROBLEM 8 Find the area under the cosine curve $y = \cos x$ from 0 to $\pi/2$. ◆

From the general derivative formulas

$$D_x \sin u = \cos u \frac{du}{dx} \qquad \text{and} \qquad D_x \cos u = -\sin u \frac{du}{dx}$$

we obtain the general integral formulas below.

Integral Formulas

For $u = u(x)$,

$$\int \sin u \, du = -\cos u + C \qquad \text{and} \qquad \int \cos u \, du = \sin u + C$$

◆ EXAMPLE 9 Find: $\int x \sin x^2 \, dx$

Solution

$$\int x \sin x^2 \, dx = \tfrac{1}{2} \int 2x \sin x^2 \, dx$$

$$= \tfrac{1}{2} \int (\sin x^2) 2x \, dx \qquad \text{Let } u = x^2\text{; then } du = 2x\,dx.$$

$$= \tfrac{1}{2} \int \sin u \, du$$

$$= -\tfrac{1}{2} \cos x^2 + C$$

Check To check, we differentiate the result to obtain the original integrand:

$$\begin{aligned} D_x(-\tfrac{1}{2}\cos x^2) &= -\tfrac{1}{2} D_x \cos x^2 \\ &= -\tfrac{1}{2}(-\sin x^2) D_x x^2 \\ &= -\tfrac{1}{2}(-\sin x^2)(2x) \\ &= x \sin x^2 \end{aligned}$$

◆

PROBLEM 9 Find: $\int \cos 20\pi t \, dt$ ◆

◆ EXAMPLE 10 Find: $\int (\sin x)^5 \cos x \, dx$

Solution This is of the form $\int u^p \, du$, where $u = \sin x$ and $du = \cos x \, dx$. Thus,

$$\int (\sin x)^5 \cos x \, dx = \frac{(\sin x)^6}{6} + C$$

◆

PROBLEM 10 Find: $\int \sqrt{\sin x} \cos x \, dx$ ◆

◆ EXAMPLE 11 Evaluate: $\int_2^{3.5} \cos x\, dx$

Solution

$$\int_2^{3.5} \cos x\, dx = \sin x \Big|_2^{3.5}$$

$$= \sin 3.5 - \sin 2 \qquad \text{Use a calculator in radian mode.}$$

$$= -0.3508 - 0.9093$$

$$= -1.2601$$ ◆

PROBLEM 11 Use a calculator to evaluate: $\int_1^{1.5} \sin x\, dx$ ◆

◆ APPLICATION

◆ EXAMPLE 12 *Total Revenue*

In Example 7 (Section 10-2), we were given the following revenue equation from the sale of ski jackets:

$$R(t) = 1.55 + 1.45 \cos \frac{\pi t}{26} \qquad 0 \le t \le 104$$

where $R(t)$ is revenue (in thousands of dollars) for a week of sales t weeks after January 1.

(A) Find the total revenue taken in over the 2 year period; that is, from $t = 0$ to $t = 104$.

(B) Find the total revenue taken in from $t = 39$ to $t = 65$.

Solutions

(A) The area under the graph of the revenue equation for the 2 year period approximates the total revenue taken in for that period:

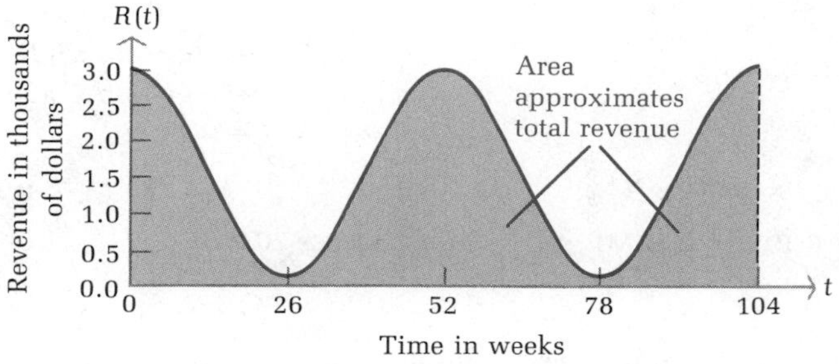

This area is given by the following definite integral:

$$\text{Total revenue} \approx \int_0^{104} \left(1.55 + 1.45 \cos \frac{\pi t}{26}\right) dt$$

$$= \left[1.55t + 1.45\left(\frac{26}{\pi}\right) \sin \frac{\pi t}{26}\right]\Big|_0^{104}$$

$$= \$161.200 \text{ thousand} \quad \text{or} \quad \$161{,}200$$

(B) The total revenue from $t = 39$ to $t = 65$ is approximated by the area under the curve from $t = 39$ to $t = 65$:

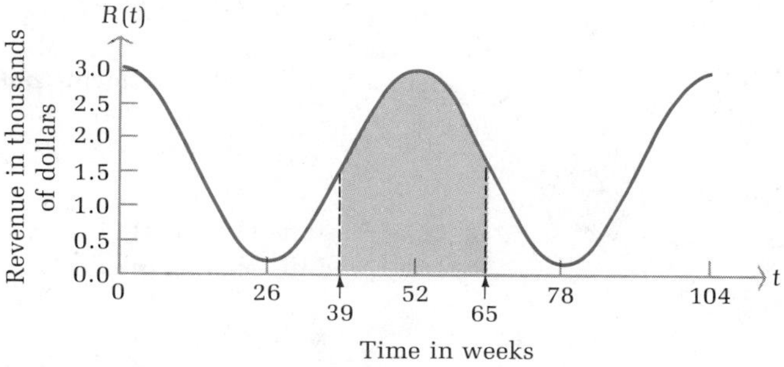

$$\text{Total revenue} \approx \int_{39}^{65} \left(1.55 + 1.45 \cos \frac{\pi t}{26}\right) dt$$

$$= \left[1.55t + 1.45\left(\frac{26}{\pi}\right) \sin \frac{\pi t}{26}\right]\Bigg|_{39}^{65}$$

$$= \$64.301 \text{ thousand} \quad \text{or} \quad \$64{,}301$$

◆

PROBLEM 12 Suppose that in Example 12 revenues from the sale of ski jackets are given approximately by

$$R(t) = 6.2 + 5.8 \cos \frac{\pi t}{6} \qquad 0 \leq t \leq 24$$

where $R(t)$ is revenue (in thousands of dollars) for a month of sales t months after January 1.

(A) Find the total revenue taken in over the 2 year period; that is, from $t = 0$ to $t = 24$.

(B) Find the total revenue taken in from $t = 4$ to $t = 8$. ◆

Answers to Matched Problems

8. 1 9. $\frac{1}{20\pi} \sin 20\pi t + C$ 10. $\frac{2}{3}(\sin x)^{3/2} + C$ 11. 0.4696

12. (A) \$148.8 thousand or \$148,800 (B) \$5.614 thousand or \$5,614

EXERCISE 10-3

Find each of the following indefinite integrals:

A

1. $\int \sin t\, dt$ 2. $\int \cos w\, dw$ 3. $\int \cos 3x\, dx$ 4. $\int \sin 2x\, dx$

5. $\int (\sin x)^{12} \cos x\, dx$ 6. $\int \sin x \cos x\, dx$

B

7. $\int \sqrt[3]{\cos x} \sin x \, dx$

8. $\int \frac{\cos x}{\sqrt{\sin x}} dx$

9. $\int x^2 \cos x^3 \, dx$

10. $\int (x+1) \sin(x^2 + 2x) \, dx$

Evaluate each of the following definite integrals:

11. $\int_0^{\pi/2} \cos x \, dx$ 12. $\int_0^{\pi/4} \cos x \, dx$ 13. $\int_{\pi/2}^{\pi} \sin x \, dx$ 14. $\int_{\pi/6}^{\pi/3} \sin x \, dx$

15. Find the shaded area under the cosine curve in the figure in the margin.

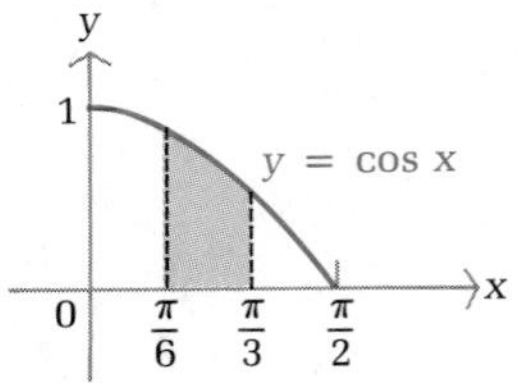

16. Find the shaded area under the sine curve in the figure below.

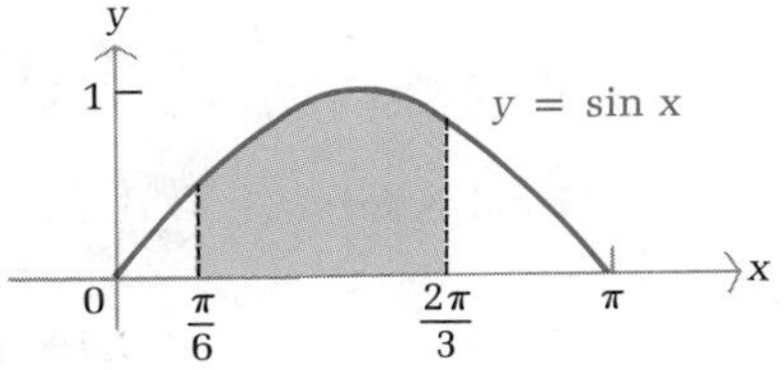

Use a calculator to evaluate the definite integrals below after performing the indefinite integration. (Remember that the limits are real numbers, so the radian mode must be used on the calculator.)

17. $\int_0^{2} \sin x \, dx$ 18. $\int_0^{0.5} \cos x \, dx$ 19. $\int_1^{2} \cos x \, dx$ 20. $\int_1^{3} \sin x \, dx$

C *Find each of the following indefinite integrals:*

21. $\int e^{\sin x} \cos x \, dx$

22. $\int e^{\cos x} \sin x \, dx$

23. $\int \frac{\cos x}{\sin x} dx$

24. $\int \frac{\sin x}{\cos x} dx$

25. $\int \tan x \, dx$

26. $\int \cot x \, dx$

APPLICATIONS

Business & Economics

27. *Seasonal business cycle.* Suppose profits on the sale of swimming suits in a department store are given approximately by

$$P(t) = 5 - 5 \cos \frac{\pi t}{26} \qquad 0 \leq t \leq 104$$

where $P(t)$ is profit (in hundreds of dollars) for a week of sales t weeks after January 1. Use definite integrals to approximate:

(A) The total profit earned during the 2 year period

(B) The total profit earned from $t = 13$ to $t = 26$

28. *Seasonal business cycle.* A soft drink company has revenues from sales over a 2 year period as given approximately by

$$R(t) = 4 - 3\cos\frac{\pi t}{6} \qquad 0 \leq t \leq 24$$

where $R(t)$ is revenue (in millions of dollars) for a month of sales t months after February 1. Use definite integrals to approximate:

(A) Total revenues taken in over the 2 year period
(B) Total revenues taken in from $t = 8$ to $t = 14$

Life Sciences

29. *Pollution.* In a large city, the amount of sulfur dioxide pollutant released into the atmosphere due to the burning of coal and oil for heating purposes is given approximately by

$$P(n) = 1 + \cos\frac{\pi n}{26} \qquad 0 \leq n \leq 104$$

where $P(n)$ is the amount of sulfur dioxide (in tons) released during the nth week after January 1.

(A) How many tons of pollutants were emitted into the atmosphere over the 2 year period?
(B) How many tons of pollutants were emitted into the atmosphere from $n = 13$ to $n = 52$?

SECTION 10-4

Chapter Review

Important Terms and Symbols

10-1 *Trigonometric Functions Review.* Angle; initial side; terminal side; vertex; degree measure; radian measure; conversion formula; standard position; graphs of sine and cosine functions; periodic function; period of a function

$$\sin x, \quad \cos x, \quad \tan x, \quad \sec x, \quad \csc x, \quad \cot x$$

10-2 *Derivatives of Trigonometric Functions.*

$$D_x \sin x = \cos x; \quad D_x \cos x = -\sin x; \quad D_x \sin u = \cos u \frac{du}{dx};$$

$$D_x \cos u = -\sin u \frac{du}{dx}$$

10-3 *Integration of Trigonometric Functions.*

$$\int \cos u\, du = \sin u + C; \quad \int \sin u\, du = -\cos u + C$$

EXERCISE 10-4 Chapter Review

Work through all the problems in this chapter review and check your answers in the back of the book. (Answers to all review problems are there.) Where weaknesses show up, review appropriate sections in the text.

A

1. Convert to radian measure in terms of π:
   (A) 30° (B) 45° (C) 60° (D) 90°
2. Evaluate without using a calculator:
   (A) $\cos \pi$ (B) $\sin 0$ (C) $\sin \frac{\pi}{2}$

Find:

3. $D_m \cos m$
4. $D_u \sin u$
5. $D_x \sin(x^2 - 2x + 1)$
6. $\int \sin 3t\, dt$

B

7. Convert to degree measure:
   (A) $\frac{\pi}{6}$ (B) $\frac{\pi}{4}$ (C) $\frac{\pi}{3}$ (D) $\frac{\pi}{2}$
8. Evaluate without using a calculator:
   (A) $\sin \frac{\pi}{6}$ (B) $\cos \frac{\pi}{4}$ (C) $\sin \frac{\pi}{3}$
9. Evaluate using a calculator:
   (A) cos 33.7 (B) sin(−118.4)

Find:

10. $D_x(x^2 - 1) \sin x$
11. $D_x(\sin x)^6$
12. $D_x \sqrt[3]{\sin x}$
13. $\int t \cos(t^2 - 1)\, dt$
14. $\int_0^{\pi} \sin u\, du$
15. $\int_0^{\pi/3} \cos x\, dx$
16. $\int_1^{2.5} \cos x\, dx$
17. Find the slope of the cosine curve $y = \cos x$ at $x = \pi/4$.
18. Find the area under the sine curve $y = \sin x$ from $x = \pi/4$ to $x = 3\pi/4$.

C

19. Convert 15° to radian measure.
20. Evaluate without using a calculator:
    (A) $\sin \frac{3\pi}{2}$ (B) $\cos \frac{5\pi}{6}$ (C) $\sin \left(\frac{-\pi}{6}\right)$

Find:

21. $D_u \tan u$

22. $D_x e^{\cos x^2}$

23. $\int e^{\sin x} \cos x \, dx$

24. $\int \tan x \, dx$

25. $\int_2^5 (5 + 2 \cos 2x) \, dx$

APPLICATIONS

Business & Economics

Problems 26–28 refer to the following: Revenues from sweater sales in a sportswear chain are given approximately by

$$R(t) = 3 + 2 \cos \frac{\pi t}{6} \qquad 0 \le t \le 24$$

where R(t) is the revenue (in thousands of dollars) for a month of sales t months after January 1.

26. (A) Find the exact values of $R(0)$, $R(2)$, $R(3)$, and $R(6)$ without using a calculator.
    (B) Use a calculator to find $R(1)$ and $R(22)$.
27. (A) What is the rate of change of revenue t months after January 1?
    (B) What is the rate of change of revenue 3 months after January 1? 10 months after January 1? 18 months after January 1?
    (C) Find all local maxima and minima for $0 < t < 24$.
    (D) Find the absolute maximum and minimum for $0 \le t \le 24$.
28. (A) Find the total revenues taken in over the 2 year period.
    (B) Find the total revenues taken in from $t = 5$ to $t = 9$.

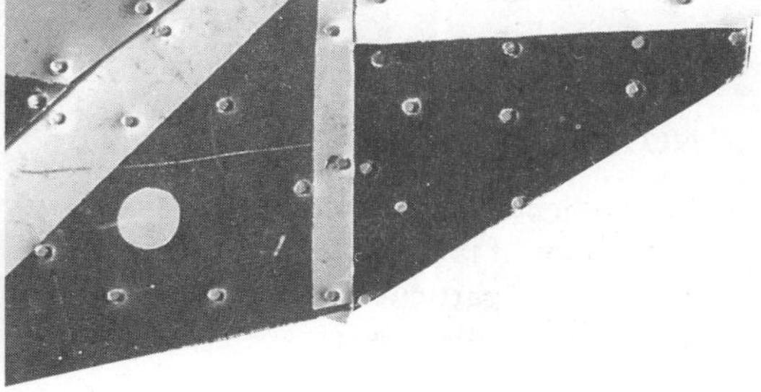

APPENDIX A

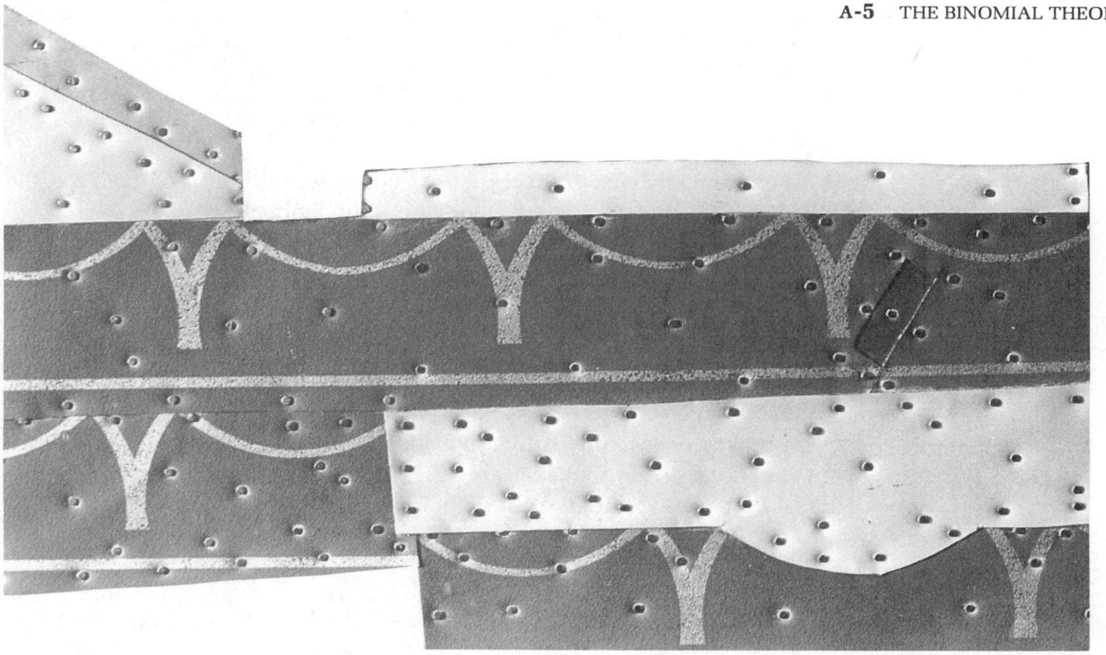

Special Topics

SECTION A-1 Sets

- SET PROPERTIES AND SET NOTATION
- SET OPERATIONS
- APPLICATION

In this section we will review a few key ideas from set theory. Set concepts and notation not only help us talk about certain mathematical ideas with greater clarity and precision, but are indispensable to a clear understanding of probability.

SET PROPERTIES AND SET NOTATION

We can think of a **set** as any collection of objects specified in such a way that we can tell whether any given object is or is not in the collection. Capital letters, such as A, B, and C, are often used to designate particular sets. Each object in a set is called a **member** or **element** of the set. Symbolically:

$a \in A$	means	"a is an element of set A"
$a \notin A$	means	"a is not an element of set A"

A set without any elements is called the **empty,** or **null, set.** For example, the set of all people over 10 feet tall is an empty set. Symbolically:

$\varnothing$	represents	"the empty, or null, set"

A set is usually described either by listing all its elements between braces { } or by enclosing a rule within braces that determines the elements of the set. Thus, if $P(x)$ is a statement about x, then

$S = \{x \mid P(x)\}$	means	"S is the set of all x such that $P(x)$ is true"

Recall that the vertical bar within the braces is read "such that." The following example illustrates the rule and listing methods of representing sets.

◆ EXAMPLE 1

	Rule	Listing
	$\{x \mid x \text{ is a weekend day}\}$	$= \{\text{Saturday, Sunday}\}$
	$\{x \mid x^2 = 4\}$	$= \{-2, 2\}$
	$\{x \mid x \text{ is an odd positive counting number}\}$	$= \{1, 3, 5, \ldots\}$ ◆

The three dots (. . .) in the last set given in Example 1 indicate that the pattern established by the first three entries continues indefinitely. The first two sets in Example 1 are **finite sets** (we intuitively know that the elements can be counted, and there is an end); the last set is an **infinite set** (we intuitively know that there is no end in counting the elements). When listing the elements in a set, we do not list an element more than once.

PROBLEM 1 Let G be the set of all numbers such that $x^2 = 9$.

(A) Denote G by the rule method. (B) Denote G by the listing method.
(C) Indicate whether the following are true or false: $3 \in G$; $9 \notin G$. ◆

If each element of a set *A* is also an element of set *B*, we say that *A* is a **subset** of *B*. For example, the set of all women students in a class is a subset of the whole class. Note that the definition allows a set to be a subset of itself. If set *A* and set *B* have exactly the same elements, then the two sets are said to be **equal.** Symbolically:

$A \subset B$	means	"*A* is a subset of *B*"
$A = B$	means	"*A* and *B* have exactly the same elements"
$A \not\subset B$	means	"*A* is not a subset of *B*"
$A \neq B$	means	"*A* and *B* do not have exactly the same elements"

It can be proved that

∅ is a subset of every set

◆ EXAMPLE 2 If $A = \{-3, -1, 1, 3\}$, $B = \{3, -3, 1, -1\}$, and $C = \{-3, -2, -1, 0, 1, 2, 3\}$, then each of the following statements is true:

$A = B$	$A \subset C$	$A \subset B$
$C \neq A$	$C \not\subset A$	$B \subset A$
$\varnothing \subset A$	$\varnothing \subset C$	$\varnothing \notin A$

◆

PROBLEM 2 Given $A = \{0, 2, 4, 6\}$, $B = \{0, 1, 2, 3, 4, 5, 6\}$, and $C = \{2, 6, 0, 4\}$, indicate whether the following relationships are true (T) or false (F):

(A) $A \subset B$ (B) $A \subset C$ (C) $A = C$
(D) $C \subset B$ (E) $B \not\subset A$ (F) $\varnothing \subset B$ ◆

◆ EXAMPLE 3 List all the subsets of the set $\{a, b, c\}$.

Solution $\{a, b, c\}, \{a, b\}, \{a, c\}, \{b, c\}, \{a\}, \{b\}, \{c\}, \varnothing$ ◆

PROBLEM 3 List all the subsets of the set $\{1, 2\}$. ◆

◆ SET OPERATIONS

The **union** of sets A and B, denoted by $A \cup B$, is the set of all elements formed by combining all the elements of A and all the elements of B into one set. Symbolically:

Union

$$A \cup B = \{x \mid x \in A \textbf{ or } x \in B\}$$

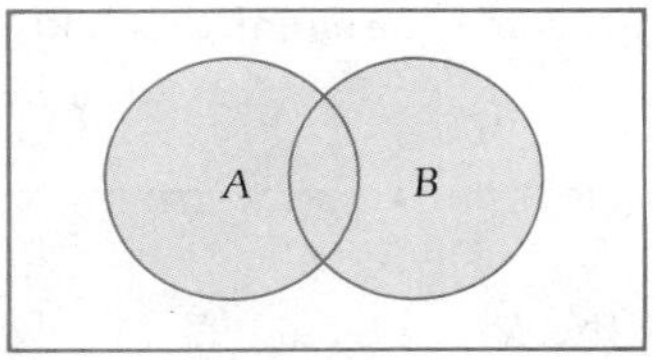

FIGURE 1
$A \cup B$ is the shaded region.

Here we use the word **or** in the way it is always used in mathematics; that is, x may be an element of set A or set B or both.

Venn diagrams are useful in visualizing set relationships. The union of two sets can be illustrated as shown in Figure 1. Note that

$$A \subset A \cup B \qquad \text{and} \qquad B \subset A \cup B$$

The **intersection** of sets A and B, denoted by $A \cap B$, is the set of elements in set A that are also in set B. Symbolically:

Intersection

$$A \cap B = \{x \mid x \in A \textbf{ and } x \in B\}$$

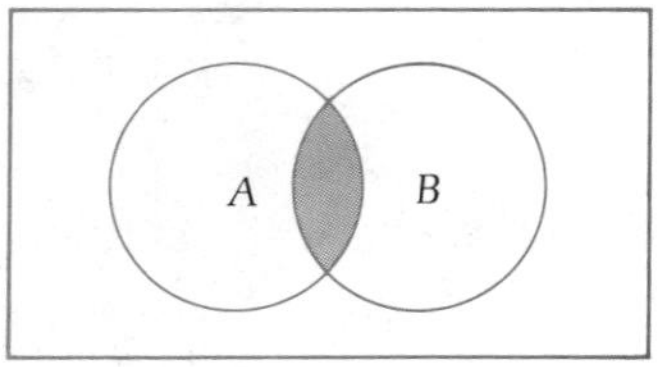

FIGURE 2
$A \cap B$ is the shaded region.

This relationship is easily visualized in the Venn diagram shown in Figure 2. Note that

$$A \cap B \subset A \qquad \text{and} \qquad A \cap B \subset B$$

If $A \cap B = \varnothing$, then the sets A and B are said to be **disjoint;** this is illustrated in Figure 3.

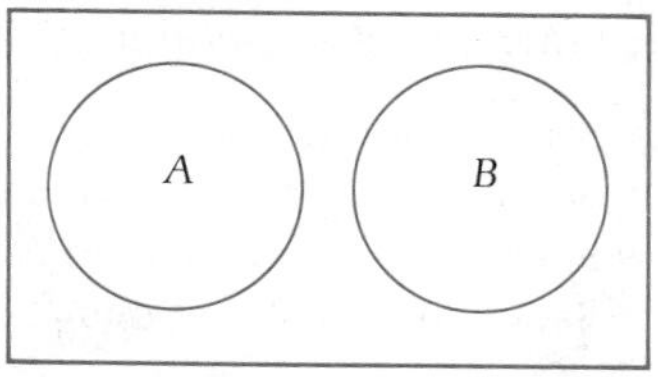

FIGURE 3
$A \cap B = \varnothing$; A and B are disjoint.

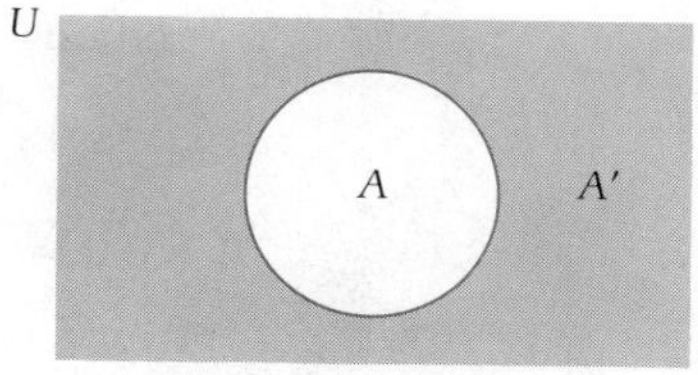

FIGURE 4
The complement of A is A'.

The set of all elements under consideration is called the **universal set *U*.** Once the universal set is determined for a particular case, all other sets under discussion must be subsets of U.

We now define one more operation on sets, called the *complement*. The **complement** of A (relative to U), denoted by A', is the set of elements in U that are not in A (see Fig. 4). Symbolically:

Complement

$$A' = \{x \in U \mid x \notin A\}$$

◆ EXAMPLE 4

If $A = \{3, 6, 9\}$, $B = \{3, 4, 5, 6, 7\}$, $C = \{4, 5, 7\}$, and $U = \{1, 2, 3, 4, 5, 6, 7, 8, 9\}$, then

$A \cup B = \{3, 4, 5, 6, 7, 9\}$
$A \cap B = \{3, 6\}$
$A \cap C = \varnothing$ — A and C are disjoint
$B' = \{1, 2, 8, 9\}$ ◆

PROBLEM 4

If $R = \{1, 2, 3, 4\}$, $S = \{1, 3, 5, 7\}$, $T = \{2, 4\}$, and $U = \{1, 2, 3, 4, 5, 6, 7, 8, 9\}$, find:

(A) $R \cup S$ (B) $R \cap S$ (C) $S \cap T$ (D) S' ◆

◆ APPLICATION

◆ EXAMPLE 5

Marketing Survey

From a survey of 100 college students, a marketing research company found that 75 students owned stereos, 45 owned cars, and 35 owned cars and stereos.

(A) How many students owned either a car or a stereo?
(B) How many students did not own either a car or a stereo?

Solutions

Venn diagrams are very useful for this type of problem. If we let

U = Set of students in sample (100)
S = Set of students who own stereos (75)
C = Set of students who own cars (45)
$S \cap C$ = Set of students who own cars and stereos (35)

then:

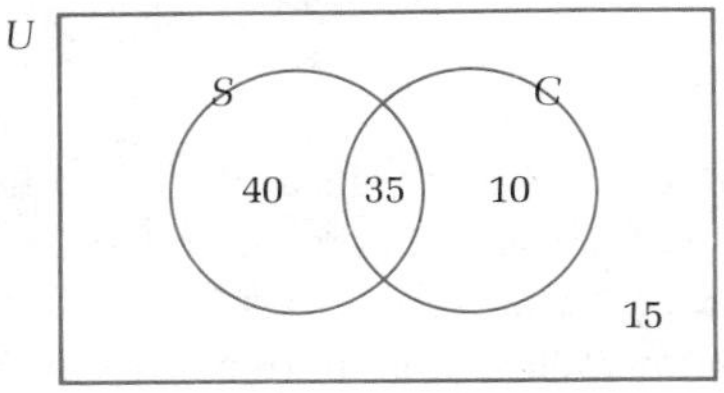

Place the number in the intersection first; then work outward:

$40 = 75 - 35$
$10 = 45 - 35$
$15 = 100 - (40 + 35 + 10)$

(A) The number of students who own either a car or a stereo is the number of students in the set $S \cup C$. You might be tempted to say that this is just the number of students in S plus the number of students in C, $75 + 45 = 120$, but this sum is larger than the sample we started with! What is wrong? We have actually counted the number in the intersection (35) twice. The correct answer, as seen in the Venn diagram, is

$$40 + 35 + 10 = 85$$

(B) The number of students who do not own either a car or a stereo is the number of students in the set $(S \cup C)'$; that is, 15. ◆

PROBLEM 5 Referring to Example 5:

(A) How many students owned a car but not a stereo?
(B) How many students did not own both a car and a stereo? ◆

Note in Example 5 and Problem 5 that the word **and** is associated with intersection and the word **or** is associated with union.

Answers to Matched Problems

1. (A) $\{x | x^2 = 9\}$ (B) $\{-3, 3\}$ (C) True; True
2. All are true 3. $\{1, 2\}, \{1\}, \{2\}, \varnothing$
4. (A) $\{1, 2, 3, 4, 5, 7\}$ (B) $\{1, 3\}$ (C) $\varnothing$ (D) $\{2, 4, 6, 8, 9\}$
5. (A) 10, the number in $S' \cap C$ (B) 65, the number in $(S \cap C)'$

EXERCISE A-1

A *Indicate true (T) or false (F).*

1. $4 \in \{2, 3, 4\}$
2. $6 \notin \{2, 3, 4\}$
3. $\{2, 3\} \subset \{2, 3, 4\}$
4. $\{3, 2, 4\} = \{2, 3, 4\}$

5. $\{3, 2, 4\} \subset \{2, 3, 4\}$
6. $\{3, 2, 4\} \in \{2, 3, 4\}$
7. $\varnothing \subset \{2, 3, 4\}$
8. $\varnothing = \{0\}$

In Problems 9–14, write the resulting set using the listing method.

9. $\{1, 3, 5\} \cup \{2, 3, 4\}$
10. $\{3, 4, 6, 7\} \cup \{3, 4, 5\}$
11. $\{1, 3, 4\} \cap \{2, 3, 4\}$
12. $\{3, 4, 6, 7\} \cap \{3, 4, 5\}$
13. $\{1, 5, 9\} \cap \{3, 4, 6, 8\}$
14. $\{6, 8, 9, 11\} \cap \{3, 4, 5, 7\}$

B *In Problems 15–20, write the resulting set using the listing method.*

15. $\{x|x - 2 = 0\}$
16. $\{x|x + 7 = 0\}$
17. $\{x|x^2 = 49\}$
18. $\{x|x^2 = 100\}$
19. {x|x is an odd number between 1 and 9, inclusive}
20. {x|x is a month starting with M}

21. For $U = \{1, 2, 3, 4, 5\}$ and $A = \{2, 3, 4\}$, find A'.
22. For $U = \{7, 8, 9, 10, 11\}$ and $A = \{7, 11\}$, find A'.

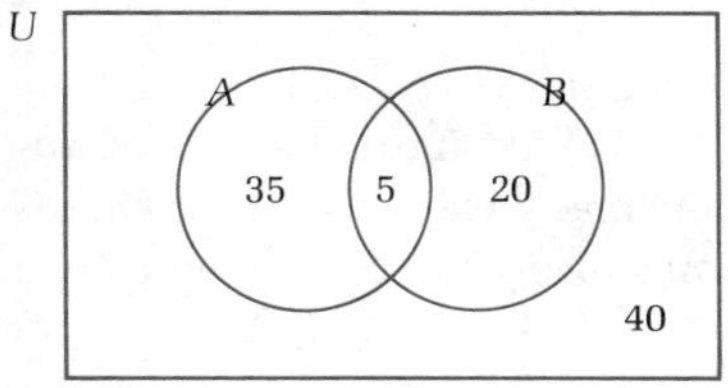

Problems 23–34 refer to the Venn diagram in the margin. How many elements are in each of the indicated sets?

23. A
24. U
25. A'
26. B'
27. $A \cup B$
28. $A \cap B$
29. $A' \cap B$
30. $A \cap B'$
31. $(A \cap B)'$
32. $(A \cup B)'$
33. $A' \cap B'$
34. U'

35. If $R = \{1, 2, 3, 4\}$ and $T = \{2, 4, 6\}$, find:

    (A) $\{x|x \in R$ **or** $x \in T\}$ (B) $R \cup T$

36. If $R = \{1, 3, 4\}$ and $T = \{2, 4, 6\}$, find:

    (A) $\{x|x \in R$ **and** $x \in T\}$ (B) $R \cap T$

37. For $P = \{1, 2, 3, 4\}$, $Q = \{2, 4, 6\}$, and $R = \{3, 4, 5, 6\}$, find $P \cup (Q \cap R)$.
38. For P, Q, and R in Problem 37, find $P \cap (Q \cup R)$.

C *Venn diagrams may be of help in Problems 39–44.*

39. If $A \cup B = B$, can we always conclude that $A \subset B$?
40. If $A \cap B = B$, can we always conclude that $B \subset A$?
41. If A and B are arbitrary sets, can we always conclude that $A \cap B \subset B$?
42. If $A \cap B = \varnothing$, can we always conclude that $B = \varnothing$?
43. If $A \subset B$ and $x \in A$, can we always conclude that $x \in B$?
44. If $A \subset B$ and $x \in B$, can we always conclude that $x \in A$?

45. How many subsets does each of the following sets have? Also, try to discover a formula in terms of n for a set with n elements.

    (A) $\{a\}$ (B) $\{a, b\}$ (C) $\{a, b, c\}$

46. How do the sets $\varnothing$, $\{\varnothing\}$, and $\{0\}$ differ from each other?

Business & Economics

Marketing survey. *Problems 47–58 refer to the following survey: A marketing survey of 1,000 car commuters found that 600 listen to the news, 500 listen to music, and 300 listen to both. Let*

N = Set of commuters in the sample who listen to news
M = Set of commuters in the sample who listen to music

Following the procedures in Example 5, find the number of commuters in each set described below.

47. $N \cup M$ **48.** $N \cap M$ **49.** $(N \cup M)'$
50. $(N \cap M)'$ **51.** $N' \cap M$ **52.** $N \cap M'$
53. Set of commuters who listen to either news or music
54. Set of commuters who listen to both news and music
55. Set of commuters who do not listen to either news or music
56. Set of commuters who do not listen to both news and music
57. Set of commuters who listen to music but not news
58. Set of commuters who listen to news but not music

59. *Committee selection.* The management of a company, a president and three vice presidents, denoted by the set $\{P, V_1, V_2, V_3\}$, wish to select a committee of 2 people from among themselves. How many ways can this committee be formed? That is, how many 2-person subsets can be formed from a set of 4 people?

60. *Voting coalition.* The management of the company in Problem 59 decides for or against certain measures as follows: The president has 2 votes and each vice president has 1 vote. Three favorable votes are needed to pass a measure. List all minimal winning coalitions; that is, list all subsets of $\{P, V_1, V_2, V_3\}$ that represent exactly 3 votes.

Life Sciences

Blood types. *When receiving a blood transfusion, a recipient must have all the antigens of the donor. A person may have one or more of the three antigens A, B, and Rh, or none at all. Eight blood types are possible, as indicated in the following Venn diagram, where U is the set of all people under consideration:*

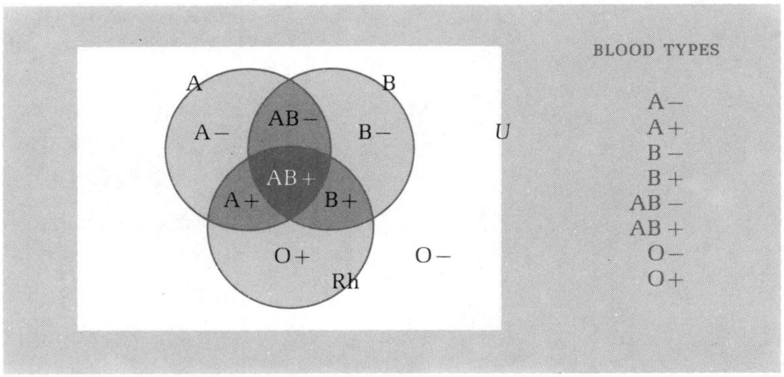

An A− person has A antigens but no B or Rh; an O+ person has Rh but neither A nor B; an AB− person has A and B antigens but no Rh; and so on.

Using the Venn diagram, indicate which of the eight blood types are included in each set.

61. $A \cap Rh$ **62.** $A \cap B$ **63.** $A \cup Rh$ **64.** $A \cup B$

65. $(A \cup B)'$ **66.** $(A \cup B \cup Rh)'$ **67.** $A' \cap B$ **68.** $Rh' \cap A$

Social Sciences

Group structures. R. D. Luce and A. D. Perry, in a study on group structure (Psychometrika, 1949, 14:95–116), used the idea of sets to formally define the notion of a clique within a group. Let G be the set of all persons in the group and let $C \subset G$. Then C is a clique provided that:

1. *C contains at least 3 elements.*
2. *For every $a, b \in C$, a* **R** *b and b* **R** *a.*
3. *For every $a \notin C$, there is at least one $b \in C$ such that a* **R̸** *b or b* **R̸** *or both.*

[Note: Interpret "a **R** *b" to mean "a relates to b," "a likes b," "a is as wealthy as b," and so on. Of course, "a* **R̸** *b" means "a does not relate to b," and so on.]*

69. Translate statement 2 into ordinary English.

70. Translate statement 3 into ordinary English.

SECTION A-2 Sequences, Series, and Summation Notation

◆ SEQUENCES

◆ SERIES AND SUMMATION NOTATION

If someone asked you to list all natural numbers that are perfect squares, you might begin by writing

1, 4, 9, 16, 25, 36

But you would soon realize that it is impossible to actually list all the perfect squares, since there are an infinite number of them. However, you could represent this collection of numbers in several different ways. One common method is to write

$$1, 4, 9, \ldots, n^2, \ldots \qquad n \in N$$

where N is the set of natural numbers. A list of numbers such as this is generally called a *sequence*. Sequences and related topics form the subject matter of this section.

◆ SEQUENCES

Consider the function f given by

$$f(n) = 2n + 1 \qquad (1)$$

where the domain of f is the set of natural numbers N. Note that

$$f(1) = 3, \quad f(2) = 5, \quad f(3) = 7, \quad \ldots$$

The function f is an example of a sequence. In general, a **sequence** is a function with domain a set of successive integers. Instead of the standard function notation used in equation (1), sequences are usually defined in terms of a special notation.

The range value $f(n)$ is usually symbolized more compactly with a symbol such as a_n. Thus, in place of equation (1), we write

$$a_n = 2n + 1$$

and the domain is understood to be the set of natural numbers unless something is said to the contrary or the context indicates otherwise. The elements in the range are called **terms of the sequence;** a_1 is the first term, a_2 is the second term, and a_n is the ***n*th term,** or **general term.**

$$\begin{aligned} a_1 &= 2(1) + 1 = 3 && \text{First term} \\ a_2 &= 2(2) + 1 = 5 && \text{Second term} \\ a_3 &= 2(3) + 1 = 7 && \text{Third term} \\ &\vdots \\ a_n &= 2n + 1 && \text{General term} \end{aligned}$$

When the terms in a sequence are written in their natural order with respect to domain values,

$$a_1, a_2, a_3, \ldots, a_n, \ldots$$

or

$$3, 5, 7, \ldots, 2n + 1, \ldots$$

the ordered list of elements is often informally referred to as a sequence. A sequence also may be represented in the abbreviated form $\{a_n\}$, where a symbol for the nth term is written within braces. For example, we could refer to the sequence $3, 5, 7, \ldots, 2n + 1, \ldots$ as the sequence $\{2n + 1\}$.

If the domain of a sequence is a finite set of successive integers, then the sequence is called a **finite sequence.** If the domain is an infinite set of successive integers, then the sequence is called an **infinite sequence.** The sequence $\{2n + 1\}$ discussed above is an infinite sequence.

◆ EXAMPLE 6 Write the first four terms of each sequence:

(A) $a_n = 3n - 2$ (B) $\left\{\frac{(-1)^n}{n}\right\}$

Solutions (A) 1, 4, 7, 10 (B) $-1, \frac{1}{2}, \frac{-1}{3}, \frac{1}{4}$ ◆

PROBLEM 6 Write the first four terms of each sequence:

(A) $a_n = -n + 3$ (B) $\left\{\frac{(-1)^n}{2^n}\right\}$ ◆

Now that we have seen how to use the general term to find the first few terms in a sequence, we consider the reverse problem. That is, can a sequence be defined just by listing the first three or four terms of the sequence? And can we then use these initial terms to find a formula for the nth term? In general, without other information, the answer to the first question is no. Many different sequences may start off with the same terms. For example, each of the following sequences starts off with the same three terms:

$$2, 4, 8, \ldots, 2^n, \ldots$$

$$2, 4, 8, \ldots, n^2 - n + 2, \ldots$$

$$2, 4, 8, \ldots, 5n + \frac{6}{n} - 9, \ldots$$

However, these are certainly different sequences. (You should verify that these sequences agree for the first three terms and differ in the fourth term by evaluating the general term for each sequence at $n = 1, 2, 3$, and 4.) Thus, simply listing the first three terms (or any other finite number of terms) does not specify a particular sequence. In fact, it can be shown that given any list of m numbers, there are an infinite number of sequences whose first m terms agree with these given numbers.

What about the second question? That is, given a few terms, can we find the general formula for at least one sequence whose first few terms agree with the given terms? The answer to this question is a qualified yes. If we can observe a simple pattern in the given terms, then we usually can construct a general term that will produce that pattern. The next example illustrates this approach.

◆ EXAMPLE 7 Find the general term of a sequence whose first four terms are:

(A) $3, 4, 5, 6, \ldots$ (B) $5, -25, 125, -625, \ldots$

Solutions

(A) Since these terms are consecutive integers, one solution is $a_n = n$, $n \geq 3$. If we want the domain of the sequence to be all natural numbers, then another solution is $b_n = n + 2$.

(B) Each of these terms can be written as the product of a power of 5 and a power of -1:

$$\begin{aligned} 5 &= (-1)^0 5^1 = a_1 \\ -25 &= (-1)^1 5^2 = a_2 \\ 125 &= (-1)^2 5^3 = a_3 \\ -625 &= (-1)^3 5^4 = a_4 \end{aligned}$$

If we choose the domain to be all natural numbers, then a solution is

$$a_n = (-1)^{n-1} 5^n$$ ◆

PROBLEM 7 Find the general term of a sequence whose first four terms are:

(A) 3, 6, 9, 12, . . . (B) 1, −2, 4, −8, . . . ◆

In general, there is usually more than one way of representing the nth term of a given sequence (see the solution of part A in Example 7). However, unless something is stated to the contrary, we assume the domain of the sequence is the set of natural numbers N.

◆ SERIES AND SUMMATION NOTATION

The sum of the terms of a sequence is called a **series.** If the sequence is finite, the corresponding series is a **finite series.** If the sequence is infinite, the corresponding series is an **infinite series.** We will consider only finite series in this section. For example,

1, 3, 5, 7, 9	Finite sequence
$1 + 3 + 5 + 7 + 9$	Finite series

Notice that we can easily evaluate this series by adding the five terms:

$$1 + 3 + 5 + 7 + 9 = 25$$

Series are often represented in a compact form called **summation notation.** Consider the following examples:

$$\sum_{k=3}^{6} k^2 = 3^2 + 4^2 + 5^2 + 6^2$$
$$= 9 + 16 + 25 + 36 = 86$$

$$\sum_{k=0}^{2} (4k + 1) = (4 \cdot 0 + 1) + (4 \cdot 1 + 1) + (4 \cdot 2 + 1)$$
$$= 1 + 5 + 9 = 15$$

In each case, the terms of the series on the right are obtained from the expression on the left by successively replacing the **summing index *k*** with integers, starting with the number indicated below the **summation sign** Σ and ending with the number that appears above Σ. The summing index may be represented by letters other than k and may start at any integer and end at any integer greater than or equal to the starting integer. Thus, if we are given the finite sequence

$$\frac{1}{2}, \frac{1}{4}, \frac{1}{8}, \ldots, \frac{1}{2^n}$$

the corresponding series is

$$\frac{1}{2} + \frac{1}{4} + \frac{1}{8} + \cdots + \frac{1}{2^n} = \sum_{j=1}^{n} \frac{1}{2^j}$$

where we have used j for the summing index.

◆ EXAMPLE 8 Write

$$\sum_{k=1}^{5} \frac{k}{k^2+1}$$

without summation notation. Do not evaluate the sum.

Solution

$$\sum_{k=1}^{5} \frac{k}{k^2+1} = \frac{1}{1^2+1} + \frac{2}{2^2+1} + \frac{3}{3^2+1} + \frac{4}{4^2+1} + \frac{5}{5^2+1}$$
$$= \frac{1}{2} + \frac{2}{5} + \frac{3}{10} + \frac{4}{17} + \frac{5}{26}$$ ◆

PROBLEM 8 Write

$$\sum_{k=1}^{5} \frac{k+1}{k}$$

without summation notation. Do not evaluate the sum. ◆

If the terms of a series are alternately positive and negative, we call the series an **alternating series.** The next example deals with the representation of such a series.

◆ EXAMPLE 9 Write the alternating series

$$\frac{1}{2} - \frac{1}{4} + \frac{1}{6} - \frac{1}{8} + \frac{1}{10} - \frac{1}{12}$$

using summation notation with:

(A) The summing index k starting at 1
(B) The summing index j starting at 0

Solutions

(A) $(-1)^{k+1}$ provides the alternation of sign, and $1/(2k)$ provides the other part of each term. Thus, we can write

$$\frac{1}{2} - \frac{1}{4} + \frac{1}{6} - \frac{1}{8} + \frac{1}{10} - \frac{1}{12} = \sum_{k=1}^{6} \frac{(-1)^{k+1}}{2k}$$

(B) $(-1)^j$ provides the alternation of sign, and $1/[2(j+1)]$ provides the other part of each term. Thus, we can write

$$\frac{1}{2} - \frac{1}{4} + \frac{1}{6} - \frac{1}{8} + \frac{1}{10} - \frac{1}{12} = \sum_{j=0}^{5} \frac{(-1)^j}{2(j+1)}$$ ◆

PROBLEM 9 Write the alternating series

$$1-\frac{1}{3}+\frac{1}{9}-\frac{1}{27}+\frac{1}{81}$$

using summation notation with:

(A) The summing index k starting at 1
(B) The summing index j starting at 0 ◆

Summation notation provides a compact notation for the sum of any list of numbers, even if the numbers are not generated by a formula. For example, suppose the results of an examination taken by a class of 10 students are given in the following list:

87, 77, 95, 83, 86, 73, 95, 68, 75, 86

If we let $a_1, a_2, a_3, \ldots, a_{10}$ represent these 10 scores, then the average test score is given by

$$\frac{1}{10}\sum_{k=1}^{10} a_k = \frac{1}{10}(87+77+95+83+86+73+95+68+75+86)$$
$$= \frac{1}{10}(825) = 82.5$$

More generally, in statistics, the **arithmetic mean** $\bar{a}$ of a list of n numbers $a_1, a_2, \ldots, a_n$ is defined as

$$\bar{a} = \frac{1}{n}\sum_{k=1}^{n} a_k$$

◆ EXAMPLE 10 Find the arithmetic mean of 3, 5, 4, 7, 4, 2, 3, and 6.

Solution

$$\bar{a} = \frac{1}{8}\sum_{k=1}^{8} a_k = \frac{1}{8}(3+5+4+7+4+2+3+6) = \frac{1}{8}(34) = 4.25$$ ◆

PROBLEM 10 Find the arithmetic mean of 9, 3, 8, 4, 3, and 6. ◆

Answers to Matched Problems

6. (A) 2, 1, 0, −1 (B) $\frac{-1}{2}, \frac{1}{4}, \frac{-1}{8}, \frac{1}{16}$
7. (A) $a_n = 3n$ (B) $a_n = (-2)^{n-1}$ 8. $2+\frac{3}{2}+\frac{4}{3}+\frac{5}{4}+\frac{6}{5}$
9. (A) $\sum_{k=1}^{5} \frac{(-1)^{k-1}}{3^{k-1}}$ (B) $\sum_{j=0}^{4} \frac{(-1)^j}{3^j}$ 10. 5.5

EXERCISE A-2

A *Write the first four terms for each sequence.*

1. $a_n = 2n + 3$
2. $a_n = 4n - 3$
3. $a_n = \frac{n+2}{n+1}$
4. $a_n = \frac{2n+1}{2n}$
5. $a_n = (-3)^{n+1}$
6. $a_n = (-\frac{1}{4})^{n-1}$

7. Write the 10th term of the sequence in Problem 1.
8. Write the 15th term of the sequence in Problem 2.
9. Write the 99th term of the sequence in Problem 3.
10. Write the 200th term of the sequence in Problem 4.

Write each series in expanded form without summation notation, and evaluate.

11. $\sum_{k=1}^{6} k$
12. $\sum_{k=1}^{5} k^2$
13. $\sum_{k=4}^{7} (2k-3)$
14. $\sum_{k=0}^{4} (-2)^k$
15. $\sum_{k=0}^{3} \frac{1}{10^k}$
16. $\sum_{k=1}^{4} \frac{1}{2^k}$

Find the arithmetic mean of each list of numbers.

17. 5, 4, 2, 1, and 6
18. 7, 9, 9, 2, and 4
19. 96, 65, 82, 74, 91, 88, 87, 91, 77, and 74
20. 100, 62, 95, 91, 82, 87, 70, 75, 87, and 82

B *Write the first five terms of each sequence.*

21. $a_n = \frac{(-1)^{n+1}}{2^n}$
22. $a_n = (-1)^n(n-1)^2$
23. $a_n = n[1 + (-1)^n]$
24. $a_n = \frac{1-(-1)^n}{n}$
25. $a_n = \left(-\frac{3}{2}\right)^{n-1}$
26. $a_n = \left(-\frac{1}{2}\right)^{n+1}$

In Problems 27–42, find the general term of a sequence whose first four terms agree with the given terms.

27. $-2, -1, 0, 1, \ldots$
28. $4, 5, 6, 7, \ldots$
29. $4, 8, 12, 16, \ldots$
30. $-3, -6, -9, -12, \ldots$
31. $\frac{1}{2}, \frac{3}{4}, \frac{5}{6}, \frac{7}{8}, \ldots$
32. $\frac{1}{2}, \frac{2}{3}, \frac{3}{4}, \frac{4}{5}, \ldots$
33. $1, -2, 3, -4, \ldots$
34. $-2, 4, -8, 16, \ldots$
35. $1, -3, 5, -7, \ldots$
36. $3, -6, 9, -12, \ldots$
37. $1, \frac{2}{5}, \frac{4}{25}, \frac{8}{125}, \ldots$
38. $\frac{4}{3}, \frac{16}{9}, \frac{64}{27}, \frac{256}{81}, \ldots$
39. $x, x^2, x^3, x^4, \ldots$
40. $1, 2x, 3x^2, 4x^3, \ldots$
41. $x, -x^3, x^5, -x^7, \ldots$
42. $x, \frac{x^2}{2}, \frac{x^3}{3}, \frac{x^4}{4}, \ldots$

Write each series in expanded form without summation notation. Do not evaluate.

43. $\sum_{k=1}^{5} (-1)^{k+1}(2k-1)^2$
44. $\sum_{k=1}^{4} \frac{(-2)^{k+1}}{2k+1}$
45. $\sum_{k=2}^{5} \frac{2^k}{2k+3}$
46. $\sum_{k=3}^{7} \frac{(-1)^k}{k^2-k}$
47. $\sum_{k=1}^{5} x^{k-1}$
48. $\sum_{k=1}^{3} \frac{1}{k} x^{k+1}$
49. $\sum_{k=0}^{4} \frac{(-1)^k x^{2k+1}}{2k+1}$
50. $\sum_{k=0}^{4} \frac{(-1)^k x^{2k}}{2k+2}$

Write each series using summation notation with:

(A) The summing index k starting at $k = 1$
(B) The summing index j starting at $j = 0$

51. $2 + 3 + 4 + 5 + 6$

52. $1^2 + 2^2 + 3^2 + 4^2$

53. $1 - \frac{1}{2} + \frac{1}{3} - \frac{1}{4}$

54. $1 - \frac{1}{3} + \frac{1}{5} - \frac{1}{7} + \frac{1}{9}$

Write each series using summation notation with the summing index k starting at $k = 1$.

55. $2 + \frac{3}{2} + \frac{4}{3} + \cdots + \frac{n+1}{n}$

56. $1 + \frac{1}{2^2} + \frac{1}{3^2} + \cdots + \frac{1}{n^2}$

57. $\frac{1}{2} - \frac{1}{4} + \frac{1}{8} - \cdots + \frac{(-1)^{n+1}}{2^n}$

58. $1 - 4 + 9 - \cdots + (-1)^{n+1}n^2$

C *Some sequences are defined by a* ***recursion formula****—that is, a formula that defines each term of the sequence in terms of one or more of the preceding terms. For example, if $\{a_n\}$ is defined by*

$$a_1 = 1 \quad \text{and} \quad a_n = 2a_{n-1} + 1 \quad \text{for } n \geq 2$$

then

$$a_2 = 2a_1 + 1 = 2 \cdot 1 + 1 = 3$$
$$a_3 = 2a_2 + 1 = 2 \cdot 3 + 1 = 7$$
$$a_4 = 2a_3 + 1 = 2 \cdot 7 + 1 = 15$$

and so on. In Problems 59–62, write the first five terms of each sequence.

59. $a_1 = 2$ and $a_n = 3a_{n-1} + 2$ for $n \geq 2$

60. $a_1 = 3$ and $a_n = 2a_{n-1} - 2$ for $n \geq 2$

61. $a_1 = 1$ and $a_n = 2a_{n-1}$ for $n \geq 2$

62. $a_1 = 1$ and $a_n = -\frac{1}{3}a_{n-1}$ for $n \geq 2$

If A is a positive real number, then the terms of the sequence defined by

$$a_1 = \frac{A}{2} \quad \text{and} \quad a_n = \frac{1}{2}\left(a_{n-1} + \frac{A}{a_{n-1}}\right) \quad \text{for } n \geq 2$$

can be used to approximate $\sqrt{A}$ to any decimal place accuracy desired. In Problems 63 and 64, compute the first four terms of this sequence for the indicated value of A, and compare the fourth term with the value of $\sqrt{A}$ obtained from a calculator.

63. $A = 2$

64. $A = 6$

SECTION A-3 Arithmetic Progressions

- ARITHMETIC PROGRESSIONS — DEFINITIONS
- SPECIAL FORMULAS
- APPLICATION

◆ ARITHMETIC PROGRESSIONS — DEFINITIONS

Consider the sequence of numbers

$$1, 4, 7, 10, 13, \ldots, 1 + 3(n - 1), \ldots$$

where each number after the first is obtained from the preceding one by adding 3 to it. This is an example of an *arithmetic progression.* In general:

Arithmetic Progression

A sequence of numbers

$$a_1, a_2, a_3, \ldots, a_n, \ldots$$

is called an **arithmetic progression** if there is a constant d, called the **common difference,** such that

$$a_n - a_{n-1} = d$$

That is,

$$a_n = a_{n-1} + d \qquad \text{for every } n > 1 \qquad (1)$$

◆ EXAMPLE 11 Which of the following can be the first four terms of an arithmetic progression, and what is its common difference?

(A) 2, 4, 8, 10, . . . (B) 3, 8, 13, 18, . . .

Solution The terms in B, with $d = 5$ ◆

PROBLEM 11 Which of the following can be the first four terms of an arithmetic progression, and what is its common difference?

(A) 15, 13, 11, 9, . . . (B) 3, 9, 27, 81, . . . ◆

◆ SPECIAL FORMULAS

Arithmetic progressions have a number of convenient properties. For example, it is easy to derive formulas for the nth term and the sum of any number of consecutive terms. To obtain a formula for the nth term of an arithmetic progression, we note that if a_1 is the first term and d is the common difference, then

$$a_2 = a_1 + d$$
$$a_3 = a_2 + d = (a_1 + d) + d = a_1 + 2d$$
$$a_4 = a_3 + d = (a_1 + 2d) + d = a_1 + 3d$$

This suggests that:

$$a_n = a_1 + (n-1)d \qquad \text{for all } n > 1 \qquad (2)$$

◆ EXAMPLE 12 Find the 21st term in the arithmetic progression: 3, 8, 13, 18, . . .

Solution Find the common difference d and use formula (2):

$$d = 5 \qquad n = 21 \qquad a_1 = 3$$

Thus

$$a_{21} = 3 + (21 - 1)5$$
$$= 103$$

◆

PROBLEM 12 Find the 51st term in the arithmetic progression: 15, 13, 11, 9, . . . ◆

We now derive two simple and very useful formulas for the sum of n consecutive terms of an arithmetic progression. Let

$$S_n = a_1 + a_2 + \cdots + a_{n-1} + a_n$$

be the sum of n terms of an arithmetic progression with common difference d. Then,

$$S_n = a_1 + (a_1 + d) + \cdots + [a_1 + (n-2)d] + [a_1 + (n-1)d]$$

Reversing the order of the sum, we obtain

$$S_n = [a_1 + (n-1)d] + [a_1 + (n-2)d] + \cdots + (a_1 + d) + a_1$$

Something interesting happens if we combine these last two equations by addition (adding corresponding terms on the right sides):

$$2S_n = [2a_1 + (n-1)d] + [2a_1 + (n-1)d] + \cdots + [2a_1 + (n-1)d] + [2a_1 + (n-1)d]$$

All the terms on the right side are the same, and there are n of them. Thus,

$$2S_n = n[2a_1 + (n-1)d]$$

and we have the following general formula:

$$S_n = \frac{n}{2}[2a_1 + (n-1)d] \qquad (3)$$

Replacing

$$[a_1 + (n-1)d] \qquad \text{in} \qquad \frac{n}{2}[a_1 + a_1 + (n-1)d]$$

by a_n from equation (2), we obtain a second useful formula for the sum:

$$S_n = \frac{n}{2}(a_1 + a_n) \qquad (4)$$

◆ EXAMPLE 13 Find the sum of the first 30 terms in the arithmetic progression: 3, 8, 13, 18, . . .

Solution Use formula (3) with $n = 30$, $a_1 = 3$, and $d = 5$:

$$S_{30} = \frac{30}{2}[2 \cdot 3 + (30-1)5] = 2{,}265$$ ◆

PROBLEM 13 Find the sum of the first 40 terms in the arithmetic progression: 15, 13, 11, 9, . . . ◆

◆ EXAMPLE 14 Find the sum of all the even numbers between 31 and 87.

Solution First, find n using equation (2):

$$a_n = a_1 + (n-1)d$$
$$86 = 32 + (n-1)2$$
$$n = 28$$

Now find s_{28} using formula (4):

$$s_n = \frac{n}{2}(a_1 + a_n)$$

$$S_{28} = \frac{28}{2}(32 + 86) = 1{,}652$$ ◆

PROBLEM 14 Find the sum of all the odd numbers between 24 and 208. ◆

◆ APPLICATION

◆ EXAMPLE 15 *Loan Repayment*

A person borrows $3,600 and agrees to repay the loan in monthly installments over a period of 3 years. The agreement is to pay 1% of the unpaid balance each month for using the money and $100 each month to reduce the loan. What is the total cost of the loan over the 3 years?

Solution Let us look at the problem relative to a time line:

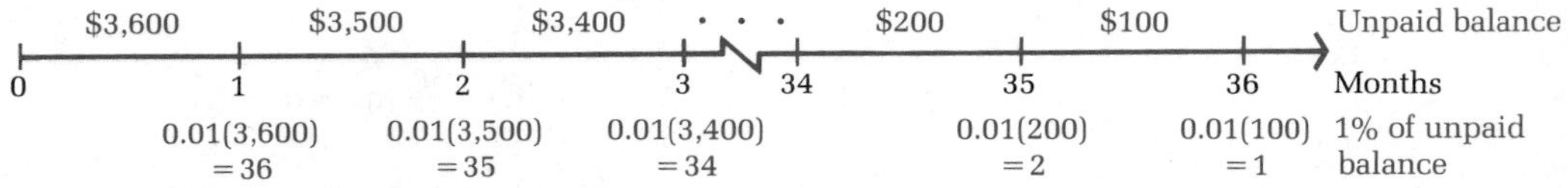

The total cost of the loan is

$$1 + 2 + \cdots + 34 + 35 + 36$$

The terms form an arithmetic progression with $n = 36$, $a_1 = 1$, and $a_{36} = 36$, so we can use formula (4):

$$S_n = \frac{n}{2}(a_1 + a_n)$$

$$S_{36} = \frac{36}{2}(1 + 36) = \$666$$

And we conclude that the total cost of the loan over the period of 3 years is $666. ◆

PROBLEM 15 Repeat Example 15 with a loan of $6,000 over a period of 5 years. ◆

Answers to Matched Problems

11. The terms in A, with $d = -2$ 12. -85 13. -960
14. 10,672 15. $1,830

EXERCISE A-3

A

1. Determine which of the following can be the first three terms of an arithmetic progression. Find the common difference d and the next two terms for those that are.
   (A) 5, 8, 11, . . . (B) 4, 8, 16, . . .
   (C) $-2, -4, -8, \ldots$ (D) $8, -2, -12, \ldots$
2. Repeat Problem 1 for:
   (A) 11, 16, 21, . . . (B) 16, 8, 4, . . .
   (C) $2, -3, -8, \ldots$ (D) $-1, -2, -4, \ldots$

Let $a_1, a_2, a_3, \ldots, a_n, \ldots$ be an arithmetic progression and S_n be the sum of the first n terms. In Problems 3–8, find the indicated quantities.

3. $a_1 = 7;\ d = 4;\ a_2 = ?;\ a_3 = ?$
4. $a_1 = -2;\ d = -3;\ a_2 = ?;\ a_3 = ?$

B

5. $a_1 = 2;\ d = 4;\ a_{21} = ?;\ S_{31} = ?$
6. $a_1 = 8;\ d = -10;\ a_{15} = ?;\ S_{23} = ?$
7. $a_1 = 18;\ a_{20} = 75;\ S_{20} = ?$
8. $a_1 = 203;\ a_{30} = 261;\ S_{30} = ?$
9. Find $f(1) + f(2) + f(3) + \cdots + f(50)$ if $f(x) = 2x - 3$.
10. Find $g(1) + g(2) + g(3) + \cdots + g(100)$ if $g(t) = 18 - 3t$.
11. Find the sum of all the odd integers between 12 and 68.
12. Find the sum of all the even integers between 23 and 97.

C

13. Show that the sum of the first n odd positive integers is n^2, using appropriate formulas from this section.
14. Show that the sum of the first n positive even integers is $n + n^2$, using formulas in this section.

APPLICATIONS

Business & Economics

15. *Salary analysis.* You are confronted with two job offers. Firm A will start you at \$24,000 per year and guarantees you a \$900 raise each year for 10 years. Firm B will start you at only \$22,000 per year, but guarantees you a \$1,300 raise each year for 10 years. Over the period of 10 years, what is the total amount each firm will pay you?
16. *Salary analysis.* In Problem 15, what would be your annual salary in each firm for the 10th year?
17. *Loan repayment.* If you borrow \$4,800 and repay the loan by paying \$200 per month to reduce the loan and 1% of the unpaid balance each month for the use of the money, what is the total cost of the loan over 24 months?
18. *Loan repayment.* Repeat Problem 17 replacing 1% with 1.5%.

SECTION A-4 Geometric Progressions

- GEOMETRIC PROGRESSIONS—DEFINITION
- SPECIAL FORMULAS
- INFINITE GEOMETRIC PROGRESSIONS

◆ GEOMETRIC PROGRESSIONS—DEFINITION

Consider the sequence of numbers

$$2, 6, 18, 54, \ldots, 2(3)^{n-1}, \ldots$$

where each number after the first is obtained from the preceding one by multiplying it by 3. This is an example of a *geometric progression*. In general:

Geometric Progression

A sequence of numbers

$$a_1, a_2, a_3, \ldots, a_n, \ldots$$

is called a **geometric progression** if there exists a nonzero constant r, called a **common ratio,** such that

$$\frac{a_n}{a_{n-1}} = r$$

That is,

$$a_n = ra_{n-1} \qquad \text{for every } n \geqslant 1 \qquad (1)$$

◆ EXAMPLE 16 Which of the following can be the first four terms of a geometric progression, and what is its common ratio?

(A) 5, 3, 1, −1, . . . (B) 1, 2, 4, 8, . . .

Solution The terms in B, with $r = 2$ ◆

PROBLEM 16 Which of the following can be the first four terms of a geometric progression, and what is its common ratio?

(A) $4, -2, 1, -\frac{1}{2}, \ldots$ (B) 2, 4, 6, 8, . . . ◆

◆ SPECIAL FORMULAS

Like arithmetic progressions, geometric progressions have several useful properties. It is easy to derive formulas for the nth term in terms of n and for the sum of any number of consecutive terms. To obtain a formula for the nth term of a geometric progression, we note that if a_1 is the first term and r is the common ratio, then

$$a_2 = ra_1$$
$$a_3 = ra_2 = r(ra_1) = r^2a_1 = a_1r^2$$
$$a_4 = ra_3 = r(r^2a_1) = r^3a_1 = a_1r^3$$

This suggests that:

$$a_n = a_1r^{n-1} \qquad \text{for all } n > 1 \qquad (2)$$

◆ EXAMPLE 17 Find the 8th term in the geometric progression: $\frac{1}{2}, \frac{1}{4}, \frac{1}{8}, \ldots$

Solution Find the common ratio r and use formula (2):

$$r = \tfrac{1}{2} \qquad n = 8 \qquad a_1 = \tfrac{1}{2}$$

Thus,

$$a_8 = (\tfrac{1}{2})(\tfrac{1}{2})^{8-1} = \tfrac{1}{256}$$ ◆

PROBLEM 17 Find the 7th term in the geometric progression: $\frac{1}{32}, -\frac{1}{16}, \frac{1}{8}, \ldots$ ◆

◆ EXAMPLE 18 If the 1st and 10th terms of a geometric progression are 2 and 4, respectively, find the common ratio r.

Solution

$$a_n = a_1 r^{n-1}$$
$$4 = 2 \cdot r^{10-1}$$
$$2 = r^9$$
$$r = 2^{1/9} \approx 1.08 \qquad \text{Use a calculator.}$$ ◆

PROBLEM 18 If the 1st and 8th terms of a geometric progression are 1,000 and 2,000, respectively, find the common ratio r. ◆

We now derive two very useful formulas for the sum of n consecutive terms of a geometric progression. Let

$$a_1, a_1r, a_1r^2, \ldots, a_1r^{n-2}, a_1r^{n-1}$$

be n terms of a geometric progression. Their sum is

$$S_n = a_1 + a_1r + a_2r^2 + \cdots + a_1r^{n-2} + a_1r^{n-1}$$

If we multiply both sides by r, we obtain

$$rS_n = a_1r + a_1r^2 + a_1r^3 + \cdots + a_1r^{n-1} + a_1r^n$$

Now combine these last two equations by subtraction to obtain

$$rS_n - S_n = (a_1r + a_1r^2 + a_1r^3 + \cdots + a_1r^{n-1} + a_1r^n) - (a_1 + a_1r + a_1r^2 + \cdots + a_1r^{n-2} + a_1r^{n-1})$$
$$(r-1)S_n = a_1r^n - a_1$$

Notice how many terms drop out on the right side. Solving for S_n, we have:

$$S_n = \frac{a_1(r^n - 1)}{r - 1} \qquad r \neq 1 \qquad (3)$$

Since $a_n = a_1 r^{n-1}$, or $ra_n = a_1 r^n$, formula (3) also can be written in the form:

$$S_n = \frac{ra_n - a_1}{r - 1} \qquad r \neq 1 \qquad (4)$$

◆ EXAMPLE 19 Find the sum of the first ten terms of the geometric progression: $1, 1.05, 1.05^2, \ldots$

Solution Use formula (3) with $a_1 = 1$, $r = 1.05$, and $n = 10$:

$$S_n = \frac{a_1(r^n - 1)}{r - 1}$$

$$S_{10} = \frac{1(1.05^{10} - 1)}{1.05 - 1}$$

$$\approx \frac{0.6289}{0.05} \approx 12.58$$

◆

PROBLEM 19 Find the sum of the first eight terms of the geometric progression: $100, 100(1.08), 100(1.08)^2, \ldots$ ◆

◆ INFINITE GEOMETRIC PROGRESSIONS

Given a geometric progression, what happens to the sum S_n of the first n terms as n increases without stopping? To answer this question, let us write formula (3) in the form

$$S_n = \frac{a_1 r^n}{r - 1} - \frac{a_1}{r - 1}$$

It is possible to show that if $|r| < 1$ (that is, $-1 < r < 1$), then r^n will approach 0 as n increases. (See what happens, for example, if you let $r = \frac{1}{2}$ and then increase n.) Thus, the first term above will approach 0 and S_n can be made as close as we please to the second term, $-a_1/(r - 1)$ [which can be written as $a_1/(1 - r)$], by taking n sufficiently large. Thus, if the common ratio r is between -1 and 1, we define the sum of an infinite geometric progression to be:

$$S_\infty = \frac{a_1}{1 - r} \qquad |r| < 1 \qquad (5)$$

If $r \leqslant -1$ or $r \geqslant 1$, then an infinite geometric progression has no sum.

◆ EXAMPLE 20

Economy Stimulation

The government has decided on a tax rebate program to stimulate the economy. Suppose you receive \$600 and you spend 80% of this, and each of the people who receive what you spend also spend 80% of what they receive, and this process continues without end. According to the **multiplier doctrine** in economics, the effect of your \$600 tax rebate on the economy is multiplied many times. What is the total amount spent if the process continues as indicated?

Solution

We need to find the sum of an infinite geometric progression with the first amount spent being $a_1 = (0.8)(\$600) = \$480$ and $r = 0.8$. Using formula (5), we obtain

$$S_\infty = \frac{a_1}{1 - r}$$

$$= \frac{\$480}{1 - 0.8} = \$2{,}400$$

Thus, assuming the process continues as indicated, we would expect the \$600 tax rebate to result in about \$2,400 of spending. ◆

PROBLEM 20

Repeat Example 20 with a tax rebate of \$1,000. ◆

Answers to Matched Problems

16. The terms in A, with $r = -\frac{1}{2}$ 17. 2 18. Approx. 1.104
19. 1,063.66 20. \$4,000

EXERCISE A-4

A

1. Determine which of the following can be the first three terms of a geometric progression. Find the common ratio r and the next two terms for those that are.

   (A) $1, -2, 4, \ldots$ (B) $7, 6, 5, \ldots$ (C) $2, 1, \frac{1}{2}, \ldots$
   (D) $2, -4, 6, \ldots$

2. Repeat Problem 1 for:

   (A) $4, -1, -6, \ldots$ (B) $15, 5, \frac{5}{3}, \ldots$ (C) $\frac{1}{4}, -\frac{1}{2}, 1, \ldots$
   (D) $\frac{1}{2}, \frac{2}{3}, \frac{3}{4}, \ldots$

Let $a_1, a_2, a_3, \ldots, a_n, \ldots$ be a geometric progression and S_n be the sum of the first n terms. In Problems 3–12, find the indicated quantities.

3. $a_1 = 3$; $r = -2$; $a_2 = ?$; $a_3 = ?$; $a_4 = ?$
4. $a_1 = 32$; $r = -\frac{1}{2}$; $a_2 = ?$; $a_3 = ?$; $a_4 = ?$
5. $a_1 = 1$; $a_7 = 729$; $r = -3$; $S_7 = ?$
6. $a_1 = 3$; $a_7 = 2{,}187$; $r = 3$; $S_7 = ?$

B

7. $a_1 = 100; \quad r = 1.08; \quad a_{10} = ?$
8. $a_1 = 240; \quad r = 1.06; \quad a_{12} = ?$
9. $a_1 = 100; \quad a_9 = 200; \quad r = ?$
10. $a_1 = 100; \quad a_{10} = 300; \quad r = ?$
11. $a_1 = 500; \quad r = 0.6; \quad S_{10} = ?; \quad S_\infty = ?$
12. $a_1 = 8{,}000; \quad r = 0.4; \quad S_{10} = ?; \quad S_\infty = ?$
13. Find the sum of each infinite geometric progression (if it exists).

    (A) 2, 4, 8, . . . (B) $2, -\frac{1}{2}, \frac{1}{8}, \ldots$
14. Repeat Problem 13 for:

    (A) 16, 4, 1, . . . (B) 1, −3, 9, . . .

C

15. Find $f(1) + f(2) + \cdots + f(10)$ if $f(x) = (\frac{1}{2})^x$.
16. Find $g(1) + g(2) + \cdots + g(10)$ if $g(x) = 2^x$.

APPLICATIONS

Business & Economics

17. *Economy stimulation.* The government, through a subsidy program, distributes $5,000,000. If we assume each individual or agency spends 70% of what is received, and 70% of this is spent, and so on, how much total increase in spending results from this government action? (Let $a_1 =$ $3,500,000.)
18. *Economy stimulation.* Repeat Problem 17 using $10,000,000 as the amount distributed and 80%.
19. *Cost-of-living adjustment.* If the cost-of-living index increased 5% for each of the past 10 years and you had a salary agreement that increased your salary by the same percentage each year, what would your present salary be if you had a salary of $20,000 per year 10 years ago? What would be your total earnings in the past 10 years? [*Hint:* $r = 1.05$]
20. *Depreciation.* In **straight-line depreciation,** an asset less its salvage value at the end of its useful life is depreciated (for tax purposes) in equal annual amounts over its useful life. Thus, a $100,000 company airplane with a salvage value of $20,000 at the end of 10 years would be depreciated at $8,000 per year for each of the 10 years.

    Since certain assets, such as airplanes, cars, and so on, depreciate more rapidly during the early years of their useful life, several methods of depreciation that take this into consideration are available to the taxpayer. One such method is called the **method of declining-balance.** The rate used cannot exceed double that used for straight-line depreciation (ignoring salvage value) and is applied to the remaining value of an asset after the previous year's depreciation has been deducted. In our airplane example, the annual rate of straight-line depreciation over the period of 10 years is 10%. Let us assume we can double this rate for the method of declining-balance. At some point before the salvage value is reached (taxpayer's choice), we must switch over to the straight-line method to depreciate the final amount of the asset.

The table illustrates the two methods of depreciation for the company airplane.

YEAR END	STRAIGHT-LINE METHOD *Amount Depreciated*	*Asset Value*	DECLINING-BALANCE METHOD *Amount Depreciated*	*Asset Value*
0	\$ 0	\$100,000	\$ 0	\$100,000
1	0.1(80,000) = 8,000	92,000	0.2(100,000) = 20,000	80,000
2	0.1(80,000) = 8,000	84,000	0.2(80,000) = 16,000	64,000
3	0.1(80,000) = 8,000	76,000	0.2(64,000) = 12,800	51,200
.	.	.	.	.
.	.	.	.	.
.	.	.	.	.
7	0.1(80,000) = 8,000	44,000	0.2(26,214) = 5,243	20,972
8	0.1(80,000) = 8,000	36,000	$\frac{972}{3}$ = 324	20,648
9	0.1(80,000) = 8,000	28,000	$\frac{972}{3}$ = 324	20,324
10	0.1(80,000) = 8,000	20,000	$\frac{972}{3}$ = 324	20,000
		Arithmetic progression		Geometric progression above dashed line

Shift to straight-line; otherwise next entry will drop below salvage value.

(A) For the declining-balance method, find the sum of the depreciation amounts above the dashed line using formula (4), and then add the entries below the line to this result.

(B) Repeat part A using formula (3).

(C) Find the asset value under declining-balance at the end of the 5th year using formula (2).

(D) Find the asset value under straight-line at the end of the 5th year using formula (2) in the preceding section.

SECTION A-5 The Binomial Theorem

- FACTORIAL
- BINOMIAL THEOREM—DEVELOPMENT

The binomial form

$$(a + b)^n$$

where n is a natural number, appears more frequently than you might expect. The coefficients in the expansion play an important role in probability studies. The *binomial formula*, which we will derive informally, enables us to expand $(a + b)^n$ directly for n any natural number. Since the formula involves *factorials*, we digress for a moment here to introduce this important concept.

◆ FACTORIAL

For n a natural number, ***n* factorial,** denoted by ***n*!**, is the product of the first n natural numbers. **Zero factorial** is defined to be 1. That is:

$$n! = n \cdot (n-1) \cdot \cdots \cdot 2 \cdot 1$$
$$1! = 1$$
$$0! = 1$$

It is also useful to note that:

$$n! = n \cdot (n-1)! \qquad n \geq 1$$

◆ EXAMPLE 21 Evaluate each:

(A) $5! = 5 \cdot 4 \cdot 3 \cdot 2 \cdot 1 = 120$ (B) $\frac{8!}{7!} = \frac{8 \cdot \cancel{7!}}{\cancel{7!}} = 8$

(C) $\frac{10!}{7!} = \frac{10 \cdot 9 \cdot 8 \cdot \cancel{7!}}{\cancel{7!}} = 720$ ◆

PROBLEM 21 Evaluate each: (A) $4!$ (B) $\frac{7!}{6!}$ (C) $\frac{8!}{5!}$ ◆

The following important formula involving factorials has applications in many areas of mathematics and statistics. We will use this formula to provide a more concise form for the expressions encountered later in this discussion.

For n and r integers satisfying $0 \leq r \leq n$,

$$C_{n,r} = \frac{n!}{r!(n-r)!}$$

◆ EXAMPLE 22 (A) $C_{9,2} = \frac{9!}{2!(9-2)!} = \frac{9!}{2!7!} = \frac{9 \cdot 8 \cdot \cancel{7!}}{2 \cdot \cancel{7!}} = 36$

(B) $C_{5,5} = \frac{5!}{5!(5-5)!} = \frac{5!}{5!0!} = \frac{5!}{5!} = 1$ ◆

PROBLEM 22 Find: (A) $C_{5,2}$ (B) $C_{6,0}$ ◆

◆ BINOMIAL THEOREM — DEVELOPMENT

Let us expand $(a + b)^n$ for several values of n to see if we can observe a pattern that leads to a general formula for the expansion for any natural number n:

$$(a + b)^1 = a + b$$
$$(a + b)^2 = a^2 + 2ab + b^2$$
$$(a + b)^3 = a^3 + 3a^2b + 3ab^2 + b^3$$
$$(a + b)^4 = a^4 + 4a^3b + 6a^2b^2 + 4ab^3 + b^4$$
$$(a + b)^5 = a^5 + 5a^4b + 10a^3b^2 + 10a^2b^3 + 5ab^4 + b^5$$

Observations

1. The expansion of $(a + b)^n$ has $(n + 1)$ terms.
2. The power of a decreases by 1 for each term as we move from left to right.
3. The power of b increases by 1 for each term as we move from left to right.
4. In each term, the sum of the powers of a and b always equals n.
5. Starting with a given term, we can get the coefficient of the next term by multiplying the coefficient of the given term by the exponent of a and dividing by the number that represents the position of the term in the series of terms. For example, in the expansion of $(a + b)^4$ above, the coefficient of the third term is found from the second term by multiplying 4 and 3, and then dividing by 2 [that is, the coefficient of the third term $= (4 \cdot 3)/2 = 6$].

We now postulate these same properties for the general case:

$$(a + b)^n = a^n + \frac{n}{1} a^{n-1}b + \frac{n(n-1)}{1 \cdot 2} a^{n-2}b^2 + \frac{n(n-1)(n-2)}{1 \cdot 2 \cdot 3} a^{n-3}b^3 + \cdots + b^n$$
$$= \frac{n!}{0!(n-0)!} a^n + \frac{n!}{1!(n-1)!} a^{n-1}b + \frac{n!}{2!(n-2)!} a^{n-2}b^2 + \frac{n!}{3!(n-3)!} a^{n-3}b^3 + \cdots + \frac{n!}{n!(n-n)!} b^n$$
$$= C_{n,0}a^n + C_{n,1}a^{n-1}b + C_{n,2}a^{n-2}b^2 + C_{n,3}a^{n-3}b^3 + \cdots + C_{n,n}b^n$$

And we are led to the formula in the binomial theorem (a formal proof requires mathematical induction, which is beyond the scope of this book):

Binomial Theorem

For all natural numbers n,

$$(a + b)^n = C_{n,0}a^n + C_{n,1}a^{n-1}b + C_{n,2}a^{n-2}b^2 + C_{n,3}a^{n-3}b^3 + \cdots + C_{n,n}b^n$$

◆ EXAMPLE 23 Use the binomial formula to expand $(u + v)^6$.

Solution

$$(u + v)^6 = C_{6,0}u^6 + C_{6,1}u^5v + C_{6,2}u^4v^2 + C_{6,3}u^3v^3 + C_{6,4}u^2v^4 + C_{6,5}uv^5 + C_{6,6}v^6$$
$$= u^6 + 6u^5v + 15u^4v^2 + 20u^3v^3 + 15u^2v^4 + 6uv^5 + v^6$$ ◆

PROBLEM 23 Use the binomial formula to expand $(x + 2)^5$. ◆

◆ EXAMPLE 24 Use the binomial formula to find the sixth term in the expansion of $(x - 1)^{18}$.

Solution

$$\text{Sixth term} = C_{18,5}x^{13}(-1)^5 = \frac{18!}{5!(18-5)!}x^{13}(-1)$$
$$= -8{,}568x^{13}$$ ◆

PROBLEM 24 Use the binomial formula to find the fourth term in the expansion of $(x - 2)^{20}$. ◆

Answers to Matched Problems

21. (A) 24 (B) 7 (C) 336 22. (A) 10 (B) 1
23. $x^5 + 5x^4 \cdot 2 + 10x^3 \cdot 2^2 + 10x^2 \cdot 2^3 + 5x \cdot 2^4 + 2^5$
$= x^5 + 10x^4 + 40x^3 + 80x^2 + 80x + 32$
24. $-9{,}120x^{17}$

EXERCISE A-5

A

Evaluate.

1. $6!$
2. $7!$
3. $\frac{10!}{9!}$
4. $\frac{20!}{19!}$
5. $\frac{12!}{9!}$
6. $\frac{10!}{6!}$
7. $\frac{5!}{2!3!}$
8. $\frac{7!}{3!4!}$
9. $\frac{6!}{5!((6-5)!}$
10. $\frac{7!}{4!(7-4)!}$
11. $\frac{20!}{3!17!}$
12. $\frac{52!}{50!2!}$

B

Evaluate

13. $C_{5,3}$
14. $C_{7,3}$
15. $C_{6,5}$
16. $C_{7,4}$
17. $C_{5,0}$
18. $C_{5,5}$
19. $C_{18,15}$
20. $C_{18,3}$

Expand each expression using the binomial formula.

21. $(a + b)^4$
22. $(m + n)^5$
23. $(x - 1)^6$
24. $(u - 2)^5$
25. $(2a - b)^5$
26. $(x - 2y)^5$

Find the indicated term in each expansion.

27. $(x - 1)^{18}$; fifth term
28. $(x - 3)^{20}$; third term
29. $(p + q)^{15}$; seventh term
30. $(p + q)^{15}$; thirteenth term
31. $(2x + y)^{12}$; eleventh term
32. $(2x + y)^{12}$; third term

C

33. Show that $C_{n,0} = C_{n,n}$.
34. Show that $C_{n,r} = C_{n,n-r}$.
35. The triangle below is called **Pascal's triangle.** Can you guess what the next two rows at the bottom are? Compare these numbers with the coefficients of binomial expansions.

$$
\begin{array}{ccccccccc}
 & & & & 1 & & & & \\
 & & & 1 & & 1 & & & \\
 & & 1 & & 2 & & 1 & & \\
 & 1 & & 3 & & 3 & & 1 & \\
1 & & 4 & & 6 & & 4 & & 1
\end{array}
$$

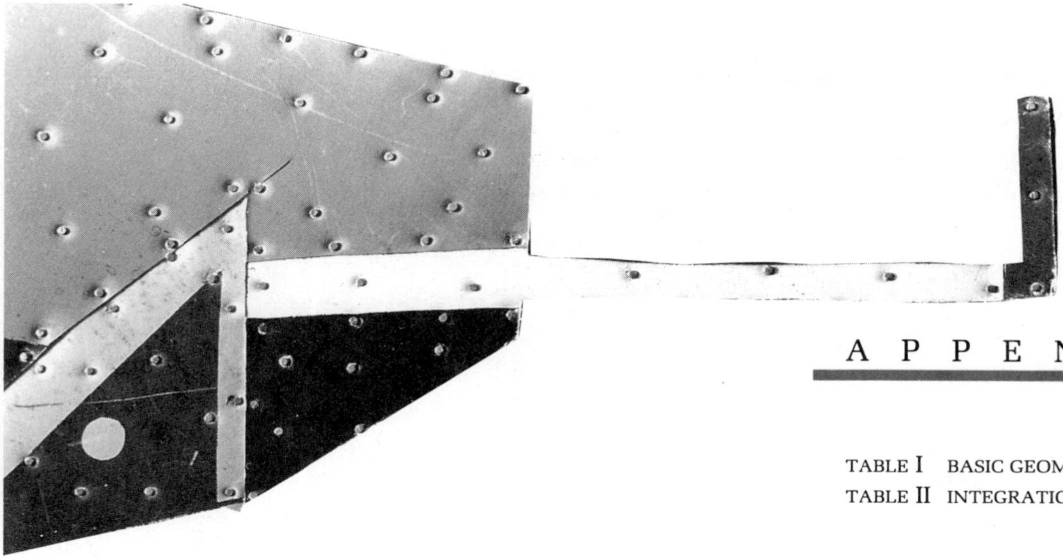

APPENDIX B

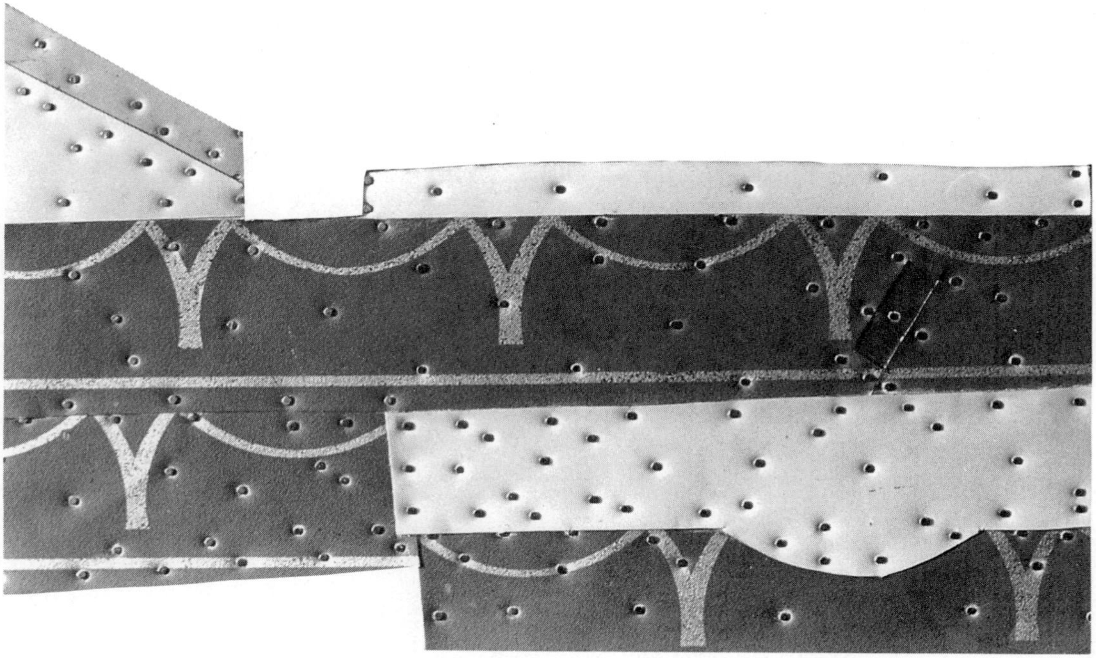

Tables

TABLE I

Basic Geometric Formulas

1. SIMILAR TRIANGLES

(A) Two triangles are similar if two angles of one triangle have the same measure as two angles of the other.
(B) If two triangles are similar, their corresponding sides are proportional:

$$\frac{a}{a'} = \frac{b}{b'} = \frac{c}{c'}$$

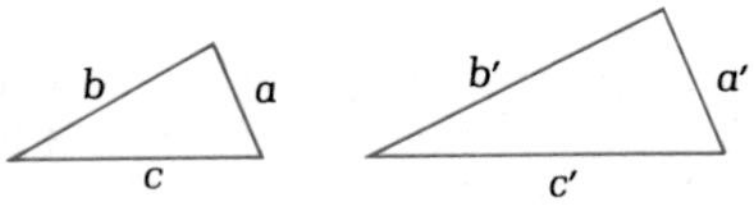

2. PYTHAGOREAN THEOREM

$c^2 = a^2 + b^2$

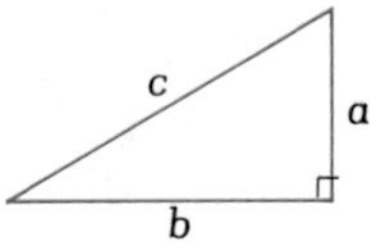

3. RECTANGLE

$A = ab$ Area
$P = 2a + 2b$ Perimeter

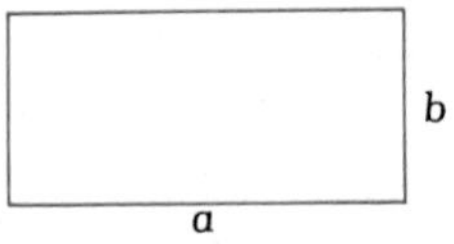

4. PARALLELOGRAM

$h =$ Height
$A = ah = ab \sin \theta$ Area
$P = 2a + 2b$ Perimeter

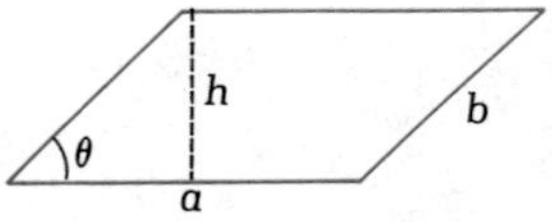

5. TRIANGLE

$h =$ Height
$A = \frac{1}{2}hc$ Area
$P = a + b + c$ Perimeter
$s = \frac{1}{2}(a + b + c)$ Semiperimeter
$A = \sqrt{s(s - a)(s - b)(s - c)}$ Area—Heron's formula

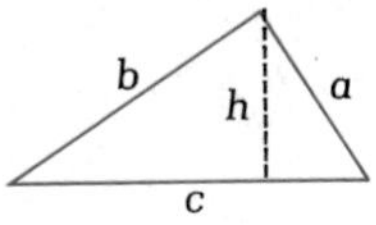

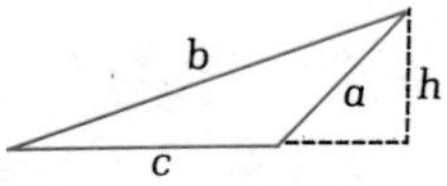

6. TRAPEZOID

Base a is parallel to base b.

$h =$ Height
$A = \frac{1}{2}(a + b)h$ Area

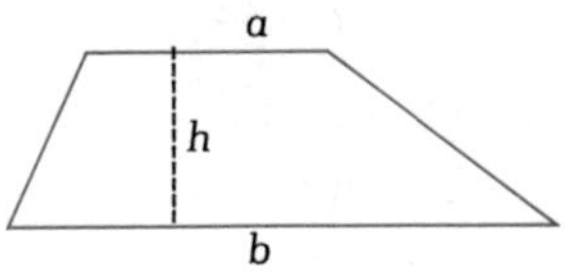

7. CIRCLE

R = Radius

D = Diameter

$D = 2R$

$A = \pi R^2 = \frac{1}{4}\pi D^2$ Area

$C = 2\pi R = \pi D$ Circumference

$\dfrac{C}{D} = \pi$ For all circles

$\pi \approx 3.141\ 59$

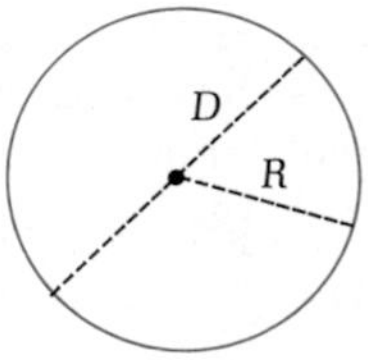

8. RECTANGULAR SOLID

$V = abc$ Volume

$T = 2ab + 2ac + 2bc$ Total surface area

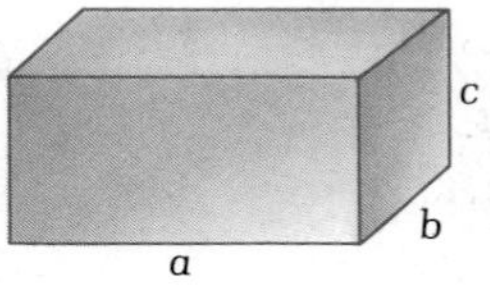

9. RIGHT CIRCULAR CYLINDER

R = Radius of base

h = Height

$V = \pi R^2 h$ Volume

$S = 2\pi Rh$ Lateral surface area

$T = 2\pi R(R + h)$ Total surface area

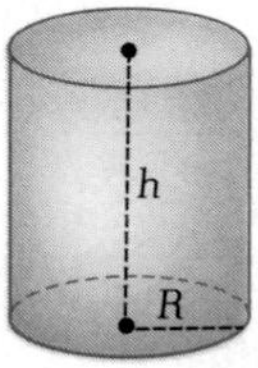

10. RIGHT CIRCULAR CONE

R = Radius of base

h = Height

s = Slant height

$V = \frac{1}{3}\pi R^2 h$ Volume

$S = \pi R s = \pi R\sqrt{R^2 + h^2}$ Lateral surface area

$T = \pi R(R + s) = \pi R(R + \sqrt{R^2 + h^2})$ Total surface area

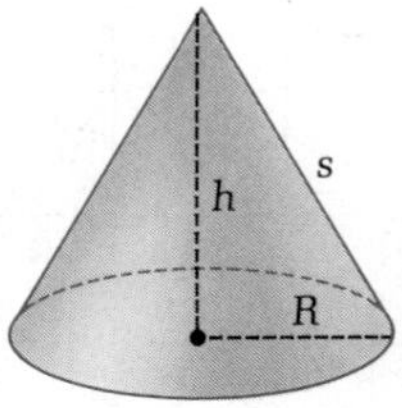

11. SPHERE

R = Radius

D = Diameter

$D = 2R$

$V = \frac{4}{3}\pi R^3 = \frac{1}{6}\pi D^3$ Volume

$S = 4\pi R^2 = \pi D^2$ Surface area

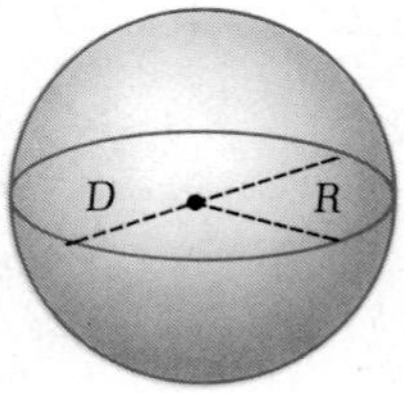

TABLE II

Integration Formulas

[*Note:* The constant of integration is omitted for each integral, but must be included in any particular application of a formula. The variable u is the variable of integration; all other symbols represent constants.]

◆ INTEGRALS INVOLVING u^n

1. $\displaystyle\int u^n\,du = \frac{u^{n+1}}{n+1}, \qquad n \neq -1$
2. $\displaystyle\int u^{-1}\,du = \int \frac{1}{u}\,du = \ln|u|$

◆ INTEGRALS INVOLVING $a + bu$; $a \neq 0$ AND $b \neq 0$

3. $\displaystyle\int \frac{1}{a+bu}\,du = \frac{1}{b}\ln|a+bu|$
4. $\displaystyle\int \frac{u}{a+bu}\,du = \frac{u}{b} - \frac{a}{b^2}\ln|a+bu|$
5. $\displaystyle\int \frac{u^2}{a+bu}\,du = \frac{(a+bu)^2}{2b^3} - \frac{2a(a+bu)}{b^3} + \frac{a^2}{b^3}\ln|a+bu|$
6. $\displaystyle\int \frac{u}{(a+bu)^2}\,du = \frac{1}{b^2}\left(\ln|a+bu| + \frac{a}{a+bu}\right)$
7. $\displaystyle\int \frac{u^2}{(a+bu)^2}\,du = \frac{(a+bu)}{b^3} - \frac{a^2}{b^3(a+bu)} - \frac{2a}{b^3}\ln|a+bu|$
8. $\displaystyle\int u(a+bu)^n\,du = \frac{(a+bu)^{n+2}}{(n+2)b^2} - \frac{a(a+bu)^{n+1}}{(n+1)b^2}, \qquad n \neq -1, -2$
9. $\displaystyle\int \frac{1}{u(a+bu)}\,du = \frac{1}{a}\ln\left|\frac{u}{a+bu}\right|$
10. $\displaystyle\int \frac{1}{u^2(a+bu)}\,du = -\frac{1}{au} + \frac{b}{a^2}\ln\left|\frac{a+bu}{u}\right|$
11. $\displaystyle\int \frac{1}{u(a+bu)^2}\,du = \frac{1}{a(a+bu)} + \frac{1}{a^2}\ln\left|\frac{u}{a+bu}\right|$
12. $\displaystyle\int \frac{1}{u^2(a+bu)^2}\,du = -\frac{a+2bu}{a^2u(a+bu)} + \frac{2b}{a^3}\ln\left|\frac{a+bu}{u}\right|$

◆ INTEGRALS INVOLVING $a^2 - u^2$, $a > 0$

13. $\displaystyle\int \frac{1}{u^2-a^2}\,du = \frac{1}{2a}\ln\left|\frac{u-a}{u+a}\right|$
14. $\displaystyle\int \frac{1}{a^2-u^2}\,du = \frac{1}{2a}\ln\left|\frac{u+a}{u-a}\right|$

◆ INTEGRALS INVOLVING $(a + bu)$ AND $(c + du)$; $b \neq 0$, $d \neq 0$, AND $ad - bc \neq 0$

15. $$\int \frac{1}{(a+bu)(c+du)}\,du = \frac{1}{ad-bc}\ln\left|\frac{c+du}{a+bu}\right|$$

16. $$\int \frac{u}{(a+bu)(c+du)}\,du = \frac{1}{ad-bc}\left(\frac{a}{b}\ln|a+bu| - \frac{c}{d}\ln|c+du|\right)$$

17. $$\int \frac{u^2}{(a+bu)(c+du)}\,du = \frac{1}{bd}\,u - \frac{1}{ad-bc}\left(\frac{a^2}{b^2}\ln|a+bu| - \frac{c^2}{d^2}\ln|c+du|\right)$$

18. $$\int \frac{1}{(a+bu)^2(c+du)}\,du = \frac{1}{ad-bc}\,\frac{1}{a+bu} + \frac{d}{(ad-bc)^2}\ln\left|\frac{c+du}{a+bu}\right|$$

19. $$\int \frac{u}{(a+bu)^2(c+du)}\,du = -\frac{a}{b(ad-bc)}\,\frac{1}{a+bu} - \frac{c}{(ad-bc)^2}\ln\left|\frac{c+du}{a+bu}\right|$$

20. $$\int \frac{a+bu}{c+du}\,du = \frac{bu}{d} + \frac{ad-bc}{d^2}\ln|c+du|$$

◆ INTEGRALS INVOLVING $\sqrt{a + bu}$, $a \neq 0$ AND $b \neq 0$

21. $$\int \sqrt{a+bu}\,du = \frac{2\sqrt{(a+bu)^3}}{3b}$$

22. $$\int u\sqrt{a+bu}\,du = \frac{2(3bu-2a)}{15b^2}\sqrt{(a+bu)^3}$$

23. $$\int u^2\sqrt{a+bu}\,du = \frac{2(15b^2u^2-12abu+8a^2)}{105b^3}\sqrt{(a+bu)^3}$$

24. $$\int \frac{1}{\sqrt{a+bu}}\,du = \frac{2\sqrt{a+bu}}{b}$$

25. $$\int \frac{u}{\sqrt{a+bu}}\,du = \frac{2(bu-2a)}{3b^2}\sqrt{a+bu}$$

26. $$\int \frac{u^2}{\sqrt{a+bu}}\,du = \frac{2(3b^2u^2-4abu+8a^2)}{15b^3}\sqrt{a+bu}$$

27. $$\int \frac{1}{u\sqrt{a+bu}}\,du = \frac{1}{\sqrt{a}}\ln\left|\frac{\sqrt{a+bu}-\sqrt{a}}{\sqrt{a+bu}+\sqrt{a}}\right|, \qquad a > 0$$

28. $$\int \frac{1}{u^2\sqrt{a+bu}}\,du = -\frac{\sqrt{a+bu}}{au} - \frac{b}{2a\sqrt{a}}\ln\left|\frac{\sqrt{a+bu}-\sqrt{a}}{\sqrt{a+bu}+\sqrt{a}}\right|, \qquad a > 0$$

◆ INTEGRALS INVOLVING $\sqrt{a^2 - u^2}$, $a > 0$

29. $$\int \frac{1}{u\sqrt{a^2-u^2}}\,du = -\frac{1}{a}\ln\left|\frac{a+\sqrt{a^2-u^2}}{u}\right|$$

30. $\displaystyle\int \frac{1}{u^2\sqrt{a^2-u^2}}\,du = -\frac{\sqrt{a^2-u^2}}{a^2u}$

31. $\displaystyle\int \frac{\sqrt{a^2-u^2}}{u}\,du = \sqrt{a^2-u^2} - a\ln\left|\frac{a+\sqrt{a^2-u^2}}{u}\right|$

◆ INTEGRALS INVOLVING $\sqrt{u^2+a^2}$, $a > 0$

32. $\displaystyle\int \sqrt{u^2+a^2}\,du = \frac{1}{2}\left(u\sqrt{u^2+a^2} + a^2\ln|u+\sqrt{u^2+a^2}|\right)$

33. $\displaystyle\int u^2\sqrt{u^2+a^2}\,du = \frac{1}{8}\left[u(2u^2+a^2)\sqrt{u^2+a^2} - a^4\ln|u+\sqrt{u^2+a^2}|\right]$

34. $\displaystyle\int \frac{\sqrt{u^2+a^2}}{u}\,du = \sqrt{u^2+a^2} - a\ln\left|\frac{a+\sqrt{u^2+a^2}}{u}\right|$

35. $\displaystyle\int \frac{\sqrt{u^2+a^2}}{u^2}\,du = -\frac{\sqrt{u^2+a^2}}{u} + \ln|u+\sqrt{u^2+a^2}|$

36. $\displaystyle\int \frac{1}{\sqrt{u^2+a^2}}\,du = \ln|u+\sqrt{u^2+a^2}|$

37. $\displaystyle\int \frac{1}{u\sqrt{u^2+a^2}}\,du = \frac{1}{a}\ln\left|\frac{u}{a+\sqrt{u^2+a^2}}\right|$

38. $\displaystyle\int \frac{u^2}{\sqrt{u^2+a^2}}\,du = \frac{1}{2}\left(u\sqrt{u^2+a^2} - a^2\ln|u+\sqrt{u^2+a^2}|\right)$

39. $\displaystyle\int \frac{1}{u^2\sqrt{u^2+a^2}}\,du = -\frac{\sqrt{u^2+a^2}}{a^2u}$

◆ INTEGRALS INVOLVING $\sqrt{u^2-a^2}$, $a > 0$

40. $\displaystyle\int \sqrt{u^2-a^2}\,du = \frac{1}{2}\left(u\sqrt{u^2-a^2} - a^2\ln|u+\sqrt{u^2-a^2}|\right)$

41. $\displaystyle\int u^2\sqrt{u^2-a^2}\,du = \frac{1}{8}\left[u(2u^2-a^2)\sqrt{u^2-a^2} - a^4\ln|u+\sqrt{u^2-a^2}|\right]$

42. $\displaystyle\int \frac{\sqrt{u^2-a^2}}{u^2}\,du = -\frac{\sqrt{u^2-a^2}}{u} + \ln|u+\sqrt{u^2-a^2}|$

43. $\displaystyle\int \frac{1}{\sqrt{u^2-a^2}}\,du = \ln|u+\sqrt{u^2-a^2}|$

44. $\displaystyle\int \frac{u^2}{\sqrt{u^2-a^2}}\,du = \frac{1}{2}\left(u\sqrt{u^2-a^2} + a^2\ln|u+\sqrt{u^2-a^2}|\right)$

45. $\displaystyle\int \frac{1}{u^2\sqrt{u^2-a^2}}\,du = \frac{\sqrt{u^2-a^2}}{a^2u}$

◆ INTEGRALS INVOLVING e^{au}, $a \neq 0$

46. $\displaystyle\int e^{au}\,du = \frac{e^{au}}{a}$

47. $$\int u^n e^{au}\,du = \frac{u^n e^{au}}{a} - \frac{n}{a}\int u^{n-1}e^{au}\,du$$

48. $$\int \frac{1}{c + de^{au}}\,du = \frac{u}{c} - \frac{1}{ac}\ln|c + de^{au}|, \quad c \neq 0$$

◆ INTEGRALS INVOLVING $\ln u$

49. $$\int \ln u\,du = u\ln u - u$$

50. $$\int \frac{\ln u}{u}\,du = \frac{1}{2}(\ln u)^2$$

51. $$\int u^n \ln u\,du = \frac{u^{n+1}}{n+1}\ln u - \frac{u^{n+1}}{(n+1)^2}, \quad n \neq -1$$

52. $$\int (\ln u)^n\,du = u(\ln u)^n - n\int (\ln u)^{n-1}\,du$$

◆ INTEGRALS INVOLVING TRIGONOMETRIC FUNCTIONS OF $au, \quad a \neq 0$

53. $$\int \sin au\,du = -\frac{1}{a}\cos au$$

54. $$\int \cos au\,du = \frac{1}{a}\sin au$$

55. $$\int \tan au\,du = -\frac{1}{a}\ln|\cos au|$$

56. $$\int \cot au\,du = \frac{1}{a}\ln|\sin au|$$

57. $$\int \sec au\,du = \frac{1}{a}\ln|\sec au + \tan au|$$

58. $$\int \csc au\,du = \frac{1}{a}\ln|\csc au - \cot au|$$

59. $$\int (\sin au)^2\,du = \frac{u}{2} - \frac{1}{4a}\sin 2au$$

60. $$\int (\cos au)^2\,du = \frac{u}{2} + \frac{1}{4a}\sin 2au$$

61. $$\int (\sin au)^n\,du = -\frac{1}{an}(\sin au)^{n-1}\cos au + \frac{n-1}{n}\int (\sin au)^{n-2}\,du, \qquad n \neq 0$$

62. $$\int (\cos au)^n\,du = \frac{1}{an}\sin au(\cos au)^{n-1} + \frac{n-1}{n}\int (\cos au)^{n-2}\,du, \qquad n \neq 0$$

Answers

CHAPTER 1

EXERCISE 1-1

1. vu **3.** $(3 + 7) + y$ **5.** $u + v$ **7.** Associative ($\cdot$) **9.** Negatives **11.** Subtraction **13.** Division **15.** Distributive **17.** Zero **19.** Associative ($\cdot$) **21.** Distributive **23.** Negatives **25.** Zero **27.** No **29.** (A) F (B) T (C) T **31.** $\sqrt{2}$ and π are two examples of infinitely many. **33.** (A) N, Z, Q, R (B) R (C) Q, R (D) Q, R **35.** (A) T (B) F, since, for example, $(8 - 4) - 2 \neq 8 - (4 - 2)$. (C) T (D) F, since, for example, $(8 \div 4) \div 2 \neq 8 \div (4 \div 2)$. **37.** $\frac{1}{11}$ **39.** 1. Commutative 2. Associative 3. Inverse 4. Identity **41.** (A) 2.166 666 666... (B) 4.582 575 69... (C) 0.437 500 000... (D) 0.261 261 261...

EXERCISE 1-2

1. 3 **3.** $x^3 + 4x^2 - 2x + 5$ **5.** $x^3 + 1$ **7.** $2x^5 + 3x^4 - 2x^3 + 11x^2 - 5x + 6$ **9.** $-5u + 2$ **11.** $6a^2 + 6a$ **13.** $a^2 - b^2$ **15.** $6x^2 - 7x - 5$ **17.** $2x^2 + xy - 6y^2$ **19.** $9y^2 - 4$ **21.** $6m^2 - mn - 35n^2$ **23.** $16m^2 - 9n^2$ **25.** $9u^2 + 24uv + 16v^2$ **27.** $a^3 - b^3$ **29.** $16x^2 + 24xy + 9y^2$ **31.** 1 **33.** $x^4 - 2x^2y^2 + y^4$ **35.** $5a^2 + 12ab - 10b^2$ **37.** $x^3 - 6x^2y + 12xy^2 - 8y^3$ **39.** $2x^2 - 2xy + 3y^2$ **41.** $8x^3 - 20x^2 + 20x + 1$ **43.** $4x^3 - 14x^2 + 8x - 6$ **45.** $m + n$ **47.** $0.09x + 0.12(10{,}000 - x) = 1{,}200 - 0.03x$ **49.** $10x + 30(3x) + 50(4{,}000 - x - 3x) = 200{,}000 - 100x$ **51.** $0.02x + 0.06(10 - x) = 0.6 - 0.04x$

EXERCISE 1-3

1. $3m^2(2m^2 - 3m - 1)$ **3.** $2uv(4u^2 - 3uv + 2v^2)$ **5.** $(7m + 5)(2m - 3)$ **7.** $(a - 4b)(3c + d)$ **9.** $(3y + 2)(y - 1)$ **11.** $(u - 5v)(u + 3v)$ **13.** Prime **15.** $(wx - y)(wx + y)$ **17.** $(3m - n)^2$ **19.** Prime **21.** $4(z - 3)(z - 4)$ **23.** $2x^2(x - 2)(x - 10)$ **25.** $x(2y - 3)^2$ **27.** $(2m - 3n)(3m + 4n)$ **29.** $uv(2u - v)(2u + v)$ **31.** $2x(x^2 - x + 4)$ **33.** $(r - t)(r^2 + rt + t^2)$ **35.** $(a + 1)(a^2 - a + 1)$ **37.** $[(x + 2) - 3y][(x + 2) + 3y]$ **39.** Prime **41.** $(6x - 6y - 1)(x - y + 4)$ **43.** $(y - 2)(y + 2)(y^2 + 1)$ **45.** $a^2(3 + ab)(9 - 3ab + a^2b^2)$

EXERCISE 1-4

1. $8d^6$ **3.** $\frac{15x^2 + 10x - 6}{180}$ **5.** $\frac{15m^2 + 14m - 6}{36m^3}$ **7.** $\frac{1}{x(x - 4)}$ **9.** $\frac{x}{x + 5}$ **11.** $\frac{x - 5}{(x - 1)^2(x + 1)}$ **13.** $\frac{2}{x - 1}$ **15.** $\frac{5}{a - 1}$ **17.** $\frac{2}{x + y}$ **19.** $\frac{7x^2 - 2x - 3}{6(x + 1)^2}$ **21.** $-\frac{1}{x}$ **23.** $\frac{-17c + 16}{15(c - 1)}$ **25.** 1 **27.** $\frac{2x + 5}{(x + 4)(x + 1)}$ **29.** $\frac{1}{x - 3}$ **31.** $\frac{1 - m}{m}$ **33.** $-cd$ **35.** $\frac{x - y}{x + y}$ **37.** 1 **39.** x

EXERCISE 1-5

1. $2/x^9$ **3.** $3w^7/2$ **5.** $2/x^3$ **7.** $1/w^5$ **9.** 5 **11.** $1/a^6$ **13.** y^6/x^{12} **15.** x **17.** $a^2\sqrt{a}$ **19.** $3x^2\sqrt{2}$ **21.** $\sqrt{m}/m$ **23.** $\sqrt{6}/3$ **25.** $\sqrt{2x}/x$ **27.** 8.23×10^{10} **29.** 7.83×10^{-1} **31.** 3.4×10^{-5} **33.** 1 **35.** 10^{14} **37.** $y^6/25x^4$ **39.** 4×10^2 **41.** $4y^3/3x^5$ **43.** $y^{12}/8x^6$ **45.** x **47.** uv **49.** $\frac{7}{4} - \frac{1}{4}x^{-3}$ **51.** $\frac{3}{4}x - x^{-1} - \frac{1}{4}x^{-3}$ **53.** $3x^4y^2z\sqrt{2y}$ **55.** $4\sqrt{3x}/x$ **57.** $\sqrt{42xy}/7y$ **59.** $2a\sqrt{3ab}/3b$ **61.** $6m^3n^3$ **63.** $2a\sqrt{3ab}/3b$ **65.** $\frac{15\sqrt{x} + 10x}{9 - 4x}$ **67.** $\frac{5\sqrt{6} - 6}{19}$ **69.** 2.4×10^{10}; 24,000,000,000 **71.** 3.125×10^4; 31,250 **73.** t^2/x^2y^{10} **75.** 4 **77.** $\frac{1}{2(x - 3)^2}$ **79.** $\frac{bc(c + b)}{c^2 + bc + b^2}$ **81.** $\frac{\sqrt{2}}{2}$ **83.** $\frac{2\sqrt{x - 2}}{x - 2}$ **85.** $\frac{1}{\sqrt{t} + \sqrt{x}}$ **87.** $\frac{1}{\sqrt{x + h} + \sqrt{x}}$

◆ EXERCISE 1-6

1. $6\sqrt[5]{x^3}$ **3.** $\sqrt[5]{(4xy^3)^2}$ **5.** $\sqrt{x^2+y^2}$ (not $x+y$) **7.** $5x^{3/4}$ **9.** $(2x^2y)^{3/5}$ **11.** $x^{1/3}+y^{1/3}$ **13.** 5 **15.** 64 **17.** -6
19. Not a rational number (not even a real number) **21.** $\frac{8}{125}$ **23.** $\frac{1}{27}$ **25.** $x^{2/5}$ **27.** m **29.** $2x/y^2$ **31.** $xy^2/2$ **33.** $2/3x^{7/12}$
35. $2y^2/3x^2$ **37.** $12x-6x^{35/4}$ **39.** $3u-13u^{1/2}v^{1/2}+4v$ **41.** $25m-n$ **43.** $9x-6x^{1/2}y^{1/2}+y$ **45.** $\frac{1}{2}x^{1/3}+x^{-1/3}$
47. $\frac{2}{3}x^{-1/4}+x^{-2/3}$ **49.** $\frac{1}{2}x^{-1/6}-\frac{1}{4}$ **51.** $2mn^2\sqrt[3]{2m}$ **53.** $2m^2n\sqrt[4]{2mn^3}$ **55.** $\sqrt[3]{x^2}$ **57.** $2a^2b\sqrt[3]{4a^2b}$ **59.** $\sqrt[4]{12x^3}/2$ **61.** $\sqrt[4]{(x-3)^3}$
63. $x\sqrt[6]{x}$ **65.** $\sqrt[6]{x^5}/x$ **67.** $\dfrac{x-2}{2(x-1)^{3/2}}$ **69.** $\dfrac{x+6}{3(x+2)^{5/3}}$ **71.** 103.2 **73.** 0.0805 **75.** 4,588

◆ EXERCISE 1-7 CHAPTER REVIEW

1. (A) $(y+z)x$ (B) $(2+x)+y$ (C) $2x+3x$ **2.** x^3+3x^2+5x-2 **3.** $x^3-3x^2-3x+22$ **4.** $3x^5+x^4-8x^3+24x^2+8x-64$
5. 3 **6.** 1 **7.** $14x^2-30x$ **8.** $9m^2-25n^2$ **9.** $6x^2-5xy-4y^2$ **10.** $4a^2-12ab+9b^2$ **11.** $(3x-2)^2$ **12.** Prime
13. $3n(2n-5)(n+1)$ **14.** $\dfrac{12a^3b-40b^2-5a}{30a^3b^2}$ **15.** $\dfrac{7x-4}{6x(x-4)}$ **16.** $\dfrac{y+2}{y(y-2)}$ **17.** u **18.** $6x^5y^{15}$ **19.** $3u^4/v^2$ **20.** 6×10^2 **21.** x^6/y^4
22. $u^{7/3}$ **23.** $3a^2/b$ **24.** $3\sqrt[5]{x^2}$ **25.** $-3(xy)^{2/3}$ **26.** $3x^2y\sqrt[3]{x^2y}$ **27.** $6x^2y^3\sqrt{xy}$ **28.** $2b\sqrt{3a}$ **29.** $\dfrac{3\sqrt{5}+5}{4}$ **30.** $\sqrt[4]{y^3}$ **31.** Subtraction
32. Commutative (+) **33.** Distributive **34.** Associative (·) **35.** Negatives **36.** Identity (+) **37.** (A) T (B) F
38. 0 and -3 are two examples of infinitely many. **39.** (A) Expressions (a) and (d) (B) None **40.** $4xy-2y^2$ **41.** $m^4-6m^2n^2+n^4$
42. $2x^3-4x^2+12x$ **43.** $x^3-6x^2y+12xy^2-8y^3$ **44.** $(x-y)(7x-y)$ **45.** Prime **46.** $3xy(2x^2+4xy-5y^2)$
47. $3(x+2y)(x^2-2xy+4y^2)$ **48.** $\dfrac{2m}{(m+2)(m-2)^2}$ **49.** $\dfrac{y^2}{x}$ **50.** $\dfrac{-ab}{a^2+ab+b^2}$ **51.** $\frac{1}{4}$ **52.** $\frac{5}{9}$ **53.** $3x^2/2y^2$ **54.** $27a^{1/6}/b^{1/2}$
55. $x+2x^{1/2}y^{1/2}+y$ **56.** $6x+7x^{1/2}y^{1/2}-3y$ **57.** 2×10^{-7} **58.** $-6x^2y^2\sqrt[5]{3x^2y}$ **59.** $x\sqrt[3]{2x^2}$ **60.** $\sqrt[5]{12x^3y^2}/2x$ **61.** $y\sqrt[3]{2x^2y}$
62. $2x-3\sqrt{xy}-5y$ **63.** $\dfrac{6x+3\sqrt{xy}}{4x-y}$ **64.** $\dfrac{4u-12\sqrt{uv}+9v}{4u-9v}$ **65.** $\dfrac{1}{\sqrt{t}+\sqrt{5}}$ **66.** $2-\frac{3}{2}x^{-1/2}$ **67.** 0 **68.** x^3+8x^2-6x+1
69. 3.213×10^6 **70.** 4.434×10^{-5} **71.** -4.541×10^{-6} **72.** 128,800 **73.** 0.3664 **74.** 1.640

CHAPTER 2

◆ EXERCISE 2-1

1. $m=5$ **3.** $x<-9$ **5.** $x\le4$ **7.** $x<-3$ or $(-\infty,-3)$ 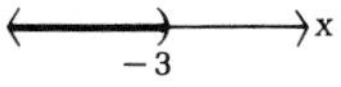 **9.** $-1\le x\le2$ or $[-1,2]$ 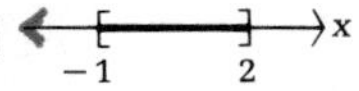 **11.** $y=8$
13. $x>-6$ **15.** $y=8$ **17.** $x=10$ **19.** $y\ge3$ **21.** $x=36$ **23.** $m<3$ **25.** $x=10$ **27.** $3\le x<7$ or $[3,7)$ 
29. $-20\le C\le20$ or $[-20,20]$ 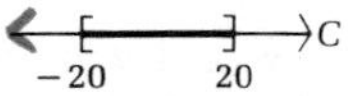 **31.** $y=\frac{3}{4}x-3$ **33.** $y=-(A/B)x+(C/B)=(-Ax+C)/B$ **35.** $C=\frac{5}{9}(F-32)$
37. $B=A/(m-n)$ **39.** $-2<x\le1$ or $(-2,1]$  **41.** 3,000 \$10 tickets; 5,000 \$6 tickets
43. \$7,200 at 10%; \$4,800 at 15% **45.** \$7,800 **47.** 5,000 **49.** 12.6 yr

◆ EXERCISE 2-2

1. ±2 **3.** $\pm\sqrt{11}$ **5.** $-2, 6$ **7.** 0, 2 **9.** $3\pm2\sqrt{3}$ **11.** $-2\pm\sqrt{2}$ **13.** 0, 2 **15.** $\pm\frac{3}{2}$ **17.** $\frac{1}{2}, -3$ **19.** $(-1\pm\sqrt{5})/2$ **21.** $(3\pm\sqrt{3})/2$
23. No real solutions **25.** $-4\pm\sqrt{11}$ **27.** $r=\sqrt{A/P}-1$ **29.** \$2 **31.** 8 ft/sec; $4\sqrt{2}$ or 5.66 ft/sec

◆ EXERCISE 2-3

1.
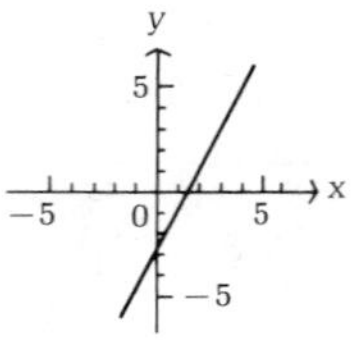

3.
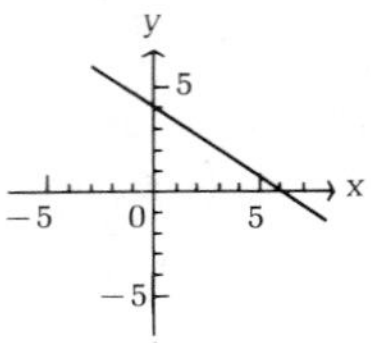

5. Slope = 2; y intercept = −3
7. Slope = $-\frac{2}{3}$; y intercept = 2
9. $y = -2x + 4$
11. $y = -\frac{3}{5}x + 3$

13.
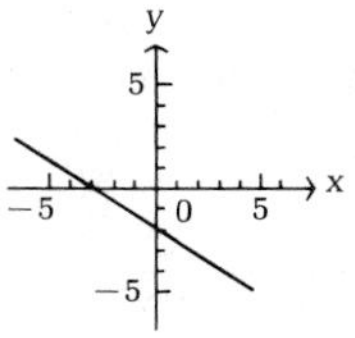

15.
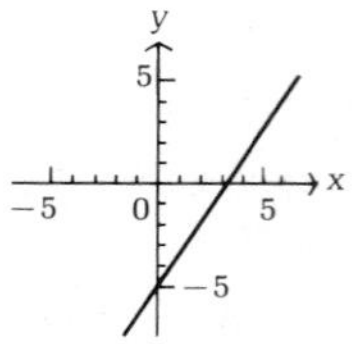

17.
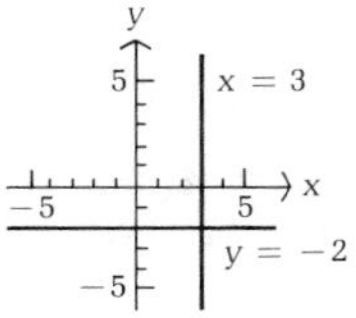

19. $y = -3x + 5$; $m = -3$
21. $y = -\frac{2}{3}x + 4$; $m = -\frac{2}{3}$
23. $y + 1 = -3(x - 4)$; $y = -3x + 11$
25. $y + 5 = \frac{2}{3}(x + 6)$; $y = \frac{2}{3}x - 1$
27. $\frac{1}{3}$
29. $-\frac{1}{5}$

31. $(y - 3) = \frac{1}{3}(x - 1)$; $x - 3y = -8$
33. $(y + 2) = -\frac{1}{5}(x + 5)$; $x + 5y = -15$
35. $x = 3$; $y = -5$
37. $x = -1$; $y = -3$
39. $y = -\frac{1}{2}x + 4$
41. (A) $y = -\frac{1}{2}x + 1$ (B) $y = 2x + 6$

43. (A) $y = \frac{1}{2}x$
(B) $y = -2x - 5$

45.
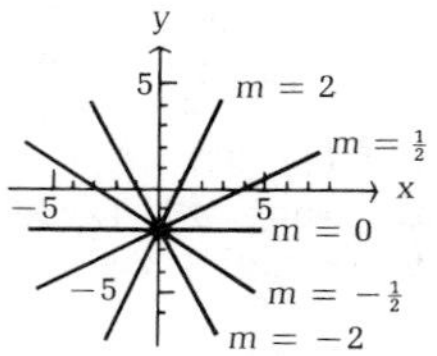

47. $x = 2$
49. $y = 3$

51. (A) \$130; \$220
(B)
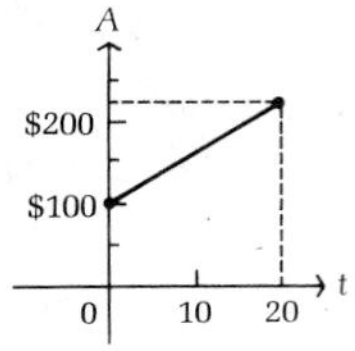

(C) 6

53. (A)
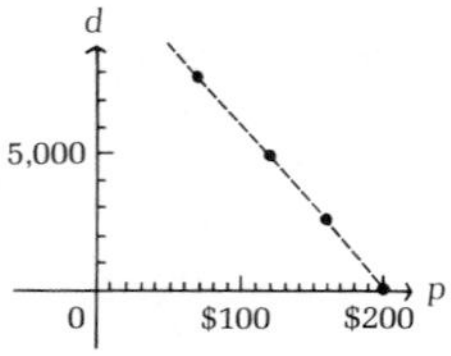

(B) $d = -60p + 12{,}000$

55. $0.2x + 0.1y = 20$
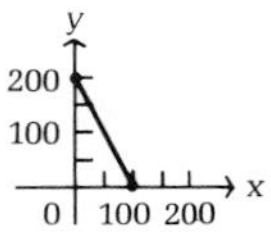

57. (A) 64 g; 35 g
(B)
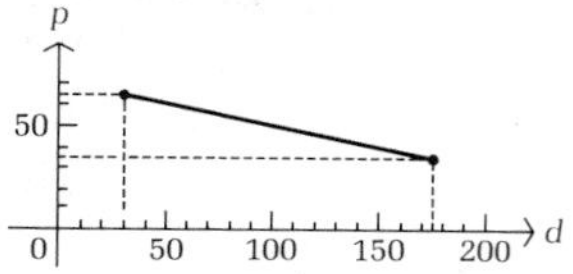

(C) $-\frac{1}{5}$

◆ EXERCISE 2-4

1. Function 3. Not a function 5. Function 7. Function; Domain = {1, 2, 4, 5}; Range = {3, 5, 7, 9} 9. Not a function 11. Function; Domain = {1, 2, 3, 4, 5, 6}; Range = {3, 4} 13. Function 15. Not a function 17. Function 19. 4 21. −5 23. −6 25. −2 27. −12 29. −1 31. −6 33. 12 35. $\frac{3}{4}$ 37. All nonnegative real numbers 39. All real numbers except $x = -3, 5$ 41. $x \geq -5$ or $[-5, \infty)$ 43. All real numbers except $x = \pm 1$ 45. All real numbers except $x = -4, 1$ 47. All real numbers 49. A function with domain R 51. A function with domain R 53. Not a function; for example, when $x = 1$, $y = \pm 3$ 55. A function with domain all real numbers except $x = 4$ 57. Not a function; for example, when $x = 4$, $y = \pm 3$ 59. 4 61. $h + 2$ 63. $h - 1$ 65. 4 67. $8a + 4h - 7$ 69. $3a^2 + 3ah + h^2$ 71. $\dfrac{1}{\sqrt{a + h} + \sqrt{a}}$ 73. $P(w) = 2w + \dfrac{50}{w}$, $w > 0$ 75. $A(\ell) = \ell(50 - \ell)$, $0 < \ell < 50$ 77. $C(x) = 96{,}000 + 80x$; \$136,000 79. $R(x) = x(200 - \frac{1}{40}x)$, $0 \leq x \leq 8{,}000$

81. (A) $V(x) = x(8 - 2x)(12 - 2x)$ (B) $0 < x < 4$ (C)

x	$V(x)$
1	60
2	64
3	36

83. $P(x) = 10x - 15 + \dfrac{270}{x}$;

x	$P(x)$
3	105
4	92.5
5	89
6	90

85. $v = \dfrac{75 - w}{15 + w}$; 1.9032 cm/sec

◆ EXERCISE 2-5

1. Slope: 2
y intercept: -4
x intercept: 2

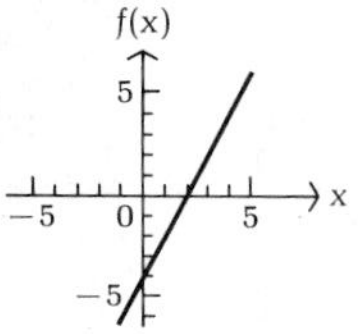

3. Slope: -2
y intercept: 4
x intercept: 2

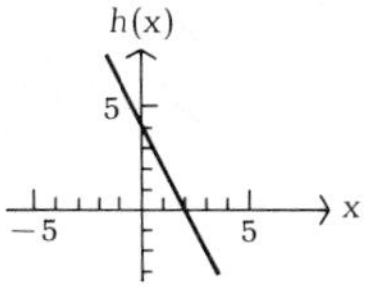

5. Slope: $-\frac{2}{3}$
y intercept: 4
x intercept: 6

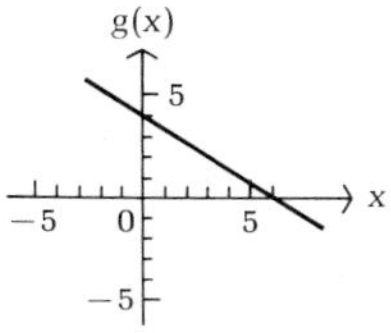

7. $f(x) = -2x + 6$
9. $f(x) = -\frac{1}{2}x + \frac{9}{2}$

11. Axis: $x = 3$
Vertex: $(3, -1)$
Min: $f(3) = -1$
y intercept: 8
x intercepts: 2, 4
Range: $[-1, \infty)$

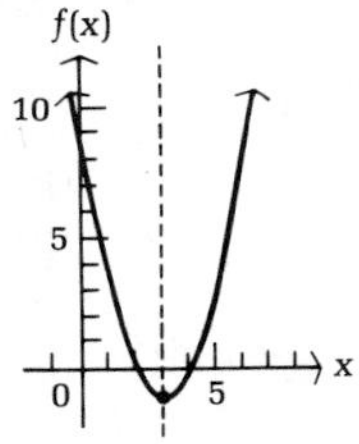

13. Axis: $x = -1$
Vertex: $(-1, 9)$
Max: $h(-1) = 9$
y intercept: 8
x intercepts: -4, 2
Range: $(-\infty, 9]$

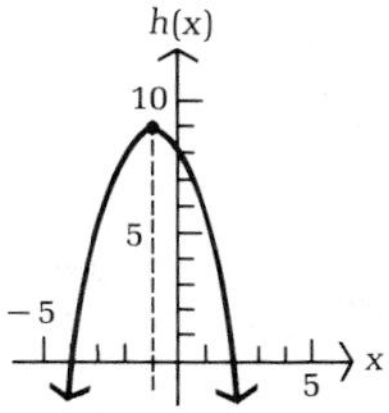

15. Axis: $x = -4$
Vertex: $(-4, 0)$
Min: $f(-4) = 0$
y intercept: 16
x intercept: -4
Range: $[0, \infty)$

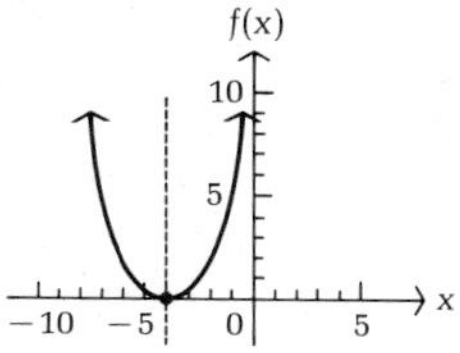

17. Axis: $u = 1$
Vertex: $(1, 3)$
Min: $f(1) = 3$
y intercept: 4
No u intercepts
Range: $[3, \infty)$

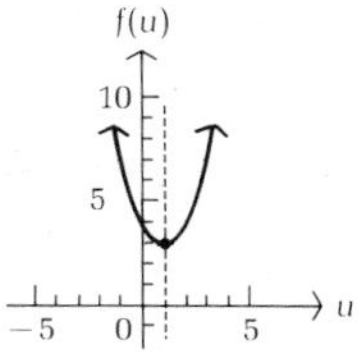

19. Axis: $x = 2$
Vertex: $(2, 6)$
Max: $h(2) = 6$
y intercept: 2
x intercepts: $2 \pm \sqrt{6}$
Range: $(-\infty, 6]$

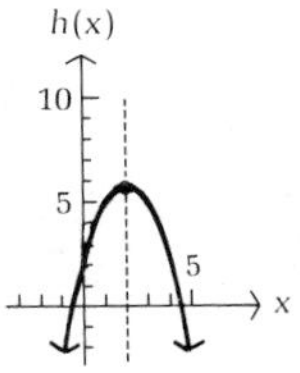

21. Axis: $x = 3$
Vertex: $(3, 9)$
Max: $f(3) = 9$
y intercept: 0
x intercepts: 0, 6
Range: $(-\infty, 9]$

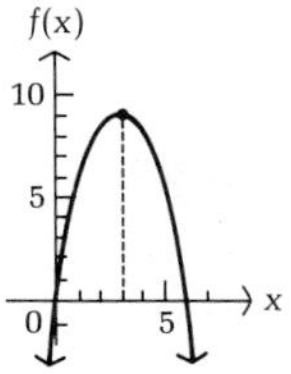

23. Axis: $s = 0$
Vertex: $(0, -4)$
Min: $F(0) = -4$
y intercept: -4
s intercepts: -2, 2
Range: $[-4, \infty)$

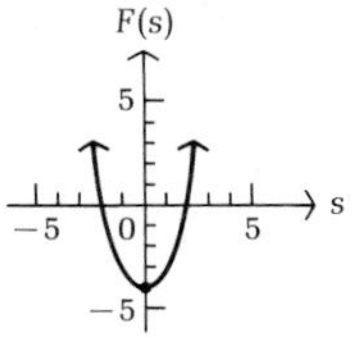

25. Axis: $x = 0$
Vertex: $(0, 4)$
Max: $F(0) = 4$
y intercept: 4
x intercepts: -2, 2
Range: $(-\infty, 4]$

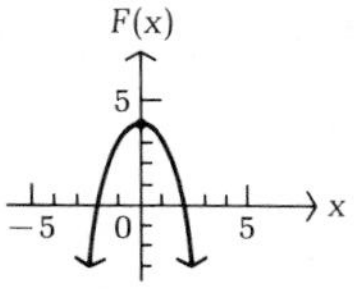

27. Domain: [0,5]
Range: {1, 3, 5}

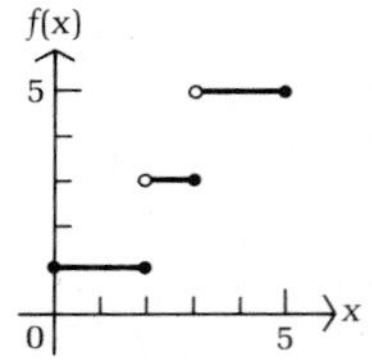

29. Domain: [−2, 2]
Range: [−2, 1]

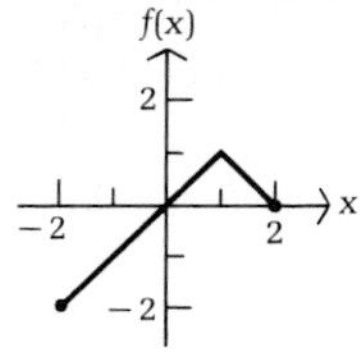

31. Domain: $(-\infty, 0) \cup (0, \infty)$
Range: $(-\infty, -2) \cup (2, \infty)$

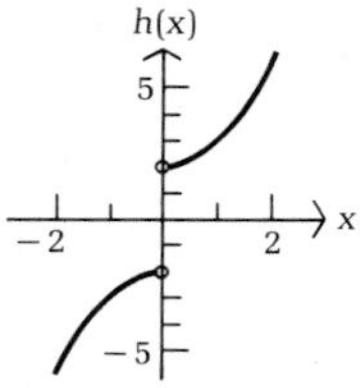

33. Domain: All real numbers
Range: [−3, 3]

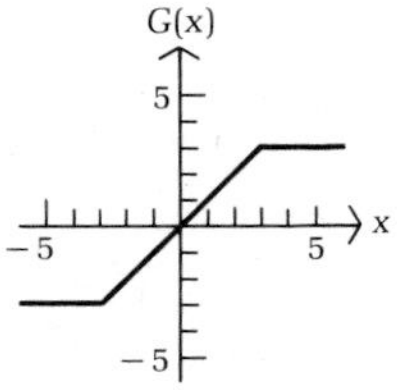

35. Axis: $x = 3.5$
Vertex: (3.5, −2.25)
Min: $f(3.5) = -2.25$
y intercept: 10
x intercepts: 2, 5
Range: $[-2.25, \infty)$

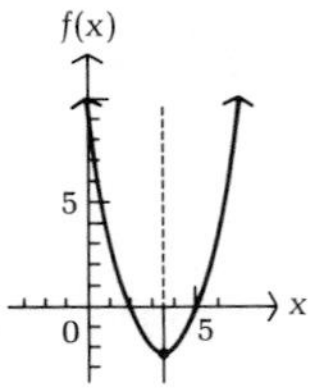

37. Axis: $x = -2.5$
Vertex: (−2.5, 8.25)
Max: $h(-2.5) = 8.25$
y intercept: 2
x intercepts: $-\frac{5}{2} \pm \frac{1}{2}\sqrt{33}$
Range: $(-\infty, 8.25]$

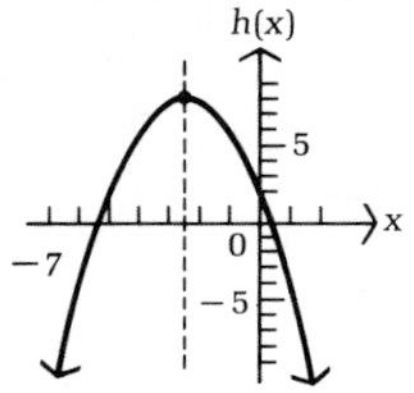

39.

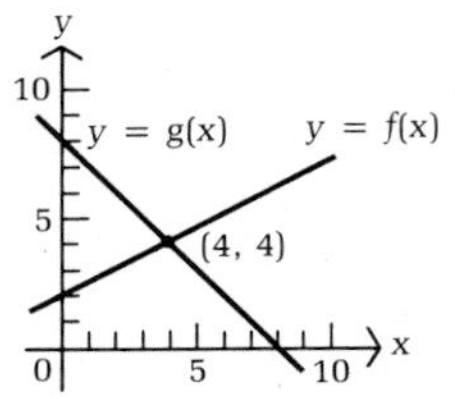

41.

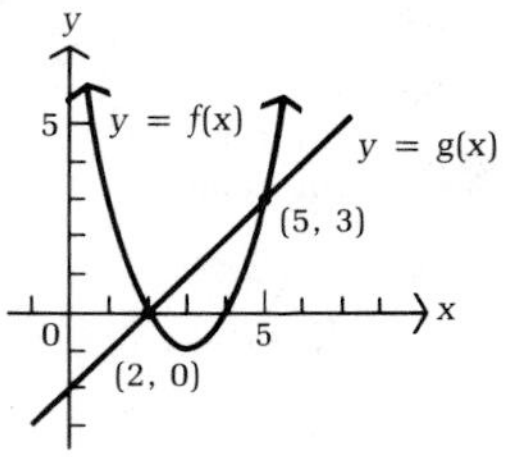

43.

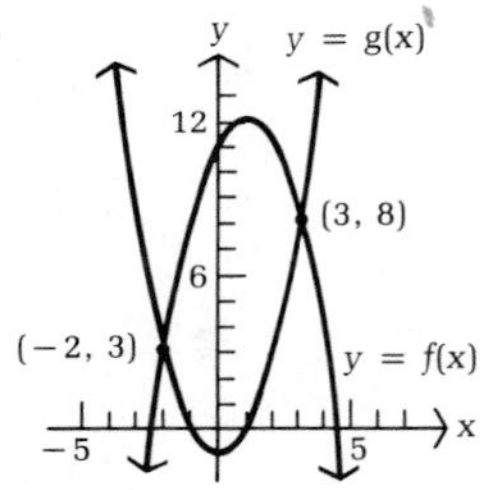

45.

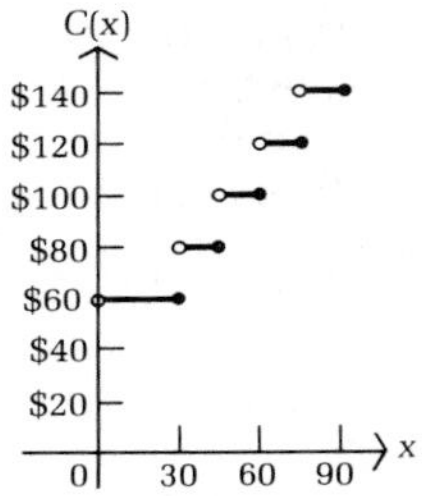

47. (A) $C = 360{,}000 - 900p$
(B) $R = xp = (9{,}000 - 30p)p = 9{,}000p - 30p^2$
(C)

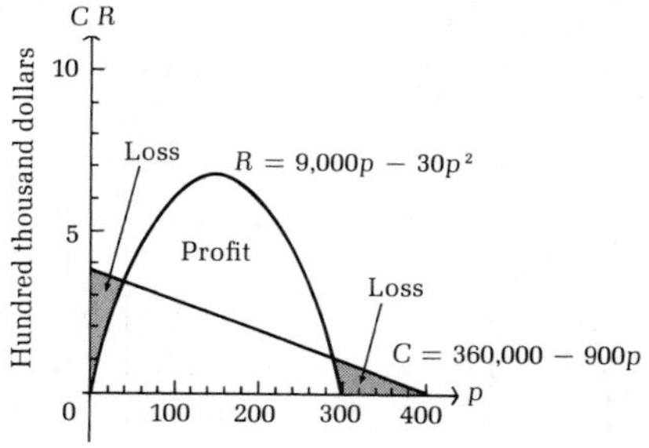

(D) \$42, \$288
(E) \$150

49. (A) $C(x) = 240{,}000 + 20x$
(B) $R(x) = 35x$
(C)

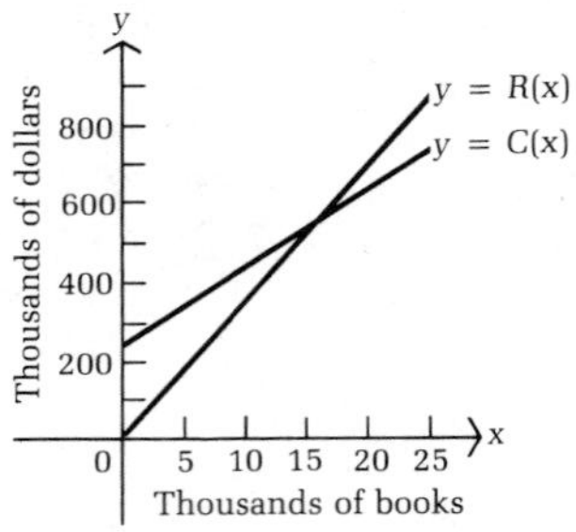

(D) 16,000 books

51.

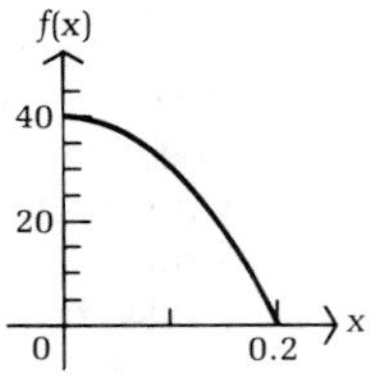

53. (A) 1 lb; 3 lb
(B)

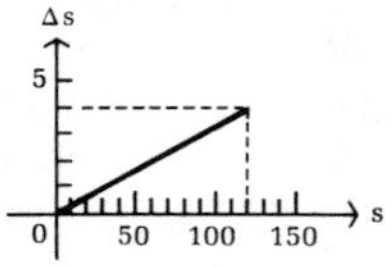

(C) $\frac{1}{30}$

◆ EXERCISE 2-6 CHAPTER REVIEW

1. $u = 36$ **2.** $x < 4$ or $(-\infty, 4)$

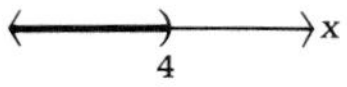

3. $x = 0, 5$ **4.**

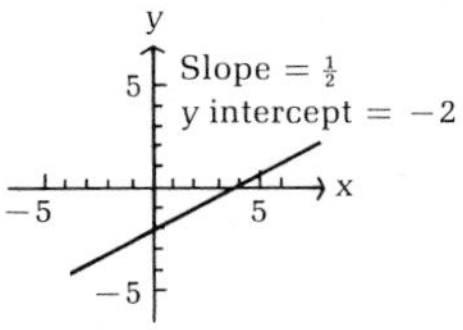

5. $y = \frac{1}{2}x + 1$ **6.**

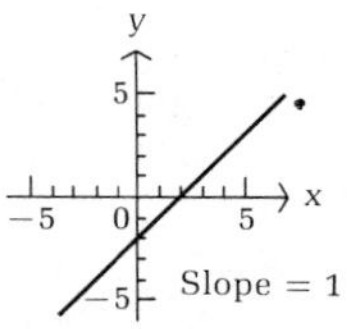

7. -2

8.

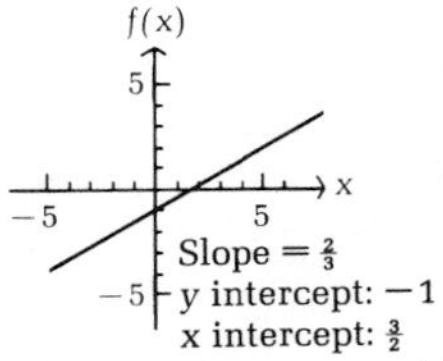

9. Min $f(x) = f(4) = -9$; vertex: $(4, -9)$ **10.** $x = 2$ **11.** $x \geq \frac{9}{2}$ or $[\frac{9}{2}, \infty)$

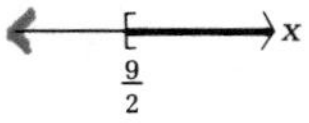

12. $2 \leq x < 12$ or $[2, 12)$

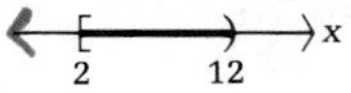

13. $y = \frac{2}{3}x - 2$ **14.** $y = 3/(x - 1)$ **15.** $x = \pm\sqrt{7}$ **16.** $x = -4, 5$ **17.** $x = (3 \pm \sqrt{17})/4$ **18.**

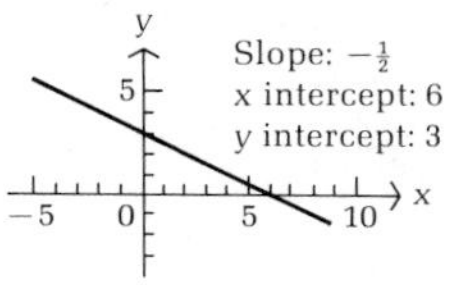

19. $x + 2y = 4$; slope $= -\frac{1}{2}$ **20.**

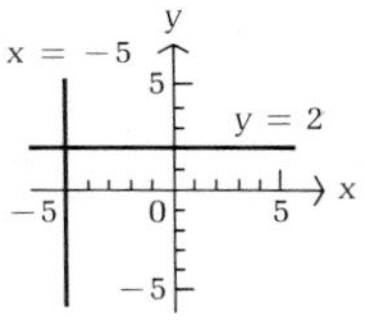

21. $y = -\frac{3}{2}x + 2$ **22.** (A) 22 (B) -36 (C) -91 (D) $-\frac{3}{4}$

23. Domain f: R; Domain g: All R except 2 **24.** 2

25. (A) A function with domain all real numbers (B) Not a function; for example, when $x = 2$, $y = \pm 3$
(C) A function with domain all real numbers except $x = -3$

26. Axis: $x = 2$
Vertex: $(2, 8)$
Max: $g(2) = 8$
y intercept: 0
x intercepts: 0, 4
Range: $(-\infty, 8]$

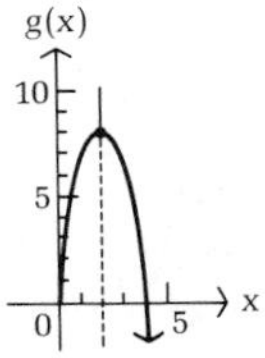

27. Domain: $[-1, 1]$; Range: $[0, 2]$

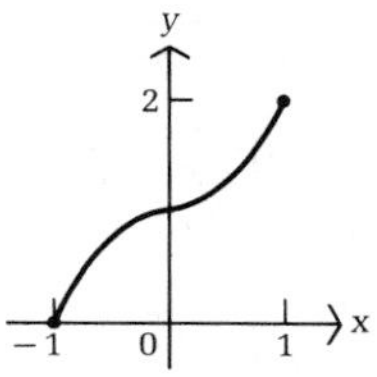

28. $x = \dfrac{-j \pm \sqrt{j^2 - 4k}}{2}$

29. $x = 4$

30. (A) $x - 2y = 8$ (B) $2x + y = 1$

31. (A) All real numbers except $x = 3$ (B) $[1, \infty)$

32. $2a + 7 + h$

33.

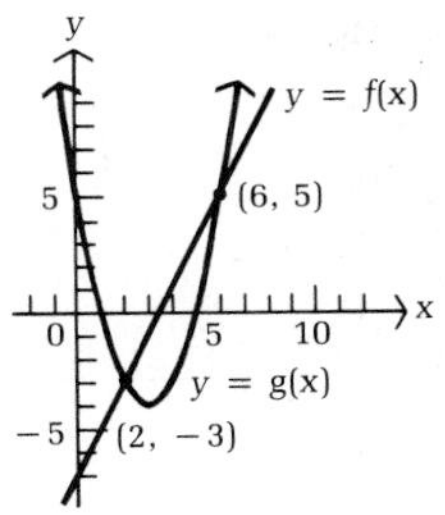

34. \$20,000 at 8%; \$40,000 at 14%

35. $\dfrac{x}{800} = \dfrac{247}{89}$; $x = \$2{,}220$

36. 10% **37.** (A) $V(t) = -1{,}250t + 12{,}000$ (B) \$5,750
38. (A) $R = \frac{8}{5}C$ (B) \$168
39. \$4.50
40. 2,400 tapes
41. (A) $R(p) = 500p - 10p^2$
(B) $C(p) = 8{,}000 - 100p$
(C)
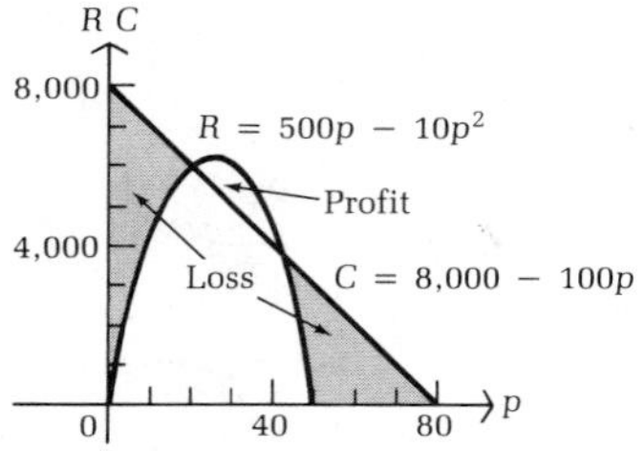

(D) \$20; \$40 (E) \$30
42. (A) $A(x) = 90x - \frac{3}{2}x^2$ (B) $0 < x < 60$ (C) $x = 30$, $y = 22.5$

CHAPTER 3

◆ EXERCISE 3-1

1.
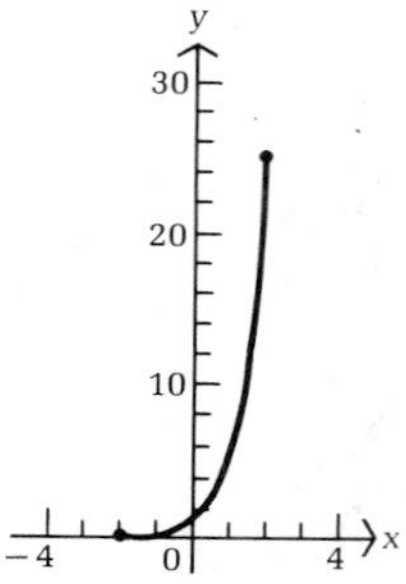
3.
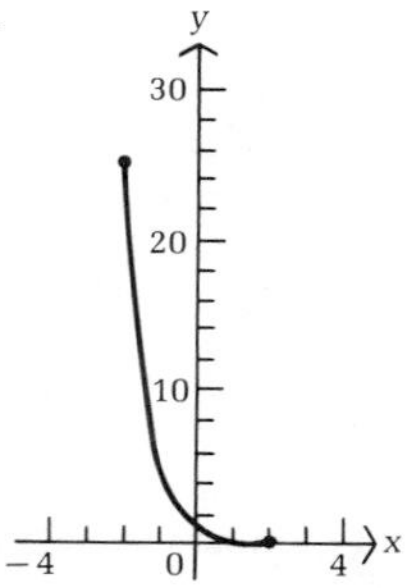
5.
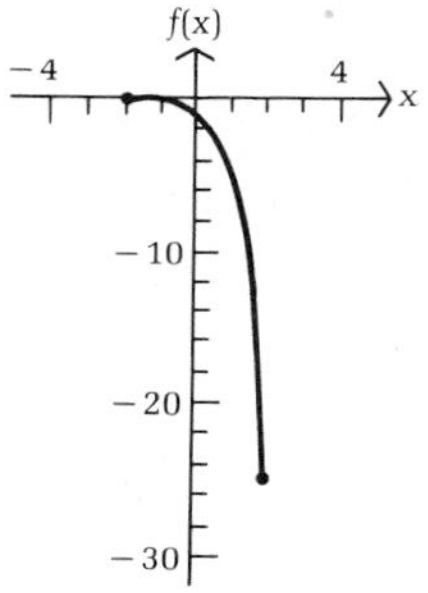
7.
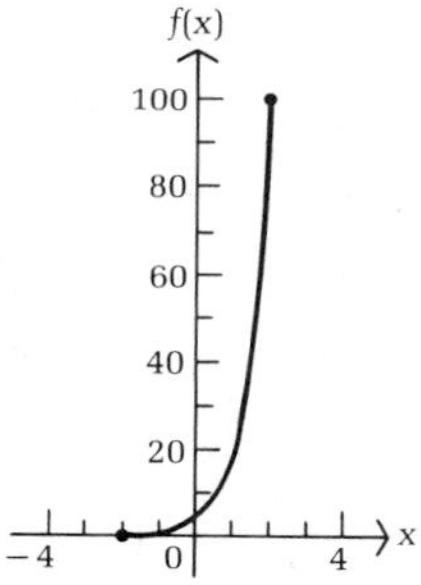
9.
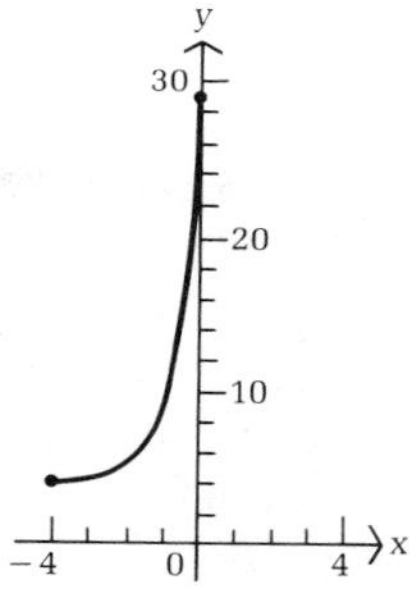

11. 4^{6xy} **13.** 5 **15.** $2^{xz}3^{yz}$ **17.** $x = 1$ **19.** $x = -1, 6$ **21.** $x = 3$ **23.** $x = 2$

25.
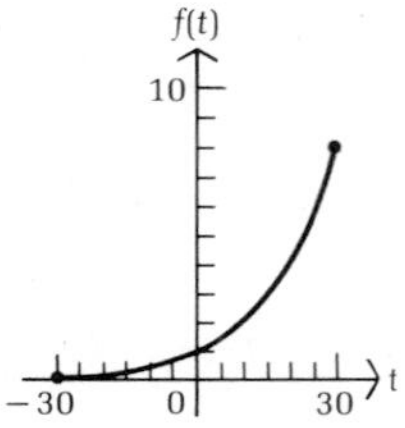
27.
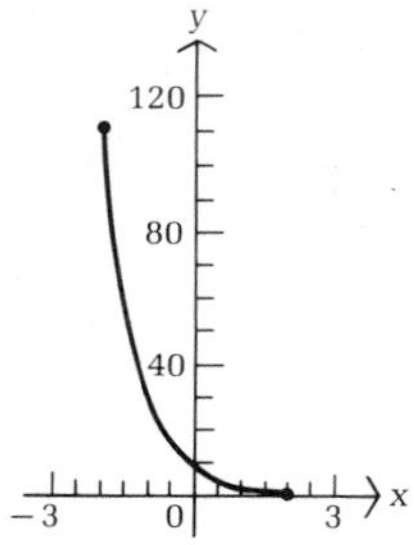
29.
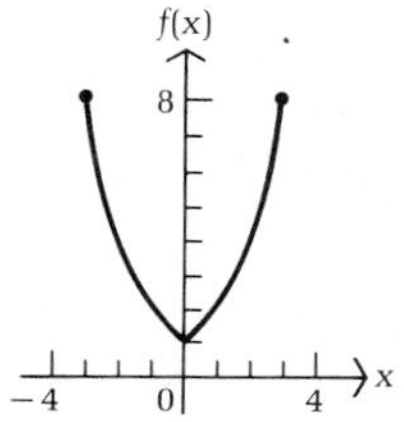
31.
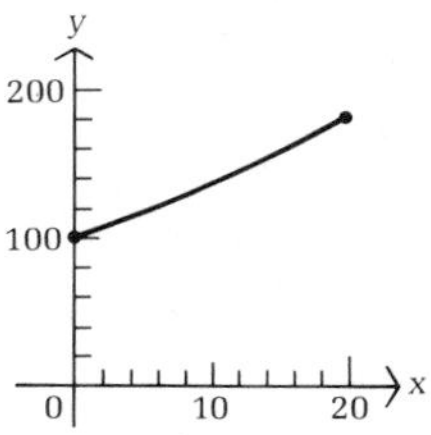
33.
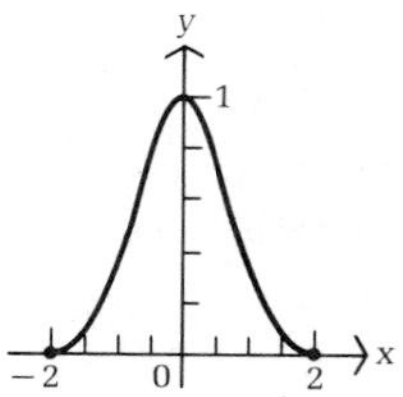

35. $3^{2x} - 3^{-2x}$ **37.** $2(3^{2x}) + 2(3^{-2x})$

39.

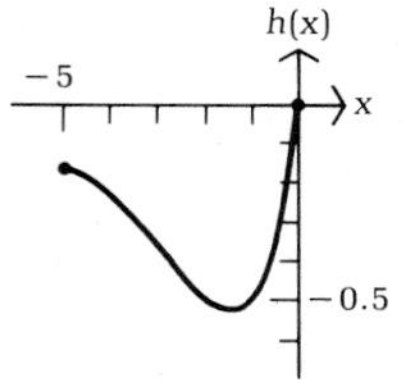

41.

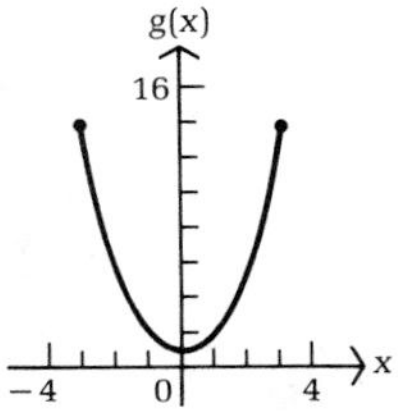

43. 0.2115
45. 0.1377
47. 15.7563
49. (A) $2,633.56 (B) $7,079.54
51. $9,217
53. (A) 33,000,000 (B) 69,000,000
55. (A) 8.49 mg (B) 0.75 mg

◆ EXERCISE 3-2

1.

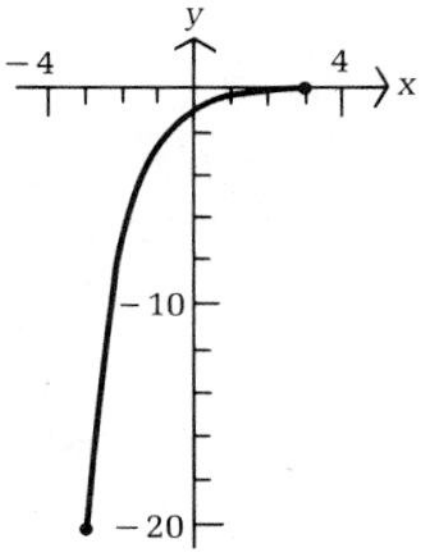

3.

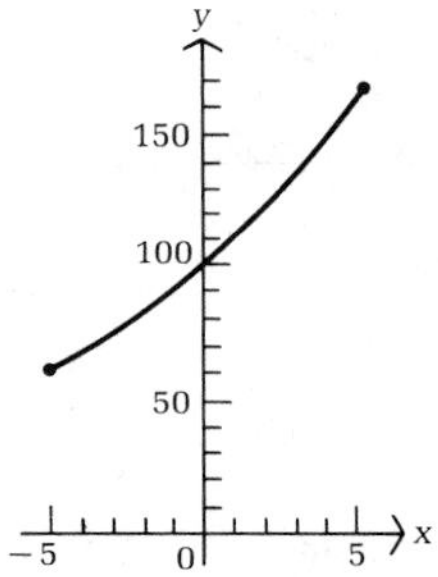

5.

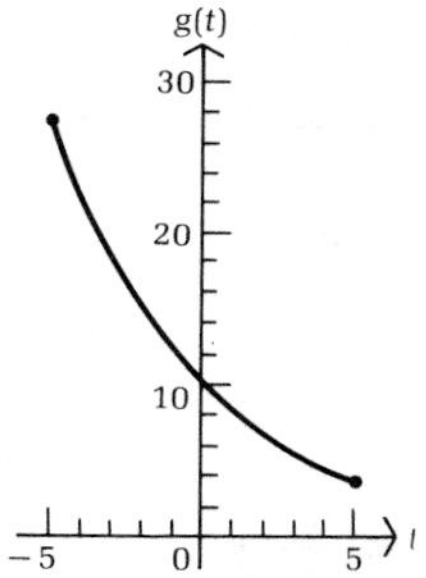

7.

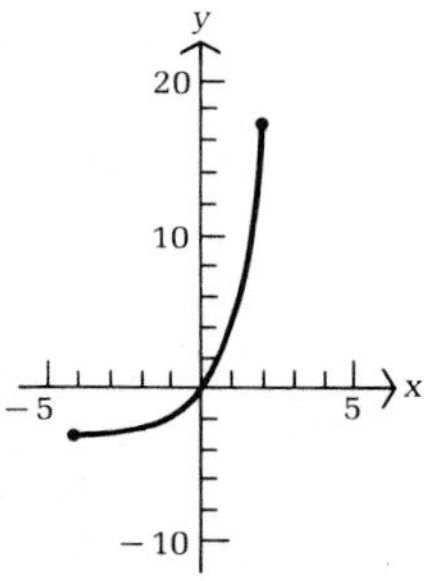

9.

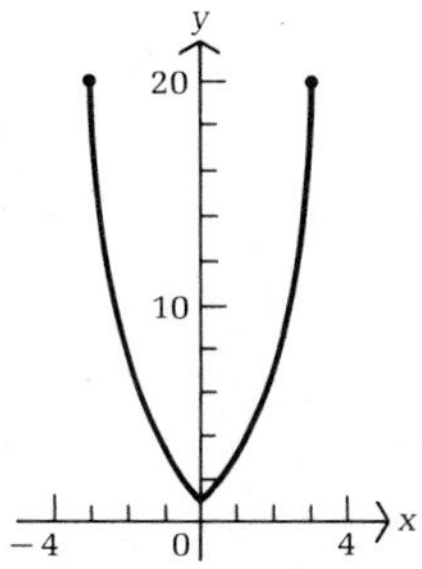

11.

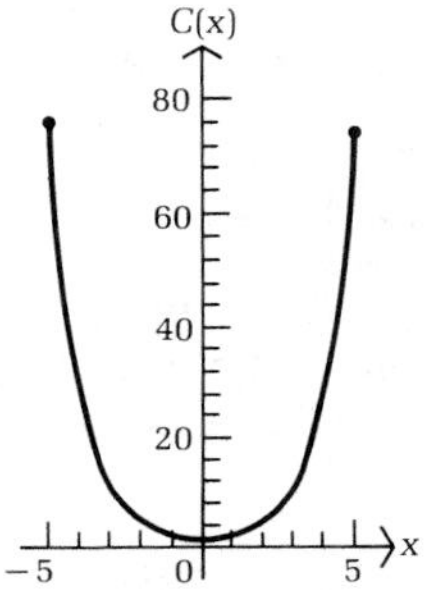

13. $e^x - e^{-x}$
15. $2/e^{2x}$
17. $x = 3$
19. $x = -3, 0$

21.

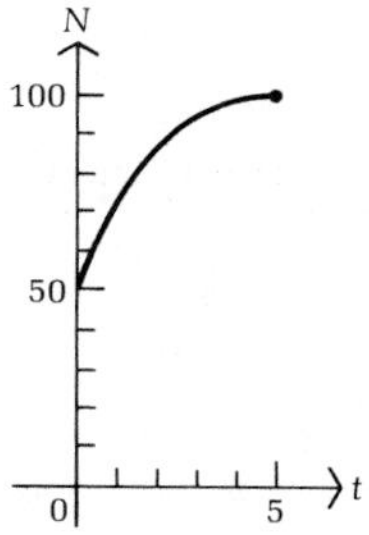

23. (A) $11,871.65 (B) $20,427.93
25. (A) $10,850.88 (B) $10,838.29
27. $28,847.49

29. N approaches 2 as t increases without bound.

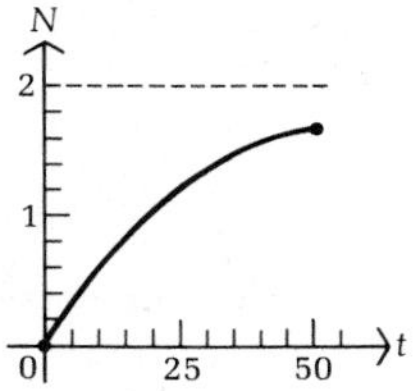

31. (A) 10% (B) 1%
33. (A) 47,699 (B) 475,757
35. 120 million

◆ EXERCISE 3-3

1. $27 = 3^3$ **3.** $10^0 = 1$ **5.** $8 = 4^{3/2}$ **7.** $\log_7 49 = 2$ **9.** $\log_4 8 = \frac{3}{2}$ **11.** $\log_b A = u$ **13.** 3 **15.** -3 **17.** 3 **19.** $\log_b P - \log_b Q$
21. $5 \log_b L$ **23.** $\log_b p - \log_b q - \log_b r - \log_b s$ **25.** $x = 9$ **27.** $y = 2$ **29.** $b = 10$ **31.** $x = 2$ **33.** $y = -2$ **35.** $b = 100$
37. $5 \log_b x - 3 \log_b y$ **39.** $\frac{1}{3} \log_b N$ **41.** $2 \log_b x + \frac{1}{3} \log_b y$ **43.** $\log_b 50 - 0.2t \log_b 2$ **45.** $\log_b P + t \log_b(1 + r)$ **47.** $\log_e 100 - 0.01t$
49. $x = 2$ **51.** $x = 8$ **53.** $x = 7$ **55.** No solution

57.

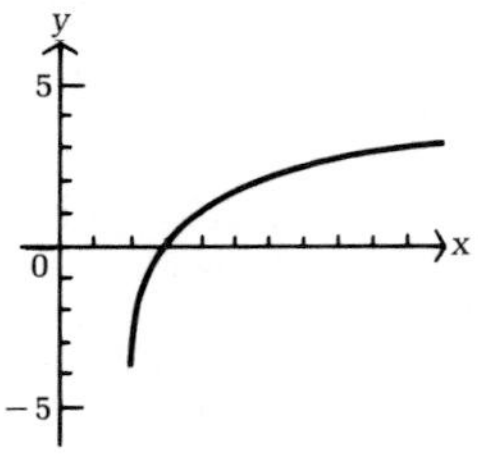

59. (A) 3.547 43 (B) −2.160 32 (C) 5.626 29 (D) −3.197 04
61. (A) 13.4431 (B) 0.0089 (C) 16.0595 (D) 0.1514
63. 4.959 **65.** 7.861 **67.** 3.301

69.

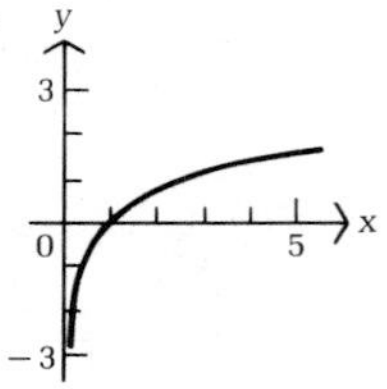

71.

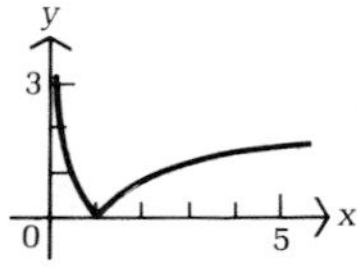

73.

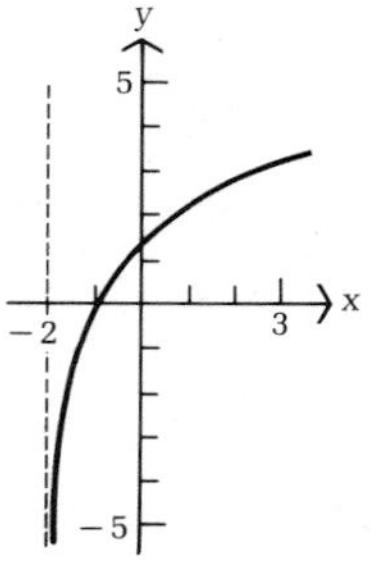

75.

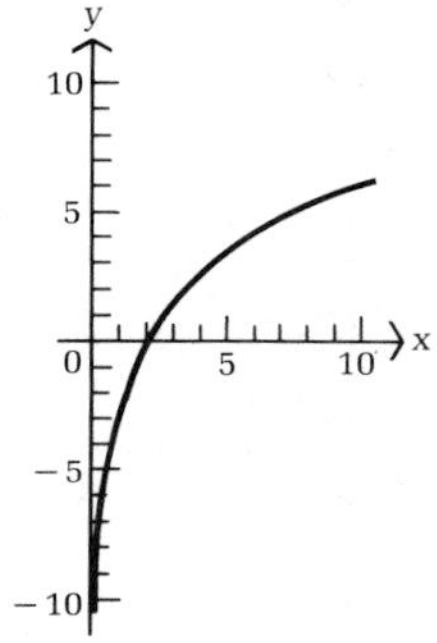

77. $\log_b 1 = 0, b > 0, b \neq 1$ **79.** $y = c \cdot 10^{0.8x}$ **81.** 12 yr **83.** $t = \dfrac{\ln 3}{\ln(1 + r)}$ **87.** Approx. 538 yr

◆ EXERCISE 3-4 CHAPTER REVIEW

1. $v = \ln u$ **2.** $y = \log x$ **3.** $M = e^N$ **4.** $u = 10^v$ **5.** 5^{2x} **6.** e^{2u^2} **7.** $x = 9$ **8.** $x = 6$ **9.** $x = 4$ **10.** $x = 2.157$ **11.** $x = 13.128$
12. $x = 1{,}273.503$ **13.** $x = 0.318$ **14.** $x = 8$ **15.** $x = 3$ **16.** $x = 3$ **17.** $x = -1, 3$ **18.** $x = 0, \frac{3}{2}$ **19.** $x = -2$ **20.** $x = \frac{1}{2}$ **21.** $x = 27$
22. $x = 13.3113$ **23.** $x = 158.7552$ **24.** $x = 0.0097$ **25.** $x = 1.4359$ **26.** $x = 1.4650$ **27.** $x = 92.1034$ **28.** $x = 9.0065$
29. $x = 2.1081$ **30.** $1 + 2e^x - e^{-x}$ **31.** $2e^{-2x} - 2$

32.

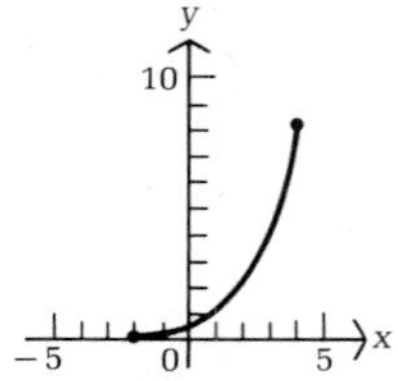

33.

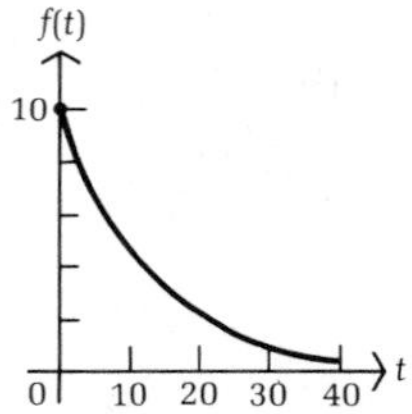

34.

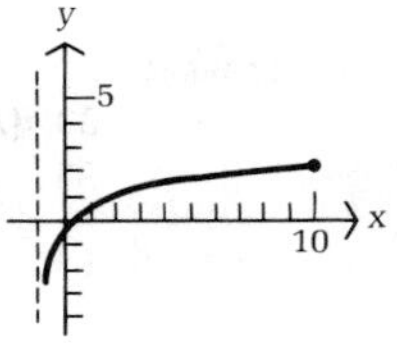

35. $x = 2$
36. $x = 2$
37. $x = 1$
38. $x = 300$
39. $y = ce^{-5t}$

40. If $\log_1 x = y$, then $1^y = x$; that is, $1 = x$ for all positive real numbers x, which is not possible. **41.** \$10,263.65 **42.** \$10,272.17
43. 8 yr **44.** 6.93 yr **45.** (A) $N = 2^{2t}$ or $N = 4^t$ (B) 15 days **46.** $k = 0.009\ 42$; 489 ft **47.** 23.4 yr **48.** 23.1 yr

CHAPTER 4

◆ EXERCISE 4-1

1. (A) 2 (B) 2 (C) Does not exist (D) 4 **3.** (A) 2 (B) 2 (C) 2 (D) Not defined **5.** $c = 0, 1$ **7.** 2, 4 **9.** All x
11. All x, except $x = 5$ **13.** All x, except $x = -2$ and $x = 3$ **15.** (A) 1 (B) 1 (C) 1 **17.** (A) 2 (B) 1 (C) Does not exist
19. (A) 1 (B) 1 (C) 1 **21.** (A) 1 (B) 1 (C) Yes **23.** (A) Does not exist (B) 1 (C) No **25.** (A) 1 (B) 3 (C) No **27.** −2

29. (A)

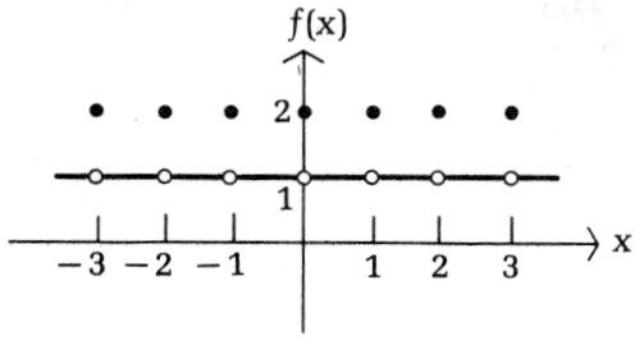

(B) 1
(C) 2
(D) No
(E) All integers

31. 5
33. 7
35. −6
37. 1
39. $(-\infty, \infty)$
41. $[5, \infty)$
43. $(-\infty, \infty)$
45. $(-\infty, 1), (1, 2), (2, \infty)$

47.

x	0.9	0.99	0.999 → 1 ← 1.001	1.01	1.1
$f(x)$	−1	−1	−1 → ? ← 1	1	1

(A) −1 (B) 1 (C) Does not exist

49.

x	0.9	0.99	0.999 → 1 ← 1.001	1.01	1.1
$f(x)$	2.71	2.97	2.997 → ? ← 3.003	3.03	3.31

(A) 3 (B) 3 (C) 3

51. Discontinuous at $x = 1$

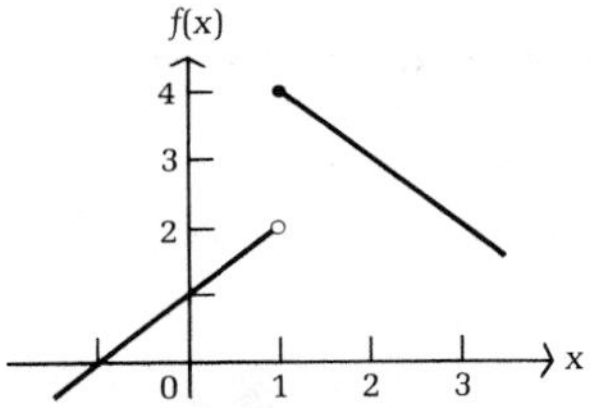

53. Continuous for all x

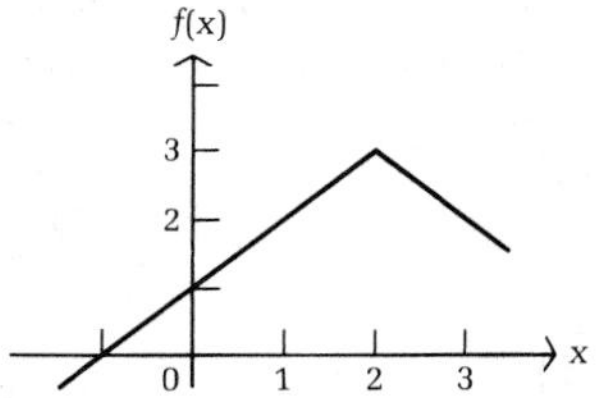

55. Discontinuous at $x = 0$

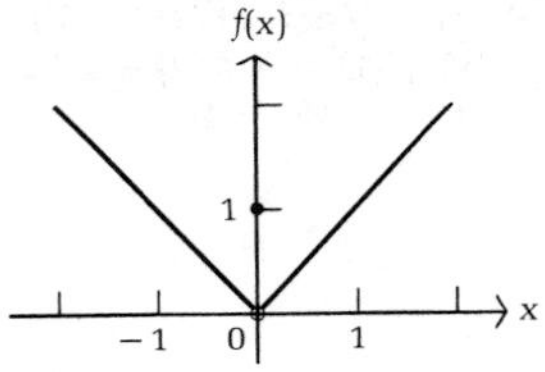

57. (A) Yes (B) Yes (C) Yes (D) Yes **59.** (A) Yes (B) No (C) Yes (D) No (E) Yes **61.** No; no

63. (A)

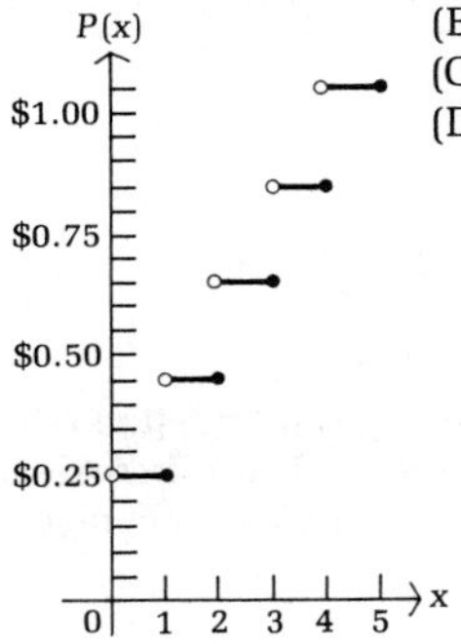

(B) $\lim_{x \to 4.5} P(x) = \$1.05$; $P(4.5) = \$1.05$
(C) $\lim_{x \to 4} P(x)$ does not exist; $P(4) = \$0.85$
(D) Continuous at $x = 4.5$; not continuous at $x = 4$

65. (A)

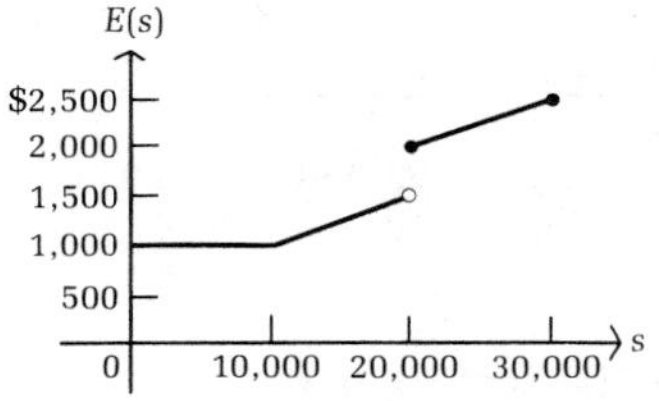

(B) $\lim_{s \to 10{,}000} E(s) = \$1{,}000$; $E(10{,}000) = \$1{,}000$
(C) $\lim_{s \to 20{,}000} E(s)$ does not exist; $E(20{,}000) = \$2{,}000$
(D) Yes; no

67. (A) t_2, t_3, t_4, t_6, t_7 (B) $\lim_{t \to t_5} N(t) = 7$; $N(t_5) = 7$ (C) $\lim_{t \to t_3} N(t)$ does not exist; $N(t_3) = 4$

◆ EXERCISE 4-2

1. -4 **3.** 36 **5.** $\frac{5}{9}$ **7.** $\sqrt{5}$ **9.** $\frac{7}{5}$ **11.** 47 **13.** -4 **15.** $\frac{5}{3}$ **17.** 243 **19.** -3 **21.** 0 **23.** 5 **25.** 2 **27.** 0 **29.** 1 **31.** $\frac{1}{2}$ **33.** 0 **35.** -1
37. -5 **39.** ∞ **41.** 2 **43.** $\frac{3}{5}$ **45.** 4 **47.** $\frac{2}{3}$ **49.** $-\infty$ **51.** ∞ **53.** Does not exist **55.** (A) 0 (B) 0 (C) $y = 0$
57. (A) ∞ (B) ∞ (C) ∞ (D) $x = -1$ **59.** 3 **61.** 4 **63.** 0 **65.** $1/(2\sqrt{2})$ **67.** Does not exist **69.** $\sqrt{3}$ **71.** $\sqrt[3]{4}$ **73.** ∞ **75.** 0 **77.** 0
79. $\frac{1}{4}$ **81.** 12 **83.** (A) $-\infty$ (B) ∞ (C) Does not exist (D) Yes **85.** $2a$ **87.** $1/(2\sqrt{a})$ **89.** (A) $\$23$ (B) $\$3.20$ (C) $\$5$ (D) $\$3$

91. (A)

COMPOUNDED	n	$A(n)$
Annually	1	$108.00
Semiannually	2	$108.16
Quarterly	4	$108.24
Monthly	12	$108.30
Weekly	52	$108.32
Daily	365	$108.33
Hourly	8,760	$108.33

(B) $\$108.33$ **93.** (A) 0.056 (B) 0.07 (C) 0.07 (D) 0
95. (A) 30 (B) 44 (C) 44 (D) 60

◆ EXERCISE 4-3

1. (A) 3 (B) 3 **3.** (A) $8 + 2h$ (B) 8 **5.** (A) 1 (B) $2 + h$ (C) 2 **7.** $f'(x) = 2$; $f'(1) = 2$, $f'(2) = 2$, $f'(3) = 2$
9. $f'(x) = -2x$; $f'(1) = -2$, $f'(2) = -4$, $f'(3) = -6$ **11.** (A) 5 (B) $3 + h$ (C) 3 (D) $y = 3x - 1$
13. (A) 5 m/sec (B) $3 + h$ m/sec (C) 3 m/sec **15.** $f'(x) = 6 - 2x$; $f'(1) = 4$, $f'(2) = 2$, $f'(3) = 0$
17. $f'(x) = 1/(2\sqrt{x})$; $f'(1) = \frac{1}{2}$, $f'(2) = 1/(2\sqrt{2})$, $f'(3) = 1/(2\sqrt{3})$ **19.** $f'(x) = 1/x^2$; $f'(1) = 1$, $f'(2) = \frac{1}{4}$, $f'(3) = \frac{1}{9}$ **21.** Yes **23.** No **25.** No
27. Yes **29.** (A) $f'(x) = 2x - 4$ (B) $-4, 0, 4$ (C)

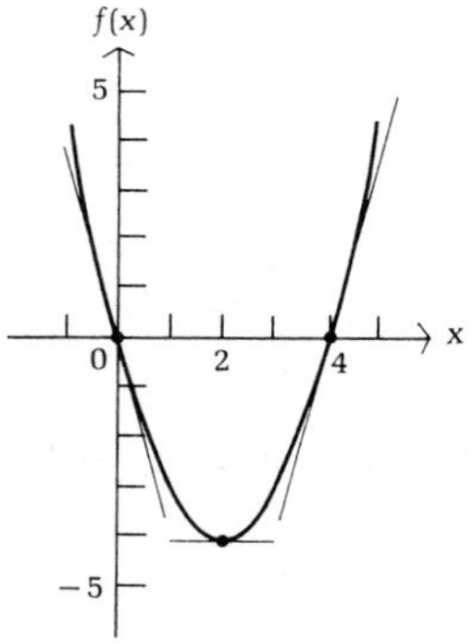

31. $v = f'(x) = 8x - 2$; 6 ft/sec, 22 ft/sec, 38 ft/sec
33. f is nondifferentiable at $x = 1$

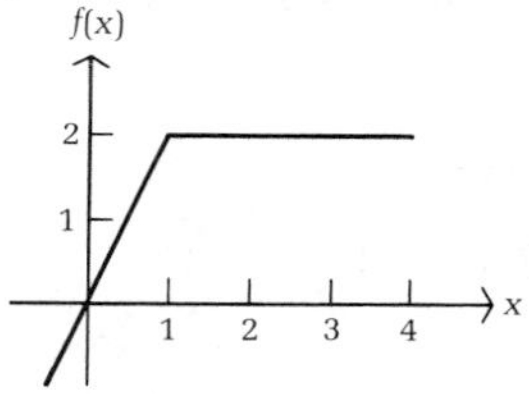

35. No
37. No
39. (A) $2 - 2x$ (B) 2 (C) -2

41. (A) 3 ($\$300$/board) (B) $4 - h$ (C) $C'(3) = 4$ ($\$400$/board) (D) $C'(x) = 10 - 2x$
(E) $C'(1) = 8$ (the rate of total cost increase at the level of production of 1 board per day is $\$800$/board); $C'(2) = 6$ (the rate of total cost increase at the level of production of 2 boards per day is $\$600$/board); $C'(3) = 4$ (the rate of total cost increase at the level of production of 3 boards per day is $\$400$/board); $C'(4) = 2$ (the rate of total cost increase at the level of production of 4 boards per day is $\$200$/board)

◆ EXERCISE 4-4

1. 0 **3.** 0 **5.** $12x^{11}$ **7.** 1 **9.** $-7x^{-8}$ **11.** $\frac{5}{2}x^{3/2}$ **13.** $-5x^{-6}$ **15.** $8x^3$ **17.** $2x^5$ **19.** $x^4/3$ **21.** $-10x^{-6}$ **23.** $-16x^{-5}$ **25.** x^{-3}
27. $-x^{-2/3}$ **29.** $4x - 3$ **31.** $15x^4 - 6x^2$ **33.** $-12x^{-5} - 4x^{-3}$ **35.** $-\frac{1}{2}x^{-2} + 2x^{-4}$ **37.** $2x^{-1/3} - \frac{5}{3}x^{-2/3}$ **39.** $-\frac{9}{5}x^{-8/5} + 3x^{-3/2}$
41. $-\frac{1}{3}x^{-4/3}$ **43.** $-6x^{-3/2} + 6x^{-3} + 1$
45. (A) $f'(x) = 6 - 2x$ (B) $f'(2) = 2$; $f'(4) = -2$ (C) $y = 2x + 4$; $y = -2x + 16$ (D) $x = 3$
47. (A) $f'(x) = 12x^3 - 12x$ (B) $f'(2) = 72$; $f'(4) = 720$ (C) $y = 72x - 127$; $y = 720x - 2{,}215$ (D) $x = -1, 0, 1$
49. (A) $v = f'(x) = 176 - 32x$ (B) $f'(0) = 176$ ft/sec; $f'(3) = 80$ ft/sec (C) $x = 5.5$ sec
51. (A) $v = f'(x) = 3x^2 - 18x + 15$ (B) $f'(0) = 15$ ft/sec; $f'(3) = -12$ ft/sec (C) $x = 1, 5$ sec **53.** $-20x^{-2}$ **55.** $2x - 3 - 10x^{-3}$

59. (A) $C'(x) = 60 - (x/2)$
(B) $C'(60) = \$30$/racket (at a production level of 60 rackets, the rate of change of total cost relative to production is \$30 per racket; thus, the cost of producing 1 more racket at this level of production is approx. \$30).
(C) \$29.75 (the marginal cost of \$30 per racket found in part B is a close approximation to this value)
(D) $C'(80) = \$20$ per racket (at a production level of 80 rackets, the rate of change of total cost relative to production is \$20 per racket; thus, the cost of producing 1 more racket at this level of production is approx. \$20)

61. (A) $N'(x) = 60 - 2x$
(B) $N'(10) = 40$ (at the \$10,000 level of advertising, there would be an approximate increase of 40 units of sales per \$1,000 increase in advertising); $N'(20) = 20$ (at the \$20,000 level of advertising, there would be an approximate increase of only 20 units of sales per \$1,000 increase in advertising); the effect of advertising levels off as the amount spent increases

63. (A) -1.37 beats/min (B) -0.58 beat/min 65. (A) 25 items/hr (B) 8.33 items/hr

◆ EXERCISE 4-5

1. $2x^3(2x) + (x^2 - 2)(6x^2) = 10x^4 - 12x^2$ 3. $(x - 3)(2) + (2x - 1)(1) = 4x - 7$ 5. $\dfrac{(x-3)(1) - x(1)}{(x-3)^2} = \dfrac{-3}{(x-3)^2}$

7. $\dfrac{(x-2)(2) - (2x+3)(1)}{(x-2)^2} = \dfrac{-7}{(x-2)^2}$ 9. $(x^2+1)(2) + (2x-3)(2x) = 6x^2 - 6x + 2$

11. $\dfrac{(2x-3)(2x) - (x^2+1)(2)}{(2x-3)^2} = \dfrac{2x^2 - 6x - 2}{(2x-3)^2}$ 13. $(x^2+2)2x + (x^2-3)2x = 4x^3 - 2x$

15. $\dfrac{(x^2-3)2x - (x^2+2)2x}{(x^2-3)^2} = \dfrac{-10x}{(x^2-3)^2}$ 17. $(2x+1)(2x-3) + (x^2-3x)(2) = 6x^2 - 10x - 3$

19. $(2x - x^2)(5) + (5x+2)(2-2x) = -15x^2 + 16x + 4$ 21. $\dfrac{(x^2+2x)(5) - (5x-3)(2x+2)}{(x^2+2x)^2} = \dfrac{-5x^2+6x+6}{(x^2+2x)^2}$

23. $\dfrac{(x^2-1)(2x-3) - (x^2-3x+1)(2x)}{(x^2-1)^2} = \dfrac{3x^2-4x+3}{(x^2-1)^2}$ 25. $f'(x) = (1+3x)(-2) + (5-2x)(3)$; $y = -11x + 29$

27. $f'(x) = \dfrac{(3x-4)(1) - (x-8)(3)}{(3x-4)^2}$; $y = 5x - 13$ 29. $f'(x) = (2x-15)(2x) + (x^2+18)(2) = 6(x-2)(x-3)$; $x = 2$, $x = 3$

31. $f'(x) = \dfrac{(x^2+1)(1) - x(2x)}{(x^2+1)^2} = \dfrac{1-x^2}{(x^2+1)^2}$; $x = -1$, $x = 1$ 33. $7x^6 - 3x^2$ 35. $-27x^{-4}$

37. $(2x^4 - 3x^3 + x)(2x-1) + (x^2 - x + 5)(8x^3 - 9x^2 + 1)$ 39. $\dfrac{(4x^2+5x-1)(6x-2) - (3x^2-2x+3)(8x+5)}{(4x^2+5x-1)^2}$

41. $9x^{1/3}(3x^2) + (x^3+5)(3x^{-2/3})$ 43. $\dfrac{(x^2-3)(2x^{-2/3}) - 6x^{1/3}(2x)}{(x^2-3)^2}$ 45. $x^{-2/3}(3x^2-4x) + (x^3-2x^2)(-\frac{2}{3}x^{-5/3})$

47. $\dfrac{(x^2+1)[(2x^2-1)(2x) + (x^2+3)(4x)] - (2x^2-1)(x^2+3)(2x)}{(x^2+1)^2}$

49. (A) $S'(t) = \dfrac{7{,}200 - 200t^2}{(t^2+36)^2}$
(B) $S(2) = 10$; $S'(2) = 4$; at $t = 2$ months, monthly sales are 10,000 and increasing at 4,000 albums per month
(C) $S(8) = 16$; $S'(8) = -0.56$; at $t = 8$ months, monthly sales are 16,000 and decreasing at 560 albums per month

51. (A) $d'(x) = \dfrac{-50{,}000(2x+10)}{(x^2+10x+25)^2} = \dfrac{-100{,}000}{(x+5)^3}$
(B) $d'(5) = -100$ radios per \$1 increase in price; $d'(10) = -30$ radios per \$1 increase in price

53. (A) $C'(t) = \dfrac{0.14 - 0.14t^2}{(t^2+1)^2}$
(B) $C'(0.5) = 0.0672$ (concentration is increasing at 0.0672 unit/hr); $C'(3) = -0.0112$ (concentration is decreasing at 0.0112 unit/hr)

55. (A) $N'(x) = \frac{(x+32)(100) - (100x+200)}{(x+32)^2} = \frac{3{,}000}{(x+32)^2}$ (B) $N'(4) = 2.31$; $N'(68) = 0.30$

◆ EXERCISE 4-6

1. $6(2x+5)^2$ **3.** $-8(5-2x)^3$ **5.** $30x(3x^2+5)^4$ **7.** $8(x^3-2x^2+2)^7(3x^2-4x)$ **9.** $(2x-5)^{-1/2}$ **11.** $-8x^3(x^4+1)^{-3}$
13. $f'(x) = 6(2x-1)^2$; $y = 6x-5$; $x = \frac{1}{2}$ **15.** $f'(x) = 2(4x-3)^{-1/2}$; $y = \frac{2}{3}x+1$; none **17.** $24x(x^2-2)^3$ **19.** $-6(x^2+3x)^{-4}(2x+3)$
21. $x(x^2+8)^{-1/2}$ **23.** $(3x+4)^{-2/3}$ **25.** $\frac{1}{2}(x^2-4x+2)^{-1/2}(2x-4) = (x-2)/(x^2-4x+2)^{1/2}$ **27.** $(-1)(2x+4)^{-2}(2) = -2/(2x+4)^2$
29. $-15x^2(x^3+4)^{-6}$ **31.** $(-1)(4x^2-4x+1)^{-2}(8x-4) = -4/(2x-1)^3$ **33.** $-2(x^2-3x)^{-3/2}(2x-3) = \frac{-2(2x-3)}{(x^2-3x)^{3/2}}$
35. $f'(x) = (4-x)^3 - 3x(4-x)^2$; $y = -16x+48$ **37.** $f'(x) = \frac{(2x-5)^3 - 6x(2x-5)^2}{(2x-5)^6}$; $y = -17x+54$
39. $f'(x) = (2x+2)^{1/2} + x(2x+2)^{-1/2}$; $y = \frac{5}{2}x - \frac{1}{2}$
41. $f'(x) = 2x(x-5)^3 + 3x^2(x-5)^2 = 5x(x-5)^2(x-2)$; $x = 0$, $x = 2$, $x = 5$
43. $f'(x) = \frac{(2x+5)^2 - 4x(2x+5)}{(2x+5)^4} = \frac{5-2x}{(2x+5)^3}$; $x = \frac{5}{2}$ **45.** $f'(x) = \frac{x-4}{\sqrt{x^2-8x+20}}$; $x = 4$
47. $18x^2(x^2+1)^2 + 3(x^2+1)^3 = 3(x^2+1)^2(7x^2+1)$ **49.** $\frac{2x^3 4(x^3-7)^3 3x^2 - (x^3-7)^4 6x^2}{4x^6} = \frac{3(x^3-7)^3(3x^3+7)}{2x^4}$
51. $(2x-3)^2[3(2x^2+1)^2(4x)] + (2x^2+1)^3[2(2x-3)(2)] = 4(2x^2+1)^2(2x-3)(8x^2-9x+1)$
53. $4x^2[\frac{1}{2}(x^2-1)^{-1/2}(2x)] + (x^2-1)^{1/2}(8x) = \frac{12x^3-8x}{\sqrt{x^2-1}}$ **55.** $\frac{(x-3)^{1/2}(2) - 2x[\frac{1}{2}(x-3)^{-1/2}]}{x-3} = \frac{x-6}{(x-3)^{3/2}}$
57. $(2x-1)^{1/2}(x^2+3)(11x^2-4x+9)$
59. (A) $C'(x) = (2x+16)^{-1/2} = 1/\sqrt{2x+16}$
(B) $C'(24) = \frac{1}{8}$ or $12.50 per calculator (the rate of change of total cost relative to production at a production level of 24 calculators is $12.50 per calculator; the cost of producing 1 more calculator at this level of production is approx. $12.50); $C'(42) = \frac{1}{10}$ or $10.00 per calculator (the rate of change of total cost relative to production at a production level of 42 calculators is $10.00 per calculator; the cost of producing 1 more calculator at this level of production is approx. $10.00)
61. $4{,}000(1+\frac{1}{12}i)^{47}$ **63.** $\frac{(4 \times 10^6)x}{(x^2-1)^{5/3}}$
65. (A) $f'(n) = n(n-2)^{-1/2} + 2(n-2)^{1/2} = \frac{3n-4}{(n-2)^{1/2}}$
(B) $f'(11) = \frac{29}{3}$ (rate of learning is $\frac{29}{3}$ units/min at the $n = 11$ level); $f'(27) = \frac{77}{5}$ (rate of learning is $\frac{77}{5}$ units/min at the $n = 27$ level)

◆ EXERCISE 4-7

1. (A) $29.50 (B) $30
3. (A) $420
(B) $\overline{C}'(500) = -0.24$; at a production level of 500 units, a unit increase in production will decrease average cost by approx. 24¢
5. (A) $R'(1{,}600) = 20$; at a production level of 1,600 units, a unit increase in production will increase revenue by approx. $20
(B) $R'(2{,}500) = -25$; at a production level of 2,500 units, a unit increase in production will decrease revenue by approx. $25
7. (A) $4.50 (B) $5
9. (A) $P'(450) = 0.5$; at a production level of 450 units, a unit increase in production will increase profit by approx. 50¢
(B) $P'(750) = -2.5$; at a production level of 750 units, a unit increase in production will decrease profit by approx. $2.50
11. (A) $1.25
(B) $\overline{P}'(150) = 0.015$; at a production level of 150 units, a unit increase in production will increase average profit by approx. 1.5¢
13. (A) $C'(x) = 60$ (B) $R(x) = 200x - (x^2/30)$ (C) $R'(x) = 200 - (x/15)$
(D) $R'(1{,}500) = 100$ (at a production level of 1,500 units, a unit increase in production will increase revenue by approx. $100); $R'(4{,}500) = -100$ (at a production level of 4,500 units, a unit increase in production will decrease revenue by approx. $100)

13. (E) Break-even points: (600, 108,000) and (3,600, 288,000)

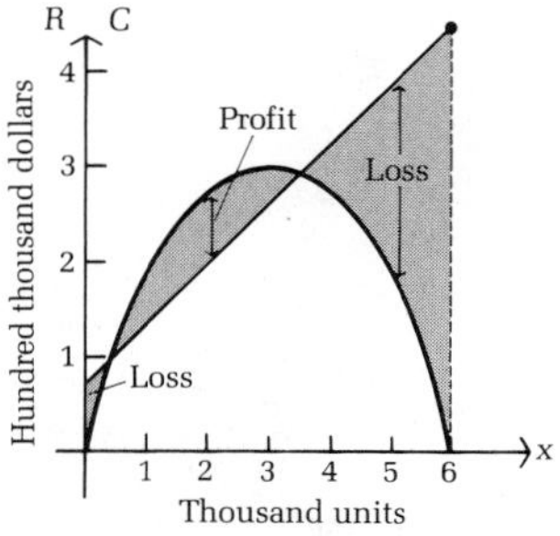

(F) $P(x) = -(x^2/30) + 140x - 72{,}000$
(G) $P'(x) = -(x/15) + 140$
(H) $P'(1{,}500) = 40$ (at a production level of 1,500 units, a unit increase in production will increase profit by approx. \$40); $P'(3{,}000) = -60$ (at a production level of 3,000 units, a unit increase in production will decrease profit by approx. \$60)

15. (A) $p = 20 - (x/50)$ (B) $R(x) = 20x - (x^2/50)$ (C) $C(x) = 4x + 1{,}400$
(D) Break-even points: (100, 1,800) and (700, 4,200)

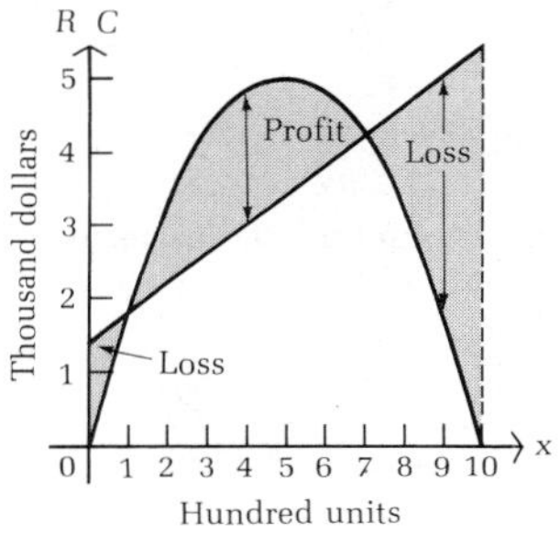

(E) $P(x) = 16x - (x^2/50) - 1{,}400$
(F) $P'(250) = 6$ (at a production level of 250 units, a unit increase in production will increase profit by approx. \$6); $P'(475) = -3$ (at a production level of 475 units, a unit increase in production will decrease profit by approx. \$3

◆ EXERCISE 4-8 CHAPTER REVIEW

1. $12x^3 - 4x$ **2.** $x^{-1/2} - 3 = \frac{1}{x^{1/2}} - 3$ **3.** 0 **4.** $-x^{-3} + x$ **5.** $(2x - 1)(3) + (3x + 2)(2) = 12x + 1$

6. $(x^2 - 1)(3x^2) + (x^3 - 3)(2x) = 5x^4 - 3x^2 - 6x$ **7.** $\frac{(x^2 + 2)2 - 2x(2x)}{(x^2 + 2)^2} = \frac{4 - 2x^2}{(x^2 + 2)^2}$ **8.** $(-1)(3x + 2)^{-2}(3) = \frac{-3}{(3x + 2)^2}$

9. $3(2x - 3)^2(2) = 6(2x - 3)^2$ **10.** $-2(x^2 + 2)^{-3}(2x) = \frac{-4x}{(x^2 + 2)^3}$ **11.** (A) Does not exist (B) 3 (C) No
12. (A) 2 (B) Not defined (C) No **13.** (A) 1 (B) 1 (C) Yes **14.** $12x^3 + 6x^{-4}$
15. $(2x^2 - 3x + 2)(2x + 2) + (x^2 + 2x - 1)(4x - 3) = 8x^3 + 3x^2 - 12x + 7$ **16.** $\frac{(x - 1)^2(2) - (2x - 3)(2)(x - 1)}{(x - 1)^4} = \frac{4 - 2x}{(x - 1)^3}$

17. $x^{-1/2} - 2x^{-3/2} = \frac{1}{\sqrt{x}} - \frac{2}{\sqrt{x^3}}$ **18.** $(x^2 - 1)[2(2x + 1)(2)] + (2x + 1)^2(2x) = 2(2x + 1)(4x^2 + x - 2)$

19. $\frac{1}{3}(x^3 - 5)^{-2/3}(3x^2) = \frac{x^2}{\sqrt[3]{(x^3 - 5)^2}}$ **20.** $-8x^{-3}$ **21.** $\frac{(2x - 3)(4)(x^2 + 2)^3(2x) - (x^2 + 2)^4(2)}{(2x - 3)^2} = \frac{2(x^2 + 2)^3(7x^2 - 12x - 2)}{(2x - 3)^2}$

22. (A) $m = f'(1) = 2$ (B) $y = 2x + 3$ **23.** (A) $m = f'(1) = 16$ (B) $y = 16x - 12$ **24.** $x = 5$ **25.** $x = -5, x = 3$
26. $x = -2, x = 2$ **27.** $x = 0, x = 3, x = \frac{15}{2}$ **28.** (A) $v = f'(x) = 32x - 4$ (B) $f'(3) = 92$ ft/sec
29. (A) $v = f'(x) = 96 - 32x$ (B) $x = 3$ sec **30.** (A) 4 (B) 6 (C) Does not exist (D) 6 (E) No
31. (A) 3 (B) 3 (C) 3 (D) 3 (E) Yes **32.** (A) ∞ (B) ∞ (C) ∞ **33.** (A) $-\infty$ (B) ∞ (C) Does not exist **34.** $(-\infty, \infty)$

35. $(-\infty, -5), (-5, \infty)$ **36.** $(-\infty, -2), (-2, 3), (3, \infty)$ **37.** $[3, \infty)$ **38.** $(-\infty, \infty)$ **39.** $\frac{2(3) - 3}{3 + 5} = \frac{3}{8}$ **40.** $2(3^2) - 3 + 1 = 16$ **41.** -1 **42.** 4

43. 4 **44.** $\frac{1}{6}$ **45.** Does not exist **46.** $1/(2\sqrt{7})$ **47.** $\sqrt{2}$ **48.** 3 **49.** ∞ **50.** $\frac{2}{3}$ **51.** 0 **52.** ∞ **53.** $2x - 1$ **54.** $1/(2\sqrt{x})$ **55.** No **56.** No **57.** No **58.** Yes

59. Discontinuous at $x = 0$

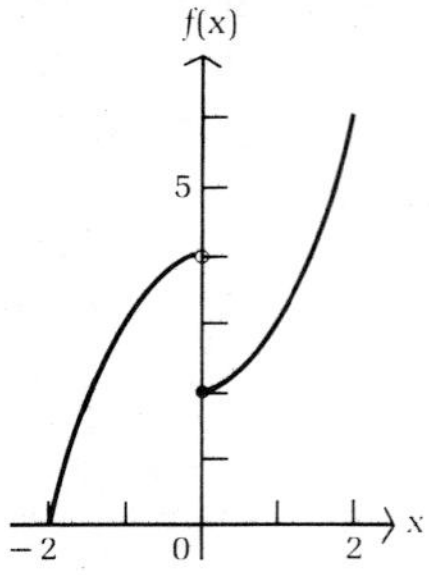

60. Continuous for all x

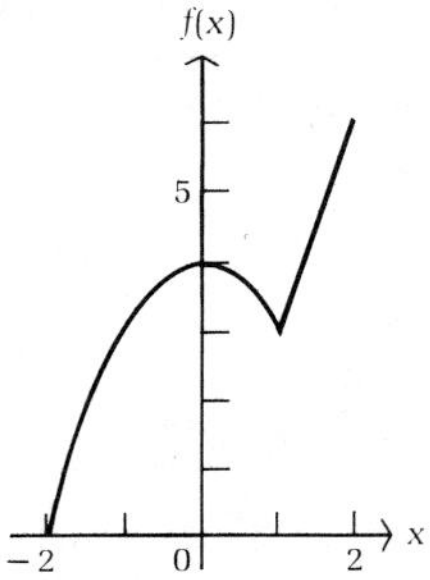

61. $7x(x-4)^3(x+3)^2$

62. $\dfrac{x^4(2x+5)}{(2x+1)^5}$

63. $\dfrac{1}{x^2\sqrt{x^2-1}}$

64. $\dfrac{4}{(x^2+4)^{3/2}}$

65. (A) 1 (B) 1 (C) 1 (D) Yes

66. (A) Yes (B) Yes (C) Yes (D) Yes

67. (A) 1 (B) -1 (C) Does not exist (D) No

68. (A) \$179.90 (B) \$180

69. (A) $C'(x) = 2$; $\overline{C}(x) = 2 + 56x^{-1}$; $\overline{C}'(x) = -56x^{-2}$

(B) $R(x) = xp = 20x - x^2$; $R'(x) = 20 - 2x$; $\overline{R}(x) = 20 - x$; $\overline{R}'(x) = -1$

(C) $P(x) = R(x) - C(x) = 18x - x^2 - 56$; $P'(x) = 18 - 2x$; $\overline{P}(x) = 18 - x - 56x^{-1}$; $\overline{P}'(x) = -1 + 56x^{-2}$

(D) Solving $R(x) = C(x)$, we find break-even points at $x = 4, 14$.

(E) $P'(7) = 4$ (increasing production increases profit); $\overline{P}'(9) = 0$ (stable); $P'(11) = -4$ (increasing production decreases profit)

(F)

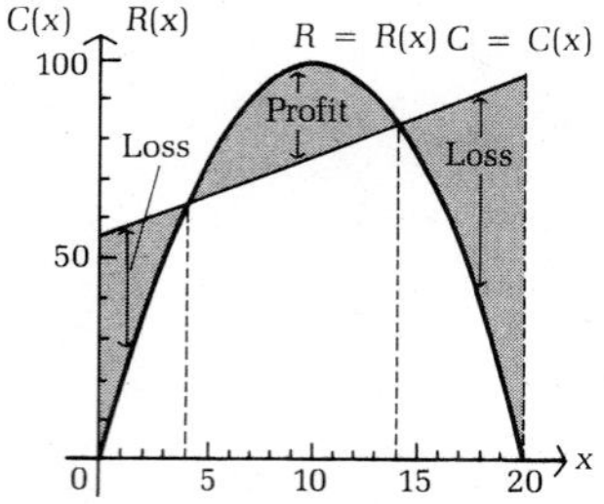

70. (A) 2 components/day (B) 3.2 components/day (C) 40 components/day

71. $C'(9) = -1$ ppm/m; $C'(99) = -0.001$ ppm/m **72.** (A) 10 items/hr (B) 5 items/hr

CHAPTER 5

◆ EXERCISE 5-1

1. (a, b); (d, f); (g, h); **3.** c, d, f **5.** b, f **7.** $-3 < x < 4$; $(-3, 4)$ **9.** $x < 3$ or $x > 7$; $(-\infty, 3) \cup (7, \infty)$

11. $-5 < x < 0$ or $x > 3$; $(-5, 0) \cup (3, \infty)$ **13.** Decreasing on $(-\infty, 8)$; increasing on $(8, \infty)$; local minimum at $x = 8$

15. Increasing on $(-\infty, 5)$; decreasing on $(5, \infty)$; local maximum at $x = 5$ **17.** Increasing for all x; no local extrema

19. Decreasing for all x; no local extrema

21. Increasing on $(-\infty, -2)$ and $(2, \infty)$; decreasing on $(-2, 2)$; local maximum at $x = -2$; local minimum at $x = 2$

23. Increasing on $(-\infty, -2)$ and $(4, \infty)$; decreasing on $(-2, 4)$; local maximum at $x = -2$; local minimum at $x = 4$

25. Increasing on $(-\infty, -1)$ and $(0, 1)$; decreasing on $(-1, 0)$ and $(1, \infty)$; local maxima at $x = -1$ and $x = 1$; local minimum at $x = 0$

27. Increasing on $(-\infty, 4)$
Decreasing on $(4, \infty)$
Horizontal tangent at $x = 4$

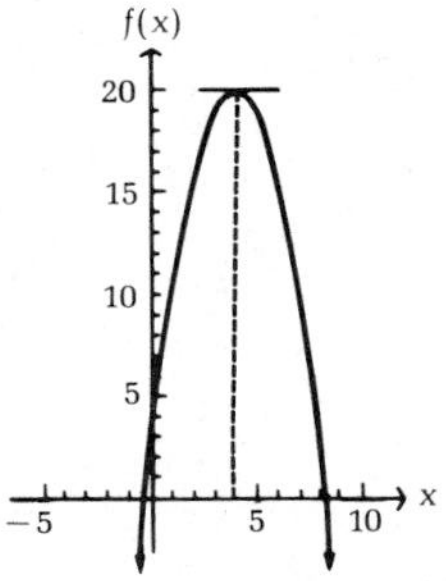

29. Increasing on $(-\infty, -1)$, $(1, \infty)$
Decreasing on $(-1, 1)$
Horizontal tangents at $x = -1, 1$

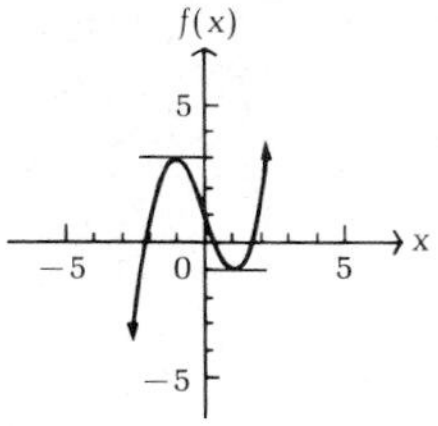

31. Decreasing for all x
Horizontal tangent at $x = 2$

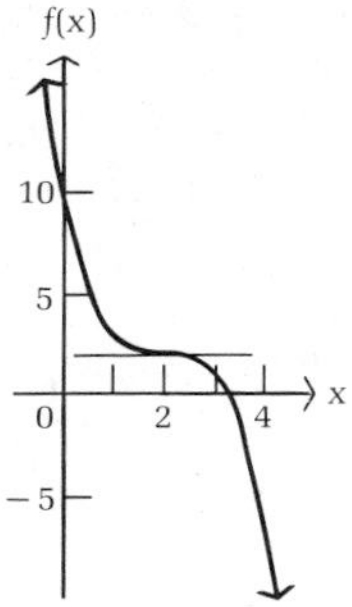

33. No critical values; increasing on $(-\infty, -2)$ and $(-2, \infty)$ no local extrema
35. Critical values: $x = -2$, $x = 2$; increasing on $(-\infty, -2)$ and $(2, \infty)$; decreasing on $(-2, 0)$ and $(0, 2)$; local maximum at $x = -2$; local minimum at $x = 2$
37. Critical value: $x = -2$; increasing on $(-2, 0)$; decreasing on $(-\infty, -2)$ and $(0, \infty)$; local minimum at $x = -2$
39. Critical values: $x = 0$, $x = 4$; increasing on $(-\infty, 0)$ and $(4, \infty)$; decreasing on $(0, 2)$ and $(2, 4)$; local maximum at $x = 0$; local minimum at $x = 4$
41. Critical values: $x = 0$, $x = 4$, $x = 6$; increasing on $(0, 4)$ and $(6, \infty)$; decreasing on $(-\infty, 0)$ and $(4, 6)$; local maximum at $x = 4$; local minima at $x = 0$ and $x = 6$
43. Critical value: $x = 2$; increasing on $(2, \infty)$; decreasing on $(-\infty, 2)$; local minimum at $x = 2$
45. Critical value: $x = 1$; increasing on $(0, 1)$; decreasing on $(1, \infty)$; local maximum at $x = 1$
47. Critical value: $x = 80$; decreasing for $0 < x < 80$; increasing for $80 < x < 150$; local minimum at $x = 80$
49. $P(x)$ is increasing over (a, b) if $P'(x) = R'(x) - C'(x) > 0$ over (a, b); that is, if $R'(x) > C'(x)$ over (a, b).
51. Critical value: $t = 1$; increasing for $0 < t < 1$; decreasing for $1 < t < 24$; local maximum at $t = 1$
53. Critical value: $t = 7$; increasing for $0 < t < 7$; decreasing for $7 < t < 24$; local maximum at $t = 7$

◆ EXERCISE 5-2

1. (a, c), (c, d), (e, g) **3.** d, e, g **5.** $6x - 4$ **7.** $40x^3$ **9.** $6x$ **11.** $24x^2(x^2 - 1) + 6(x^2 - 1)^2 = 6(x^2 - 1)(5x^2 - 1)$ **13.** $6x^{-3} + 12x^{-4}$
15. $f(2) = -2$ is a local minimum **17.** $f(-1) = 2$ is a local maximum; $f(2) = -25$ is a local minimum **19.** No local extrema
21. $f(-2) = -6$ is a local minimum; $f(0) = 10$ is a local maximum; $f(2) = -6$ is a local minimum **23.** $f(0) = 2$ is a local minimum
25. $f(-4) = -8$ is a local maximum; $f(4) = 8$ is a local minimum **27.** Concave upward for all x; no inflection points
29. Concave upward on $(6, \infty)$; concave downward on $(-\infty, 6)$; inflection point at $x = 6$
31. Concave upward on $(-\infty, -2)$ and $(2, \infty)$; concave downward on $(-2, 2)$; inflection points at $x = -2$ and $x = 2$
33. Concave upward on $(0, 2)$; concave downward on $(-\infty, 0)$ and $(2, \infty)$; inflection points at $x = 0$ and $x = 2$

35. Local maximum at $x = 0$
Local minimum at $x = 4$
Inflection point at $x = 2$

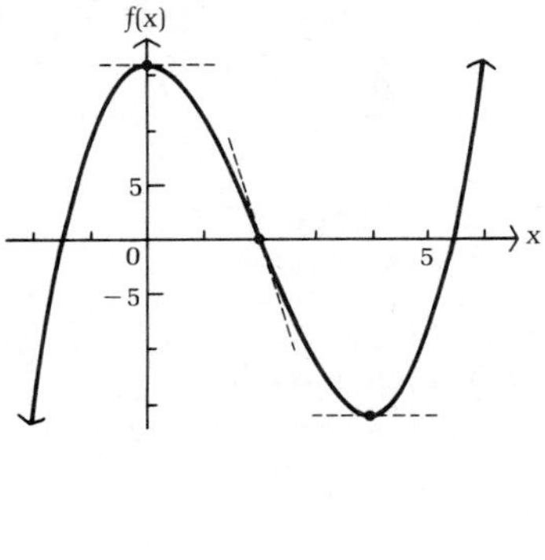

37. Inflection point at $x = 0$

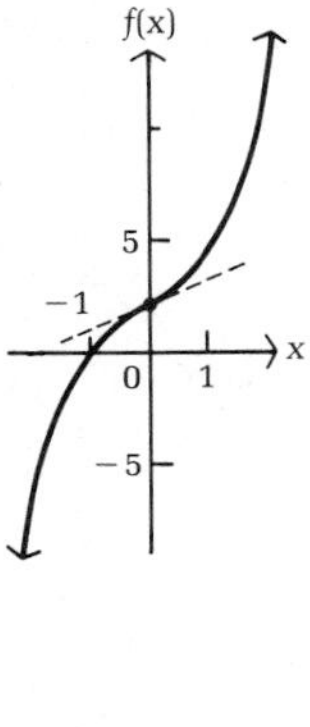

39. Inflection point at $x = 2$

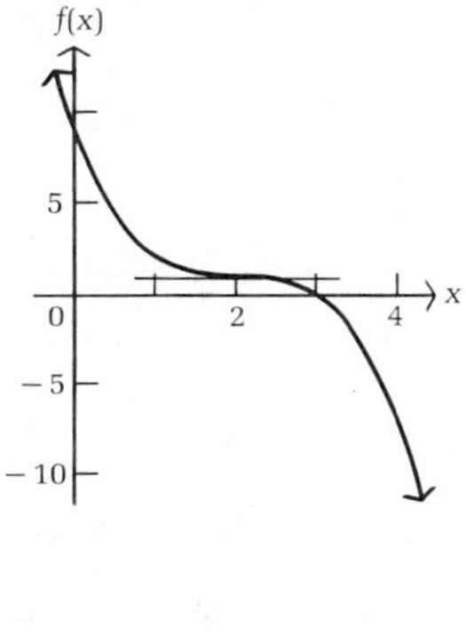

41. Local maximum at $x = -2$
Local minimum at $x = 2$
Inflection point at $x = 0$

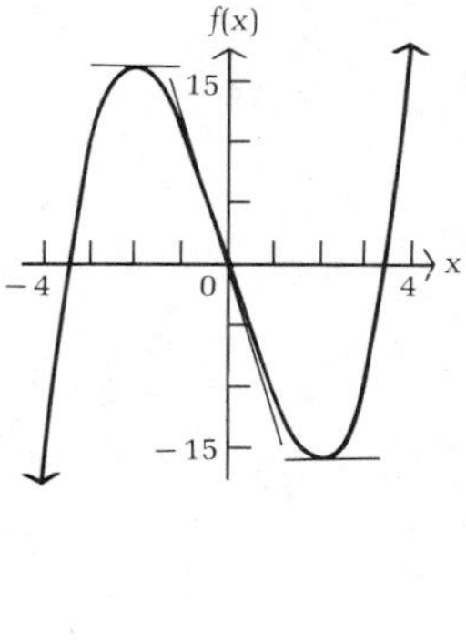

43. Inflection points at $x = -2$ and $x = 2$ **45.** Inflection points at $x = -6$, $x = 0$, and $x = 6$
47. (A) Local maximum at $x = 60$ (B) Concave downward on the whole interval $(0, 80)$

49. (A) Increasing on $(10, 25)$; decreasing on $(25, 40)$
(B) Inflection point at $x = 25$
(C)

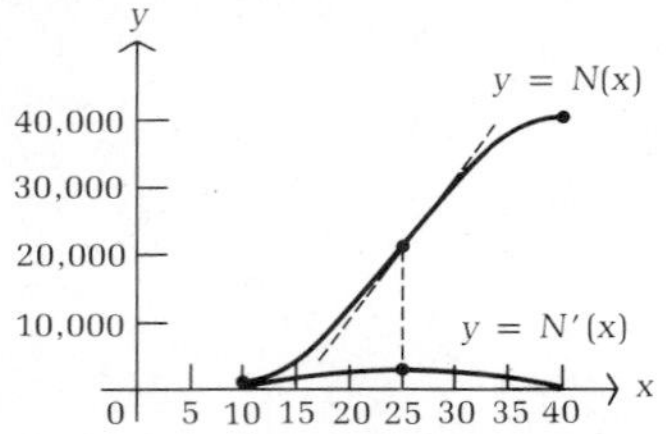

(D) Max $N'(x) = N'(25) = 2{,}025$

51. (A) Increasing on $(0, 10)$ decreasing on $(10, 20)$
(B) Inflection point at $t = 10$
(C)

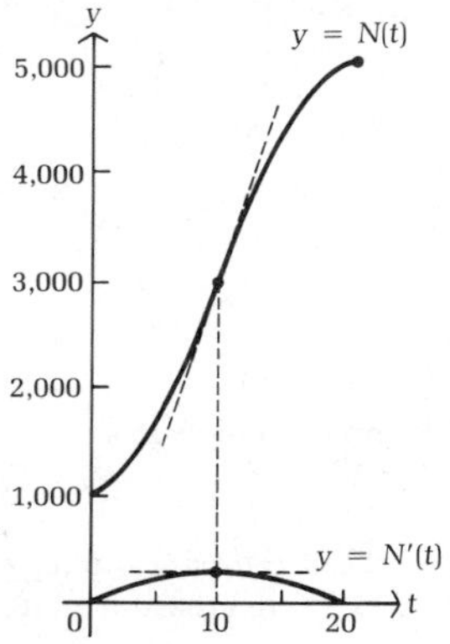

(D) $N'(10) = 300$

53. (A) Increasing on $(5, \infty)$; decreasing on $(0, 5)$
(B) Inflection point at $n = 5$

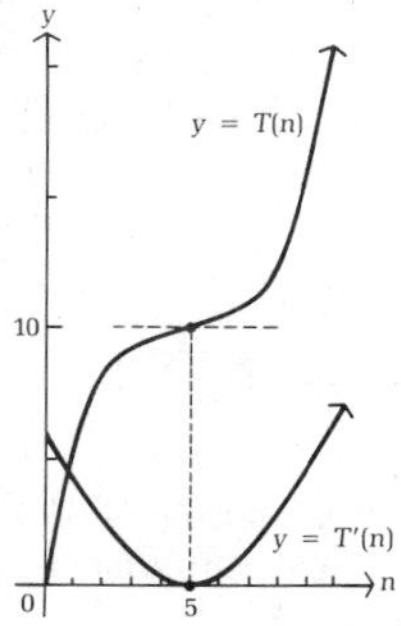

(C) $T'(5) = 0$

◆ EXERCISE 5-3

1. (b, d), $(d, 0)$, (g, ∞) **3.** $x = 0$ **5.** (a, d), (e, h) **7.** $x = a$, $x = h$ **9.** $x = d$, $x = e$
11. Horizontal asymptote: $y = 2$; vertical asymptote: $x = -2$
13. Horizontal asymptote: $y = 1$; vertical asymptotes: $x = -1$ and $x = 1$ **15.** No horizontal or vertical asymptotes
17. Horizontal asymptote: $y = 0$; no vertical asymptotes **19.** No horizontal asymptote; vertical asymptote: $x = 3$

21. Domain: All real numbers
y intercept: 5
x intercepts: 1, 5
Decreasing on $(-\infty, 3)$
Increasing on $(3, \infty)$
Local minimum at $x = 3$
Concave upward on $(-\infty, \infty)$

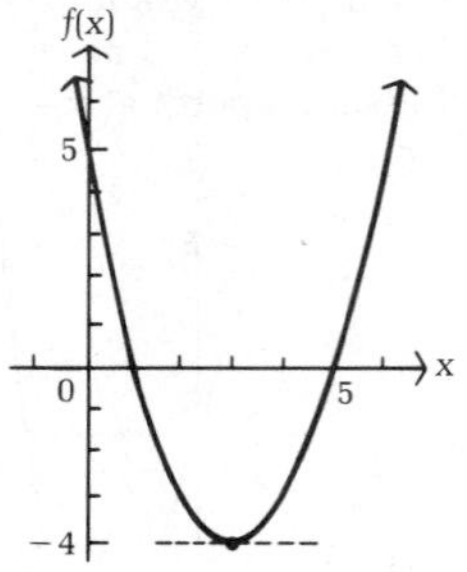

23. Domain: All real numbers
y intercept: 0
x intercepts: 0, 6
Increasing on $(-\infty, 0)$ and $(4, \infty)$
Decreasing on $(0, 4)$
Local maximum at $x = 0$
Local minimum at $x = 4$
Concave upward on $(2, \infty)$
Concave downward on $(-\infty, 2)$
Inflection point at $x = 2$

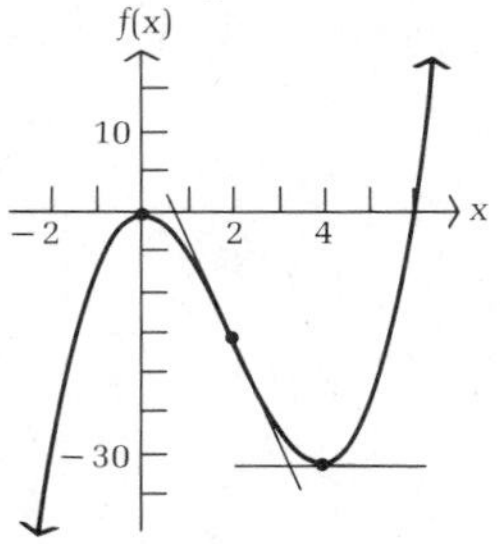

25. Domain: All real numbers
y intercept: 16
x intercepts: -4, 2
Increasing on $(-\infty, -2)$ and $(2, \infty)$
Decreasing on $(-2, 2)$
Local maximum at $x = -2$
Local minimum at $x = 2$
Concave upward on $(0, \infty)$
Concave downward on $(-\infty, 0)$
Inflection point at $x = 0$

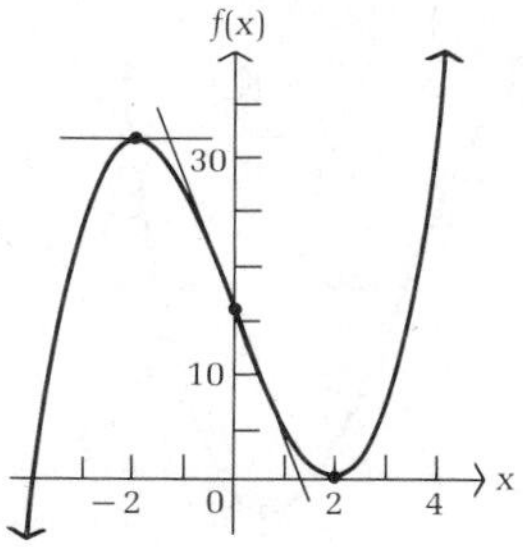

27. Domain: All real numbers
y intercept: 0
x intercepts: 0, 4
Increasing on $(-\infty, 3)$
Decreasing on $(3, \infty)$
Local maximum at $x = 3$
Concave upward on $(0, 2)$
Concave downward on $(-\infty, 0)$ and $(2, \infty)$
Inflection points at $x = 0$ and $x = 2$

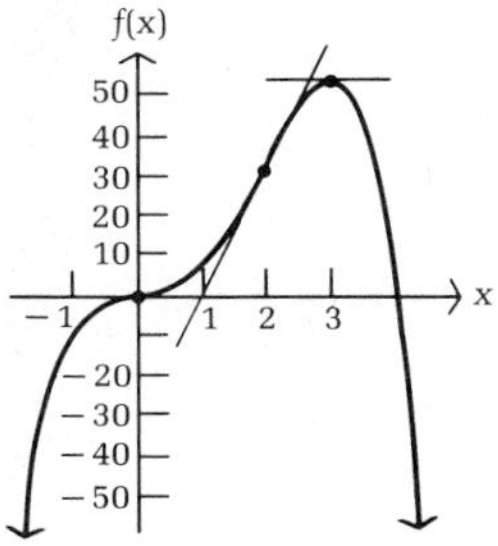

29. Domain: All real numbers except 3
y intercept: -1
x intercept: -3
Horizontal asymptote: $y = 1$
Vertical asymptote: $x = 3$
Decreasing on $(-\infty, 3)$ and $(3, \infty)$
Concave upward on $(3, \infty)$
Concave downward on $(-\infty, 3)$

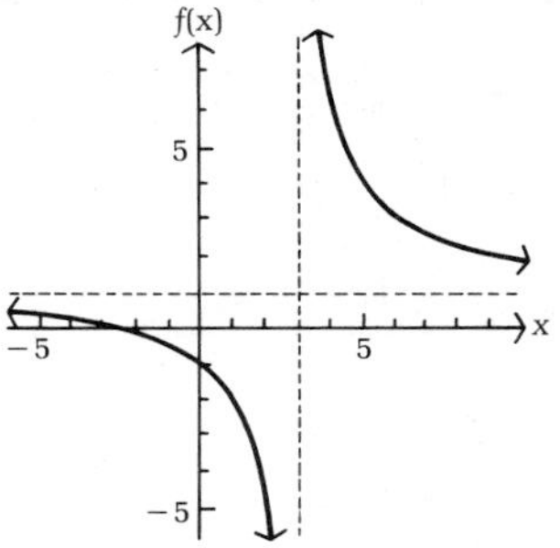

31. Domain: All real numbers except 2
y intercept: 0
x intercept: 0
Horizontal asymptote: $y = 1$
Vertical asymptote: $x = 2$
Decreasing on $(-\infty, 2)$ and $(2, \infty)$
Concave downward on $(-\infty, 2)$
Concave upward on $(2, \infty)$

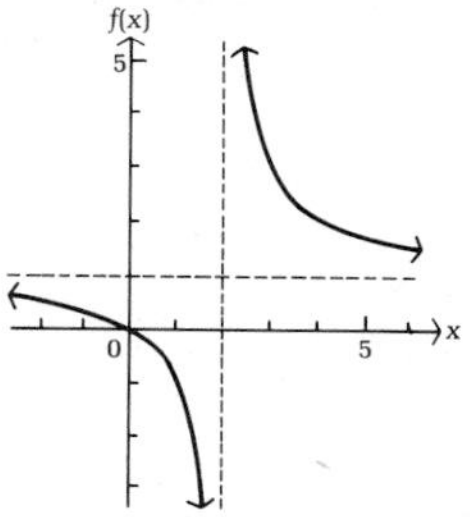

33. Domain: All real numbers except 0
Vertical asymptote: $x = 0$
Increasing on $(-\infty, -1)$ and $(1, \infty)$
Decreasing on $(-1, 0)$ and $(0, 1)$
Local maximum at $x = -1$
Local minimum at $x = 1$
Concave upward on $(0, \infty)$
Concave downward on $(-\infty, 0)$

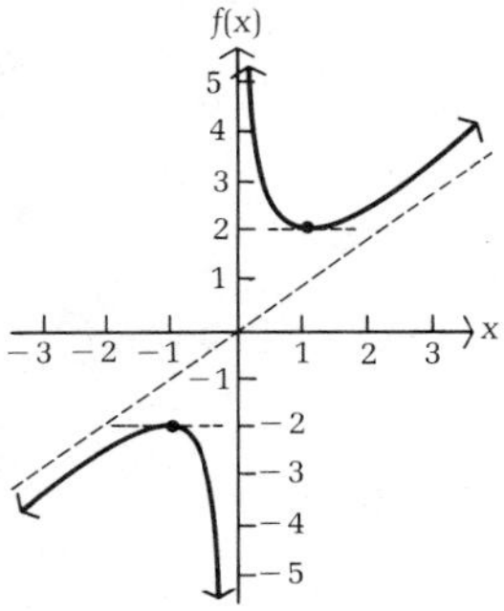

35. Domain: All real numbers
y intercept: 0
x intercepts: $-1, 0, 1$
Increasing on $(-\infty, -\sqrt{3}/3)$ and $(\sqrt{3}/3, \infty)$
Decreasing on $(-\sqrt{3}/3, \sqrt{3}/3)$
Local maximum at $x = -\sqrt{3}/3$
Local minimum at $x = \sqrt{3}/3$
Concave downward on $(-\infty, 0)$
Concave upward on $(0, \infty)$
Inflection point at $x = 0$

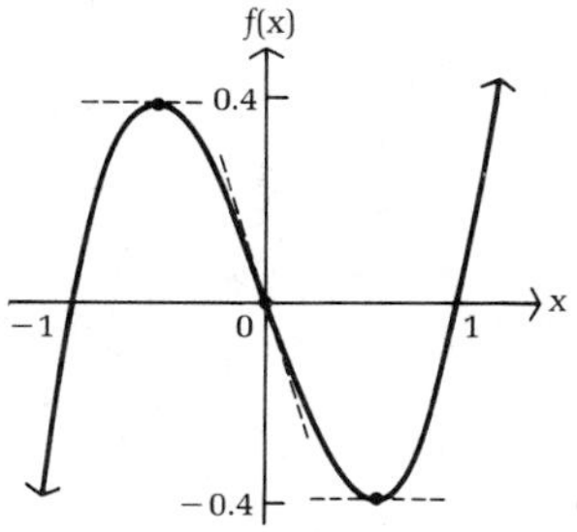

37. Domain: All real numbers
y intercept: 27
x intercepts: $-3, 3$
Increasing on $(-\infty, -\sqrt{3})$ and $(0, \sqrt{3})$
Decreasing on $(-\sqrt{3}, 0)$ and $(\sqrt{3}, \infty)$
Local maxima at $x = -\sqrt{3}$ and $x = \sqrt{3}$
Local minimum at $x = 0$
Concave upward on $(-1, 1)$
Concave downward on $(-\infty, -1)$ and $(1, \infty)$
Inflection points at $x = -1$ and $x = 1$

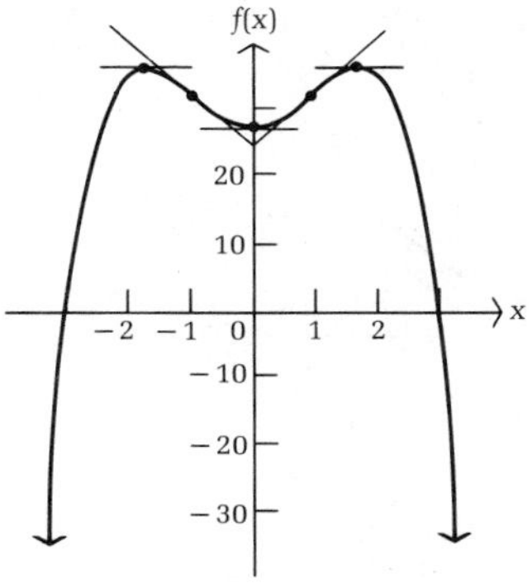

39. Domain: All real numbers
y intercept: 16
x intercepts: $-2, 2$
Decreasing on $(-\infty, -2)$ and $(0, 2)$
Increasing on $(-2, 0)$ and $(2, \infty)$
Local minima at $x = -2$ and $x = 2$
Local maximum at $x = 0$
Concave upward on $(-\infty, -2\sqrt{3}/3)$ and $(2\sqrt{3}/3, \infty)$
Concave downward on $(-2\sqrt{3}/3, 2\sqrt{3}/3)$
Inflection points at $x = -2\sqrt{3}/3$ and $x = 2\sqrt{3}/3$

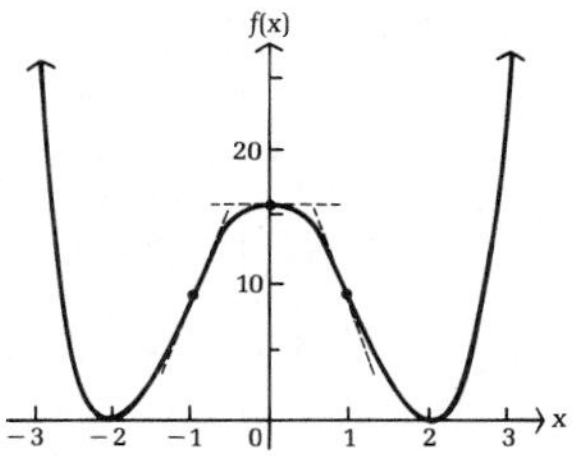

41. Domain: All real numbers
y intercept: 0
x intercepts: 0, 1.5
Decreasing on $(-\infty, 0)$ and $(0, 1.25)$
Increasing on $(1.25, \infty)$
Local minimum at $x = 1.25$
Concave upward on $(-\infty, 0)$ and $(1, \infty)$
Concave downward on $(0, 1)$
Inflection points at $x = 0$ and $x = 1$

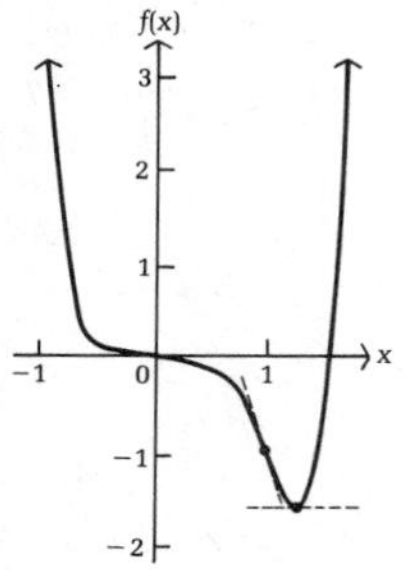

43. Domain: All real numbers except ± 2
y intercept: 0
x intercept: 0
Horizontal asymptote: $y = 0$
Vertical asymptotes: $x = -2$, $x = 2$
Decreasing on $(-\infty, -2)$, $(-2, 2)$, and $(2, \infty)$
Concave upward on $(-2, 0)$ and $(2, \infty)$
Concave downward on $(-\infty, -2)$ and $(0, 2)$
Inflection point at $x = 0$

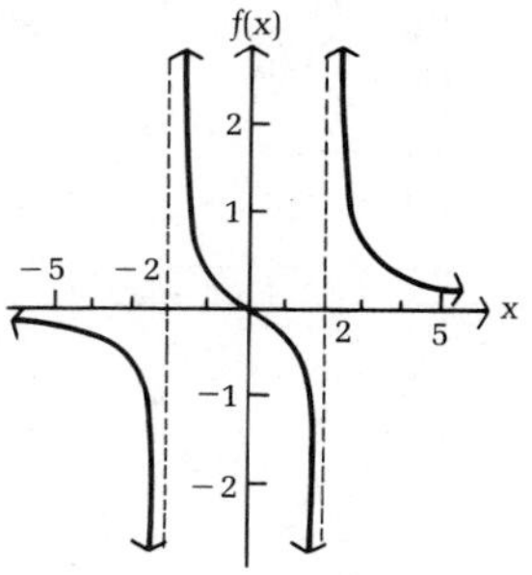

45. Domain: All real numbers
y intercept: 1
Horizontal asymptote: $y = 0$
Increasing on $(-\infty, 0)$
Decreasing on $(0, \infty)$
Local maximum at $x = 0$
Concave upward on $(-\infty, -\sqrt{3}/3)$ and $(\sqrt{3}/3, \infty)$
Concave downward on $(-\sqrt{3}/3, \sqrt{3}/3)$
Inflection points at $x = -\sqrt{3}/3$ and $x = \sqrt{3}/3$

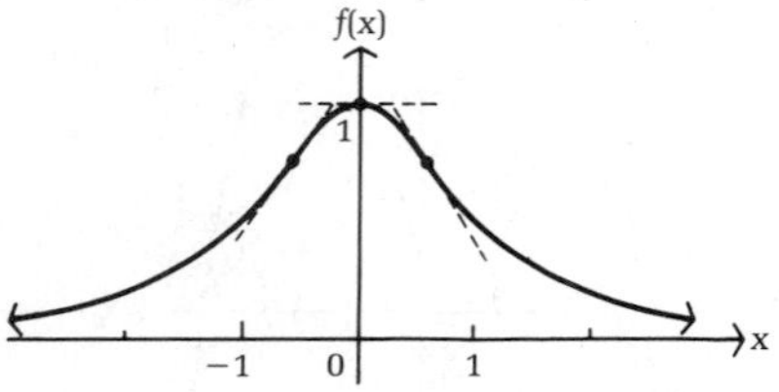

47.

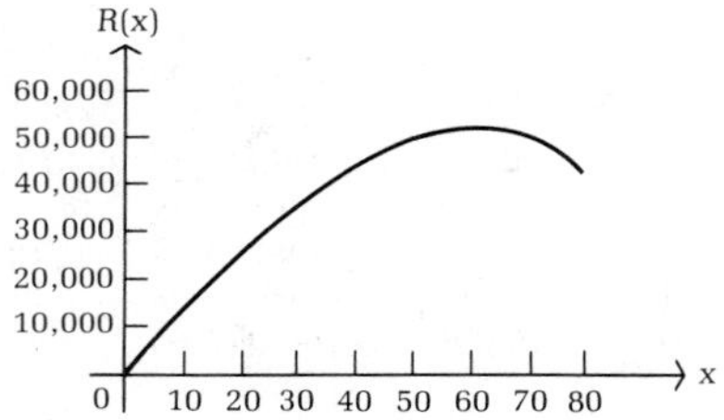

49. (A) Increasing on (0, 1)
(B) Concave upward on (0, 1)
(C) $x = 1$ is a vertical asymptote
(D) The origin is both an x and a y intercept
(E)

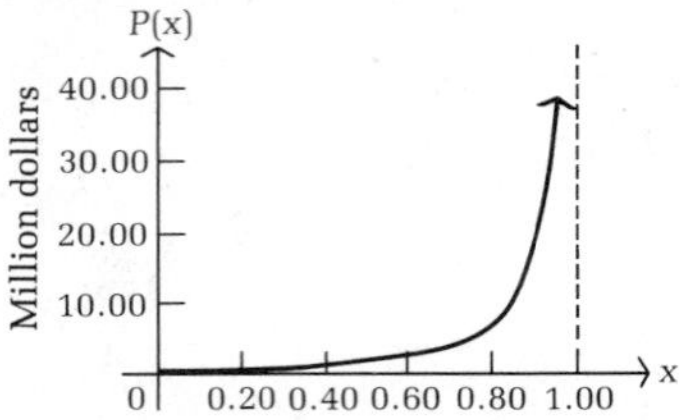

51. (A) $\overline{C}(n) = \dfrac{3,200}{n} + 250 + 50n$
(B)

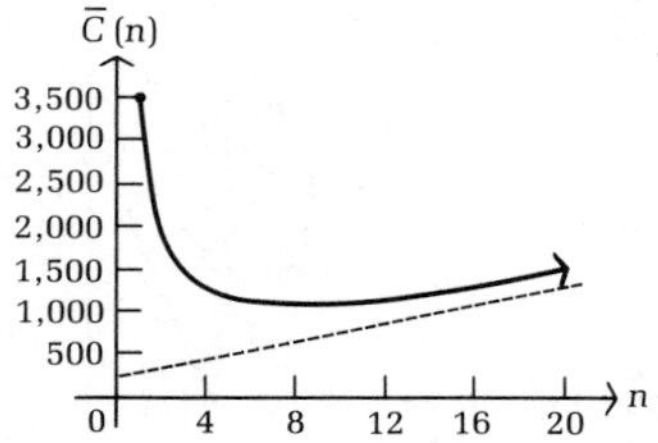

(C) 8 yr

53. (A)

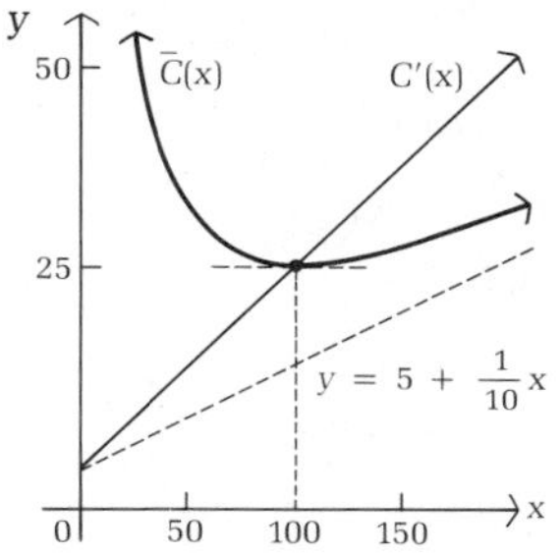

(B) 25 at $x = 100$

55.

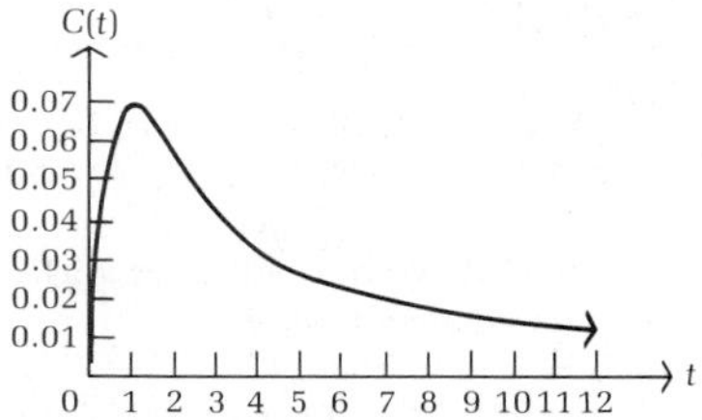

57.

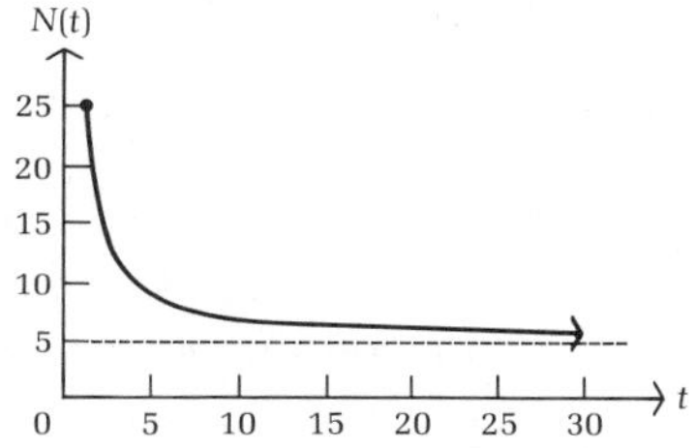

◆ EXERCISE 5-4

1. Min $f(x) = f(2) = 1$; no maximum **3.** Max $f(x) = f(4) = 26$; no minimum **5.** No absolute extrema exist. **7.** Max $f(x) = f(2) = 16$
9. Min $f(x) = f(2) = 14$
11. (A) Max $f(x) = f(5) = 14$; Min $f(x) = f(-1) = -22$ (B) Max $f(x) = f(1) = -2$; Min $f(x) = f(-1) = -22$
(C) Max $f(x) = f(5) = 14$; Min $f(x) = f(3) = -6$
13. (A) Max $f(x) = f(0) = 126$; Min $f(x) = f(2) = -26$ (B) Max $f(x) = f(7) = 49$; Min $f(x) = f(2) = -26$
(C) Max $f(x) = f(6) = 6$; Min $f(x) = f(3) = -15$
15. Exactly in half **17.** 15 and -15 **19.** A square of side 25 cm; maximum area = 625 cm^2 **21.** 3,000 pairs; \$3.00/pair
23. \$35; \$6,125 **25.** 40 trees; 1,600 lb **27.** $(10 - 2\sqrt{7})/3 = 1.57$ in. squares
29. 20 ft by 40 ft (with the expensive side being one of the short sides) **31.** 10,000 books in 5 printings
33. (A) $x = 5.1$ mi (B) $x = 10$ mi **35.** 4 days; 20 bacteria/cm^3 **37.** 50 mice per order **39.** 1 month; 2 ft **41.** 4 yr from now

◆ EXERCISE 5-5

1. $\Delta x = 3$; $\Delta y = 45$; $\Delta y/\Delta x = 15$ **3.** 12 **5.** 12 **7.** $dy = (24x - 3x^2)\,dx$ **9.** $dy = \left(2x - \frac{x^2}{3}\right) dx$ **11.** $dy = -\frac{295}{x^{3/2}}\,dx$

13. (A) $12 + 3\,\Delta x$ (B) 12 **15.** $dy = 6(2x+1)^2\,dx$ **17.** $dy = \frac{9 - x^2}{(x^2+9)^2}\,dx$ **19.** $dy = 1.4$; $\Delta y = 1.44$ **21.** $dy = -3$; $\Delta y = -3\frac{1}{3}$

23. 120 $in.^3$ **25.** $dy = \frac{6x - 2}{3(3x^2 - 2x + 1)^{2/3}}\,dx$ **27.** $dy = 3.9$; $\Delta y = 3.83$ **29.** 40 unit increase; 20 unit increase **31.** $-\$6$; \$4

33. -1.37/min; -0.58/min **35.** 1.26 mm^2 **37.** 3 wpm **39.** (A) 2,100 increase (B) 4,800 increase (C) 2,100 increase

1. $(a, c_1), (c_3, c_6)$ **2.** $(c_1, c_3), (c_6, b)$ **3.** $(a, c_2), (c_4, c_5), (c_7, b)$ **4.** c_3 **5.** c_6 **6.** c_1, c_3, c_5 **7.** c_6 **8.** c_2, c_4, c_5, c_7
9. $f''(x) = 12x^2 + 30x$ **10.** $y'' = 8/x^3$ **11.** $dy = (3x^2 + 4)\,dx$ **12.** $dy = 18x(3x^2 - 7)^2\,dx$ **13.** $-3 < x < 4$; $(-3, 4)$
14. $-3 < x < 0$ or $x > 5$; $(-3, 0) \cup (5, \infty)$
15. (A) All real numbers (B) y intercept: 0; x intercepts: 0, 9 (C) No horizontal or vertical asymptotes
16. (A) 3, 9 (B) 3, 9 (C) Increasing on $(-\infty, 3)$ and $(9, \infty)$; decreasing on $(3, 9)$
(D) Local maximum at $x = 3$; local minimum at $x = 9$
17. (A) Concave downward on $(-\infty, 6)$; concave upward on $(6, \infty)$ (B) Inflection point at $x = 6$
18.

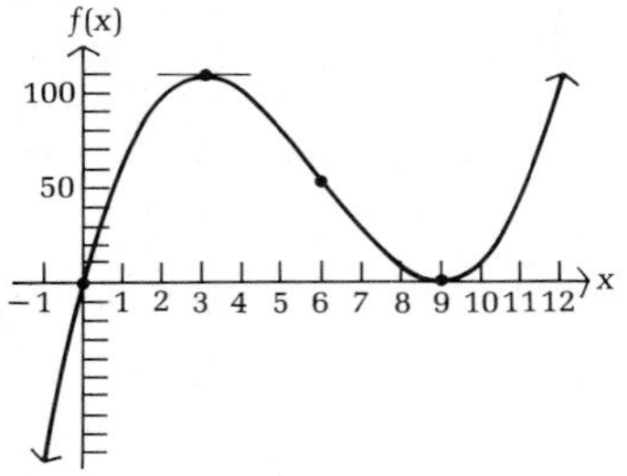

19. (A) All real numbers except -2 (B) y intercept: 0; x intercept: 0 (C) Horizontal asymptote: $y = 3$; vertical asymptote: $x = -2$
20. (A) None (B) -2 (C) Increasing on $(-\infty, -2)$ and $(-2, \infty)$ (D) None
21. (A) Concave upward on $(-\infty, -2)$; concave downward on $(-2, \infty)$ (B) No inflection points
22.

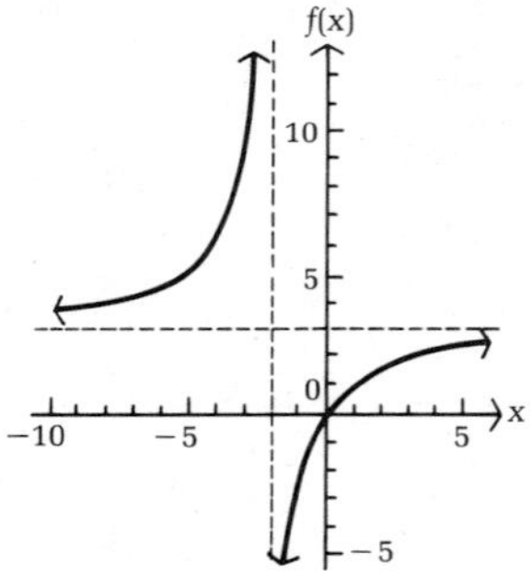

23. Min $f(x) = f(2) = -4$; Max $f(x) = f(5) = 77$ **24.** Min $f(x) = f(2) = 8$ **25.** Horizontal asymptote: $y = 0$; no vertical asymptotes
26. No horizontal asymptotes; vertical asymptotes: $x = -3$ and $x = 3$ **27.** $dy = 7.3$; $\Delta y = 7.45$
28. Max $f'(x) = f'(2) = 12$

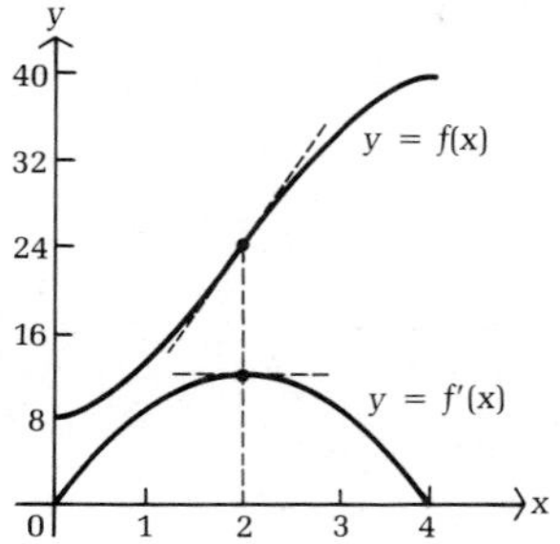

29. Each number is 20; minimum sum is 40
30. Domain: All real numbers
y intercept: -3
x intercepts: $-3, 1$
No vertical or horizontal asymptotes
Increasing on $(-2, \infty)$
Decreasing on $(-\infty, -2)$
Local minimum at $x = -2$
Concave upward on $(-\infty, -1)$ and $(1, \infty)$
Concave downward on $(-1, 1)$
Inflection points at $x = -1$ and $x = 1$

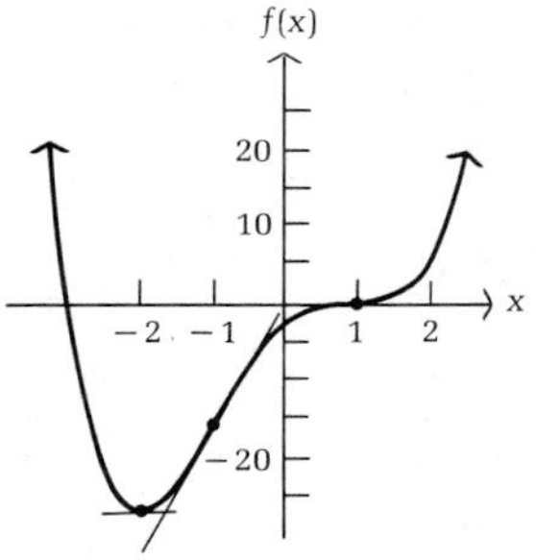

31. $dy = -0.0031$; $\Delta y = -0.0031$
32. Max $P(x) = P(3{,}000) = \$175{,}000$

33. Min $\overline{C}(x) = \overline{C}(200) = 50$

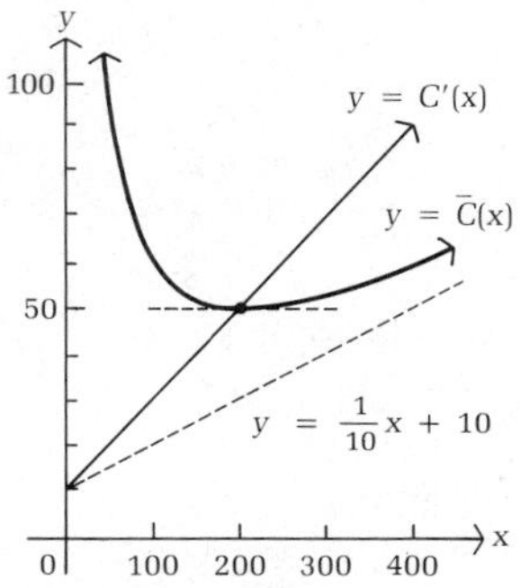

34. \$49; \$6,724
35. 12 orders/yr
36. \$110
37. 3 days
38. 2 yr from now

CHAPTER 6

◆ EXERCISE 6-1

1. \$1,221.40; \$1,648.72; \$2,225.54 **3.** 11.55 **5.** 10.99 **7.** 0.14

9.

n	$[1 + (1/n)]^n$
10	2.593 74
100	2.704 81
1,000	2.716 92
10,000	2.718 15
100,000	2.718 27
1,000,000	2.718 28
10,000,000	2.718 28
↓	↓
∞	$e = 2.718\ 281\ 828\ 459 \ldots$

11. \$55,463.90 **13.** \$9,931.71 **15.** $r = \frac{1}{4} \ln 1.5 \approx 0.1014$ or 10.14%

17. (A)

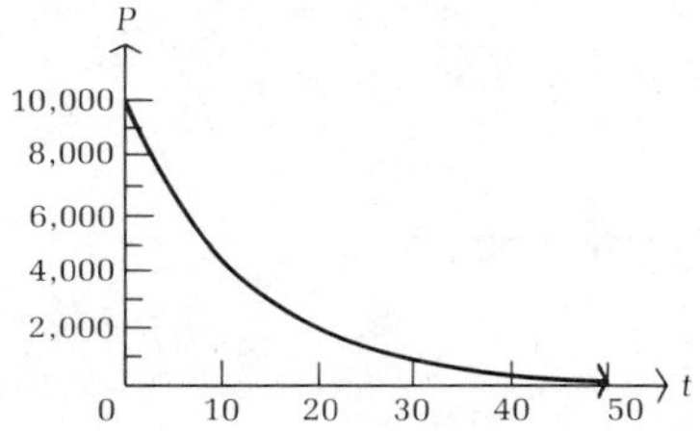

(B) $\lim_{t \to \infty} 10{,}000e^{-0.08t} = 0$

19. 2.77 yr **21.** 13.86% **23.** $A = Pe^{rt}$; $2P = Pe^{rt}$; $e^{rt} = 2$; $\ln e^{rt} = \ln 2$; $rt = \ln 2$; $t = (\ln 2)/r$ **25.** 34.66 yr **27.** 3.47%
29. $t = -(\ln 0.5)/0.000\ 433\ 2 \approx 1{,}600$ yr **31.** $r = (\ln 0.5)/30 \approx -0.0231$ **33.** Approx. 521 yr

◆ EXERCISE 6-2

1. $6e^x - \frac{7}{x}$ **3.** $2ex^{e-1} + 3e^x$ **5.** $\frac{5}{x}$ **7.** $\frac{2 \ln x}{x}$ **9.** $x^3 + 4x^3 \ln x = x^3(1 + 4 \ln x)$ **11.** $x^3e^x + 3x^2e^x = x^2e^x(x + 3)$

13. $\frac{(x^2 + 9)e^x - 2xe^x}{(x^2 + 9)^2} = \frac{e^x(x^2 - 2x + 9)}{(x^2 + 9)^2}$ **15.** $\frac{x^3 - 4x^3 \ln x}{x^8} = \frac{1 - 4 \ln x}{x^5}$ **17.** $3(x + 2)^2 \ln x + \frac{(x + 2)^3}{x} = (x + 2)^2\left(3 \ln x + \frac{x + 2}{x}\right)$

19. $(x + 1)^3e^x + 3(x + 1)^2e^x = (x + 1)^2e^x(x + 4)$ **21.** $\frac{2xe^x - (x^2 + 1)e^x}{(e^x)^2} = \frac{2x - x^2 - 1}{e^x}$ **23.** $(\ln x)^3 + 3(\ln x)^2 = (\ln x)^2(\ln x + 3)$

25. $-15e^x(4 - 5e^x)^2$ **27.** $\frac{1}{2x\sqrt{1 + \ln x}}$ **29.** xe^x **31.** $4x \ln x$ **33.** $y = ex$ **35.** $y = \frac{1}{e}x$

37. Max $f(x) = f(e^3) = e^3 \approx 20.086$ **39.** Min $f(x) = f(1) = e \approx 2.718$ **41.** Max $f(x) = f(e^{1/2}) = 2e^{-1/2} \approx 1.213$

43. Domain: All real numbers
y intercept: 0
x intercept: 0
Horizontal asymptote: $y = 1$
Decreasing on $(-\infty, \infty)$
Concave downward on $(-\infty, \infty)$

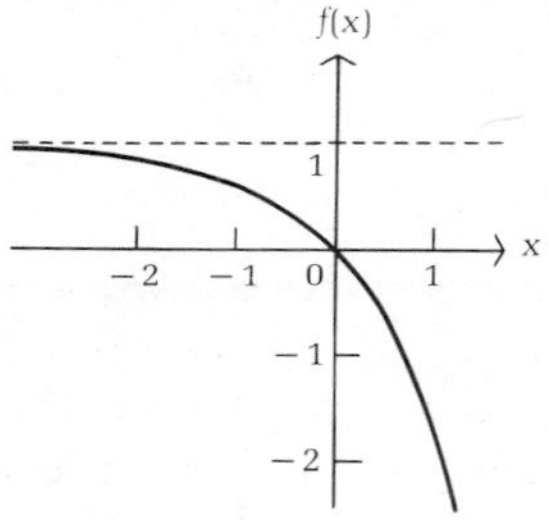

45. Domain: $(0, \infty)$
Vertical asymptote: $x = 0$
Increasing on $(1, \infty)$
Decreasing on $(0, 1)$
Local minimum at $x = 1$
Concave upward on $(0, \infty)$

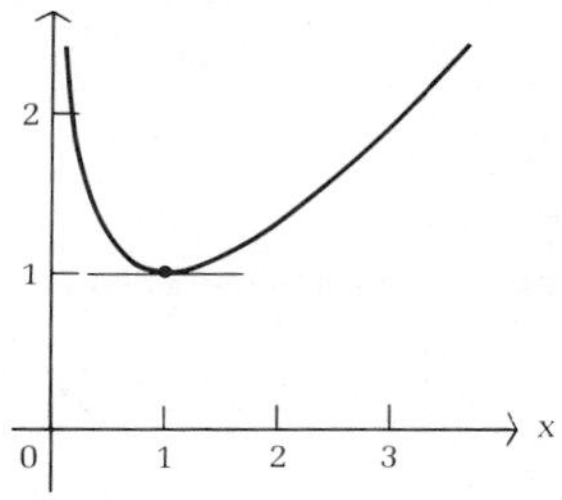

47. Domain: All real numbers
y intercept: 3
x intercept: 3
Horizontal asymptote: $y = 0$
Increasing on $(-\infty, 2)$
Decreasing on $(2, \infty)$
Local maximum at $x = 2$
Concave upward on $(-\infty, 1)$
Concave downward on $(1, \infty)$
Inflection point at $x = 1$

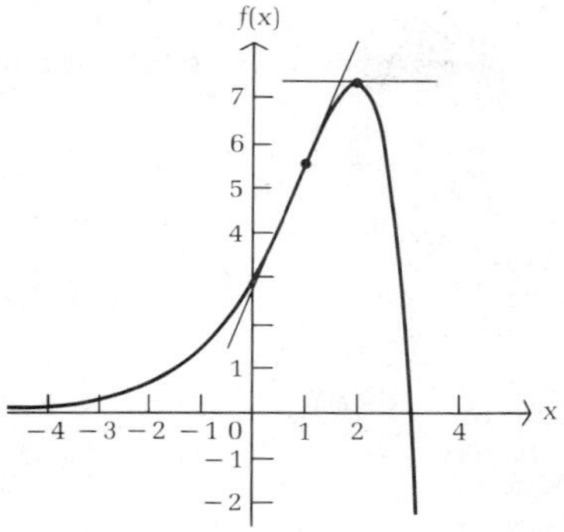

49. Domain: $(0, \infty)$
x intercept: 1
Increasing on $(e^{-1/2}, \infty)$
Decreasing on $(0, e^{-1/2})$
Local minimum at $x = e^{-1/2}$
Concave upward on $(e^{-3/2}, \infty)$
Concave downward on $(0, e^{-3/2})$
Inflection point at $x = e^{-3/2}$

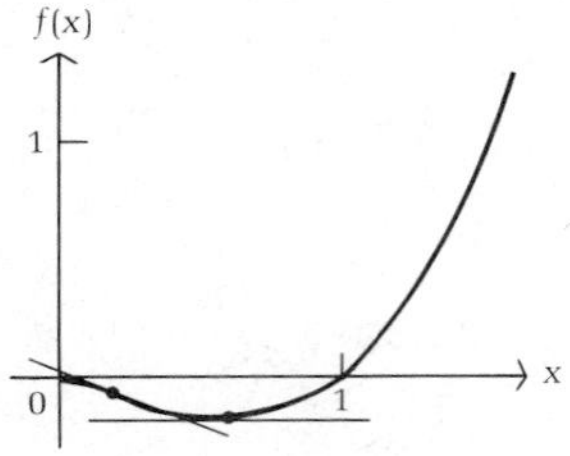

51. $p = \$2$ **53.** Min $\overline{C}(x) = \overline{C}(e^7) \approx \$99.91$ **55.** (A) At \$3.68 each, the maximum revenue will be \$3,680/wk (in the test city).
(B)

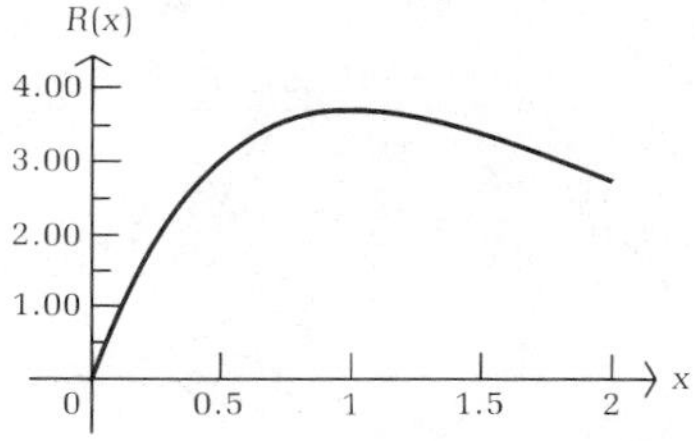

57. At the 40 lb weight level, blood pressure would increase at the rate of 0.44 mm of mercury/lb of weight gain. At the 90 lb weight level, blood pressure would increase at the rate of 0.19 mm of mercury/lb of weight gain.

59. (A) After 1 hr, the concentration is decreasing at the rate of 1.60 mg/ml/hr; after 4 hr, the concentration is decreasing at the rate of 0.08 mg/ml/hr.

(B)

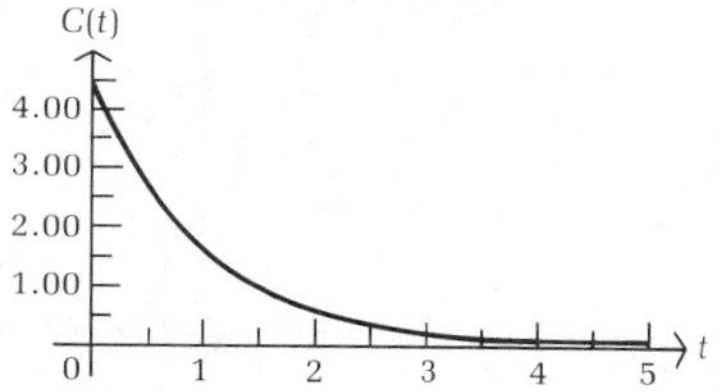

61. $dR/dS = k/S$

◆ EXERCISE 6-3

1. $y = u^3$; $u = 2x + 5$ **3.** $y = \ln u$; $u = 2x^2 + 7$ **5.** $y = e^u$; $u = x^2 - 2$ **7.** $y = (2 + e^x)^2$; $dy/dx = 2e^x(2 + e^x)$

9. $y = e^{2-x^4}$; $dy/dx = -4x^3e^{2-x^4}$ **11.** $y = \ln(4x^5 - 7)$; $\dfrac{dy}{dx} = \dfrac{20x^4}{4x^5 - 7}$ **13.** $\dfrac{1}{x - 3}$ **15.** $\dfrac{-2}{3 - 2t}$ **17.** $6e^{2x}$ **19.** $-8e^{-4t}$ **21.** $-3e^{-0.03x}$

23. $\dfrac{4}{x + 1}$ **25.** $4e^{2x} - 3e^x$ **27.** $(6x - 2)e^{3x^2-2x}$ **29.** $\dfrac{2t + 3}{t^2 + 3t}$ **31.** $\dfrac{x}{x^2 + 1}$

33. $\dfrac{4[\ln(t^2 + 1)]^3(2t)}{t^2 + 1} = \dfrac{8t[\ln(t^2 + 1)]^3}{t^2 + 1}$ **35.** $4(e^{2x} - 1)^3(2e^{2x}) = 8e^{2x}(e^{2x} - 1)^3$ **37.** $\dfrac{(x^2 + 1)(2e^{2x}) - e^{2x}(2x)}{(x^2 + 1)^2} = \dfrac{2e^{2x}(x^2 - x + 1)}{(x^2 + 1)^2}$

39. $(x^2 + 1)(-e^{-x}) + e^{-x}(2x) = e^{-x}(2x - x^2 - 1)$ **41.** $\dfrac{e^{-x}}{x} - e^{-x}\ln x$ **43.** $\dfrac{-2x}{(1 + x^2)[\ln(1 + x^2)]^2}$ **45.** $\dfrac{-2x}{3(1 - x^2)[\ln(1 - x^2)]^{2/3}}$

47. Domain: $(-\infty, \infty)$
y intercept: 0
x intercept: 0
Horizontal asymptote: $y = 1$
Increasing on $(-\infty, \infty)$
Concave downward on $(-\infty, \infty)$

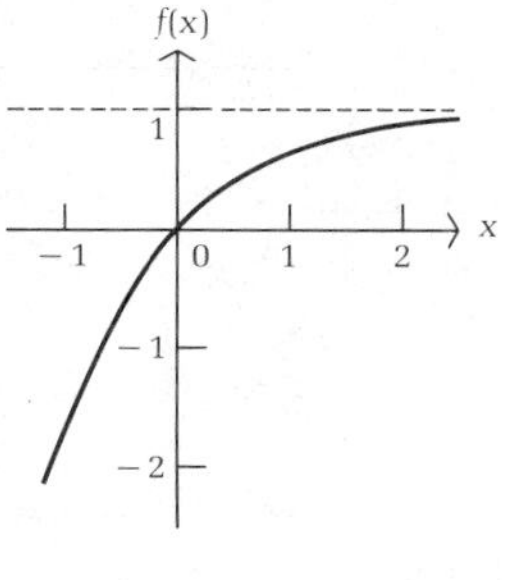

49. Domain: $(-\infty, 1)$
y intercept: 0
x intercept: 0
Vertical asymptote: $x = 1$
Decreasing on $(-\infty, 1)$
Concave downward on $(-\infty, 1)$

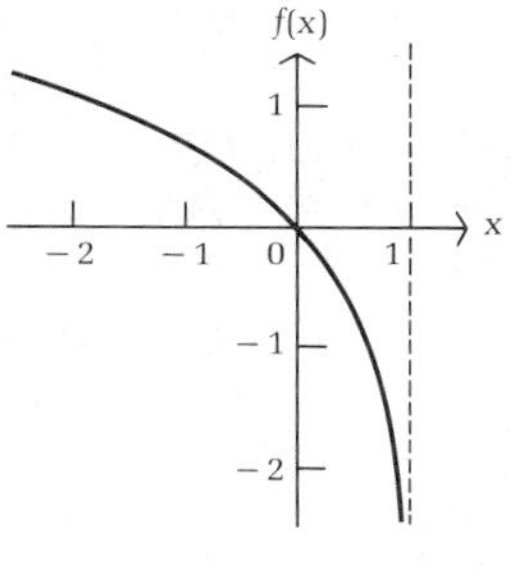

51. Domain: $(-\infty, \infty)$
y intercept: 1
Horizontal asymptote: $y = 0$
Increasing on $(-\infty, 0)$
Decreasing on $(0, \infty)$
Local maximum at $x = 0$
Concave upward on $(-\infty, -1)$ and $(1, \infty)$
Concave downward on $(-1, 1)$
Inflection points at $x = -1$ and $x = 1$

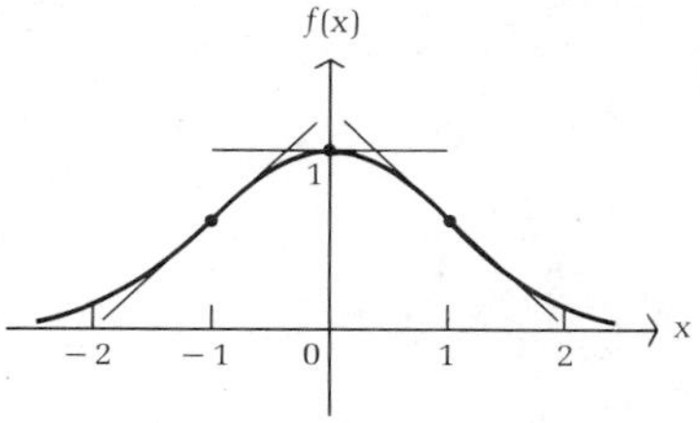

53. $y = 1 + [\ln(2 + e^x)]^2$; $\dfrac{dy}{dx} = \dfrac{2e^x \ln(2 + e^x)}{2 + e^x}$ **55.** $\dfrac{1}{\ln 2}\left(\dfrac{6x}{3x^2 - 1}\right)$ **57.** $(2x + 1)(10^{x^2+x})(\ln 10)$

61. A maximum revenue of \$735.80 is realized at a production level of 20 units at \$36.79 each.

63. −\$27,145/yr; −\$18,196/yr; −\$11,036/yr **65.** (A) 23 days; \$26,685; about 50%
(B)

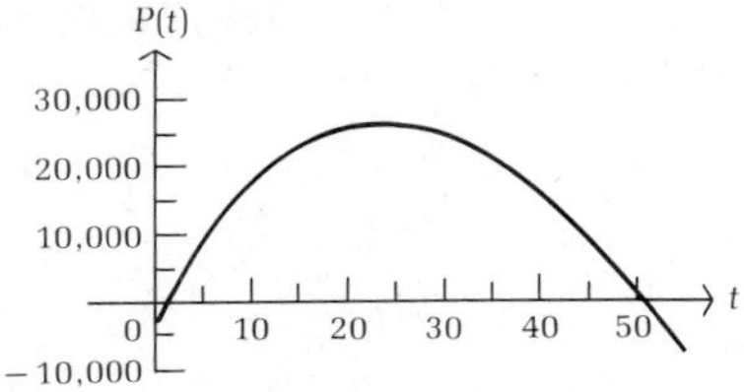

67. 2.27 mm of mercury/yr; 0.81 mm of mercury/yr; 0.41 mm of mercury/yr
69. $A'(t) = 2(\ln 2)5{,}000e^{2t\ln 2} = 10{,}000(\ln 2)2^{2t}$; $A'(1) = 27{,}726$ bacteria/hr (rate of change at the end of the first hour); $A'(5) = 7{,}097{,}827$ bacteria/hr (rate of change at the end of the fifth hour)
71.

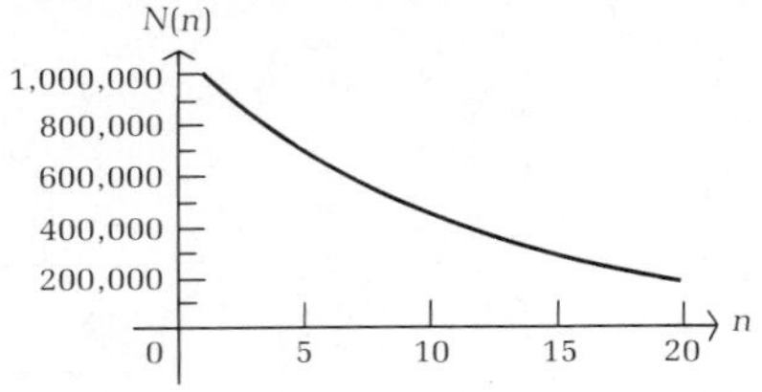

◆ EXERCISE 6-4

1. $y' = 6x$; 6 **3.** $y' = \dfrac{3x}{y}$; 3 **5.** $y' = \dfrac{1}{2y+1}$; $\frac{1}{3}$ **7.** $y' = -\dfrac{y}{x}$; $-\frac{3}{2}$ **9.** $y' = -\dfrac{2y}{2x+1}$; 4 **11.** $y' = \dfrac{6-2y}{x}$; −1 **13.** $y' = \dfrac{2x}{e^y - 2y}$; 2
15. $y' = \dfrac{3x^2y}{y+1}$; $\frac{3}{2}$ **17.** $y' = \dfrac{6x^2y - y\ln y}{x+2y}$; 2 **19.** $x' = \dfrac{2tx - 3t^2}{2x - t^2}$; 8 **21.** $y = -x + 5$ **23.** $y = \frac{2}{5}x - \frac{12}{5}$; $y = \frac{3}{5}x + \frac{12}{5}$
25. $y' = \dfrac{1}{3(1+y)^2 + 1}$; $\frac{1}{13}$ **27.** $y' = \dfrac{3(x-2y)^2}{6(x-2y)^2 + 4y}$; $\frac{3}{10}$ **29.** $y' = \dfrac{3x^2(7+y^2)^{1/2}}{y}$; 16 **31.** $y' = \dfrac{y}{2xy^2 - x}$; 1 **33.** $p' = \dfrac{1}{2p-2}$
35. $p' = -\dfrac{\sqrt{10{,}000 - p^2}}{p}$ **37.** $\dfrac{dL}{dV} = \dfrac{-(L+m)}{V+n}$

◆ EXERCISE 6-5

1. 240 **3.** $\frac{9}{4}$ **5.** $\frac{1}{2}$ **7.** Decreasing at 9 units/sec **9.** Approx. −3.03 ft/sec **11.** $dA/dt \approx 126$ ft²/sec **13.** 3,768 cc/min
15. 6 lb/in.²/hr **17.** $-\frac{9}{4}$ ft/sec **19.** $\frac{20}{3}$ ft/sec **21.** (A) $dC/dt = \$15{,}000$/wk (B) $dR/dt = -\$50{,}000$/wk (C) $dP/dt = -\$65{,}000$/wk
23. Approx. 100 ft³/min

◆ EXERCISE 6-6 CHAPTER REVIEW

1. \$3,136.62; \$4,919.21; \$12,099.29 **2.** $\dfrac{2}{x} + 3e^x$ **3.** $2e^{2x-3}$ **4.** $\dfrac{2}{2x+7}$ **5.** (A) $y = \ln(3 + e^x)$ (B) $\dfrac{dy}{dx} = \dfrac{e^x}{3+e^x}$

6. $y' = \dfrac{9x^2}{4y}$; $\frac{9}{8}$ **7.** $dy/dt = 216$

8. Domain: All real numbers
y intercept: 100
Horizontal asymptote: $y = 0$
Decreasing on $(-\infty, \infty)$
Concave upward on $(-\infty, \infty)$

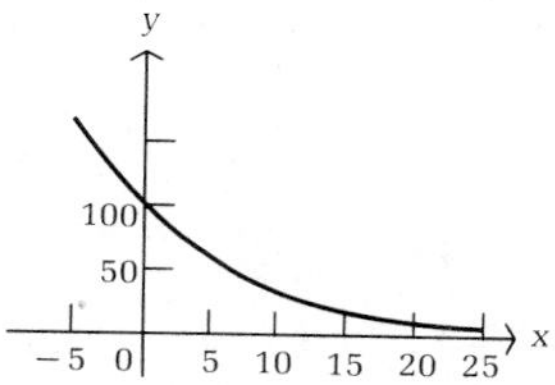

9. $\dfrac{7[(\ln z)^6 + 1]}{z}$ 10. $x^5(1 + 6 \ln x)$

11. $\dfrac{e^x(x - 6)}{x^7}$ 12. $\dfrac{6x^2 - 3}{2x^3 - 3x}$

13. $(3x^2 - 2x)e^{x^3 - x^2}$ 14. $\dfrac{1 - 2x \ln 5x}{xe^{2x}}$

15. $y = -x + 2$; $y = -ex + 1$

16. $y' = \dfrac{3y - 2x}{8y - 3x}$; $\frac{8}{19}$ 17. $x' = \dfrac{4tx}{3x^2 - 2t^2}$; -4 18. $y' = \dfrac{1}{e^y + 2y}$; 1 19. $y' = \dfrac{2xy}{1 + 2y^2}$; $\frac{2}{3}$ 20. $dy/dt = -2$ units/sec

21. 0.27 ft/sec 22. $dR/dt = 1/\pi \approx 0.318$ in./min 23. Max $f(x) = f(e^{4.5}) = 2e^{4.5} \approx 180.03$ 24. Max $f(x) = f(0.5) = 5e^{-1} \approx 1.84$

25. Domain: All real numbers
y intercept: 0
x intercept: 0
Horizontal asymptote: $y = 5$
Increasing on $(-\infty, \infty)$
Concave downward on $(-\infty, \infty)$

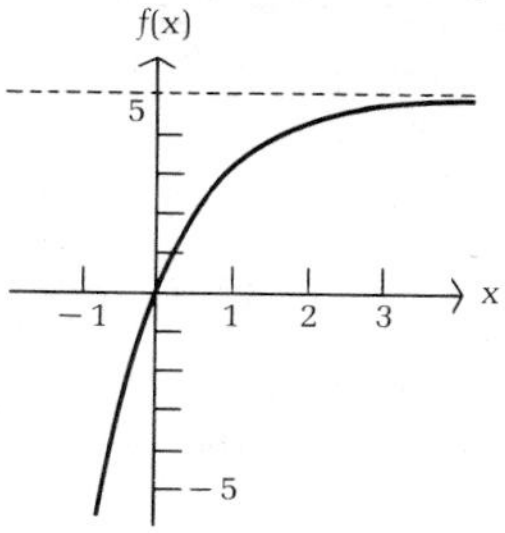

26. Domain: $(0, \infty)$
x intercept: 1
Increasing on $(e^{-1/3}, \infty)$
Decreasing on $(0, e^{-1/3})$
Local minimum at $x = e^{-1/3}$
Concave upward on $(e^{-5/6}, \infty)$
Concave downward on $(0, e^{-5/6})$
Inflection point at $x = e^{-5/6}$

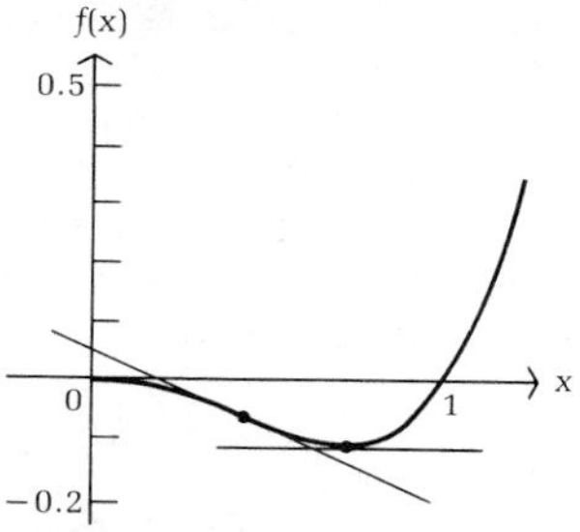

27. (A) $y = [\ln(4 - e^x)]^3$ (B) $\dfrac{dy}{dx} = \dfrac{-3e^x[\ln(4 - e^x)]^2}{4 - e^x}$

28. $2x(5^{x^2-1})(\ln 5)$

29. $\left(\dfrac{1}{\ln 5}\right)\dfrac{2x - 1}{x^2 - x}$

30. $\dfrac{2x + 1}{2(x^2 + x)\sqrt{\ln(x^2 + x)}}$

31. $y' = \dfrac{2x - e^{xy}y}{xe^{xy} - 1}$; 0

32. (A) 15 yr (B) 13.9 yr

33. $A'(t) = 10e^{0.1t}$; $A'(1) = \$11.05$/yr; $A'(10) = \$27.18$/yr 34. $R'(x) = (1{,}000 - 20x)e^{-0.02x}$

35. A maximum revenue of $18,394 is realized at a production level of 50 units at $367.88 each.

36.

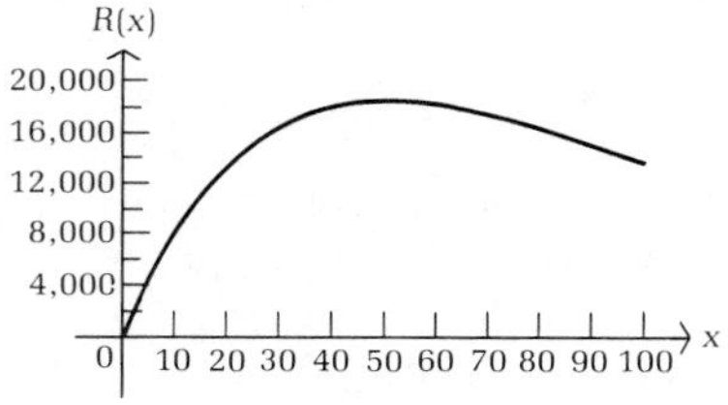

37. Min $\overline{C}(x) = \overline{C}(e^5) \approx \$49.66$

38. $p' = \dfrac{-(5{,}000 - 2p^3)^{1/2}}{3p^2}$

39. $\dfrac{dR}{dt} = \$110$/day

40. -1.111 mg/ml/hr; -0.335 mg/ml/hr

41. $\dfrac{dR}{dt} = -\dfrac{3}{2\pi} \approx 0.477$ ml/day

42. (A) Increasing at the rate of 2.68 units/day at the end of 1 day of training; increasing at the rate of 0.54 unit/day after 5 days of training

(B)

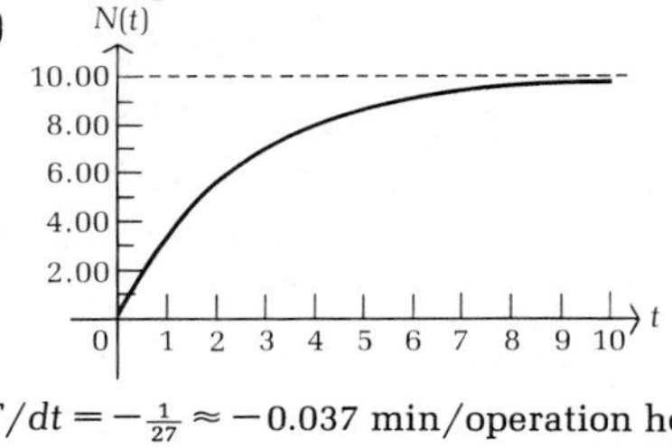

43. $dT/dt = -\frac{1}{27} \approx -0.037$ min/operation hour

CHAPTER 7

◆ EXERCISE 7-1

1. $7x + C$ **3.** $(x^7/7) + C$ **5.** $2t^4 + C$ **7.** $u^2 + u + C$ **9.** $x^3 + x^2 - 5x + C$ **11.** $(s^5/5) - \frac{4}{3}s^6 + C$ **13.** $3e^t + C$ **15.** $2\ln|z| + C$
17. $y = 40x^5 + C$ **19.** $P = 24x - 3x^2 + C$ **21.** $y = \frac{1}{3}u^6 - u^3 - u + C$ **23.** $y = e^x + 3x + C$ **25.** $x = 5\ln|t| + t + C$ **27.** $4x^{3/2} + C$
29. $-4x^{-2} + C$ **31.** $2\sqrt{u} + C$ **33.** $-(x^{-2}/8) + C$ **35.** $-(u^{-4}/8) + C$ **37.** $x^3 + 2x^{-1} + C$ **39.** $2x^5 + 2x^{-4} - 2x + C$
41. $2x^{3/2} + 4x^{1/2} + C$ **43.** $\frac{3}{5}x^{5/3} + 2x^{-2} + C$ **45.** $(e^x/4) - (3x^2/8) + C$ **47.** $-z^{-2} - z^{-1} + \ln|z| + C$ **49.** $y = x^2 - 3x + 5$
51. $C(x) = 2x^3 - 2x^2 + 3{,}000$ **53.** $x = 40\sqrt{t}$ **55.** $y = -2x^{-1} + 3\ln|x| - x + 3$ **57.** $x = 4e^t - 2t - 3$ **59.** $y = 2x^2 - 3x + 1$
61. $x^2 + x^{-1} + C$ **63.** $\frac{1}{2}x^2 + x^{-2} + C$ **65.** $e^x - 2\ln|x| + C$ **67.** $M = t + t^{-1} + \frac{3}{4}$ **69.** $y = 3x^{5/3} + 3x^{2/3} - 6$ **71.** $p(x) = 10x^{-1} + 10$
73. $P(x) = 50x - 0.02x^2$; $P(100) = \$4{,}800$ **75.** $R(x) = 100x - (x^2/10)$; $p = 100 - (x/10)$; $p = \$30$
77. $S(t) = 2{,}000 - 15t^{5/3}$; $80^{3/5} \approx 14$ months **79.** $L(x) = 4{,}800x^{1/2}$; $L(25) = 24{,}000$ labor-hours **81.** $W(h) = 0.0005h^3$; $W(70) = 171.5$ lb
83. 19,400

◆ EXERCISE 7-2

1. $\frac{1}{6}(x^2 - 4)^6 + C$ **3.** $e^{4x} + C$ **5.** $\ln|2t + 3| + C$ **7.** $\frac{1}{24}(3x - 2)^8 + C$ **9.** $\frac{1}{16}(x^2 + 3)^8 + C$ **11.** $-20e^{-0.5t} + C$ **13.** $\frac{1}{10}\ln|10x + 7| + C$
15. $\frac{1}{4}e^{2x^2} + C$ **17.** $\frac{1}{3}\ln|x^3 + 4| + C$ **19.** $-\frac{1}{18}(3t^2 + 1)^{-3} + C$ **21.** $\frac{1}{3}(4 - x^3)^{-1} + C$ **23.** $\frac{2}{5}(x + 4)^{5/2} - \frac{8}{3}(x + 4)^{3/2} + C$
25. $\frac{2}{3}(x - 3)^{3/2} + 6(x - 3)^{1/2} + C$ **27.** $\frac{1}{11}(x - 4)^{11} + \frac{2}{5}(x - 4)^{10} + C$ **29.** $\frac{1}{8}(1 + e^{2x})^4 + C$ **31.** $\frac{1}{2}\ln|4 + 2x + x^2| + C$ **33.** $e^{x^2+x+1} + C$
35. $\frac{1}{4}(e^x - 2x)^4 + C$ **37.** $-\frac{1}{12}(x^4 + 2x^2 + 1)^{-3} + C$ **39.** $\frac{1}{9}(3x^2 + 7)^{3/2} + C$ **41.** $\frac{1}{8}x^8 + \frac{4}{5}x^5 + 2x^2 + C$ **43.** $\frac{1}{9}(x^3 + 2)^3 + C$
45. $\frac{1}{4}(2x^4 + 3)^{1/2} + C$ **47.** $\frac{1}{4}(\ln x)^4 + C$ **49.** $e^{-1/x} + C$ **51.** $x = \frac{1}{3}(t^3 + 5)^7 + C$ **53.** $y = 3(t^2 - 4)^{1/2} + C$ **55.** $p = -(e^x - e^{-x})^{-1} + C$
59. $p(x) = 2{,}000/(3x + 50)$; 250 bottles **61.** $C(x) = 12x + 500\ln(x + 1) + 2{,}000$; $\overline{C}(1{,}000) = \$17.45$
63. $S(t) = 10t + 100e^{-0.1t} - 100$, $0 \leq t \leq 24$; $S(12) \approx \$50$ million **65.** $Q(t) = 100\ln(t + 1) + 5t$, $0 \leq t \leq 20$; $Q(9) \approx 275$ thousand barrels
67. $W(t) = 2e^{0.1t}$; $W(8) \approx 4.45$ g **69.** $N(t) = 5{,}000 - 1{,}000\ln(1 + t^2)$; $N(10) \approx 385$ bacteria/ml
71. $N(t) = 100 - 60e^{-0.1t}$, $0 \leq t \leq 15$; $N(15) \approx 87$ words/min **73.** $E(t) = 12{,}000 - 10{,}000(t + 1)^{-1/2}$; $E(15) = 9{,}500$ students

◆ EXERCISE 7-3

1. $A = 1{,}000e^{0.08t}$ **3.** $A = 8{,}000e^{0.06t}$ **5.** $p(x) = 100e^{-0.05x}$ **7.** $N = L(1 - e^{-0.051t})$ **9.** $I = I_0e^{-0.00942x}$; $x \approx 74$ ft
11. $Q = 3e^{-0.04t}$; $Q(10) = 2.01$ ml **13.** $-0.023\ 117$ **15.** Approx. 24,200 yr **17.** 104 times; 67 times
19. (A) 7 people; 353 people (B) 400

◆ EXERCISE 7-4

1. 5 **3.** 5 **5.** 2 **7.** 48 **9.** $-\frac{7}{3} \approx -2.333$ **11.** 2 **13.** $\frac{1}{2}(e^2 - 1)$ **15.** $2\ln 3.5$ **17.** -2 **19.** 14 **21.** $5^6 = 15{,}625$ **23.** $\ln 4$
25. $20(e^{0.25} - e^{-0.5}) \approx 13.55$ **27.** $\frac{56}{3} \approx 18.667$ **29.** $\frac{28}{3} \approx 9.333$ **31.** $\frac{1}{6}[(e^2 - 2)^3 - 1]$ **33.** $-3 - \ln 2$ **35.** $\frac{1}{6}(15^{3/2} - 5^{3/2})$
37. $\frac{1}{2}(\ln 2 - \ln 3)$ **39.** 0 **41.** $\int_0^5 500(t - 12)\,dt = -\$23{,}750$; $\int_5^{10} 500(t - 12)\,dt = -\$11{,}250$
43. Useful life $= \sqrt{\ln 55} \approx 2$ yr; Total profit $= \frac{51}{22} - \frac{5}{2}e^{-4} \approx 2.272$ or \$2,272 **45.** 4,800 labor-hours
47. $100\ln 11 + 50 \approx 290$ thousand barrels; $100\ln 21 - 100\ln 11 + 50 \approx 115$ thousand barrels
49. $20 + 100e^{-1.2} \approx \$50$ million; $120 + 100e^{-2.4} - 100e^{-1.2} \approx \$99$ million **51.** 134 billion ft^3
53. $2e^{0.8} - 2 \approx 2.45$ g; $2e^{1.6} - 2e^{0.8} \approx 5.45$ g
55. An increase of $60 - 60e^{-0.5} \approx 24$ words/min; an increase of $60e^{-0.5} - 60e^{-1} \approx 14$ words/min; an increase of $60e^{-1} - 60e^{-1.5} \approx 9$ words/min

◆ EXERCISE 7-5

1. 16 **3.** 7 **5.** $\frac{7}{3}$ **7.** 9 **9.** $e^2 - e^{-1}$ **11.** $-\ln 0.5$ **13.** 15 **15.** 32 **17.** 36 **19.** 9 **21.** $\frac{5}{2}$ **23.** $2e + \ln 2 - 2e^{0.5}$ **25.** $\frac{23}{3}$ **27.** 17 **29.** $\frac{28}{3}$
31. $\frac{4}{3}$ **33.** $\frac{343}{3}$ **35.** 8 **37.** $\frac{407}{4}$
39. Total production from the end of the 5th year to the end of the 10th year is $50 + 100\ln 20 - 100\ln 15 \approx 79$ thousand barrels.

41. Total profit over the 5 year useful life of the game is $20 - 30e^{-1.5} \approx 13.306$ or \$13,306.
43. 1935: 0.412; 1947: 0.231; income is more equally distributed in 1947.
45. 1963: 0.818; 1983: 0.846; total assets are less equally distributed in 1983.
47. Total weight gain during the first 10 hr is $3e - 3 \approx 5.15$ g.
49. Average number of words learned during the second 2 hr is $15 \ln 4 - 15 \ln 2 \approx 10$.

◆ EXERCISE 7-6 CHAPTER REVIEW

1. $t^3 - t^2 + C$ **2.** 12 **3.** $-3t^{-1} - 3t + C$ **4.** $\frac{15}{2}$ **5.** $-2e^{-0.5x} + C$ **6.** $2 \ln 5$ **7.** $y = f(x) = x^3 - 2x + 4$ **8.** 12 **9.** $\frac{1}{8}(6x - 5)^{4/3} + C$
10. 2 **11.** $-2x^{-1} - e^{x^2} + C$ **12.** $(20^{3/2} - 8)/3$ **13.** $-\frac{1}{2}e^{-2x} + \ln|x| + C$ **14.** $-500(e^{-0.2} - 1) \approx 90.63$ **15.** $y = f(x) = 3 \ln|x| + x^{-1} + 4$
16. $y = 3x^2 + x - 4$ **17.** $\frac{64}{3}$ **18.** $\frac{1}{2} \ln 10$ **19.** 0.45 **20.** $\frac{1}{48}(2x^4 + 5)^6 + C$ **21.** $-\ln(e^{-x} + 3) + C$ **22.** $-(e^x + 2)^{-1} + C$
23. $\frac{1}{3}(\ln x)^3 + C$ **24.** $\frac{1}{8}x^8 - \frac{2}{5}x^5 + \frac{1}{2}x^2 + C$ **25.** $\frac{2}{3}(6 - x)^{3/2} - 12(6 - x)^{1/2} + C$ **26.** $\frac{1,234}{15} \approx 82.267$ **27.** $\frac{64}{15}$ **28.** $y = 3e^{x^3} - 1$
29. $N = 800e^{0.06t}$ **30.** $\frac{46}{3}$ **31.** $P(x) = 100x - 0.01x^2$; $P(10) = \$999$ **32.** $\int_0^{15} (60 - 4t)\,dt = 450$ thousand barrels
33. $\int_{10}^{40} \left(150 - \frac{x}{10}\right) dx = \$4,425$ **34.** Useful life $= 10 \ln \frac{20}{3} \approx 19$ yr; total profit $= 143 - 200e^{-1.9} \approx 113.086$ or \$113,086
35. $S(t) = 50 - 50e^{-0.08t}$; $50 - 50e^{-0.96} \approx \$31$ million; $-(\ln 0.2)/0.8 \approx 20$ months
36. Current: 0.3; projected: 0.2; income will be more equally distributed 10 yr from now. **37.** 1 cm^2 **38.** 800 gal
39. (A) $226e^{0.11} \approx 252$ million (B) $\frac{\ln 2}{0.011} \approx 63$ yr **40.** $\frac{-\ln 0.04}{0.000\ 123\ 8} \approx 26,000$ yr
41. $N(t) = 95 - 70e^{-0.1t}$; $N(15) \approx 79$ words/min

CHAPTER 8

◆ EXERCISE 8-1

1. (A) 120 (B) 124 **3.** (A) 123 (B) 124 **5.** (A) -4 (B) -5.33 **7.** (A) -5 (B) -5.33 **9.** 23.4 **11.** 55.3 **13.** 250 **15.** 2
17. $\frac{45}{28} \approx 1.61$ **19.** $2(1 - e^{-2}) \approx 1.73$ **21.** 0.791 **23.** 1.650 **25.** 0.748 **27.** 0.747 **29.** $[f(b) - f(a)]/(b - a)$
31. (A) $I = -200t + 600$ (B) $\frac{1}{3} \int_0^3 (-200t + 600)\,dt = 300$ **33.** \$16,000 **35.** (A) \$420 (B) \$135,000 **37.** \$149.18; \$122.96
39. $50e^{0.6} - 50e^{0.4} - 10 \approx \$6.51$
41. $(40 + 100e^{-0.7})/7 \approx 12.8$ or 12,800 hamburgers; $(140 + 100e^{-1.4} - 100e^{-0.7})/7 \approx 16.4$ or 16,400 hamburgers **43.** \$10,000
45. 3,120,000 ft^2 **47.** 10°C **49.** 1.1 liters **51.** $0.6 \ln 2 + 0.1 \approx 0.516$; $(4.2 \ln 625 + 2.4 - 4.2 \ln 49)/24 \approx 0.546$

◆ EXERCISE 8-2

1. \$12,500 **3.** $8,000(e^{0.15} - 1) \approx \$1,295$ **5.** $10,000(e^{0.8} - 1) \approx \$12,255$ **7.** $12,500(e^{0.4} - e^{-0.08}) \approx \$7,109$
9. $25,000(e^{0.36} - 1) - 9,000 \approx \$1,833$ **11.** \$625,000 **13.** \$9,500 **15.** $\bar{p} = 24$; $CS = \$3,380$; $PS = \$1,690$
17. $\bar{p} \approx 49$; $CS = 55,990 - 80,000e^{-0.49} \approx \$6,980$; $PS = 54,010 - 30,000e^{0.49} \approx \$5,041$ **19.** \$2,000
21. Clothing store: $FV = 120,000(e^{0.5} - 1) \approx \$77,847$; computer store: $FV = 200,000(e^{0.5} - e^{0.25}) \approx \$72,939$; the clothing store is the better investment.
23. Bond: $FV = 10,000e^{0.4} \approx \$14,918$; business: $FV = 25,000(e^{0.4} - 1) \approx \$12,296$; the bond is the better investment.
25. \$46,283 **27.** $\frac{k}{r}(e^{rT} - 1)$

◆ EXERCISE 8-3

1. $\frac{1}{3}xe^{3x} - \frac{1}{9}e^{3x} + C$ **3.** $\frac{x^3}{3}\ln x - \frac{x^3}{9} + C$ **5.** $-xe^{-x} - e^{-x} + C$ **7.** $\frac{1}{2}e^{x^2} + C$
9. $(xe^x - 4e^x)|_0^1 = -3e + 4 \approx -4.1548$ **11.** $(x \ln 2x - x)|_1^3 = (3 \ln 6 - 3) - (\ln 2 - 1) \approx 2.6821$ **13.** $\ln(x^2 + 1) + C$ **15.** $(\ln x)^2/2 + C$
17. $\frac{2}{3}x^{3/2} \ln x - \frac{4}{9}x^{3/2} + C$ **19.** $(x^2 - 2x + 2)e^x + C$ **21.** $\frac{xe^{ax}}{a} - \frac{e^{ax}}{a^2} + C$ **23.** $\left(-\frac{\ln x}{x} - \frac{1}{x}\right)\Big|_1^e = -\frac{2}{e} + 1 \approx 0.2642$
25. $6 \ln 6 - 4 \ln 4 - 2 \approx 3.205$ **27.** $xe^{x-2} - e^{x-2} + C$ **29.** $\frac{1}{2}(1 + x^2)\ln(1 + x^2) - \frac{1}{2}(1 + x^2) + C$ **31.** $(1 + e^x)\ln(1 + e^x) - (1 + e^x) + C$
33. $x(\ln x)^2 - 2x \ln x + 2x + C$ **35.** $x(\ln x)^3 - 3x(\ln x)^2 + 6x \ln x - 6x + C$ **37.** $(e^2 - 3)/2$ **39.** $2e^2 - 3$
41. $P(t) = t^2 + te^{-t} + e^{-t} - 1$ **43.** $31{,}250 - 18{,}750e^{0.4} \approx \$3{,}278$ **45.** $1 - 2e^{-1} \approx 0.264$ **47.** $(10 - 2 \ln 6)/3 \approx 2.1388$ppm **49.** 20,980

◆ EXERCISE 8-4

1. $\ln\left|\frac{x}{1 + x}\right| + C$ **3.** $\frac{1}{3 + x} + 2 \ln\left|\frac{5 + 2x}{3 + x}\right| + C$ **5.** $\frac{2(x - 32)}{3}\sqrt{16 + x} + C$ **7.** $\frac{1}{2}\ln\left|\frac{2}{2 + \sqrt{x^2 + 4}}\right| + C$
9. $\frac{1}{3}x^3 \ln x - \frac{1}{9}x^3 + C$ **11.** $9 \ln \frac{3}{2} - 2 \approx 1.6492$ **13.** $\frac{1}{2} \ln \frac{12}{5} \approx 0.4377$ **15.** $\ln 3 \approx 1.0986$
17. $-\frac{\sqrt{4x^2 + 1}}{x} + 2 \ln|2x + \sqrt{4x^2 + 1}| + C$ **19.** $\frac{1}{2} \ln|x^2 + \sqrt{x^4 - 16}| + C$ **21.** $\frac{1}{6}(x^3\sqrt{x^6 + 4} + 4 \ln|x^3 + \sqrt{x^6 + 4}|) + C$
23. $-\frac{\sqrt{4 - x^4}}{8x^2} + C$ **25.** $\frac{1}{5} \ln\left|\frac{3 + 4e^x}{2 + e^x}\right| + C$ **27.** $\frac{2}{3}(\ln x - 8)\sqrt{4 + \ln x} + C$ **29.** $\frac{1}{5}x^2e^{5x} - \frac{2}{25}xe^{5x} + \frac{2}{125}e^{5x} + C$
31. $-x^3e^{-x} - 3x^2e^{-x} - 6xe^{-x} - 6e^{-x} + C$ **33.** $x(\ln x)^3 - 3x(\ln x)^2 + 6x \ln x - 6x + C$ **35.** $\frac{64}{3}$ **37.** $\frac{1}{2} \ln \frac{9}{5} \approx 0.2939$
39. $\frac{1}{2} \ln|x^2 + 2x| + C$ **41.** $\frac{2}{3} \ln|3 + x| + \frac{1}{3} \ln|x| + C$ **43.** $3{,}000 + 1{,}500 \ln \frac{1}{3} \approx \$1{,}352$ **45.** $100{,}000e - 250{,}000 \approx \$21{,}828$
47. $\frac{19}{135} \approx 0.1407$ **49.** $S(t) = 1 + t - \frac{1}{1 + t} - 2 \ln|1 + t|$; $24.96 - 2 \ln 25 \approx \$18.5$ million **51.** $100 \ln 3 \approx 110$ ft **53.** $60 \ln 5 \approx 97$ items

◆ EXERCISE 8-5 CHAPTER REVIEW

1. $\frac{1}{4}xe^{4x} - \frac{1}{16}e^{4x} + C$ **2.** $\frac{1}{2}x^2 \ln x - \frac{1}{4}x^2 + C$ **3.** $\frac{1}{1 + x} + \ln\left|\frac{x}{1 + x}\right| + C$
4. $-\frac{\sqrt{1 + x}}{x} - \frac{1}{2}\ln\left|\frac{\sqrt{1 + x} - 1}{\sqrt{1 + x} + 1}\right| + C$ **5.** 44 **6.** 30.8 **7.** 7 **8.** 1 **9.** $\frac{15}{2} - 8 \ln 8 + 8 \ln 4 \approx 1.955$
10. $\frac{1}{6}(3x\sqrt{9x^2 - 49} - 49 \ln|3x + \sqrt{9x^2 - 49}|) + C$ **11.** $-2te^{-0.5t} - 4e^{-0.5t} + C$ **12.** $\frac{1}{3}x^3 \ln x - \frac{1}{9}x^3 + C$ **13.** $x - \ln|1 + 2e^x| + C$
14. $2e \ln 2 \approx 3.768$ **15.** 2.317 **16.** $\frac{13}{2}$ **17.** $\frac{1}{3}(\ln x)^3 + C$ **18.** $\frac{1}{2}x^2(\ln x)^2 - \frac{1}{2}x^2 \ln x + \frac{1}{4}x^2 + C$ **19.** $\sqrt{x^2 - 36} + C$
20. $\frac{1}{2} \ln|x^2 + \sqrt{x^4 - 36}| + C$ **21.** $50 \ln 10 - 42 \ln 6 - 24 \approx 15.875$ **22.** $x(\ln x)^2 - 2x \ln x + 2x + C$ **23.** 109 items
24. $16e^{2.5} - 16e^2 - 8 \approx \$68.70$ **25.** (A) $50{,}000(e^{0.25} - 1) \approx \$14{,}201$ (B) $25{,}000(e^{0.75} - e^{0.25}) \approx \$20{,}824$ (C) \$6,623
26. (A) \$1,000 (B) \$800 (C) $\bar{p} = 40$; $CS = \$2{,}250$; $PS = \$2{,}700$ **27.** 3,374,000 barrels **28.** $120 \ln 2 - 120 \ln 1.5 - 30 \approx 4.522$ ml
29. 6.5 ppm **30.** $25 \ln|3 + \sqrt{10}| \approx 45.461$ or 45,461

CHAPTER 9

◆ EXERCISE 9-1

1. 10 **3.** 1 **5.** 0 **7.** 1 **9.** 6 **11.** 150 **13.** 16π **15.** 791 **17.** 0.192 **19.** 118 **21.** $100e^{0.8} \approx 222.55$ **23.** $2x + h$ **25.** $2y^2$
27. $E(0, 0, 3)$; $F(2, 0, 3)$ **29.** \$4,400; \$6,000; \$7,100 **31.** $R(p, q) = -5p^2 + 6pq - 4q^2 + 200p + 300q$; $R(2, 3) = \$1{,}280$; $R(3, 2) = \$1{,}175$
33. 30,065 units **35.** \$272,615.08 **37.** $T(70, 47) \approx 29$ min; $T(60, 27) = 33$ min **39.** $C(6, 8) = 75$; $C(8.1, 9) = 90$
41. $Q(12, 10) = 120$; $Q(10, 12) \approx 83$

◆ EXERCISE 9-2

1. 3 **3.** 2 **5.** $-4xy$ **7.** -6 **9.** $10xy^3$ **11.** 60 **13.** $2x - 2y + 6$ **15.** 6 **17.** -2 **19.** 2 **21.** $2e^{2x+3y}$ **23.** $6e^{2x+3y}$ **25.** $6e^2$ **27.** $4e^3$
29. $f_x(x, y) = 6x(x^2 - y^3)^2$; $f_y(x, y) = -9y^2(x^2 - y^3)^2$ **31.** $f_x(x, y) = 24xy(3x^2y - 1)^3$; $f_y(x, y) = 12x^2(3x^2y - 1)^3$
33. $f_x(x, y) = 2x/(x^2 + y^2)$; $f_y(x, y) = 2y/(x^2 + y^2)$ **35.** $f_x(x, y) = y^4e^{xy^2}$; $f_y(x, y) = 2xy^3e^{xy^2} + 2ye^{xy^2}$
37. $f_x(x, y) = 4xy^2/(x^2 + y^2)^2$; $f_y(x, y) = -4x^2y/(x^2 + y^2)^2$ **39.** $f_{xx}(x, y) = 2y^2 + 6x$; $f_{xy}(x, y) = 4xy = f_{yx}(x, y)$; $f_{yy}(x, y) = 2x^2$
41. $f_{xx}(x, y) = -2y/x^3$; $f_{xy}(x, y) = (-1/y^2) + (1/x^2) = f_{yx}(x, y)$; $f_{yy}(x, y) = 2x/y^3$
43. $f_{xx}(x, y) = (2y + xy^2)e^{xy}$; $f_{xy}(x, y) = (2x + x^2y)e^{xy} = f_{yx}(x, y)$; $f_{yy}(x, y) = x^3e^{xy}$ **45.** $x = 2$ and $y = 4$
47. $f_{xx}(x, y) + f_{yy}(x, y) = (2y^2 - 2x^2)/(x^2 + y^2)^2 + (2x^2 - 2y^2)/(x^2 + y^2)^2 = 0$ **49.** (A) $2x$ (B) $4y$
51. $P_x(1{,}200, 1{,}800) = 24$: profit will increase approx. \$24 per unit increase in production of type *A* calculators at the (1,200, 1,800) output level; $P_y(1{,}200, 1{,}800) = -48$: profit will decrease approx. \$48 per unit increase in production of type *B* calculators at the (1,200, 1,800) output level
53. $\partial x/\partial p = -5$: a \$1 increase in the price of brand *A* will decrease the demand for brand *A* by 5 lb at any price level (p, q); $\partial y/\partial p = 2$: a \$1 increase in the price of brand *A* will increase the demand for brand *B* by 2 lb at any price level (p, q)
55. (A) $f_x(x, y) = 7.5x^{-0.25}y^{0.25}$; $f_y(x, y) = 2.5x^{0.75}y^{-0.75}$
(B) Marginal productivity of labor $= f_x(600, 100) \approx 4.79$; Marginal productivity of capital $= f_y(600, 100) \approx 9.58$ (C) Capital
57. Competitive **59.** Complementary
61. (A) $f_w(w, h) = 6.65w^{-0.575}h^{0.725}$; $f_h(w, h) = 11.34w^{0.425}h^{-0.275}$
(B) $f_w(65, 57) = 11.31$: for a 65 lb child 57 in. tall, the rate of change in surface area is 11.31 in.^{2} for each pound gained in weight (height is held fixed); $f_h(65, 57) = 21.99$: for a 65 lb child 57 in. tall, the rate of change in surface area is 21.99 in.^{2} for each inch gained in height (weight is held fixed)
63. $C_W(6, 8) = 12.5$: index increases approx. 12.5 units for 1 in. increase in width of head (length held fixed) when $W = 6$ and $L = 8$; $C_L(6, 8) = -9.38$: index decreases approx. 9.38 units for 1 in. increase in length (width held fixed) when $W = 6$ and $L = 8$

◆ EXERCISE 9-3

1. $f(-2, 0) = 10$ is a local maximum **3.** $f(-1, 3) = 4$ is a local minimum **5.** f has a saddle point at $(3, -2)$
7. $f(3, 2) = 33$ is a local maximum **9.** $f(2, 2) = 8$ is a local minimum **11.** f has a saddle point at $(0, 0)$
13. f has a saddle point at $(0, 0)$; $f(1, 1) = -1$ is a local minimum
15. f has a saddle point at $(0, 0)$; $f(3, 18) = -162$ and $f(-3, -18) = -162$ are local minima
17. The test fails at $(0, 0)$; f has saddle points at $(2, 2)$ and $(2, -2)$ **19.** 2,000 type *A* and 4,000 type *B*; Max $P = P(2, 4) = \$15$ million

21. (A)

p	q	x	y
\$10	\$12	56	16
\$11	\$11	6	56

(B) A maximum weekly profit of \$288 is realized for $p = \$10$ and $q = \$12$. **23.** $P(x, y) = P(4, 2)$ **25.** 8 by 4 by 2 in. **27.** 20 by 20 by 40 in.

◆ EXERCISE 9-4

1. Max $f(x, y) = f(3, 3) = 18$ **3.** Min $f(x, y) = f(3, 4) = 25$
5. Max $f(x, y) = f(3, 3) = f(-3, -3) = 18$; Min $f(x, y) = f(3, -3) = f(-3, 3) = -18$
7. Maximum product is 25 when each number is 5 **9.** Min $f(x, y, z) = f(-4, 2, -6) = 56$
11. Max $f(x, y, z) = f(2, 2, 2) = 6$; Min $f(x, y, z) = f(-2, -2, -2) = -6$
13. 60 of model *A* and 30 of model *B* will yield a minimum cost of \$32,400 per week
15. (A) 8,000 units of labor and 1,000 units of capital; Max $N(x, y) = N(8{,}000, 1{,}000) \approx 263{,}902$ units
(B) Marginal productivity of money ≈ 0.6598; increase in production $\approx 32{,}990$ units
17. 8 by 8 by $\frac{8}{3}$ in. **19.** $x = 50$ ft and $y = 200$ ft; maximum area is 10,000 ft^2

◆ EXERCISE 9-5

1.
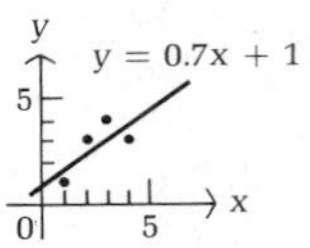

3.
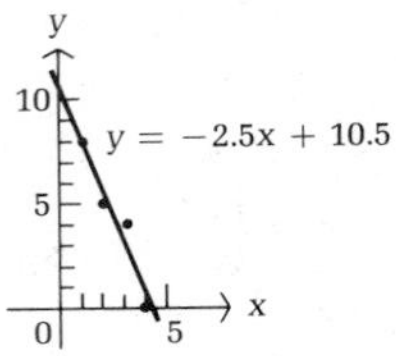

5.
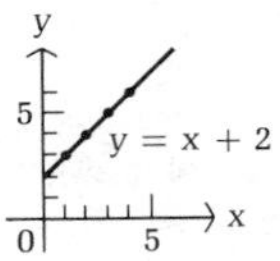

7. $y = 2.12x + 10.8$; $y = 63.8$ when $x = 25$
9. $y = -1.2x + 12.6$; $y = 10.2$ when $x = 2$
11. $y = -1.53x + 26.67$; $y = 14.4$ when $x = 8$

13. $y = 0.75x^2 - 3.45x + 4.75$

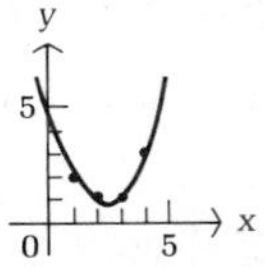

17. (A) $y = 743x + 10{,}012$ (B) \$17,442
19. (A) $y = -0.48x + 4.38$ (B) \$6.56/bottle
21. (A) $P = -0.66T + 48.8$ (B) 11.18 beats/min
23. (A) $D = -3.1A + 54.6$ (B) 45%

◆ EXERCISE 9-6

1. (A) $3x^2y^4 + C(x)$ (B) $3x^2$ **3.** (A) $2x^2 + 6xy + 5x + E(y)$ (B) $35 + 30y$ **5.** (A) $\sqrt{y + x^2} + E(y)$ (B) $\sqrt{y+4} - \sqrt{y}$ **7.** 9 **9.** 330

11. $(56 - 20\sqrt{5})/3$ **13.** 16 **15.** 49 **17.** $\frac{1}{8}\int_1^5\int_{-1}^1 (x+y)^2\,dy\,dx = \frac{32}{3}$ **19.** $\frac{1}{15}\int_1^4\int_2^7 (x/y)\,dy\,dx = \frac{1}{2}\ln\frac{7}{2} \approx 0.6264$

21. $\frac{4}{3}$ cubic units **23.** $\frac{32}{3}$ cubic units **25.** $\int_0^1\int_1^2 xe^{xy}\,dy\,dx = \frac{1}{2} + \frac{1}{2}e^2 - e$ **27.** $\int_0^1\int_{-1}^1 \frac{2y + 3xy^2}{1 + x^2}\,dy\,dx = \ln 2$

29. $\frac{1}{0.4}\int_{0.6}^{0.8}\int_5^7 \frac{y}{1-x}\,dy\,dx = 30\ln 2 \approx \$20.8$ billion

31. $\frac{1}{10}\int_{10}^{20}\int_1^2 x^{0.75}y^{0.25}\,dy\,dx = \frac{8}{175}(2^{1.25} - 1)(20^{1.75} - 10^{1.75}) \approx 8.375$ or 8,375 units

33. $\frac{1}{192}\int_{-8}^{8}\int_{-6}^{6} [10 - \frac{1}{10}(x^2 + y^2)]\,dy\,dx = \frac{20}{3}$ insects/ft² **35.** $\frac{1}{8}\int_{-2}^{2}\int_{-1}^{1} [100 - 15(x^2 + y^2)]\,dy\,dx = 75$ ppm

37. $\frac{1}{10{,}000}\int_{2{,}000}^{3{,}000}\int_{50}^{60} 0.000\,013\,3xy^2\,dy\,dx \approx 100.86$ ft **39.** $\frac{1}{16}\int_8^{16}\int_{10}^{12} 100\frac{x}{y}\,dy\,dx = 600\ln 1.2 \approx 109.4$

◆ EXERCISE 9-7 CHAPTER REVIEW

1. $f(5, 10) = 2{,}900$; $f_x(x, y) = 40$; $f_y(x, y) = 70$ **2.** $\partial^2 z/\partial x^2 = 6xy^2$; $\partial^2 z/\partial x\,\partial y = 6x^2y$ **3.** $2xy^3 + 2y^2 + C(x)$ **4.** $3x^2y^2 + 4xy + E(y)$
5. 1 **6.** $f(2, 3) = 7$; $f_y(x, y) = -2x + 2y + 3$; $f_y(2, 3) = 5$ **7.** $(-8)(-6) - (4)^2 = 32$ **8.** $(1, 3, -\frac{1}{2})$, $(-1, -3, \frac{1}{2})$
9. $y = -1.5x + 15.5$; $y = 0.5$ when $x = 10$ **10.** 18 **11.** $f_x(x, y) = 2xe^{x^2+2y}$; $f_y(x, y) = 2e^{x^2+2y}$; $f_{xy}(x, y) = 4xe^{x^2+2y}$
12. $f_x(x, y) = 10x(x^2 + y^2)^4$; $f_{xy}(x, y) = 80xy(x^2 + y^2)^3$ **13.** $f(2, 3) = -25$ is a local minimum; f has a saddle point at $(-2, 3)$
14. Max $f(x, y) = f(6, 4) = 24$ **15.** Min $f(x, y, z) = f(2, 1, 2) = 9$ **16.** $y = \frac{116}{165}x + \frac{100}{3}$ **17.** $\frac{27}{5}$ **18.** 4 cubic units
19. (A) $P_x(1, 3) = 8$; profit will increase \$8,000 for 100 unit increase in product A if production of product B is held fixed at an output level of (1, 3)
(B) For 200 units of A and 300 units of B, $P(2, 3) = \$100$ thousand is a local maximum.
20. 8 by 6 by 2 in. **21.** $y = 0.63x + 1.33$; profit in sixth year is \$5.11 million
22. (A) Marginal productivity of labor ≈ 8.37; marginal productivity of capital ≈ 1.67; management should encourage increased use of labor.

22. (B) 80 units of labor and 40 units of capital; Max $N(x, y) = N(80, 40) \approx 696$ units; marginal productivity of money ≈ 0.0696; increase in production ≈ 139 units

(C) $\frac{1}{1{,}000}\int_{50}^{100}\int_{20}^{40} 10x^{0.8}y^{0.2}\,dy\,dx = \frac{(40^{1.2} - 20^{1.2})(100^{1.8} - 50^{1.8})}{216} = 621$ items

23. $T_x(70, 17) = -0.924$ min/ft increase in depth when $V = 70$ ft³ and $x = 17$ ft **24.** 36 ppm **25.** 50,000
26. $y = \frac{1}{2}x + 48$; $y = 68$ when $x = 40$

CHAPTER 10

◆ EXERCISE 10-1

1. $\pi/3$ radians **3.** $\pi/4$ radian **5.** $\pi/6$ radian **7.** II **9.** IV **11.** I **13.** 1 **15.** 0 **17.** -1 **19.** 60° **21.** 45° **23.** 30° **25.** $\sqrt{3}/2$
27. $\frac{1}{2}$ **29.** $-\sqrt{3}/2$ **31.** 0.1411 **33.** -0.6840 **35.** 0.7970 **37.** $3\pi/20$ radian **39.** 15° **41.** 1 **43.** 2 **45.** $1/\sqrt{3}$ or $\sqrt{3}/3$
49. (A) $P(13) = 5$, $P(26) = 10$, $P(39) = 5$, $P(52) = 0$ (B) $P(30) \approx 9.43$, $P(100) \approx 0.57$
51. (A) $V(0) = 0.10$, $V(1) = 0.45$, $V(2) = 0.80$, $V(3) = 0.45$, $V(7) = 0.45$ (B) $V(3.5) \approx 0.20$, $V(5.7) \approx 0.76$ **53.** (A) $-5.6°$ (B) $-4.7°$

◆ EXERCISE 10-2

1. $-\sin t$ **3.** $3x^2 \cos x^3$ **5.** $t \cos t + \sin t$ **7.** $(\cos x)^2 - (\sin x)^2$ **9.** $5(\sin x)^4 \cos x$ **11.** $\frac{\cos x}{2\sqrt{\sin x}}$ **13.** $-\frac{x^{-1/2}}{2}\sin\sqrt{x} = \frac{-\sin\sqrt{x}}{2\sqrt{x}}$

15. $f'\left(\frac{\pi}{6}\right) = \cos\frac{\pi}{6} = \frac{\sqrt{3}}{2}$ **17.** $\frac{(\cos x)^2 + (\sin x)^2}{(\cos x)^2} = \frac{1}{(\cos x)^2} = (\sec x)^2$ **19.** $\frac{x\cos\sqrt{x^2 - 1}}{\sqrt{x^2 - 1}}$ **21.** $2e^x \cos x$

23. (A) $P'(t) = \frac{5\pi}{26}\sin\frac{\pi t}{26}$, $0 \le t \le 104$

(B) $P'(8) = \$0.50$ hundred or \$50/week; $P'(26) = \$0$/week; $P'(50) = -\$0.14$ hundred or $-\$14$/week

(C)

t	$P(t)$	
26	\$1,000	Local maximum
52	\$0	Local minimum
78	\$1,000	Local maximum

(D)

t	$P(t)$	
0	\$0	Absolute minimum
26	\$1,000	Absolute maximum
52	\$0	Absolute minimum
78	\$1,000	Absolute maximum
104	\$0	Absolute minimum

25. (A) $V'(t) = \frac{0.35\pi}{2}\sin\frac{\pi t}{2}$, $0 \le t \le 8$

(B) $V'(3) = -0.55$ liter/sec; $V'(4) = 0.00$ liter/sec; $V'(5) = 0.55$ liter/sec

(C)

t	$V(t)$	
2	0.80	Local maximum
4	0.10	Local minimum
6	0.80	Local maximum

(D)

t	$V(t)$	
0	0.10	Absolute minimum
2	0.80	Absolute maximum
4	0.10	Absolute minimum
6	0.80	Absolute maximum
8	0.10	Absolute minimum

◆ EXERCISE 10-3

1. $-\cos t + C$ **3.** $\frac{1}{3}\sin 3x + C$ **5.** $\frac{1}{13}(\sin x)^{13} + C$ **7.** $-\frac{3}{4}(\cos x)^{4/3} + C$ **9.** $\frac{1}{3}\sin x^3 + C$ **11.** 1 **13.** 1 **15.** $\sqrt{3}/2 - \frac{1}{2} \approx 0.366$
17. 1.4161 **19.** 0.0678 **21.** $e^{\sin x} + C$ **23.** $\ln|\sin x| + C$ **25.** $-\ln|\cos x| + C$
27. (A) \$520 hundred or \$52,000 (B) \$106.38 hundred or \$10,638 **29.** (A) 104 tons (B) 31 tons

◆ EXERCISE 10-4 CHAPTER REVIEW

1. (A) $\pi/6$ (B) $\pi/4$ (C) $\pi/3$ (D) $\pi/2$ **2.** (A) -1 (B) 0 (C) 1 **3.** $-\sin m$ **4.** $\cos u$ **5.** $(2x-2)\cos(x^2-2x+1)$
6. $-\frac{1}{3}\cos 3t + C$ **7.** (A) 30° (B) 45° (C) 60° (D) 90° **8.** (A) $\frac{1}{2}$ (B) $\sqrt{2}/2$ (C) $\sqrt{3}/2$ **9.** (A) -0.6543 (B) 0.8308
10. $(x^2-1)\cos x + 2x\sin x$ **11.** $6(\sin x)^5\cos x$ **12.** $(\cos x)/[3(\sin x)^{2/3}]$ **13.** $\frac{1}{2}\sin(t^2-1) + C$ **14.** 2 **15.** $\sqrt{3}/2$ **16.** -0.243
17. $-\sqrt{2}/2$ **18.** $\sqrt{2}$ **19.** $\pi/12$ **20.** (A) -1 (B) $-\sqrt{3}/2$ (C) $-\frac{1}{2}$ **21.** $1/(\cos u)^2 = (\sec u)^2$ **22.** $-2x(\sin x^2)e^{\cos x^2}$ **23.** $e^{\sin x} + C$
24. $-\ln|\cos x| + C$ **25.** 15.2128 **26.** (A) $R(0) = 5$, $R(2) = 4$, $R(3) = 3$, $R(6) = 1$ (B) $R(1) = 4{,}732$, $R(22) = 4$

27. (A) $R'(t) = -\dfrac{\pi}{3}\sin\dfrac{\pi t}{6},\ 0 \le t \le 24$

(B) $R'(3) = -\$1.047$ thousand or $-\$1{,}047$/month; $R'(10) = \$0.907$ thousand or \$907/month; $R'(18) = \$0.000$ thousand

(C)

t	$P(t)$	
6	\$1,000	Local minimum
12	\$5,000	Local maximum
18	\$1,000	Local minimum

(D)

t	$P(t)$	
0	\$5,000	Absolute maximum
6	\$1,000	Absolute minimum
12	\$5,000	Absolute maximum
18	\$1,000	Absolute minimum
24	\$5,000	Absolute maximum

28. (A) \$72 thousand or \$72,000 (B) \$6.270 thousand or \$6,270

APPENDIX A

◆ EXERCISE A-1

1. T **3.** T **5.** T **7.** T **9.** {1, 2, 3, 4, 5} **11.** {3, 4} **13.** ∅ **15.** {2} **17.** {−7, 7} **19.** {1, 3, 5, 7, 9} **21.** $A' = \{1, 5\}$ **23.** 40 **25.** 60
27. 60 **29.** 20 **31.** 95 **33.** 40 **35.** (A) {1, 2, 3, 4, 6} (B) {1, 2, 3, 4, 6} **37.** {1, 2, 3, 4, 6} **39.** Yes **41.** Yes **43.** Yes
45. (A) 2 (B) 4 (C) 8; 2^n **47.** 800 **49.** 200 **51.** 200 **53.** 800 **55.** 200 **57.** 200 **59.** 6 **61.** A+, AB+
63. A−, A+, B+, AB−, AB+, O+ **65.** O+, O− **67.** B−, B+ **69.** Everybody in the clique relates to each other.

◆ EXERCISE A-2

1. 5, 7, 9, 11 **3.** $\frac{3}{2}, \frac{4}{3}, \frac{5}{4}, \frac{6}{5}$ **5.** $9, -27, 81, -243$ **7.** 23 **9.** $\frac{101}{100}$ **11.** $1+2+3+4+5+6 = 21$ **13.** $5+7+9+11 = 32$
15. $1 + \frac{1}{10} + \frac{1}{100} + \frac{1}{1{,}000} = \frac{1{,}111}{1{,}000}$ **17.** 3.6 **19.** 82.5 **21.** $\frac{1}{2}, -\frac{1}{4}, \frac{1}{8}, -\frac{1}{16}, \frac{1}{32}$ **23.** 0, 4, 0, 8, 0 **25.** $1, -\frac{3}{2}, \frac{9}{4}, -\frac{27}{8}, \frac{81}{16}$ **27.** $a_n = n - 3$
29. $a_n = 4n$ **31.** $a_n = (2n-1)/2n$ **33.** $a_n = (-1)^{n+1}n$ **35.** $a_n = (-1)^{n+1}(2n-1)$ **37.** $a_n = (\frac{2}{5})^{n-1}$ **39.** $a_n = x^n$ **41.** $a_n = (-1)^{n+1}x^{2n-1}$

43. $1 - 9 + 25 - 49 + 81$ **45.** $\frac{4}{7} + \frac{8}{9} + \frac{16}{11} + \frac{32}{13}$ **47.** $1 + x + x^2 + x^3 + x^4$ **49.** $x - \dfrac{x^3}{3} + \dfrac{x^5}{5} - \dfrac{x^7}{7} + \dfrac{x^9}{9}$

51. (A) $\sum_{k=1}^{5}(k+1)$ (B) $\sum_{j=0}^{4}(j+2)$ **53.** (A) $\sum_{k=1}^{4}\dfrac{(-1)^{k+1}}{k}$ (B) $\sum_{j=0}^{3}\dfrac{(-1)^j}{j+1}$ **55.** $\sum_{k=1}^{n}\dfrac{k+1}{k}$ **57.** $\sum_{k=1}^{n}\dfrac{(-1)^{k+1}}{2^k}$ **59.** 2, 8, 26, 80, 242

61. 1, 2, 4, 8, 16 **63.** $1, \frac{3}{2}, \frac{17}{12}, \frac{577}{408}$; $a_4 = \frac{577}{408} \approx 1.414\ 216$, $\sqrt{2} \approx 1.414\ 214$

◆ EXERCISE A-3

1. (A) $d = 3$; 14, 17 (B) Not an arithmetic progression (C) Not an arithmetic progression (D) $d = -10$; -22, -32
3. $a_2 = 11$; $a_3 = 15$ **5.** $a_{21} = 82$; $S_{31} = 1{,}922$ **7.** $S_{20} = 930$ **9.** 2,400 **11.** 1,120
13. Use $a_1 = 1$ and $d = 2$ in $S_n = (n/2)[2a_1 + (n-1)d]$. **15.** Firm *A*: \$280,500; firm *B*: \$278,500
17. \$48 + \$46 + · · · + \$4 + \$2 = \$600

◆ EXERCISE A-4

1. (A) $r = -2$; $a_4 = -8$, $a_5 = 16$ (B) Not a geometric progression (C) $r = \frac{1}{2}$; $a_4 = \frac{1}{4}$, $a_5 = \frac{1}{8}$ (D) Not a geometric progression
3. $a_2 = -6$; $a_3 = 12$; $a_4 = -24$ **5.** $S_7 = 547$ **7.** $a_{10} = 199.90$ **9.** $r = 1.09$ **11.** $S_{10} = 1{,}242$; $S_\infty = 1{,}250$
13. (A) Does not exist (B) $S_\infty = \frac{8}{5} = 1.6$ **15.** 0.999 **17.** About \$11,670,000 **19.** \$31,027; \$251,558

◆ EXERCISE A-5

1. 720 **3.** 10 **5.** 1,320 **7.** 10 **9.** 6 **11.** 1,140 **13.** 10 **15.** 6 **17.** 1 **19.** 816
21. $C_{4,0}a^4 + C_{4,1}a^3b + C_{4,2}a^2b^2 + C_{4,3}ab^3 + C_{4,4}b^4 = a^4 + 4a^3b + 6a^2b^2 + 4ab^3 + b^4$ **23.** $x^6 - 6x^5 + 15x^4 - 20x^3 + 15x^2 - 6x + 1$
25. $32a^5 - 80a^4b + 80a^3b^2 - 40a^2b^3 + 10ab^4 - b^5$ **27.** $3{,}060x^{14}$ **29.** $5{,}005p^9q^6$ **31.** $264x^2y^{10}$ **33.** $C_{n,0} = \dfrac{n!}{0!n!} = 1$; $C_{n,n} = \dfrac{n!}{n!0!} = 1$
35. 1 5 10 10 5 1; 1 6 15 20 15 6 1

Index

◆ BUSINESS & ECONOMICS APPLICATIONS